Texts in Applied Mathematics 37

Texts in Applied Mathematics

1. *Sirovich:* Introduction to Applied Mathematics.
2. *Wiggins:* Introduction to Applied Nonlinear Dynamical Systems and Chaos.
3. *Hale/Koçak:* Dynamics and Bifurcations.
4. *Chorin/Marsden:* A Mathematical Introduction to Fluid Mechanics, Third Edition.
5. *Hubbard/West:* Differential Equations: A Dynamical Systems Approach: Ordinary Differential Equations.
6. *Sontag:* Mathematical Control Theory: Deterministic Finite Dimensional Systems Second Edition.
7. *Perko:* Differential Equations and Dynamical Systems, Third Edition.
8. *Seaborn:* Hypergeometric Functions and Their Applications.
9. *Pipkin:* A Course on Integral Equations.
10. *Hoppensteadt/Peskin:* Modeling and Simulation in Medicine and the Life Sciences, Second Edition.
11. *Braun:* Differential Equations and Their Applications, Fourth Edition.
12. *Stoer/Bulirsch:* Introduction to Numerical Analysis, Third Edition.
13. *Renardy/Rogers:* An Introduction to Partial Differential Equations.
14. *Banks:* Growth and Diffusion Phenomena: Mathematical Frameworks and Applications.
15. *Brenner/Scott:* The Mathematical Theory of Finite Element Methods, Second Edition.
16. *Van de Velde:* Concurrent Scientific Computing.
17. *Marsden/Ratiu:* Introduction to Mechanics and Symmetry, Second Edition.
18. *Hubbard/West:* Differential Equations: A Dynamical Systems Approach: Higher-Dimensional Systems.
19. *Kaplan/Glass:* Understanding Nonlinear Dynamics.
20. *Holmes:* Introduction to Perturbation Methods.
21. *Curtain/Zwart:* An Introduction to Infinite-Dimensional Linear Systems Theory.
22. *Thomas:* Numerical Partial Differential Equations: Finite Difference Methods.
23. *Taylor:* Partial Differential Equations: Basic Theory.
24. *Merkin:* Introduction to the Theory of Stability of Motion.
25. *Naber:* Topology, Geometry, and Gauge Fields: Foundations.
26. *Polderman/Willems:* Introduction to Mathematical Systems Theory: A Behavioral Approach.
27. *Reddy:* Introductory Functional Analysis: with Applications to Boundary Value Problems and Finite Elements.
28. *Gustafson/Wilcox:* Analytical and Computational Methods of Advanced Engineering Mathematics.

(*continued after index*)

Alfio Quarteroni Riccardo Sacco Fausto Saleri

Numerical Mathematics

Second Edition

With 135 Figures and 45 Tables

Springer

Alfio Quarteroni
SB-IACS-CMS, EPFL
1015 Lausanne, Switzerland
and
Dipartimento di Matematica-MOX
Politecnico di Milano
Piazza Leonardo da Vinci, 32
20133 Milano, Italy
E-mail: alfio.quarteroni@epfl.ch

Riccardo Sacco
Dipartimento di Matematica
Politecnico di Milano
Piazza Leonardo da Vinci, 32
20133 Milano, Italy
E-mail: riccardo.sacco@polimi.it

Fausto Saleri
Dipartimento di Matematica–MOX
Politecnico di Milano
Piazza Leonardo da Vinci, 32
20133 Milano, Italy
E-mail: fausto.saleri@polimi.it

Mathematics Subject Classification (2000): 15-01, 34-01, 35-01, 65-01

ISBN 978-3-642-07101-0   e-ISBN 978-3-540-49809-4

Springer is a part of Springer Science+Business Media.

springer.com

Softcover reprint of the hardcover 2nd edition 2007

Cover design: design & production GmbH, Heidelberg

Preface

Numerical mathematics is the branch of mathematics that proposes, develops, analyzes and applies methods from scientific computing to several fields including analysis, linear algebra, geometry, approximation theory, functional equations, optimization and differential equations. Other disciplines such as physics, the natural and biological sciences, engineering, and economics and the financial sciences frequently give rise to problems that need scientific computing for their solutions.

As such, numerical mathematics is the crossroad of several disciplines of great relevance in modern applied sciences, and can become a crucial tool for their qualitative and quantitative analysis. This role is also emphasized by the continual development of computers and algorithms, which make it possible nowadays, using scientific computing, to tackle problems of such a large size that real-life phenomena can be simulated providing accurate responses at affordable computational cost.

The corresponding spread of numerical software represents an enrichment for the scientific community. However, the user has to make the correct choice of the method (or the algorithm) which best suits the problem at hand. As a matter of fact, no black-box methods or algorithms exist that can effectively and accurately solve all kinds of problems.

One of the purposes of this book is to provide the mathematical foundations of numerical methods, to analyze their basic theoretical properties (stability, accuracy, computational complexity), and demonstrate their performances on examples and counterexamples which outline their pros and cons. This is done using the MATLAB® [1] software environment. This choice satisfies the two fundamental needs of user-friendliness and wide-spread diffusion, making it available on virtually every computer.

Every chapter is supplied with examples, exercises and applications of the discussed theory to the solution of real-life problems. The reader is thus in the ideal condition for acquiring the theoretical knowledge that is required to

[1] MATLAB is a trademark of The MathWorks, Inc.

make the right choice among the numerical methodologies and make use of the related computer programs.

This book is primarily addressed to undergraduate students, with particular focus on the degree courses in Engineering, Mathematics, Physics and Computer Science. The attention which is paid to the applications and the related development of software makes it valuable also for graduate students, researchers and users of scientific computing in the most widespread professional fields.

The content of the volume is organized into four Parts and 13 chapters.

Part I comprises two chapters in which we review basic linear algebra and introduce the general concepts of consistency, stability and convergence of a numerical method as well as the basic elements of computer arithmetic.

Part II is on numerical linear algebra, and is devoted to the solution of linear systems (Chapters 3 and 4) and eigenvalues and eigenvectors computation (Chapter 5).

We continue with Part III where we face several issues about functions and their approximation. Specifically, we are interested in the solution of nonlinear equations (Chapter 6), solution of nonlinear systems and optimization problems (Chapter 7), polynomial approximation (Chapter 8) and numerical integration (Chapter 9).

Part IV, which demands a mathematical background, is concerned with approximation, integration and transforms based on orthogonal polynomials (Chapter 10), solution of initial value problems (Chapter 11), boundary value problems (Chapter 12) and initial-boundary value problems for parabolic and hyperbolic equations (Chapter 13).
Part I provides the indispensable background. Each of the remaining Parts has a size and a content that make it well suited for a semester course.

A guideline index to the use of the numerous MATLAB programs developed in the book is reported at the end of the volume. These programs are also available at the web site address:

http://www1.mate.polimi.it/~calnum/programs.html.

For the reader's ease, any code is accompanied by a brief description of its input/output parameters.

We express our thanks to the staff at Springer-Verlag New York for their expert guidance and assistance with editorial aspects, as well as to Dr. Martin Peters from Springer-Verlag Heidelberg and Dr. Francesca Bonadei from Springer-Italia for their advice and friendly collaboration all along this project.

We gratefully thank Professors L. Gastaldi and A. Valli for their useful comments on Chapters 12 and 13.

We also wish to express our gratitude to our families for their forbearance and understanding, and dedicate this book to them.

Lausanne, Milan
January 2000

Alfio Quarteroni
Riccardo Sacco
Fausto Saleri

Preface to the Second Edition

This second edition is characterized by a thourough overall revision.

Regarding the styling of the book, we have improved the readibility of pictures, tables and program headings.

Regarding the scientific contents, we have introduced several changes in the chapter on iterative methods for the solution of linear systems as well as in the chapter on polynomial approximation of functions and data.

Lausanne, Milan
September 2006

Alfio Quarteroni
Riccardo Sacco
Fausto Saleri

Contents

Part I Getting Started

Part III Around Functions and Functionals

Part I

Getting Started

1

Foundations of Matrix Analysis

In this chapter we recall the basic elements of linear algebra which will be employed in the remainder of the text. For most of the proofs as well as for the details, the reader is referred to [Bra75], [Nob69], [Hal58]. Further results on eigenvalues can be found in [Hou75] and [Wil65].

1.1 Vector Spaces

Definition 1.1 A *vector space* over the numeric field K ($K = \mathbb{R}$ or $K = \mathbb{C}$) is a nonempty set V, whose elements are called *vectors* and in which two operations are defined, called *addition* and *scalar multiplication*, that enjoy the following properties:

1. addition is commutative and associative;
2. there exists an element $\mathbf{0} \in V$ (the *zero vector* or *null vector*) such that $\mathbf{v} + \mathbf{0} = \mathbf{v}$ for each $\mathbf{v} \in V$;
3. $0 \cdot \mathbf{v} = \mathbf{0}$, $1 \cdot \mathbf{v} = \mathbf{v}$, for each $\mathbf{v} \in V$, where 0 and 1 are respectively the zero and the unity of K;
4. for each element $\mathbf{v} \in V$ there exists its opposite, $-\mathbf{v}$, in V such that $\mathbf{v} + (-\mathbf{v}) = \mathbf{0}$;
5. the following distributive properties hold

$$\forall \alpha \in K, \ \forall \mathbf{v}, \mathbf{w} \in V, \ \alpha(\mathbf{v} + \mathbf{w}) = \alpha\mathbf{v} + \alpha\mathbf{w},$$

$$\forall \alpha, \beta \in K, \ \forall \mathbf{v} \in V, \ (\alpha + \beta)\mathbf{v} = \alpha\mathbf{v} + \beta\mathbf{v};$$

6. the following associative property holds

$$\forall \alpha, \beta \in K, \ \forall \mathbf{v} \in V, \ (\alpha\beta)\mathbf{v} = \alpha(\beta\mathbf{v}).$$

■

Example 1.1 Remarkable instances of vector spaces are:
- $V = \mathbb{R}^n$ (respectively $V = \mathbb{C}^n$): the set of the n-tuples of real (respectively complex) numbers, $n \geq 1$;
- $V = \mathbb{P}_n$: the set of polynomials $p_n(x) = \sum_{k=0}^{n} a_k x^k$ with real (or complex) coefficients a_k having degree less than or equal to n, $n \geq 0$;
- $V = C^p([a, b])$: the set of real (or complex)-valued functions which are continuous on $[a, b]$ up to their p-th derivative, $0 \leq p < \infty$. •

Definition 1.2 We say that a nonempty part W of V is a *vector subspace* of V iff W is a vector space over K. ■

Example 1.2 The vector space $\mathbb{P}_n$ is a vector subspace of $C^\infty(\mathbb{R})$, which is the space of infinite continuously differentiable functions on the real line. A trivial subspace of any vector space is the one containing only the zero vector. •

In particular, the set W of the linear combinations of a system of p vectors of V, $\{\mathbf{v}_1, \ldots, \mathbf{v}_p\}$, is a vector subspace of V, called the *generated subspace* or *span* of the vector system, and is denoted by

$$\begin{aligned} W &= \text{span}\{\mathbf{v}_1, \ldots, \mathbf{v}_p\} \\ &= \{\mathbf{v} = \alpha_1\mathbf{v}_1 + \ldots + \alpha_p\mathbf{v}_p \quad \text{with } \alpha_i \in K,\ i = 1, \ldots, p\}. \end{aligned}$$

The system $\{\mathbf{v}_1, \ldots, \mathbf{v}_p\}$ is called a system of *generators* for W.
If $W_1, \ldots, W_m$ are vector subspaces of V, then the set

$$S = \{\mathbf{w} : \ \mathbf{w} = \mathbf{v}_1 + \ldots + \mathbf{v}_m \text{ with } \mathbf{v}_i \in W_i,\ i = 1, \ldots, m\}$$

is also a vector subspace of V. We say that S is the *direct sum* of the subspaces W_i if any element $\mathbf{s} \in S$ admits a unique representation of the form $\mathbf{s} = \mathbf{v}_1 + \ldots + \mathbf{v}_m$ with $\mathbf{v}_i \in W_i$ and $i = 1, \ldots, m$. In such a case, we shall write $S = W_1 \oplus \ldots \oplus W_m$.

Definition 1.3 A system of vectors $\{\mathbf{v}_1, \ldots, \mathbf{v}_m\}$ of a vector space V is called *linearly independent* if the relation

$$\alpha_1\mathbf{v}_1 + \alpha_2\mathbf{v}_2 + \ldots + \alpha_m\mathbf{v}_m = \mathbf{0}$$

with $\alpha_1, \alpha_2, \ldots, \alpha_m \in K$ implies that $\alpha_1 = \alpha_2 = \ldots = \alpha_m = 0$. Otherwise, the system will be called *linearly dependent.* ■

We call a *basis* of V any system of linearly independent generators of V. If $\{\mathbf{u}_1, \ldots, \mathbf{u}_n\}$ is a basis of V, the expression $\mathbf{v} = v_1\mathbf{u}_1 + \ldots + v_n\mathbf{u}_n$ is called the *decomposition* of $\mathbf{v}$ with respect to the basis and the scalars $v_1, \ldots, v_n \in K$ are the *components* of $\mathbf{v}$ with respect to the given basis. Moreover, the following property holds.

Property 1.1 *Let V be a vector space which admits a basis of n vectors. Then every system of linearly independent vectors of V has at most n elements and any other basis of V has n elements. The number n is called the dimension of V and we write $dim(V) = n$.*
If, instead, for any n there always exist n linearly independent vectors of V, the vector space is called infinite dimensional.

Example 1.3 For any integer p the space $C^p([a,b])$ is infinite dimensional. The spaces $\mathbb{R}^n$ and $\mathbb{C}^n$ have dimension equal to n. The usual basis for $\mathbb{R}^n$ is the set of *unit vectors* $\{\mathbf{e}_1, \ldots, \mathbf{e}_n\}$ where $(\mathbf{e}_i)_j = \delta_{ij}$ for $i, j = 1, \ldots n$, where δ_{ij} denotes the *Kronecker symbol* equal to 0 if $i \neq j$ and 1 if $i = j$. This choice is of course not the only one that is possible (see Exercise 2). •

1.2 Matrices

Let m and n be two positive integers. We call a *matrix* having m rows and n columns, or a matrix $m \times n$, or a matrix (m,n), with elements in K, a set of mn scalars $a_{ij} \in K$, with $i = 1, \ldots, m$ and $j = 1, \ldots n$, represented in the following rectangular array

$$\mathrm{A} = \begin{bmatrix} a_{11} & a_{12} & \ldots & a_{1n} \\ a_{21} & a_{22} & \ldots & a_{2n} \\ \vdots & \vdots & & \vdots \\ a_{m1} & a_{m2} & \ldots & a_{mn} \end{bmatrix}. \tag{1.1}$$

When $K = \mathbb{R}$ or $K = \mathbb{C}$ we shall respectively write $\mathrm{A} \in \mathbb{R}^{m \times n}$ or $\mathrm{A} \in \mathbb{C}^{m \times n}$, to explicitly outline the numerical fields which the elements of A belong to. Capital letters will be used to denote the matrices, while the lower case letters corresponding to those upper case letters will denote the matrix entries.
We shall abbreviate (1.1) as $\mathrm{A} = (a_{ij})$ with $i = 1, \ldots, m$ and $j = 1, \ldots n$. The index i is called row index, while j is the column index. The set $(a_{i1}, a_{i2}, \ldots, a_{in})$ is called the *i-th row* of A; likewise, $(a_{1j}, a_{2j}, \ldots, a_{mj})$ is the *j-th column* of A.

If $n = m$ the matrix is called *squared* or having order n and the set of the entries $(a_{11}, a_{22}, \ldots, a_{nn})$ is called its *main diagonal*.

A matrix having one row or one column is called a *row vector* or *column vector* respectively. Unless otherwise specified, we shall always assume that a vector is a column vector. In the case $n = m = 1$, the matrix will simply denote a scalar of K.
Sometimes it turns out to be useful to distinguish within a matrix the set made up by specified rows and columns. This prompts us to introduce the following definition.

Definition 1.4 Let A be a matrix $m \times n$. Let $1 \leq i_1 < i_2 < \ldots < i_k \leq m$ and $1 \leq j_1 < j_2 < \ldots < j_l \leq n$ two sets of contiguous indexes. The matrix $\mathrm{S}(k \times l)$

of entries $s_{pq} = a_{i_p j_q}$ with $p = 1, \ldots, k$, $q = 1, \ldots, l$ is called a *submatrix* of A. If $k = l$ and $i_r = j_r$ for $r = 1, \ldots, k$, S is called a *principal submatrix* of A. ■

Definition 1.5 A matrix $A(m \times n)$ is called *block partitioned* or said to be *partitioned into submatrices* if

$$A = \begin{bmatrix} A_{11} & A_{12} & \ldots & A_{1l} \\ A_{21} & A_{22} & \ldots & A_{2l} \\ \vdots & \vdots & \ddots & \vdots \\ A_{k1} & A_{k2} & \ldots & A_{kl} \end{bmatrix},$$

where A_{ij} are submatrices of A. ■

Among the possible partitions of A, we recall in particular the partition by columns

$$A = (\mathbf{a}_1, \ \mathbf{a}_2, \ \ldots, \mathbf{a}_n),$$

$\mathbf{a}_i$ being the i-th column vector of A. In a similar way the partition by rows of A can be defined. To fix the notations, if A is a matrix $m \times n$, we shall denote by

$$A(i_1 : i_2, j_1 : j_2) = (a_{ij}) \ i_1 \leq i \leq i_2, \ j_1 \leq j \leq j_2$$

the submatrix of A of size $(i_2 - i_1 + 1) \times (j_2 - j_1 + 1)$ that lies between the rows i_1 and i_2 and the columns j_1 and j_2. Likewise, if $\mathbf{v}$ is a vector of size n, we shall denote by $\mathbf{v}(i_1 : i_2)$ the vector of size $i_2 - i_1 + 1$ made up by the i_1-th to the i_2-th components of $\mathbf{v}$.

These notations are convenient in view of programming the algorithms that will be presented throughout the volume in the MATLAB language.

1.3 Operations with Matrices

Let $A = (a_{ij})$ and $B = (b_{ij})$ be two matrices $m \times n$ over K. We say that A is *equal* to B, if $a_{ij} = b_{ij}$ for $i = 1, \ldots, m$, $j = 1, \ldots, n$. Moreover, we define the following operations:

- *matrix sum*: the matrix sum is the matrix $A + B = (a_{ij} + b_{ij})$. The neutral element in a matrix sum is the *null matrix*, still denoted by 0 and made up only by null entries;
- *matrix multiplication by a scalar*: the multiplication of A by $\lambda \in K$, is a matrix $\lambda A = (\lambda a_{ij})$;
- *matrix product*: the product of two matrices A and B of sizes (m, p) and (p, n) respectively, is a matrix $C(m, n)$ whose entries are $c_{ij} = \sum_{k=1}^{p} a_{ik} b_{kj}$, for $i = 1, \ldots, m$, $j = 1, \ldots, n$.

The matrix product is associative and distributive with respect to the matrix sum, but it is not in general commutative. The square matrices for which the property AB = BA holds, will be called *commutative.*

In the case of square matrices, the neutral element in the matrix product is a square matrix of order n called the *unit matrix of order* n or, more frequently, the *identity matrix* given by $\mathrm{I}_n = (\delta_{ij})$. The identity matrix is, by definition, the only matrix $n \times n$ such that $\mathrm{AI}_n = \mathrm{I}_n\mathrm{A} = \mathrm{A}$ for all square matrices A. In the following we shall omit the subscript n unless it is strictly necessary. The identity matrix is a special instance of a *diagonal matrix* of order n, that is, a square matrix of the type $\mathrm{D} = (d_{ii}\delta_{ij})$. We will use in the following the notation $\mathrm{D} = \mathrm{diag}(d_{11}, d_{22}, \cdots, d_{nn})$.

Finally, if A is a square matrix of order n and p is an integer, we define A^p as the product of A with itself iterated p times. We let $\mathrm{A}^0 = \mathrm{I}$.

Let us now address the so-called *elementary row operations* that can be performed on a matrix. They consist of:

- multiplying the i-th row of a matrix by a scalar α; this operation is equivalent to pre-multiplying A by the matrix $\mathrm{D} = \mathrm{diag}(1, \ldots, 1, \alpha, 1, \ldots, 1)$, where α occupies the i-th position;
- exchanging the i-th and j-th rows of a matrix; this can be done by pre-multiplying A by the matrix $\mathrm{P}^{(i,j)}$ of elements

$$p_{rs}^{(i,j)} = \begin{cases} 1 & \text{if } r = s = 1, \ldots, i-1, i+1, \ldots, j-1, j+1, \ldots n, \\ 1 & \text{if } r = j, s = i \text{ or } r = i, s = j, \\ 0 & \text{otherwise.} \end{cases} \quad (1.2)$$

  Matrices like (1.2) are called *elementary permutation matrices.* The product of elementary permutation matrices is called a *permutation matrix*, and it performs the row exchanges associated with each elementary permutation matrix. In practice, a permutation matrix is a reordering by rows of the identity matrix;
- adding α times the j-th row of a matrix to its i-th row. This operation can also be performed by pre-multiplying A by the matrix $\mathrm{I} + \mathrm{N}_\alpha^{(i,j)}$, where $\mathrm{N}_\alpha^{(i,j)}$ is a matrix having null entries except the one in position i, j whose value is α.

1.3.1 Inverse of a Matrix

Definition 1.6 A square matrix A of order n is called *invertible* (or *regular* or *nonsingular*) if there exists a square matrix B of order n such that A B = B A = I. B is called the *inverse matrix* of A and is denoted by A^{-1}. A matrix which is not invertible is called *singular.* ■

If A is invertible its inverse is also invertible, with $(\mathrm{A}^{-1})^{-1} = \mathrm{A}$. Moreover, if A and B are two invertible matrices of order n, their product AB is also invertible, with $(\mathrm{A\,B})^{-1} = \mathrm{B}^{-1}\mathrm{A}^{-1}$. The following property holds.

Property 1.2 *A square matrix is invertible iff its column vectors are linearly independent.*

Definition 1.7 We call the *transpose* of a matrix $A \in \mathbb{R}^{m\times n}$ the matrix $n \times m$, denoted by A^T, that is obtained by exchanging the rows of A with the columns of A. ■

Clearly, $(A^T)^T = A$, $(A+B)^T = A^T + B^T$, $(AB)^T = B^T A^T$ and $(\alpha A)^T = \alpha A^T$ $\forall \alpha \in \mathbb{R}$. If A is invertible, then also $(A^T)^{-1} = (A^{-1})^T = A^{-T}$.

Definition 1.8 Let $A \in \mathbb{C}^{m\times n}$; the matrix $B = A^H \in \mathbb{C}^{n\times m}$ is called the *conjugate transpose* (or *adjoint*) of A if $b_{ij} = \bar{a}_{ji}$, where $\bar{a}_{ji}$ is the complex conjugate of a_{ji}. ■

In analogy with the case of the real matrices, it turns out that $(A+B)^H = A^H + B^H$, $(AB)^H = B^H A^H$ and $(\alpha A)^H = \bar{\alpha} A^H$ $\forall \alpha \in \mathbb{C}$.

Definition 1.9 A matrix $A \in \mathbb{R}^{n\times n}$ is called *symmetric* if $A = A^T$, while it is *antisymmetric* if $A = -A^T$. Finally, it is called *orthogonal* if $A^T A = AA^T = I$, that is $A^{-1} = A^T$. ■

Permutation matrices are orthogonal and the same is true for their products.

Definition 1.10 A matrix $A \in \mathbb{C}^{n\times n}$ is called *hermitian* or *self-adjoint* if $A^T = \bar{A}$, that is, if $A^H = A$, while it is called *unitary* if $A^H A = AA^H = I$. Finally, if $AA^H = A^H A$, A is called *normal*. ■

As a consequence, a unitary matrix is one such that $A^{-1} = A^H$.
Of course, a unitary matrix is also normal, but it is not in general hermitian. For instance, the matrix of the Example 1.4 is unitary, although not symmetric (if $s \neq 0$). We finally notice that the diagonal entries of an hermitian matrix must necessarily be real (see also Exercise 5).

1.3.2 Matrices and Linear Mappings

Definition 1.11 A *linear map* from $\mathbb{C}^n$ into $\mathbb{C}^m$ is a function $f : \mathbb{C}^n \longrightarrow \mathbb{C}^m$ such that $f(\alpha \mathbf{x} + \beta \mathbf{y}) = \alpha f(\mathbf{x}) + \beta f(\mathbf{y})$, $\forall \alpha, \beta \in K$ and $\forall \mathbf{x}, \mathbf{y} \in \mathbb{C}^n$. ■

The following result links matrices and linear maps.

Property 1.3 *Let* $f : \mathbb{C}^n \longrightarrow \mathbb{C}^m$ *be a linear map. Then, there exists a unique matrix* $A_f \in \mathbb{C}^{m\times n}$ *such that*

$$f(\mathbf{x}) = A_f \mathbf{x} \qquad \forall \mathbf{x} \in \mathbb{C}^n. \tag{1.3}$$

Conversely, if $A_f \in \mathbb{C}^{m\times n}$ *then the function defined in* (1.3) *is a linear map from* $\mathbb{C}^n$ *into* $\mathbb{C}^m$.

Example 1.4 An important example of a linear map is the counterclockwise *rotation* by an angle ϑ in the plane (x_1, x_2). The matrix associated with such a map is given by

$$\mathrm{G}(\vartheta) = \begin{bmatrix} c & -s \\ s & c \end{bmatrix}, \quad c = \cos(\vartheta), \; s = \sin(\vartheta)$$

and it is called a *rotation matrix*. •

1.3.3 Operations with Block-Partitioned Matrices

All the operations that have been previously introduced can be extended to the case of a block-partitioned matrix A, provided that the size of each single block is such that any single matrix operation is well-defined.

Indeed, the following result can be shown (see, e.g., [Ste73]).

Property 1.4 *Let* A *and* B *be the block matrices*

$$\mathrm{A} = \begin{bmatrix} \mathrm{A}_{11} & \ldots & \mathrm{A}_{1l} \\ \vdots & \ddots & \vdots \\ \mathrm{A}_{k1} & \ldots & \mathrm{A}_{kl} \end{bmatrix}, \mathrm{B} = \begin{bmatrix} \mathrm{B}_{11} & \ldots & \mathrm{B}_{1n} \\ \vdots & \ddots & \vdots \\ \mathrm{B}_{m1} & \ldots & \mathrm{B}_{mn} \end{bmatrix},$$

where A_{ij} *and* B_{ij} *are matrices* $(k_i \times l_j)$ *and* $(m_i \times n_j)$*. Then we have*

1.

$$\lambda\mathrm{A} = \begin{bmatrix} \lambda\mathrm{A}_{11} & \ldots & \lambda\mathrm{A}_{1l} \\ \vdots & \ddots & \vdots \\ \lambda\mathrm{A}_{k1} & \ldots & \lambda\mathrm{A}_{kl} \end{bmatrix}, \quad \lambda \in \mathbb{C}; \; \mathrm{A}^T = \begin{bmatrix} \mathrm{A}_{11}^T & \ldots & \mathrm{A}_{k1}^T \\ \vdots & \ddots & \vdots \\ \mathrm{A}_{1l}^T & \ldots & \mathrm{A}_{kl}^T \end{bmatrix};$$

2. if $k = m$*,* $l = n$*,* $m_i = k_i$ *and* $n_j = l_j$*, then*

$$\mathrm{A} + \mathrm{B} = \begin{bmatrix} \mathrm{A}_{11} + \mathrm{B}_{11} & \ldots & \mathrm{A}_{1l} + \mathrm{B}_{1l} \\ \vdots & \ddots & \vdots \\ \mathrm{A}_{k1} + \mathrm{B}_{k1} & \ldots & \mathrm{A}_{kl} + \mathrm{B}_{kl} \end{bmatrix};$$

3. if $l = m$*,* $l_i = m_i$ *and* $k_i = n_i$*, then, letting* $C_{ij} = \sum_{s=1}^{m} \mathrm{A}_{is}\mathrm{B}_{sj}$*,*

$$\mathrm{AB} = \begin{bmatrix} \mathrm{C}_{11} & \ldots & \mathrm{C}_{1l} \\ \vdots & \ddots & \vdots \\ \mathrm{C}_{k1} & \ldots & \mathrm{C}_{kl} \end{bmatrix}.$$

1.4 Trace and Determinant of a Matrix

Let us consider a square matrix A of order n. The *trace* of a matrix is the sum of the diagonal entries of A, that is $\mathrm{tr(A)} = \sum_{i=1}^{n} a_{ii}$.

We call the *determinant* of A the scalar defined through the following formula

$$\det(\mathrm{A}) = \sum_{\boldsymbol{\pi} \in P} \mathrm{sign}(\boldsymbol{\pi}) a_{1\pi_1} a_{2\pi_2} \ldots a_{n\pi_n},$$

where $P = \left\{\boldsymbol{\pi} = (\pi_1, \ldots, \pi_n)^T\right\}$ is the set of the $n!$ vectors that are obtained by permuting the index vector $\mathbf{i} = (1, \ldots, n)^T$ and $\mathrm{sign}(\boldsymbol{\pi})$ equal to 1 (respectively, -1) if an even (respectively, odd) number of exchanges is needed to obtain $\boldsymbol{\pi}$ from $\mathbf{i}$.

The following properties hold

$$\det(\mathrm{A}) = \det(\mathrm{A}^T),\ \det(\mathrm{AB}) = \det(\mathrm{A})\det(\mathrm{B}),\ \det(\mathrm{A}^{-1}) = 1/\det(\mathrm{A}),$$

$$\det(\mathrm{A}^H) = \overline{\det(\mathrm{A})},\ \det(\alpha \mathrm{A}) = \alpha^n \det(\mathrm{A}),\ \forall \alpha \in K.$$

Moreover, if two rows or columns of a matrix coincide, the determinant vanishes, while exchanging two rows (or two columns) produces a change of sign in the determinant. Of course, the determinant of a diagonal matrix is the product of the diagonal entries.

Denoting by A_{ij} the matrix of order $n-1$ obtained from A by eliminating the i-th row and the j-th column, we call the *complementary minor* associated with the entry a_{ij} the determinant of the matrix A_{ij}. We call the *k-th principal (dominating) minor* of A, d_k, the determinant of the principal submatrix of order k, $\mathrm{A}_k = \mathrm{A}(1:k, 1:k)$. If we denote by $\Delta_{ij} = (-1)^{i+j}\det(\mathrm{A}_{ij})$ the *cofactor* of the entry a_{ij}, the actual computation of the determinant of A can be performed using the following recursive relation

$$\det(\mathrm{A}) = \begin{cases} a_{11} & \text{if } n = 1, \\ \\ \displaystyle\sum_{j=1}^{n} \Delta_{ij} a_{ij}, & \text{for } n > 1, \end{cases} \tag{1.4}$$

which is known as the *Laplace rule*. If A is a square invertible matrix of order n, then

$$\mathrm{A}^{-1} = \frac{1}{\det(\mathrm{A})}\mathrm{C},$$

where C is the matrix having entries Δ_{ji}, $i, j = 1, \ldots, n$.

As a consequence, a square matrix is invertible iff its determinant is nonvanishing. In the case of nonsingular diagonal matrices the inverse is still a diagonal matrix having entries given by the reciprocals of the diagonal entries of the matrix.

Every *orthogonal matrix* is invertible, its inverse is given by A^T, moreover $\det(\mathrm{A}) = \pm 1$.

1.5 Rank and Kernel of a Matrix

Let A be a rectangular matrix $m \times n$. We call the *determinant of order q (with $q \geq 1$) extracted from matrix* A, the determinant of any square matrix of order q obtained from A by eliminating $m - q$ rows and $n - q$ columns.

Definition 1.12 The *rank* of A (denoted by rank(A)) is the maximum order of the nonvanishing determinants extracted from A. A matrix has *complete or full rank* if $\mathrm{rank}(\mathrm{A}) = \min(m,n)$. ■

Notice that the rank of A represents the maximum number of linearly independent column vectors of A that is, the dimension of the *range* of A, defined as

$$\mathrm{range}(\mathrm{A}) = \{\mathbf{y} \in \mathbb{R}^m : \ \mathbf{y} = \mathrm{A}\mathbf{x} \text{ for } \mathbf{x} \in \mathbb{R}^n\}. \tag{1.5}$$

Rigorously speaking, one should distinguish between the column rank of A and the row rank of A, the latter being the maximum number of linearly independent row vectors of A. Nevertheless, it can be shown that the row rank and column rank do actually coincide.

The *kernel* of A is defined as the subspace

$$\ker(\mathrm{A}) = \{\mathbf{x} \in \mathbb{R}^n : \ \mathrm{A}\mathbf{x} = \mathbf{0}\}.$$

The following relations hold:

1. $\mathrm{rank}(\mathrm{A}) = \mathrm{rank}(\mathrm{A}^T)$ (if $\mathrm{A} \in \mathbb{C}^{m \times n}$, $\mathrm{rank}(\mathrm{A}) = \mathrm{rank}(\mathrm{A}^H)$);
2. $\mathrm{rank}(\mathrm{A}) + \dim(\ker(\mathrm{A})) = n$.

In general, $\dim(\ker(\mathrm{A})) \neq \dim(\ker(\mathrm{A}^T))$. If A is a nonsingular square matrix, then $\mathrm{rank}(\mathrm{A}) = n$ and $\dim(\ker(\mathrm{A})) = 0$.

Example 1.5 Let

$$\mathrm{A} = \begin{bmatrix} 1 & 1 & 0 \\ 1 & -1 & 1 \end{bmatrix}.$$

Then, $\mathrm{rank}(\mathrm{A}) = 2$, $\dim(\ker(\mathrm{A})) = 1$ and $\dim(\ker(\mathrm{A}^T)) = 0$. •

We finally notice that for a matrix $A \in \mathbb{C}^{n\times n}$ the following properties are equivalent:

1. A is nonsingular;
2. $\det(A) \neq 0$;
3. $\ker(A) = \{\mathbf{0}\}$;
4. $\mathrm{rank}(A) = n$;
5. A has linearly independent rows and columns.

1.6 Special Matrices

1.6.1 Block Diagonal Matrices

These are matrices of the form $D = \mathrm{diag}(D_1, \ldots, D_n)$, where D_i are square matrices with $i = 1, \ldots, n$. Clearly, each single diagonal block can be of different size. We shall say that a block diagonal matrix has size n if n is the number of its diagonal blocks. The determinant of a block diagonal matrix is given by the product of the determinants of the single diagonal blocks.

1.6.2 Trapezoidal and Triangular Matrices

A matrix $A(m \times n)$ is called *upper trapezoidal* if $a_{ij} = 0$ for $i > j$, while it is *lower trapezoidal* if $a_{ij} = 0$ for $i < j$. The name is due to the fact that, in the case of upper trapezoidal matrices, with $m < n$, the nonzero entries of the matrix form a trapezoid.

A *triangular matrix* is a square trapezoidal matrix of order n of the form

$$L = \begin{bmatrix} l_{11} & 0 & \ldots & 0 \\ l_{21} & l_{22} & \ldots & 0 \\ \vdots & \vdots & & \vdots \\ l_{n1} & l_{n2} & \ldots & l_{nn} \end{bmatrix} \text{ or } U = \begin{bmatrix} u_{11} & u_{12} & \ldots & u_{1n} \\ 0 & u_{22} & \ldots & u_{2n} \\ \vdots & \vdots & & \vdots \\ 0 & 0 & \ldots & u_{nn} \end{bmatrix}.$$

The matrix L is called *lower triangular* while U is *upper triangular*.
Let us recall some algebraic properties of triangular matrices that are easy to check.

- The determinant of a triangular matrix is the product of the diagonal entries;
- the inverse of a lower (respectively, upper) triangular matrix is still lower (respectively, upper) triangular;
- the product of two lower triangular (respectively, upper trapezoidal) matrices is still lower triangular (respectively, upper trapezoidal);
- if we call *unit triangular matrix* a triangular matrix that has diagonal entries equal to 1, then, the product of lower (respectively, upper) unit triangular matrices is still lower (respectively, upper) unit triangular.

1.6.3 Banded Matrices

The matrices introduced in the previous section are a special instance of banded matrices. Indeed, we say that a matrix $A \in \mathbb{R}^{m\times n}$ (or in $\mathbb{C}^{m\times n}$) has *lower band p* if $a_{ij} = 0$ when $i > j + p$ and *upper band q* if $a_{ij} = 0$ when $j > i + q$. Diagonal matrices are banded matrices for which $p = q = 0$, while trapezoidal matrices have $p = m - 1$, $q = 0$ (lower trapezoidal), $p = 0$, $q = n - 1$ (upper trapezoidal).

Other banded matrices of relevant interest are the *tridiagonal matrices* for which $p = q = 1$ and the *upper bidiagonal* ($p = 0$, $q = 1$) or *lower bidiagonal* ($p = 1$, $q = 0$). In the following, $\text{tridiag}_n(\mathbf{b}, \mathbf{d}, \mathbf{c})$ will denote the triadiagonal matrix of size n having respectively on the lower and upper principal diagonals the vectors $\mathbf{b} = (b_1, \ldots, b_{n-1})^T$ and $\mathbf{c} = (c_1, \ldots, c_{n-1})^T$, and on the principal diagonal the vector $\mathbf{d} = (d_1, \ldots, d_n)^T$. If $b_i = \beta$, $d_i = \delta$ and $c_i = \gamma$, β, δ and γ being given constants, the matrix will be denoted by $\text{tridiag}_n(\beta, \delta, \gamma)$.

We also mention the so-called *lower Hessenberg matrices* ($p = m - 1$, $q = 1$) and *upper Hessenberg matrices* ($p = 1$, $q = n - 1$) that have the following structure

$$\mathrm{H} = \begin{bmatrix} h_{11} & h_{12} & & 0 \\ h_{21} & h_{22} & \ddots & \\ \vdots & & \ddots & h_{m-1n} \\ h_{m1} & \ldots & \ldots & h_{mn} \end{bmatrix} \text{ or } \mathrm{H} = \begin{bmatrix} h_{11} & h_{12} & \ldots & h_{1n} \\ h_{21} & h_{22} & & h_{2n} \\ & \ddots & \ddots & \vdots \\ 0 & & h_{mn-1} & h_{mn} \end{bmatrix}.$$

Matrices of similar shape can obviously be set up in the block-like format.

1.7 Eigenvalues and Eigenvectors

Let A be a square matrix of order n with real or complex entries; the number $\lambda \in \mathbb{C}$ is called an *eigenvalue* of A if there exists a nonnull vector $\mathbf{x} \in \mathbb{C}^n$ such that $\mathrm{A}\mathbf{x} = \lambda\mathbf{x}$. The vector $\mathbf{x}$ is the *eigenvector* associated with the eigenvalue λ and the set of the eigenvalues of A is called the *spectrum* of A, denoted by $\sigma(\mathrm{A})$. We say that $\mathbf{x}$ and $\mathbf{y}$ are respectively a *right eigenvector* and a *left eigenvector* of A, associated with the eigenvalue λ, if

$$\mathrm{A}\mathbf{x} = \lambda\mathbf{x}, \; \mathbf{y}^H\mathrm{A} = \lambda\mathbf{y}^H.$$

The eigenvalue λ corresponding to the eigenvector $\mathbf{x}$ can be determined by computing the *Rayleigh quotient* $\lambda = \mathbf{x}^H\mathrm{A}\mathbf{x}/(\mathbf{x}^H\mathbf{x})$. The number λ is the solution of the *characteristic equation*

$$p_{\mathrm{A}}(\lambda) = \det(\mathrm{A} - \lambda\mathrm{I}) = 0,$$

where $p_{\mathrm{A}}(\lambda)$ is the *characteristic polynomial.* Since this latter is a polynomial of degree n with respect to λ, there certainly exist n eigenvalues of A not necessarily distinct. The following properties can be proved

$$\det(\mathrm{A}) = \prod_{i=1}^{n} \lambda_i, \ \mathrm{tr}(\mathrm{A}) = \sum_{i=1}^{n} \lambda_i, \tag{1.6}$$

and since $\det(\mathrm{A}^T - \lambda \mathrm{I}) = \det((\mathrm{A} - \lambda \mathrm{I})^T) = \det(\mathrm{A} - \lambda \mathrm{I})$ one concludes that $\sigma(\mathrm{A}) = \sigma(\mathrm{A}^T)$ and, in an analogous way, that $\sigma(\mathrm{A}^H) = \sigma(\bar{\mathrm{A}})$.

From the first relation in (1.6) it can be concluded that a matrix is singular iff it has at least one null eigenvalue, since $p_\mathrm{A}(0) = \det(\mathrm{A}) = \Pi_{i=1}^n \lambda_i$.

Secondly, if A has real entries, $p_\mathrm{A}(\lambda)$ turns out to be a real-coefficient polynomial so that complex eigenvalues of A shall necessarily occur in complex conjugate pairs.

Finally, due to the Cayley-Hamilton Theorem if $p_\mathrm{A}(\lambda)$ is the characteristic polynomial of A, then $p_\mathrm{A}(\mathrm{A}) = 0$, where $p_\mathrm{A}(\mathrm{A})$ denotes a matrix polynomial (for the proof see, e.g., [Axe94], p. 51).

The maximum module of the eigenvalues of A is called the *spectral radius* of A and is denoted by

$$\rho(\mathrm{A}) = \max_{\lambda \in \sigma(\mathrm{A})} |\lambda|. \tag{1.7}$$

Characterizing the eigenvalues of a matrix as the roots of a polynomial implies in particular that λ is an eigenvalue of $\mathrm{A} \in \mathbb{C}^{n \times n}$ iff $\bar{\lambda}$ is an eigenvalue of A^H. An immediate consequence is that $\rho(\mathrm{A}) = \rho(\mathrm{A}^H)$. Moreover, $\forall \mathrm{A} \in \mathbb{C}^{n \times n}$, $\forall \alpha \in \mathbb{C}$, $\rho(\alpha \mathrm{A}) = |\alpha| \rho(\mathrm{A})$, and $\rho(\mathrm{A}^k) = [\rho(\mathrm{A})]^k \ \forall k \in \mathbb{N}$.

Finally, assume that A is a block triangular matrix

$$\mathrm{A} = \begin{bmatrix} \mathrm{A}_{11} & \mathrm{A}_{12} & \ldots & \mathrm{A}_{1k} \\ 0 & \mathrm{A}_{22} & \ldots & \mathrm{A}_{2k} \\ \vdots & & \ddots & \vdots \\ 0 & \ldots & 0 & \mathrm{A}_{kk} \end{bmatrix}.$$

As $p_\mathrm{A}(\lambda) = p_{\mathrm{A}_{11}}(\lambda) p_{\mathrm{A}_{22}}(\lambda) \cdots p_{\mathrm{A}_{kk}}(\lambda)$, the spectrum of A is given by the union of the spectra of each single diagonal block. As a consequence, if A is triangular, the eigenvalues of A are its diagonal entries.

For each eigenvalue λ of a matrix A the set of the eigenvectors associated with λ, together with the null vector, identifies a subspace of $\mathbb{C}^n$ which is called the *eigenspace* associated with λ and corresponds by definition to ker(A-λI). The dimension of the eigenspace is

$$\dim\left[\ker(\mathrm{A} - \lambda \mathrm{I})\right] = n - \mathrm{rank}(\mathrm{A} - \lambda \mathrm{I}),$$

and is called *geometric multiplicity* of the eigenvalue λ. It can never be greater than the *algebraic multiplicity* of λ, which is the multiplicity of λ as a root of the characteristic polynomial. Eigenvalues having geometric multiplicity strictly less than the algebraic one are called *defective*. A matrix having at least one defective eigenvalue is called *defective*.

The eigenspace associated with an eigenvalue of a matrix A is invariant with respect to A in the sense of the following definition.

Definition 1.13 A subspace S in $\mathbb{C}^n$ is called *invariant* with respect to a square matrix A if $AS \subset S$, where AS is the transformed of S through A. ■

1.8 Similarity Transformations

Definition 1.14 Let C be a square nonsingular matrix having the same order as the matrix A. We say that the matrices A and $C^{-1}AC$ are *similar*, and the transformation from A to $C^{-1}AC$ is called a *similarity transformation*. Moreover, we say that the two matrices are *unitarily similar* if C is unitary. ■

Two similar matrices share the same spectrum and the same characteristic polynomial. Indeed, it is easy to check that if $(\lambda, \mathbf{x})$ is an eigenvalue-eigenvector pair of A, $(\lambda, C^{-1}\mathbf{x})$ is the same for the matrix $C^{-1}AC$ since

$$(C^{-1}AC)C^{-1}\mathbf{x} = C^{-1}A\mathbf{x} = \lambda C^{-1}\mathbf{x}.$$

We notice in particular that the product matrices AB and BA, with $A \in \mathbb{C}^{n\times m}$ and $B \in \mathbb{C}^{m\times n}$, are not similar but satisfy the following property (see [Hac94], p.18, Theorem 2.4.6)

$$\sigma(AB)\setminus\{0\} = \sigma(BA)\setminus\{0\},$$

that is, AB and BA share the same spectrum apart from null eigenvalues so that $\rho(AB) = \rho(BA)$.

The use of similarity transformations aims at reducing the complexity of the problem of evaluating the eigenvalues of a matrix. Indeed, if a given matrix could be transformed into a similar matrix in diagonal or triangular form, the computation of the eigenvalues would be immediate. The main result in this direction is the following theorem (for the proof, see [Dem97], Theorem 4.2).

Property 1.5 (Schur decomposition) *Given* $A \in \mathbb{C}^{n\times n}$, *there exists* U *unitary such that*

$$U^{-1}AU = U^H AU = \begin{bmatrix} \lambda_1 & b_{12} & \ldots & b_{1n} \\ 0 & \lambda_2 & & b_{2n} \\ \vdots & & \ddots & \vdots \\ 0 & \ldots & 0 & \lambda_n \end{bmatrix} = T,$$

where λ_i *are the eigenvalues of* A.

It thus turns out that every matrix A is unitarily similar to an upper triangular matrix. The matrices T and U are not necessarily unique [Hac94]. The Schur decomposition theorem gives rise to several important results; among them, we recall:

1. every hermitian matrix is *unitarily similar* to a diagonal real matrix, that is, when A is hermitian every Schur decomposition of A is diagonal. In such an event, since

$$\mathrm{U}^{-1}\mathrm{AU} = \Lambda = \mathrm{diag}(\lambda_1, \ldots, \lambda_n),$$

   it turns out that $\mathrm{AU} = \mathrm{U}\Lambda$, that is, $\mathbf{A}\mathbf{u}_i = \lambda_i \mathbf{u}_i$ for $i = 1, \ldots, n$ so that the column vectors of U are the eigenvectors of A. Moreover, since the eigenvectors are orthogonal two by two, it turns out that an hermitian matrix has a system of orthonormal eigenvectors that generates the whole space $\mathbb{C}^n$. Finally, it can be shown that a matrix A of order n is similar to a diagonal matrix D iff the eigenvectors of A form a basis for $\mathbb{C}^n$ [Axe94];
2. a matrix $\mathrm{A} \in \mathbb{C}^{n\times n}$ is normal iff it is unitarily similar to a diagonal matrix. As a consequence, a normal matrix $\mathrm{A} \in \mathbb{C}^{n\times n}$ admits the following *spectral decomposition*: $\mathrm{A} = \mathrm{U}\Lambda\mathrm{U}^H = \sum_{i=1}^{n} \lambda_i \mathbf{u}_i \mathbf{u}_i^H$ being U unitary and Λ diagonal [SS90];
3. let A and B be two normal and commutative matrices; then, the generic eigenvalue μ_i of A+B is given by the sum $\lambda_i + \xi_i$, where λ_i and ξ_i are the eigenvalues of A and B associated with the same eigenvector.

There are, of course, nonsymmetric matrices that are similar to diagonal matrices, but these are not unitarily similar (see, e.g., Exercise 7).
The Schur decomposition can be improved as follows (for the proof see, e.g., [Str80], [God66]).

Property 1.6 (Canonical Jordan Form) *Let* A *be any square matrix. Then, there exists a nonsingular matrix* X *which transforms* A *into a block diagonal matrix* J *such that*

$$\mathrm{X}^{-1}\mathrm{AX} = \mathrm{J} = \mathrm{diag}\left(\mathrm{J}_{k_1}(\lambda_1), \mathrm{J}_{k_2}(\lambda_2), \ldots, \mathrm{J}_{k_l}(\lambda_l)\right),$$

which is called canonical Jordan form, λ_j *being the eigenvalues of* A *and* $\mathrm{J}_k(\lambda) \in \mathbb{C}^{k\times k}$ *a Jordan block of the form* $\mathrm{J}_1(\lambda) = \lambda$ *if* $k = 1$ *and*

$$\mathrm{J}_k(\lambda) = \begin{bmatrix} \lambda & 1 & 0 & \ldots & 0 \\ 0 & \lambda & 1 & \cdots & \vdots \\ \vdots & \ddots & \ddots & 1 & 0 \\ \vdots & & \ddots & \lambda & 1 \\ 0 & \ldots & \ldots & 0 & \lambda \end{bmatrix}, \qquad \textit{for } k > 1.$$

If an eigenvalue is defective, the size of the corresponding Jordan block is greater than one. Therefore, the canonical Jordan form tells us that a matrix can be diagonalized by a similarity transformation iff it is nondefective. For this reason, the nondefective matrices are called *diagonalizable*. In particular, normal matrices are diagonalizable.

Partitioning X by columns, $X = (\mathbf{x}_1, \ldots, \mathbf{x}_n)$, it can be seen that the k_i vectors associated with the Jordan block $J_{k_i}(\lambda_i)$ satisfy the following recursive relation

$$\begin{aligned} &A\mathbf{x}_l = \lambda_i \mathbf{x}_l, \qquad l = \sum_{j=1}^{i-1} m_j + 1, \\ &A\mathbf{x}_j = \lambda_i \mathbf{x}_j + \mathbf{x}_{j-1},\ j = l+1, \ldots, l-1+k_i, \ \text{if } k_i \neq 1. \end{aligned} \qquad (1.8)$$

The vectors $\mathbf{x}_i$ are called *principal vectors* or *generalized eigenvectors* of A.

Example 1.6 Let us consider the following matrix

$$A = \begin{bmatrix} 7/4 & 3/4 & -1/4 & -1/4 & -1/4 & 1/4 \\ 0 & 2 & 0 & 0 & 0 & 0 \\ -1/2 & -1/2 & 5/2 & 1/2 & -1/2 & 1/2 \\ -1/2 & -1/2 & -1/2 & 5/2 & 1/2 & 1/2 \\ -1/4 & -1/4 & -1/4 & -1/4 & 11/4 & 1/4 \\ -3/2 & -1/2 & -1/2 & 1/2 & 1/2 & 7/2 \end{bmatrix}.$$

The Jordan canonical form of A and its associated matrix X are given by

$$J = \begin{bmatrix} 2 & 1 & 0 & 0 & 0 & 0 \\ 0 & 2 & 0 & 0 & 0 & 0 \\ 0 & 0 & 3 & 1 & 0 & 0 \\ 0 & 0 & 0 & 3 & 1 & 0 \\ 0 & 0 & 0 & 0 & 3 & 0 \\ 0 & 0 & 0 & 0 & 0 & 2 \end{bmatrix}, \quad X = \begin{bmatrix} 1 & 0 & 0 & 0 & 0 & 1 \\ 0 & 1 & 0 & 0 & 0 & 1 \\ 0 & 0 & 1 & 0 & 0 & 1 \\ 0 & 0 & 0 & 1 & 0 & 1 \\ 0 & 0 & 0 & 0 & 1 & 1 \\ 1 & 1 & 1 & 1 & 1 & 1 \end{bmatrix}.$$

Notice that two different Jordan blocks are related to the same eigenvalue ($\lambda = 2$). It is easy to check property (1.8). Consider, for example, the Jordan block associated with the eigenvalue $\lambda_2 = 3$; we have

$$\begin{aligned} A\mathbf{x}_3 &= [0\ 0\ 3\ 0\ 0\ 3]^T = 3\,[0\ 0\ 1\ 0\ 0\ 1]^T = \lambda_2 \mathbf{x}_3, \\ A\mathbf{x}_4 &= [0\ 0\ 1\ 3\ 0\ 4]^T = 3\,[0\ 0\ 0\ 1\ 0\ 1]^T + [0\ 0\ 1\ 0\ 0\ 1]^T = \lambda_2 \mathbf{x}_4 + \mathbf{x}_3, \\ A\mathbf{x}_5 &= [0\ 0\ 0\ 1\ 3\ 4]^T = 3\,[0\ 0\ 0\ 0\ 1\ 1]^T + [0\ 0\ 0\ 1\ 0\ 1]^T = \lambda_2 \mathbf{x}_5 + \mathbf{x}_4. \end{aligned}$$

•

1.9 The Singular Value Decomposition (SVD)

Any matrix can be reduced in diagonal form by a suitable pre and post-multiplication by unitary matrices. Precisely, the following result holds.

Property 1.7 *Let* $A \in \mathbb{C}^{m\times n}$. *There exist two unitary matrices* $U \in \mathbb{C}^{m\times m}$ *and* $V \in \mathbb{C}^{n\times n}$ *such that*

$$U^H A V = \Sigma = \text{diag}(\sigma_1, \ldots, \sigma_p) \in \mathbb{R}^{m\times n} \qquad \textit{with } p = \min(m, n) \qquad (1.9)$$

and $\sigma_1 \geq \ldots \geq \sigma_p \geq 0$. *Formula* (1.9) *is called Singular Value Decomposition or (SVD) of* A *and the numbers* σ_i *(or* $\sigma_i(A)$*) are called singular values of* A.

If A is a real-valued matrix, U and V will also be real-valued and in (1.9) U^T must be written instead of U^H. The following characterization of the singular values holds

$$\sigma_i(A) = \sqrt{\lambda_i(A^H A)}, \quad i = 1, \dots, p. \tag{1.10}$$

Indeed, from (1.9) it follows that $A = U\Sigma V^H$, $A^H = V\Sigma^H U^H$ so that, U and V being unitary, $A^H A = V\Sigma^H \Sigma V^H$, that is, $\lambda_i(A^H A) = \lambda_i(\Sigma^H \Sigma) = (\sigma_i(A))^2$. Since AA^H and $A^H A$ are hermitian matrices, the columns of U, called the *left singular vectors* of A, turn out to be the eigenvectors of AA^H (see Section 1.8) and, therefore, they are not uniquely defined. The same holds for the columns of V, which are the *right singular vectors* of A.

Relation (1.10) implies that if $A \in \mathbb{C}^{n\times n}$ is hermitian with eigenvalues given by $\lambda_1, \lambda_2, \dots, \lambda_n$, then the singular values of A coincide with the modules of the eigenvalues of A. Indeed because $AA^H = A^2$, $\sigma_i = \sqrt{\lambda_i^2} = |\lambda_i|$ for $i = 1, \dots, n$. As far as the rank is concerned, if

$$\sigma_1 \geq \dots \geq \sigma_r > \sigma_{r+1} = \dots = \sigma_p = 0,$$

then the rank of A is r, the kernel of A is the span of the column vectors of V, $\{\mathbf{v}_{r+1}, \dots, \mathbf{v}_n\}$, and the range of A is the span of the column vectors of U, $\{\mathbf{u}_1, \dots, \mathbf{u}_r\}$.

Definition 1.15 Suppose that $A \in \mathbb{C}^{m\times n}$ has rank equal to r and that it admits a SVD of the type $U^H AV = \Sigma$. The matrix $A^\dagger = V\Sigma^\dagger U^H$ is called the *Moore-Penrose pseudo-inverse* matrix, being

$$\Sigma^\dagger = \mathrm{diag}\left(\frac{1}{\sigma_1}, \dots, \frac{1}{\sigma_r}, 0, \dots, 0\right). \tag{1.11}$$

■

The matrix $A^\dagger$ is also called the *generalized inverse* of A (see Exercise 13). Indeed, if $\mathrm{rank}(A) = n < m$, then $A^\dagger = (A^T A)^{-1} A^T$, while if $n = m = \mathrm{rank}(A)$, $A^\dagger = A^{-1}$. For further properties of $A^\dagger$, see also Exercise 12.

1.10 Scalar Product and Norms in Vector Spaces

Very often, to quantify errors or measure distances one needs to compute the magnitude of a vector or a matrix. For that purpose we introduce in this section the concept of a vector norm and, in the following one, of a matrix norm. We refer the reader to [Ste73], [SS90] and [Axe94] for the proofs of the properties that are reported hereafter.

Definition 1.16 A *scalar product* on a vector space V defined over K is any map $(\cdot, \cdot)$ acting from $V \times V$ into K which enjoys the following properties:

1. it is linear with respect to the vectors of V, that is
$$(\gamma\mathbf{x} + \lambda\mathbf{z}, \mathbf{y}) = \gamma(\mathbf{x}, \mathbf{y}) + \lambda(\mathbf{z}, \mathbf{y}), \ \forall\mathbf{x}, \mathbf{y}, \mathbf{z} \in V, \ \forall\gamma, \lambda \in K;$$
2. it is *hermitian*, that is, $(\mathbf{y}, \mathbf{x}) = \overline{(\mathbf{x}, \mathbf{y})}$, $\forall\mathbf{x}, \mathbf{y} \in V$;
3. it is *positive definite*, that is, $(\mathbf{x}, \mathbf{x}) > 0$, $\forall\mathbf{x} \neq \mathbf{0}$ (in other words, $(\mathbf{x}, \mathbf{x}) \geq 0$, and $(\mathbf{x}, \mathbf{x}) = 0$ if and only if $\mathbf{x} = \mathbf{0}$).

■

In the case $V = \mathbb{C}^n$ (or $\mathbb{R}^n$), an example is provided by the classical Euclidean scalar product given by
$$(\mathbf{x}, \mathbf{y}) = \mathbf{y}^H\mathbf{x} = \sum_{i=1}^{n} x_i\bar{y}_i,$$
where $\bar{z}$ denotes the complex conjugate of z.

Moreover, for any given square matrix A of order n and for any $\mathbf{x}$, $\mathbf{y} \in \mathbb{C}^n$ the following relation holds
$$(\mathrm{A}\mathbf{x}, \mathbf{y}) = (\mathbf{x}, \mathrm{A}^H\mathbf{y}). \tag{1.12}$$
In particular, since for any matrix $\mathrm{Q} \in \mathbb{C}^{n\times n}$, $(\mathrm{Q}\mathbf{x}, \mathrm{Q}\mathbf{y}) = (\mathbf{x}, \mathrm{Q}^H\mathrm{Q}\mathbf{y})$, one gets

Property 1.8 *Unitary matrices preserve the Euclidean scalar product, that is,* $(\mathrm{Q}\mathbf{x}, \mathrm{Q}\mathbf{y}) = (\mathbf{x}, \mathbf{y})$ *for any unitary matrix* Q *and for any pair of vectors* $\mathbf{x}$ *and* $\mathbf{y}$.

Definition 1.17 Let V be a vector space over K. We say that the map $\|\cdot\|$ from V into $\mathbb{R}$ is a *norm* on V if the following axioms are satisfied:

1. (*i*) $\|\mathbf{v}\| \geq 0$ $\forall\mathbf{v} \in V$ and (*ii*) $\|\mathbf{v}\| = 0$ if and only if $\mathbf{v} = \mathbf{0}$;
2. $\|\alpha\mathbf{v}\| = |\alpha|\|\mathbf{v}\|$ $\forall\alpha \in K, \forall\mathbf{v} \in V$ (homogeneity property);
3. $\|\mathbf{v} + \mathbf{w}\| \leq \|\mathbf{v}\| + \|\mathbf{w}\|$ $\forall\mathbf{v}, \mathbf{w} \in V$ (triangular inequality),

where $|\alpha|$ denotes the absolute value of α if $K = \mathbb{R}$, the module of α if $K = \mathbb{C}$. ■

The pair $(V, \|\cdot\|)$ is called a *normed space*. We shall distinguish among norms by a suitable subscript at the margin of the double bar symbol. In the case the map $|\cdot|$ from V into $\mathbb{R}$ enjoys only the properties 1(*i*), 2 and 3 we shall call such a map a *seminorm*. Finally, we shall call a *unit vector* any vector of V having unit norm.
An example of a normed space is $\mathbb{R}^n$, equipped for instance by the *p-norm* (or *Hölder norm*); this latter is defined for a vector $\mathbf{x}$ of components $\{x_i\}$ as
$$\|\mathbf{x}\|_p = \left(\sum_{i=1}^{n}|x_i|^p\right)^{1/p}, \qquad \text{for } 1 \leq p < \infty. \tag{1.13}$$

Notice that the limit as p goes to infinity of $\|\mathbf{x}\|_p$ exists, is finite, and equals the maximum module of the components of $\mathbf{x}$. Such a limit defines in turn a norm, called the *infinity norm* (or *maximum norm*), given by

$$\|\mathbf{x}\|_\infty = \max_{1\leq i\leq n} |x_i|.$$

When $p = 2$, from (1.13) the standard definition of *Euclidean norm* is recovered

$$\|\mathbf{x}\|_2 = (\mathbf{x},\mathbf{x})^{1/2} = \left(\sum_{i=1}^{n} |x_i|^2\right)^{1/2} = \left(\mathbf{x}^T\mathbf{x}\right)^{1/2},$$

for which the following property holds.

Property 1.9 (Cauchy-Schwarz inequality) *For any pair* $\mathbf{x},\mathbf{y} \in \mathbb{R}^n$,

$$|(\mathbf{x},\mathbf{y})| = |\mathbf{x}^T\mathbf{y}| \leq \|\mathbf{x}\|_2\ \|\mathbf{y}\|_2, \tag{1.14}$$

where strict equality holds iff $\mathbf{y} = \alpha\mathbf{x}$ *for some* $\alpha \in \mathbb{R}$.

We recall that the scalar product in $\mathbb{R}^n$ can be related to the p-norms introduced over $\mathbb{R}^n$ in (1.13) by the *Hölder inequality*

$$|(\mathbf{x},\mathbf{y})| \leq \|\mathbf{x}\|_p \|\mathbf{y}\|_q, \ \text{with}\ \frac{1}{p} + \frac{1}{q} = 1.$$

In the case where V is a finite-dimensional space the following property holds (for a sketch of the proof, see Exercise 14).

Property 1.10 *Any vector norm* $\|\cdot\|$ *defined on* V *is a continuous function of its argument, namely,* $\forall\varepsilon > 0$, $\exists C > 0$ *such that if* $\|\mathbf{x} - \widehat{\mathbf{x}}\| \leq \varepsilon$ *then* $|\ \|\mathbf{x}\| - \|\widehat{\mathbf{x}}\|\ | \leq C\varepsilon$, *for any* $\mathbf{x}$, $\widehat{\mathbf{x}} \in V$.

New norms can be easily built using the following result.

Property 1.11 *Let* $\|\cdot\|$ *be a norm of* $\mathbb{R}^n$ *and* $\mathrm{A} \in \mathbb{R}^{n\times n}$ *be a matrix with* n *linearly independent columns. Then, the function* $\|\cdot\|_{\mathrm{A}^2}$ *acting from* $\mathbb{R}^n$ *into* $\mathbb{R}$ *defined as*

$$\|\mathbf{x}\|_{\mathrm{A}^2} = \|\mathrm{A}\mathbf{x}\| \qquad \forall\mathbf{x} \in \mathbb{R}^n,$$

is a norm of $\mathbb{R}^n$.

Two vectors $\mathbf{x}$, $\mathbf{y}$ in V are said to be *orthogonal* if $(\mathbf{x},\mathbf{y}) = 0$. This statement has an immediate geometric interpretation when $V = \mathbb{R}^2$ since in such a case

$$(\mathbf{x},\mathbf{y}) = \|\mathbf{x}\|_2\|\mathbf{y}\|_2 \cos(\vartheta),$$

Table 1.1. Equivalence constants for the main norms of $\mathbb{R}^n$

c_{pq}	$q=1$	$q=2$	$q=\infty$
$p=1$	1	1	1
$p=2$	$n^{-1/2}$	1	1
$p=\infty$	n^{-1}	$n^{-1/2}$	1

C_{pq}	$q=1$	$q=2$	$q=\infty$
$p=1$	1	$n^{1/2}$	n
$p=2$	1	1	$n^{1/2}$
$p=\infty$	1	1	1

where ϑ is the angle between the vectors $\mathbf{x}$ and $\mathbf{y}$. As a consequence, if $(\mathbf{x},\mathbf{y}) = 0$ then ϑ is a right angle and the two vectors are orthogonal in the geometric sense.

Definition 1.18 Two norms $\|\cdot\|_p$ and $\|\cdot\|_q$ on V are *equivalent* if there exist two positive constants c_{pq} and C_{pq} such that

$$c_{pq}\|\mathbf{x}\|_q \le \|\mathbf{x}\|_p \le C_{pq}\|\mathbf{x}\|_q \;\forall \mathbf{x} \in V.$$

■

In a finite-dimensional normed space all norms are equivalent. In particular, if $V = \mathbb{R}^n$ it can be shown that for the p-norms, with $p = 1$, 2, and ∞, the constants c_{pq} and C_{pq} take the value reported in Table 1.1.

In this book we shall often deal with sequences of vectors and with their *convergence*. For this purpose, we recall that a sequence of vectors $\{\mathbf{x}^{(k)}\}$ in a vector space V having finite dimension n, converges to a vector $\mathbf{x}$, and we write $\lim_{k\to\infty} \mathbf{x}^{(k)} = \mathbf{x}$ if

$$\lim_{k\to\infty} x_i^{(k)} = x_i,\; i = 1,\ldots,n, \tag{1.15}$$

where $x_i^{(k)}$ and x_i are the components of the corresponding vectors with respect to a basis of V. If $V = \mathbb{R}^n$, due to the uniqueness of the limit of a sequence of real numbers, (1.15) implies also the uniqueness of the limit, if existing, of a sequence of vectors.

We further notice that in a finite-dimensional space all the norms are topologically equivalent in the sense of convergence, namely, given a sequence of vectors $\mathbf{x}^{(k)}$, we have that

$$|||\mathbf{x}^{(k)}||| \to 0 \;\Leftrightarrow\; \|\mathbf{x}^{(k)}\| \to 0 \text{ if } k \to \infty,$$

where $|||\cdot|||$ and $\|\cdot\|$ are any two vector norms. As a consequence, we can establish the following link between norms and limits.

Property 1.12 *Let $\|\cdot\|$ be a norm in a finite dimensional space V. Then*

$$\lim_{k\to\infty} \mathbf{x}^{(k)} = \mathbf{x} \Leftrightarrow \lim_{k\to\infty} \|\mathbf{x} - \mathbf{x}^{(k)}\| = 0,$$

where $\mathbf{x} \in V$ and $\{\mathbf{x}^{(k)}\}$ is a sequence of elements of V.

1.11 Matrix Norms

Definition 1.19 A *matrix norm* is a mapping $\|\cdot\| : \mathbb{R}^{m\times n} \to \mathbb{R}$ such that:

1. $\|\mathrm{A}\| \geq 0\ \forall \mathrm{A} \in \mathbb{R}^{m\times n}$ and $\|\mathrm{A}\| = 0$ if and only if $\mathrm{A} = 0$;
2. $\|\alpha\mathrm{A}\| = |\alpha|\|\mathrm{A}\|\quad \forall\alpha \in \mathbb{R}, \forall \mathrm{A} \in \mathbb{R}^{m\times n}$ (homogeneity);
3. $\|\mathrm{A} + \mathrm{B}\| \leq \|\mathrm{A}\| + \|\mathrm{B}\|\quad \forall \mathrm{A}, \mathrm{B} \in \mathbb{R}^{m\times n}$ (triangular inequality).

■

Unless otherwise specified we shall employ the same symbol $\|\cdot\|$, to denote matrix norms and vector norms.

We can better characterize the matrix norms by introducing the concepts of compatible norm and norm induced by a vector norm.

Definition 1.20 We say that a matrix norm $\|\cdot\|$ is *compatible* or *consistent* with a vector norm $\|\cdot\|$ if

$$\|\mathrm{A}\mathbf{x}\| \leq \|\mathrm{A}\|\ \|\mathbf{x}\|, \qquad \forall \mathbf{x} \in \mathbb{R}^n. \tag{1.16}$$

More generally, given three norms, all denoted by $\|\cdot\|$, albeit defined on $\mathbb{R}^m$, $\mathbb{R}^n$ and $\mathbb{R}^{m\times n}$, respectively, we say that they are consistent if $\forall \mathbf{x} \in \mathbb{R}^n$, $\mathrm{A}\mathbf{x} = \mathbf{y} \in \mathbb{R}^m$, $\mathrm{A} \in \mathbb{R}^{m\times n}$, we have that $\|\mathbf{y}\| \leq \|\mathrm{A}\|\ \|\mathbf{x}\|$. ■

In order to single out matrix norms of practical interest, the following property is in general required

Definition 1.21 We say that a matrix norm $\|\cdot\|$ is *sub-multiplicative* if $\forall \mathrm{A} \in \mathbb{R}^{n\times m}$, $\forall \mathrm{B} \in \mathbb{R}^{m\times q}$

$$\|\mathrm{AB}\| \leq \|\mathrm{A}\|\ \|\mathrm{B}\|. \tag{1.17}$$

■

This property is not satisfied by any matrix norm. For example (taken from [GL89]), the norm $\|\mathrm{A}\|_\Delta = \max|a_{ij}|$ for $i = 1, \ldots, n$, $j = 1, \ldots, m$ does not satisfy (1.17) if applied to the matrices

$$\mathrm{A} = \mathrm{B} = \begin{bmatrix} 1\ 1 \\ 1\ 1 \end{bmatrix},$$

since $2 = \|\mathrm{AB}\|_\Delta > \|\mathrm{A}\|_\Delta\|\mathrm{B}\|_\Delta = 1$.
Notice that, given a certain sub-multiplicative matrix norm $\|\cdot\|_\alpha$, there always exists a consistent vector norm. For instance, given any fixed vector $\mathbf{y} \neq \mathbf{0}$ in $\mathbb{C}^n$, it suffices to define the consistent vector norm as

$$\|\mathbf{x}\| = \|\mathbf{x}\mathbf{y}^H\|_\alpha \qquad \mathbf{x} \in \mathbb{C}^n.$$

As a consequence, in the case of sub-multiplicative matrix norms it is no longer necessary to explicitly specify the vector norm with respect to the matrix norm is consistent.

Example 1.7 The norm

$$\|A\|_F = \sqrt{\sum_{i,j=1}^{n} |a_{ij}|^2} = \sqrt{\mathrm{tr}(AA^H)} \tag{1.18}$$

is a matrix norm called the *Frobenius norm* (or *Euclidean norm* in $\mathbb{C}^{n^2}$) and is compatible with the Euclidean vector norm $\|\cdot\|_2$. Indeed,

$$\|\mathbf{Ax}\|_2^2 = \sum_{i=1}^{n} \left| \sum_{j=1}^{n} a_{ij} x_j \right|^2 \leq \sum_{i=1}^{n} \left(\sum_{j=1}^{n} |a_{ij}|^2 \sum_{j=1}^{n} |x_j|^2 \right) = \|A\|_F^2 \|\mathbf{x}\|_2^2.$$

Notice that for such a norm $\|I_n\|_F = \sqrt{n}$. •

In view of the definition of a natural norm, we recall the following theorem.

Theorem 1.1 *Let* $\|\cdot\|$ *be a vector norm. The function*

$$\|A\| = \sup_{\mathbf{x} \neq \mathbf{0}} \frac{\|\mathbf{Ax}\|}{\|\mathbf{x}\|} \tag{1.19}$$

is a matrix norm called induced matrix norm or natural matrix norm.

Proof. We start by noticing that (1.19) is equivalent to

$$\|A\| = \sup_{\|\mathbf{x}\|=1} \|\mathbf{Ax}\|. \tag{1.20}$$

Indeed, one can define for any $\mathbf{x} \neq \mathbf{0}$ the unit vector $\mathbf{u} = \mathbf{x}/\|\mathbf{x}\|$, so that (1.19) becomes

$$\|A\| = \sup_{\|\mathbf{u}\|=1} \|\mathbf{Au}\| = \|\mathbf{Aw}\| \qquad \text{with } \|\mathbf{w}\| = 1.$$

This being taken as given, let us check that (1.19) (or, equivalently, (1.20)) is actually a norm, making direct use of Definition 1.19.

1. If $\|\mathbf{Ax}\| \geq 0$, then it follows that $\|A\| = \sup_{\|\mathbf{x}\|=1} \|\mathbf{Ax}\| \geq 0$. Moreover

$$\|A\| = \sup_{\mathbf{x} \neq \mathbf{0}} \frac{\|\mathbf{Ax}\|}{\|\mathbf{x}\|} = 0 \Leftrightarrow \|\mathbf{Ax}\| = 0 \ \forall \mathbf{x} \neq \mathbf{0},$$

and $A\mathbf{x} = \mathbf{0}\ \forall \mathbf{x} \neq \mathbf{0}$ if and only if A=0; therefore $\|A\| = 0 \Leftrightarrow A = 0$.
2. Given a scalar α,

$$\|\alpha A\| = \sup_{\|\mathbf{x}\|=1} \|\alpha \mathbf{Ax}\| = |\alpha| \sup_{\|\mathbf{x}\|=1} \|\mathbf{Ax}\| = |\alpha|\ \|A\|.$$

3. Finally, triangular inequality holds. Indeed, by definition of supremum, if $\mathbf{x} \neq \mathbf{0}$ then

$$\frac{\|\mathbf{Ax}\|}{\|\mathbf{x}\|} \leq \|A\| \Rightarrow \|\mathbf{Ax}\| \leq \|A\| \|\mathbf{x}\|,$$

so that, taking $\mathbf{x}$ with unit norm, one gets

$$\|(A + B)\mathbf{x}\| \leq \|\mathbf{Ax}\| + \|\mathbf{Bx}\| \leq \|A\| + \|B\|,$$

from which it follows that $\|A + B\| = \sup_{\|\mathbf{x}\|=1} \|(A + B)\mathbf{x}\| \leq \|A\| + \|B\|$.

◇

Relevant instances of induced matrix norms are the so-called *p-norms* defined as

$$\|A\|_p = \sup_{\mathbf{x}\neq \mathbf{0}} \frac{\|A\mathbf{x}\|_p}{\|\mathbf{x}\|_p}.$$

The 1-norm and the infinity norm are easily computable since

$$\|A\|_1 = \max_{j=1,\ldots,n} \sum_{i=1}^{m} |a_{ij}|, \ \|A\|_\infty = \max_{i=1,\ldots,m} \sum_{j=1}^{n} |a_{ij}|,$$

and they are called the *column sum norm* and the *row sum norm*, respectively.

Moreover, we have $\|A\|_1 = \|A^T\|_\infty$ and, if A is self-adjoint or real symmetric, $\|A\|_1 = \|A\|_\infty$.

A special discussion is deserved by the *2-norm* or *spectral norm* for which the following theorem holds.

Theorem 1.2 *Let $\sigma_1(A)$ be the largest singular value of* A. *Then*

$$\|A\|_2 = \sqrt{\rho(A^H A)} = \sqrt{\rho(AA^H)} = \sigma_1(A). \tag{1.21}$$

In particular, if A *is hermitian (or real and symmetric), then*

$$\|A\|_2 = \rho(A), \tag{1.22}$$

while, if A *is unitary,* $\|A\|_2 = 1$.

Proof. Since $A^H A$ is hermitian, there exists a unitary matrix U such that

$$U^H A^H A U = \text{diag}(\mu_1, \ldots, \mu_n),$$

where μ_i are the (positive) eigenvalues of $A^H A$. Let $\mathbf{y} = U^H \mathbf{x}$, then

$$\|A\|_2 = \sup_{\mathbf{x}\neq\mathbf{0}} \sqrt{\frac{(A^H A\mathbf{x}, \mathbf{x})}{(\mathbf{x}, \mathbf{x})}} = \sup_{\mathbf{y}\neq\mathbf{0}} \sqrt{\frac{(U^H A^H A U\mathbf{y}, \mathbf{y})}{(\mathbf{y}, \mathbf{y})}}$$

$$= \sup_{\mathbf{y}\neq\mathbf{0}} \sqrt{\sum_{i=1}^{n} \mu_i |y_i|^2 / \sum_{i=1}^{n} |y_i|^2} = \sqrt{\max_{i=1,\ldots,n} |\mu_i|},$$

from which (1.21) follows, thanks to (1.10).

If A is hermitian, the same considerations as above apply directly to A.

Finally, if A is unitary, we have

$$\|A\mathbf{x}\|_2^2 = (A\mathbf{x}, A\mathbf{x}) = (\mathbf{x}, A^H A\mathbf{x}) = \|\mathbf{x}\|_2^2,$$

so that $\|A\|_2 = 1$. ◇

As a consequence, the computation of $\|A\|_2$ is much more expensive than that of $\|A\|_\infty$ or $\|A\|_1$. However, if only an estimate of $\|A\|_2$ is required, the following relations can be profitably employed in the case of square matrices

$$\max_{i,j}|a_{ij}| \leq \|A\|_2 \leq n \max_{i,j}|a_{ij}|,$$
$$\frac{1}{\sqrt{n}}\|A\|_\infty \leq \|A\|_2 \leq \sqrt{n}\|A\|_\infty,$$
$$\frac{1}{\sqrt{n}}\|A\|_1 \leq \|A\|_2 \leq \sqrt{n}\|A\|_1,$$
$$\|A\|_2 \leq \sqrt{\|A\|_1 \, \|A\|_\infty}.$$

For other estimates of similar type we refer to Exercise 17. Moreover, if A is normal then $\|A\|_2 \leq \|A\|_p$ for any n and all $p \geq 2$.

Theorem 1.3 *Let $|||\cdot|||$ be a matrix norm induced by a vector norm $\|\cdot\|$. Then, the following relations hold:*

1. $\|Ax\| \leq |||A||| \, \|x\|$, *that is, $|||\cdot|||$ is a norm compatible with $\|\cdot\|$;*
2. $|||I||| = 1$;
3. $|||AB||| \leq |||A||| \, |||B|||$, *that is, $|||\cdot|||$ is sub-multiplicative.*

Proof. Part 1 of the theorem is already contained in the proof of Theorem 1.1, while part 2 follows from the fact that $|||I||| = \sup_{x \neq 0}\|Ix\|/\|x\| = 1$. Part 3 is simple to check. ◇

Notice that the p-norms are sub-multiplicative. Moreover, we remark that the sub-multiplicativity property by itself would only allow us to conclude that $|||I||| \geq 1$. Indeed, $|||I||| = |||I \cdot I||| \leq |||I|||^2$.

1.11.1 Relation between Norms and the Spectral Radius of a Matrix

We next recall some results that relate the spectral radius of a matrix to matrix norms and that will be widely employed in Chapter 4.

Theorem 1.4 *Let $\|\cdot\|$ be a consistent matrix norm; then*

$$\rho(A) \leq \|A\| \qquad \forall A \in \mathbb{C}^{n\times n}.$$

Proof. Let λ be an eigenvalue of A and $\mathbf{v} \neq \mathbf{0}$ an associated eigenvector. As a consequence, since $\|\cdot\|$ is consistent, we have

$$|\lambda| \, \|\mathbf{v}\| = \|\lambda\mathbf{v}\| = \|A\mathbf{v}\| \leq \|A\| \, \|\mathbf{v}\|,$$

so that $|\lambda| \leq \|A\|$. ◇

More precisely, the following property holds (see for the proof [IK66], p. 12, Theorem 3).

Property 1.13 *Let $A \in \mathbb{C}^{n\times n}$ and $\varepsilon > 0$. Then, there exists an induced matrix norm $\|\cdot\|_{A,\varepsilon}$ (depending on ε) such that*

$$\|A\|_{A,\varepsilon} \leq \rho(A) + \varepsilon.$$

As a result, having fixed an arbitrarily small tolerance, there always exists a matrix norm which is arbitrarily close to the spectral radius of A, namely

$$\rho(\mathrm{A}) = \inf_{\|\cdot\|} \|\mathrm{A}\|, \tag{1.23}$$

the infimum being taken on the set of all the consistent norms.

For the sake of clarity, we notice that the spectral radius is a sub-multiplicative seminorm, since it is not true that $\rho(\mathrm{A}) = 0$ iff $\mathrm{A} = 0$. As an example, any triangular matrix with null diagonal entries clearly has spectral radius equal to zero. Moreover, we have the following result.

Property 1.14 *Let* A *be a square matrix and let* $\|\cdot\|$ *be a consistent norm. Then*

$$\lim_{m\to\infty} \|\mathrm{A}^m\|^{1/m} = \rho(\mathrm{A}).$$

1.11.2 Sequences and Series of Matrices

A sequence of matrices $\left\{\mathrm{A}^{(k)}\right\} \in \mathbb{R}^{n\times n}$ is said to *converge* to a matrix $\mathrm{A} \in \mathbb{R}^{n\times n}$ if

$$\lim_{k\to\infty} \|\mathrm{A}^{(k)} - \mathrm{A}\| = 0.$$

The choice of the norm does not influence the result since in $\mathbb{R}^{n\times n}$ all norms are equivalent. In particular, when studying the convergence of iterative methods for solving linear systems (see Chapter 4), one is interested in the so-called *convergent matrices* for which

$$\lim_{k\to\infty} \mathrm{A}^k = 0,$$

0 being the null matrix. The following theorem holds.

Theorem 1.5 *Let* A *be a square matrix; then*

$$\lim_{k\to\infty} \mathrm{A}^k = 0 \Leftrightarrow \rho(\mathrm{A}) < 1. \tag{1.24}$$

Moreover, the geometric series $\sum_{k=0}^{\infty} \mathrm{A}^k$ *is convergent iff* $\rho(\mathrm{A}) < 1$. *In such a case*

$$\sum_{k=0}^{\infty} \mathrm{A}^k = (\mathrm{I} - \mathrm{A})^{-1}. \tag{1.25}$$

As a result, if $\rho(\mathrm{A}) < 1$ *the matrix* $\mathrm{I} - \mathrm{A}$ *is invertible and the following inequalities hold*

$$\frac{1}{1 + \|\mathrm{A}\|} \le \|(\mathrm{I} - \mathrm{A})^{-1}\| \le \frac{1}{1 - \|\mathrm{A}\|}, \tag{1.26}$$

where $\|\cdot\|$ *is an induced matrix norm such that* $\|\mathrm{A}\| < 1$.

Proof. Let us prove (1.24). Let $\rho(\mathrm{A}) < 1$, then $\exists\varepsilon > 0$ such that $\rho(\mathrm{A}) < 1 - \varepsilon$ and thus, thanks to Property 1.13, there exists an induced matrix norm $\|\cdot\|$ such that $\|\mathrm{A}\| \le \rho(\mathrm{A}) + \varepsilon < 1$. From the fact that $\|\mathrm{A}^k\| \le \|\mathrm{A}\|^k < 1$ and from the definition of convergence it turns out that as $k \to \infty$ the sequence $\{\mathrm{A}^k\}$ tends to zero. Conversely, assume that $\lim_{k\to\infty} \mathrm{A}^k = 0$ and let λ denote an eigenvalue of A. Then, $\mathrm{A}^k\mathbf{x} = \lambda^k\mathbf{x}$, being $\mathbf{x}(\neq\mathbf{0})$ an eigenvector associated with λ, so that $\lim_{k\to\infty} \lambda^k = 0$. As a consequence, $|\lambda| < 1$ and because this is true for a generic eigenvalue one gets $\rho(\mathrm{A}) < 1$ as desired. Relation (1.25) can be obtained noting first that the eigenvalues of I−A are given by $1-\lambda(\mathrm{A})$, $\lambda(\mathrm{A})$ being the generic eigenvalue of A. On the other hand, since $\rho(\mathrm{A}) < 1$, we deduce that I−A is nonsingular. Then, from the identity

$$(\mathrm{I} - \mathrm{A})(\mathrm{I} + \mathrm{A} + \ldots + \mathrm{A}^n) = (\mathrm{I} - \mathrm{A}^{n+1})$$

and taking the limit for n tending to infinity the thesis follows since

$$(\mathrm{I} - \mathrm{A})\sum_{k=0}^{\infty} \mathrm{A}^k = \mathrm{I}.$$

Finally, thanks to Theorem 1.3, the equality $\|\mathrm{I}\| = 1$ holds, so that

$$1 = \|\mathrm{I}\| \le \|\mathrm{I} - \mathrm{A}\| \; \|(\mathrm{I} - \mathrm{A})^{-1}\| \le (1 + \|\mathrm{A}\|) \, \|(\mathrm{I} - \mathrm{A})^{-1}\|,$$

giving the first inequality in (1.26). As for the second part, noting that $\mathrm{I} = \mathrm{I} - \mathrm{A} + \mathrm{A}$ and multiplying both sides on the right by $(\mathrm{I} - \mathrm{A})^{-1}$, one gets $(\mathrm{I} - \mathrm{A})^{-1} = \mathrm{I} + \mathrm{A}(\mathrm{I} - \mathrm{A})^{-1}$. Passing to the norms, we obtain

$$\|(\mathrm{I} - \mathrm{A})^{-1}\| \le 1 + \|\mathrm{A}\| \; \|(\mathrm{I} - \mathrm{A})^{-1}\|,$$

and thus the second inequality, since $\|\mathrm{A}\| < 1$. ◇

Remark 1.1 The assumption that there exists an induced matrix norm such that $\|\mathrm{A}\| < 1$ is justified by Property 1.13, recalling that A is convergent and, therefore, $\rho(\mathrm{A}) < 1$. ■

Notice that (1.25) suggests an algorithm to approximate the inverse of a matrix by a truncated series expansion.

1.12 Positive Definite, Diagonally Dominant and M-matrices

Definition 1.22 A matrix $\mathrm{A} \in \mathbb{C}^{n\times n}$ is *positive definite in* $\mathbb{C}^n$ if the number $(\mathrm{A}\mathbf{x}, \mathbf{x})$ is real and positive $\forall\mathbf{x} \in \mathbb{C}^n$, $\mathbf{x} \neq \mathbf{0}$. A matrix $\mathrm{A} \in \mathbb{R}^{n\times n}$ is *positive definite in* $\mathbb{R}^n$ if $(\mathrm{A}\mathbf{x}, \mathbf{x}) > 0$ $\forall\mathbf{x} \in \mathbb{R}^n$, $\mathbf{x} \neq \mathbf{0}$. If the strict inequality is substituted by the weak one ($\ge$) the matrix is called *positive semi-definite*. ■

Example 1.8 Matrices that are positive definite in $\mathbb{R}^n$ are not necessarily symmetric. An instance is provided by matrices of the form

$$A = \begin{bmatrix} 2 & \alpha \\ -2-\alpha & 2 \end{bmatrix} \tag{1.27}$$

for $\alpha \neq -1$. Indeed, for any nonnull vector $\mathbf{x} = (x_1, x_2)^T$ in $\mathbb{R}^2$

$$(A\mathbf{x}, \mathbf{x}) = 2(x_1^2 + x_2^2 - x_1 x_2) > 0.$$

Notice that A is *not* positive definite in $\mathbb{C}^2$. Indeed, if we take a complex vector $\mathbf{x}$ we find out that the number $(A\mathbf{x}, \mathbf{x})$ is not real-valued in general. •

Definition 1.23 Let $A \in \mathbb{R}^{n\times n}$. The matrices

$$A_S = \frac{1}{2}(A + A^T), \; A_{SS} = \frac{1}{2}(A - A^T)$$

are respectively called the *symmetric part* and the *skew-symmetric part* of A. Obviously, $A = A_S + A_{SS}$. If $A \in \mathbb{C}^{n\times n}$, the definitions modify as follows: $A_S = \frac{1}{2}(A + A^H)$ and $A_{SS} = \frac{1}{2}(A - A^H)$. ■

The following property holds

Property 1.15 *A real matrix* A *of order* n *is positive definite iff its symmetric part* A_S *is positive definite.*

Indeed, it suffices to notice that, due to (1.12) and the definition of A_{SS}, $\mathbf{x}^T A_{SS} \mathbf{x} = 0 \; \forall \mathbf{x} \in \mathbb{R}^n$. For instance, the matrix in (1.27) has a positive definite symmetric part, since

$$A_S = \frac{1}{2}(A + A^T) = \begin{bmatrix} 2 & -1 \\ -1 & 2 \end{bmatrix}.$$

This holds more generally (for the proof see [Axe94]).

Property 1.16 *Let* $A \in \mathbb{C}^{n\times n}$ *(respectively,* $A \in \mathbb{R}^{n\times n}$*); if* $(A\mathbf{x}, \mathbf{x})$ *is real-valued* $\forall \mathbf{x} \in \mathbb{C}^n$*, then* A *is hermitian (respectively, symmetric).*

An immediate consequence of the above results is that matrices that are positive definite in $\mathbb{C}^n$ do satisfy the following characterizing property.

Property 1.17 *A square matrix* A *of order* n *is positive definite in* $\mathbb{C}^n$ *iff it is hermitian and has positive eigenvalues. Thus, a positive definite matrix is nonsingular.*

In the case of positive definite real matrices in $\mathbb{R}^n$, results more specific than those presented so far hold only if the matrix is also symmetric (this is the reason why many textbooks deal only with symmetric positive definite matrices). In particular

Property 1.18 *Let* $\mathrm{A} \in \mathbb{R}^{n\times n}$ *be symmetric. Then,* A *is positive definite iff one of the following properties is satisfied:*

1. $(\mathrm{A}\mathbf{x}, \mathbf{x}) > 0 \ \forall \mathbf{x} \neq \mathbf{0}$ *with* $\mathbf{x} \in \mathbb{R}^n$;
2. *the eigenvalues of the principal submatrices of* A *are all positive;*
3. *the dominant principal minors of* A *are all positive (Sylvester criterion);*
4. *there exists a nonsingular matrix* H *such that* $\mathrm{A} = \mathrm{H}^T\mathrm{H}$.

All the diagonal entries of a positive definite matrix are positive. Indeed, if $\mathbf{e}_i$ is the i-th vector of the canonical basis of $\mathbb{R}^n$, then $\mathbf{e}_i^T \mathrm{A}\mathbf{e}_i = a_{ii} > 0$.

Moreover, it can be shown that if A is symmetric positive definite, the entry with the largest module must be a diagonal entry (these last two properties are therefore necessary conditions for a matrix to be positive definite).

We finally notice that if A is symmetric positive definite and $\mathrm{A}^{1/2}$ is the only positive definite matrix that is a solution of the matrix equation $\mathrm{X}^2 = \mathrm{A}$, the norm

$$\|\mathbf{x}\|_{\mathrm{A}} = \|\mathrm{A}^{1/2}\mathbf{x}\|_2 = (\mathrm{A}\mathbf{x}, \mathbf{x})^{1/2} \tag{1.28}$$

defines a vector norm, called the *energy norm* of the vector $\mathbf{x}$. Related to the energy norm is the *energy scalar product* given by $(\mathbf{x}, \mathbf{y})_{\mathrm{A}} = (\mathrm{A}\mathbf{x}, \mathbf{y})$.

Definition 1.24 A matrix $\mathrm{A} \in \mathbb{R}^{n\times n}$ is called *diagonally dominant by rows* if

$$|a_{ii}| \geq \sum_{j=1, j\neq i}^{n} |a_{ij}|, \text{ with } i = 1, \ldots, n,$$

while it is called *diagonally dominant by columns* if

$$|a_{ii}| \geq \sum_{j=1, j\neq i}^{n} |a_{ji}|, \text{ with } i = 1, \ldots, n.$$

If the inequalities above hold in a strict sense, A is called *strictly diagonally dominant* (by rows or by columns, respectively). ■

A strictly diagonally dominant matrix that is symmetric with positive diagonal entries is also positive definite.

Definition 1.25 A nonsingular matrix $\mathrm{A} \in \mathbb{R}^{n\times n}$ is an *M-matrix* if $a_{ij} \leq 0$ for $i \neq j$ and if all the entries of its inverse are nonnegative. ■

M-matrices enjoy the so-called *discrete maximum principle*, that is, if A is an M-matrix and $\mathrm{A}\mathbf{x} \leq \mathbf{0}$, then $\mathbf{x} \leq \mathbf{0}$ (where the inequalities are meant componentwise). In this connection, the following result can be useful.

Property 1.19 (M-criterion) *Let a matrix* A *satisfy* $a_{ij} \leq 0$ *for* $i \neq j$. *Then* A *is an M-matrix if and only if there exists a vector* $\mathbf{w} > 0$ *such that* $\mathrm{A}\mathbf{w} > 0$.

Finally, M-matrices are related to strictly diagonally dominant matrices by the following property.

Property 1.20 *A matrix* $A \in \mathbb{R}^{n\times n}$ *that is strictly diagonally dominant by rows and whose entries satisfy the relations* $a_{ij} \leq 0$ *for* $i \neq j$ *and* $a_{ii} > 0$*, is an M-matrix.*

For further results about M-matrices, see for instance [Axe94] and [Var62].

1.13 Exercises

1. Let W_1 and W_2 be two subspaces of $\mathbb{R}^n$. Prove that if $V = W_1 \oplus W_2$, then $\dim(V) = \dim(W_1) + \dim(W_2)$, while in general
$$\dim(W_1 + W_2) = \dim(W_1) + \dim(W_2) - \dim(W_1 \cap W_2).$$
[*Hint* : Consider a basis for $W_1 \cap W_2$ and first extend it to W_1, then to W_2, verifying that the basis formed by the set of the obtained vectors is a basis for the sum space.]
2. Check that the following set of vectors
$$\mathbf{v}_i = \left(x_1^{i-1}, x_2^{i-1}, \ldots, x_n^{i-1}\right),\ i = 1, 2, \ldots, n,$$
forms a basis for $\mathbb{R}^n$, $x_1, \ldots, x_n$ being a set of n distinct points of $\mathbb{R}$.
3. Exhibit an example showing that the product of two symmetric matrices may be nonsymmetric.
4. Let B be a skew-symmetric matrix, namely, $B^T = -B$. Let $A = (I+B)(I-B)^{-1}$ and show that $A^{-1} = A^T$.
5. A matrix $A \in \mathbb{C}^{n\times n}$ is called *skew-hermitian* if $A^H = -A$. Show that the diagonal entries of A must be purely imaginary numbers.
6. Let A, B and A+B be invertible matrices of order n. Show that also $A^{-1} + B^{-1}$ is nonsingular and that
$$\left(A^{-1} + B^{-1}\right)^{-1} = A\,(A+B)^{-1}\,B = B\,(A+B)^{-1}\,A.$$
[*Solution* : $\left(A^{-1} + B^{-1}\right)^{-1} = A\left(I + B^{-1}A\right)^{-1} = A\,(B+A)^{-1}\,B$. The second equality is proved similarly by factoring out B and A, respectively from left and right.]
7. Given the nonsymmetric real matrix
$$A = \begin{bmatrix} 0 & 1 & 1 \\ 1 & 0 & -1 \\ -1 & -1 & 0 \end{bmatrix},$$
check that it is similar to the diagonal matrix $D = \mathrm{diag}(1, 0, -1)$ and find its eigenvectors. Is this matrix normal?
[*Solution* : the matrix is not normal.]

8. Let A be a square matrix of order n. Check that if $P(\mathrm{A}) = \sum_{k=0}^{n} c_k \mathrm{A}^k$ and $\lambda(\mathrm{A})$ are the eigenvalues of A, then the eigenvalues of $P(\mathrm{A})$ are given by $\lambda(P(\mathrm{A})) = P(\lambda(\mathrm{A}))$. In particular, prove that $\rho(\mathrm{A}^2) = [\rho(\mathrm{A})]^2$.
9. Prove that a matrix of order n having n distinct eigenvalues cannot be defective. Moreover, prove that a normal matrix cannot be defective.
10. *Commutativity of matrix product.* Show that if A and B are square matrices that share the same set of eigenvectors, then AB = BA. Prove, by a counterexample, that the converse is false.
11. Let A be a normal matrix whose eigenvalues are $\lambda_1, \ldots, \lambda_n$. Show that the singular values of A are $|\lambda_1|, \ldots, |\lambda_n|$.
12. Let $\mathrm{A} \in \mathbb{C}^{m\times n}$ with $\mathrm{rank}(\mathrm{A}) = n$. Show that $\mathrm{A}^\dagger = (\mathrm{A}^T\mathrm{A})^{-1}\mathrm{A}^T$ enjoys the following properties:

$$(1)\ \mathrm{A}^\dagger\mathrm{A} = \mathrm{I}_n;\ (2)\ \mathrm{A}^\dagger\mathrm{A}\mathrm{A}^\dagger = \mathrm{A}^\dagger,\ \mathrm{A}\mathrm{A}^\dagger\mathrm{A} = \mathrm{A};\ (3)\ \text{if } m = n,\ \mathrm{A}^\dagger = \mathrm{A}^{-1}.$$

13. Show that the Moore-Penrose pseudo-inverse matrix $\mathrm{A}^\dagger$ is the only matrix that minimizes the functional

$$\min_{\mathrm{X}\in\mathbb{C}^{n\times m}} \|\mathrm{AX} - \mathrm{I}_m\|_{\mathrm{F}},$$

where $\|\cdot\|_{\mathrm{F}}$ is the Frobenius norm.
14. Prove Property 1.10.
[*Solution* : For any $\mathbf{x}, \widehat{\mathbf{x}} \in V$ show that $|\ \|\mathbf{x}\| - \|\widehat{\mathbf{x}}\|\ | \leq \|\mathbf{x} - \widehat{\mathbf{x}}\|$. Assuming that $\dim(V) = n$ and expanding the vector $\mathbf{w} = \mathbf{x} - \widehat{\mathbf{x}}$ on a basis of V, show that $\|\mathbf{w}\| \leq C\|\mathbf{w}\|_\infty$, from which the thesis follows by imposing in the first obtained inequality that $\|\mathbf{w}\|_\infty \leq \varepsilon$.]
15. Prove Property 1.11 in the case $\mathrm{A} \in \mathbb{R}^{n\times m}$ with m linearly independent columns. [*Hint* : First show that $\|\cdot\|_{\mathrm{A}}$ fulfills all the properties characterizing a norm: positiveness (A has linearly independent columns, thus if $\mathbf{x} \neq \mathbf{0}$, then $\mathrm{A}\mathbf{x} \neq \mathbf{0}$, which proves the thesis), homogeneity and triangular inequality.]
16. Show that for a rectangular matrix $\mathrm{A} \in \mathbb{R}^{m\times n}$

$$\|\mathrm{A}\|_{\mathrm{F}}^2 = \sigma_1^2 + \ldots + \sigma_p^2,$$

where p is the minimum between m and n, σ_i are the singular values of A and $\|\cdot\|_{\mathrm{F}}$ is the Frobenius norm.
17. Assuming $p, q = 1, 2, \infty, F$, recover the following table of equivalence constants c_{pq} such that $\forall \mathrm{A} \in \mathbb{R}^{n\times n}$, $\|\mathrm{A}\|_p \leq c_{pq}\|\mathrm{A}\|_q$.

c_{pq}	$q = 1$	$q = 2$	$q = \infty$	$q = F$
$p = 1$	1	$\sqrt{n}$	n	$\sqrt{n}$
$p = 2$	$\sqrt{n}$	1	$\sqrt{n}$	1
$p = \infty$	n	$\sqrt{n}$	1	$\sqrt{n}$
$p = F$	$\sqrt{n}$	$\sqrt{n}$	$\sqrt{n}$	1

18. A matrix norm for which $\|\mathrm{A}\| = \|\ |\mathrm{A}|\ \|$ is called *absolute norm*, having denoted by $|\mathrm{A}|$ the matrix of the absolute values of the entries of A. Prove that $\|\cdot\|_1$, $\|\cdot\|_\infty$ and $\|\cdot\|_F$ are absolute norms, while $\|\cdot\|_2$ is not. Show that for this latter

$$\frac{1}{\sqrt{n}}\|\mathrm{A}\|_2 \leq \|\ |\mathrm{A}|\ \|_2 \leq \sqrt{n}\|\mathrm{A}\|_2.$$

2

Principles of Numerical Mathematics

The basic concepts of consistency, stability and convergence of a numerical method will be introduced in a very general context in the first part of the chapter: they provide the common framework for the analysis of any method considered henceforth. The second part of the chapter deals with the computer finite representation of real numbers and the analysis of error propagation in machine operations.

2.1 Well-posedness and Condition Number of a Problem

Consider the following problem: find x such that

$$F(x, d) = 0, \tag{2.1}$$

where d is the set of data which the solution depends on and F is the functional relation between x and d. According to the kind of problem that is represented in (2.1), the variables x and d may be real numbers, vectors or functions. Typically, (2.1) is called a *direct* problem if F and d are given and x is the unknown, *inverse* problem if F and x are known and d is the unknown, *identification* problem when x and d are given while the functional relation F is the unknown (these latter problems will not be covered in this volume).

Problem (2.1) is *well posed* if it admits a *unique* solution x which *depends with continuity on the data*. We shall use the terms *well posed* and *stable* in an interchanging manner and we shall deal henceforth only with well-posed problems.

A problem which does not enjoy the property above is called *ill posed* or *unstable* and before undertaking its numerical solution it has to be regularized, that is, it must be suitably transformed into a well-posed problem (see, for instance [Mor84]). Indeed, it is not appropriate to pretend the numerical method can cure the pathologies of an intrinsically ill-posed problem.

Example 2.1 A simple instance of an ill-posed problem is finding the number of real roots of a polynomial. For example, the polynomial $p(x) = x^4 - x^2$

$(2a-1)+a(a-1)$ exhibits a discontinuous variation of the number of real roots as a continuously varies in the real field. We have, indeed, 4 real roots if $a \geq 1$, 2 if $a \in [0,1)$ while no real roots exist if $a < 0$. •

Let D be the set of admissible data, i.e. the set of the values of d in correspondance of which problem (2.1) admits a unique solution. Continuous dependence on the data means that small perturbations on the data d of D yield "small" changes in the solution x. Precisely, let $d \in D$ and denote by δd a perturbation admissible in the sense that $d + \delta d \in D$ and by δx the corresponding change in the solution, in such a way that

$$F(x + \delta x, d + \delta d) = 0. \tag{2.2}$$

Then, we require that

$$\begin{aligned} &\exists \eta_0 = \eta_0(d) > 0,\ \exists K_0 = K_0(d) \text{ such that} \\ &\text{if } \|\delta d\| \leq \eta_0 \text{ then } \|\delta x\| \leq K_0 \|\delta d\|. \end{aligned} \tag{2.3}$$

The norms used for the data and for the solution may not coincide, whenever d and x represent variables of different kinds.

Remark 2.1 The property of continuous dependence on the data could have been stated in the following alternative way, which is more akin to the classical form of Analysis

$$\forall \varepsilon > 0\ \exists \delta = \delta(\varepsilon) \text{ such that if } \|\delta d\| \leq \delta \text{ then } \|\delta x\| \leq \varepsilon.$$

The form (2.3) is however more suitable to express in the following the concept of *numerical stability*, that is, the property that small perturbations on the data yield perturbations of the same order on the solution. ■

With the aim of making the stability analysis more quantitative, we introduce the following definition.

Definition 2.1 For problem (2.1) we define the *relative condition number* to be

$$K(d) = \sup \left\{ \frac{\|\delta x\|/\|x\|}{\|\delta d\|/\|d\|},\ \delta d \neq 0,\ d + \delta d \in D \right\}. \tag{2.4}$$

Whenever $d = 0$ or $x = 0$, it is necessary to introduce the *absolute condition number*, given by

$$K_{abs}(d) = \sup \left\{ \frac{\|\delta x\|}{\|\delta d\|},\ \delta d \neq 0,\ d + \delta d \in D \right\}. \tag{2.5}$$

■

Problem (2.1) is called *ill-conditioned* if $K(d)$ is "big" for any admissible datum d (the precise meaning of "small" and "big" is going to change depending on the considered problem).

The property of a problem of being well-conditioned is independent of the numerical method that is being used to solve it. In fact, it is possible to generate stable as well as unstable numerical schemes for solving well-conditioned problems. The concept of stability for an algorithm or for a numerical method is analogous to that used for problem (2.1) and will be made precise in the next section.

Remark 2.2 (Ill-posed problems) Even in the case in which the condition number does not exist (formally, it is infinite), it is not necessarily true that the problem is ill-posed. In fact there exist well posed problems (for instance, the search of multiple roots of algebraic equations, see Example 2.2) for which the condition number is infinite, but such that they can be reformulated in equivalent problems (that is, having the same solutions) with a finite condition number. ■

If problem (2.1) admits a unique solution, then there necessarily exists a mapping G, that we call *resolvent*, between the sets of the data and of the solutions, such that

$$x = G(d), \text{ that is } F(G(d), d) = 0. \tag{2.6}$$

According to this definition, (2.2) yields $x + \delta x = G(d + \delta d)$. Assuming that G is differentiable in d and denoting formally by $G'(d)$ its derivative with respect to d (if $G : \mathbb{R}^n \to \mathbb{R}^m$, $G'(d)$ will be the Jacobian matrix of G evaluated at the vector d), a Taylor's expansion of G truncated at first order ensures that

$$G(d + \delta d) - G(d) = G'(d)\delta d + o(\|\delta d\|) \qquad \text{for } \delta d \to 0,$$

where $\|\cdot\|$ is a suitable vector norm and $o(\cdot)$ is the classical infinitesimal symbol denoting an infinitesimal term of higher order with respect to its argument. Neglecting the infinitesimal of higher order with respect to $\|\delta d\|$, from (2.4) and (2.5) we respectively deduce that

$$K(d) \simeq \|G'(d)\| \frac{\|d\|}{\|G(d)\|}, \qquad K_{abs}(d) \simeq \|G'(d)\|, \tag{2.7}$$

where the symbol $\|\cdot\|$, when applied to a matrix, denotes the induced matrix norm (1.19) associated with the vector norm introduced above. The estimates in (2.7) are of great practical usefulness in the analysis of problems in the form (2.6), as shown in the forthcoming examples.

Example 2.2 (Algebraic equations of second degree) The solutions to the algebraic equation $x^2 - 2px + 1 = 0$, with $p \geq 1$, are $x_\pm = p \pm \sqrt{p^2 - 1}$. In this case, $F(x, p) = x^2 - 2px + 1$, the datum d is the coefficient p, while x is the vector of components $\{x_+, x_-\}$. As for the condition number, we notice that (2.6) holds by taking $G : \mathbb{R} \to \mathbb{R}^2$, $G(p) = \{x_+, x_-\}$. Letting $G_\pm(p) = x_\pm$, it follows that $G'_\pm(p) = 1 \pm p/\sqrt{p^2 - 1}$. Using (2.7) with $\|\cdot\| = \|\cdot\|_2$ we get

$$K(p) \simeq \frac{|p|}{\sqrt{p^2 - 1}}, \qquad p > 1. \tag{2.8}$$

From (2.8) it turns out that in the case of separated roots (say, if $p \geq \sqrt{2}$) problem $F(x, p) = 0$ is well conditioned. The behavior dramatically changes in the case of multiple roots, that is when $p = 1$. First of all, one notices that the function $G_\pm(p) = p \pm \sqrt{p^2 - 1}$ is no longer differentiable for $p = 1$, which makes (2.8) meaningless. On the other hand, equation (2.8) shows that, for p close to 1, the problem at hand is *ill conditioned*. However, the problem is not *ill posed*. Indeed, following Remark 2.2, it is possible to reformulate it in an equivalent manner as $F(x, t) = x^2 - ((1 + t^2)/t)x + 1 = 0$, with $t = p + \sqrt{p^2 - 1}$, whose roots $x_- = t$ and $x_+ = 1/t$ coincide for $t = 1$. The change of parameter thus removes the singularity that is present in the former representation of the roots as functions of p. The two roots $x_- = x_-(t)$ and $x_+ = x_+(t)$ are now indeed regular functions of t in the neighborhood of $t = 1$ and evaluating the condition number by (2.7) yields $K(t) \simeq 1$ for any value of t. The transformed problem is thus well conditioned. •

Example 2.3 (Systems of linear equations) Consider the linear system $\mathrm{A}\mathbf{x} = \mathbf{b}$, where $\mathbf{x}$ and $\mathbf{b}$ are two vectors in $\mathbb{R}^n$, while A is the matrix $(n \times n)$ of the real coefficients of the system. Suppose that A is nonsingular; in such a case x is the unknown solution $\mathbf{x}$, while the data d are the right-hand side $\mathbf{b}$ and the matrix A, that is, $d = \{b_i, a_{ij}, 1 \leq i, j \leq n\}$.

Suppose now that we perturb only the right-hand side $\mathbf{b}$. We have $d = \mathbf{b}$, $\mathbf{x} = G(\mathbf{b}) = \mathrm{A}^{-1}\mathbf{b}$ so that, $G'(\mathbf{b}) = \mathrm{A}^{-1}$, and (2.7) yields

$$K(d) \simeq \frac{\|\mathrm{A}^{-1}\| \, \|\mathbf{b}\|}{\|\mathrm{A}^{-1}\mathbf{b}\|} = \frac{\|\mathrm{A}\mathbf{x}\|}{\|\mathbf{x}\|}\|\mathrm{A}^{-1}\| \leq \|\mathrm{A}\| \, \|\mathrm{A}^{-1}\| = K(\mathrm{A}), \tag{2.9}$$

where $K(\mathrm{A})$ is the condition number of matrix A (see Sect. 3.1.1) and the use of a consistent matrix norm is understood. Therefore, if A is well conditioned, solving the linear system $\mathrm{A}\mathbf{x}=\mathbf{b}$ is a stable problem with respect to perturbations of the right-hand side $\mathbf{b}$. Stability with respect to perturbations on the entries of A will be analyzed in Sect. 3.10. •

Example 2.4 (Nonlinear equations) Let $f : \mathbb{R} \to \mathbb{R}$ be a function of class C^1 and consider the nonlinear equation

$$F(x, d) = f(x) = \varphi(x) - d = 0,$$

where $\varphi : \mathbb{R} \to \mathbb{R}$ is a suitable function and $d \in \mathbb{R}$ a datum (possibly equal to zero). The problem is well defined only if φ is invertible in a neighborhood of d: in such a case, indeed, $x = \varphi^{-1}(d)$ and the resolvent is $G = \varphi^{-1}$. Since $(\varphi^{-1})'(d) = [\varphi'(x)]^{-1}$, the first relation in (2.7) yields, for $d \neq 0$,

$$K(d) \simeq |[\varphi'(x)]^{-1}|\frac{|d|}{|x|}, \tag{2.10}$$

while if $d = 0$ or $x = 0$ we have

$$K_{abs}(d) \simeq |[\varphi'(x)]^{-1}|. \tag{2.11}$$

The problem is thus ill posed if x is a multiple root of $\varphi(x) - d$; it is ill conditioned when $\varphi'(x)$ is "small", well conditioned when $\varphi'(x)$ is "large". We shall further address this subject in Secttion6.1. •

In view of (2.7), the quantity $\|G'(d)\|$ is an approximation of $K_{abs}(d)$ and is sometimes called *first order absolute condition number.* This latter represents the limit of the Lipschitz constant of G (see Section 11.1) as the perturbation on the data tends to zero.
Such a number does not always provide a sound estimate of the condition number $K_{abs}(d)$. This happens, for instance, when G' vanishes at a point whilst G is nonnull in a neighborhood of the same point. For example, take $x = G(d) = \cos(d) - 1$ for $d \in (-\pi/2, \pi/2)$; we have $G'(0) = 0$, while $K_{abs}(0) = 2/\pi$.

2.2 Stability of Numerical Methods

We shall henceforth suppose the problem (2.1) to be well posed. A numerical method for the approximate solution of (2.1) will consist, in general, of a sequence of approximate problems

$$F_n(x_n, d_n) = 0 \qquad n \geq 1 \tag{2.12}$$

depending on a certain parameter n (to be defined case by case). The understood expectation is that $x_n \to x$ as $n \to \infty$, i.e. that the numerical solution converges to the exact solution. For that, it is necessary that $d_n \to d$ and that F_n "approximates" F, as $n \to \infty$. Precisely, if the datum d of problem (2.1) is admissible for F_n, we say that (2.12) is *consistent* if

$$F_n(x, d) = F_n(x, d) - F(x, d) \to 0 \text{ for } n \to \infty, \tag{2.13}$$

where x is the solution to problem (2.1) corresponding to the datum d.

The meaning of this definition will be made precise in the next chapters for any single class of considered problems.

A method is said to be *strongly consistent* if $F_n(x, d) = 0$ for *any* value of n and not only for $n \to \infty$.

In some cases (e.g., when iterative methods are used) problem (2.12) could take the following form

$$F_n(x_n, x_{n-1}, \ldots, x_{n-q}, d_n) = 0 \qquad n \geq q, \tag{2.14}$$

where $x_0, x_1, \ldots, x_{q-1}$ are given. In such a case, the property of strong consistency becomes $F_n(x, x, \ldots, x, d) = 0$ for all $n \geq q$.

Example 2.5 Let us consider the following iterative method (known as Newton's method and discussed in Section 6.2.2) for approximating a simple root α of a function $f : \mathbb{R} \rightarrow \mathbb{R}$,

$$\text{given } x_0, \quad x_n = x_{n-1} - \frac{f(x_{n-1})}{f'(x_{n-1})}, \qquad n \geq 1. \tag{2.15}$$

The method (2.15) can be written in the form (2.14) by setting $F_n(x_n, x_{n-1}, f) = x_n - x_{n-1} + f(x_{n-1})/f'(x_{n-1})$ and is strongly consistent since $F_n(\alpha, \alpha, f) = 0$ for all $n \geq 1$.

Consider now the following numerical method (known as the composite midpoint rule discussed in Section 9.2) for approximating $x = \int_a^b f(t)\,dt$,

$$x_n = H \sum_{k=1}^{n} f\left(\frac{t_k + t_{k+1}}{2}\right), \qquad n \geq 1,$$

where $H = (b - a)/n$ and $t_k = a + (k - 1)H$, $k = 1, \ldots, n + 1$. This method is consistent; it is also strongly consistent provided that f is a piecewise linear polynomial.

More generally, all numerical methods obtained from the mathematical problem by truncation of limit operations (such as integrals, derivatives, series, ...) are *not* strongly consistent. •

Recalling what has been previously stated about problem (2.1), in order for the numerical method to be *well posed* (or *stable*) we require that for any fixed n, there exists a unique solution x_n corresponding to the datum d_n, that the computation of x_n as a function of d_n is unique and, furthermore, that x_n depends continuously on the data. More precisely, let d_n be an arbitrary element of D_n, where D_n is the set of all admissible data for (2.12). Let δd_n be a perturbation admissible in the sense that $d_n + \delta d_n \in D_n$, and let δx_n denote the corresponding perturbation on the solution, that is

$$F_n(x_n + \delta x_n, d_n + \delta d_n) = 0.$$

Then we require that

$$\begin{aligned} &\exists \eta_0 = \eta_0(d_n) > 0,\ \exists K_0 = K_0(d_n) \text{ such that} \\ &\text{if } \|\delta d_n\| \leq \eta_0 \text{ then } \|\delta x_n\| \leq K_0 \|\delta d_n\|. \end{aligned} \tag{2.16}$$

As done in (2.4), we introduce for each problem in the sequence (2.12) the quantities

$$\begin{aligned} K_n(d_n) &= \sup \left\{ \frac{\|\delta x_n\| / \|x_n\|}{\|\delta d_n\| / \|d_n\|}, \delta d_n \neq 0,\ d_n + \delta d_n \in D_n \right\}, \\ K_{abs,n}(d_n) &= \sup \left\{ \frac{\|\delta x_n\|}{\|\delta d_n\|},\ \delta d_n \neq 0,\ d_n + \delta d_n \in D_n \right\}. \end{aligned} \tag{2.17}$$

The numerical method is said to be well conditioned if $K_n(d_n)$ is "small" for any admissible datum d_n, ill conditioned otherwise. As in (2.6), let us consider the case where, for each n, the functional relation (2.12) defines a mapping G_n between the sets of the numerical data and the solutions

$$x_n = G_n(d_n), \quad \text{that is } F_n(G_n(d_n), d_n) = 0. \tag{2.18}$$

Assuming that G_n is differentiable, we can obtain from (2.17)

$$K_n(d_n) \simeq \|G'_n(d_n)\| \frac{\|d_n\|}{\|G_n(d_n)\|}, \qquad K_{abs,n}(d_n) \simeq \|G'_n(d_n)\|. \tag{2.19}$$

We observe that, in the case where the sets of admissible data in problems (2.1) and (2.12) coincide, we can use in (2.16) and (2.17) the quantity d instead of d_n. In such a case, we can define the *relative and absolute asymptotic condition number* corresponding to the datum d as follows

$$K^{num}(d) = \lim_{k\to\infty} \sup_{n\geq k} K_n(d), \qquad K^{num}_{abs}(d) = \lim_{k\to\infty} \sup_{n\geq k} K_{abs,n}(d).$$

Example 2.6 (Sum and subtraction) The function $f : \mathbb{R}^2 \to \mathbb{R}$, $f(a,b) = a+b$, is a linear mapping whose gradient is the vector $f'(a,b) = (1,1)^T$. Using the vector norm $\|\cdot\|_1$ defined in (1.13) yields $K(a,b) \simeq (|a|+|b|)/(|a+b|)$, from which it follows that summing two numbers of the same sign is a well conditioned operation, being $K(a,b) \simeq 1$. On the other hand, subtracting two numbers almost equal is ill conditioned, since $|a+b| \ll |a|+|b|$. This fact, already pointed out in Example 2.2, leads to the *cancellation of significant digits* whenever numbers can be represented using only a finite number of digits (as in *floating-point* arithmetic, see Sect. 2.5). •

Example 2.7 Consider again the problem of computing the roots of a polynomial of second degree analyzed in Example 2.2. When $p > 1$ (separated roots), such a problem is well conditioned. However, we generate an *unstable* algorithm if we evaluate the root x_- by the formula $x_- = p - \sqrt{p^2-1}$. This formula is indeed subject to errors due to *numerical cancellation* of significant digits (see Sect. 2.4) that are introduced by the finite arithmetic of the computer. A possible remedy to this trouble consists of computing $x_+ = p + \sqrt{p^2-1}$ at first, then $x_- = 1/x_+$. Alternatively, one can solve $F(x,p) = x^2 - 2px + 1 = 0$ using Newton's method (proposed in Example 2.5), which reads:

$$\text{given } x_0, \; x_n = x_{n-1} - (x_{n-1}^2 - 2px_{n-1} + 1)/(2x_{n-1} - 2p) = f_n(p), \; n \geq 1.$$

Applying (2.19) for $p > 1$ yields $K_n(p) \simeq |p|/|x_n - p|$. To compute $K^{num}(p)$ we notice that, in the case when the algorithm converges, the solution x_n would converge to one of the roots x_+ or x_-; therefore, $|x_n - p| \to \sqrt{p^2-1}$ and thus $K_n(p) \to K^{num}(p) \simeq |p|/\sqrt{p^2-1}$, in perfect agreement with the value (2.8) of the condition number of the exact problem.

We can conclude that Newton's method for the search of simple roots of a second order algebraic equation is ill conditioned if $|p|$ is very close to 1, while it is well conditioned in the other cases. •

The final goal of numerical approximation is, of course, to build, through numerical problems of the type (2.12), solutions x_n that "get closer" to the solution of problem (2.1) as much as n gets larger. This concept is made precise in the next definition.

Definition 2.2 The numerical method (2.12) is *convergent* iff

$$\begin{aligned} &\forall \varepsilon > 0 \; \exists n_0 = n_0(\varepsilon), \; \exists \delta = \delta(n_0, \varepsilon) > 0 \text{ such that} \\ &\forall n > n_0(\varepsilon), \; \forall \delta d_n : \|\delta d_n\| \le \delta \qquad \Rightarrow \|x(d) - x_n(d + \delta d_n)\| \le \varepsilon, \end{aligned} \tag{2.20}$$

where d is an admissible datum for the problem (2.1), $x(d)$ is the corresponding solution and $x_n(d + \delta d_n)$ is the solution of the numerical problem (2.12) with datum $d + \delta d_n$. ■

To verify the implication (2.20) it suffices to check that under the same assumptions

$$\|x(d + \delta d_n) - x_n(d + \delta d_n)\| \le \frac{\varepsilon}{2}. \tag{2.21}$$

Indeed, thanks to (2.3) we have

$$\begin{aligned} \|x(d) - x_n(d + \delta d_n)\| &\le \|x(d) - x(d + \delta d_n)\| \\ &+ \|x(d + \delta d_n) - x_n(d + \delta d_n)\| \le K_0 \|\delta d_n\| + \tfrac{\varepsilon}{2}. \end{aligned}$$

Choosing $\delta = \min\{\eta_0, \varepsilon/(2K_0)\}$ one obtains (2.20).

Measures of the convergence of x_n to x are given by the *absolute error* or the *relative error*, respectively defined as

$$E(x_n) = |x - x_n|, \qquad E_{rel}(x_n) = \frac{|x - x_n|}{|x|} \quad (\text{if } x \ne 0). \tag{2.22}$$

In the cases where x and x_n are matrix or vector quantities, in addition to the definitions in (2.22) (where the absolute values are substituted by suitable norms) it is sometimes useful to introduce the *relative error by component* defined as

$$E^c_{rel}(x_n) = \max_{i,j} \frac{|(x - x_n)_{ij}|}{|x_{ij}|}. \tag{2.23}$$

2.2.1 Relations between Stability and Convergence

The concepts of stability and convergence are strongly connected.
First of all, if problem (2.1) is well posed, a *necessary* condition in order for the numerical problem (2.12) to be convergent is that it is stable.

Let us thus assume that the method is convergent, that is, (2.20) holds for an arbitrary $\varepsilon > 0$. We have

$$\begin{aligned}\|\delta x_n\| = \|x_n(d+\delta d_n) - x_n(d)\| &\le \|x_n(d) - x(d)\| \\ &+\|x(d) - x(d+\delta d_n)\| + \|x(d+\delta d_n) - x_n(d+\delta d_n)\| \\ &\le K(\delta(n_0,\varepsilon),d)\|\delta d_n\| + \varepsilon,\end{aligned} \qquad (2.24)$$

having used (2.3) and (2.21) twice. Choosing now δd_n such that $\|\delta d_n\| \le \eta_0$, we deduce that $\|\delta x_n\|/\|\delta d_n\|$ can be bounded by $K_0 = K(\delta(n_0,\varepsilon),d) + 1$, provided that $\varepsilon \le \|\delta d_n\|$, so that the method is stable. Thus, we are interested in stable numerical methods since only these can be convergent.

The stability of a numerical method becomes a *sufficient* condition for the numerical problem (2.12) to converge if this latter is also consistent with problem (2.1). Indeed, under these assumptions we have

$$\begin{aligned}\|x(d+\delta d_n) - x_n(d+\delta d_n)\| &\le \|x(d+\delta d_n) - x(d)\| \\ &+\|x(d) - x_n(d)\| + \|x_n(d) - x_n(d+\delta d_n)\|.\end{aligned}$$

Thanks to (2.3), the first term at right-hand side can be bounded by $\|\delta d_n\|$ (up to a multiplicative constant independent of δd_n). A similar bound holds for the third term, due to the stability property (2.16). Finally, concerning the remaining term, if F_n is differentiable with respect to the variable x, an expansion in a Taylor series gives

$$F_n(x(d),d) - F_n(x_n(d),d) = \frac{\partial F_n}{\partial x}|_{(\overline{x},d)}(x(d) - x_n(d)),$$

for a suitable $\overline{x}$ "between" $x(d)$ and $x_n(d)$. Assuming also that $\partial F_n/\partial x$ is invertible, we get

$$x(d) - x_n(d) = \left(\frac{\partial F_n}{\partial x}\right)^{-1}_{|(\overline{x},d)} [F_n(x(d),d) - F_n(x_n(d),d)]. \qquad (2.25)$$

On the other hand, replacing $F_n(x_n(d),d)$ with $F(x(d),d)$ (since both terms are equal to zero) and passing to the norms, we find

$$\|x(d) - x_n(d)\| \le \left\| \left(\frac{\partial F_n}{\partial x}\right)^{-1}_{|(\overline{x},d)} \right\| \|F_n(x(d),d) - F(x(d),d)\|.$$

Thanks to (2.13) we can thus conclude that $\|x(d) - x_n(d)\| \to 0$ for $n \to \infty$. The result that has just been proved, although stated in qualitative terms, is a milestone in numerical analysis, known as *equivalence theorem* (or Lax-Richtmyer theorem): "*for a consistent numerical method, stability is equivalent to convergence*". A rigorous proof of this theorem is available in [Dah56] for the case of linear Cauchy problems, or in [Lax65] and in [RM67] for linear well-posed initial value problems.

2.3 A priori and a posteriori Analysis

The stability analysis of a numerical method can be carried out following different strategies:

1. *forward analysis*, which provides a bound to the variations $\|\delta x_n\|$ on the solution due to both perturbations in the data and to errors that are intrinsic to the numerical method;
2. *backward analysis*, which aims at estimating the perturbations that should be "impressed" to the data of a given problem in order to obtain the results actually computed under the assumption of working in exact arithmetic. Equivalently, given a certain computed solution $\widehat{x}_n$, backward analysis looks for the perturbations δd_n on the data such that $F_n(\widehat{x}_n, d_n + \delta d_n) = 0$. Notice that, when performing such an estimate, *no* account at all is taken into the way $\widehat{x}_n$ has been obtained (that is, which method has been employed to generate it).

Forward and backward analyses are two different instances of the so called *a priori analysis*. This latter can be applied to investigate not only the stability of a numerical method, but also its convergence. In this case it is referred to as *a priori error analysis*, which can again be performed using either a forward or a backward technique.

A priori error analysis is distincted from the so called *a posteriori error analysis*, which aims at producing an estimate of the error on the grounds of quantities that are actually computed by a specific numerical method. Typically, denoting by $\widehat{x}_n$ the computed numerical solution, approximation to the solution x of problem (2.1), the a posteriori error analysis aims at evaluating the error $x - \widehat{x}_n$ as a function of the residual $r_n = F(\widehat{x}_n, d)$ by means of constants that are called *stability factors* (see [EEHJ96]).

Example 2.8 For the sake of illustration, consider the problem of finding the zeros $\alpha_1, \ldots, \alpha_n$ of a polynomial $p_n(x) = \sum_{k=0}^{n} a_k x^k$ of degree n.

Denoting by $\tilde{p}_n(x) = \sum_{k=0}^{n} \tilde{a}_k x^k$ a perturbed polynomial whose zeros are $\tilde{\alpha}_i$, forward analysis aims at estimating the error between two corresponding zeros α_i and $\tilde{\alpha}_i$, in terms of the variations on the coefficients $a_k - \tilde{a}_k$, $k = 0, 1, \ldots, n$.

On the other hand, let $\{\hat{\alpha}_i\}$ be the approximate zeros of p_n (computed somehow). Backward analysis provides an estimate of the perturbations δa_k which should be impressed to the coefficients so that $\sum_{k=0}^{n}(a_k + \delta a_k)\hat{\alpha}_i^k = 0$, for a fixed $\hat{\alpha}_i$. The goal of a posteriori error analysis would rather be to provide an estimate of the error $\alpha_i - \hat{\alpha}_i$ as a function of the residual value $p_n(\hat{\alpha}_i)$.

This analysis will be carried out in Section 6.1. •

Example 2.9 Consider the linear system $\mathrm{A}\mathbf{x}=\mathbf{b}$, where $\mathrm{A} \in \mathbb{R}^{n \times n}$ is a nonsingular matrix.

For the perturbed system $\tilde{\mathrm{A}}\tilde{\mathbf{x}} = \tilde{\mathbf{b}}$, forward analysis provides an estimate of the error $\mathbf{x} - \tilde{\mathbf{x}}$ in terms of $\mathrm{A} - \tilde{\mathrm{A}}$ and $\mathbf{b} - \tilde{\mathbf{b}}$, while backward analysis estimates the perturbations $\delta\mathrm{A} = (\delta a_{ij})$ and $\delta\mathbf{b} = (\delta b_i)$ which should be impressed to the entries of A and **b** in order to get $(\mathrm{A} + \delta\mathrm{A})\widehat{\mathbf{x}}_n = \mathbf{b} + \delta\mathbf{b}$, $\widehat{\mathbf{x}}_n$ being the solution of the linear system (computed somehow). Finally, a posteriori error analysis looks for an estimate of the error $\mathbf{x} - \widehat{\mathbf{x}}_n$ as a function of the residual $\mathbf{r}_n = \mathbf{b} - \mathrm{A}\widehat{\mathbf{x}}_n$.

We will develop this analysis in Section 3.1. •

It is important to point out the role played by the a posteriori analysis in devising strategies for *adaptive error control.* These strategies, by suitably changing the discretization parameters (for instance, the spacing between nodes in the numerical integration of a function or a differential equation), employ the a posteriori analysis in order to ensure that the error does not exceed a fixed tolerance.

A numerical method that makes use of an adaptive error control is called *adaptive numerical method.* In practice, a method of this kind applies in the computational process the idea of *feedback*, by activating on the grounds of a computed solution a convergence test which ensures the control of error within a fixed tolerance. In case the convergence test fails, a suitable strategy for modifying the discretization parameters is automatically adopted in order to enhance the accuracy of the solution to be newly computed, and the overall procedure is iterated until the convergence check is passed.

2.4 Sources of Error in Computational Models

Whenever the numerical problem (2.12) is an approximation to the mathematical problem (2.1) and this latter is in turn a model of a physical problem (which will be shortly denoted by PP), we shall say that (2.12) is a *computational model* for PP.

In this process the global error, denoted by e, is expressed by the difference between the actually computed solution, $\widehat{x}_n$, and the physical solution, x_{ph}, of which x provides a model. The global error e can thus be interpreted as being the sum of the error e_m of the mathematical model, given by $x - x_{ph}$, and the error e_c of the computational model, $\widehat{x}_n - x$, that is $e = e_m + e_c$ (see Figure 2.1).

The error e_m will in turn take into account the error of the mathematical model in strict sense (that is, the extent at which the functional equation (2.1) does realistically describe the problem PP) and the error on the data (that is, how much accurately does d provide a measure of the real physical data). In the same way, e_c turns out to be the combination of the numerical discretization error $e_n = x_n - x$, the error e_a introduced by the numerical algorithm and the *roundoff* error introduced by the computer during the actual solution of problem (2.12) (see Sect. 2.5).

In general, we can thus outline the following sources of error:

1. errors due to the model, that can be controlled by a proper choice of the mathematical model;

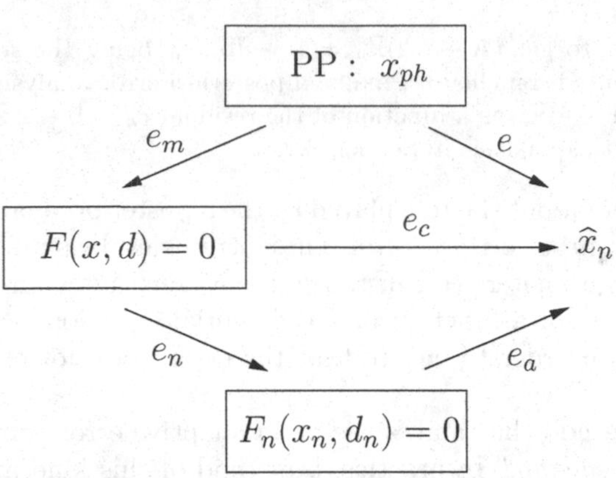

Fig. 2.1. Errors in computational models

2. errors in the data, that can be reduced by enhancing the accuracy in the measurement of the data themselves;
3. truncation errors, arising from having replaced in the numerical model limits by operations that involve a finite number of steps;
4. rounding errors.

The errors at the items 3. and 4. give rise to the *computational error*. A numerical method will thus be convergent if this error can be made arbitrarily small by increasing the computational effort. Of course, convergence is the primary, albeit not unique, goal of a numerical method, the others being *accuracy*, *reliability* and *efficiency*.

Accuracy means that the errors are small with respect to a fixed tolerance. It is usually quantified by the order of infinitesimal of the error e_n with respect to the discretization characteristic parameter (for instance the largest grid spacing between the discretization nodes). By the way, we notice that machine *precision* does not limit, on theoretical grounds, the accuracy.

Reliability means it is likely that the global error can be guaranteed to be below a certain tolerance. Of course, a numerical model can be considered to be reliable only if suitably *tested*, that is, successfully applied to several test cases.

Efficiency means that the computational complexity that is needed to control the error (that is, the amount of operations and the size of the memory required) is as small as possible.

Having encountered the term *algorithm* several times in this section, we cannot refrain from providing an intuitive description of it. By *algorithm* we mean a directive that indicates, through elementary operations, all the passages that are needed to solve a specific problem. An algorithm can in turn contain sub-algorithms and must have the feature of terminating after a finite number of elementary operations. As a consequence, the executor of the algorithm (machine or human being) must find within the algorithm itself all the instructions to completely solve the problem at hand (provided that the necessary resources for its execution are available).

For instance, the statement that a polynomial of second degree surely admits two roots in the complex plane does not characterize an algorithm, whereas the formula yielding the roots *is* an algorithm (provided that the sub-algorithms needed to correctly execute all the operations have been defined in turn).

Finally, the *complexity of an algorithm* is a measure of its executing time. Calculating the complexity of an algorithm is therefore a part of the analysis of the efficiency of a numerical method. Since several algorithms, with different complexities, can be employed to solve the same problem P, it is useful to introduce the concept of *complexity of a problem*, this latter meaning the complexity of the algorithm that has minimum complexity among those solving P. The complexity of a problem is typically measured by a parameter directly associated with P. For instance, in the case of the product of two square matrices, the computational complexity can be expressed as a function of a power of the matrix size n (see, [Str69]).

2.5 Machine Representation of Numbers

Any machine operation is affected by *rounding errors* or *roundoff.* They are due to the fact that on a computer only a finite subset of the set of real numbers can be represented. In this section, after recalling the positional notation of real numbers, we introduce their machine representation.

2.5.1 The Positional System

Let a base $\beta \in \mathbb{N}$ be fixed with $\beta \geq 2$, and let x be a real number with a finite number of digits x_k with $0 \leq x_k < \beta$ for $k = -m, \ldots, n$. The notation (conventionally adopted)

$$x_\beta = (-1)^s \left[x_n x_{n-1} \ldots x_1 x_0 . x_{-1} x_{-2} \ldots x_{-m}\right], \; x_n \neq 0 \qquad (2.26)$$

is called the *positional representation* of x with respect to the base β. The point between x_0 and x_{-1} is called decimal point if the base is 10, binary point if the base is 2, while s depends on the sign of x ($s = 0$ if x is positive, 1 if negative). Relation (2.26) actually means

$$x_\beta = (-1)^s \left(\sum_{k=-m}^{n} x_k \beta^k \right).$$

Example 2.10 The conventional writing $x_{10} = 425.33$ denotes the number $x = 4 \cdot 10^2 + 2 \cdot 10 + 5 + 3 \cdot 10^{-1} + 3 \cdot 10^{-2}$, while $x_6 = 425.33$ would denote the real number $x = 4 \cdot 6^2 + 2 \cdot 6 + 5 + 3 \cdot 6^{-1} + 3 \cdot 6^{-2}$. A rational number can of course have a finite number of digits in a base and an infinite number of digits in another base. For example, the fraction 1/3 has infinite digits in base 10, being $x_{10} = 0.\bar{3}$, while it has only one digit in base 3, being $x_3 = 0.1$. •

Any real number can be approximated by numbers having a finite representation. Indeed, having fixed the base β, the following property holds

$$\forall \varepsilon > 0,\ \forall x_\beta \in \mathbb{R},\ \exists y_\beta \in \mathbb{R} \text{ such that } |y_\beta - x_\beta| < \varepsilon,$$

where y_β has finite positional representation.
In fact, given the positive number $x_\beta = x_n x_{n-1} \ldots x_0 . x_{-1} \ldots x_{-m} \ldots$ with a number of digits, finite or infinite, for any $r \geq 1$ one can build two numbers

$$x_\beta^{(l)} = \sum_{k=0}^{r-1} x_{n-k} \beta^{n-k},\ x_\beta^{(u)} = x_\beta^{(l)} + \beta^{n-r+1},$$

having r digits, such that $x_\beta^{(l)} < x_\beta < x_\beta^{(u)}$ and $x_\beta^{(u)} - x_\beta^{(l)} = \beta^{n-r+1}$. If r is chosen in such a way that $\beta^{n-r+1} < \varepsilon$, then taking y_β equal to $x_\beta^{(l)}$ or $x_\beta^{(u)}$ yields the desired inequality. This result legitimates the computer representation of real numbers (and thus by a finite number of digits).

Although theoretically speaking all the bases are equivalent, in the computational practice three are the bases generally employed: base 2 or binary, base 10 or decimal (the most natural) and base 16 or hexadecimal. Almost all modern computers use base 2, apart from a few which traditionally employ base 16. In what follows, we will assume that β is an even integer.

In the binary representation, digits reduce to the two symbols `0` and `1`, called *bits* (*binary digits*), while in the hexadecimal case the symbols used for the representation of the digits are `0,1,...,9,A,B,C,D,E,F`. Clearly, the smaller the adopted base, the longer the string of characters needed to represent the same number.

To simplify notations, we shall write x instead of x_β, leaving the base β understood.

2.5.2 The Floating-point Number System

Assume a given computer has N memory positions in which to store any number. The most natural way to make use of these positions in the representation of a real number x different from zero is to fix one of them for its sign, $N-k-1$ for the integer digits and k for the digits beyond the point, in such a way that

$$x = (-1)^s \cdot [a_{N-2} a_{N-3} \ldots a_k \, . \, a_{k-1} \ldots a_0], \tag{2.27}$$

s being equal to 1 or 0. Notice that one memory position is equivalent to one bit storage only when $\beta = 2$. The set of numbers of this kind is called *fixed-point system*. Equation (2.27) stands for

$$x = (-1)^s \cdot \beta^{-k} \sum_{j=0}^{N-2} a_j \beta^j \qquad (2.28)$$

and therefore this representation amounts to fixing a scaling factor for all the representable numbers.

The use of fixed point strongly limits the value of the minimum and maximum numbers that can be represented on the computer, unless a very large number N of memory positions is employed. This drawback can be easily overcome if the scaling in (2.28) is allowed to be varying. In such a case, given a non vanishing real number x, its *floating-point* representation is given by

$$x = (-1)^s \cdot (0.a_1 a_2 \dots a_t) \cdot \beta^e = (-1)^s \cdot m \cdot \beta^{e-t}, \qquad (2.29)$$

where $t \in \mathbb{N}$ is the number of allowed significant digits a_i (with $0 \le a_i \le \beta - 1$), $m = a_1 a_2 \dots a_t$ an integer number called *mantissa* such that $0 \le m \le \beta^t - 1$ and e an integer number called *exponent.* Clearly, the exponent can vary within a finite interval of admissible values: we let $L \le e \le U$ (typically $L < 0$ and $U > 0$). The N memory positions are now distributed among the sign (one position), the significant digits (t positions) and the digits for the exponent (the remaining $N - t - 1$ positions). The number zero has a separate representation.

Typically, on the computer there are two formats available for the *floating-point* number representation: *single* and *double precision.* In the case of binary representation, these formats correspond in the standard version to the representation with $N = 32$ *bits* (single precision)

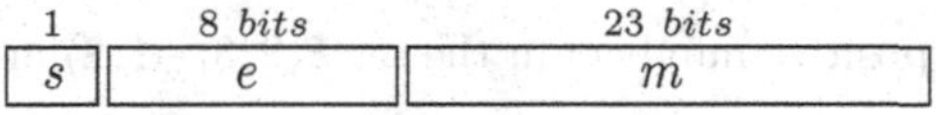

and with $N = 64$ *bits* (double precision)

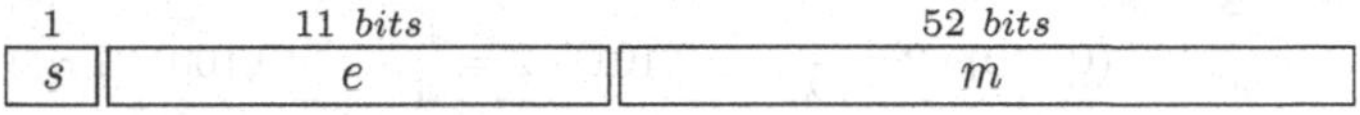

Let us denote by

$$\mathbb{F}(\beta, t, L, U) = \{0\} \cup \left\{ x \in \mathbb{R} : \ x = (-1)^s \beta^e \sum_{i=1}^{t} a_i \beta^{-i} \right\}$$

the set of *floating-point numbers* with t significant digits, base $\beta \ge 2$, $0 \le a_i \le \beta - 1$, and range (L,U) with $L \le e \le U$.

In order to enforce *uniqueness in a number representation,* it is typically assumed that $a_1 \ne 0$ and $m \ge \beta^{t-1}$. In such an event a_1 is called the principal significant digit, while a_t is the last significant digit and the representation of x is called *normalized.* The mantissa m is now varying between β^{t-1} and $\beta^t - 1$.

For instance, in the case $\beta = 10$, $t = 4$, $L = -1$ and $U = 4$, without the assumption that $a_1 \neq 0$, the number 1 would admit the following representations

$$0.1000 \cdot 10^1,\ 0.0100 \cdot 10^2,\ 0.0010 \cdot 10^3,\ 0.0001 \cdot 10^4.$$

To always have uniqueness in the representation, it is assumed that also the number zero has its own sign (typically $s = 0$ is assumed).
It can be immediately noticed that if $x \in \mathbb{F}(\beta, t, L, U)$ then also $-x \in \mathbb{F}(\beta, t, L, U)$. Moreover, the following lower and upper bounds hold for the absolute value of x

$$x_{min} = \beta^{L-1} \le |x| \le \beta^U(1 - \beta^{-t}) = x_{max}. \tag{2.30}$$

The cardinality of $\mathbb{F}(\beta, t, L, U)$ (henceforth shortly denoted by $\mathbb{F}$) is

$$card\ \mathbb{F} = 2(\beta - 1)\beta^{t-1}(U - L + 1) + 1.$$

From (2.30) it turns out that it is not possible to represent any number (apart from zero) whose absolute value is less than x_{min}. This latter limitation can be overcome by completing $\mathbb{F}$ by the set $\mathbb{F}_D$ of the *floating-point de-normalized* numbers obtained by removing the assumption that a_1 is non null, only for the numbers that are referred to the minimum exponent L. In such a way the uniqueness in the representation is not lost and it is possible to generate numbers that have mantissa between 1 and $\beta^{t-1} - 1$ and belong to the interval $(-\beta^{L-1}, \beta^{L-1})$. The smallest number in this set has absolute value equal to β^{L-t}.

Example 2.11 The positive numbers in the set $\mathbb{F}(2, 3, -1, 2)$ are

$$(0.111) \cdot 2^2 = \frac{7}{2}, \quad (0.110) \cdot 2^2 = 3, \quad (0.101) \cdot 2^2 = \frac{5}{2}, \quad (0.100) \cdot 2^2 = 2,$$

$$(0.111) \cdot 2 = \frac{7}{4}, \quad (0.110) \cdot 2 = \frac{3}{2}, \quad (0.101) \cdot 2 = \frac{5}{4}, \quad (0.100) \cdot 2 = 1,$$

$$(0.111) = \frac{7}{8}, \quad (0.110) = \frac{3}{4}, \quad (0.101) = \frac{5}{8}, \quad (0.100) = \frac{1}{2},$$

$$(0.111) \cdot 2^{-1} = \frac{7}{16}, (0.110) \cdot 2^{-1} = \frac{3}{8}, (0.101) \cdot 2^{-1} = \frac{5}{16}, (0.100) \cdot 2^{-1} = \frac{1}{4}.$$

They are included between $x_{min} = \beta^{L-1} = 2^{-2} = 1/4$ and $x_{max} = \beta^U(1 - \beta^{-t}) = 2^2(1-2^{-3}) = 7/2$. As a whole, we have $(\beta-1)\beta^{t-1}(U-L+1) = (2-1)2^{3-1}(2+1+1) = 16$ strictly positive numbers. Their opposites must be added to them, as well as the number zero. We notice that when $\beta = 2$, the first significant digit in the normalized representation is necessarily equal to 1 and thus it may not be stored in the computer (in such an event, we call it *hidden bit*).

When considering also the positive de-normalized numbers, we should complete the above set by adding the following numbers

$$(.011)_2 \cdot 2^{-1} = \frac{3}{16}, \ (.010)_2 \cdot 2^{-1} = \frac{1}{8}, \ (.001)_2 \cdot 2^{-1} = \frac{1}{16}.$$

According to what previously stated, the smallest de-normalized number is $\beta^{L-t} = 2^{-1-3} = 1/16$. •

2.5.3 Distribution of Floating-point Numbers

The *floating-point* numbers are not equally spaced along the real line, but they get dense close to the smallest representable number. It can be checked that the spacing between a number $x \in \mathbb{F}$ and its next nearest $y \in \mathbb{F}$, where both x and y are assumed to be non null, is at least $\beta^{-1}\epsilon_M|x|$ and at most $\epsilon_M|x|$, being $\epsilon_M = \beta^{1-t}$ the *machine epsilon*. This latter represents the distance between the number 1 and the nearest *floating-point* number, and therefore it is the smallest number of $\mathbb{F}$ such that $1 + \epsilon_M > 1$.
Having instead fixed an interval of the form $[\beta^e, \beta^{e+1}]$, the numbers of $\mathbb{F}$ that belong to such an interval are equally spaced and have distance equal to β^{e-t}. Decreasing (or increasing) by one the exponent gives rise to a decrement (or increment) of a factor β of the distance between consecutive numbers.

Unlike the absolute distance, the relative distance between two consecutive numbers has a periodic behavior which depends only on the mantissa m. Indeed, denoting by $(-1)^s m(x)\beta^{e-t}$ one of the two numbers, the distance Δx from the successive one is equal to $(-1)^s \beta^{e-t}$, which implies that the relative distance is

$$\frac{\Delta x}{x} = \frac{(-1)^s \beta^{e-t}}{(-1)^s m(x)\beta^{e-t}} = \frac{1}{m(x)}. \tag{2.31}$$

Within the interval $[\beta^e, \beta^{e+1}]$, the ratio in (2.31) is decreasing as x increases since in the normalized representation the mantissa varies from β^{t-1} to $\beta^t - 1$ (not included). However, as soon as $x = \beta^{e+1}$, the relative distance gets back to the value β^{-t+1} and starts decreasing on the successive intervals, as shown in Figure 2.2. This oscillatory phenomenon is called *wobbling precision* and the greater the base β, the more pronounced the effect. This is another reason why small bases are preferably employed in computers.

2.5.4 IEC/IEEE Arithmetic

The possibility of building sets of *floating-point* numbers that differ in base, number of significant digits and *range* of the exponent has prompted in the past the development, for almost any computer, of a particular system $\mathbb{F}$. In order to avoid this proliferation of numerical systems, a *standard* has been fixed that is nowadays almost universally accepted. This standard was developed in 1985 by the Institute of Electrical and Electronics Engineers (shortly, IEEE) and was approved in 1989 by the International Electronical Commission (IEC) as the international standard IEC559 and it is now known by this

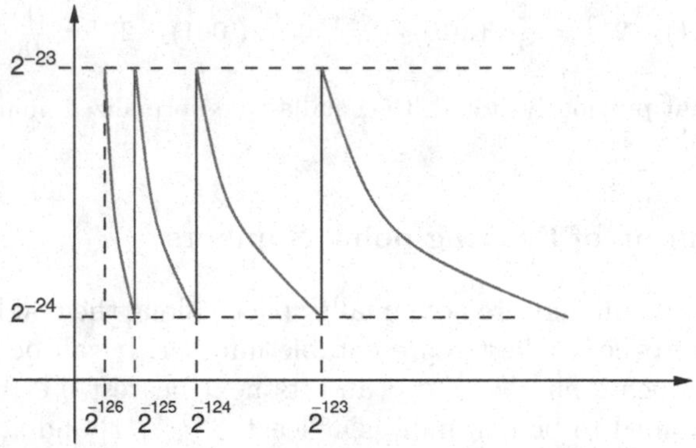

Fig. 2.2. Variation of relative distance for the set of numbers $\mathbb{F}(2, 24, -125, 128)$ IEC/IEEE in single precision

Table 2.1. Lower or upper limits in the standard IEC559 for the extended format of *floating-point* numbers

	single	double		single	double
N	≥ 43 *bits*	≥ 79 *bits*	t	≥ 32	≥ 64
L	≤ -1021	≤ 16381	U	≥ 1024	≥ 16384

Table 2.2. IEC559 codings of some exceptional values

value	exponent	mantissa
± 0	$L-1$	0
$\pm\infty$	$U+1$	0
NaN	$U+1$	$\neq 0$

name (IEC is an organization analogue to the International Standardization Organization (ISO) in the field of electronics). The standard IEC559 endorses two formats for the *floating-point* numbers: a *basic format*, made by the system $\mathbb{F}(2, 24, -125, 128)$ for the single precision, and by $\mathbb{F}(2, 53, -1021, 1024)$ for the double precision, both including the de-normalized numbers, and an *extended format*, for which only the main limitations are fixed (see Table 2.1). Almost all the computers nowadays satisfy the requirements above. We summarize in Table 2.2 the special codings that are used in IEC559 to deal with the values ± 0, $\pm\infty$ and with the so-called *non numbers* (shortly, NaN, that is *not a number*), which correspond for instance to $0/0$ or to other exceptional operations.

2.5.5 Rounding of a Real Number in its Machine Representation

The fact that on any computer only a subset $\mathbb{F}(\beta, t, L, U)$ of $\mathbb{R}$ is actually available poses several practical problems, first of all the representation in $\mathbb{F}$

of *any* given real number. To this concern, notice that, even if x and y were two numbers in $\mathbb{F}$, the result of an operation on them does not necessarily belong to $\mathbb{F}$. Therefore, we must define an arithmetic also on $\mathbb{F}$.

The simplest approach to solve the first problem consists of rounding $x \in \mathbb{R}$ in such a way that the rounded number belongs to $\mathbb{F}$. Among all the possible rounding operations, let us consider the following one. Given $x \in \mathbb{R}$ in the normalized positional notation let us substitute x by its representant $fl(x)$ in $\mathbb{F}$, defined as

$$fl(x) = (-1)^s(0.\,a_1a_2\ldots\widetilde{a}_t)\cdot\beta^e, \quad \tilde{a}_t = \begin{cases} a_t & \text{if } a_{t+1} < \beta/2, \\ a_t+1 & \text{if } a_{t+1} \geq \beta/2. \end{cases} \tag{2.32}$$

The mapping $fl : \mathbb{R} \to \mathbb{F}$ is the most commonly used and is called *rounding* (in the *chopping* one would take more trivially $\widetilde{a}_t = a_t$). Clearly, $fl(x) = x$ if $x \in \mathbb{F}$ and moreover $fl(x) \leq fl(y)$ if $x \leq y\ \forall x, y \in \mathbb{R}$ (monotonicity property).

Remark 2.3 (Overflow and underflow) Everything written so far holds only for the numbers that in (2.29) have exponent e within the *range* of $\mathbb{F}$. If, indeed, $x \in (-\infty, -x_{max}) \cup (x_{max}, \infty)$ the value $fl(x)$ is not defined, while if $x \in (-x_{min}, x_{min})$ the operation of *rounding* is defined anyway (even in absence of de-normalized numbers). In the first case, if x is the result of an operation on numbers of $\mathbb{F}$, we speak about *overflow*, in the second case about *underflow* (or *graceful underflow* if de-normalized numbers are accounted for). The *overflow* is handled by the system through an interrupt of the executing program. ■

Apart from exceptional situations, we can easily quantify the error, absolute and relative, that is made by substituting $fl(x)$ for x. The following result can be shown (see for instance [Hig96], Theorem 2.2).

Property 2.1 *If $x \in \mathbb{R}$ is such that $x_{min} \leq |x| \leq x_{max}$, then*

$$fl(x) = x(1+\delta) \text{ with } |\delta| \leq \mathrm{u} \tag{2.33}$$

where

$$\mathrm{u} = \frac{1}{2}\beta^{1-t} = \frac{1}{2}\epsilon_M \tag{2.34}$$

is the so-called roundoff unit (or machine precision).

As a consequence of (2.33), the following bound holds for the relative error

$$E_{rel}(x) = \frac{|x - fl(x)|}{|x|} \leq \mathrm{u}, \tag{2.35}$$

while, for the absolute error, one gets

$$E(x) = |x - fl(x)| \leq \beta^{e-t}|(a_1 \ldots a_t.a_{t+1}\ldots) - (a_1 \ldots \tilde{a}_t)|.$$

From (2.32), it follows that

$$|(a_1 \dots a_t . a_{t+1} \dots) - (a_1 \dots \tilde{a}_t)| \leq \beta^{-1} \frac{\beta}{2},$$

from which

$$E(x) \leq \frac{1}{2} \beta^{-t+e}.$$

Remark 2.4 In the MATLAB environment it is possible to know immediately the value of ϵ_M, which is given by the system variable eps. ■

2.5.6 Machine Floating-point Operations

As previously stated, it is necessary to define on the set of machine numbers an arithmetic which is analogous, as far as possible, to the arithmetic in $\mathbb{R}$. Thus, given any arithmetic operation $\circ : \mathbb{R} \times \mathbb{R} \to \mathbb{R}$ on two operands in $\mathbb{R}$ (the symbol $\circ$ may denote sum, subtraction, multiplication or division), we shall denote by $\boxed{\circ}$ the corresponding machine operation

$$\boxed{\circ} : \mathbb{R} \times \mathbb{R} \to \mathbb{F}, \qquad x \boxed{\circ}\, y = fl(fl(x) \circ fl(y)).$$

From the properties of *floating-point* numbers one could expect that for the operations on two operands, whenever well defined, the following property holds: $\forall x, y \in \mathbb{F}$, $\exists \delta \in \mathbb{R}$ such that

$$x \boxed{\circ}\, y = (x \circ y)(1 + \delta) \qquad \text{with } |\delta| \leq \mathrm{u}. \tag{2.36}$$

In order for (2.36) to be satisfied when $\circ$ is the operator of subtraction, it will require an additional assumption on the structure of the numbers in $\mathbb{F}$, that is the presence of the so-called round digit (which is addressed at the end of this section). In particular, when $\circ$ is the sum operator, it follows that for all $x, y \in \mathbb{F}$ (see Exercise 10)

$$\frac{|x \boxed{+}\, y - (x + y)|}{|x + y|} \leq \mathrm{u}(1 + \mathrm{u}) \frac{|x| + |y|}{|x + y|} + \mathrm{u}, \tag{2.37}$$

so that the relative error associated with every machine operation will be small, unless $x + y$ is not small by itself. An aside comment is deserved by the case of the sum of two numbers close in module, but opposite in sign. In fact, in such a case $x + y$ can be quite small, this generating the so-called *cancellation errors* (as evidenced in Example 2.6).

It is important to notice that, together with properties of standard arithmetic that are preserved when passing to *floating-point* arithmetic (like, for instance, the commutativity of the sum of two addends, or the product of two factors), other properties are lost. An example is given by the associativity of sum: it can indeed be shown (see Exercise 11) that in general

Table 2.3. Results for some exceptional operations

exception	examples	result
non valid operation	$0/0$, $0 \cdot \infty$	NaN
$overflow$		$\pm\infty$
division by zero	$1/0$	$\pm\infty$
$underflow$		subnormal numbers

$$x \boxed{+} (y \boxed{+} z) \neq (x \boxed{+} y) \boxed{+} z.$$

We shall denote by *flop* the single elementary *floating-point* operation (sum, subtraction, multiplication or division) (the reader is warned that in some texts flop identifies an operation of the form $a + b \cdot c$). According to the previous convention, a scalar product between two vectors of length n will require $2n - 1$ flops, a product matrix-vector $2(m-1)n$ flops if the matrix is $n \times m$ and finally, a product matrix-matrix $2(r-1)mn$ flops if the two matrices are $m \times r$ and $r \times n$ respectively.

Remark 2.5 (IEC559 arithmetic) The IEC559 standard also defines a closed arithmetic on $\mathbb{F}$, this meaning that any operation on it produces a result that can be represented within the system itself, although not necessarily being expected from a pure mathematical standpoint. As an example, in Table 2.3 we report the results that are obtained in exceptional situations. The presence of a NaN (*Not a Number*) in a sequence of operations automatically implies that the result is a NaN. General acceptance of this standard is still ongoing. ■

We mention that not all the *floating-point* systems satisfy (2.36). One of the main reasons is the absence of the *round digit* in subtraction, that is, an extra-bit that gets into action on the mantissa level when the subtraction between two *floating-point* numbers is performed. To demonstrate the importance of the round digit, let us consider the following example with a system $\mathbb{F}$ having $\beta = 10$ and $t = 2$. Let us subtract 1 and 0.99. We have

$$\begin{array}{ll} 10^1 \cdot 0.1 & 10^1 \cdot 0.10 \\ 10^0 \cdot 0.99 \Rightarrow & \underline{10^1 \cdot 0.09} \\ & 10^1 \cdot 0.01 \longrightarrow \boxed{10^0 \cdot 0.10} \end{array}$$

that is, the result differs from the exact one by a factor 10. If we now execute the same subtraction using the round digit, we obtain the exact result. Indeed

$$\begin{array}{ll} 10^1 \cdot 0.1 & 10^1 \cdot 0.10 \\ 10^0 \cdot 0.99 \Rightarrow & \underline{10^1 \cdot 0.09\boxed{9}} \\ & 10^1 \cdot 0.00\boxed{1} \longrightarrow \boxed{10^0 \cdot 0.01} \end{array}$$

In fact, it can be shown that addition and subtraction, if executed without round digit, do not satisfy the property

$$fl(x \pm y) = (x \pm y)(1 + \delta) \text{ with } |\delta| \leq \mathbf{u},$$

but the following one

$$fl(x \pm y) = x(1 + \alpha) \pm y(1 + \beta) \text{ with } |\alpha| + |\beta| \leq \mathbf{u}.$$

An arithmetic for which this latter event happens is called *aberrant*. In some computers the round digit does not exist, most of the care being spent on velocity in the computation. Nowadays, however, the trend is to use even two round digits (see [HP94] for technical details about the subject).

2.6 Exercises

1. Use (2.7) to compute the condition number $K(d)$ of the following expressions
$$(1)\ x - a^d = 0,\ a > 0\ (2)\ d - x + 1 = 0,$$
d being the datum, a a parameter and x the "unknown".
[*Solution* : (1) $K(d) \simeq |d||\log a|$, (2) $K(d) = |d|/|d+1|$.]
2. Study the well posedness and the conditioning in the infinity norm of the following problem as a function of the datum d: find x and y such that
$$\begin{cases} x + dy = 1, \\ dx + y = 0. \end{cases}$$
[*Solution* : the given problem is a linear system whose matrix is $\mathrm{A} = \begin{bmatrix} 1 & d \\ d & 1 \end{bmatrix}$.
It is well-posed if A is nonsingular, i.e., if $d \neq \pm 1$. In such a case, $K_\infty(\mathrm{A}) = |(|d|+1)/(|d|-1)|$.]
3. Study the conditioning of the solving formula $x_{\pm} = -p \pm \sqrt{p^2+q}$ for the second degree equation $x^2 + 2px - q$ with respect to changes in the parameters p and q separately.
[*Solution* : $K(p) = |p|/\sqrt{p^2+q}$, $K(q) = |q|/(2|x_{\pm}|\sqrt{p^2+q})$.]
4. Consider the following Cauchy problem
$$\begin{cases} x'(t) = x_0 e^{at}\left(a\cos(t) - \sin(t)\right), t > 0, \\ x(0) = x_0, \end{cases} \tag{2.38}$$
whose solution is $x(t) = x_0 e^{at}\cos(t)$ (a is a given real number). Study the conditioning of (2.38) with respect to the choice of the initial datum and check that on unbounded intervals it is well conditioned if $a < 0$, while it is ill conditioned if $a > 0$.
[*Hint* : consider the definition of $K_{abs}(a)$.]
5. Let $\widehat{x} \neq 0$ be an approximation of a nonnull quantity x. Find the relation between the relative error $\epsilon = |x - \widehat{x}|/|x|$ and $\tilde{E} = |x - \widehat{x}|/|\widehat{x}|$.
6. Determine all the elements of the set $\mathbb{F} = (10, 6, -9, 9)$, in both normalized and de-normalized cases.

7. Consider the set of the de-normalized numbers $\mathbb{F}_D$ and study the behavior of the absolute distance and of the relative distance between two of these numbers. Does the *wobbling precision* effect arise again?
[*Hint* : for these numbers, uniformity in the relative density is lost. As a consequence, the absolute distance remains constant (equal to β^{L-t}), while the relative one rapidly grows as x tends to zero.]
8. What is the value of 0^0 in IEEE arithmetic?
[*Solution* : ideally, the outcome should be NaN. In practice, IEEE systems recover the value 1. A motivation of this result can be found in [Gol91].]
9. Show that, due to cancellation errors, the following sequence

$$I_0 = \log\frac{6}{5},\ I_k + 5I_{k-1} = \frac{1}{k}, \quad k = 1, 2, \ldots, n, \tag{2.39}$$

is not well suited to finite arithmetic computations of the integral $I_n = \int_0^1 \frac{x^n}{x+5}dx$ when n is sufficiently large, although it works in infinite arithmetic.
[*Hint* : consider the initial perturbed datum $\tilde{I}_0 = I_0 + \mu_0$ and study the propagation of the error μ_0 within (2.39).]
10. Prove (2.37).
[*Solution* : notice that

$$\frac{|x \boxed{+} y - (x+y)|}{|x+y|} \leq \frac{|x \boxed{+} y - (fl(x) + fl(y))|}{|x+y|} + \frac{|fl(x) - x + fl(y) - y|}{|x+y|}.$$

Then, use (2.36) and (2.35).]
11. Given $x, y, z \in \mathbb{F}$ with $x+y$, $y+z$, $x+y+z$ that fall into the *range* of $\mathbb{F}$, show that

$$|(x \boxed{+} y) \boxed{+} z - (x+y+z)| \leq C_1 \simeq (2|x+y| + |z|)\mathbf{u}$$
$$|x \boxed{+} (y \boxed{+} z) - (x+y+z)| \leq C_2 \simeq (|x| + 2|y+z|)\mathbf{u}.$$

12. Which among the following approximations of π,

$$\pi = 4\left(1 - \frac{1}{3} + \frac{1}{5} - \frac{1}{7} + \frac{1}{9} - \ldots\right),$$
$$\pi = 6\left(0.5 + \frac{(0.5)^3}{2\cdot 3} + \frac{3(0.5)^5}{2\cdot 4\cdot 5} + \frac{3\cdot 5(0.5)^7}{2\cdot 4\cdot 6\cdot 7} + \ldots\right) \tag{2.40}$$

better limits the propagation of rounding errors? Compare using MATLAB the obtained results as a function of the number of the terms in each sum in (2.40).
13. Analyze the stability, with respect to propagation of rounding errors, of the following two MATLAB codes to evaluate $f(x) = (e^x - 1)/x$ for $|x| \ll 1$

```
% Algorithm 1
if x == 0
  f = 1;
else
  f = (exp(x) - 1) / x;
end
```

```
% Algorithm 2
y = exp (x);
if y == 1
  f = 1;
else
  f = (y - 1) / log (y);
end
```

[*Solution* : the first algorithm is inaccurate due to cancellation errors, while the second one (in presence of round digit) is stable and accurate.]

14. In binary arithmetic one can show [Dek71] that the rounding error in the sum of two numbers a and b, with $a \geq b$, can be computed as

$$((a \boxed{+} b) \boxed{-} a) \boxed{-} b).$$

Based on this property, a method has been proposed, called *Kahan compensated sum*, to compute the sum of n addends a_i in such a way that the rounding errors are compensated. In practice, letting the initial rounding error $e_1 = 0$ and $s_1 = a_1$, at the i-th step, with $i \geq 2$, the algorithm evaluates $y_i = x_i - e_{i-1}$, the sum is updated setting $s_i = s_{i-1} + y_i$ and the new rounding error is computed as $e_i = (s_i - s_{i-1}) - y_i$. Implement this algorithm in MATLAB and check its accuracy by evaluating again the second expression in (2.40).

15. The area $A(T)$ of a triangle T with sides a, b and c, can be computed using the following formula

$$A(T) = \sqrt{p(p-a)(p-b)(p-c)},$$

where p is half the perimeter of T. Show that in the case of strongly deformed triangles ($a \simeq b + c$), this formula lacks accuracy and check this experimentally.

Part II

Numerical Linear Algebra

3

Direct Methods for the Solution of Linear Systems

A system of m linear equations in n unknowns consists of a set of algebraic relations of the form

$$\sum_{j=1}^{n} a_{ij}x_j = b_i, \; i = 1, \ldots, m, \tag{3.1}$$

where x_j are the unknowns, a_{ij} are the coefficients of the system and b_i are the components of the right hand side. System (3.1) can be more conveniently written in matrix form as

$$\mathrm{A}\mathbf{x} = \mathbf{b}, \tag{3.2}$$

where we have denoted by $\mathrm{A} = (a_{ij}) \in \mathbb{C}^{m\times n}$ the coefficient matrix, by $\mathbf{b}=(b_i) \in \mathbb{C}^m$ the right side vector and by $\mathbf{x}=(x_i) \in \mathbb{C}^n$ the unknown vector, respectively. We call a *solution* of (3.2) any n-tuple of values x_i which satisfies (3.1).

In this chapter we shall be mainly dealing with real-valued square systems of order n, that is, systems of the form (3.2) with $\mathrm{A} \in \mathbb{R}^{n\times n}$ and $\mathbf{b} \in \mathbb{R}^n$. In such cases existence and uniqueness of the solution of (3.2) are ensured if one of the following (equivalent) hypotheses holds:

1. A is invertible;
2. rank(A)=n;
3. the homogeneous system $\mathrm{A}\mathbf{x}=\mathbf{0}$ admits only the null solution.

The solution of system (3.2) is formally provided by *Cramer's rule*

$$x_j = \frac{\Delta_j}{\det(\mathrm{A})}, \qquad j = 1, \ldots, n, \tag{3.3}$$

where Δ_j is the determinant of the matrix obtained by substituting the j-th column of A with the right hand side $\mathbf{b}$. This formula is, however, of little practical use. Indeed, if the determinants are evaluated by the recursive relation

(1.4), the computational effort of Cramer's rule is of the order of $(n+1)!$ flops and therefore turns out to be unacceptable even for small dimensions of A (for instance, a computer able to perform 10^9 flops per second would take $9.6 \cdot 10^{47}$ years to solve a linear system of only 50 equations).
For this reason, numerical methods that are alternatives to Cramer's rule have been developed. They are called *direct* methods if they yield the solution of the system in a finite number of steps, *iterative* if they require (theoretically) an infinite number of steps. Iterative methods will be addressed in the next chapter. We notice from now on that the choice between a direct and an iterative method does not depend only on the theoretical efficiency of the scheme, but also on the particular type of matrix, on memory storage requirements and, finally, on the architecture of the computer.

3.1 Stability Analysis of Linear Systems

Solving a linear system by a numerical method invariably leads to the introduction of rounding errors. Only using stable numerical methods can keep away the propagation of such errors from polluting the accuracy of the solution. In this section two aspects of stability analysis will be addressed.

Firstly, we will analyze the sensitivity of the solution of (3.2) to changes in the data A and $\mathbf{b}$ (forward a priori analysis). Secondly, assuming that an approximate solution $\widehat{\mathbf{x}}$ of (3.2) is available, we shall quantify the perturbations on the data A and $\mathbf{b}$ in order for $\widehat{\mathbf{x}}$ to be the exact solution of a perturbed system (backward a priori analysis). The size of these perturbations will in turn allow us to measure the accuracy of the computed solution $\widehat{\mathbf{x}}$ by the use of a posteriori analysis.

3.1.1 The Condition Number of a Matrix

The *condition number* of a matrix $\mathrm{A} \in \mathbb{C}^{n\times n}$ is defined as

$$K(\mathrm{A}) = \|\mathrm{A}\| \, \|\mathrm{A}^{-1}\|, \qquad (3.4)$$

where $\|\cdot\|$ is an induced matrix norm. In general $K(\mathrm{A})$ depends on the choice of the norm; this will be made clear by introducing a subscript into the notation, for instance, $K_\infty(\mathrm{A}) = \|\mathrm{A}\|_\infty \, \|\mathrm{A}^{-1}\|_\infty$. More generally, $K_p(\mathrm{A})$ will denote the condition number of A in the p-norm. Remarkable instances are $p=1$, $p=2$ and $p=\infty$ (we refer to Exercise 1 for the relations among $K_1(\mathrm{A})$, $K_2(\mathrm{A})$ and $K_\infty(\mathrm{A})$).

As already noticed in Example 2.3, an increase in the condition number produces a higher sensitivity of the solution of the linear system to changes in the data. Let us start by noticing that $K(\mathrm{A}) \geq 1$ since

$$1 = \|\mathrm{A}\mathrm{A}^{-1}\| \leq \|\mathrm{A}\| \, \|\mathrm{A}^{-1}\| = K(\mathrm{A}).$$

Moreover, $K(\mathrm{A}^{-1}) = K(\mathrm{A})$ and $\forall \alpha \in \mathbb{C}$ with $\alpha \neq 0$, $K(\alpha\mathrm{A}) = K(\mathrm{A})$. Finally, if A is orthogonal, $K_2(\mathrm{A}) = 1$ since $\|\mathrm{A}\|_2 = \sqrt{\rho(\mathrm{A}^T\mathrm{A})} = \sqrt{\rho(\mathrm{I})} = 1$ and $\mathrm{A}^{-1} = \mathrm{A}^T$. The condition number of a singular matrix is set equal to infinity.

For $p = 2$, $K_2(\mathrm{A})$ can be characterized as follows. Starting from (1.21), it can be proved that

$$K_2(\mathrm{A}) = \|\mathrm{A}\|_2 \; \|\mathrm{A}^{-1}\|_2 = \frac{\sigma_1(\mathrm{A})}{\sigma_n(\mathrm{A})},$$

where $\sigma_1(\mathrm{A})$ and $\sigma_n(\mathrm{A})$ are the maximum and minimum singular values of A (see Property 1.7). As a consequence, in the case of symmetric positive definite matrices we have

$$K_2(\mathrm{A}) = \frac{\lambda_{max}}{\lambda_{min}} = \rho(\mathrm{A})\rho(\mathrm{A}^{-1}), \tag{3.5}$$

where λ_{max} and λ_{min} are the maximum and minimum eigenvalues of A. To check (3.5), notice that

$$\|\mathrm{A}\|_2 = \sqrt{\rho(\mathrm{A}^T\mathrm{A})} = \sqrt{\rho(\mathrm{A}^2)} = \sqrt{\lambda_{max}^2} = \lambda_{max}.$$

Moreover, since $\lambda(\mathrm{A}^{-1}) = 1/\lambda(\mathrm{A})$, one gets $\|\mathrm{A}^{-1}\|_2 = 1/\lambda_{min}$ from which (3.5) follows. For that reason, $K_2(\mathrm{A})$ is called *spectral condition number.*

Remark 3.1 Define the relative distance of $\mathrm{A} \in \mathbb{C}^{n\times n}$ from the set of singular matrices with respect to the p-norm by

$$\mathrm{dist}_p(\mathrm{A}) = \min\left\{ \frac{\|\delta\mathrm{A}\|_p}{\|\mathrm{A}\|_p} : \; \mathrm{A} + \delta\mathrm{A} \text{ is singular} \right\}.$$

It can then be shown that ([Kah66], [Gas83])

$$\mathrm{dist}_p(\mathrm{A}) = \frac{1}{K_p(\mathrm{A})}. \tag{3.6}$$

Equation (3.6) suggests that a matrix A with a high condition number can behave like a singular matrix of the form A+δA. In other words, null perturbations in the right hand side do not necessarily yield nonvanishing changes in the solution since, if A+δA is singular, the homogeneous system $(\mathrm{A}+\delta\mathrm{A})\mathbf{z} = \mathbf{0}$ does no longer admit only the null solution. Notice that if the following condition holds

$$\|\mathrm{A}^{-1}\|_p\|\delta\mathrm{A}\|_p < 1. \tag{3.7}$$

then the matrix A+δA is nonsingular (see, e.g., [Atk89], Theorem 7.12). ■

Relation (3.6) seems to suggest that a natural candidate for measuring the ill-conditioning of a matrix is its determinant, since from (3.3) one is prompted to conclude that small determinants mean nearly-singular matrices. However this conclusion is wrong, as there exist examples of matrices with small (respectively, high) determinants and small (respectively, high) condition numbers (see Exercise 2).

3.1.2 Forward a priori Analysis

In this section we introduce a measure of the sensitivity of the system to changes in the data. These changes will be interpreted in Section 3.10 as being the effects of rounding errors induced by the numerical method used to solve the system. For a more comprehensive analysis of the subject we refer to [Dat95], [GL89], [Ste73] and [Var62].

Due to rounding errors, a numerical method for solving (3.2) does not provide the exact solution but only an approximate one, which satisfies a perturbed system. In other words, a numerical method yields an (exact) solution $\mathbf{x} + \boldsymbol{\delta}\mathbf{x}$ of the perturbed system

$$(\mathrm{A} + \delta\mathrm{A})(\mathbf{x} + \boldsymbol{\delta}\mathbf{x}) = \mathbf{b} + \boldsymbol{\delta}\mathbf{b}. \tag{3.8}$$

The next result provides an estimate of $\boldsymbol{\delta}\mathbf{x}$ in terms of $\delta\mathrm{A}$ and $\boldsymbol{\delta}\mathbf{b}$.

Theorem 3.1 *Let* $\mathrm{A} \in \mathbb{R}^{n\times n}$ *be a nonsingular matrix and* $\delta\mathrm{A} \in \mathbb{R}^{n\times n}$ *be such that* (3.7) *is satisfied for an induced matrix norm* $\|\cdot\|$. *Then, if* $\mathbf{x} \in \mathbb{R}^n$ *is the solution of* $\mathrm{A}\mathbf{x}=\mathbf{b}$ *with* $\mathbf{b} \in \mathbb{R}^n$ $(\mathbf{b} \neq \mathbf{0})$ *and* $\boldsymbol{\delta}\mathbf{x} \in \mathbb{R}^n$ *satisfies* (3.8) *for* $\boldsymbol{\delta}\mathbf{b} \in \mathbb{R}^n$,

$$\frac{\|\boldsymbol{\delta}\mathbf{x}\|}{\|\mathbf{x}\|} \leq \frac{K(\mathrm{A})}{1 - K(\mathrm{A})\|\delta\mathrm{A}\|/\|\mathrm{A}\|}\left(\frac{\|\boldsymbol{\delta}\mathbf{b}\|}{\|\mathbf{b}\|} + \frac{\|\delta\mathrm{A}\|}{\|\mathrm{A}\|}\right). \tag{3.9}$$

Proof. From (3.7) it follows that the matrix $\mathrm{A}^{-1}\delta\mathrm{A}$ has norm less than 1. Then, due to Theorem 1.5, $\mathrm{I} + \mathrm{A}^{-1}\delta\mathrm{A}$ is invertible and from (1.26) it follows that

$$\|(\mathrm{I} + \mathrm{A}^{-1}\delta\mathrm{A})^{-1}\| \leq \frac{1}{1 - \|\mathrm{A}^{-1}\delta\mathrm{A}\|} \leq \frac{1}{1 - \|\mathrm{A}^{-1}\|\ \|\delta\mathrm{A}\|}. \tag{3.10}$$

On the other hand, solving for $\boldsymbol{\delta}\mathbf{x}$ in (3.8) and recalling that $\mathrm{A}\mathbf{x} = \mathbf{b}$, one gets

$$\boldsymbol{\delta}\mathbf{x} = (\mathrm{I} + \mathrm{A}^{-1}\delta\mathrm{A})^{-1}\mathrm{A}^{-1}(\boldsymbol{\delta}\mathbf{b} - \delta\mathrm{A}\mathbf{x}),$$

from which, passing to the norms and using (3.10), it follows that

$$\|\boldsymbol{\delta}\mathbf{x}\| \leq \frac{\|\mathrm{A}^{-1}\|}{1 - \|\mathrm{A}^{-1}\|\ \|\delta\mathrm{A}\|}\left(\|\boldsymbol{\delta}\mathbf{b}\| + \|\delta\mathrm{A}\|\ \|\mathbf{x}\|\right).$$

Finally, dividing both sides by $\|\mathbf{x}\|$ (which is nonzero since $\mathbf{b} \neq \mathbf{0}$ and A is nonsingular) and noticing that $\|\mathbf{x}\| \geq \|\mathbf{b}\|/\|\mathrm{A}\|$, the result follows. ◇

Well-conditioning alone is not enough to yield an accurate solution of the linear system. It is indeed crucial, as pointed out in Chapter 2, to resort to stable algorithms. Conversely, ill-conditioning does not necessarily exclude that for particular choices of the right side $\mathbf{b}$ the overall conditioning of the system is good (see Exercise 4).
A particular case of Theorem 3.1 is the following.

Theorem 3.2 *Assume that the conditions of Theorem* 3.1 *hold and let* $\delta\mathrm{A} = 0$. *Then*

$$\frac{1}{K(\mathrm{A})}\frac{\|\boldsymbol{\delta}\mathbf{b}\|}{\|\mathbf{b}\|} \leq \frac{\|\boldsymbol{\delta}\mathbf{x}\|}{\|\mathbf{x}\|} \leq K(\mathrm{A})\frac{\|\boldsymbol{\delta}\mathbf{b}\|}{\|\mathbf{b}\|}. \tag{3.11}$$

Proof. We will prove only the first inequality since the second one directly follows from (3.9). Relation $\boldsymbol{\delta}\mathbf{x} = \mathrm{A}^{-1}\boldsymbol{\delta}\mathbf{b}$ yields $\|\boldsymbol{\delta}\mathbf{b}\| \leq \|\mathrm{A}\|\,\|\boldsymbol{\delta}\mathbf{x}\|$. Multiplying both sides by $\|\mathbf{x}\|$ and recalling that $\|\mathbf{x}\| \leq \|\mathrm{A}^{-1}\|\,\|\mathbf{b}\|$ it follows that $\|\mathbf{x}\|\,\|\boldsymbol{\delta}\mathbf{b}\| \leq K(\mathrm{A})\|\mathbf{b}\|\,\|\boldsymbol{\delta}\mathbf{x}\|$, which is the desired inequality. ◇

In order to employ the inequalities (3.9) and (3.11) in the analysis of propagation of rounding errors in the case of direct methods, $\|\delta\mathrm{A}\|$ and $\|\boldsymbol{\delta}\mathbf{b}\|$ should be bounded in terms of the dimension of the system and of the characteristics of the *floating-point* arithmetic that is being used.

It is indeed reasonable to expect that the perturbations induced by a method for solving a linear system are such that $\|\delta\mathrm{A}\| \leq \gamma\|\mathrm{A}\|$ and $\|\boldsymbol{\delta}\mathbf{b}\| \leq \gamma\|\mathbf{b}\|$, γ being a positive number that depends on the roundoff unit $\mathbf{u}$ (for example, we shall assume henceforth that $\gamma = \beta^{1-t}$, where β is the base and t is the number of digits of the mantissa of the floating-point system $\mathbb{F}$). In such a case (3.9) can be completed by the following theorem.

Theorem 3.3 *Assume that* $\|\delta\mathrm{A}\| \leq \gamma\|\mathrm{A}\|$, $\|\boldsymbol{\delta}\mathbf{b}\| \leq \gamma\|\mathbf{b}\|$ *with* $\gamma \in \mathbb{R}^+$ *and* $\delta\mathrm{A} \in \mathbb{R}^{n\times n}$, $\boldsymbol{\delta}\mathbf{b} \in \mathbb{R}^n$. *Then, if* $\gamma K(\mathrm{A}) < 1$ *the following inequalities hold*

$$\frac{\|\mathbf{x}+\boldsymbol{\delta}\mathbf{x}\|}{\|\mathbf{x}\|} \leq \frac{1+\gamma K(\mathrm{A})}{1-\gamma K(\mathrm{A})}, \tag{3.12}$$

$$\frac{\|\boldsymbol{\delta}\mathbf{x}\|}{\|\mathbf{x}\|} \leq \frac{2\gamma}{1-\gamma K(\mathrm{A})}K(\mathrm{A}). \tag{3.13}$$

Proof. From (3.8) it follows that $(\mathrm{I}+\mathrm{A}^{-1}\delta\mathrm{A})(\mathbf{x}+\boldsymbol{\delta}\mathbf{x}) = \mathbf{x}+\mathrm{A}^{-1}\boldsymbol{\delta}\mathbf{b}$. Moreover, since $\gamma K(\mathrm{A}) < 1$ and $\|\delta\mathrm{A}\| \leq \gamma\|\mathrm{A}\|$ it turns out that $\mathrm{I}+\mathrm{A}^{-1}\delta\mathrm{A}$ is nonsingular. Taking the inverse of such a matrix and passing to the norms we get $\|\mathbf{x}+\boldsymbol{\delta}\mathbf{x}\| \leq \|(\mathrm{I}+\mathrm{A}^{-1}\delta\mathrm{A})^{-1}\|\left(\|\mathbf{x}\|+\gamma\|\mathrm{A}^{-1}\|\,\|\mathbf{b}\|\right)$. From Theorem 1.5 it then follows that

$$\|\mathbf{x}+\boldsymbol{\delta}\mathbf{x}\| \leq \frac{1}{1-\|\mathrm{A}^{-1}\delta\mathrm{A}\|}\left(\|\mathbf{x}\|+\gamma\|\mathrm{A}^{-1}\|\,\|\mathbf{b}\|\right),$$

which implies (3.12), since $\|\mathrm{A}^{-1}\delta\mathrm{A}\| \leq \gamma K(\mathrm{A})$ and $\|\mathbf{b}\| \leq \|\mathrm{A}\|\,\|\mathbf{x}\|$.
Let us prove (3.13). Subtracting (3.2) from (3.8) it follows that

$$\mathrm{A}\boldsymbol{\delta}\mathbf{x} = -\delta\mathrm{A}(\mathbf{x}+\boldsymbol{\delta}\mathbf{x})+\boldsymbol{\delta}\mathbf{b}.$$

Inverting A and passing to the norms, the following inequality is obtained

$$\begin{aligned}\|\boldsymbol{\delta}\mathbf{x}\| &\leq \|\mathrm{A}^{-1}\delta\mathrm{A}\|\,\|\mathbf{x}+\boldsymbol{\delta}\mathbf{x}\|+\|\mathrm{A}^{-1}\|\,\|\boldsymbol{\delta}\mathbf{b}\| \\ &\leq \gamma K(\mathrm{A})\|\mathbf{x}+\boldsymbol{\delta}\mathbf{x}\|+\gamma\|\mathrm{A}^{-1}\|\,\|\mathbf{b}\|.\end{aligned} \tag{3.14}$$

Dividing both sides by $\|\mathbf{x}\|$ and using the triangular inequality $\|\mathbf{x}+\boldsymbol{\delta}\mathbf{x}\| \leq \|\boldsymbol{\delta}\mathbf{x}\| + \|\mathbf{x}\|$, we finally get (3.13). ◇

Remarkable instances of perturbations $\delta\mathrm{A}$ and $\boldsymbol{\delta}\mathbf{b}$ are those for which $|\delta\mathrm{A}| \leq \gamma|\mathrm{A}|$ and $|\boldsymbol{\delta}\mathbf{b}| \leq \gamma|\mathbf{b}|$ with $\gamma \geq 0$. Hereafter, the *absolute value notation* $\mathrm{B} = |\mathrm{A}|$ denotes the matrix $n \times n$ having entries $b_{ij} = |a_{ij}|$ with $i, j = 1, \ldots, n$ and the inequality $\mathrm{C} \leq \mathrm{D}$, with $\mathrm{C}, \mathrm{D} \in \mathbb{R}^{m\times n}$ has the following meaning

$$c_{ij} \leq d_{ij} \text{ for } i = 1, \ldots, m, \; j = 1, \ldots, n.$$

If $\|\cdot\|_\infty$ is considered, from (3.14) it follows that

$$\begin{aligned}\frac{\|\boldsymbol{\delta}\mathbf{x}\|_\infty}{\|\mathbf{x}\|_\infty} &\leq \gamma \frac{\|\,|\mathrm{A}^{-1}|\,|\mathrm{A}|\,|\mathbf{x}| + |\mathrm{A}^{-1}|\,|\mathbf{b}|\,\|_\infty}{(1-\gamma\|\,|\mathrm{A}^{-1}|\,|\mathrm{A}|\,\|_\infty)\|\mathbf{x}\|_\infty} \\ &\leq \frac{2\gamma}{1-\gamma\|\,|\mathrm{A}^{-1}|\,|\mathrm{A}|\,\|_\infty}\|\,|\mathrm{A}^{-1}|\,|\mathrm{A}|\,\|_\infty.\end{aligned} \tag{3.15}$$

Estimate (3.15) is generally too pessimistic; however, the following *componentwise error estimates* of $\boldsymbol{\delta}\mathbf{x}$ can be derived from (3.15)

$$\begin{aligned}&|\delta x_i| \leq \gamma|\mathbf{r}_{(i)}^T|\,|\mathrm{A}|\,|\mathbf{x}+\boldsymbol{\delta}\mathbf{x}|, \; i = 1, \ldots, n \text{ if } \boldsymbol{\delta}\mathbf{b} = \mathbf{0}, \\ &\frac{|\delta x_i|}{|x_i|} \leq \gamma\frac{|\mathbf{r}_{(i)}^T|\,|\mathbf{b}|}{|\mathbf{r}_{(i)}^T\mathbf{b}|}, \qquad i = 1, \ldots, n \text{ if } \delta\mathrm{A} = 0,\end{aligned} \tag{3.16}$$

being $\mathbf{r}_{(i)}^T$ the row vector $\mathbf{e}_i^T\mathrm{A}^{-1}$. Estimates (3.16) are more stringent than (3.15), as can be seen in Example 3.1. The first inequality in (3.16) can be used when the perturbed solution $\mathbf{x}+\boldsymbol{\delta}\mathbf{x}$ is known, being henceforth $\mathbf{x}+\boldsymbol{\delta}\mathbf{x}$ the solution computed by a numerical method.

In the case where $|\mathrm{A}^{-1}|\,|\mathbf{b}| = |\mathbf{x}|$, the parameter γ in (3.15) is equal to 1. For such systems the components of the solution are insensitive to perturbations to the right side. A slightly worse situation occurs when A is a triangular M-matrix and $\mathbf{b}$ has positive entries. In such a case γ is bounded by $2n-1$, since

$$|\mathbf{r}_{(i)}^T|\,|\mathrm{A}|\,|\mathbf{x}| \leq (2n-1)|x_i|.$$

For further details on the subject we refer to [Ske79], [CI95] and [Hig89]. Results linking componentwise estimates to normwise estimates through the so-called *hypernorms* can be found in [ADR92].

Example 3.1 Consider the linear system $\mathrm{A}\mathbf{x}=\mathbf{b}$ with

$$\mathrm{A} = \begin{bmatrix} \alpha & \frac{1}{\alpha} \\ 0 & \frac{1}{\alpha} \end{bmatrix}, \; \mathbf{b} = \begin{bmatrix} \alpha^2 + \frac{1}{\alpha} \\ \frac{1}{\alpha} \end{bmatrix}$$

which has solution $\mathbf{x} = [\alpha, 1]^T$, where $0 < \alpha < 1$. Let us compare the results obtained using (3.15) and (3.16). From

$$|\mathrm{A}^{-1}|\ |\mathrm{A}|\ |\mathbf{x}| = |\mathrm{A}^{-1}|\ |\mathbf{b}| = \left[\alpha + \frac{2}{\alpha^2}, 1\right]^T \tag{3.17}$$

it follows that the supremum of (3.17) is unbounded as $\alpha \to 0$, exactly as it happens in the case of $\|\mathrm{A}\|_\infty$. On the other hand, the amplification factor of the error in (3.16) is bounded. Indeed, the component of the maximum absolute value, x_2, of the solution, satisfies $|\mathbf{r}_{(2)}^T|\ |\mathrm{A}|\ |\mathbf{x}|/|x_2| = 1$. •

3.1.3 Backward a priori Analysis

The numerical methods that we shall consider in the following do not require the explicit computation of the inverse of A to solve $\mathrm{A}\mathbf{x}=\mathbf{b}$. However, we can always assume that they yield an approximate solution of the form $\widehat{\mathbf{x}} = \mathrm{C}\mathbf{b}$, where the matrix C, due to rounding errors, is an approximation of A^{-1}. In practice, C is very seldom constructed; in case this should happen, the following result yields an estimate of the error that is made substituting C for A^{-1} (see [IK66], Chapter 2, Theorem 7).

Property 3.1 *Let* $\mathrm{R} = \mathrm{AC} - \mathrm{I}$; *if* $\|\mathrm{R}\| < 1$, *then* A *and* C *are nonsingular and*

$$\|\mathrm{A}^{-1}\| \leq \frac{\|\mathrm{C}\|}{1 - \|\mathrm{R}\|}, \quad \frac{\|\mathrm{R}\|}{\|\mathrm{A}\|} \leq \|\mathrm{C} - \mathrm{A}^{-1}\| \leq \frac{\|\mathrm{C}\|\ \|\mathrm{R}\|}{1 - \|\mathrm{R}\|}. \tag{3.18}$$

In the frame of backward a priori analysis we can interpret C as being the inverse of $\mathrm{A} + \delta\mathrm{A}$ (for a suitable unknown $\delta\mathrm{A}$). We are thus assuming that $\mathrm{C}(\mathrm{A} + \delta\mathrm{A}) = \mathrm{I}$. This yields

$$\delta\mathrm{A} = \mathrm{C}^{-1} - \mathrm{A} = -(\mathrm{AC} - \mathrm{I})\mathrm{C}^{-1} = -\mathrm{R}\mathrm{C}^{-1}$$

and, as a consequence, if $\|\mathrm{R}\| < 1$ it turns out that

$$\|\delta\mathrm{A}\| \leq \frac{\|\mathrm{R}\|\ \|\mathrm{A}\|}{1 - \|\mathrm{R}\|}, \tag{3.19}$$

having used the first inequality in (3.18), where A is assumed to be an approximation of the inverse of C (notice that the roles of C and A can be interchanged).

3.1.4 A posteriori Analysis

Having approximated the inverse of A by a matrix C turns into having an approximation of the solution of the linear system (3.2). Let us denote by $\mathbf{y}$ a known approximate solution. The aim of the a posteriori analysis is to relate the (unknown) error $\mathbf{e} = \mathbf{y} - \mathbf{x}$ to quantities that can be computed using $\mathbf{y}$ and C.

The starting point of the analysis relies on the fact that the *residual vector* $\mathbf{r} = \mathbf{b} - \mathrm{A}\mathbf{y}$ is in general nonzero, since $\mathbf{y}$ is just an approximation to the unknown exact solution. The residual can be related to the error through Property 3.1 as follows. We have $\mathbf{e} = \mathrm{A}^{-1}(\mathrm{A}\mathbf{y} - \mathbf{b}) = -\mathrm{A}^{-1}\mathbf{r}$ and thus, if $\|\mathrm{R}\| < 1$ then

$$\|\mathbf{e}\| \leq \frac{\|\mathbf{r}\| \, \|\mathrm{C}\|}{1 - \|\mathrm{R}\|}. \tag{3.20}$$

Notice that the estimate does not necessarily require $\mathbf{y}$ to coincide with the solution $\widehat{\mathbf{x}} = \mathrm{C}\mathbf{b}$ of the backward *a priori* analysis. One could therefore think of computing C only for the purpose of using the estimate (3.20) (for instance, in the case where (3.2) is solved through the Gauss elimination method, one can compute C a posteriori using the LU factorization of A, see Sections 3.3 and 3.3.1).

We conclude by noticing that if $\boldsymbol{\delta}\mathbf{b}$ is interpreted in (3.11) as being the residual of the computed solution $\mathbf{y} = \mathbf{x} + \boldsymbol{\delta}\mathbf{x}$, it also follows that

$$\frac{\|\mathbf{e}\|}{\|\mathbf{x}\|} \leq K(\mathrm{A}) \frac{\|\mathbf{r}\|}{\|\mathbf{b}\|}. \tag{3.21}$$

The estimate (3.21) is not used in practice since the computed residual is affected by rounding errors. A more significant estimate (in the $\|\cdot\|_\infty$ norm) is obtained letting $\widehat{\mathbf{r}} = fl(\mathbf{b} - \mathrm{A}\mathbf{y})$ and assuming that $\widehat{\mathbf{r}} = \mathbf{r} + \boldsymbol{\delta}\mathbf{r}$ with $|\boldsymbol{\delta}\mathbf{r}| \leq \gamma_{n+1}(|\mathrm{A}| \, |\mathbf{y}| + |\mathbf{b}|)$, where $\gamma_{n+1} = (n+1)\mathrm{u}/(1-(n+1)\mathrm{u}) > 0$, from which we have

$$\frac{\|\mathbf{e}\|_\infty}{\|\mathbf{y}\|_\infty} \leq \frac{\| \, |\mathrm{A}^{-1}|(|\widehat{\mathbf{r}}| + \gamma_{n+1}(|\mathrm{A}||\mathbf{y}| + |\mathbf{b}|))\|_\infty}{\|\mathbf{y}\|_\infty}.$$

Formulae like this last one are implemented in the library for linear algebra LAPACK (see [ABB+92]).

3.2 Solution of Triangular Systems

Consider the nonsingular 3×3 lower triangular system

$$\begin{bmatrix} l_{11} & 0 & 0 \\ l_{21} & l_{22} & 0 \\ l_{31} & l_{32} & l_{33} \end{bmatrix} \begin{bmatrix} x_1 \\ x_2 \\ x_3 \end{bmatrix} = \begin{bmatrix} b_1 \\ b_2 \\ b_3 \end{bmatrix}.$$

Since the matrix is nonsingular, its diagonal entries l_{ii}, $i = 1, 2, 3$, are nonvanishing, hence we can solve sequentially for the unknown values x_i, $i = 1, 2, 3$, as follows

$$x_1 = b_1/l_{11},$$
$$x_2 = (b_2 - l_{21}x_1)/l_{22},$$
$$x_3 = (b_3 - l_{31}x_1 - l_{32}x_2)/l_{33}.$$

This algorithm can be extended to systems $n \times n$ and is called *forward substitution*. In the case of a system $\mathrm{L}\mathbf{x}=\mathbf{b}$, with L being a nonsingular lower triangular matrix of order n ($n \geq 2$), the method takes the form

$$x_1 = \frac{b_1}{l_{11}},$$
$$x_i = \frac{1}{l_{ii}}\left(b_i - \sum_{j=1}^{i-1} l_{ij}x_j\right), \; i = 2, \ldots, n. \tag{3.22}$$

The number of multiplications and divisions to execute the algorithm is equal to $n(n+1)/2$, while the number of sums and subtractions is $n(n-1)/2$. The global operation count for (3.22) is thus n^2 flops.

Similar conclusions can be drawn for a linear system $\mathrm{U}\mathbf{x}=\mathbf{b}$, where U is a nonsingular upper triangular matrix of order n ($n \geq 2$). In this case the algorithm is called *backward substitution* and in the general case can be written as

$$x_n = \frac{b_n}{u_{nn}},$$
$$x_i = \frac{1}{u_{ii}}\left(b_i - \sum_{j=i+1}^{n} u_{ij}x_j\right), \; i = n-1, \ldots, 1. \tag{3.23}$$

Its computational cost is still n^2 flops.

3.2.1 Implementation of Substitution Methods

Each i-th step of algorithm (3.22) requires performing the scalar product between the row vector $\mathrm{L}(i, 1 : i-1)$ (this notation denoting the vector extracted from matrix L taking the elements of the i-th row from the first to the (i-1)-th column) and the column vector $\mathbf{x}(1 : i-1)$. The access to matrix L is thus by row; for that reason, the forward substitution algorithm, when implemented in the form above, is called *row-oriented.*

Its coding is reported in Program 1.

Program 1 - forwardrow : Forward substitution: row-oriented version

```
function [x]=forwardrow(L,b)
% FORWARDROW forward substitution: row oriented version.
% X=FORWARDROW(L,B) solves the lower triangular system L*X=B with the
% forward substitution method in the row-oriented version.
[n,m]=size(L);
if n ~= m, error('Only square systems'); end
if min(abs(diag(L))) == 0, error('The system is singular'); end
```

```
x(1,1) = b(1)/L(1,1);
for i = 2:n
   x (i,1) = (b(i)-L(i,1:i-1)*x(1:i-1,1))/L(i,i);
end
return
```

To obtain a *column-oriented* version of the same algorithm, we take advantage of the fact that i-th component of the vector $\mathbf{x}$, once computed, can be conveniently eliminated from the system.

An implementation of such a procedure, where the solution $\mathbf{x}$ is overwritten on the right vector $\mathbf{b}$, is reported in Program 2.

Program 2 - forwardcol : Forward substitution: column-oriented version

```
function [b]=forwardcol(L,b)
% FORWARDCOL forward substitution: column oriented version.
% X=FORWARDCOL(L,B) solves the lower triangular system L*X=B with the
% forward substitution method in the column-oriented version.
[n,m]=size(L);
if n ~= m, error('Only square systems'); end
if min(abs(diag(L))) == 0, error('The system is singular'); end
for j=1:n-1
   b(j)= b(j)/L(j,j); b(j+1:n)=b(j+1:n)-b(j)*L(j+1:n,j);
end
b(n) = b(n)/L(n,n);
return
```

Implementing the same algorithm by a row-oriented rather than a column-oriented approach, might dramatically change its performance (but of course, not the solution). The choice of the form of implementation must therefore be subordinated to the specific hardware that is used.

Similar considerations hold for the backward substitution method, presented in (3.23) in its row-oriented version.
In Program 3 only the column-oriented version of the algorithm is coded. As usual, the vector $\mathbf{x}$ is overwritten on $\mathbf{b}$.

Program 3 - backwardcol : Backward substitution: column-oriented version

```
function [b]=backwardcol(U,b)
% BACKWARDCOL backward substitution: column oriented version.
% X=BACKWARDCOL(U,B) solves the upper triangular system U*X=B with the
% backward substitution method in the column-oriented version.
[n,m]=size(U);
if n ~= m, error('Only square systems'); end
if min(abs(diag(U))) == 0, error('The system is singular'); end
for j = n:-1:2,
   b(j)=b(j)/U(j,j); b(1:j-1)=b(1:j-1)-b(j)*U(1:j-1,j);
end
```

```
b(1) = b(1)/U(1,1);
return
```

When large triangular systems must be solved, only the triangular portion of the matrix should be stored leading to considerable saving of memory resources.

3.2.2 Rounding Error Analysis

The analysis developed so far has not accounted for the presence of rounding errors. When including these, the forward and backward substitution algorithms no longer yield the exact solutions to the systems $\mathbf{L}\mathbf{x}=\mathbf{b}$ and $\mathbf{U}\mathbf{x}=\mathbf{b}$, but rather provide approximate solutions $\widehat{\mathbf{x}}$ that can be regarded as being *exact* solutions to the perturbed systems

$$(\mathrm{L} + \delta\mathrm{L})\widehat{\mathbf{x}} = \mathbf{b}, \ (\mathrm{U} + \delta\mathrm{U})\widehat{\mathbf{x}} = \mathbf{b},$$

where $\delta\mathrm{L} = (\delta l_{ij})$ and $\delta\mathrm{U} = (\delta u_{ij})$ are perturbation matrices. In order to apply the estimates (3.9) carried out in Section 3.1.2, we must provide estimates of the perturbation matrices, δL and δU, as a function of the entries of L and U, of their size and of the characteristics of the floating-point arithmetic. For this purpose, it can be shown that

$$|\delta\mathrm{T}| \leq \frac{n\mathrm{u}}{1 - n\mathrm{u}}|\mathrm{T}|, \tag{3.24}$$

where T is equal to L or U, $\mathrm{u} = \frac{1}{2}\beta^{1-t}$ is the *roundoff* unit defined in (2.34). Clearly, if $n\mathrm{u} < 1$ from (3.24) it turns out that, using a Taylor expansion, $|\delta\mathrm{T}| \leq n\mathrm{u}|\mathrm{T}| + \mathcal{O}(\mathrm{u}^2)$. Moreover, from (3.24) and (3.9) it follows that, if $n\mathrm{u}K(\mathrm{T}) < 1$, then

$$\frac{\|\mathbf{x} - \widehat{\mathbf{x}}\|}{\|\mathbf{x}\|} \leq \frac{n\mathrm{u}K(\mathrm{T})}{1 - n\mathrm{u}K(\mathrm{T})} = n\mathrm{u}K(\mathrm{T}) + \mathcal{O}(\mathrm{u}^2) \tag{3.25}$$

for the norms $\|\cdot\|_1$, $\|\cdot\|_\infty$ and the Frobenius norm. If u is sufficiently small (as typically happens), the perturbations introduced by the rounding errors in the solution of a triangular system can thus be neglected. As a consequence, the accuracy of the solution computed by the forward or backward substitution algorithm is generally very high.

These results can be improved by introducing some additional assumptions on the entries of L or U. In particular, if the entries of U are such that $|u_{ii}| \geq |u_{ij}|$ for any $j > i$, then

$$|x_i - \widehat{x}_i| \leq 2^{n-i+1}\frac{n\mathrm{u}}{1 - n\mathrm{u}}\max_{j\geq i}|\widehat{x}_j|, \qquad 1 \leq i \leq n.$$

The same result holds if T=L, provided that $|l_{ii}| \geq |l_{ij}|$ for any $j < i$, or if L and U are diagonally dominant. The previous estimates will be employed in Sections 3.3.1 and 3.4.2.

For the proofs of the results reported so far, see [FM67], [Hig89] and [Hig88].

3.2.3 Inverse of a Triangular Matrix

The algorithm (3.23) can be employed to explicitly compute the inverse of an upper triangular matrix. Indeed, given an upper triangular matrix U, the column vectors $\mathbf{v}_i$ of the inverse $V=(\mathbf{v}_1, \ldots, \mathbf{v}_n)$ of U satisfy the following linear systems

$$\mathrm{U}\mathbf{v}_i = \mathbf{e}_i, \; i = 1, \ldots, n, \tag{3.26}$$

where $\{\mathbf{e}_i\}$ is the canonical basis of $\mathbb{R}^n$ (defined in Example 1.3). Solving for $\mathbf{v}_i$ thus requires the application of algorithm (3.23) n times to (3.26).

This procedure is quite inefficient since at least half the entries of the inverse of U are null. Let us take advantage of this as follows. Denote by $\mathbf{v}'_k = (v'_{1k}, \ldots, v'_{kk})^T$ the vector of size k such that

$$\mathrm{U}^{(k)}\mathbf{v}'_k = \mathbf{l}_k \; k = 1, \ldots, n, \tag{3.27}$$

where $\mathrm{U}^{(k)}$ is the principal submatrix of U of order k and $\mathbf{l}_k$ the vector of $\mathbb{R}^k$ having null entries, except the first one which is equal to 1. Systems (3.27) are upper triangular, but have order k and can be again solved using the method (3.23). We end up with the following inversion algorithm for upper triangular matrices: for $k = n, n-1, \ldots, 1$ compute

$$\begin{aligned} v'_{kk} &= u_{kk}^{-1}, \\ v'_{ik} &= -u_{ii}^{-1} \sum_{j=i+1}^{k} u_{ij} v'_{jk}, \text{ for } i = k-1, k-2, \ldots, 1. \end{aligned} \tag{3.28}$$

At the end of this procedure the vectors $\mathbf{v}'_k$ furnish the nonvanishing entries of the columns of U^{-1}. The algorithm requires about $n^3/3 + (3/4)n^2$ flops. Once again, due to rounding errors, the algorithm (3.28) no longer yields the exact solution, but an approximation of it. The error that is introduced can be estimated using the backward a priori analysis carried out in Section 3.1.3.

A similar procedure can be constructed from (3.22) to compute the inverse of a lower triangular system.

3.3 The Gaussian Elimination Method (GEM) and LU Factorization

The Gaussian elimination method aims at reducing the system $\mathrm{A}\mathbf{x}=\mathbf{b}$ to an equivalent system (that is, having the same solution) of the form $\mathrm{U}\mathbf{x}=\widehat{\mathbf{b}}$, where U is an upper triangular matrix and $\widehat{\mathbf{b}}$ is an updated right side vector. This latter system can then be solved by the backward substitution method. Let us denote the original system by $\mathrm{A}^{(1)}\mathbf{x} = \mathbf{b}^{(1)}$. During the reduction procedure we basically employ the property which states that replacing one of the equations by the difference between this equation and another one multiplied by a nonnull constant yields an equivalent system (i.e., one with the same solution).

Thus, consider a nonsingular matrix $\mathrm{A} \in \mathbb{R}^{n\times n}$, and suppose that the diagonal entry a_{11} is nonvanishing. Introducing the *multipliers*

$$m_{i1} = \frac{a_{i1}^{(1)}}{a_{11}^{(1)}}, \, i = 2, 3, \ldots, n,$$

where $a_{ij}^{(1)}$ denote the elements of $\mathrm{A}^{(1)}$, it is possible to eliminate the unknown x_1 from the rows other than the first one by simply subtracting from row i, with $i = 2, \ldots, n$, the first row multiplied by m_{i1} and doing the same on the right side. If we now define

$$a_{ij}^{(2)} = a_{ij}^{(1)} - m_{i1} a_{1j}^{(1)}, \, i, j = 2, \ldots, n,$$

$$b_i^{(2)} = b_i^{(1)} - m_{i1} b_1^{(1)}, \quad i = 2, \ldots, n,$$

where $b_i^{(1)}$ denote the components of $\mathbf{b}^{(1)}$, we get a new system of the form

$$\begin{bmatrix} a_{11}^{(1)} & a_{12}^{(1)} & \ldots & a_{1n}^{(1)} \\ 0 & a_{22}^{(2)} & \ldots & a_{2n}^{(2)} \\ \vdots & \vdots & & \vdots \\ 0 & a_{n2}^{(2)} & \ldots & a_{nn}^{(2)} \end{bmatrix} \begin{bmatrix} x_1 \\ x_2 \\ \vdots \\ x_n \end{bmatrix} = \begin{bmatrix} b_1^{(1)} \\ b_2^{(2)} \\ \vdots \\ b_n^{(2)} \end{bmatrix},$$

which we denote by $\mathrm{A}^{(2)}\mathbf{x} = \mathbf{b}^{(2)}$, that is equivalent to the starting one. Similarly, we can transform the system in such a way that the unknown x_2 is eliminated from rows $3, \ldots, n$. In general, we end up with the finite sequence of systems

$$\mathrm{A}^{(k)}\mathbf{x} = \mathbf{b}^{(k)}, \, 1 \le k \le n, \tag{3.29}$$

where, for $k \geq 2$, matrix $\mathrm{A}^{(k)}$ takes the following form

$$\mathrm{A}^{(k)} = \begin{bmatrix} a_{11}^{(1)} & a_{12}^{(1)} & \ldots & \ldots & \ldots & a_{1n}^{(1)} \\ 0 & a_{22}^{(2)} & & & & a_{2n}^{(2)} \\ \vdots & & \ddots & & & \vdots \\ 0 & \ldots & 0 & a_{kk}^{(k)} & \ldots & a_{kn}^{(k)} \\ \vdots & & \vdots & \vdots & & \vdots \\ 0 & \ldots & 0 & a_{nk}^{(k)} & \ldots & a_{nn}^{(k)} \end{bmatrix},$$

having assumed that $a_{ii}^{(i)} \neq 0$ for $i = 1, \ldots, k-1$. It is clear that for $k = n$ we obtain the upper triangular system $\mathrm{A}^{(n)}\mathbf{x} = \mathbf{b}^{(n)}$

$$\begin{bmatrix} a_{11}^{(1)} & a_{12}^{(1)} & \dots & \dots & a_{1n}^{(1)} \\ 0 & a_{22}^{(2)} & & & a_{2n}^{(2)} \\ \vdots & & \ddots & & \vdots \\ 0 & & & \ddots & \vdots \\ 0 & & & & a_{nn}^{(n)} \end{bmatrix} \begin{bmatrix} x_1 \\ x_2 \\ \vdots \\ \vdots \\ x_n \end{bmatrix} = \begin{bmatrix} b_1^{(1)} \\ b_2^{(2)} \\ \vdots \\ \vdots \\ b_n^{(n)} \end{bmatrix}.$$

Consistently with the notations that have been previously introduced, we denote by U the upper triangular matrix $\mathrm{A}^{(n)}$. The entries $a_{kk}^{(k)}$ are called *pivots* and must obviously be nonnull for $k = 1, \dots, n-1$.

In order to highlight the formulae which transform the k-th system into the $k+1$-th one, for $k = 1, \dots, n-1$ we assume that $a_{kk}^{(k)} \neq 0$ and define the multiplier

$$m_{ik} = \frac{a_{ik}^{(k)}}{a_{kk}^{(k)}}, \; i = k+1, \dots, n. \tag{3.30}$$

Then we let

$$\begin{aligned} a_{ij}^{(k+1)} &= a_{ij}^{(k)} - m_{ik} a_{kj}^{(k)}, \; i, j = k+1, \dots, n \\ b_i^{(k+1)} &= b_i^{(k)} - m_{ik} b_k^{(k)}, \quad i = k+1, \dots, n. \end{aligned} \tag{3.31}$$

Example 3.2 Let us use GEM to solve the following system

$$(\mathrm{A}^{(1)}\mathbf{x} = \mathbf{b}^{(1)}) \begin{cases} x_1 + \frac{1}{2}x_2 + \frac{1}{3}x_3 = \frac{11}{6} \\ \frac{1}{2}x_1 + \frac{1}{3}x_2 + \frac{1}{4}x_3 = \frac{13}{12} \\ \frac{1}{3}x_1 + \frac{1}{4}x_2 + \frac{1}{5}x_3 = \frac{47}{60} \end{cases},$$

which admits the solution $\mathbf{x}=[1,\ 1,\ 1]^T$. At the first step we compute the multipliers $m_{21} = 1/2$ and $m_{31} = 1/3$, and subtract from the second and third equation of the system the first row multiplied by m_{21} and m_{31}, respectively. We obtain the equivalent system

$$(\mathrm{A}^{(2)}\mathbf{x} = \mathbf{b}^{(2)}) \begin{cases} x_1 + \frac{1}{2}x_2 + \frac{1}{3}x_3 = \frac{11}{6} \\ 0 + \frac{1}{12}x_2 + \frac{1}{12}x_3 = \frac{1}{6} \\ 0 + \frac{1}{12}x_2 + \frac{4}{45}x_3 = \frac{31}{180} \end{cases}.$$

If we now subtract the second row multiplied by $m_{32} = 1$ from the third one, we end up with the upper triangular system

$$(\mathrm{A}^{(3)}\mathbf{x} = \mathbf{b}^{(3)}) \begin{cases} x_1 + \frac{1}{2}x_2 + \frac{1}{3}x_3 = \frac{11}{6} \\ 0 + \frac{1}{12}x_2 + \frac{1}{12}x_3 = \frac{1}{6} \\ 0 + \quad 0 + \frac{1}{180}x_3 = \frac{1}{180} \end{cases},$$

from which we immediately compute $x_3 = 1$ and then, by back substitution, the remaining unknowns $x_1 = x_2 = 1$. •

Remark 3.2 The matrix in Example 3.2 is called the *Hilbert matrix* of order 3. In the general $n \times n$ case, its entries are

$$h_{ij} = 1/(i+j-1), \qquad i,j = 1,\ldots,n. \tag{3.32}$$

As we shall see later on, this matrix provides the paradigm of an ill-conditioned matrix. ■

To complete Gaussian elimination $2(n-1)n(n+1)/3 + n(n-1)$ flops are required, plus n^2 flops to backsolve the triangular system $\mathrm{U}\,\mathbf{x} = \mathbf{b}^{(n)}$. Therefore, about $(2n^3/3 + 2n^2)$ flops are needed to solve the linear system using GEM. Neglecting the lower order terms, we can state that the Gaussian elimination process has a cost of $2n^3/3$ flops.

As previously noticed, GEM terminates safely iff the pivotal elements $a_{kk}^{(k)}$, for $k = 1,\ldots,n-1$, are nonvanishing. Unfortunately, having nonnull diagonal entries in A is not enough to prevent zero pivots to arise during the elimination process. For example, matrix A in (3.33) is nonsingular and has nonzero diagonal entries

$$\mathrm{A} = \begin{bmatrix} 1 & 2 & 3 \\ 2 & 4 & 5 \\ 7 & 8 & 9 \end{bmatrix}, \; \mathrm{A}^{(2)} = \begin{bmatrix} 1 & 2 & 3 \\ 0 & \boxed{0} & -1 \\ 0 & -6 & -12 \end{bmatrix}. \tag{3.33}$$

Nevertheless, when GEM is applied, it is interrupted at the second step since $a_{22}^{(2)} = 0$.

More restrictive conditions on A are thus needed to ensure the applicability of the method. We shall see in Section 3.3.1 that if the leading dominating minors d_i of A are nonzero for $i = 1,\ldots,n-1$, then the corresponding pivotal entries $a_{ii}^{(i)}$ must necessarily be nonvanishing. We recall that d_i is the determinant of A_i, the i-th principal submatrix made by the first i rows and columns of A. The matrix in the previous example does not satisfy this condition, having $d_1 = 1$ and $d_2 = 0$.

Classes of matrices exist such that GEM can be always safely employed in its basic form (3.31). Among them, we recall the following ones:

1. matrices *diagonally dominant by rows*;
2. matrices *diagonally dominant by columns*. In such a case one can even show that the multipliers are in module less than or equal to 1 (see Property 3.2);
3. matrices *symmetric and positive definite* (see Theorem 3.6).

For a rigorous derivation of these results, we refer to the forthcoming sections.

3.3.1 GEM as a Factorization Method

In this section we show how GEM is equivalent to performing a factorization of the matrix A into the product of two matrices, A=LU, with U=$\mathrm{A}^{(n)}$. Since L and U depend only on A and not on the right hand side, the same factorization can be reused when solving several linear systems having the same matrix A but different right hand side **b**, with a considerable reduction of the operation count (indeed, the main computational effort, about $2n^3/3$ flops, is spent in the elimination procedure).

Let us go back to Example 3.2 concerning the Hilbert matrix H_3. In practice, to pass from $A^{(1)}=H_3$ to the matrix $A^{(2)}$ at the second step, we have multiplied the system by the matrix

$$M_1 = \begin{bmatrix} 1 & 0 & 0 \\ -\frac{1}{2} & 1 & 0 \\ -\frac{1}{3} & 0 & 1 \end{bmatrix} = \begin{bmatrix} 1 & 0 & 0 \\ -m_{21} & 1 & 0 \\ -m_{31} & 0 & 1 \end{bmatrix}.$$

Indeed,

$$M_1 A = M_1 A^{(1)} = \begin{bmatrix} 1 & \frac{1}{2} & \frac{1}{3} \\ 0 & \frac{1}{12} & \frac{1}{12} \\ 0 & \frac{1}{12} & \frac{4}{45} \end{bmatrix} = A^{(2)}.$$

Similarly, to perform the second (and last) step of GEM, we must multiply $A^{(2)}$ by the matrix

$$M_2 = \begin{bmatrix} 1 & 0 & 0 \\ 0 & 1 & 0 \\ 0 & -1 & 1 \end{bmatrix} = \begin{bmatrix} 1 & 0 & 0 \\ 0 & 1 & 0 \\ 0 & -m_{32} & 1 \end{bmatrix},$$

where $A^{(3)} = M_2 A^{(2)}$. Therefore

$$M_2 M_1 A = A^{(3)} = U. \tag{3.34}$$

On the other hand, matrices M_1 and M_2 are lower triangular, their product is still lower triangular, as is their inverse; thus, from (3.34) one gets

$$A = (M_2 M_1)^{-1} U = LU,$$

which is the desired factorization of A.

This identity can be generalized as follows. Setting

$$\mathbf{m}_k = [0, \ldots, 0, m_{k+1,k}, \ldots, m_{n,k}]^T \in \mathbb{R}^n,$$

and defining

$$
\mathrm{M}_k = \begin{bmatrix} 1 \ldots & 0 & 0 \ldots 0 \\ \vdots \ \ddots & \vdots & \vdots \quad \vdots \\ 0 & 1 & 0 \quad 0 \\ 0 & -m_{k+1,k} & 1 \quad 0 \\ \vdots \ \vdots & \vdots & \vdots \ \ddots \ \vdots \\ 0 \ldots & -m_{n,k} & 0 \ldots 1 \end{bmatrix} = \mathrm{I}_n - \mathbf{m}_k \mathbf{e}_k^T
$$

as the k-th *Gaussian transformation matrix*, one finds out that

$$
(\mathrm{M}_k)_{ip} = \delta_{ip} - (\mathbf{m}_k \mathbf{e}_k^T)_{ip} = \delta_{ip} - m_{ik}\delta_{kp}, \qquad i,p = 1,\ldots,n.
$$

On the other hand, from (3.31) we have that

$$
a_{ij}^{(k+1)} = a_{ij}^{(k)} - m_{ik}\delta_{kk}a_{kj}^{(k)} = \sum_{p=1}^{n} (\delta_{ip} - m_{ik}\delta_{kp}) a_{pj}^{(k)}, \qquad i,j = k+1,\ldots,n,
$$

or, equivalently,

$$
\mathrm{A}^{(k+1)} = \mathrm{M}_k \mathrm{A}^{(k)}. \tag{3.35}
$$

As a consequence, at the end of the elimination process the matrices M_k, with $k = 1,\ldots,n-1$, and the matrix U have been generated such that

$$
\mathrm{M}_{n-1}\mathrm{M}_{n-2}\cdots\mathrm{M}_1\mathrm{A} = \mathrm{U}.
$$

The matrices M_k are unit lower triangular with inverse given by

$$
\mathrm{M}_k^{-1} = 2\mathrm{I}_n - \mathrm{M}_k = \mathrm{I}_n + \mathbf{m}_k\mathbf{e}_k^T, \tag{3.36}
$$

while the matrix $(\mathbf{m}_i\mathbf{e}_i^T)(\mathbf{m}_j\mathbf{e}_j^T)$ is equal to the null matrix if $i \neq j$. As a consequence, we have

$$
\begin{aligned}
\mathrm{A} &= \mathrm{M}_1^{-1}\mathrm{M}_2^{-1}\cdots\mathrm{M}_{n-1}^{-1}\mathrm{U} \\
&= (\mathrm{I}_n + \mathbf{m}_1\mathbf{e}_1^T)(\mathrm{I}_n + \mathbf{m}_2\mathbf{e}_2^T)\cdots(\mathrm{I}_n + \mathbf{m}_{n-1}\mathbf{e}_{n-1}^T)\mathrm{U} \\
&= \left(\mathrm{I}_n + \sum_{i=1}^{n-1}\mathbf{m}_i\mathbf{e}_i^T\right)\mathrm{U} \\
&= \begin{bmatrix} 1 & 0 & \ldots & \ldots & 0 \\ m_{21} & 1 & & & \vdots \\ \vdots & m_{32} & \ddots & & \vdots \\ \vdots & \vdots & & \ddots & 0 \\ m_{n1} & m_{n2} & \ldots & m_{n,n-1} & 1 \end{bmatrix} \mathrm{U}.
\end{aligned} \tag{3.37}
$$

Defining $L = (M_{n-1}M_{n-2}\cdots M_1)^{-1} = M_1^{-1}\cdots M_{n-1}^{-1}$, it follows that

$$A = LU.$$

We notice that, due to (3.37), the subdiagonal entries of L are the multipliers m_{ik} produced by GEM, while the diagonal entries are equal to one.

Once the matrices L and U have been computed, solving the linear system consists only of solving successively the two triangular systems

$$\mathbf{L}\mathbf{y} = \mathbf{b},$$

$$\mathbf{U}\mathbf{x} = \mathbf{y}.$$

The computational cost of the factorization process is obviously the same as that required by GEM.

The following result establishes a link between the leading dominant minors of a matrix and its LU factorization induced by GEM.

Theorem 3.4 *Let $A \in \mathbb{R}^{n\times n}$. The LU factorization of A with $l_{ii} = 1$ for $i = 1,\ldots,n$ exists and is unique iff the principal submatrices A_i of A of order $i = 1,\ldots,n-1$ are nonsingular.*

Proof. The existence of the LU factorization can be proved following the steps of the GEM. Here we prefer to pursue an alternative approach, which allows for proving at the same time both existence and uniqueness and that will be used again in later sections.

Let us assume that the principal submatrices A_i of A are nonsingular for $i = 1,\ldots,n-1$ and prove, by induction on i, that under this hypothesis the LU factorization of $A(= A_n)$ with $l_{ii} = 1$ for $i = 1,\ldots,n$, exists and is unique.

The property is obviously true if $i = 1$. Assume therefore that there exists an unique LU factorization of A_{i-1} of the form $A_{i-1} = L^{(i-1)}U^{(i-1)}$ with $l_{kk}^{(i-1)} = 1$ for $k = 1,\ldots,i-1$, and show that there exists an unique factorization also for A_i. We partition A_i by block matrices as

$$A_i = \begin{bmatrix} A_{i-1} & \mathbf{c} \\ \mathbf{d}^T & a_{ii} \end{bmatrix},$$

and look for a factorization of A_i of the form

$$A_i = L^{(i)}U^{(i)} = \begin{bmatrix} L^{(i-1)} & \mathbf{0} \\ \mathbf{l}^T & 1 \end{bmatrix} \begin{bmatrix} U^{(i-1)} & \mathbf{u} \\ \mathbf{0}^T & u_{ii} \end{bmatrix}, \tag{3.38}$$

having also partitioned by blocks the factors $L^{(i)}$ and $U^{(i)}$. Computing the product of these two factors and equating by blocks the elements of A_i, it turns out that the vectors $\mathbf{l}$ and $\mathbf{u}$ are the solutions to the linear systems $L^{(i-1)}\mathbf{u} = \mathbf{c}$, $\mathbf{l}^TU^{(i-1)} = \mathbf{d}^T$.

On the other hand, since $0 \neq \det(A_{i-1}) = \det(L^{(i-1)})\det(U^{(i-1)})$, the matrices $L^{(i-1)}$ and $U^{(i-1)}$ are nonsingular and, as a result, $\mathbf{u}$ and $\mathbf{l}$ exist and are unique.

Thus, there exists a unique factorization of A_i, where u_{ii} is the unique solution of the equation $u_{ii} = a_{ii} - \mathbf{l}^T\mathbf{u}$. This completes the induction step of the proof.

It now remains to prove that, if the factorization at hand exists and is unique, then the first $n-1$ principal submatrices of A must be nonsingular. We shall distinguish the case where A is singular and when it is nonsingular.

Let us start from the second one and assume that the LU factorization of A with $l_{ii} = 1$ for $i = 1, \dots, n$, exists and is unique. Then, due to (3.38), we have $A_i = L^{(i)}U^{(i)}$ for $i = 1, \dots, n$. Thus

$$\det(A_i) = \det(L^{(i)})\det(U^{(i)}) = \det(U^{(i)}) = u_{11}u_{22}\cdots u_{ii}, \qquad (3.39)$$

from which, taking $i = n$ and A nonsingular, we obtain $u_{11}u_{22}\cdots u_{nn} \neq 0$, and thus, necessarily, $\det(A_i) = u_{11}u_{22}\cdots u_{ii} \neq 0$ for $i = 1, \dots, n-1$.

Now let A be a singular matrix and assume that (at least) one diagonal entry of U is equal to zero. Denote by u_{kk} the null entry of U with minimum index k. Thanks to (3.38), the factorization can be computed without troubles until the $k+1$-th step. From that step on, since the matrix $U^{(k)}$ is singular, existence and uniqueness of the vector $\mathbf{l}^T$ are certainly lost, and, thus, the same holds for the uniqueness of the factorization. In order for this not to occur before the process has factorized the whole matrix A, the u_{kk} entries must all be nonzero up to the index $k = n-1$ included, and thus, due to (3.39), all the principal submatrices A_k must be nonsingular for $k = 1, \dots, n-1$. ◇

From the above theorem we conclude that, if an A_i, with $i = 1, \dots, n-1$, is singular, then the factorization may either not exist or not be unique.

Example 3.3 Consider the matrices

$$B = \begin{bmatrix} 1 & 2 \\ 1 & 2 \end{bmatrix}, \quad C = \begin{bmatrix} 0 & 1 \\ 1 & 0 \end{bmatrix}, \quad D = \begin{bmatrix} 0 & 1 \\ 0 & 2 \end{bmatrix}.$$

According to Theorem 3.4, the singular matrix B, having nonsingular leading minor $B_1 = 1$, admits a unique LU factorization. The remaining two examples outline that, if the assumptions of the theorem are not fulfilled, the factorization may fail to exist or be unique.

Actually, the nonsingular matrix C, with C_1 singular, does not admit any factorization, while the (singular) matrix D, with D_1 singular, admits an infinite number of factorizations of the form $D = L_\beta U_\beta$, with

$$L_\beta = \begin{bmatrix} 1 & 0 \\ \beta & 1 \end{bmatrix}, \quad U_\beta = \begin{bmatrix} 0 & 1 \\ 0 & 2-\beta \end{bmatrix}, \quad \forall \beta \in \mathbb{R}.$$

•

In the case where the LU factorization is unique, we point out that, because $\det(A) = \det(LU) = \det(L)\det(U) = \det(U)$, the determinant of A is given by

$$\det(A) = u_{11} \cdots u_{nn}.$$

Let us now recall the following property (referring for its proof to [GL89] or [Hig96]).

Property 3.2 *If* A *is a matrix diagonally dominant by rows or by columns, then the LU factorization of* A *exists and is unique. In particular, if* A *is diagonally dominant by columns, then* $|l_{ij}| \leq 1 \; \forall i, j$.

In the proof of Theorem 3.4 we exploited the fact the the diagonal entries of L are equal to 1. In a similar manner, we could have fixed to 1 the diagonal entries of the upper triangular matrix U, obtaining a variant of GEM that will be considered in Section 3.3.4.

The freedom in setting up either the diagonal entries of L or those of U, implies that several LU factorizations exist which can be obtained one from the other by multiplication with a suitable diagonal matrix (see Section 3.4.1).

3.3.2 The Effect of Rounding Errors

If rounding errors are taken into account, the factorization process induced by GEM yields two matrices, $\widehat{\mathrm{L}}$ and $\widehat{\mathrm{U}}$, such that $\widehat{\mathrm{L}}\widehat{\mathrm{U}} = \mathrm{A} + \delta\mathrm{A}$, $\delta\mathrm{A}$ being a perturbation matrix. The size of such a perturbation can be estimated by

$$|\delta\mathrm{A}| \leq \frac{n\mathrm{u}}{1 - n\mathrm{u}} |\widehat{\mathrm{L}}| \, |\widehat{\mathrm{U}}|, \tag{3.40}$$

where u is the *roundoff* unit, under the assumption that $n\mathrm{u} < 1$. (For the proof of this result we refer to [Hig89].) From (3.40) it is seen that the presence of small pivotal entries can make the right side of the inequality virtually unbounded, with a consequent loss of control on the size of the perturbation matrix $\delta\mathrm{A}$. The interest is thus in finding out estimates like (3.40) of the form

$$|\delta\mathrm{A}| \leq g(\mathrm{u}) |\mathrm{A}|,$$

where $g(\mathrm{u})$ is a suitable positive function of u. For instance, assuming that $\widehat{\mathrm{L}}$ and $\widehat{\mathrm{U}}$ have nonnegative entries, then since $|\widehat{\mathrm{L}}| \, |\widehat{\mathrm{U}}| = |\widehat{\mathrm{L}}\widehat{\mathrm{U}}|$ one gets

$$|\widehat{\mathrm{L}}| \, |\widehat{\mathrm{U}}| = |\widehat{\mathrm{L}}\widehat{\mathrm{U}}| = |\mathrm{A} + \delta\mathrm{A}| \leq |\mathrm{A}| + |\delta\mathrm{A}| \leq |\mathrm{A}| + \frac{n\mathrm{u}}{1 - n\mathrm{u}} |\widehat{\mathrm{L}}| \, |\widehat{\mathrm{U}}|, \tag{3.41}$$

from which the desired bound is achieved by taking $g(\mathrm{u}) = n\mathrm{u}/(1 - 2n\mathrm{u})$, with $n\mathrm{u} < 1/2$.

The technique of pivoting, examined in Section 3.5, keeps the size of the pivotal entries under control and makes it possible to obtain estimates like (3.41) for any matrix.

3.3.3 Implementation of LU Factorization

Since L is a lower triangular matrix with diagonal entries equal to 1 and U is upper triangular, it is possible (and convenient) to store the LU factorization directly in the same memory area that is occupied by the matrix A. More precisely, U is stored in the upper triangular part of A (including the diagonal), whilst L occupies the lower triangular portion of A (the diagonal entries of L are not stored since they are implicitly assumed to be 1).

A coding of the algorithm is reported in Program 4. The output matrix A contains the overwritten LU factorization.

Program 4 - lukji : LU factorization of matrix A: *kji* version

```
function [A]=lukji(A)
% LUKJI LU factorization of a matrix A in the kji version
%  Y=LUKJI(A): U is stored in the upper triangular part of Y and L is stored
%  in the strict lower triangular part of Y.
[n,m]=size(A);
if n ~= m, error('Only square systems'); end
for k=1:n-1
   if A(k,k)==0; error('Null pivot element'); end
   A(k+1:n,k)=A(k+1:n,k)/A(k,k);
   for j=k+1:n
      i=[k+1:n]; A(i,j)=A(i,j)-A(i,k)*A(k,j);
   end
end
return
```

This implementation of the factorization algorithm is commonly referred to as the kji version, due to the order in which the cycles are executed. In a more appropriate notation, it is called the $SAXPY - kji$ version, due to the fact that the basic operation of the algorithm, which consists of multiplying a scalar A by a vector $\mathbf{X}$, summing another vector $\mathbf{Y}$ and then storing the result, is usually called SAXPY (i.e. *S*calar *A X P*lus *Y*).

The factorization can of course be executed by following a different order. In general, the forms in which the cycle on index i precedes the cycle on j are called *row-oriented*, whilst the others are called *column-oriented*. As usual, this terminology refers to the fact that the matrix is accessed by rows or by columns.

An example of LU factorization, jki version and column-oriented, is given in Program 5. This version is commonly called $GAXPY - jki$, since the basic operation (a product matrix-vector), is called GAXPY which stands for Generalized sAXPY (see for further details [DGK84]). In the GAXPY operation the scalar A of the SAXPY operation is replaced by a matrix.

Program 5 - lujki : LU factorization of matrix A: *jki* version

```
function [A]=lujki(A)
% LUJKI LU factorization of a matrix A in the jki version
% Y=LUJKI(A): U is stored in the upper triangular part of Y and L is stored
%  in the strict lower triangular part of Y.
[n,m]=size(A);
if n ~= m, error('Only square systems'); end
for j=1:n
```

```
  if A(j,j)==0; error('Null pivot element'); end
  for k=1:j-1
     i=[k+1:n]; A(i,j)=A(i,j)-A(i,k)*A(k,j);
  end
  i=[j+1:n];  A(i,j)=A(i,j)/A(j,j);
end
return
```

3.3.4 Compact Forms of Factorization

Remarkable variants of LU factorization are the Crout factorization and Doolittle factorization, and are known also as *compact forms* of the Gauss elimination method. This name is due to the fact that these approaches require less intermediate results than the standard GEM to generate the factorization of A.

Computing the LU factorization of A is formally equivalent to solving the following nonlinear system of n^2 equations

$$a_{ij} = \sum_{r=1}^{\min(i,j)} l_{ir} u_{rj}, \tag{3.42}$$

the unknowns being the $n^2 + n$ coefficients of the triangular matrices L and U. If we arbitrarily set n coefficients to 1, for example the diagonal entries of L or U, we end up with the Doolittle and Crout methods, respectively, which provide an efficient way to solve system (3.42).

In fact, supposing that the first $k-1$ columns of L and rows of U are available and setting $l_{kk} = 1$ (Doolittle method), the following equations are obtained from (3.42)

$$\begin{aligned} a_{kj} &= \sum_{r=1}^{k-1} l_{kr} u_{rj} + \boxed{u_{kj}}, \quad j = k, \ldots, n, \\ a_{ik} &= \sum_{r=1}^{k-1} l_{ir} u_{rk} + \boxed{l_{ik}}\, u_{kk}, \; i = k+1, \ldots, n. \end{aligned}$$

Note that these equations can be solved in a *sequential* way with respect to the boxed variables u_{kj} and l_{ik}. From the Doolittle compact method we thus obtain first the k-th row of U and then the k-th column of L, as follows: for $k = 1, \ldots, n$

$$\begin{aligned} u_{kj} &= a_{kj} - \sum_{r=1}^{k-1} l_{kr} u_{rj}, \qquad j = k, \ldots, n, \\ l_{ik} &= \frac{1}{u_{kk}} \left(a_{ik} - \sum_{r=1}^{k-1} l_{ir} u_{rk} \right), \; i = k+1, \ldots, n. \end{aligned} \tag{3.43}$$

The Crout factorization is generated similarly, computing first the k-th column of L and then the k-th row of U: for $k = 1, \ldots, n$

$$l_{ik} = a_{ik} - \sum_{r=1}^{k-1} l_{ir} u_{rk}, \qquad i = k, \ldots, n,$$

$$u_{kj} = \frac{1}{l_{kk}} \left(a_{kj} - \sum_{r=1}^{k-1} l_{kr} u_{rj} \right), \; j = k+1, \ldots, n,$$

where we set $u_{kk} = 1$. Recalling the notations introduced above, the Doolittle factorization is nothing but the ijk version of GEM.

We provide in Program 6 the implementation of the Doolittle scheme. Notice that now the main computation is a dot product, so this scheme is also known as the $DOT - ijk$ version of GEM.

Program 6 - luijk : LU factorization of the matrix A: ijk version

```
function [A]=luijk(A)
% LUIJK LU factorization of a matrix A in the ijk version
% Y=LUIJK(A): U is stored in the upper triangular part of Y and L is stored
%   in the strict lower triangular part of Y.
[n,m]=size(A);
if n ~= m, error('Only square systems'); end
for i=1:n
   for j=2:i
      if A(j,j)==0; error('Null pivot element'); end
      A(i,j-1)=A(i,j-1)/A(j-1,j-1);
      k=[1:j-1];   A(i,j)=A(i,j)-A(i,k)*A(k,j);
   end
   k=[1:i-1];
   for j=i+1:n
      A(i,j)=A(i,j)-A(i,k)*A(k,j);
   end
end
return
```

3.4 Other Types of Factorization

We now address factorizations suitable for symmetric and rectangular matrices.

3.4.1 LDM^T Factorization

It is possible to devise other types of factorizations of A. Specifically, we will address some variants where the factorization of A is of the form

$$A = LDM^T,$$

where L, M^T and D are lower triangular, upper triangular and diagonal matrices, respectively.

After the construction of this factorization, the resolution of the system can be carried out solving first the lower triangular system $\mathbf{L}\mathbf{y}=\mathbf{b}$, then the diagonal one $\mathbf{D}\mathbf{z}=\mathbf{y}$, and finally the upper triangular system $\mathbf{M}^T\mathbf{x}=\mathbf{z}$, with a cost of $n^2 + n$ flops. In the symmetric case, we obtain M = L and the LDL^T factorization can be computed with half the cost (see Section 3.4.2).

The LDM^T factorization enjoys a property analogous to the one in Theorem 3.4 for the LU factorization. In particular, the following result holds.

Theorem 3.5 *If all the principal minors of a matrix* $A \in \mathbb{R}^{n\times n}$ *are nonzero then there exist a unique diagonal matrix* D*, a unique unit lower triangular matrix* L *and a unique unit upper triangular matrix* M^T*, such that* $A = LDM^T$.

Proof. By Theorem 3.4 we already know that there exists a unique LU factorization of A with $l_{ii} = 1$ for $i = 1, \ldots, n$. If we set the diagonal entries of D equal to u_{ii} (nonzero because U is nonsingular), then $A = LU = LD(D^{-1}U)$. Upon defining $M^T = D^{-1}U$, the existence of the LDM^T factorization follows, where $D^{-1}U$ is a unit upper triangular matrix. The uniqueness of the LDM^T factorization is a consequence of the uniqueness of the LU factorization. ◇

The above proof shows that, since the diagonal entries of D coincide with those of U, we could compute L, M^T and D starting from the LU factorization of A. It suffices to compute M^T as $D^{-1}U$. Nevertheless, this algorithm has the same cost as the standard LU factorization. Likewise, it is also possible to compute the three matrices of the factorization by enforcing the identity $A=LDM^T$ entry by entry.

3.4.2 Symmetric and Positive Definite Matrices: The Cholesky Factorization

As already pointed out, the factorization LDM^T simplifies considerably when A is symmetric because in such a case M=L, yielding the so-called LDL^T factorization. The computational cost halves, with respect to the LU factorization, to about $(n^3/3)$ flops.

As an example, the Hilbert matrix of order 3 admits the following LDL^T factorization

$$H_3 = \begin{bmatrix} 1 & \frac{1}{2} & \frac{1}{3} \\ \frac{1}{2} & \frac{1}{3} & \frac{1}{4} \\ \frac{1}{3} & \frac{1}{4} & \frac{1}{5} \end{bmatrix} = \begin{bmatrix} 1 & 0 & 0 \\ \frac{1}{2} & 1 & 0 \\ \frac{1}{3} & 1 & 1 \end{bmatrix} \begin{bmatrix} 1 & 0 & 0 \\ 0 & \frac{1}{12} & 0 \\ 0 & 0 & \frac{1}{180} \end{bmatrix} \begin{bmatrix} 1 & \frac{1}{2} & \frac{1}{3} \\ 0 & 1 & 1 \\ 0 & 0 & 1 \end{bmatrix}.$$

In the case that A is also positive definite, the diagonal entries of D in the LDL^T factorization are positive. Moreover, we have the following result.

Theorem 3.6 *Let* $A \in \mathbb{R}^{n\times n}$ *be a symmetric and positive definite matrix. Then, there exists a unique upper triangular matrix* H *with positive diagonal entries such that*

$$A = H^T H. \tag{3.44}$$

This factorization is called Cholesky factorization and the entries h_{ij} *of* H^T *can be computed as follows:* $h_{11} = \sqrt{a_{11}}$ *and, for* $i = 2, \ldots, n$,

$$\begin{aligned} h_{ij} &= \left(a_{ij} - \sum_{k=1}^{j-1} h_{ik}h_{jk}\right) / h_{jj},\ j = 1, \ldots, i-1, \\ h_{ii} &= \left(a_{ii} - \sum_{k=1}^{i-1} h_{ik}^2\right)^{1/2}. \end{aligned} \tag{3.45}$$

Proof. Let us prove the theorem proceeding by induction on the size i of the matrix (as done in Theorem 3.4), recalling that if $A_i \in \mathbb{R}^{i\times i}$ is symmetric positive definite, then all its principal submatrices enjoy the same property.

For $i = 1$ the result is obviously true. Thus, suppose that it holds for $i-1$ and prove that it also holds for i. There exists an upper triangular matrix H_{i-1} such that $A_{i-1} = H_{i-1}^T H_{i-1}$. Let us partition A_i as

$$A_i = \begin{bmatrix} A_{i-1} & \mathbf{v} \\ \mathbf{v}^T & \alpha \end{bmatrix},$$

with $\alpha \in \mathbb{R}^+$, $\mathbf{v} \in \mathbb{R}^{i-1}$ and look for a factorization of A_i of the form

$$A_i = H_i^T H_i = \begin{bmatrix} H_{i-1}^T & \mathbf{0} \\ \mathbf{h}^T & \beta \end{bmatrix} \begin{bmatrix} H_{i-1} & \mathbf{h} \\ \mathbf{0}^T & \beta \end{bmatrix}.$$

Enforcing the equality with the entries of A_i yields the equations $H_{i-1}^T \mathbf{h} = \mathbf{v}$ and $\mathbf{h}^T\mathbf{h} + \beta^2 = \alpha$. The vector $\mathbf{h}$ is thus uniquely determined, since H_{i-1}^T is nonsingular. As for β, due to the properties of determinants

$$0 < \det(A_i) = \det(H_i^T)\det(H_i) = \beta^2(\det(H_{i-1}))^2,$$

we can conclude that it must be a real number. As a result, $\beta = \sqrt{\alpha - \mathbf{h}^T\mathbf{h}}$ is the desired diagonal entry and this concludes the inductive argument.

Let us now prove formulae (3.45). The fact that $h_{11} = \sqrt{a_{11}}$ is an immediate consequence of the induction argument for $i = 1$. In the case of a generic i, relations $(3.45)_1$ are the forward substitution formulae for the solution of the linear system $H_{i-1}^T\mathbf{h} = \mathbf{v} = (a_{1i}, a_{2i}, \ldots, a_{i-1,i})^T$, while formulae $(3.45)_2$ state that $\beta = \sqrt{\alpha - \mathbf{h}^T\mathbf{h}}$, where $\alpha = a_{ii}$. ◇

The algorithm which implements (3.45) requires about $(n^3/3)$ flops and it turns out to be stable with respect to the propagation of rounding errors. It can indeed be shown that the upper triangular matrix $\tilde{H}$ is such that $\tilde{H}^T\tilde{H} = A + \delta A$, where δA is a pertubation matrix such that $\|\delta A\|_2 \le 8n(n+1)\mathbf{u}\|A\|_2$, when the rounding errors are considered and assuming that $2n(n+1)\mathrm{u} \leq 1-(n+1)\mathrm{u}$ (see [Wil68]).

Also, for the Cholesky factorization it is possible to overwrite the matrix H^T in the lower triangular portion of A, without any further memory storage. By doing so, both A and the factorization are preserved, noting that A is stored in the upper triangular section since it is symmetric and that its diagonal entries can be computed as $a_{11} = h_{11}^2$, $a_{ii} = h_{ii}^2 + \sum_{k=1}^{i-1} h_{ik}^2$, $i = 2, \ldots, n$. An example of implementation of the Cholesky factorization is coded in Program 7.

Program 7 - chol2 : Cholesky factorization

```
function [A]=chol2(A)
% CHOL2 Cholesky factorization of a s.p.d. matrix A.
% R=CHOL2(A) produces an upper triangular matrix R such that R'*R=A.
[n,m]=size(A);
if n ~= m, error('Only square systems'); end
for k=1:n-1
   if A(k,k) ¡= 0, error('Null or negative pivot element'); end
   A(k,k)=sqrt(A(k,k));  A(k+1:n,k)=A(k+1:n,k)/A(k,k);
   for j=k+1:n,  A(j:n,j)=A(j:n,j)-A(j:n,k)*A(j,k);  end
end
A(n,n)=sqrt(A(n,n));
A = tril(A); A=A';
return
```

3.4.3 Rectangular Matrices: The QR Factorization

Definition 3.1 A matrix $\mathrm{A} \in \mathbb{R}^{m\times n}$, with $m \geq n$, admits a *QR factorization* if there exist an orthogonal matrix $\mathrm{Q} \in \mathbb{R}^{m\times m}$ and an upper trapezoidal matrix $\mathrm{R} \in \mathbb{R}^{m\times n}$ with null rows from the $n+1$-th one on, such that

$$\mathrm{A} = \mathrm{QR}. \tag{3.46}$$

■

This factorization can be constructed either using suitable transformation matrices (Givens or Householder matrices, see Section 5.6.1) or using the Gram-Schmidt orthogonalization algorithm discussed below.

It is also possible to generate a reduced version of the QR factorization (3.46), as stated in the following result.

Property 3.3 *Let* $\mathrm{A} \in \mathbb{R}^{m\times n}$ *be a matrix of rank* n *for which a QR factorization is known. Then there exists a unique factorization of* A *of the form*

$$\mathrm{A} = \widetilde{\mathrm{Q}}\widetilde{\mathrm{R}} \tag{3.47}$$

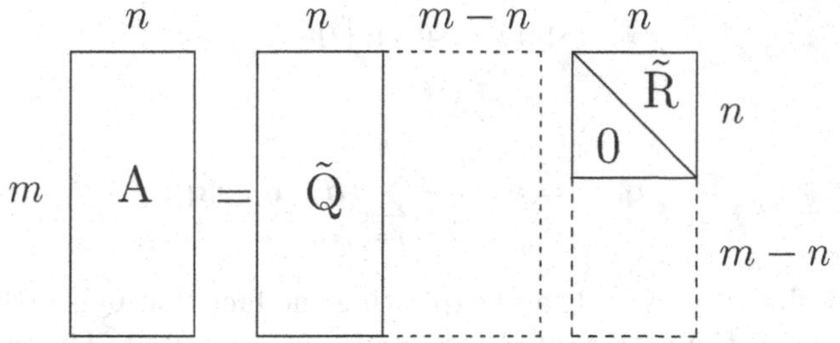

Fig. 3.1. The reduced factorization. The matrices of the QR factorization are drawn in dashed lines

where $\widetilde{\mathrm{Q}}$ *and* $\widetilde{\mathrm{R}}$ *are submatrices of* Q *and* R *given respectively by*

$$\widetilde{\mathrm{Q}} = \mathrm{Q}(1:m, 1:n),\ \widetilde{\mathrm{R}} = \mathrm{R}(1:n, 1:n). \tag{3.48}$$

Moreover, $\widetilde{\mathrm{Q}}$ *has orthonormal vector columns and* $\widetilde{\mathrm{R}}$ *is upper triangular and coincides with the Cholesky factor* H *of the symmetric positive definite matrix* $\mathrm{A}^T\mathrm{A}$, *that is,* $\mathrm{A}^T\mathrm{A} = \widetilde{\mathrm{R}}^T\widetilde{\mathrm{R}}$.

If A has rank n (i.e., full rank), then the column vectors of $\tilde{\mathrm{Q}}$ form an orthonormal basis for the vector space range(A) (defined in (1.5)). As a consequence, constructing the QR factorization can also be interpreted as a procedure for generating an orthonormal basis for a given set of vectors. If A has rank $r < n$, the QR factorization does not necessarily yield an orthonormal basis for range(A). However, one can obtain a factorization of the form

$$\mathrm{Q}^T\mathrm{AP} = \begin{bmatrix} \mathrm{R}_{11} & \mathrm{R}_{12} \\ 0 & 0 \end{bmatrix},$$

where Q is orthogonal, P is a permutation matrix and R_{11} is a nonsingular upper triangular matrix of order r.

In general, when using the QR factorization, we shall always refer to its reduced form (3.47) as it finds a remarkable application in the solution of overdetermined systems (see Section 3.13).

The matrix factors $\tilde{\mathrm{Q}}$ and $\tilde{\mathrm{R}}$ in (3.47) can be computed using the *Gram-Schmidt orthogonalization.* Starting from a set of linearly independent vectors, $\mathbf{x}_1, \ldots, \mathbf{x}_n$, this algorithm generates a new set of mutually orthogonal vectors, $\mathbf{q}_1, \ldots, \mathbf{q}_n$, given by

$$\begin{aligned} &\mathbf{q}_1 = \mathbf{x}_1, \\ &\mathbf{q}_{k+1} = \mathbf{x}_{k+1} - \sum_{i=1}^{k} \frac{(\mathbf{q}_i, \mathbf{x}_{k+1})}{(\mathbf{q}_i, \mathbf{q}_i)} \mathbf{q}_i, \qquad k = 1, \ldots, n-1. \end{aligned} \tag{3.49}$$

Denoting by $\mathbf{a}_1, \ldots, \mathbf{a}_n$ the column vectors of A, we set $\tilde{\mathbf{q}}_1 = \mathbf{a}_1/\|\mathbf{a}_1\|_2$ and, for $k = 1, \ldots, n-1$, compute the column vectors of $\tilde{\mathrm{Q}}$ as

$$\tilde{\mathbf{q}}_{k+1} = \mathbf{q}_{k+1}/\|\mathbf{q}_{k+1}\|_2,$$

where

$$\mathbf{q}_{k+1} = \mathbf{a}_{k+1} - \sum_{j=1}^{k} (\tilde{\mathbf{q}}_j, \mathbf{a}_{k+1})\tilde{\mathbf{q}}_j.$$

Next, imposing that $\mathrm{A}=\tilde{\mathrm{Q}}\tilde{\mathrm{R}}$ and exploiting the fact that $\tilde{\mathrm{Q}}$ is orthogonal (that is, $\tilde{\mathrm{Q}}^{-1} = \tilde{\mathrm{Q}}^T$), the entries of $\tilde{\mathrm{R}}$ can easily be computed. The overall computational cost of the algorithm is of the order of mn^2 flops.

It is also worth noting that if A has full rank, the matrix $\mathrm{A}^T\mathrm{A}$ is symmetric and positive definite (see Section 1.9) and thus it admits a unique Cholesky factorization of the form $\mathrm{H}^T\mathrm{H}$. On the other hand, since the orthogonality of $\tilde{\mathrm{Q}}$ implies

$$\mathrm{H}^T\mathrm{H} = \mathrm{A}^T\mathrm{A} = \tilde{\mathrm{R}}^T\tilde{\mathrm{Q}}^T\tilde{\mathrm{Q}}\tilde{\mathrm{R}} = \tilde{\mathrm{R}}^T\tilde{\mathrm{R}},$$

we conclude that $\tilde{\mathrm{R}}$ is actually the Cholesky factor H of $\mathrm{A}^T\mathrm{A}$. Thus, the diagonal entries of $\tilde{\mathrm{R}}$ are all nonzero only if A has full rank.
The Gram-Schmidt method is of little practical use since the generated vectors lose their linear independence due to rounding errors. Indeed, in *floating-point* arithmetic the algorithm produces very small values of $\|\mathbf{q}_{k+1}\|_2$ and $\tilde{r}_{kk}$ with a consequent numerical instability and loss of orthogonality for matrix $\tilde{\mathrm{Q}}$ (see Example 3.4).

These drawbacks suggest employing a more stable version, known as *modified Gram-Schmidt method.* At the beginning of the $k+1$-th step, the projections of the vector $\mathbf{a}_{k+1}$ along the vectors $\tilde{\mathbf{q}}_1, \ldots, \tilde{\mathbf{q}}_k$ are progressively subtracted from $\mathbf{a}_{k+1}$. On the resulting vector, the orthogonalization step is then carried out. In practice, after computing $(\tilde{\mathbf{q}}_1, \mathbf{a}_{k+1})\tilde{\mathbf{q}}_1$ at the $k+1$-th step, this vector is immediately subtracted from $\mathbf{a}_{k+1}$. As an example, one lets

$$\mathbf{a}_{k+1}^{(1)} = \mathbf{a}_{k+1} - (\tilde{\mathbf{q}}_1, \mathbf{a}_{k+1})\tilde{\mathbf{q}}_1.$$

This new vector $\mathbf{a}_{k+1}^{(1)}$ is projected along the direction of $\tilde{\mathbf{q}}_2$ and the obtained projection is subtracted from $\mathbf{a}_{k+1}^{(1)}$, yielding

$$\mathbf{a}_{k+1}^{(2)} = \mathbf{a}_{k+1}^{(1)} - (\tilde{\mathbf{q}}_2, \mathbf{a}_{k+1}^{(1)})\tilde{\mathbf{q}}_2$$

and so on, until $\mathbf{a}_{k+1}^{(k)}$ is computed.

It can be checked that $\mathbf{a}_{k+1}^{(k)}$ coincides with the corresponding vector $\mathbf{q}_{k+1}$ in the standard Gram-Schmidt process, since, due to the orthogonality of vectors $\tilde{\mathbf{q}}_1, \tilde{\mathbf{q}}_2, \ldots, \tilde{\mathbf{q}}_k$,

$$\begin{aligned}\mathbf{a}_{k+1}^{(k)} &= \mathbf{a}_{k+1} - (\tilde{\mathbf{q}}_1, \mathbf{a}_{k+1})\tilde{\mathbf{q}}_1 - (\tilde{\mathbf{q}}_2, \mathbf{a}_{k+1} - (\tilde{\mathbf{q}}_1, \mathbf{a}_{k+1})\tilde{\mathbf{q}}_1)\,\tilde{\mathbf{q}}_2 + \ldots \\ &= \mathbf{a}_{k+1} - \sum_{j=1}^{k} (\tilde{\mathbf{q}}_j, \mathbf{a}_{k+1})\tilde{\mathbf{q}}_j.\end{aligned}$$

Program 8 implements the modified Gram-Schmidt method. Notice that it is not possible to overwrite the computed QR factorization on the matrix A. In general, the matrix $\widetilde{R}$ is overwritten on A, whilst $\widetilde{Q}$ is stored separately. The computational cost of the modified Gram-Schmidt method has the order of $2mn^2$ flops.

Program 8 - modgrams : Modified Gram-Schmidt method

```
function [Q,R]=modgrams(A)
% MODGRAMS QR factorization of a matrix A.
% [Q,R]=MODGRAMS(A) produces an upper trapezoidal matrix R and an orthogonal
% matrix Q such that Q*R=A.
[m,n]=size(A);
Q=zeros(m,n);  Q(1:m,1) = A(1:m,1);  R=zeros(n);  R(1,1)=1;
for k = 1:n
  R(k,k) = norm(A(1:m,k));
  Q(1:m,k) = A(1:m,k)/R(k,k);
  j=[k+1:n];
  R(k,j) = Q (1:m,k)'*A(1:m,j);
  A(1:m,j) = A (1:m,j)-Q(1:m,k)*R(k,j);
end
return
```

Example 3.4 Let us consider the Hilbert matrix H_4 of order 4 (see (3.32)). The matrix $\tilde{Q}$, generated by the standard Gram-Schmidt algorithm, is orthogonal up to the order of 10^{-10}, being

$$I - \tilde{Q}^T\tilde{Q} = 10^{-10}\begin{bmatrix} 0.0000 & -0.0000 & 0.0001 & -0.0041 \\ -0.0000 & 0 & 0.0004 & -0.0099 \\ 0.0001 & 0.0004 & 0 & -0.4785 \\ -0.0041 & -0.0099 & -0.4785 & 0 \end{bmatrix}$$

and $\|I - \tilde{Q}^T\tilde{Q}\|_\infty = 4.9247 \cdot 10^{-11}$. Using the modified Gram-Schmidt method, we would obtain

$$I - \tilde{Q}^T\tilde{Q} = 10^{-12}\begin{bmatrix} 0.0001 & -0.0005 & 0.0069 & -0.2853 \\ -0.0005 & 0 & -0.0023 & 0.0213 \\ 0.0069 & -0.0023 & 0.0002 & -0.0103 \\ -0.2853 & 0.0213 & -0.0103 & 0 \end{bmatrix}$$

and this time $\|I - \tilde{Q}^T\tilde{Q}\|_\infty = 3.1686 \cdot 10^{-13}$.

An improved result can be obtained using, instead of Program 8, the intrinsic function `qr` of MATLAB. This function can be properly employed to generate both the factorization (3.46) as well as its reduced version (3.47). •

3.5 Pivoting

As previously pointed out, the GEM process breaks down as soon as a zero pivotal entry is computed. In such an event, one needs to resort to the so-called *pivoting* technique, which amounts to exchanging rows (or columns) of the system in such a way that nonvanishing pivots are obtained.

Example 3.5 Let us go back to matrix (3.33) for which GEM furnishes at the second step a zero pivotal element. By simply exchanging the second row with the third one, we can execute one step further of the elimination method, finding a nonzero pivot. The generated system is equivalent to the original one and it can be noticed that it is already in upper triangular form. Indeed

$$\mathrm{A}^{(2)} = \begin{bmatrix} 1 & 2 & 3 \\ 0 & -6 & -12 \\ 0 & 0 & -1 \end{bmatrix} = \mathrm{U},$$

while the transformation matrices are given by

$$\mathrm{M}_1 = \begin{bmatrix} 1 & 0 & 0 \\ -2 & 1 & 0 \\ -7 & 0 & 1 \end{bmatrix}, \ \mathrm{M}_2 = \begin{bmatrix} 1 & 0 & 0 \\ 0 & 1 & 0 \\ 0 & 0 & 1 \end{bmatrix}.$$

From an algebraic standpoint, a *permutation* of the rows of A has been performed. In fact, it now no longer holds that $\mathrm{A}=\mathrm{M}_1^{-1}\mathrm{M}_2^{-1}\mathrm{U}$, but rather $\mathrm{A}=\mathrm{M}_1^{-1}\boxed{\mathrm{P}}\mathrm{M}_2^{-1}\mathrm{U}$, P being the permutation matrix

$$\mathrm{P} = \begin{bmatrix} 1 & 0 & 0 \\ 0 & 0 & 1 \\ 0 & 1 & 0 \end{bmatrix}. \tag{3.50}$$

•

The pivoting strategy adopted in Example 3.5 can be generalized by looking, at each step k of the elimination procedure, for a nonzero pivotal entry by searching within the entries of the subcolumn $\mathrm{A}^{(k)}(k:n,k)$. For that reason, it is called *partial pivoting* (by rows).

From (3.30) it can be seen that a large value of m_{ik} (generated for example by a small value of the pivot $a_{kk}^{(k)}$) might amplify the rounding errors affecting the entries $a_{kj}^{(k)}$. Therefore, in order to ensure a better stability, the pivotal element is chosen as the largest entry (in module) of the column $\mathrm{A}^{(k)}(k:n,k)$ and partial pivoting is generally performed at every step of the elimination procedure, even if not strictly necessary (that is, even if nonzero pivotal entries are found).

Alternatively, the searching process could have been extended to the whole submatrix $\mathrm{A}^{(k)}(k:n,k:n)$, ending up with a *complete pivoting* (see Figure 3.2). Notice, however, that while partial pivoting requires an additional cost of about n^2 searches, complete pivoting needs about $2n^3/3$, with a considerable increase of the computational cost of GEM.

Example 3.6 Let us consider the linear system $\mathrm{A}\mathbf{x} = \mathbf{b}$ with

$$\mathrm{A} = \begin{bmatrix} 10^{-13} & 1 \\ 1 & 1 \end{bmatrix}$$

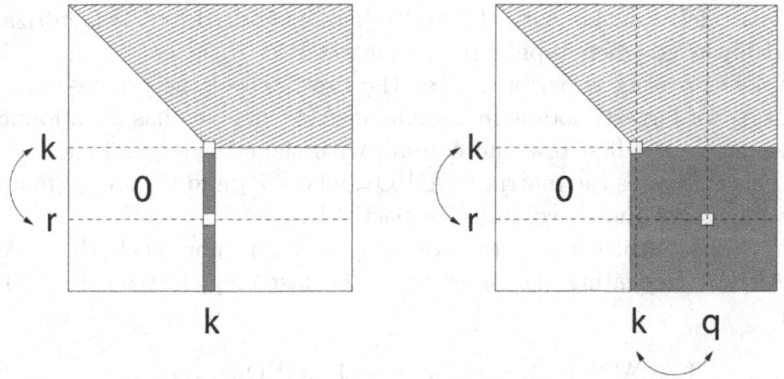

Fig. 3.2. Partial pivoting by row (*left*) or complete pivoting (*right*). Darker areas of the matrix are those involved in the searching for the pivotal entry

and where $\mathbf{b}$ is chosen in such a way that $\mathbf{x} = (1,1)^T$ is the exact solution. Suppose we use base 2 and 16 significant digits. GEM without pivoting would give $\mathbf{x}_{GEM} = [0.99920072216264, 1]^T$, while GEM plus partial pivoting furnishes the exact solution up to the 16-th digit. •

Let us analyze how partial pivoting affects the LU factorization induced by GEM. At the first step of GEM with partial pivoting, after finding out the entry a_{r1} of maximum module in the first column, the elementary permutation matrix P_1 which exchanges the first row with the r-th row is constructed (if $r = 1$, P_1 is the identity matrix). Next, the first Gaussian transformation matrix M_1 is generated and we set $\mathrm{A}^{(2)} = \mathrm{M}_1\mathrm{P}_1\mathrm{A}^{(1)}$. A similar approach is now taken on $\mathrm{A}^{(2)}$, searching for a new permutation matrix P_2 and a new matrix M_2 such that

$$\mathrm{A}^{(3)} = \mathrm{M}_2\mathrm{P}_2\mathrm{A}^{(2)} = \mathrm{M}_2\mathrm{P}_2\mathrm{M}_1\mathrm{P}_1\mathrm{A}^{(1)}.$$

Executing all the elimination steps, the resulting upper triangular matrix U is now given by

$$\mathrm{U} = \mathrm{A}^{(n)} = \mathrm{M}_{n-1}\mathrm{P}_{n-1}\cdots\mathrm{M}_1\mathrm{P}_1\mathrm{A}^{(1)}. \tag{3.51}$$

Letting $\mathrm{M} = \mathrm{M}_{n-1}\mathrm{P}_{n-1}\cdots\mathrm{M}_1\mathrm{P}_1$ and $\mathrm{P} = \mathrm{P}_{n-1}\cdots\mathrm{P}_1$, we obtain that U=MA and, thus, $\mathrm{U} = (\mathrm{MP}^{-1})\mathrm{PA}$. It can easily be checked that the matrix $\mathrm{L} = \mathrm{PM}^{-1}$ is unit lower triangular, so that the LU factorization reads

$$\mathrm{PA} = \mathrm{LU}. \tag{3.52}$$

One should not be worried by the presence of the inverse of M, since $\mathrm{M}^{-1} = \mathrm{P}_1^{-1}\mathrm{M}_1^{-1}\cdots\mathrm{P}_{n-1}^{-1}\mathrm{M}_{n-1}^{-1}$ and $\mathrm{P}_i^{-1} = \mathrm{P}_i^T$, while $\mathrm{M}_i^{-1} = 2\mathrm{I}_n - \mathrm{M}_i$.

Once L, U and P are available, solving the initial linear system amounts to solving the triangular systems $\mathrm{L}\mathbf{y} = \mathrm{P}\mathbf{b}$ and $\mathrm{U}\mathbf{x} = \mathbf{y}$. Notice that the entries

of the matrix L coincide with the multipliers computed by LU factorization, without pivoting, when applied to the matrix PA.
If complete pivoting is performed, at the first step of the process, once the element a_{qr} of largest module in submatrix $\mathrm{A}(1:n, 1:n)$ has been found, we must exchange the first row and column with the q-th row and the r-th column. This generates the matrix $\mathrm{P}_1\mathrm{A}^{(1)}\mathrm{Q}_1$, where P_1 and Q_1 are permutation matrices by rows and by columns, respectively.

As a consequence, the action of matrix M_1 is now such that $\mathrm{A}^{(2)} = \mathrm{M}_1\mathrm{P}_1\mathrm{A}^{(1)}\mathrm{Q}_1$. Repeating the process, at the last step, instead of (3.51) we obtain

$$\mathrm{U} = \mathrm{A}^{(n)} = \mathrm{M}_{n-1}\mathrm{P}_{n-1}\cdots\mathrm{M}_1\mathrm{P}_1\mathrm{A}^{(1)}\mathrm{Q}_1\cdots\mathrm{Q}_{n-1}.$$

In the case of complete pivoting the LU factorization becomes

$$\mathrm{PAQ} = \mathrm{LU},$$

where $\mathrm{Q} = \mathrm{Q}_1\cdots\mathrm{Q}_{n-1}$ is a permutation matrix accounting for all permutations that have been operated. By construction, matrix L is still lower triangular, with module entries less than or equal to 1. As happens in partial pivoting, the entries of L are the multipliers produced by the LU factorization process without pivoting, when applied to the matrix PAQ.

Program 9 is an implementation of the LU factorization with complete pivoting. For an efficient computer implementation of the LU factorization with partial pivoting, we refer to the MATLAB intrinsic function `lu`.

Program 9 - LUpivtot : LU factorization with complete pivoting

```
function [L,U,P,Q]=LUpivtot(A)
%LUPIVTOT LU factorization with complete pivoting
% [L,U,P,Q]=LUPIVTOT(A) returns unit lower triangular matrix L, upper
% triangular matrix U and permutation matrices P and Q so that P*A*Q=L*U.
[n,m]=size(A);
if n ~= m, error('Only square systems'); end
P=eye(n); Q=P; Minv=P; I=eye(n);
for k=1:n-1
   [Pk,Qk]=pivot(A,k,n,I); A=Pk*A*Qk;
   [Mk,Mkinv]=MGauss(A,k,n);
   A=Mk*A;   P=Pk*P;    Q=Q*Qk;
   Minv=Minv*Pk*Mkinv;
end
U=triu(A); L=P*Minv;
return

function [Mk,Mkinv]=MGauss(A,k,n)
Mk=eye(n);
i=[k+1:n];
Mk(i,k)=-A(i,k)/A(k,k);
```

```
Mkinv=2*eye(n)-Mk;
return

function [Pk,Qk]=pivot(A,k,n,I)
[y,i]=max(abs(A(k:n,k:n)));
[piv,jpiv]=max(y);
ipiv=i(jpiv);
jpiv=jpiv+k-1;
ipiv=ipiv+k-1;
Pk=I; Pk(ipiv,ipiv)=0; Pk(k,k)=0; Pk(k,ipiv)=1; Pk(ipiv,k)=1;
Qk=I; Qk(jpiv,jpiv)=0; Qk(k,k)=0; Qk(k,jpiv)=1; Qk(jpiv,k)=1;
return
```

Remark 3.3 The presence of large pivotal entries is not in itself sufficient to guarantee accurate solutions, as demonstrated by the following example (taken from [JM92]). For the linear system

$$\begin{bmatrix} -4000 & 2000 & 2000 \\ 2000 & 0.78125 & 0 \\ 2000 & 0 & 0 \end{bmatrix} \begin{bmatrix} x_1 \\ x_2 \\ x_3 \end{bmatrix} = \begin{bmatrix} 400 \\ 1.3816 \\ 1.9273 \end{bmatrix},$$

at the first step the pivotal entry coincides with the diagonal entry -4000 itself. However, executing GEM with 8 digits on such a matrix yields the solution

$$\widehat{\mathbf{x}} = [0.00096365, \ -0.698496, \ 0.90042329]^T,$$

whose first component drastically differs from that of the exact solution $\mathbf{x} = [1.9273, \ -0.698496, \ 0.9004233]^T$. The cause of this behaviour should be ascribed to the wide variations among the system coefficients. Such cases can be remedied by a suitable *scaling* of the matrix (see Section 3.12.1). ■

Remark 3.4 (Pivoting for symmetric matrices) As already noticed, pivoting is not strictly necessary if A is symmetric and positive definite. A separate comment is deserved when A is symmetric but not positive definite, since pivoting could destroy the symmetry of the matrix. This can be avoided by employing a complete pivoting of the form PAP^T, even though this pivoting can only turn out into a reordering of the *diagonal* entries of A. As a consequence, the presence on the diagonal of A of small entries might inhibit the advantages of the pivoting. To deal with matrices of this kind, special algorithms are needed (like the Parlett-Reid method [PR70] or the Aasen method [Aas71]) for whose description we refer to [GL89], and to [JM92] for the case of sparse matrices. ■

3.6 Computing the Inverse of a Matrix

The explicit computation of the inverse of a matrix can be carried out using the LU factorization as follows. Denoting by X the inverse of a nonsingular matrix A$\in \mathbb{R}^{n\times n}$, the column vectors of X are the solutions to the linear systems $\mathrm{A}\mathbf{x}_i = \mathbf{e}_i$, for $i = 1, \ldots, n$.

Supposing that PA=LU, where P is the partial pivoting permutation matrix, we must solve $2n$ triangular systems of the form

$$\mathrm{L}\mathbf{y}_i = \mathrm{P}\mathbf{e}_i,\ \mathrm{U}\mathbf{x}_i = \mathbf{y}_i,\ i = 1, \ldots, n,$$

i.e., a succession of linear systems having the same coefficient matrix but different right hand sides. The computation of the inverse of a matrix is a costly procedure which can sometimes be even less stable than GEM (see [Hig88]).

An alternative approach for computing the inverse of A is provided by the *Faddev* or *Leverrier formula*, which, letting B_0=I, recursively computes

$$\alpha_k = \frac{1}{k}\mathrm{tr}(\mathrm{AB}_{k-1}),\ \mathrm{B}_k = -\mathrm{AB}_{k-1} + \alpha_k \mathrm{I},\ k = 1, 2, \ldots, n.$$

Since $\mathrm{B}_n = 0$, if $\alpha_n \neq 0$ we get

$$\mathrm{A}^{-1} = \frac{1}{\alpha_n}\mathrm{B}_{n-1},$$

and the computational cost of the method for a full matrix is equal to $(n-1)n^3$ flops (for further details see [FF63], [Bar89]).

3.7 Banded Systems

Discretization methods for boundary value problems often lead to solving linear systems with matrices having banded, block or sparse forms. Exploiting the structure of the matrix allows for a dramatic reduction in the computational costs of the factorization and of the substitution algorithms. In the present and forthcoming sections, we shall address special variants of GEM or LU factorization that are properly devised for dealing with matrices of this kind. For the proofs and a more comprehensive treatment, we refer to [GL89] and [Hig88] for banded or block matrices, while we refer to [JM92], [GL81] and [Saa96] for sparse matrices and the techniques for their storage.

The main result for banded matrices is the following.

Property 3.4 *Let* A$\in \mathbb{R}^{n\times n}$. *Suppose that there exists a* LU *factorization of* A. *If* A *has upper bandwidth q and lower bandwidth p, then* L *has lower bandwidth p and* U *has upper bandwidth q.*

In particular, notice that the same memory area used for A is enough to also store its LU factorization. Consider, indeed, that a matrix A having upper bandwidth q and lower bandwidth p is usually stored in a matrix B $(p + q + 1) \times n$, assuming that

$$b_{i-j+q+1,j} = a_{ij}$$

for all the indices i, j that fall into the band of the matrix. For instance, in the case of the tridiagonal matrix A=tridiag$_5(-1, 2, -1)$ (where $q = p = 1$), the compact storage reads

$$\mathrm{B} = \begin{bmatrix} 0 & -1 & -1 & -1 & -1 \\ 2 & 2 & 2 & 2 & 2 \\ -1 & -1 & -1 & -1 & 0 \end{bmatrix}.$$

The same format can be used for storing the factorization LU of A. It is clear that this storage format can be quite inconvenient in the case where only a few bands of the matrix are large. In the limit, if only one column and one row were full, we would have $p = q = n$ and thus B would be a full matrix with a lot of zero entries.

Finally, we notice that the inverse of a banded matrix is generally full (as happens for the matrix A considered above).

3.7.1 Tridiagonal Matrices

Consider the particular case of a linear system with nonsingular tridiagonal matrix A given by

$$\mathrm{A} = \begin{bmatrix} a_1 & c_1 & & 0 \\ b_2 & a_2 & \ddots & \\ & \ddots & & c_{n-1} \\ 0 & & b_n & a_n \end{bmatrix}.$$

In such an event, the matrices L and U of the LU factorization of A are bidiagonal matrices of the form

$$\mathrm{L} = \begin{bmatrix} 1 & & 0 \\ \beta_2 & 1 & \\ & \ddots & \ddots & \\ 0 & & \beta_n & 1 \end{bmatrix}, \mathrm{U} = \begin{bmatrix} \alpha_1 & c_1 & & 0 \\ & \alpha_2 & \ddots & \\ & & \ddots & c_{n-1} \\ 0 & & & \alpha_n \end{bmatrix}.$$

The coefficients α_i and β_i can easily be computed by the following relations

$$\alpha_1 = a_1, \; \beta_i = \frac{b_i}{\alpha_{i-1}}, \; \alpha_i = a_i - \beta_i c_{i-1}, \; i = 2, \ldots, n. \tag{3.53}$$

This is known as the *Thomas algorithm* and can be regarded as a particular instance of the Doolittle factorization, without pivoting. When one is not interested in storing the coefficients of the original matrix, the entries α_i and β_i can be overwritten on A.

The Thomas algorithm can also be extended to solve the whole tridiagonal system $\mathbf{Ax} = \mathbf{f}$. This amounts to solving two bidiagonal systems $\mathbf{Ly} = \mathbf{f}$ and $\mathbf{Ux} = \mathbf{y}$, for which the following formulae hold:

$$(\mathbf{Ly} = \mathbf{f}) \; y_1 = f_1, \, y_i = f_i - \beta_i y_{i-1}, \, i = 2, \dots, n, \qquad (3.54)$$

$$(\mathbf{Ux} = \mathbf{y}) \; x_n = \frac{y_n}{\alpha_n}, \, x_i = (y_i - c_i x_{i+1}) / \alpha_i, \, i = n-1, \dots, 1. \qquad (3.55)$$

The algorithm requires only $8n - 7$ flops: precisely, $3(n-1)$ flops for the factorization (3.53) and $5n - 4$ flops for the substitution procedure (3.54)-(3.55).

As for the stability of the method, if A is a nonsingular tridiagonal matrix and $\widehat{\mathrm{L}}$ and $\widehat{\mathrm{U}}$ are the factors actually computed, then

$$|\delta \mathrm{A}| \le (4\mathrm{u} + 3\mathrm{u}^2 + \mathrm{u}^3)|\widehat{\mathrm{L}}| \, |\widehat{\mathrm{U}}|,$$

where δA is implicitly defined by the relation $\mathrm{A} + \delta\mathrm{A} = \widehat{\mathrm{L}}\widehat{\mathrm{U}}$ while $\mathbf{u}$ is the *roundoff* unit. In particular, if A is also symmetric and positive definite or it is an M-matrix, we have

$$|\delta \mathrm{A}| \le \frac{4\mathrm{u} + 3\mathrm{u}^2 + \mathrm{u}^3}{1 - \mathrm{u}} |\mathrm{A}|,$$

which implies the stability of the factorization procedure in such cases. A similar result holds even if A is diagonally dominant.

3.7.2 Implementation Issues

An implementation of the LU factorization for banded matrices is shown in Program 10.

Program 10 - luband : LU factorization for a banded matrix

```
function [A]=luband(A,p,q)
%LUBAND LU factorization for a banded matrix
%  Y=LUBAND(A,P,Q): U is stored in the upper triangular part of Y and L is stored
%  in the strict lower triangular part of Y for a banded matrix A
%  with an upper bandwidth Q and a lower bandwidth P.
[n,m]=size(A);
if n ~= m, error('Only square systems'); end
for k = 1:n-1
   for i = k+1:min(k+p,n), A(i,k)=A(i,k)/A(k,k);  end
   for j = k+1:min(k+q,n)
      i = [k+1:min(k+p,n)];
      A(i,j)=A(i,j)-A(i,k)*A(k,j);
   end
end
return
```

In the case where $n \gg p$ and $n \gg q$, this algorithm approximately takes $2npq$ flops, with a considerable saving with respect to the case in which A is a full matrix.

Similarly, *ad hoc* versions of the substitution methods can be devised (see Programs 11 and 12). Their costs are, respectively, of the order of $2np$ flops and $2nq$ flops, always assuming that $n \gg p$ and $n \gg q$.

Program 11 - forwband : Forward substitution for a banded matrix L

```
function [b]=forwband (L,p,b)
%FORWBAND forward substitution for a banded matrix
% X=FORWBAND(L,P,B) solves the lower triangular system L*X=B
% where L is a matrix with lower bandwidth P.
[n,m]=size(L);
if n ~= m, error('Only square systems'); end
for j = 1:n
   i=[j+1:min(j+p,n)]; b(i) = b(i) - L(i,j)*b(j);
end
return
```

Program 12 - backband : Backward substitution for a banded matrix U

```
function [b]=backband (U,q,b)
%BACKBAND forward substitution for a banded matrix
% X=BACKBAND(U,Q,B) solves the upper triangular system U*X=B
% where U is a matrix with upper bandwidth Q.
[n,m]=size(U);
if n ~= m, error('Only square systems'); end
for j=n:-1:1
   b (j) = b (j) / U (j,j);
   i = [max(1,j-q):j-1]; b(i)=b(i)-U(i,j)*b(j);
end
return
```

The programs assume that the whole matrix is stored (including also the zero entries).

Concerning the tridiagonal case, the Thomas algorithm can be implemented in several ways. In particular, when implementing it on computers where divisions are more costly than multiplications, it is possible (and convenient) to devise a version of the algorithm without divisions in (3.55), by resorting to the following form of the factorization

$$A = LDM^T =$$

$$\begin{bmatrix} \gamma_1^{-1} & 0 & & 0 \\ b_2 & \gamma_2^{-1} & \ddots & \\ & \ddots & \ddots & 0 \\ 0 & & b_n & \gamma_n^{-1} \end{bmatrix} \begin{bmatrix} \gamma_1 & & & 0 \\ & \gamma_2 & & \\ & & \ddots & \\ 0 & & & \gamma_n \end{bmatrix} \begin{bmatrix} \gamma_1^{-1} & c_1 & & 0 \\ 0 & \gamma_2^{-1} & \ddots & \\ & \ddots & \ddots & c_{n-1} \\ 0 & & 0 & \gamma_n^{-1} \end{bmatrix}.$$

The coefficients γ_i can be recursively computed by the formulae

$$\gamma_i = (a_i - b_i\gamma_{i-1}c_{i-1})^{-1}, \, i = 1, \ldots, n,$$

where $\gamma_0 = 0$, $b_1 = 0$ and $c_n = 0$ have been assumed. The forward and backward substitution algorithms respectively read:

$$\begin{aligned} &(\mathbf{Ly} = \mathbf{f}) \;\; y_1 = \gamma_1 f_1, \, y_i = \gamma_i(f_i - b_i y_{i-1}), \, i = 2, \ldots, n, \\ &(\mathbf{Ux} = \mathbf{y}) \;\; x_n = y_n \quad x_i = y_i - \gamma_i c_i x_{i+1}, \quad i = n-1, \ldots, 1. \end{aligned} \tag{3.56}$$

In Program 13 we show an implementation of the Thomas algorithm in the form (3.56), without divisions. The input vectors a, b and c contain the coefficients of the tridiagonal matrix $\{a_i\}$, $\{b_i\}$ and $\{c_i\}$, respectively, while the vector f contains the components f_i of the right-hand side $\mathbf{f}$.

Program 13 - modthomas : Thomas algorithm, modified version

```
function [x] = modthomas (a,b,c,f)
%MODTHOMAS modified version of the Thomas algorithm
% X=MODTHOMAS(A,B,C,F) solves the system T*X=F where T
% is the tridiagonal matrix T=tridiag(B,A,C).
n=length(a);
b=[0; b];
c=[c; 0];
gamma(1)=1/a(1);
for i=2:n
   gamma(i)=1/(a(i)-b(i)*gamma(i-1)*c(i-1));
end
y(1)=gamma(1)*f (1);
for i =2:n
   y(i)=gamma(i)*(f(i)-b(i)*y(i-1));
end
x(n,1)=y(n);
for i=n-1:-1:1
   x(i,1)=y(i)-gamma(i)*c(i)*x(i+1,1);
end
return
```

3.8 Block Systems

In this section we deal with the LU factorization of block-partitioned matrices, where each block can possibly be of a different size. Our aim is twofold: optimizing the storage occupation by suitably exploiting the structure of the matrix and reducing the computational cost of the solution of the system.

3.8.1 Block LU Factorization

Let $A \in \mathbb{R}^{n\times n}$ be the following block partitioned matrix

$$A = \begin{bmatrix} A_{11} & A_{12} \\ A_{21} & A_{22} \end{bmatrix},$$

where $A_{11} \in \mathbb{R}^{r\times r}$ is a nonsingular square matrix whose factorization $L_{11}D_1R_{11}$ is known, while $A_{22} \in \mathbb{R}^{(n-r)\times(n-r)}$. In such a case it is possible to factorize A using only the LU factorization of the block A_{11}. Indeed, it is true that

$$\begin{bmatrix} A_{11} & A_{12} \\ A_{21} & A_{22} \end{bmatrix} = \begin{bmatrix} L_{11} & 0 \\ L_{21} & I_{n-r} \end{bmatrix} \begin{bmatrix} D_1 & 0 \\ 0 & \Delta_2 \end{bmatrix} \begin{bmatrix} R_{11} & R_{12} \\ 0 & I_{n-r} \end{bmatrix},$$

where

$$L_{21} = A_{21}R_{11}^{-1}D_1^{-1}, \; R_{12} = D_1^{-1}L_{11}^{-1}A_{12},$$
$$\Delta_2 = A_{22} - L_{21}D_1R_{12}.$$

If necessary, the reduction procedure can be repeated on the matrix Δ_2, thus obtaining a block-version of the LU factorization.

If A_{11} were a scalar, the above approach would reduce by one the size of the factorization of a given matrix. Applying iteratively this method yields an alternative way of performing the Gauss elimination.

We also notice that the proof of Theorem 3.4 can be extended to the case of block matrices, obtaining the following result.

Theorem 3.7 *Let* $A \in \mathbb{R}^{n\times n}$ *be partitioned in* $m \times m$ *blocks* A_{ij} *with* $i, j = 1, \ldots, m$. A *admits a unique LU block factorization (with* L *having unit diagonal entries) iff the* $m-1$ *dominant principal block minors of* A *are nonzero.*

Since the block factorization is an equivalent formulation of the standard LU factorization of A, the stability analysis carried out for the latter holds for its block-version as well. Improved results concerning the efficient use in block algorithms of fast forms of matrix-matrix product are dealt with in [Hig88]. In the forthcoming section we focus solely on block-tridiagonal matrices.

3.8.2 Inverse of a Block-partitioned Matrix

The inverse of a block matrix can be constructed using the LU factorization introduced in the previous section. A remarkable application is when A is a block matrix of the form

$$A = C + UBV,$$

where C is a block matrix that is "easy" to invert (for instance, when C is given by the diagonal blocks of A), while U, B and V take into account the connections between the diagonal blocks. In such an event A can be inverted by using the *Sherman-Morrison* or *Woodbury* formula

$$\mathrm{A}^{-1} = (\mathrm{C}+\mathrm{UBV})^{-1} = \mathrm{C}^{-1} - \mathrm{C}^{-1}\mathrm{U}\left(\mathrm{I}+\mathrm{BVC}^{-1}\mathrm{U}\right)^{-1}\mathrm{BVC}^{-1}, \quad (3.57)$$

having assumed that C and $\mathrm{I}+\mathrm{BVC}^{-1}\mathrm{U}$ are two nonsingular matrices. This formula has several practical and theoretical applications, and is particularly effective if connections between blocks are of modest relevance.

3.8.3 Block Tridiagonal Systems

Consider block tridiagonal systems of the form

$$\mathrm{A}_n\mathbf{x} = \begin{bmatrix} \mathrm{A}_{11} & \mathrm{A}_{12} & & 0 \\ \mathrm{A}_{21} & \mathrm{A}_{22} & \ddots & \\ & \ddots & \ddots & \mathrm{A}_{n-1,n} \\ 0 & & \mathrm{A}_{n,n-1} & \mathrm{A}_{nn} \end{bmatrix} \begin{bmatrix} \mathbf{x}_1 \\ \vdots \\ \vdots \\ \mathbf{x}_n \end{bmatrix} = \begin{bmatrix} \mathbf{b}_1 \\ \vdots \\ \vdots \\ \mathbf{b}_n \end{bmatrix}, \quad (3.58)$$

where A_{ij} are matrices of order $n_i \times n_j$ and $\mathbf{x}_i$ and $\mathbf{b}_i$ are column vectors of size n_i, for $i, j = 1, \ldots, n$. We assume that the diagonal blocks are squared, although not necessarily of the same size. For $k = 1, \ldots, n$, set

$$\mathrm{A}_k = \begin{bmatrix} \mathrm{I}_{n_1} & & & 0 \\ \mathrm{L}_1 & \mathrm{I}_{n_2} & & \\ & \ddots & \ddots & \\ 0 & & \mathrm{L}_{k-1} & \mathrm{I}_{n_k} \end{bmatrix} \begin{bmatrix} \mathrm{U}_1 & \mathrm{A}_{12} & & 0 \\ & \mathrm{U}_2 & \ddots & \\ & & \ddots & \mathrm{A}_{k-1,k} \\ 0 & & & \mathrm{U}_k \end{bmatrix}.$$

Equating for $k = n$ the matrix above with the corresponding blocks of A_n, it turns out that $\mathrm{U}_1 = \mathrm{A}_{11}$, while the remaining blocks can be obtained solving sequentially, for $i = 2, \ldots, n$, the systems $\mathrm{L}_{i-1}\mathrm{U}_{i-1} = \mathrm{A}_{i,i-1}$ for the columns of L and computing $\mathrm{U}_i = \mathrm{A}_{ii} - \mathrm{L}_{i-1}\mathrm{A}_{i-1,i}$.
This procedure is well defined only if all the matrices U_i are nonsingular, which is the case if, for instance, the matrices $\mathrm{A}_1, \ldots, \mathrm{A}_n$ are nonsingular. As an alternative, one could resort to factorization methods for banded matrices, even if this requires the storage of a large number of zero entries (unless a suitable reordering of the rows of the matrix is performed).

A remarkable instance is when the matrix is *block tridiagonal* and *symmetric*, with symmetric and positive definite blocks. In such a case (3.58) takes the form

$$\begin{bmatrix} \mathrm{A}_{11} & \mathrm{A}_{21}^T & & 0 \\ \mathrm{A}_{21} & \mathrm{A}_{22} & \ddots & \\ & \ddots & \ddots & \mathrm{A}_{n,n-1}^T \\ 0 & & \mathrm{A}_{n,n-1} & \mathrm{A}_{nn} \end{bmatrix} \begin{bmatrix} \mathbf{x}_1 \\ \vdots \\ \vdots \\ \mathbf{x}_n \end{bmatrix} = \begin{bmatrix} \mathbf{b}_1 \\ \vdots \\ \vdots \\ \mathbf{b}_n \end{bmatrix}.$$

Here we consider an extension to the block case of the Thomas algorithm, which aims at transforming A into a block bidiagonal matrix. To this purpose, we first have to eliminate the block corresponding to matrix A_{21}. Assume that the Cholesky factorization of A_{11} is available and denote by H_{11} the Cholesky factor. If we multiply the first row of the block system by H_{11}^{-T}, we find

$$H_{11}\mathbf{x}_1 + H_{11}^{-T}A_{21}^T\mathbf{x}_2 = H_{11}^{-T}\mathbf{b}_1.$$

Letting $H_{21} = H_{11}^{-T}A_{21}^T$ and $\mathbf{c}_1 = H_{11}^{-T}\mathbf{b}_1$, it follows that $A_{21} = H_{21}^T H_{11}$ and thus the first two rows of the system are

$$\begin{aligned} H_{11}\mathbf{x}_1 + H_{21}\mathbf{x}_2 &= \mathbf{c}_1, \\ H_{21}^T H_{11}\mathbf{x}_1 + A_{22}\mathbf{x}_2 + A_{32}^T\mathbf{x}_3 &= \mathbf{b}_2. \end{aligned}$$

As a consequence, multiplying the first row by H_{21}^T and subtracting it from the second one, the unknown $\mathbf{x}_1$ is eliminated and the following equivalent equation is obtained

$$A_{22}^{(1)}\mathbf{x}_2 + A_{32}^T\mathbf{x}_3 = \mathbf{b}_2 - H_{21}\mathbf{c}_1,$$

with $A_{22}^{(1)} = A_{22} - H_{21}^T H_{21}$. At this point, the factorization of $A_{22}^{(1)}$ is carried out and the unknown $\mathbf{x}_2$ is eliminated from the third row of the system, and the same is repeated for the remaining rows of the system. At the end of the procedure, which requires solving $(n-1)\sum_{j=1}^{n-1} n_j$ linear systems to compute the matrices $H_{i+1,i}$, $i = 1, \ldots, n-1$, we end up with the following block bidiagonal system

$$\begin{bmatrix} H_{11} & H_{21} & & 0 \\ & H_{22} & \ddots & \\ & & \ddots & H_{n,n-1} \\ 0 & & & H_{nn} \end{bmatrix} \begin{bmatrix} \mathbf{x}_1 \\ \vdots \\ \vdots \\ \mathbf{x}_n \end{bmatrix} = \begin{bmatrix} \mathbf{c}_1 \\ \vdots \\ \vdots \\ \mathbf{c}_n \end{bmatrix},$$

which can be solved with a (block) back substitution method. If all blocks have the same size p, then the number of multiplications required by the algorithm is about $(7/6)(n-1)p^3$ flops (assuming both p and n very large).

3.9 Sparse Matrices

In this section we briefly address the numerical solution of linear *sparse* systems, that is, systems where the matrix A$\in \mathbb{R}^{n\times n}$ has a number of nonzero entries of the order of n (and not n^2). We call a *pattern* of a sparse matrix the set of its nonzero coefficients.

Banded matrices with sufficiently small bands are sparse matrices. Obviously, for a sparse matrix the matrix structure itself is redundant and it can be more conveniently substituted by a vector-like structure by means of *matrix compacting techniques*, like the banded matrix format discussed in Section 3.7.

For sake of convenience, we associate with a sparse matrix A an *oriented graph* G(A). A graph is a pair $(\mathcal{V}, \mathcal{X})$ where $\mathcal{V}$ is a set of p points and $\mathcal{X}$ is a set of q ordered pairs of elements of $\mathcal{V}$ that are linked by a line. The elements of $\mathcal{V}$ are called the *vertices* of the graph, while the connection lines are called the *paths* of the graph.

The graph G(A) associated with a matrix A$\in \mathbb{R}^{m\times n}$ can be constructed by identifying the vertices with the set of the indices from 1 to the maximum between m and n and supposing that a path exists which connects two vertices i and j if $a_{ij} \neq 0$ and is directed from i to j, for $i = 1, \ldots, m$ and $j = 1, \ldots, n$. For a diagonal entry $a_{ii} \neq 0$, the path joining the vertex i with itself is called a *loop*. Since an orientation is associated with each side, the graph is called oriented (or finite directed). As an example, Figure 3.3 displays the pattern of a symmetric and sparse 12×12 matrix, together with its associated graph.

As previously noticed, during the factorization procedure, nonzero entries can be generated in memory positions that correspond to zero entries in the starting matrix. This action is referred to as *fill-in*. Figure 3.4 shows the effect of fill-in on the sparse matrix whose pattern is shown in Figure 3.3. Since use of *pivoting* in the factorization process makes things even more complicated, we shall only consider the case of symmetric positive definite matrices for which pivoting is not necessary.

A first remarkable result concerns the amount of fill-in. Let $m_i(\mathrm{A}) = i - \min\{j < i : \ a_{ij} \neq 0\}$ and denote by $\mathcal{E}(\mathrm{A})$ the *convex hull* of A, given by

$$\mathcal{E}(\mathrm{A}) = \{(i,j) : \ 0 < i - j \leq m_i(\mathrm{A})\} . \tag{3.59}$$

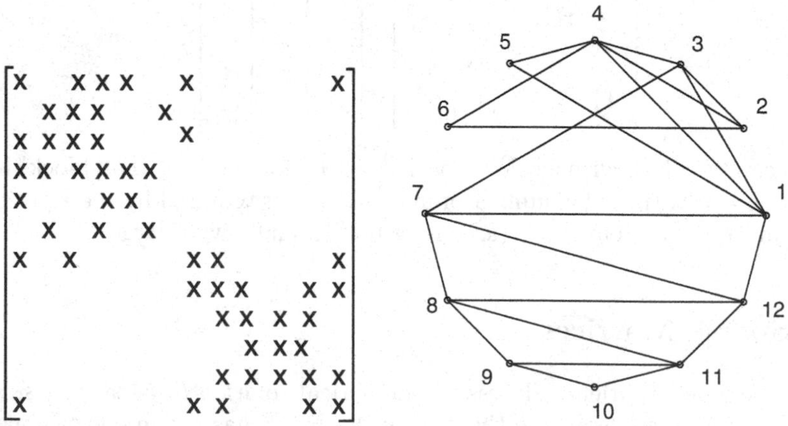

Fig. 3.3. *Pattern* of a symmetric sparse matrix (*left*) and of its associated graph (*right*). For the sake of clarity, the loops have not been drawn; moreover, since the matrix is symmetric, only one of the two sides associated with each $a_{ij} \neq 0$ has been reported

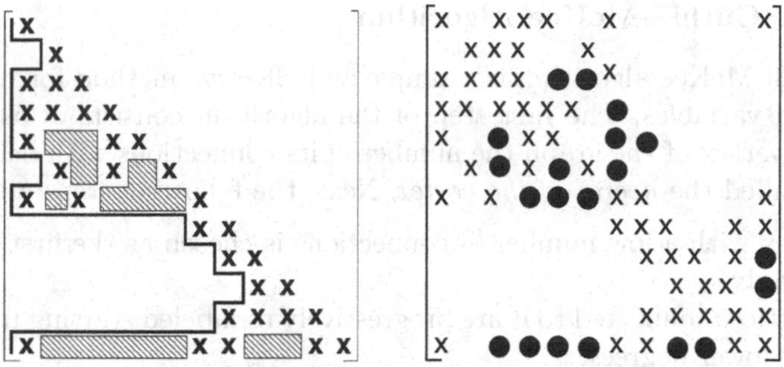

Fig. 3.4. The shaded regions in the left figure show the areas of the matrix that can be affected by fill-in, for the matrix considered in Figure 3.3. Solid lines denote the boundary of $\mathcal{E}(\mathrm{A})$. The right figure displays the factors that have been actually computed. Black dots denote the elements of A that were originarily equal to zero

For a symmetric positive definite matrix, we have

$$\mathcal{E}(\mathrm{A}) = \mathcal{E}(\mathrm{H} + \mathrm{H}^T), \tag{3.60}$$

where H is the Cholesky factor, so that fill-in is confined within the convex hull of A (see Figure 3.4). Moreover, if we denote by $l_k(\mathrm{A})$ the number of active rows at the k-th step of the factorization (i.e., the number of rows of A with $i > k$ and $a_{ik} \neq 0$), the computational cost of the factorization process is

$$\frac{1}{2}\sum_{k=1}^{n} l_k(\mathrm{A})\left(l_k(\mathrm{A}) + 3\right) \qquad \text{flops}, \tag{3.61}$$

having accounted for all the nonzero entries of the convex hull. Confinement of fill-in within $\mathcal{E}(\mathrm{A})$ ensures that the LU factorization of A can be stored without extra memory areas simply by storing all the entries of $\mathcal{E}(\mathrm{A})$ (including the null elements). However, such a procedure might still be highly inefficient due to the large number of zero entries in the hull (see Exercise 11).

On the other hand, from (3.60) one gets that the reduction in the convex hull reflects a reduction of fill-in, and in turn, due to (3.61), of the number of operations needed to perform the factorization. For this reason several strategies for reordering the graph of the matrix have been devised. Among them, we recall the Cuthill-McKee method, which will be addressed in the next section.

An alternative consists of decomposing the matrix into sparse submatrices, with the aim of reducing the original problem to the solution of subproblems of reduced size, where matrices can be stored in full format. This approach leads to submatrix decomposition methods which will be addressed in Section 3.9.2.

3.9.1 The Cuthill-McKee Algorithm

The Cuthill-McKee algorithm is a simple and effective method for reordering the system variables. The first step of the algorithm consists of associating with each vertex of the graph the number of its connections with neighboring vertices, called the *degree of the vertex*. Next, the following steps are taken:

1. a vertex with a low number of connections is chosen as the first vertex of the graph;
2. the vertices connected to it are progressively re-labeled starting from those having lower degrees;
3. the procedure is repeated starting from the vertices connected to the second vertex in the updated list. The nodes already re-labeled are ignored. Then, a third new vertex is considered, and so on, until all the vertices have been explored.

The usual way to improve the efficiency of the algorithm is based on the so-called *reverse* form of the Cuthill-McKee method. This consists of executing the Cuthill-McKee algorithm described above where, at the end, the i-th vertex is moved into the $n-i+1$-th position of the list, n being the number of nodes in the graph. Figure 3.5 reports, for comparison, the graphs obtained using the direct and reverse Cuthill-McKee reordering in the case of the matrix pattern represented in Figure 3.3, while in Figure 3.6 the factors L and U are compared. Notice the absence of fill-in when the reverse Cuthill-McKee method is used.

Remark 3.5 For an efficient solution of linear systems with sparse matrices, we mention the public domain libraries SPARSKIT [Saa90], UMFPACK [DD95] and the MATLAB `sparfun` package. ∎

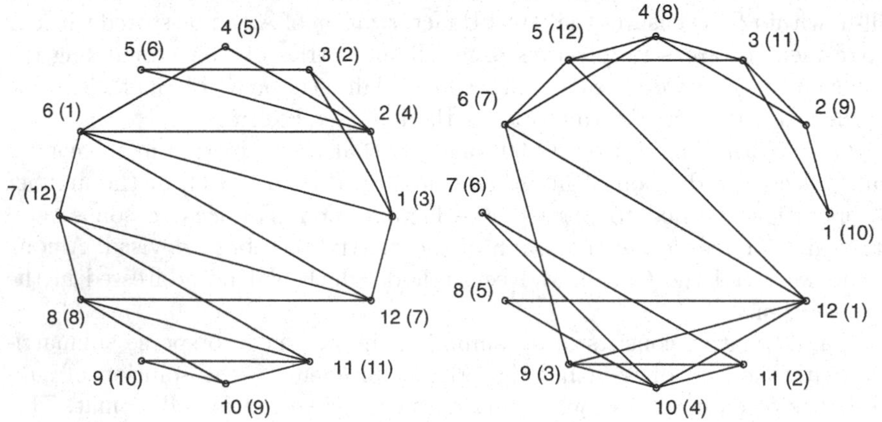

Fig. 3.5. Reordered graphs using the direct (*left*) and reverse (*right*) Cuthill-McKee algorithm. The label of each vertex, before reordering is performed, is reported in braces

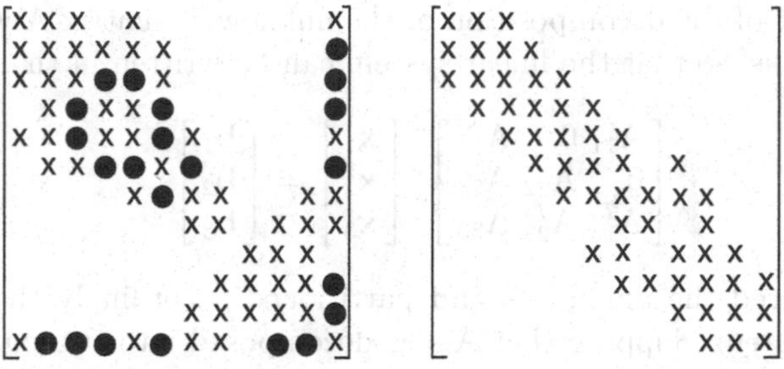

Fig. 3.6. Factors L and U after the direct (*left*) and reverse (*right*) Cuthill-McKee reordering. In the second case, fill-in is absent

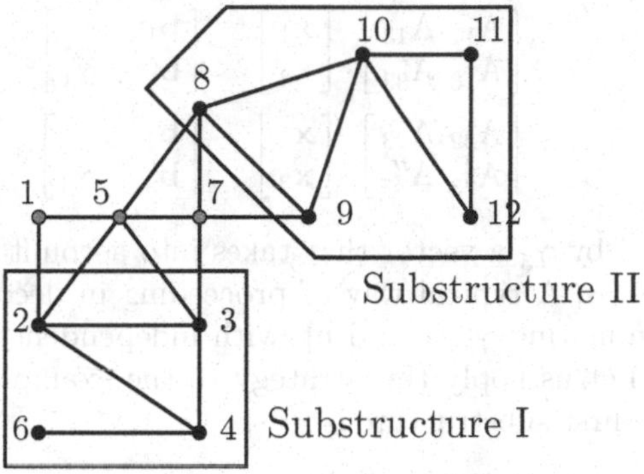

Fig. 3.7. Decomposition into two substructures

3.9.2 Decomposition into Substructures

These methods have been developed in the framework of numerical approximation of partial differential equations. Their basic strategy consists of splitting the solution of the original linear system into subsystems of smaller size which are almost independent from each other and can be easily interpreted as a reordering technique.

We describe the methods on a special example, referring for a more comprehensive presentation to [BSG96]. Consider the linear system $A\mathbf{x}=\mathbf{b}$, where A is a symmetric positive definite matrix whose pattern is shown in Figure 3.3. To help develop an intuitive understanding of the method, we draw the graph of A in the form as in Figure 3.7.

We then partition the graph of A into the two subgraphs (or substructures) identified in the figure and denote by $\mathbf{x}_k$, $k = 1, 2$, the vectors of the unknowns relative to the nodes that belong to the interior of the k-th substructure. We also denote by $\mathbf{x}_3$ the vector of the unknowns that lie along the interface between the two substructures. Referring to the decomposition in Figure 3.7, we have $\mathbf{x}_1 = [2,\ 3,\ 4,\ 6]^T$, $\mathbf{x}_2 = [8,\ 9,\ 10,\ 11,\ 12]^T$ and $\mathbf{x}_3 = [1,\ 5,\ 7]^T$.

As a result of the decomposition of the unknowns, matrix A will be partitioned in blocks, so that the linear system can be written in the form

$$\begin{bmatrix} A_{11} & 0 & A_{13} \\ 0 & A_{22} & A_{23} \\ A_{13}^T & A_{23}^T & A_{33} \end{bmatrix} \begin{bmatrix} \mathbf{x}_1 \\ \mathbf{x}_2 \\ \mathbf{x}_3 \end{bmatrix} = \begin{bmatrix} \mathbf{b}_1 \\ \mathbf{b}_2 \\ \mathbf{b}_3 \end{bmatrix},$$

having reordered the unknowns and partitioned accordingly the right hand side of the system. Suppose that A_{33} is decomposed into two parts, A'_{33} and A''_{33}, which represent the contributions to A_{33} of each substructure. Similarly, let the right hand side $\mathbf{b}_3$ be decomposed as $\mathbf{b}'_3 + \mathbf{b}''_3$. The original linear system is now equivalent to the following pair

$$\begin{bmatrix} A_{11} & A_{13} \\ A_{13}^T & A'_{33} \end{bmatrix} \begin{bmatrix} \mathbf{x}_1 \\ \mathbf{x}_3 \end{bmatrix} = \begin{bmatrix} \mathbf{b}_1 \\ \mathbf{b}'_3 + \boldsymbol{\gamma}_3 \end{bmatrix},$$

$$\begin{bmatrix} A_{22} & A_{23} \\ A_{23}^T & A''_{33} \end{bmatrix} \begin{bmatrix} \mathbf{x}_2 \\ \mathbf{x}_3 \end{bmatrix} = \begin{bmatrix} \mathbf{b}_2 \\ \mathbf{b}''_3 - \boldsymbol{\gamma}_3 \end{bmatrix},$$

having denoted by $\boldsymbol{\gamma}_3$ a vector that takes into account the coupling between the substructures. A typical way of proceeding in decomposition techniques consists of eliminating $\boldsymbol{\gamma}_3$ to end up with independent systems, one for each substructure. Let us apply this strategy to the example at hand. The linear system for the first substructure is

$$\begin{bmatrix} A_{11} & A_{13} \\ A_{13}^T & A'_{33} \end{bmatrix} \begin{bmatrix} \mathbf{x}_1 \\ \mathbf{x}_3 \end{bmatrix} = \begin{bmatrix} \mathbf{b}_1 \\ \mathbf{b}'_3 + \boldsymbol{\gamma}_3 \end{bmatrix}. \tag{3.62}$$

Let us now factorize A_{11} as $H_{11}^T H_{11}$ and proceed with the reduction method already described in Section 3.8.3 for block tridiagonal matrices. We obtain the system

$$\begin{bmatrix} H_{11} & H_{21} \\ 0 & A'_{33} - H_{21}^T H_{21} \end{bmatrix} \begin{bmatrix} \mathbf{x}_1 \\ \mathbf{x}_3 \end{bmatrix} = \begin{bmatrix} \mathbf{c}_1 \\ \mathbf{b}'_3 + \boldsymbol{\gamma}_3 - H_{21}^T \mathbf{c}_1 \end{bmatrix},$$

where $H_{21} = H_{11}^{-T} A_{13}$ and $\mathbf{c}_1 = H_{11}^{-T} \mathbf{b}_1$. The second equation of this system yields $\boldsymbol{\gamma}_3$ explicitly as

$$\boldsymbol{\gamma}_3 = \left(A'_{33} - H_{21}^T H_{21}\right) \mathbf{x}_3 - \mathbf{b}'_3 + H_{21}^T \mathbf{c}_1.$$

Substituting this equation into the system for the second substructure, one ends up with a system only in the unknowns $\mathbf{x}_2$ and $\mathbf{x}_3$

$$\begin{bmatrix} A_{22} & A_{23} \\ A_{23}^T & A'''_{33} \end{bmatrix} \begin{bmatrix} \mathbf{x}_2 \\ \mathbf{x}_3 \end{bmatrix} = \begin{bmatrix} \mathbf{b}_2 \\ \mathbf{b}'''_3 \end{bmatrix}, \tag{3.63}$$

where $A'''_{33} = A_{33} - H_{21}^T H_{21}$ and $\mathbf{b}'''_3 = \mathbf{b}_3 - H_{21}^T \mathbf{c}_1$. Once (3.63) has been solved, it will be possible, by backsubstitution into (3.62), to compute also $\mathbf{x}_1$.

The technique described above can be easily extended to the case of several substructures and its efficiency will increase the more the substructures are mutually independent. It reproduces *in nuce* the so-called *frontal method* (introduced by Irons [Iro70]), which is quite popular in the solution of finite element systems (for an implementation, we refer to the UMFPACK library [DD95]).

Remark 3.6 (The Schur complement) An approach that is dual to the above method consists of reducing the starting system to a system acting only on the interface unknowns $\mathbf{x}_3$, passing through the assembling of the Schur complement of matrix A, defined in the 3×3 case at hand as

$$S = A_{33} - A_{13}^T A_{11}^{-1} A_{13} - A_{23}^T A_{22}^{-1} A_{23}.$$

The original problem is thus equivalent to the system

$$S\mathbf{x}_3 = \mathbf{b}_3 - A_{13}^T A_{11}^{-1} \mathbf{b}_1 - A_{23}^T A_{22}^{-1} \mathbf{b}_2.$$

This system is full (even if the matrices A_{ij} were sparse) and can be solved using either a direct or an iterative method, provided that a suitable preconditioner is available. Once $\mathbf{x}_3$ has been computed, one can get $\mathbf{x}_1$ and $\mathbf{x}_2$ by solving two systems of reduced size, whose matrices are A_{11} and A_{22}, respectively.

We also notice that if the block matrix A is symmetric and positive definite, then the linear system on the Schur complement S is no more ill-conditioned than the original system on A, since

$$K_2(S) \leq K_2(A)$$

(for a proof, see Lemma 3.12, [Axe94]. See also [CM94] and [QV99]). ■

3.9.3 Nested Dissection

This is a renumbering technique quite similar to substructuring. In practice, it consists of repeating the decomposition process several times at each substructure level, until the size of each single block is made sufficiently small. In Figure 3.8 a possible *nested dissection* is shown in the case of the matrix considered in the previous section. Once the subdivision procedure has been completed, the vertices are renumbered starting with the nodes belonging to the latest substructuring level and moving progressively up to the first level. In the example at hand, the new node ordering is 11, 9, 7, 6, 12, 8, 4, 2, 1, 5, 3.

This procedure is particularly effective if the problem has a large size and the substructures have few connections between them or exhibit a repetitive pattern [Geo73].

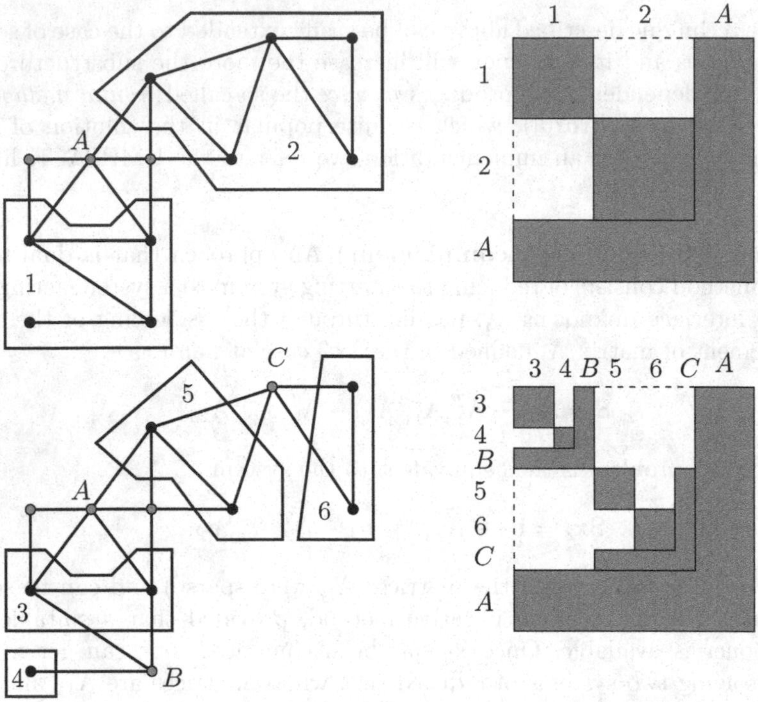

Fig. 3.8. Two steps of *nested dissection*. Graph partitioning (*left*) and matrix reordering (*right*)

3.10 Accuracy of the Solution Achieved Using GEM

Let us analyze the effects of rounding errors on the accuracy of the solution yielded by GEM. Suppose that A and **b** are a matrix and a vector of *floating-point* numbers. Denoting by $\widehat{\mathrm{L}}$ and $\widehat{\mathrm{U}}$, respectively, the matrices of the LU factorization induced by GEM and computed in *floating-point* arithmetic, the solution $\widehat{\mathbf{x}}$ yielded by GEM can be regarded as being the solution (in exact arithmetic) of the perturbed system $(\mathrm{A}+\delta\mathrm{A})\widehat{\mathbf{x}} = \mathbf{b}$, where $\delta\mathrm{A}$ is a perturbation matrix such that

$$|\delta\mathrm{A}| \leq n\mathrm{u}\left(3|\mathrm{A}| + 5|\widehat{\mathrm{L}}||\widehat{\mathrm{U}}|\right) + \mathcal{O}(\mathrm{u}^2), \tag{3.64}$$

where u is the *roundoff* unit and the matrix absolute value notation has been used (see [GL89], Section 3.4.6). As a consequence, the entries of $\delta\mathrm{A}$ will be small in size if the entries of $\widehat{\mathrm{L}}$ and $\widehat{\mathrm{U}}$ are small. Using partial pivoting allows for bounding below 1 the module of the entries of $\widehat{\mathrm{L}}$ in such a way that, passing to the infinity norm and noting that $\|\widehat{\mathrm{L}}\|_\infty \leq n$, the estimate (3.64) becomes

$$\|\delta\mathrm{A}\|_\infty \leq n\mathrm{u}\left(3\|\mathrm{A}\|_\infty + 5n\|\widehat{\mathrm{U}}\|_\infty\right) + \mathcal{O}(\mathrm{u}^2). \tag{3.65}$$

The bound for $\|\delta \mathrm{A}\|_\infty$ in (3.65) is of practical use only if it is possible to provide an estimate for $\|\widehat{\mathrm{U}}\|_\infty$. With this aim, backward analysis can be carried out introducing the so-called *growth factor*

$$\rho_n = \frac{\max\limits_{i,j,k}|\widehat{a}_{ij}^{(k)}|}{\max\limits_{i,j}|a_{ij}|}. \tag{3.66}$$

Taking advantage of the fact that $|\widehat{u}_{ij}| \le \rho_n \max\limits_{i,j}|a_{ij}|$, the following result due to Wilkinson can be drawn from (3.65),

$$\|\delta \mathrm{A}\|_\infty \le 8\mathrm{u}n^3\rho_n\|\mathrm{A}\|_\infty + \mathcal{O}(\mathrm{u}^2). \tag{3.67}$$

The growth factor can be bounded by 2^{n-1} and, although in most of the cases it is of the order of 10, there exist matrices for which the inequality in (3.67) becomes an equality (see, for instance, Exercise 5). For some special classes of matrices, a sharp bound for ρ_n can be found:

1. for banded matrices with upper and lower bands equal to p, $\rho_n \le 2^{2p-1} - (p-1)2^{p-2}$. As a consequence, in the tridiagonal case one gets $\rho_n \le 2$;
2. for Hessenberg matrices, $\rho_n \le n$;
3. for symmetric positive definite matrices, $\rho_n = 1$;
4. for matrices strictly diagonally dominant by columns, $\rho_n \le 2$.

To achieve better stability when using GEM for arbitrary matrices, resorting to complete pivoting would seem to be mandatory, since it ensures that $\rho_n \le n^{1/2}\left(2 \cdot 3^{1/2} \cdot \ldots \cdot n^{1/(n-1)}\right)^{1/2}$. Indeed, this growth is slower than 2^{n-1} as n increases.

However, apart from very special instances, GEM with only partial pivoting exhibits acceptable growth factors. This make it the most commonly employed method in the computational practice.

Example 3.7 Consider the linear system (3.2) with

$$\mathrm{A} = \begin{bmatrix} \varepsilon & 1 \\ 1 & 0 \end{bmatrix}, \ \mathbf{b} = \begin{bmatrix} 1+\varepsilon \\ 1 \end{bmatrix}, \tag{3.68}$$

which admits the exact solution $\mathbf{x}=\mathbf{1}$ for any value of ε. The matrix is well-conditioned, having $K_\infty(\mathrm{A}) = (1+\varepsilon)^2$. Attempting to solve the system for $\varepsilon = 10^{-15}$ by the LU factorization with 16 significant digits, and using the Programs 5, 2 and 3, yields the solution $\widehat{\mathbf{x}} = [0.8881784197001253, 1.000000000000000]^T$, with an error greater than 11% on the first component. Some insight into the causes of the inaccuracy of the computed solution can be drawn from (3.64). Indeed this latter does not provide a uniformly small bound for all the entries of matrix $\delta\mathrm{A}$, rather

$$|\delta \mathrm{A}| \le \begin{bmatrix} 3.55 \cdot 10^{-30} & 1.33 \cdot 10^{-15} \\ 1.33 \cdot 10^{-15} & \boxed{2.22} \end{bmatrix}.$$

Notice that the entries of the corresponding matrices $\widehat{\mathrm{L}}$ and $\widehat{\mathrm{U}}$ are quite large in module. Conversely, resorting to GEM with partial or complete pivoting yields the exact solution of the system (see Exercise 6). •

Let us now address the role of the condition number in the error analysis for GEM. GEM yields a solution $\widehat{\mathbf{x}}$ that is typically characterized by having a small residual $\widehat{\mathbf{r}} = \mathbf{b} - \mathrm{A}\widehat{\mathbf{x}}$ (see [GL89]). This feature, however, does not ensure that the error $\mathbf{x} - \widehat{\mathbf{x}}$ is small when $K(\mathrm{A}) \gg 1$ (see Example 3.8). In fact, if $\boldsymbol{\delta}\mathbf{b}$ in (3.11) is regarded as being the residual, then

$$\frac{\|\mathbf{x} - \widehat{\mathbf{x}}\|}{\|\mathbf{x}\|} \leq K(\mathrm{A})\|\widehat{\mathbf{r}}\| \frac{1}{\|\mathrm{A}\|\|\mathbf{x}\|} \leq K(\mathrm{A})\frac{\|\widehat{\mathbf{r}}\|}{\|\mathbf{b}\|}.$$

This result will be applied to devise methods, based on the a posteriori analysis, for improving the accuracy of the solution of GEM (see Section 3.12).

Example 3.8 Consider the linear system $\mathrm{A}\mathbf{x} = \mathbf{b}$ with

$$\mathrm{A} = \begin{bmatrix} 1 & 1.0001 \\ 1.0001 & 1 \end{bmatrix}, \; \mathbf{b} = \begin{bmatrix} 1 \\ 1 \end{bmatrix},$$

which admits the solution $\mathbf{x} = [0.499975\ldots, 0.499975\ldots]^T$. Assuming as an approximate solution the vector $\widehat{\mathbf{x}} = [-4.499775, 5.5002249]^T$, one finds the residual $\widehat{\mathbf{r}} \simeq [-0.001, 0]^T$, which is small although $\widehat{\mathbf{x}}$ is quite different from the exact solution. The reason for this is due to the ill-conditioning of matrix A. Indeed in this case $K_\infty(\mathrm{A}) = 20001$. •

An estimate of the number of exact significant digits of a numerical solution of a linear system can be given as follows. From (3.13), letting $\gamma = \mathrm{u}$ and assuming that $\mathrm{u}K_\infty(\mathrm{A}) \leq 1/2$ we get

$$\frac{\|\boldsymbol{\delta}\mathbf{x}\|_\infty}{\|\mathbf{x}\|_\infty} \leq \frac{2\mathrm{u}K_\infty(\mathrm{A})}{1 - \mathrm{u}K_\infty(\mathrm{A})} \leq 4\mathrm{u}K_\infty(\mathrm{A}).$$

As a consequence

$$\frac{\|\widehat{\mathbf{x}} - \mathbf{x}\|_\infty}{\|\mathbf{x}\|_\infty} \simeq \mathrm{u}K_\infty(\mathrm{A}). \tag{3.69}$$

Assuming that $\mathrm{u} \simeq \beta^{-t}$ and $K_\infty(\mathrm{A}) \simeq \beta^m$, one gets that the solution $\widehat{\mathbf{x}}$ computed by GEM will have at least $t - m$ exact digits, t being the number of digits available for the mantissa. In other words, the ill-conditioning of a system depends both on the capability of the *floating-point* arithmetic that is being used and on the accuracy that is required in the solution.

3.11 An Approximate Computation of $K(\mathrm{A})$

Suppose that the linear system (3.2) has been solved by a factorization method. To determine the accuracy of the computed solution, the analysis carried out in Section 3.10 can be used if an estimate of the condition number

$K(\mathrm{A})$ of A, which we denote by $\widehat{K}(\mathrm{A})$, is available. Indeed, although evaluating $\|\mathrm{A}\|$ can be an easy task if a suitable norm is chosen (for instance, $\|\cdot\|_1$ or $\|\cdot\|_\infty$), it is by no means reasonable (or computationally convenient) to compute A^{-1} if the only purpose is to evaluate $\|\mathrm{A}^{-1}\|$. For this reason, we describe in this section a procedure (proposed in [CMSW79]) that approximates $\|\mathrm{A}^{-1}\|$ with a computational cost of the order of n^2 flops.

The basic idea of the algorithm is as follows: $\forall \mathbf{d} \in \mathbb{R}^n$ with $\mathbf{d} \neq \mathbf{0}$, thanks to the definition of matrix norm, $\|\mathrm{A}^{-1}\| \geq \|\mathbf{y}\|/\|\mathbf{d}\| = \gamma(\mathbf{d})$ with $\mathrm{A}\mathbf{y} = \mathbf{d}$. Thus, we look for $\mathbf{d}$ in such a way that $\gamma(\mathbf{d})$ is as large as possible and assume the obtained value as an estimate of $\|\mathrm{A}^{-1}\|$.

For the method to be effective, the selection of $\mathbf{d}$ is crucial. To explain how to do this, we start by assuming that the QR factorization of A has been computed and that $K_2(\mathrm{A})$ is to be approximated. In such an event, since $K_2(\mathrm{A}) = K_2(\mathrm{R})$ due to Property 1.8, it suffices to estimate $\|\mathrm{R}^{-1}\|_2$ instead of $\|\mathrm{A}^{-1}\|_2$. Considerations related to the SVD of R induce approximating $\|\mathrm{R}^{-1}\|_2$ by the following algorithm:

compute the vectors $\mathbf{x}$ and $\mathbf{y}$, solutions to the systems

$$\mathrm{R}^T\mathbf{x} = \mathbf{d}, \ \mathrm{R}\mathbf{y} = \mathbf{x}, \tag{3.70}$$

then estimate $\|\mathrm{R}^{-1}\|_2$ by the ratio $\gamma_2 = \|\mathbf{y}\|_2/\|\mathbf{x}\|_2$. The vector $\mathbf{d}$ appearing in (3.70) should be determined in such a way that γ_2 is as close as possible to the value actually attained by $\|\mathrm{R}^{-1}\|_2$. It can be shown that, except in very special cases, γ_2 provides for any choice of $\mathbf{d}$ a reasonable (although not very accurate) estimate of $\|\mathrm{R}^{-1}\|_2$ (see Exercise 15). As a consequence, a proper selection of $\mathbf{d}$ can encourage this natural trend.

Before going on, it is worth noting that computing $K_2(\mathrm{R})$ is not an easy matter even if an estimate of $\|\mathrm{R}^{-1}\|_2$ is available. Indeed, it would remain to compute $\|\mathrm{R}\|_2 = \sqrt{\rho(\mathrm{R}^T\mathrm{R})}$. To overcome this difficulty, we consider henceforth $K_1(\mathrm{R})$ instead of $K_2(\mathrm{R})$ since $\|\mathrm{R}\|_1$ is easily computable. Then, heuristics allows us to assume that the ratio $\gamma_1 = \|\mathbf{y}\|_1/\|\mathbf{x}\|_1$ is an estimate of $\|\mathrm{R}^{-1}\|_1$, exactly as γ_2 is an estimate of $\|\mathrm{R}^{-1}\|_2$.

Let us now deal with the choice of $\mathbf{d}$. Since $\mathrm{R}^T\mathbf{x} = \mathbf{d}$, the generic component x_k of $\mathbf{x}$ can be formally related to $x_1, \ldots, x_{k-1}$ through the formulae of forward substitution as

$$\begin{aligned} &r_{11}x_0 = d_1, \\ &r_{kk}x_k = d_k - (r_{1k}x_1 + \ldots + r_{k-1,k}x_{k-1}), \ k \geq 1. \end{aligned} \tag{3.71}$$

Assume that the components of $\mathbf{d}$ are of the form $d_k = \pm\theta_k$, where θ_k are random numbers and set arbitrarily $d_1 = \theta_1$. Then, $x_1 = \theta_1/r_{11}$ is completely determined, while $x_2 = (d_2 - r_{12}x_1)/r_{22}$ depends on the sign of d_2. We set the sign of d_2 as the opposite of $r_{12}x_1$ in such a way to make $\|\mathbf{x}(1:2)\|_1 = |x_1| + |x_2|$, for a fixed x_1, the largest possible. Once x_2 is known, we compute x_3 following the same criterion, and so on, until x_n.

This approach sets the sign of each component of $\mathbf{d}$ and yields a vector $\mathbf{x}$ with a presumably large $\|\cdot\|_1$. However, it can fail since it is based on the idea (which is in general not true) that maximizing $\|\mathbf{x}\|_1$ can be done by selecting at each step k in (3.71) the component x_k which guarantees the maximum increase of $\|\mathbf{x}(1:k-1)\|_1$ (without accounting for the fact that all the components are related).

Therefore, we need to modify the method by including a sort of "look-ahead" strategy, which accounts for the way of choosing d_k affects all later values x_i, with $i > k$, still to be computed. Concerning this point, we notice that for a generic row i of the system it is always possible to compute at step k the vector $\mathbf{p}^{(k-1)}$ with components

$$p_i^{(k-1)} = 0, \qquad i = 1, \ldots, k-1,$$
$$p_i^{(k-1)} = r_{1i}x_1 + \ldots + r_{k-1,i}x_{k-1}, \; i = k, \ldots, n.$$

Thus $x_k = (\pm\theta_k - p_k^{(k-1)})/r_{kk}$. We denote the two possible values of x_k by x_k^+ and x_k^-. The choice between them is now taken not only accounting for which of the two most increases $\|\mathbf{x}(1:k)\|_1$, but also evaluating the increase of $\|\mathbf{p}^{(k)}\|_1$. This second contribution accounts for the effect of the choice of d_k on the components that are still to be computed. We can include both criteria in a unique test. Denoting by

$$p_i^{(k)^+} = 0, \qquad p_i^{(k)^-} = 0, \qquad i = 1, \ldots, k,$$
$$p_i^{(k)^+} = p_i^{(k-1)} + r_{ki}x_k^+, \; p_i^{(k)^-} = p_i^{(k-1)} + r_{ki}x_k^-, \; i = k+1, \ldots, n,$$

the components of the vectors $\mathbf{p}^{(k)^+}$ and $\mathbf{p}^{(k)^-}$ respectively, we set each k-th step $d_k = +\theta_k$ or $d_k = -\theta_k$ according to whether $|r_{kk}x_k^+| + \|\mathbf{p}^{(k)^+}\|_1$ is greater or less than $|r_{kk}x_k^-| + \|\mathbf{p}^{(k)^-}\|_1$.

Under this choice $\mathbf{d}$ is completely determined and the same holds for $\mathbf{x}$. Now, solving the system $\mathrm{R}\mathbf{y} = \mathbf{x}$, we are warranted that $\|\mathbf{y}\|_1/\|\mathbf{x}\|_1$ is a reliable approximation to $\|\mathrm{R}^{-1}\|_1$, so that we can set $\widehat{K}_1(\mathrm{A}) = \|\mathrm{R}\|_1\|\mathbf{y}\|_1/\|\mathbf{x}\|_1$.

In practice the PA=LU factorization introduced in Section 3.5 is usually available. Based on the previous considerations and on some heuristics, an analogous procedure to that shown above can be conveniently employed to approximate $\|\mathrm{A}^{-1}\|_1$. Precisely, instead of systems (3.70), we must now solve

$$(\mathrm{LU})^T\mathbf{x} = \mathbf{d}, \; \mathrm{LU}\mathbf{y} = \mathbf{x}.$$

We set $\|\mathbf{y}\|_1/\|\mathbf{x}\|_1$ as the approximation of $\|\mathrm{A}^{-1}\|_1$ and, consequently, we define $\widehat{K}_1(\mathrm{A})$. The strategy for selecting $\mathbf{d}$ can be the same as before; indeed, solving $(\mathrm{LU})^T\mathbf{x} = \mathbf{d}$ amounts to solving

$$\mathrm{U}^T\mathbf{z} = \mathbf{d}, \; \mathrm{L}^T\mathbf{x} = \mathbf{z}, \tag{3.72}$$

and thus, since U^T is lower triangular, we can proceed as in the previous case. A remarkable difference concerns the computation of **x**. Indeed, while the matrix R^T in the second system of (3.70) has the same condition number as R, the second system in (3.72) has a matrix L^T which could be even more ill-conditioned than U^T. If this were the case, solving for **x** could lead to an inaccurate outcome, thus making the whole process useless.

Fortunately, resorting to partial pivoting prevents this circumstance from occurring, ensuring that any ill-condition in A is reflected in a corresponding ill-condition in U. Moreover, picking θ_k randomly between 1/2 and 1 guarantees accurate results even in the special cases where L turns out to be ill-conditioned.

The algorithm presented below is implemented in the LINPACK library [BDMS79] and in the MATLAB function `rcond`. This function, in order to avoid rounding errors, returns as output parameter the reciprocal of $\widehat{K}_1(\mathrm{A})$. A more accurate estimator, described in [Hig88], is implemented in the MATLAB function `condest`.

Program 14 implements the approximate evaluation of K_1 for a matrix A of generic form. The input parameters are the size `n` of the matrix `A`, the matrix `A`, the factors `L`, `U` of its PA=LU factorization and the vector `theta` containing the random numbers θ_k, for $k = 1, \ldots, n$.

Program 14 - condest2 : Algorithm for the approximation of $K_1(\mathrm{A})$

```
function [k1]=condest2(A,L,U,theta)
%CONDEST2 Condition number
% K1=CONDEST2(A,L,U,THETA) returns an approximation of the condition
% number of a matrix A. L and U are the factor of the LU factorization of A.
% THETA contains random numbers.
[n,m]=size(A);
if n ~= m, error('Only square matrices'); end
p = zeros(1,n);
for k=1:n
   zplus=(theta(k)-p(k))/U(k,k);  zminu=(-theta(k)-p(k))/U(k,k);
   splus=abs(theta(k)-p(k));        sminu=abs(-theta(k)-p(k));
   for i=k+1:n
      splus=splus+abs(p(i)+U(k,i)*zplus);
      sminu=sminu+abs(p(i)+U(k,i)*zminu);
   end
   if splus >= sminu,  z(k)=zplus;  else,  z(k)=zminu;  end
   i=[k+1:n];  p(i)=p(i)+U(k,i)*z(k);
end
z = z';
x = backwardcol(L',z);
w = forwardcol(L,x);
y = backwardcol(U,w);
k1=norm(A,1)*norm(y,1)/norm(x,1);
return
```

Example 3.9 Let us consider the Hilbert matrix H_4. Its condition number $K_1(H_4)$, computed using the MATLAB function `invhilb` which returns the exact inverse of H_4, is $2.8375 \cdot 10^4$. Running Program 14 with `theta`$=[1,1,1,1]^T$ gives the reasonable estimate $\widehat{K}_1(H_4) = 2.1509 \cdot 10^4$ (which is the same as the output of `rcond`), while the function `condest` returns the exact result. •

3.12 Improving the Accuracy of GEM

As previously noted if the matrix of the system is ill-conditioned, the solution generated by GEM could be inaccurate even though its residual is small. In this section, we mention two techniques for improving the accuracy of the solution computed by GEM.

3.12.1 Scaling

If the entries of A vary greatly in size, it is likely that during the elimination process large entries are summed to small entries, with a consequent onset of rounding errors. A remedy consists of performing a *scaling* of the matrix A before the elimination is carried out.

Example 3.10 Consider again the matrix A of Remark 3.3. Multiplying it on the right and on the left with matrix D=diag(0.0005, 1, 1), we obtain the scaled matrix

$$\tilde{A} = DAD = \begin{bmatrix} -0.001 & 1 & 1 \\ 1 & 0.78125 & 0 \\ 1 & 0 & 0 \end{bmatrix}.$$

Applying GEM to the scaled system $\tilde{A}\tilde{\mathbf{x}} = D\mathbf{b} = [0.2,\ 1.3816,\ 1.9273]^T$, we get the correct solution $\mathbf{x} = D\tilde{\mathbf{x}}$. •

Row scaling of A amounts to finding a diagonal nonsingular matrix D_1 such that the diagonal entries of $D_1 A$ are of the same size. The linear system $A\mathbf{x} = \mathbf{b}$ transforms into

$$D_1 A\mathbf{x} = D_1 \mathbf{b}.$$

When both rows and columns of A are to be scaled, the *scaled* version of (3.2) becomes

$$(D_1 A D_2)\mathbf{y} = D_1 \mathbf{b} \ \text{ with } \mathbf{y} = D_2^{-1}\mathbf{x},$$

having also assumed that D_2 is invertible. Matrix D_1 scales the equations, while D_2 scales the unknowns. Notice that, to prevent rounding errors, the *scaling* matrices are chosen in the form

$$D_1 = \text{diag}(\beta^{r_1}, \ldots, \beta^{r_n}),\ D_2 = \text{diag}(\beta^{c_1}, \ldots, \beta^{c_n}),$$

where β is the base of the used *floating-point* arithmetic and the exponents $r_1, \ldots, r_n$, $c_1, \ldots, c_n$ must be determined. It can be shown that

$$\frac{\|D_2^{-1}(\widehat{\mathbf{x}} - \mathbf{x})\|_\infty}{\|D_2^{-1}\mathbf{x}\|_\infty} \simeq \mathrm{u}K_\infty(D_1AD_2).$$

Therefore, *scaling* will be effective if $K_\infty(D_1AD_2)$ is much less than $K_\infty(A)$. Finding convenient matrices D_1 and D_2 is not in general an easy matter.

A strategy consists, for instance, of picking up D_1 and D_2 in such a way that $\|D_1AD_2\|_\infty$ and $\|D_1AD_2\|_1$ belong to the interval $[1/\beta, 1]$, where β is the base of the used *floating-point* arithmetic (see [McK62] for a detailed analysis in the case of the Crout factorization).

Remark 3.7 (The Skeel condition number) The *Skeel condition number*, defined as $\mathrm{cond}(A) = \|\, |A^{-1}|\, |A|\, \|_\infty$, is the supremum over the set $\mathbf{x} \in \mathbb{R}^n$, with $\mathbf{x} \neq \mathbf{0}$, of the numbers

$$\mathrm{cond}(A, \mathbf{x}) = \frac{\|\, |A^{-1}|\, |A|\, |\mathbf{x}|\, \|_\infty}{\|\mathbf{x}\|_\infty}.$$

Unlike what happens for $K(A)$, cond(A,**x**) is invariant with respect to a *scaling* by rows of A, that is, to transformations of A of the form DA, where D is a nonsingular diagonal matrix. As a consequence, cond(A) provides a sound indication of the ill-conditioning of a matrix, irrespectively of any possible row diagonal *scaling*. ■

3.12.2 Iterative Refinement

Iterative refinement is a technique for improving the accuracy of a solution yielded by a direct method. Suppose that the linear system (3.2) has been solved by means of LU factorization (with partial or complete pivoting), and denote by $\mathbf{x}^{(0)}$ the computed solution. Having fixed an error tolerance, *tol*, the iterative refinement performs as follows: for $i = 0, 1, \ldots$, until convergence:

1. compute the residual $\mathbf{r}^{(i)} = \mathbf{b} - A\mathbf{x}^{(i)}$;
2. solve the linear system $A\mathbf{z} = \mathbf{r}^{(i)}$ using the LU factorization of A;
3. update the solution setting $\mathbf{x}^{(i+1)} = \mathbf{x}^{(i)} + \mathbf{z}$;
4. if $\|\mathbf{z}\|/\|\mathbf{x}^{(i+1)}\| < tol$, then terminate the process returning the solution $\mathbf{x}^{(i+1)}$. Otherwise, the algorithm restarts at step 1.

In absence of rounding errors, the process would stop at the first step, yielding the exact solution. The convergence properties of the method can be improved by computing the residual $\mathbf{r}^{(i)}$ in double precision, while computing the other quantities in single precision. We call this procedure *mixed-precision iterative refinement* (shortly, MPR), as compared to *fixed-precision iterative refinement* (FPR).

It can be shown that, if $\| \, |A^{-1}| \, |\widehat{L}| \, |\widehat{U}| \, \|_\infty$ is sufficiently small, then at each step i of the algorithm, the relative error $\|\mathbf{x} - \mathbf{x}^{(i)}\|_\infty / \|\mathbf{x}\|_\infty$ is reduced by a factor ρ, which is given by

$$\rho \simeq 2\; n \; \mathrm{cond}(A, \mathbf{x})\mathrm{u} \; (\mathrm{FPR}),$$

$$\rho \simeq \mathrm{u} \qquad\qquad\qquad (\mathrm{MPR}),$$

where ρ is independent of the condition number of A in the case of MPR. Slow convergence of FPR is a clear indication of the ill-conditioning of the matrix, as it can be shown that, if p is the number of iterations for the method to converge, then $K_\infty(A) \simeq \beta^{t(1-1/p)}$.

Even if performed in fixed precision, iterative refinement is worth using since it improves the overall stability of any direct method for solving the system. We refer to [Ric81], [Ske80], [JW77] [Ste73], [Wil63] and [CMSW79] for an overview of this subject.

3.13 Undetermined Systems

We have seen that the solution of the linear system $A\mathbf{x}=\mathbf{b}$ exists and is unique if $n = m$ and A is nonsingular. In this section we give a meaning to the solution of a linear system both in the *overdetermined* case, where $m > n$, and in the *underdetermined* case, corresponding to $m < n$. We notice that an undetermined system generally has no solution unless the right side $\mathbf{b}$ is an element of range(A).
For a detailed presentation, we refer to [LH74], [GL89] and [Bjö88].

Given $A \in \mathbb{R}^{m\times n}$ with $m \geq n$, $\mathbf{b} \in \mathbb{R}^m$, we say that $\mathbf{x}^* \in \mathbb{R}^n$ is a solution of the linear system $A\mathbf{x}=\mathbf{b}$ *in the least-squares sense* if

$$\Phi(\mathbf{x}^*) = \|A\mathbf{x}^* - \mathbf{b}\|_2^2 \leq \min_{\mathbf{x}\in\mathbb{R}^n} \|A\mathbf{x} - \mathbf{b}\|_2^2 = \min_{\mathbf{x}\in\mathbb{R}^n} \Phi(\mathbf{x}). \qquad (3.73)$$

The problem thus consists of minimizing the Euclidean norm of the residual. The solution of (3.73) can be found by imposing the condition that the gradient of the function Φ in (3.73) must be equal to zero at $\mathbf{x}^*$. From

$$\Phi(\mathbf{x}) = (A\mathbf{x} - \mathbf{b})^T (A\mathbf{x} - \mathbf{b}) = \mathbf{x}^T A^T A\mathbf{x} - 2\mathbf{x}^T A^T \mathbf{b} + \mathbf{b}^T\mathbf{b},$$

we find that

$$\nabla\Phi(\mathbf{x}^*) = 2A^T A\mathbf{x}^* - 2A^T\mathbf{b} = 0,$$

from which it follows that $\mathbf{x}^*$ must be the solution of the square system

$$A^T A\mathbf{x}^* = A^T\mathbf{b} \qquad (3.74)$$

known as the system of *normal equations*. The system is nonsingular if A has *full rank* and in such a case the least-squares solution exists and is unique.

We notice that $B = A^T A$ is a symmetric and positive definite matrix. Thus, in order to solve the normal equations, one could first compute the Cholesky factorization $B = H^T H$ and then solve the two systems $H^T \mathbf{y} = A^T \mathbf{b}$ and $H\mathbf{x}^* = \mathbf{y}$. However, due to roundoff errors, the computation of $A^T A$ may be affected by a loss of significant digits, with a consequent loss of positive definiteness or nonsingularity of the matrix, as happens in the following example (implemented in MATLAB) where for a matrix A with full rank, the corresponding matrix $fl(A^T A)$ turns out to be singular

$$A = \begin{bmatrix} 1 & 1 \\ 2^{-27} & 0 \\ 0 & 2^{-27} \end{bmatrix}, \quad fl(A^T A) = \begin{bmatrix} 1 & 1 \\ 1 & 1 \end{bmatrix}.$$

Therefore, in the case of ill-conditioned matrices it is more convenient to utilize the QR factorization introduced in Section 3.4.3. Indeed, the following result holds.

Theorem 3.8 *Let $A \in \mathbb{R}^{m\times n}$, with $m \geq n$, be a full rank matrix. Then the unique solution of* (3.73) *is given by*

$$\mathbf{x}^* = \tilde{R}^{-1}\tilde{Q}^T\mathbf{b}, \tag{3.75}$$

where $\tilde{R} \in \mathbb{R}^{n\times n}$ and $\tilde{Q} \in \mathbb{R}^{m\times n}$ are the matrices defined in (3.48) *starting from the QR factorization of* A. *Moreover, the minimum of Φ is given by*

$$\Phi(\mathbf{x}^*) = \sum_{i=n+1}^{m} [(Q^T\mathbf{b})_i]^2.$$

Proof. The QR factorization of A exists and is unique since A has full rank. Thus, there exist two matrices, $Q \in \mathbb{R}^{m\times m}$ and $R \in \mathbb{R}^{m\times n}$ such that A=QR, where Q is orthogonal. Since orthogonal matrices preserve the Euclidean scalar product (see Property 1.8), it follows that

$$\|A\mathbf{x} - \mathbf{b}\|_2^2 = \|R\mathbf{x} - Q^T\mathbf{b}\|_2^2.$$

Recalling that R is upper trapezoidal, we have

$$\|R\mathbf{x} - Q^T\mathbf{b}\|_2^2 = \|\tilde{R}\mathbf{x} - \tilde{Q}^T\mathbf{b}\|_2^2 + \sum_{i=n+1}^{m} [(Q^T\mathbf{b})_i]^2,$$

so that the minimum is achieved when $\mathbf{x} = \mathbf{x}^*$. ◇

For more details about the analysis of the computational cost of the algorithm (which depends on the actual implementation of the QR factorization), as well as for results about its stability, we refer the reader to the texts quoted at the beginning of the section.

If A does not have full rank, the solution techniques above fail, since in this case if $\mathbf{x}^*$ is a solution to (3.73), the vector $\mathbf{x}^* + \mathbf{z}$, with $\mathbf{z} \in \ker(A)$, is a solution too. We must therefore introduce a further constraint to enforce the uniqueness of the solution. Typically, one requires that $\mathbf{x}^*$ has minimal Euclidean norm, so that the least-squares problem can be formulated as:

$$\begin{aligned}&\text{find } \mathbf{x}^* \in \mathbb{R}^n \text{ with minimal Euclidean norm such that}\\ &\|\mathrm{A}\mathbf{x}^* - \mathbf{b}\|_2^2 \leq \min_{\mathbf{x}\in\mathbb{R}^n} \|\mathrm{A}\mathbf{x} - \mathbf{b}\|_2^2.\end{aligned} \tag{3.76}$$

This problem is consistent with (3.73) if A has full rank, since in this case (3.73) has a unique solution which necessarily must have minimal Euclidean norm.

The tool for solving (3.76) is the singular value decomposition (or SVD, see Section 1.9), for which the following theorem holds.

Theorem 3.9 *Let* $\mathrm{A} \in \mathbb{R}^{m\times n}$ *with SVD given by* $\mathrm{A} = \mathrm{U}\Sigma\mathrm{V}^T$. *Then the unique solution to* (3.76) *is*

$$\mathbf{x}^* = \mathrm{A}^\dagger \mathbf{b}, \tag{3.77}$$

where $\mathrm{A}^\dagger$ *is the pseudo-inverse of* A *introduced in Definition 1.15.*

Proof. Using the SVD of A, problem (3.76) is equivalent to finding $\mathbf{w} = \mathrm{V}^T\mathbf{x}$ such that $\mathbf{w}$ has minimal Euclidean norm and

$$\|\Sigma\mathbf{w} - \mathrm{U}^T\mathbf{b}\|_2^2 \leq \|\Sigma\mathbf{y} - \mathrm{U}^T\mathbf{b}\|_2^2, \quad \forall \mathbf{y} \in \mathbb{R}^n.$$

If r is the number of nonzero singular values σ_i of A, then

$$\|\Sigma\mathbf{w} - \mathrm{U}^T\mathbf{b}\|_2^2 = \sum_{i=1}^{r} \left(\sigma_i w_i - (\mathrm{U}^T\mathbf{b})_i\right)^2 + \sum_{i=r+1}^{m} \left((\mathrm{U}^T\mathbf{b})_i\right)^2,$$

which is minimum if $w_i = (\mathrm{U}^T\mathbf{b})_i/\sigma_i$ for $i = 1, \ldots, r$. Moreover, it is clear that among the vectors $\mathbf{w}$ of $\mathbb{R}^n$ having the first r components fixed, the one with minimal Euclidean norm has the remaining $n-r$ components equal to zero. Thus the solution vector is $\mathbf{w}^* = \Sigma^\dagger\mathrm{U}^T\mathbf{b}$, that is, $\mathbf{x}^* = \mathrm{V}\Sigma^\dagger\mathrm{U}^T\mathbf{b} = \mathrm{A}^\dagger\mathbf{b}$, where $\Sigma^\dagger$ is the diagonal matrix defined in (1.11). ◇

As for the stability of problem (3.76), we point out that if the matrix A does not have full rank, the solution $\mathbf{x}^*$ is not necessarily a continuous function of the data, so that small changes on these latter might produce large variations in $\mathbf{x}^*$. An example of this is shown below.

Example 3.11 Consider the system $\mathrm{A}\mathbf{x} = \mathbf{b}$ with

$$\mathrm{A} = \begin{bmatrix} 1 & 0 \\ 0 & 0 \\ 0 & 0 \end{bmatrix}, \ \mathbf{b} = \begin{bmatrix} 1 \\ 2 \\ 3 \end{bmatrix}, \ \mathrm{rank}(\mathrm{A}) = 1.$$

Using the MATLAB function `svd` we can compute the SVD of A. Then computing the pseudo-inverse, one finds the solution vector $\mathbf{x}^* = (1,\ 0)^T$. If we perturb the null entry a_{22}, with the value 10^{-12}, the perturbed matrix has (full) rank 2 and the solution (which is unique in the sense of (3.73)) is now given by $\widehat{\mathbf{x}}^* = \left(1,\ 2\cdot 10^{12}\right)^T$. •

We refer the reader to Section 5.8.3 for the approximate computation of the SVD of a matrix.

In the case of underdetermined systems, for which $m < n$, if A has full rank the QR factorization can still be used. In particular, when applied to the transpose matrix A^T, the method yields the solution of minimal Euclidean norm. If, instead, the matrix has not full rank, one must resort to SVD.

Remark 3.8 If $m = n$ (square system), both SVD and QR factorization can be used to solve the linear system $A\mathbf{x}=\mathbf{b}$, as alternatives to GEM. Even though these algorithms require a number of flops far superior to GEM (SVD, for instance, requires $12n^3$ flops), they turn out to be more accurate when the system is ill-conditioned and nearly singular. ■

Example 3.12 Compute the solution to the linear system $H_{15}\mathbf{x}=\mathbf{b}$, where H_{15} is the Hilbert matrix of order 15 (see (3.32)) and the right-hand side is chosen in such a way that the exact solution is the unit vector $\mathbf{x} = \mathbf{1}$. Using GEM with partial pivoting yields a solution affected by a relative error larger than 100%. A solution of much better quality is obtained by passing through the computation of the pseudo-inverse, where the entries in Σ that are less than 10^{-13} are set equal to zero. •

3.14 Applications

In this section we present two problems, suggested by structural mechanics and grid generation in finite element analysis, whose solutions require solving large linear systems.

3.14.1 Nodal Analysis of a Structured Frame

Let us consider a structured frame which is made by rectilinear beams connected among them through hinges (referred to as the *nodes*) and suitably constrained to the ground. External loads are assumed to be applied at the nodes of the frame and for any beam in the frame the internal actions amount to a unique force of constant strength and directed as the beam itself. If the normal stress acting on the beam is a traction we assume that it has positive sign, otherwise the action has negative sign. Structured frames are frequently employed as covering structures for large size public buildings like exhibition stands, railway stations or airport halls.

To determine the internal actions in the frame, that are the unknowns of the mathematical problem, a *nodal analysis* is used (see [Zie77]): the equilibrium with respect to translation is imposed at every node of the frame yielding a sparse and large-size linear system. The resulting matrix has a sparsity pattern which depends on the numbering of the unknowns and that can strongly affect the computational effort of the LU factorization due to fill-in. We will show that the fill-in can be dramatically reduced by a suitable reordering of the unknowns.

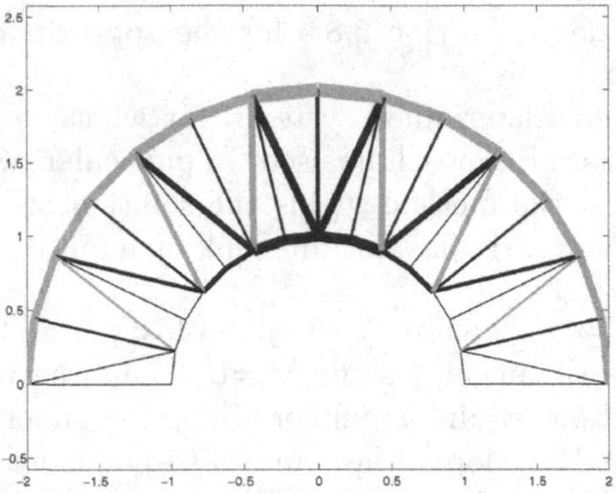

Fig. 3.9. A structured frame loaded at the point $(0, 1)$

The structure shown in Figure 3.9 is arc-shaped and is symmetric with respect to the origin. The radii r and R of the inner and outer circles are equal to 1 and 2, respectively. An external vertical load of unit size directed downwards is applied at $(0, 1)$ while the frame is constrained to ground through a hinge at $(-(r + R), 0)$ and a bogie at $(r + R, 0)$. To generate the structure we have partitioned the half unit circle in n_θ uniform slices, resulting in a total number of $n = 2(n_\theta + 1)$ nodes and a matrix size of $m = 2n$. The structure in Figure 3.9 has $n_\theta = 7$ and the unknowns are numbered following a counterclockwise labeling of the beams starting from the node at $(1, 0)$.

We have represented the structure along with the internal actions computed by solving the nodal equilibrium equations where the width of the beams is proportional to the strength of the computed action. Black is used to identify tractions whereas gray is associated with compressions. As expected the maximum traction stress is attained at the node where the external load is applied.

We show in Figure 3.10 the sparsity pattern of matrix A (*left*) and that of the L-factor of its LU factorization with partial pivoting (*right*) in the case $n_\theta = 40$ which corresponds to a size of 164×164. Notice the large fill-in effect arising in the lower part of L which results in an increase of the nonzero entries from 645 (before the factorization) to 1946 (after the factorization).

In view of the solution of the linear system by a direct method, the increase of the nonzero entries demands for a suitable reordering of the unknowns. For this purpose we use the MATLAB function `symrcm` which implements the symmetric reverse Cuthill-McKee algorithm described in Section 3.9.1. The sparsity pattern, after reordering, is shown in Figure 3.11 (*left*) while the L-factor of the LU factorization of the reordered matrix is shown in Figure 3.11 (*right*). The results indicate that the reordering procedure has "scattered" the sparsity pattern throughout the matrix with a relatively modest increase of the nonzero entries from 645 to 1040.

The effectiveness of the symmetric reverse Cuthill-McKee reordering procedure is demonstrated in Figure 3.12 which shows the number of nonzero

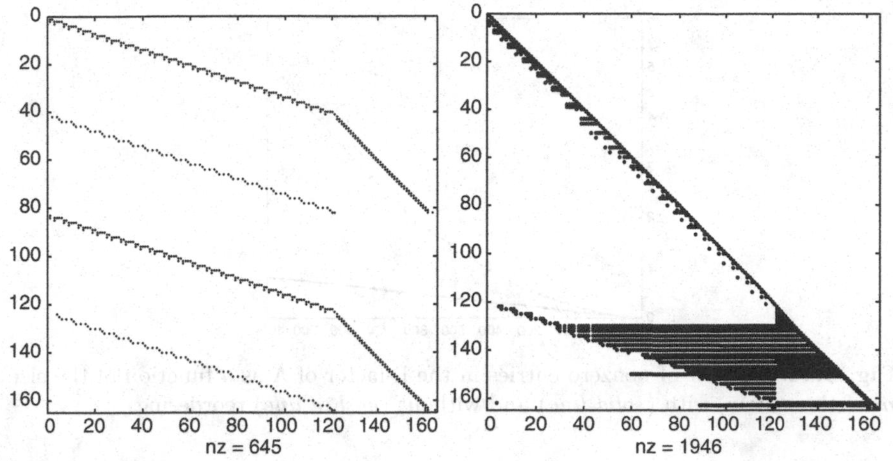

Fig. 3.10. Sparsity pattern of matrix A (*left*) and of the L-factor of the LU factorization with partial pivoting (*right*) in the case $n_\theta = 40$

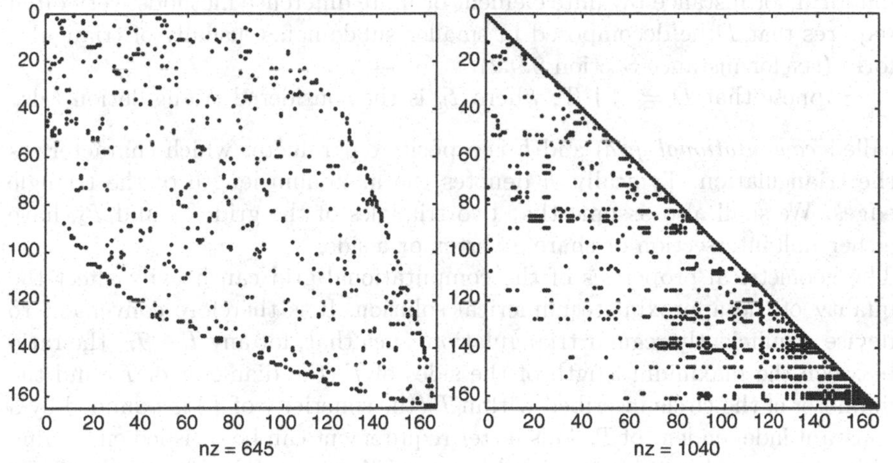

Fig. 3.11. Sparsity pattern of matrix A (*left*) after a reordering with the symmetric reverse Cuthill-McKee algorithm and the L-factor of the LU factorization of the reordered matrix with partial pivoting (*right*) in the case $n_\theta = 40$

entries **nz** in the L-factor of A as a function of the size m of the matrix (represented on the x-axis). In the reordered case (solid line) a linear increase of **nz** with m can be clearly appreciated at the expense of a dramatic fill-in growing with m if no reordering is performed (dashed line).

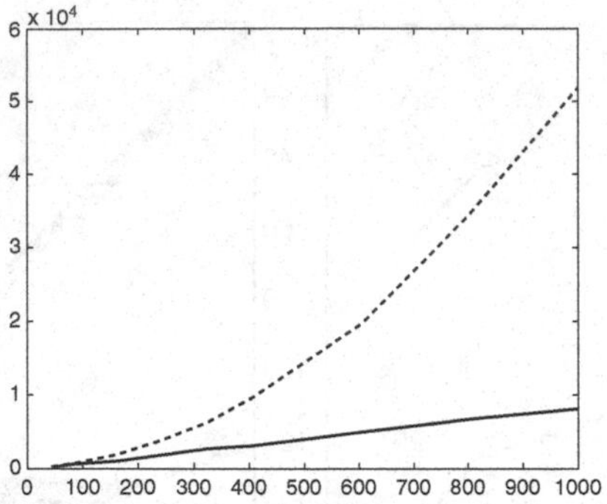

Fig. 3.12. Number of nonzero entries in the L-factor of A as a function of the size m of the matrix, with (*solid line*) and without (*dashed line*) reordering

3.14.2 Regularization of a Triangular Grid

The numerical solution of a problem in a two-dimensional domain D of polygonal form, for instance by finite element or finite difference methods, very often requires that D be decomposed in smaller subdomains, usually of triangular form (see for instance Section 9.9.2).

Suppose that $\overline{D} = \bigcup_{T \in \mathcal{T}_h} T$, where $\mathcal{T}_h$ is the considered triangulation (also called *computational grid*) and h is a positive parameter which characterizes the triangulation. Typically, h denotes the maximum length of the triangle edges. We shall also assume that two triangles of the grid, T_1 and T_2, have either null intersection or share a vertex or a side.
The geometrical properties of the computational grid can heavily affect the quality of the approximate numerical solution. It is therefore convenient to devise a sufficiently regular triangulation, such that, for any $T \in \mathcal{T}_h$, the ratio between the maximum length of the sides of T (the diameter of T) and the diameter of the circle inscribed within T (the sphericity of T) is bounded by a constant independent of T. This latter requirement can be satisfied employing a *regularization* procedure, applied to an existing grid. We refer to [Ver96] for further details on this subject.
Let us assume that $\mathcal{T}_h$ contains N_T triangles and N vertices, of which N_b, lying on the boundary ∂D of D, are kept fixed and having coordinates $\mathbf{x}_i^{(\partial D)} = [x_i^{(\partial D)}, y_i^{(\partial D)}]^T$. We denote by $\mathcal{N}_h$ the set of grid nodes, excluding the boundary nodes, and for each node $\mathbf{x}_i = (x_i, y_i)^T \in \mathcal{N}_h$, let $\mathcal{P}_i$ and $\mathcal{Z}_i$ respectively be the set of triangles $T \in \mathcal{T}_h$ sharing $\mathbf{x}_i$ (called the *patch* of $\mathbf{x}_i$) and the set of nodes of $\mathcal{P}_i$ except node $\mathbf{x}_i$ itself (see Figure 3.13, right). We let $n_i = \dim(\mathcal{Z}_i)$.

The regularization procedure consists of moving the generic node $\mathbf{x}_i$ to a new position which is determined by the center of gravity of the polygon generated by joining the nodes of $\mathcal{Z}_i$, and for that reason it is called a *barycentric regularization*. The effect of such a procedure is to force all the

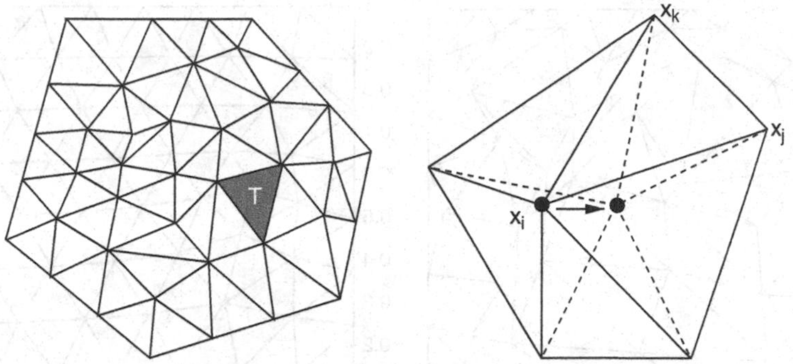

Fig. 3.13. An example of a decomposition into triangles of a polygonal domain D (*left*), and the effect of the barycentric regularization on a patch of triangles (*right*). The newly generated grid is plotted in dashed line

triangles that belong to the interior of the domain to assume a shape that is as regular as possible (in the limit, each triangle should be equilateral). In practice, we let

$$\mathbf{x}_i = \left(\sum_{\mathbf{x}_j \in \mathcal{Z}_i} \mathbf{x}_j \right) / n_i, \qquad \forall \mathbf{x}_i \in \mathcal{N}_h, \qquad \mathbf{x}_i = \mathbf{x}_i^{(\partial D)} \quad \text{if } \mathbf{x}_i \in \partial D.$$

Two systems must then be solved, one for the x-components $\{x_i\}$ and the other for the y-components $\{y_i\}$. Denoting by z_i the generic unknown, the i-th row of the system, in the case of internal nodes, reads

$$n_i z_i - \sum_{z_j \subset \mathcal{Z}_i} z_j = 0, \quad \forall i \in \mathcal{N}_h, \tag{3.78}$$

while for the boundary nodes the identities $z_i = z_i^{(\partial D)}$ hold. Equations (3.78) yield a system of the form $\mathrm{A}\mathbf{z} = \mathbf{b}$, where A is a symmetric and positive definite matrix of order $N - N_b$ which can be shown to be an M-matrix (see Section 1.12). This property ensures that the new grid coordinates satisfy minimum and maximum discrete principles, that is, they take a value which is between the minimum and the maximum values attained on the boundary.

Let us apply the regularization technique to the triangulation of the unit square in Figure 3.14, which is affected by a severe non uniformity of the triangle size. The grid consists of $N_T = 112$ triangles and $N = 73$ vertices, of which $N_b = 32$ are on the boundary. The size of each of the two linear systems (3.78) is thus equal to 41 and their solution is carried out by the LU factorization of matrix A in its original form and using its sparse format, obtained using the Cuthill-McKee inverse reordering algorithm described in Section 3.9.1.

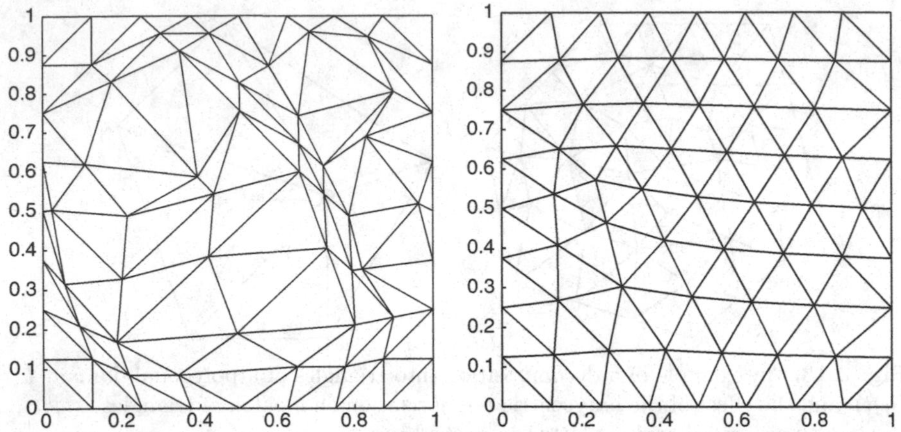

Fig. 3.14. Triangulation before (*left*) and after (*right*) the regularization

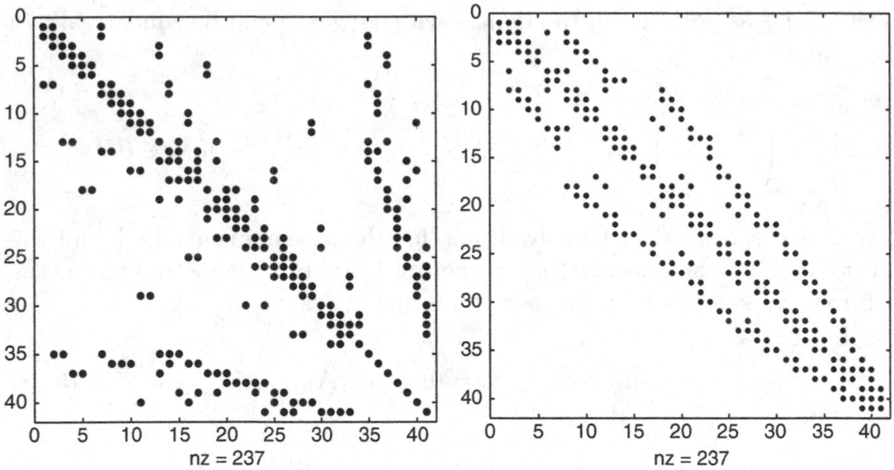

Fig. 3.15. Sparsity *patterns* of matrix A without and with reordering (*left* and *right*, respectively)

In Figure 3.15 the sparsity *patterns* of A are displayed, without and with reordering; the integer nz = 237 denotes the number of nonzero entries in the matrix. Notice that in the second case there is a decrease in the bandwidth of the matrix, to which corresponds a large reduction in the operation count from 61623 to 5552. The final configuration of the grid is displayed in Figure 3.14 (*right*), which clearly shows the effectiveness of the regularization procedure.

3.15 Exercises

1. For any square matrix A$\in \mathbb{R}^{n\times n}$, prove the following relations

$$\frac{1}{n}K_2(\mathrm{A}) \leq K_1(\mathrm{A}) \leq nK_2(\mathrm{A}), \ \frac{1}{n}K_\infty(\mathrm{A}) \leq K_2(\mathrm{A}) \leq nK_\infty(\mathrm{A}),$$
$$\frac{1}{n^2}K_1(\mathrm{A}) \leq K_\infty(\mathrm{A}) \leq n^2 K_1(\mathrm{A}).$$

They allow us to conclude that if a matrix is ill-conditioned in a certain norm it remains so even in another norm, up to a factor depending on n.

2. Check that the matrix B $\in \mathbb{R}^{n\times n}$: $b_{ii} = 1$, $b_{ij} = -1$ if $i < j$, $b_{ij} = 0$ if $i > j$, has determinant equal to 1, yet $K_\infty(\mathrm{B})$ is large (equal to $n2^{n-1}$).
3. Prove that $K(\mathrm{AB}) \leq K(\mathrm{A})K(\mathrm{B})$, for any two square nonsingular matrices A,B$\in \mathbb{R}^{n\times n}$.
4. Given the matrix A $\in \mathbb{R}^{2\times 2}$, $a_{11} = a_{22} = 1$, $a_{12} = \gamma$, $a_{21} = 0$, check that for $\gamma \geq 0$, $K_\infty(\mathrm{A}) = K_1(\mathrm{A}) = (1+\gamma)^2$. Next, consider the linear system A$\mathbf{x} = \mathbf{b}$ where $\mathbf{b}$ is such that $\mathbf{x} = (1-\gamma, 1)^T$ is the solution. Find a bound for $\|\boldsymbol{\delta}\mathbf{x}\|_\infty/\|\mathbf{x}\|_\infty$ in terms of $\|\boldsymbol{\delta}\mathbf{b}\|_\infty/\|\mathbf{b}\|_\infty$ when $\boldsymbol{\delta}\mathbf{b} = (\delta_1, \delta_2)^T$. Is the problem well- or ill-conditioned?
5. Consider the matrix A $\in \mathbb{R}^{n\times n}$, with entries $a_{ij} = 1$ if $i = j$ or $j = n$, $a_{ij} = -1$ if $i > j$, zero otherwise. Show that A admits an LU factorization, with $|l_{ij}| \leq 1$ and $u_{nn} = 2^{n-1}$.
6. Consider matrix (3.68) in Example 3.7. Prove that the matrices $\widehat{\mathrm{L}}$ and $\widehat{\mathrm{U}}$ have entries very large in module. Check that using GEM with complete pivoting yields the exact solution.
7. Devise a variant of GEM that transforms a nonsingular matrix A $\in \mathbb{R}^{n\times n}$ directly into a diagonal matrix D. This process is commonly known as the *Gauss-Jordan method.* Find the Gauss-Jordan transformation matrices G_i, $i = 1, \ldots, n$, such that $\mathrm{G}_n \cdots \mathrm{G}_1\mathrm{A} = \mathrm{D}$.
8. Let A be a sparse matrix of order n. Prove that the computational cost of the LU factorization of A is given by (3.61). Prove also that it is always less than

$$\frac{1}{2}\sum_{k=1}^{n} m_k(\mathrm{A})\left(m_k(\mathrm{A})+3\right).$$

9. Prove that, if A is a symmetric and positive definite matrix, solving the linear system A$\mathbf{x} = \mathbf{b}$ amounts to computing $\mathbf{x}= \sum_{i=1}^n (c_i/\lambda_i)\mathbf{v}_i$, where λ_i are the eigenvalues of A and $\mathbf{v}_i$ are the corresponding eigenvectors.
10. (From [JM92]). Consider the following linear system

$$\begin{bmatrix} 1001 & 1000 \\ 1000 & 1001 \end{bmatrix} \begin{bmatrix} x_1 \\ x_2 \end{bmatrix} = \begin{bmatrix} b_1 \\ b_2 \end{bmatrix}.$$

Using Exercise 9, explain why, when $\mathbf{b} = [2001,\ 2001]^T$, a small change $\boldsymbol{\delta}\mathbf{b} = [1, 0]^T$ produces large variations in the solution, while, conversely, when $\mathbf{b} = [1,\ -1]^T$, a small variation $\boldsymbol{\delta}\mathbf{x} = [0.001, 0]^T$ in the solution induces a large change in $\mathbf{b}$.
[*Hint* : expand the right hand side on the basis of the eigenvectors of the matrix.]

11. Characterize the fill-in for a matrix $A \in \mathbb{R}^{n\times n}$ having nonzero entries only on the main diagonal and on the first column and last row. Propose a permutation that minimizes the fill-in.
[*Hint* : it suffices to exchange the first row and the first column with the last row and the last column, respectively.]
12. Consider the linear system $H_n\mathbf{x} = \mathbf{b}$, where H_n is the Hilbert matrix of order n. Estimate, as a function of n, the maximum number of significant digits that are expected when solving the system by GEM.
13. Given the vectors

$$\mathbf{v}_1 = [1,\ 1,\ 1,\ -1]^T,\ \mathbf{v}_2 = [2,\ -1,\ -1,\ 1]^T,$$
$$\mathbf{v}_3 = [0,\ 3,\ 3,\ -3]^T,\ \mathbf{v}_4 = [-1,\ 2,\ 2,\ 1]^T,$$

generate an orthonormal system using the Gram-Schmidt algorithm, in either its standard and modified versions, and compare the obtained results. What is the dimension of the space generated by the given vectors?
14. Prove that if A=QR then

$$\frac{1}{n}K_1(A) \le K_1(R) \le nK_1(A),$$

while $K_2(A) = K_2(R)$.
15. Let $A \in \mathbb{R}^{n\times n}$ be a nonsingular matrix. Determine the conditions under which the ratio $\|\mathbf{y}\|_2/\|\mathbf{x}\|_2$, with $\mathbf{x}$ and $\mathbf{y}$ as in (3.70), approximates $\|A^{-1}\|_2$.
[*Solution* : let $U\Sigma V^T$ be the singular value decomposition of A. Denote by $\mathbf{u}_i$, $\mathbf{v}_i$ the column vectors of U and V, respectively, and expand the vector $\mathbf{d}$ in (3.70) on the basis spanned by $\{\mathbf{v}_i\}$. Then $\mathbf{d} = \sum_{i=1}^n \tilde{d}_i\mathbf{v}_i$ and, from (3.70), $\mathbf{x} = \sum_{i=1}^n (\tilde{d}_i/\sigma_i)\mathbf{u}_i$, $\mathbf{y} = \sum_{i=1}^n (\tilde{d}_i/\sigma_i^2)\mathbf{v}_i$, having denoted the singular values of A by $\sigma_1, \ldots, \sigma_n$.
The ratio

$$\|\mathbf{y}\|_2/\|\mathbf{x}\|_2 = \left[\sum_{i=1}^n (\tilde{d}_i/\sigma_i^2)^2 / \sum_{i=1}^n (\tilde{d}_i/\sigma_i)^2\right]^{1/2}$$

is about equal to $\sigma_n^{-1} = \|A^{-1}\|_2$ if: (*i*) $\mathbf{y}$ has a relevant component in the direction of $\mathbf{v}_n$ (i.e., if $\tilde{d}_n$ is not excessively small), and (*ii*) the ratio $\tilde{d}_n/\sigma_n$ is not negligible with respect to the ratios $\tilde{d}_i/\sigma_i$ for $i = 1, \ldots, n-1$. This last circumstance certainly occurs if A is ill-conditioned in the $\|\cdot\|_2$-norm since $\sigma_n \ll \sigma_1$.]

4

Iterative Methods for Solving Linear Systems

Iterative methods formally yield the solution $\mathbf{x}$ of a linear system after an infinite number of steps. At each step they require the computation of the residual of the system. In the case of a full matrix, their computational cost is therefore of the order of n^2 operations for each iteration, to be compared with an overall cost of the order of $\frac{2}{3}n^3$ operations needed by direct methods. Iterative methods can therefore become competitive with direct methods provided the number of iterations that are required to converge (within a prescribed tolerance) is either independent of n or scales sublinearly with respect to n.

In the case of large sparse matrices, as discussed in Section 3.9, direct methods may be unconvenient due to the dramatic fill-in, although extremely efficient direct solvers can be devised on sparse matrices featuring special structures like, for example, those encountered in the approximation of partial differential equations (see Chapters 12 and 13).

Finally, we notice that, when A is ill-conditioned, a combined use of direct and iterative methods is made possible by preconditioning techniques that will be addressed in Section 4.3.2.

4.1 On the Convergence of Iterative Methods

The basic idea of iterative methods is to construct a sequence of vectors $\mathbf{x}^{(k)}$ that enjoy the property of *convergence*

$$\mathbf{x} = \lim_{k\to\infty} \mathbf{x}^{(k)}, \tag{4.1}$$

where $\mathbf{x}$ is the solution to (3.2). In practice, the iterative process is stopped at the minimum value of n such that $\|\mathbf{x}^{(n)} - \mathbf{x}\| < \varepsilon$, where ε is a fixed tolerance and $\|\cdot\|$ is any convenient vector norm. However, since the exact solution is obviously not available, it is necessary to introduce suitable stopping criteria to monitor the convergence of the iteration (see Section 4.6).

To start with, we consider iterative methods of the form

$$\text{given } \mathbf{x}^{(0)},\ \mathbf{x}^{(k+1)} = \mathrm{B}\mathbf{x}^{(k)} + \mathbf{f}, \quad k \geq 0, \tag{4.2}$$

having denoted by B an $n \times n$ square matrix called the *iteration matrix* and by **f** a vector that is obtained from the right hand side **b**.

Definition 4.1 An iterative method of the form (4.2) is said to be *consistent* with (3.2) if **f** and B are such that $\mathbf{x} = \mathrm{B}\mathbf{x} + \mathbf{f}$. Equivalently,

$$\mathbf{f} = (\mathrm{I} - \mathrm{B})\mathrm{A}^{-1}\mathbf{b}.$$

■

Having denoted by

$$\mathbf{e}^{(k)} = \mathbf{x}^{(k)} - \mathbf{x} \tag{4.3}$$

the error at the k-th step of the iteration, the condition for convergence (4.1) amounts to requiring that $\lim_{k\to\infty} \mathbf{e}^{(k)} = \mathbf{0}$ for any choice of the initial datum $\mathbf{x}^{(0)}$ (often called the *initial guess*).

Consistency alone does not suffice to ensure the convergence of the iterative method (4.2), as shown in the following example.

Example 4.1 To solve the linear system $2\mathrm{I}\mathbf{x} = \mathbf{b}$, consider the iterative method

$$\mathbf{x}^{(k+1)} = -\mathbf{x}^{(k)} + \mathbf{b},$$

which is obviously consistent. This scheme is not convergent for any choice of the initial guess. If, for instance, $\mathbf{x}^{(0)} = \mathbf{0}$, the method generates the sequence $\mathbf{x}^{(2k)} = \mathbf{0}$, $\mathbf{x}^{(2k+1)} = \mathbf{b}$, $k = 0, 1, \ldots$.

On the other hand, if $\mathbf{x}^{(0)} = \frac{1}{2}\mathbf{b}$ the method is convergent. •

Theorem 4.1 *Let* (4.2) *be a consistent method. Then, the sequence of vectors* $\{\mathbf{x}^{(k)}\}$ *converges to the solution of* (3.2) *for any choice of* $\mathbf{x}^{(0)}$ *iff* $\rho(\mathrm{B}) < 1$.

Proof. From (4.3) and the consistency assumption, the recursive relation $\mathbf{e}^{(k+1)} = \mathrm{B}\mathbf{e}^{(k)}$ is obtained. Therefore,

$$\mathbf{e}^{(k)} = \mathrm{B}^k\mathbf{e}^{(0)}, \qquad \forall k = 0, 1, \ldots \tag{4.4}$$

Thus, thanks to Theorem 1.5, it follows that $\lim_{k\to\infty} \mathrm{B}^k\mathbf{e}^{(0)} = \mathbf{0}$ for any $\mathbf{e}^{(0)}$ iff $\rho(\mathrm{B}) < 1$.

Conversely, suppose that $\rho(\mathrm{B}) > 1$, then there exists at least one eigenvalue $\lambda(\mathrm{B})$ with module greater than 1. Let $\mathbf{e}^{(0)}$ be an eigenvector associated with λ; then $\mathrm{B}\mathbf{e}^{(0)} = \lambda\mathbf{e}^{(0)}$ and, therefore, $\mathbf{e}^{(k)} = \lambda^k\mathbf{e}^{(0)}$. As a consequence, $\mathbf{e}^{(k)}$ cannot tend to $\mathbf{0}$ as $k \to \infty$, since $|\lambda| > 1$. ◇

From (1.23) and Theorem 1.4 it follows that a sufficient condition for convergence to hold is that $\|\mathrm{B}\| < 1$, for any consistent matrix norm. It is reasonable to expect that the convergence is faster when $\rho(\mathrm{B})$ is smaller so that an estimate of $\rho(\mathrm{B})$ might provide a sound indication of the convergence of the algorithm. Other remarkable quantities in convergence analysis are contained in the following definition.

Definition 4.2 Let B be the iteration matrix. We call:

1. $\|\mathrm{B}^m\|$ the *convergence factor* after m steps of the iteration;
2. $\|\mathrm{B}^m\|^{1/m}$ the *average convergence factor* after m steps;
3. $R_m(\mathrm{B}) = -\frac{1}{m}\log\|\mathrm{B}^m\|$ the *average convergence rate* after m steps.

■

These quantities are too expensive to compute since they require evaluating B^m. Therefore, it is usually preferred to estimate the *asymptotic convergence rate*, which is defined as

$$R(\mathrm{B}) = \lim_{k\to\infty} R_k(\mathrm{B}) = -\log\rho(\mathrm{B}), \tag{4.5}$$

where Property 1.14 has been accounted for. In particular, if B were symmetric, we would have

$$R_m(\mathrm{B}) = -\frac{1}{m}\log\|\mathrm{B}^m\|_2 = -\log\rho(\mathrm{B}).$$

In the case of nonsymmetric matrices, $\rho(\mathrm{B})$ sometimes provides an overoptimistic estimate of $\|\mathrm{B}^m\|^{1/m}$ (see [Axe94], Section 5.1). Indeed, although $\rho(\mathrm{B}) < 1$, the convergence to zero of the sequence $\|\mathrm{B}^m\|$ might be nonmonotone (see Exercise 1). We finally notice that, due to (4.5), $\rho(\mathrm{B})$ is the *asymptotic convergence factor*. Criteria for estimating the quantities defined so far will be addressed in Section 4.6.

Remark 4.1 The iterations introduced in (4.2) are a special instance of iterative methods of the form

$$\mathbf{x}^{(0)} = \mathbf{f}_0(\mathrm{A}, \mathbf{b}),$$

$$\mathbf{x}^{(n+1)} = \mathbf{f}_{n+1}(\mathbf{x}^{(n)}, \mathbf{x}^{(n-1)}, \ldots, \mathbf{x}^{(n-m)}, \mathrm{A}, \mathbf{b}), \text{ for } n \geq m,$$

where $\mathbf{f}_i$ and $\mathbf{x}^{(m)}, \ldots, \mathbf{x}^{(1)}$ are given functions and vectors, respectively. The number of steps which the current iteration depends on is called the *order of the method.* If the functions $\mathbf{f}_i$ are independent of the step index i, the method is called *stationary*, otherwise it is *nonstationary*. Finally, if $\mathbf{f}_i$ depends linearly on $\mathbf{x}^{(0)}, \ldots, \mathbf{x}^{(m)}$, the method is called *linear*, otherwise it is *nonlinear.*

In the light of these definitions, the methods considered so far are therefore *stationary linear iterative methods of first order.* In Section 4.3, examples of nonstationary linear methods will be provided. ■

4.2 Linear Iterative Methods

A general technique to devise consistent linear iterative methods is based on an additive *splitting* of the matrix A of the form A=P−N, where P and N are two suitable matrices and P is nonsingular. For reasons that will be clear in the later sections, P is called *preconditioning matrix* or *preconditioner*.

Precisely, given $\mathbf{x}^{(0)}$, one can compute $\mathbf{x}^{(k)}$ for $k \geq 1$, solving the systems

$$\mathrm{P}\mathbf{x}^{(k+1)} = \mathrm{N}\mathbf{x}^{(k)} + \mathbf{b}, \quad k \geq 0. \tag{4.6}$$

The iteration matrix of method (4.6) is $\mathrm{B} = \mathrm{P}^{-1}\mathrm{N}$, while $\mathbf{f} = \mathrm{P}^{-1}\mathbf{b}$. Alternatively, (4.6) can be written in the form

$$\mathbf{x}^{(k+1)} = \mathbf{x}^{(k)} + \mathrm{P}^{-1}\mathbf{r}^{(k)}, \tag{4.7}$$

where

$$\mathbf{r}^{(k)} = \mathbf{b} - \mathrm{A}\mathbf{x}^{(k)} \tag{4.8}$$

denotes the *residual* vector at step k. Relation (4.7) outlines the fact that a linear system, with coefficient matrix P, must be solved to update the solution at step $k+1$. Thus P, besides being nonsingular, ought to be easily invertible, in order to keep the overall computational cost low. (Notice that, if P were equal to A and N=0, method (4.7) would converge in one iteration, but at the same cost of a direct method).

Let us mention two results that ensure convergence of the iteration (4.7), provided suitable conditions on the splitting of A are fulfilled (for their proof, we refer to [Hac94]).

Property 4.1 *Let* $\mathrm{A} = \mathrm{P} - \mathrm{N}$, *with* A *and* P *symmetric and positive definite. If the matrix* $2\mathrm{P} - \mathrm{A}$ *is positive definite, then the iterative method defined in* (4.7) *is convergent for any choice of the initial datum* $\mathbf{x}^{(0)}$ *and*

$$\rho(\mathrm{B}) = \|\mathrm{B}\|_{\mathrm{A}} = \|\mathrm{B}\|_{\mathrm{P}} < 1.$$

Moreover, the convergence of the iteration is monotone with respect to the norms $\|\cdot\|_{\mathrm{P}}$ *and* $\|\cdot\|_{\mathrm{A}}$ *(i.e.,* $\|\mathbf{e}^{(k+1)}\|_{\mathrm{P}} < \|\mathbf{e}^{(k)}\|_{\mathrm{P}}$ *and* $\|\mathbf{e}^{(k+1)}\|_{\mathrm{A}} < \|\mathbf{e}^{(k)}\|_{\mathrm{A}}$ $k = 0, 1, \ldots$*).*

Property 4.2 *Let* $\mathrm{A} = \mathrm{P} - \mathrm{N}$ *with* A *being symmetric and positive definite. If the matrix* $\mathrm{P} + \mathrm{P}^T - \mathrm{A}$ *is positive definite, then* P *is invertible, the iterative method defined in* (4.7) *is monotonically convergent with respect to norm* $\|\cdot\|_{\mathrm{A}}$ *and* $\rho(\mathrm{B}) \leq \|\mathrm{B}\|_{\mathrm{A}} < 1$.

4.2.1 Jacobi, Gauss-Seidel and Relaxation Methods

In this section we consider some classical linear iterative methods.

If the diagonal entries of A are nonzero, we can single out in each equation the corresponding unknown, obtaining the equivalent linear system

$$x_i = \frac{1}{a_{ii}}\left[b_i - \sum_{\substack{j=1\\j\neq i}}^{n} a_{ij}x_j\right], \qquad i = 1,\ldots,n. \tag{4.9}$$

In the Jacobi method, once an arbitrarily initial guess $\mathbf{x}^{(0)}$ has been chosen, $\mathbf{x}^{(k+1)}$ is computed by the formulae

$$x_i^{(k+1)} = \frac{1}{a_{ii}}\left[b_i - \sum_{\substack{j=1\\j\neq i}}^{n} a_{ij}x_j^{(k)}\right], \ i = 1,\ldots,n. \tag{4.10}$$

This amounts to performing the following splitting for A

$$\mathrm{P} = \mathrm{D},\ \mathrm{N} = \mathrm{D} - \mathrm{A} = \mathrm{E} + \mathrm{F},$$

where D is the diagonal matrix of the diagonal entries of A, E is the lower triangular matrix of entries $e_{ij} = -a_{ij}$ if $i > j$, $e_{ij} = 0$ if $i \leq j$, and F is the upper triangular matrix of entries $f_{ij} = -a_{ij}$ if $j > i$, $f_{ij} = 0$ if $j \leq i$. As a consequence, $\mathrm{A} = \mathrm{D} - (\mathrm{E} + \mathrm{F})$.

The iteration matrix of the Jacobi method is thus given by

$$\mathrm{B}_J = \mathrm{D}^{-1}(\mathrm{E} + \mathrm{F}) = \mathrm{I} - \mathrm{D}^{-1}\mathrm{A}. \tag{4.11}$$

A generalization of the Jacobi method is the over-relaxation method (or JOR), in which, having introduced a relaxation parameter ω, (4.10) is replaced by

$$x_i^{(k+1)} = \frac{\omega}{a_{ii}}\left[b_i - \sum_{\substack{j=1\\j\neq i}}^{n} a_{ij}x_j^{(k)}\right] + (1-\omega)x_i^{(k)}, \qquad i = 1,\ldots,n.$$

The corresponding iteration matrix is

$$\mathrm{B}_{J_\omega} = \omega\mathrm{B}_J + (1-\omega)\mathrm{I}. \tag{4.12}$$

In the form (4.7), the JOR method corresponds to

$$\mathbf{x}^{(k+1)} = \mathbf{x}^{(k)} + \omega\mathrm{D}^{-1}\mathbf{r}^{(k)}.$$

This method is consistent for any $\omega \neq 0$ and for $\omega = 1$ it coincides with the Jacobi method.

The Gauss-Seidel method differs from the Jacobi method in the fact that at the $k+1$-th step the available values of $x_i^{(k+1)}$ are being used to update the solution, so that, instead of (4.10), one has

$$x_i^{(k+1)} = \frac{1}{a_{ii}} \left[b_i - \sum_{j=1}^{i-1} a_{ij} x_j^{(k+1)} - \sum_{j=i+1}^{n} a_{ij} x_j^{(k)} \right], \; i = 1, \ldots, n. \tag{4.13}$$

This method amounts to performing the following splitting for A

$$\mathrm{P} = \mathrm{D} - \mathrm{E}, \; \mathrm{N} = \mathrm{F},$$

and the associated iteration matrix is

$$\mathrm{B}_{GS} = (\mathrm{D} - \mathrm{E})^{-1}\mathrm{F}. \tag{4.14}$$

Starting from Gauss-Seidel method, in analogy to what was done for Jacobi iterations, we introduce the successive over-relaxation method (or SOR method)

$$x_i^{(k+1)} = \frac{\omega}{a_{ii}} \left[b_i - \sum_{j=1}^{i-1} a_{ij} x_j^{(k+1)} - \sum_{j=i+1}^{n} a_{ij} x_j^{(k)} \right] + (1-\omega) x_i^{(k)}, \tag{4.15}$$

for $i = 1, \ldots, n$. The method (4.15) can be written in vector form as

$$(\mathrm{I} - \omega \mathrm{D}^{-1}\mathrm{E})\mathbf{x}^{(k+1)} = [(1-\omega)\mathrm{I} + \omega \mathrm{D}^{-1}\mathrm{F}]\mathbf{x}^{(k)} + \omega \mathrm{D}^{-1}\mathbf{b}, \tag{4.16}$$

from which the iteration matrix is

$$\mathrm{B}(\omega) = (\mathrm{I} - \omega \mathrm{D}^{-1}\mathrm{E})^{-1}[(1-\omega)\mathrm{I} + \omega \mathrm{D}^{-1}\mathrm{F}]. \tag{4.17}$$

Multiplying by D both sides of (4.16) and recalling that $\mathrm{A} = \mathrm{D} - (\mathrm{E} + \mathrm{F})$ yields the following form (4.7) of the SOR method

$$\mathbf{x}^{(k+1)} = \mathbf{x}^{(k)} + \left(\frac{1}{\omega}\mathrm{D} - \mathrm{E} \right)^{-1} \mathbf{r}^{(k)}.$$

It is consistent for any $\omega \neq 0$ and for $\omega = 1$ it coincides with Gauss-Seidel method. In particular, if $\omega \in (0, 1)$ the method is called under-relaxation, while if $\omega > 1$ it is called over-relaxation.

4.2.2 Convergence Results for Jacobi and Gauss-Seidel Methods

There exist special classes of matrices for which it is possible to state a priori some convergence results for the methods examined in the previous section. The first result in this direction is the following.

Theorem 4.2 *If* A *is a strictly diagonally dominant matrix by rows, the Jacobi and Gauss-Seidel methods are convergent.*

Proof. Let us prove the part of the theorem concerning the Jacobi method, while for the Gauss-Seidel method we refer to [Axe94]. Since A is strictly diagonally dominant by rows, $|a_{ii}| > \sum_{j=1}^{n} |a_{ij}|$ for $j \neq i$ and $i = 1, \ldots, n$. As a consequence, $\|\mathrm{B}_J\|_\infty = \max_{i=1,\ldots,n} \sum_{j=1, j\neq i}^{n} |a_{ij}|/|a_{ii}| < 1$, so that the Jacobi method is convergent. ◇

Theorem 4.3 *If* A *and* 2D − A *are symmetric and positive definite matrices, then the Jacobi method is convergent and* $\rho(\mathrm{B}_J) = \|\mathrm{B}_J\|_\mathrm{A} = \|\mathrm{B}_J\|_\mathrm{D}$.

Proof. The theorem follows from Property 4.1 taking P=D. ◇

In the case of the JOR method, the assumption on 2D − A can be removed, yielding the following result.

Theorem 4.4 *If* A *is symmetric positive definite, then the JOR method is convergent if* $0 < \omega < 2/\rho(\mathrm{D}^{-1}\mathrm{A})$.

Proof. The result immediately follows from (4.12) and noting that A has real positive eigenvalues. ◇

Concerning the Gauss-Seidel method, the following result holds.

Theorem 4.5 *If* A *is symmetric positive definite, the Gauss-Seidel method is monotonically convergent with respect to the norm* $\|\cdot\|_\mathrm{A}$.

Proof. We can apply Property 4.2 to the matrix P=D−E, upon checking that $\mathrm{P} + \mathrm{P}^T - \mathrm{A}$ is positive definite. Indeed

$$\mathrm{P} + \mathrm{P}^T - \mathrm{A} = 2\mathrm{D} - \mathrm{E} - \mathrm{F} - \mathrm{A} = \mathrm{D},$$

having observed that $(\mathrm{D} - \mathrm{E})^T = \mathrm{D} - \mathrm{F}$. We conclude by noticing that D is positive definite, since it is the diagonal of A. ◇

Finally, if A is tridiagonal (or block tridiagonal), it can be shown that

$$\rho(\mathrm{B}_{GS}) = \rho^2(\mathrm{B}_J) \tag{4.18}$$

(see [You71] for the proof). From (4.18) we can conclude that both methods converge or fail to converge at the same time. In the former case, the Gauss-Seidel method is more rapidly convergent than the Jacobi method, and the asymptotic convergence rate of the Gauss-Seidel method is twice than that of the Jacobi method. In particular, if A is tridiagonal and symmetric positive definite, Theorem 4.5 implies convergence of the Gauss-Seidel method, and (4.18) ensures convergence also for the Jacobi method.

Relation (4.18) holds even if A enjoys the following *A-property*.

Definition 4.3 A *consistently ordered* matrix $\mathrm{M} \in \mathbb{R}^{n\times n}$ (that is, a matrix such that $\alpha\mathrm{D}^{-1}\mathrm{E} + \alpha^{-1}\mathrm{D}^{-1}\mathrm{F}$, for $\alpha \neq 0$, has eigenvalues that do not depend on α, where $\mathrm{M} = \mathrm{D} - \mathrm{E} - \mathrm{F}$, $\mathrm{D} = \mathrm{diag}(m_{11}, \ldots, m_{nn})$, E and F are strictly lower and upper triangular matrices, respectively) enjoys the A-property if it can be partitioned in the 2×2 block form

$$\mathrm{M} = \begin{bmatrix} \tilde{\mathrm{D}}_1 & \mathrm{M}_{12} \\ \mathrm{M}_{21} & \tilde{\mathrm{D}}_2 \end{bmatrix},$$

where $\tilde{\mathrm{D}}_1$ and $\tilde{\mathrm{D}}_2$ are diagonal matrices. ∎

When dealing with general matrices, no a priori conclusions on the convergence properties of the Jacobi and Gauss-Seidel methods can be drawn, as shown in Example 4.2.

Example 4.2 Consider the 3×3 linear systems of the form $\mathrm{A}_i\mathbf{x} = \mathbf{b}_i$, where $\mathbf{b}_i$ is always taken in such a way that the solution of the system is the unit vector, and the matrices A_i are

$$\mathrm{A}_1 = \begin{bmatrix} 3 & 0 & 4 \\ 7 & 4 & 2 \\ -1 & 1 & 2 \end{bmatrix}, \qquad \mathrm{A}_2 = \begin{bmatrix} -3 & 3 & -6 \\ -4 & 7 & -8 \\ 5 & 7 & -9 \end{bmatrix},$$

$$\mathrm{A}_3 = \begin{bmatrix} 4 & 1 & 1 \\ 2 & -9 & 0 \\ 0 & -8 & -6 \end{bmatrix}, \qquad \mathrm{A}_4 = \begin{bmatrix} 7 & 6 & 9 \\ 4 & 5 & -4 \\ -7 & -3 & 8 \end{bmatrix}.$$

It can be checked that the Jacobi method does fail to converge for A_1 ($\rho(\mathrm{B}_J) = 1.33$), while the Gauss-Seidel scheme is convergent. Conversely, in the case of A_2, the Jacobi method is convergent, while the Gauss-Seidel method fails to converge ($\rho(\mathrm{B}_{GS}) = 1.\bar{1}$). In the remaining two cases, the Jacobi method is more slowly convergent than the Gauss-Seidel method for matrix A_3 ($\rho(\mathrm{B}_J) = 0.44$ against $\rho(\mathrm{B}_{GS}) = 0.018$), and the converse is true for A_4 ($\rho(\mathrm{B}_J) = 0.64$ while $\rho(\mathrm{B}_{GS}) = 0.77$). •

We conclude the section with the following result.

Theorem 4.6 *If the Jacobi method is convergent, then the JOR method converges if* $0 < \omega \leq 1$.

Proof. From (4.12) we obtain that the eigenvalues of B_{J_ω} are

$$\mu_k = \omega\lambda_k + 1 - \omega, \qquad k = 1, \ldots, n,$$

where λ_k are the eigenvalues of B_J. Then, recalling the Euler formula for the representation of a complex number, we let $\lambda_k = r_k e^{i\theta_k}$ and get

$$|\mu_k|^2 = \omega^2 r_k^2 + 2\omega r_k \cos(\theta_k)(1-\omega) + (1-\omega)^2 \leq (\omega r_k + 1 - \omega)^2,$$

which is less than 1 if $0 < \omega \leq 1$. ◇

4.2.3 Convergence Results for the Relaxation Method

The following result provides a necessary condition on ω in order the SOR method to be convergent.

Theorem 4.7 *For any $\omega \in \mathbb{R}$ we have $\rho(\mathrm{B}(\omega)) \geq |\omega - 1|$; therefore, the SOR method fails to converge if $\omega \leq 0$ or $\omega \geq 2$.*

Proof. If $\{\lambda_i\}$ denote the eigenvalues of the SOR iteration matrix, then

$$\left|\prod_{i=1}^{n} \lambda_i\right| = \left|\det\left[(1-\omega)\mathrm{I} + \omega \mathrm{D}^{-1}\mathrm{F}\right]\right| = |1-\omega|^n.$$

Therefore, at least one eigenvalue λ_i must exist such that $|\lambda_i| \geq |1-\omega|$ and thus, in order for convergence to hold, we must have $|1-\omega| < 1$, that is $0 < \omega < 2$. $\diamond$

Assuming that A is symmetric and positive definite, the condition $0 < \omega < 2$, besides being necessary, becomes also sufficient for convergence. Indeed the following result holds (for the proof, see [Hac94]).

Property 4.3 (Ostrowski) *If* A *is symmetric and positive definite, then the SOR method is convergent iff $0 < \omega < 2$. Moreover, its convergence is monotone with respect to $\|\cdot\|_{\mathrm{A}}$.*

Finally, if A is *strictly diagonally dominant* by rows, the SOR method converges if $0 < \omega \leq 1$.

The results above show that the SOR method is more or less rapidly convergent, depending on the choice of the relaxation parameter ω. The question of how to determine the value ω_{opt} for which the convergence rate is the highest possible can be given a satisfactory answer only in special cases (see, for instance, [Axe94], [You71], [Var62] or [Wac66]). Here we limit ourselves to quoting the following result (whose proof is in [Axe94]).

Property 4.4 *If the matrix* A *enjoys the A-property and if B_J has real eigenvalues, then the SOR method converges for any choice of $\mathbf{x}^{(0)}$ iff $\rho(\mathrm{B}_J) < 1$ and $0 < \omega < 2$. Moreover,*

$$\omega_{opt} = \frac{2}{1 + \sqrt{1 - \rho^2(\mathrm{B}_J)}} \tag{4.19}$$

and the corresponding asymptotic convergence factor is

$$\rho(\mathrm{B}(\omega_{opt})) = \frac{1 - \sqrt{1 - \rho^2(\mathrm{B}_J)}}{1 + \sqrt{1 - \rho^2(\mathrm{B}_J)}}.$$

4.2.4 A priori Forward Analysis

In the previous analysis we have neglected the rounding errors. However, as shown in the following example (taken from [HW76]), they can dramatically affect the convergence rate of the iterative method.

Example 4.3 Let A be a lower bidiagonal matrix of order 100 with entries $a_{ii} = 1.5$ and $a_{i,i-1} = 1$, and let $\mathbf{b} \in \mathbb{R}^{100}$ be the right-side with $b_i = 2.5$. The exact solution of the system $\mathrm{A}\mathbf{x} = \mathbf{b}$ has components $x_i = 1-(-2/3)^i$. The SOR method with $\omega = 1.5$ should be convergent, working in exact arithmetic, since $\rho(\mathrm{B}(1.5)) = 0.5$ (far below one). However, running Program 16 with $\mathbf{x}^{(0)} = fl(\mathbf{x}) + \epsilon_M$, which is extremely close to the exact value, the sequence $\mathbf{x}^{(k)}$ diverges and after 100 iterations the algorithm yields a solution with $\|\mathbf{x}^{(100)}\|_\infty = 10^{13}$. The flaw is due to rounding error propagation and must not be ascribed to a possible ill-conditioning of the matrix since $K_\infty(\mathrm{A}) \simeq 5$. •

To account for rounding errors, let us denote by $\widehat{\mathbf{x}}^{(k+1)}$ the solution (in finite arithmetic) generated by an iterative method of the form (4.6) after k steps. Due to rounding errors, $\widehat{\mathbf{x}}^{(k+1)}$ can be regarded as the exact solution to the problem

$$\mathrm{P}\widehat{\mathbf{x}}^{(k+1)} = \mathrm{N}\widehat{\mathbf{x}}^{(k)} + \mathbf{b} - \boldsymbol{\zeta}_k, \tag{4.20}$$

with

$$\boldsymbol{\zeta}_k = \delta\mathrm{P}_{k+1}\widehat{\mathbf{x}}^{(k+1)} - \mathbf{g}_k.$$

The matrix $\delta\mathrm{P}_{k+1}$ accounts for the rounding errors in the solution of (4.6), while the vector $\mathbf{g}_k$ includes the errors made in the evaluation of $\mathrm{N}\widehat{\mathbf{x}}^{(k)} + \mathbf{b}$. From (4.20), we obtain

$$\widehat{\mathbf{x}}^{(k+1)} = \mathrm{B}^{k+1}\mathbf{x}^{(0)} + \sum_{j=0}^{k} \mathrm{B}^j\mathrm{P}^{-1}(\mathbf{b} - \boldsymbol{\zeta}_{k-j})$$

and for the absolute error $\widehat{\mathbf{e}}^{(k+1)} = \mathbf{x} - \widehat{\mathbf{x}}^{(k+1)}$

$$\widehat{\mathbf{e}}^{(k+1)} = \mathrm{B}^{k+1}\mathbf{e}^{(0)} + \sum_{j=0}^{k} \mathrm{B}^j\mathrm{P}^{-1}\boldsymbol{\zeta}_{k-j}.$$

The first term represents the error that is made by the iterative method in exact arithmetic; if the method is convergent, this error is negligible for sufficiently large values of k. The second term refers instead to rounding error propagation; its analysis is quite technical and is carried out, for instance, in [Hig88] in the case of Jacobi, Gauss-Seidel and SOR methods.

4.2.5 Block Matrices

The methods of the previous sections are also referred to as *point* (or *line*) iterative methods, since they act on single entries of matrix A. It is possible to devise *block* versions of the algorithms, provided that D denotes the block diagonal matrix whose entries are the $m \times m$ diagonal blocks of matrix A (see Section 1.6).

The *block Jacobi method* is obtained taking again P=D and N=D-A. The method is well-defined only if the diagonal blocks of D are nonsingular. If A is decomposed in $p \times p$ square blocks, the block Jacobi method is

$$\mathrm{A}_{ii}\mathbf{x}_i^{(k+1)} = \mathbf{b}_i - \sum_{\substack{j=1 \\ j\neq i}}^{p} \mathrm{A}_{ij}\mathbf{x}_j^{(k)}, \; i = 1, \ldots, p,$$

having also decomposed the solution vector and the right side in blocks of size p, denoted by $\mathbf{x}_i$ and $\mathbf{b}_i$, respectively. As a result, at each step, the block Jacobi method requires solving p linear systems of matrices A_{ii}. Theorem 4.3 is still valid, provided that D is substituted by the corresponding block diagonal matrix.

In a similar manner, the block Gauss-Seidel and block SOR methods can be introduced.

4.2.6 Symmetric Form of the Gauss-Seidel and SOR Methods

Even if A is a symmetric matrix, the Gauss-Seidel and SOR methods generate iteration matrices that are not necessarily symmetric. For that, we introduce in this section a technique that allows for symmetrizing these schemes. The final aim is to provide an approach for generating symmetric preconditioners (see Section 4.3.2).

Firstly, let us remark that an analogue of the Gauss-Seidel method can be constructed, by simply exchanging E with F. The following iteration can thus be defined, called the *backward Gauss-Seidel method*

$$(\mathrm{D} - \mathrm{F})\mathbf{x}^{(k+1)} = \mathrm{E}\mathbf{x}^{(k)} + \mathbf{b},$$

with iteration matrix given by $\mathrm{B}_{GSb} = (\mathrm{D} - \mathrm{F})^{-1}\mathrm{E}$.
The *symmetric Gauss-Seidel method* is obtained by combining an iteration of Gauss-Seidel method with an iteration of backward Gauss-Seidel method. Precisely, the k-th iteration of the symmetric Gauss-Seidel method is

$$(\mathrm{D} - \mathrm{E})\mathbf{x}^{(k+1/2)} = \mathrm{F}\mathbf{x}^{(k)} + \mathbf{b}, \qquad (\mathrm{D} - \mathrm{F})\mathbf{x}^{(k+1)} = \mathrm{E}\mathbf{x}^{(k+1/2)} + \mathbf{b}.$$

Eliminating $\mathbf{x}^{(k+1/2)}$, the following scheme is obtained

$$\begin{aligned}
\mathbf{x}^{(k+1)} &= \mathrm{B}_{SGS}\mathbf{x}^{(k)} + \mathbf{b}_{SGS}, \\
\mathrm{B}_{SGS} &= (\mathrm{D} - \mathrm{F})^{-1}\mathrm{E}(\mathrm{D} - \mathrm{E})^{-1}\mathrm{F}, \\
\mathbf{b}_{SGS} &= (\mathrm{D} - \mathrm{F})^{-1}[\mathrm{E}(\mathrm{D} - \mathrm{E})^{-1} + \mathrm{I}]\mathbf{b}.
\end{aligned} \tag{4.21}$$

The preconditioning matrix associated with (4.21) is

$$\mathrm{P}_{SGS} = (\mathrm{D} - \mathrm{E})\mathrm{D}^{-1}(\mathrm{D} - \mathrm{F}).$$

The following result can be proved (see [Hac94]).

Property 4.5 *If* A *is a symmetric positive definite matrix, the symmetric Gauss-Seidel method is convergent, and, moreover,* B_{SGS} *is symmetric positive definite.*

In a similar manner, defining the backward SOR method

$$(D - \omega F)\mathbf{x}^{(k+1)} = [\omega E + (1-\omega)D]\,\mathbf{x}^{(k)} + \omega\mathbf{b},$$

and combining it with a step of SOR method, the following *symmetric SOR method* or *SSOR*, is obtained

$$\mathbf{x}^{(k+1)} = B_s(\omega)\mathbf{x}^{(k)} + \mathbf{b}_\omega,$$

where

$$B_s(\omega) = (D - \omega F)^{-1}(\omega E + (1-\omega)D)(D - \omega E)^{-1}(\omega F + (1-\omega)D),$$

$$\mathbf{b}_\omega = \omega(2-\omega)(D - \omega F)^{-1}D(D - \omega E)^{-1}\mathbf{b}.$$

The preconditioning matrix of this scheme is

$$P_{SSOR}(\omega) = \left(\frac{1}{\omega}D - E\right)\frac{\omega}{2-\omega}D^{-1}\left(\frac{1}{\omega}D - F\right). \qquad (4.22)$$

If A is symmetric and positive definite, the SSOR method is convergent if $0 < \omega < 2$ (see [Hac94] for the proof). Typically, the SSOR method with an optimal choice of the relaxation parameter converges more slowly than the corresponding SOR method. However, the value of $\rho(B_s(\omega))$ is less sensitive to a choice of ω around the optimal value (in this respect, see the behavior of the spectral radii of the two iteration matrices in Figure 4.1). For this reason, the optimal value of ω that is chosen in the case of SSOR method is usually the same used for the SOR method (for further details, we refer to [You71]).

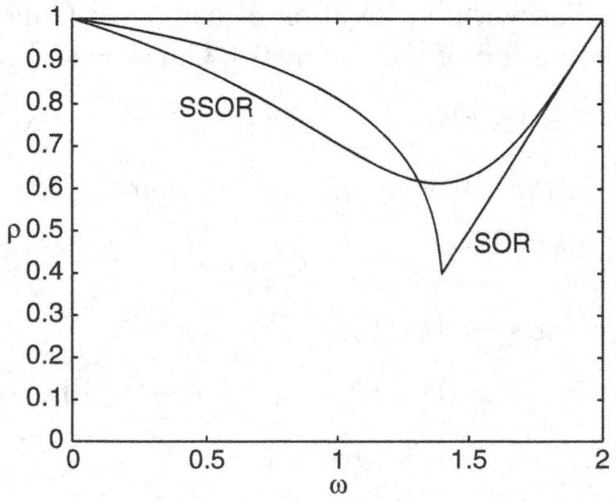

Fig. 4.1. Spectral radius of the iteration matrix of SOR and SSOR methods, as a function of the relaxation parameter ω for the matrix $\mathrm{tridiag}_{10}(-1, 2, -1)$

4.2.7 Implementation Issues

We provide the programs implementing the Jacobi and Gauss-Seidel methods in their point form and with relaxation.

In Program 15 the JOR method is implemented (the Jacobi method is obtained as a special case setting `omega` = 1). The stopping test monitors the Euclidean norm of the residual at each iteration, normalized to the value of the initial residual.
Notice that each component `x(i)` of the solution vector can be computed independently; this method can thus be easily parallelized.

Program 15 - jor : JOR method

```
function [x,iter]=jor(A,b,x0,nmax,tol,omega)
%JOR JOR method
% [X,ITER]=JOR(A,B,X0,NMAX,TOL,OMEGA) attempts to solve the system
% A*X=B with the JOR method. TOL specifies the tolerance of the method.
% NMAX specifies the maximum number of iterations. X0 specifies the initial
% guess. OMEGA is the relaxation parameter. ITER is the iteration number at
% which X is computed.
[n,m]=size(A);
if n ~= m, error('Only square systems'); end
iter=0;
r = b-A*x0; r0=norm(r); err=norm(r); x=x0;
while err > tol & iter < nmax
   iter = iter + 1;
   for i=1:n
      s = 0;
      for j = 1:i-1, s=s+A(i,j)*x(j);   end
      for j = i+1:n, s=s+A(i,j)*x(j);   end
      xnew(i,1)=omega*(b(i)-s)/A(i,i)+(1-omega)*x(i);
   end
   x=xnew;  r=b-A*x;  err=norm(r)/r0;
end
return
```

Program 16 implements the SOR method. Taking `omega=1` yields the Gauss-Seidel method.
Unlike the Jacobi method, this scheme is fully sequential. However, it can be efficiently implemented without storing the solution of the previous step, with a saving of memory storage.

Program 16 - sor : SOR method

```
function [x,iter]=sor(A,b,x0,nmax,tol,omega)
%SOR SOR method
% [X,ITER]=SOR(A,B,X0,NMAX,TOL,OMEGA) attempts to solve the system
% A*X=B with the SOR method. TOL specifies the tolerance of the method.
```

```
% NMAX specifies the maximum number of iterations. X0 specifies the initial
% guess. OMEGA is the relaxation parameter. ITER is the iteration number at
% which X is computed.
[n,m]=size(A);
if n ~= m, error('Only square systems'); end
iter=0; r=b-A*x0; r0=norm(r); err=norm(r); xold=x0;
while err > tol & iter < nmax
   iter = iter + 1;
   for i=1:n
      s=0;
      for j = 1:i-1, s=s+A(i,j)*x(j); end
      for j = i+1:n, s=s+A(i,j)*xold(j); end
      x(i,1)=omega*(b(i)-s)/A(i,i)+(1-omega)*xold(i);
   end
   xold=x;  r=b-A*x; err=norm(r)/r0;
end
return
```

4.3 Stationary and Nonstationary Iterative Methods

Denote by

$$\mathrm{R_P} = \mathrm{I} - \mathrm{P}^{-1}\mathrm{A}$$

the iteration matrix associated with (4.7). Proceeding as in the case of relaxation methods, (4.7) can be generalized introducing a relaxation (or acceleration) parameter α. This leads to the following *stationary Richardson method*

$$\mathbf{x}^{(k+1)} = \mathbf{x}^{(k)} + \alpha \mathrm{P}^{-1}\mathbf{r}^{(k)}, \qquad k \geq 0. \tag{4.23}$$

More generally, allowing α to depend on the iteration index, the *nonstationary Richardson method* or *semi-iterative method* is given by

$$\mathbf{x}^{(k+1)} = \mathbf{x}^{(k)} + \alpha_k \mathrm{P}^{-1}\mathbf{r}^{(k)}, \qquad k \geq 0. \tag{4.24}$$

The iteration matrix at the k-th step for (4.24) (depending on k) is

$$\mathrm{R}_{\alpha_k} = \mathrm{I} - \alpha_k \mathrm{P}^{-1}\mathrm{A},$$

with $\alpha_k = \alpha$ in the stationary case. If $\mathrm{P} = \mathrm{I}$, the family of methods (4.24) will be called *nonpreconditioned.* The Jacobi and Gauss-Seidel methods can be regarded as stationary Richardson methods with $\mathrm{P} = \mathrm{D}$ and $\mathrm{P} = \mathrm{D} - \mathrm{E}$, respectively (and $\alpha = 1$ in both cases).

We can rewrite (4.24) (and, thus, also (4.23)) in a form of greater interest for computation. Letting $\mathbf{z}^{(k)} = \mathrm{P}^{-1}\mathbf{r}^{(k)}$ (the so-called *preconditioned residual*), we get $\mathbf{x}^{(k+1)} = \mathbf{x}^{(k)} + \alpha_k \mathbf{z}^{(k)}$ and $\mathbf{r}^{(k+1)} = \mathbf{b} - \mathrm{A}\mathbf{x}^{(k+1)} = \mathbf{r}^{(k)} - \alpha_k \mathrm{A}\mathbf{z}^{(k)}$.

To summarize, a nonstationary Richardson method requires at each $k+1$-th step the following operations:

$$\begin{aligned}
&\text{solve the linear system } \mathrm{P}\mathbf{z}^{(k)} = \mathbf{r}^{(k)},\\
&\text{compute the acceleration parameter } \alpha_k,\\
&\text{update the solution } \mathbf{x}^{(k+1)} = \mathbf{x}^{(k)} + \alpha_k \mathbf{z}^{(k)},\\
&\text{update the residual } \mathbf{r}^{(k+1)} = \mathbf{r}^{(k)} - \alpha_k \mathrm{A}\mathbf{z}^{(k)}.
\end{aligned} \qquad (4.25)$$

4.3.1 Convergence Analysis of the Richardson Method

Let us first consider the stationary Richardson methods for which $\alpha_k = \alpha$ for $k \geq 0$. The following convergence result holds.

Theorem 4.8 *For any nonsingular matrix* P, *the stationary Richardson method* (4.23) *is convergent iff*

$$\frac{2\mathrm{Re}\lambda_i}{\alpha|\lambda_i|^2} > 1 \ \forall i = 1, \ldots, n, \qquad (4.26)$$

where $\lambda_i \in \mathbb{C}$ *are the eigenvalues of* $\mathrm{P}^{-1}\mathrm{A}$.

Proof. Let us apply Theorem 4.1 to the iteration matrix $\mathrm{R}_\alpha = \mathrm{I} - \alpha\mathrm{P}^{-1}\mathrm{A}$. The condition $|1 - \alpha\lambda_i| < 1$ for $i = 1, \ldots, n$ yields the inequality

$$(1 - \alpha\mathrm{Re}\lambda_i)^2 + \alpha^2(\mathrm{Im}\lambda_i)^2 < 1,$$

from which (4.26) immediately follows. ◇

Let us notice that, if the sign of the real parts of the eigenvalues of $\mathrm{P}^{-1}\mathrm{A}$ is not constant, the stationary Richardson method *cannot* converge.
More specific results can be obtained provided that suitable assumptions are made on the spectrum of $\mathrm{P}^{-1}\mathrm{A}$.

Theorem 4.9 *Assume that* P *is a nonsingular matrix and that* $\mathrm{P}^{-1}\mathrm{A}$ *has positive real eigenvalues, ordered in such a way that* $\lambda_1 \geq \lambda_2 \geq \ldots \geq \lambda_n > 0$. *Then, the stationary Richardson method* (4.23) *is convergent iff* $0 < \alpha < 2/\lambda_1$. *Moreover, letting*

$$\alpha_{opt} = \frac{2}{\lambda_1 + \lambda_n}, \qquad (4.27)$$

the spectral radius of the iteration matrix R_α *is minimum if* $\alpha = \alpha_{opt}$, *with*

$$\rho_{opt} = \min_\alpha [\rho(\mathrm{R}_\alpha)] = \frac{\lambda_1 - \lambda_n}{\lambda_1 + \lambda_n}. \qquad (4.28)$$

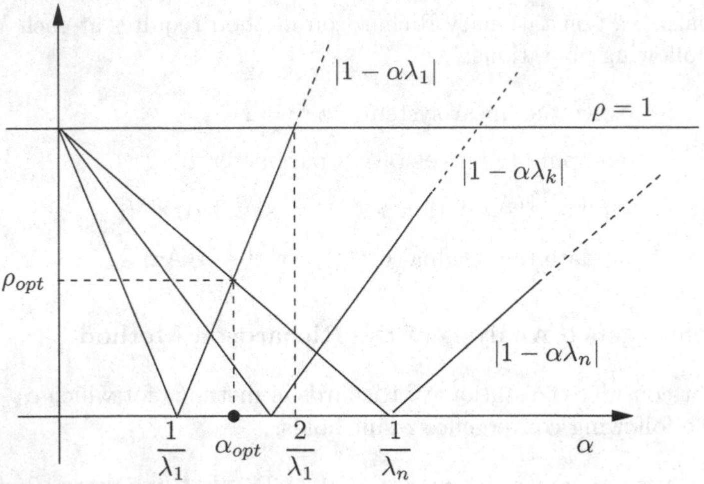

Fig. 4.2. Spectral radius of R_α as a function of the eigenvalues of $P^{-1}A$

Proof. The eigenvalues of R_α are given by $\lambda_i(R_\alpha) = 1 - \alpha\lambda_i$, so that (4.23) is convergent iff $|\lambda_i(R_\alpha)| < 1$ for $i = 1, \ldots, n$, that is, if $0 < \alpha < 2/\lambda_1$. It follows (see Figure 4.2) that $\rho(R_\alpha)$ is minimum when $1 - \alpha\lambda_n = \alpha\lambda_1 - 1$, that is, for $\alpha = 2/(\lambda_1 + \lambda_n)$, which furnishes the desired value for α_{opt}. By substitution, the desired value of ρ_{opt} is obtained. ◇

If $P^{-1}A$ is symmetric positive definite, it can be shown that the convergence of the Richardson method is monotone with respect to either $\|\cdot\|_2$ and $\|\cdot\|_A$. In such a case, using (4.28), we can also relate ρ_{opt} to $K_2(P^{-1}A)$ as follows

$$\rho_{opt} = \frac{K_2(P^{-1}A) - 1}{K_2(P^{-1}A) + 1}, \; \alpha_{opt} = \frac{2\|A^{-1}P\|_2}{K_2(P^{-1}A) + 1}. \tag{4.29}$$

The choice of a suitable preconditioner P is, therefore, of paramount importance for improving the convergence of a Richardson method. Of course, such a choice should also account for the need of keeping the computational effort as low as possible. In Section 4.3.2, some preconditioners of common use in practice will be described.

Corollary 1 *Assume that* A *is a symmetric positive definite matrix with eigenvalues* $\lambda_1 \geq \lambda_2 \geq \ldots \geq \lambda_n$. *Then, if* $0 < \alpha < 2/\lambda_1$, *the nonpreconditioned stationary Richardson method is convergent and*

$$\|\mathbf{e}^{(k+1)}\|_A \leq \rho(R_\alpha)\|\mathbf{e}^{(k)}\|_A, \quad k \geq 0. \tag{4.30}$$

The same result holds for the preconditioned Richardson method, provided that the matrices P, A *and* $P^{-1}A$ *are symmetric positive definite.*

Proof. The convergence is a consequence of Theorem 4.8. Moreover, we notice that

$$\|\mathbf{e}^{(k+1)}\|_A = \|R_\alpha\mathbf{e}^{(k)}\|_A = \|A^{1/2}R_\alpha\mathbf{e}^{(k)}\|_2 \leq \|A^{1/2}R_\alpha A^{-1/2}\|_2\|A^{1/2}\mathbf{e}^{(k)}\|_2.$$

The matrix R_α is symmetric positive definite and is similar to $A^{1/2}R_\alpha A^{-1/2}$. Therefore,

$$\|A^{1/2}R_\alpha A^{-1/2}\|_2 = \rho(R_\alpha).$$

The result (4.30) follows by noting that $\|A^{1/2}\mathbf{e}^{(k)}\|_2 = \|\mathbf{e}^{(k)}\|_A$. A similar proof can be carried out in the preconditioned case, provided we replace A with $P^{-1}A$. ◇

Finally, the inequality (4.30) holds even if only P and A are symmetric positive definite (for the proof, see [QV94], Chapter 2).

4.3.2 Preconditioning Matrices

All the methods introduced in the previous sections can be cast in the form (4.2), so that they can be regarded as being methods for solving the system

$$(I - B)\mathbf{x} = \mathbf{f} = P^{-1}\mathbf{b}.$$

On the other hand, since $B = P^{-1}N$, system (3.2) can be equivalently reformulated as

$$P^{-1}A\mathbf{x} = P^{-1}\mathbf{b}. \tag{4.31}$$

The latter is the *preconditioned system*, being P the *preconditioning matrix* or *left preconditioner*. *Right* and *centered* preconditioners can be introduced as well, if system (3.2) is transformed, respectively, as

$$AP^{-1}\mathbf{y} = \mathbf{b},\ \mathbf{y} = P\mathbf{x},$$

or

$$P_L^{-1}AP_R^{-1}\mathbf{y} = P_L^{-1}\mathbf{b},\ \mathbf{y} = P_R\mathbf{x}.$$

There are *point preconditioners* and *block preconditioners*, depending on whether they are applied to the single entries of A or to the blocks of a partition of A. The iterative methods considered so far correspond to fixed-point iterations on a left-preconditioned system. As stressed by (4.25), computing the inverse of P is not mandatory; actually, the role of P is to "precondition" the residual $\mathbf{r}^{(k)}$ through the solution of the additional system $P\mathbf{z}^{(k)} = \mathbf{r}^{(k)}$.

Since the preconditioner acts on the spectral radius of the iteration matrix, it would be useful to pick up, for a given linear system, an *optimal preconditioner*, i.e., a preconditioner which is able to make the number of iterations required for convergence independent of the size of the system. Notice that the choice P=A is optimal but, trivially, "inefficient"; some alternatives of greater computational interest will be examined below.

There is not a general roadmap to devise optimal preconditioners. However, an established "rule of thumb" is that P is a good preconditioner for A if $P^{-1}A$ is near to being a normal matrix and if its eigenvalues are clustered within a sufficiently small region of the complex field. The choice of a preconditioner must also be guided by practical considerations, noticeably, its computational cost and its memory requirements.

Preconditioners can be divided into two main categories: algebraic and functional preconditioners, the difference being that the algebraic preconditioners are independent of the problem that originated the system to be solved, and are actually constructed via algebraic procedures, while the functional preconditioners take advantage of the knowledge of the problem and are constructed as a function of it. In addition to the preconditioners already introduced in Section 4.2.6, we give a description of other algebraic preconditioners of common use.

1. *Diagonal preconditioners*: choosing P as the diagonal of A is generally effective if A is symmetric positive definite. A usual choice in the nonsymmetric case is to set

$$p_{ii} = \left(\sum_{j=1}^{n} a_{ij}^2 \right)^{1/2}.$$

   Block diagonal preconditioners can be constructed in a similar manner. We remark that devising an optimal diagonal preconditioner is far from being trivial, as previously noticed in Section 3.12.1 when dealing with the scaling of a matrix.
2. *Incomplete LU factorization* (shortly ILU) and *Incomplete Cholesky factorization* (shortly IC).
   An incomplete factorization of A is a process that computes $P = L_{in}U_{in}$, where L_{in} is a lower triangular matrix and U_{in} is an upper triangular matrix. These matrices are approximations of the *exact* matrices L, U of the LU factorization of A and are chosen in such a way that the residual matrix $R = A - L_{in}U_{in}$ satisfies some prescribed requirements, such as having zero entries in specified locations.
   For a given matrix M, the L-part (U-part) of M will mean henceforth the lower (upper) triangular part of M. Moreover, we assume that the factorization process can be carried out without resorting to pivoting.
   The basic approach to incomplete factorization, consists of requiring the approximate factors L_{in} and U_{in} to have the same sparsity pattern as the L-part and U-part of A, respectively. A general algorithm for constructing an incomplete factorization is to perform Gauss elimination as follows: at each step k, compute $m_{ik} = a_{ik}^{(k)}/a_{kk}^{(k)}$ only if $a_{ik} \neq 0$ for $i = k+1,\ldots,n$. Then, compute for $j = k+1,\ldots,n$ $a_{ij}^{(k+1)}$ only if $a_{ij} \neq 0$. This algorithm is implemented in Program 17, where the matrices L_{in} and U_{in} are progressively overwritten onto the L-part and U-part of A.

Program 17 - basicILU : Incomplete LU factorization

```
function [A] = basicILU(A)
%BASICILU Incomplete LU factorization.
```

```
% Y=BASICILU(A): U is stored in the upper triangular part of Y and L is stored
% in the strict lower triangular part of Y. The factors L and U have the
% same sparsity as that of the matrix A.
[n,m]=size(A);
if n ~= m, error('Only square matrices'); end
for k=1:n-1
   for i=k+1:n,
      if A(i,k) ~= 0
         if A(k,k) == 0, error('Null pivot element'); end
         A(i,k)=A(i,k)/A(k,k);
         for j=k+1:n
            if A(i,j) ~= 0
               A(i,j)=A(i,j)-A(i,k)*A(k,j);
            end
         end
      end
   end
end
return
```

We notice that having L_{in} and U_{in} with the same patterns as the L and U-parts of A, respectively, does not necessarily imply that R has the same sparsity pattern as A, but guarantees that $r_{ij} = 0$ if $a_{ij} \neq 0$, as is shown in Figure 4.3.

The resulting incomplete factorization is known as ILU(0), where "0" means that no fill-in has been introduced in the factorization process. An alternative strategy might be to fix the structure of L_{in} and U_{in}

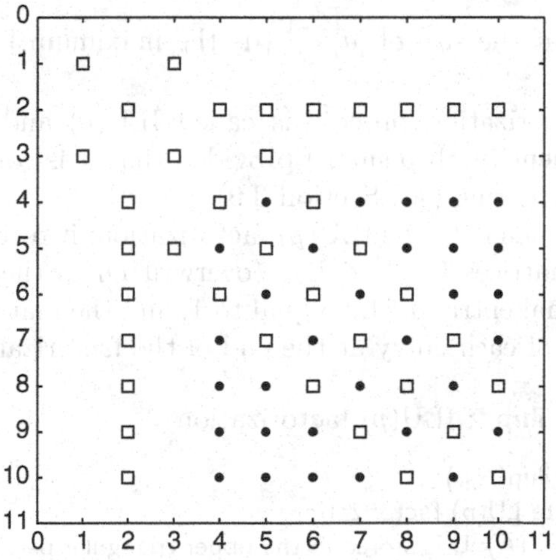

Fig. 4.3. The sparsity pattern of the original matrix A is represented by the squares, while the pattern of $R = A - L_{in}U_{in}$, computed by Program 17, is drawn by the bullets

irrespectively of that of A, in such a way that some computational criteria are satisfied (for example, that the incomplete factors have the simplest possible structure).

The accuracy of the ILU(0) factorization can obviously be improved by allowing some fill-in to arise, and thus, by accepting nonzero entries in the factorization whereas A has elements equal to zero. To this purpose, it is convenient to introduce a function, which we call *fill-in level*, that is associated with each entry of A and that is being modified during the factorization process. If the fill-in level of an element is greater than an admissible value $p \in \mathbb{N}$, the corresponding entry in U_{in} or L_{in} is set equal to zero.

Let us explain how this procedure works, assuming that the matrices L_{in} and U_{in} are progressively overwritten to A (as happens in Program 4). The fill-in level of an entry $a_{ij}^{(k)}$ is denoted by lev_{ij}, where the dependence on k is understood, and it should provide a reasonable estimate of the size of the entry during the factorization process. Actually, we are assuming that if $lev_{ij} = q$ then $|a_{ij}| \simeq \delta^q$ with $\delta \in (0,1)$, so that q is greater when $|a_{ij}^{(k)}|$ is smaller.

At the starting step of the procedure, the level of the nonzero entries of A and of the diagonal entries is set equal to 0, while the level of the null entries is set equal to infinity. For any row $i = 2, \ldots, n$, the following operations are performed: if $lev_{ik} \leq p$, $k = 1, \ldots, i-1$, the entry m_{ik} of L_{in} and the entries $a_{ij}^{(k+1)}$ of U_{in}, $j = i+1, \ldots, n$, are updated. Moreover, if $a_{ij}^{(k+1)} \neq 0$ the value lev_{ij} is updated as being the minimum between the available value of lev_{ij} and $lev_{ik} + lev_{kj} + 1$. The reason of this choice is that $|a_{ij}^{(k+1)}| = |a_{ij}^{(k)} - m_{ik} a_{kj}^{(k)}| \simeq |\delta^{lev_{ij}} - \delta^{lev_{ik}+lev_{kj}+1}|$, so that one can assume that the size of $|a_{ij}^{(k+1)}|$ is the maximum between $\delta^{lev_{ij}}$ and $\delta^{lev_{ik}+lev_{kj}+1}$.

The above factorization process is called ILU(p) and turns out to be extremely efficient (with p small) provided that it is coupled with a suitable matrix reordering (see Section 3.9).

Program 18 implements the ILU(p) factorization; it returns in output the approximate matrices L_{in} and U_{in} (overwritten to the input matrix `a`), with the diagonal entries of L_{in} equal to 1, and the matrix `lev` containing the fill-in level of each entry at the end of the factorization.

Program 18 - ilup : ILU(p) factorization

```
function [A,lev]=ilup(A,p)
%ILUP Incomplete LU(p) factorization.
%  [Y,LEV]=ILUP(A): U is stored in the upper triangular part of Y and L is stored
%  in the strict lower triangular part of Y. The factors L and U
%  have a fill-in level P. LEV contains the fill-in level of
%  each entry at the end of the factorization.
```

```
[n,m]=size(A);
if n ~= m, error('Only square matrices'); end
lev=Inf*ones(n,n);
i=(A~=0);
lev(i)=0,
for i=2:n
  for k=1:i-1
    if lev(i,k) <= p
      if A(k,k)==0, error('Null pivot element'); end
      A(i,k)=A(i,k)/A(k,k);
      for j=k+1:n
        A(i,j)=A(i,j)-A(i,k)*A(k,j);
        if A(i,j) ~= 0
          lev(i,j)=min(lev(i,j),lev(i,k)+lev(k,j)+1);
        end
      end
    end
  end
  for j=1:n, if lev(i,j) > p, A(i,j) = 0; end, end
end
return
```

Example 4.4 Consider the matrix $A \in \mathbb{R}^{46\times 46}$ associated with the finite difference approximation of the Laplace operator $\Delta\cdot = \frac{\partial^2\cdot}{\partial x^2} + \frac{\partial^2\cdot}{\partial y^2}$ (see Section 12.6). This matrix can be generated with the following MATLAB commands: `G=numgrid('B',10); A=delsq(G)` and corresponds to the discretization of the differential operator on a domain having the shape of the exterior of a butterfly and included in the square $[-1,1]^2$. The number of nonzero entries of A is 174. Figure 4.4 shows the pattern of matrix A (drawn by the bullets) and the entries in the pattern added by the ILU(1) and ILU(2) factorizations due to fill-in (denoted by the squares and the triangles, respectively). Notice that these entries are all contained within the envelope of A since no pivoting has been performed. •

The ILU(p) process can be carried out without knowing the actual values of the entries of A, but only working on their fill-in levels. Therefore, we can distinguish between a *symbolic factorization* (the generation of the levels) and an *actual factorization* (the computation of the entries of ILU(p) starting from the informations contained in the level function). The scheme is thus particularly effective when several linear systems must be solved, with matrices having the same structure but different entries. On the other hand, for certain classes of matrices, the fill-in level does not always provide a sound indication of the *actual* size attained by the entries. In such cases, it is better to monitor the size of the entries of R by neglecting each time the entries that are too small. For instance, one can drop out the entries $a_{ij}^{(k+1)}$ such that

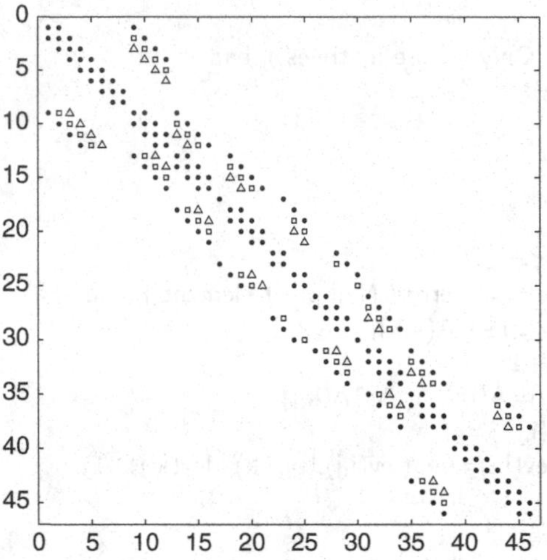

Fig. 4.4. Pattern of the matrix A in Example 4.4 (*bullets*); entries added by the ILU(1) and ILU(2) factorizations (*squares* and *triangles*, respectively)

$$|a_{ij}^{(k+1)}| \le c|a_{ii}^{(k+1)} a_{jj}^{(k+1)}|^{1/2}, \qquad i,j = 1,\ldots,n,$$

with $0 < c < 1$ (see [Axe94]).

In the strategies considered so far, the entries of the matrix that are dropped out can no longer be recovered in the incomplete factorization process. Some remedies exist for this drawback: for instance, at the end of each k-th step of the factorization, one can sum, row by row, the discarded entries to the diagonal entries of U_{in}. By doing so, an incomplete factorization known as MILU (Modified ILU) is obtained, which enjoys the property of being exact with respect to the constant vectors, i.e., such that $R\mathbf{1} = \mathbf{0}$ (see [Axe94] for other formulations). In the practice, this simple trick provides, for a wide class of matrices, a better preconditioner than obtained with the ILU method. In the case of symmetric positive definite matrices one can resort to the Modified Incomplete Cholesky Factorization (MICh).

We conclude by mentioning the ILUT factorization, which collects the features of ILU(p) and MILU. This factorization can also include partial pivoting by columns with a slight increase of the computational cost. For an efficient implementation of incomplete factorizations, we refer to the MATLAB function `luinc` in the toolbox `sparfun`.

The existence of the ILU factorization is not guaranteed for all nonsingular matrices (see for an example [Elm86]) and the process stops if zero pivotal entries arise. Existence theorems can be proved if A is an M-matrix [MdV77] or diagonally dominant [Man80]. It is worth noting that sometimes the ILU factorization turns out to be more stable than the complete LU factorization [GM83].

3. *Polynomial preconditioners*: the preconditioning matrix is defined as

$$P^{-1} = p(A),$$

where p is a polynomial in A, usually of low degree.
A remarkable example is given by Neumann polynomial preconditioners. Letting $A = D - C$, we have $A = (I - CD^{-1})D$, from which

$$A^{-1} = D^{-1}(I - CD^{-1})^{-1} = D^{-1}(I + CD^{-1} + (CD^{-1})^2 + \ldots).$$

A preconditioner can then be obtained by truncating the series above at a certain power p. This method is actually effective only if $\rho(CD^{-1}) < 1$, which is the necessary condition in order the series to be convergent.
4. *Least-squares preconditioners*: A^{-1} is approximated by a least-squares polynomial $p_s(A)$ (see Section 3.13). Since the aim is to make matrix $I - P^{-1}A$ as close as possible to the null matrix, the least-squares approximant $p_s(A)$ is chosen in such a way that the function $\varphi(x) = 1 - p_s(x)x$ is minimized. This preconditioning technique works effectively only if A is symmetric and positive definite.

For further results on preconditioners, see [dV89] and [Axe94].

Example 4.5 Consider the matrix $A \in \mathbb{R}^{324\times 324}$ associated with the finite difference approximation of the Laplace operator on the square $[-1,1]^2$. This matrix can be generated with the following MATLAB commands: `G=numgrid('N',20); A=delsq(G)`. The condition number of the matrix is $K_2(A) = 211.3$. In Table 4.1 we show the values of $K_2(P^{-1}A)$ computed using the ILU(p) and Neumann preconditioners, with $p = 0, 1, 2, 3$. In the last case D is the diagonal part of A. •

Remark 4.2 Let A and P be real symmetric matrices of order n, with P positive definite. The eigenvalues of the preconditioned matrix $P^{-1}A$ are solutions of the algebraic equation

$$\mathbf{Ax} = \lambda \mathbf{Px}, \tag{4.32}$$

where $\mathbf{x}$ is an eigenvector associated with the eigenvalue λ. Equation (4.32) is an example of *generalized eigenvalue problem* (see Section 5.9 for a thorough

Table 4.1. Spectral condition numbers of the preconditioned matrix A of Example 4.5 as a function of p

p	ILU(p)	Neumann
0	22.3	211.3
1	12	36.91
2	8.6	48.55
3	5.6	18.7

discussion) and the eigenvalue λ can be computed through the following generalized Rayleigh quotient

$$\lambda = \frac{(\mathrm{A}\mathbf{x}, \mathbf{x})}{(\mathrm{P}\mathbf{x}, \mathbf{x})}.$$

Applying the Courant-Fisher Theorem (see Section 5.11) yields

$$\frac{\lambda_{min}(\mathrm{A})}{\lambda_{max}(\mathrm{P})} \leq \lambda \leq \frac{\lambda_{max}(\mathrm{A})}{\lambda_{min}(\mathrm{P})}. \tag{4.33}$$

Relation (4.33) provides a lower and upper bound for the eigenvalues of the preconditioned matrix as a function of the extremal eigenvalues of A and P, and therefore it can be profitably used to estimate the condition number of $\mathrm{P}^{-1}\mathrm{A}$. ■

4.3.3 The Gradient Method

The expression of the optimal parameter that has been provided in Theorem 4.9 is of limited usefulness in practical computations, since it requires the knowledge of the extremal eigenvalues of the matrix $\mathrm{P}^{-1}\mathrm{A}$. In the special case of symmetric and positive definite matrices, however, the optimal acceleration parameter can be *dynamically* computed at each step k as follows.

We first notice that, for such matrices, solving system (3.2) is equivalent to finding the minimizer $\mathbf{x} \in \mathbb{R}^n$ of the quadratic form

$$\Phi(\mathbf{y}) = \frac{1}{2}\mathbf{y}^T\mathrm{A}\mathbf{y} - \mathbf{y}^T\mathbf{b},$$

which is called the *energy of system* (3.2). Indeed, the gradient of Φ is given by

$$\nabla\Phi(\mathbf{y}) = \frac{1}{2}(\mathrm{A}^T + \mathrm{A})\mathbf{y} - \mathbf{b} = \mathrm{A}\mathbf{y} - \mathbf{b}. \tag{4.34}$$

As a consequence, if $\nabla\Phi(\mathbf{x}) = \mathbf{0}$ then $\mathbf{x}$ is a solution of the original system. Conversely, if $\mathbf{x}$ is a solution, then

$$\Phi(\mathbf{y}) = \Phi(\mathbf{x} + (\mathbf{y} - \mathbf{x})) = \Phi(\mathbf{x}) + \frac{1}{2}(\mathbf{y} - \mathbf{x})^T\mathrm{A}(\mathbf{y} - \mathbf{x}), \qquad \forall \mathbf{y} \in \mathbb{R}^n$$

and thus, $\Phi(\mathbf{y}) > \Phi(\mathbf{x})$ if $\mathbf{y} \neq \mathbf{x}$, i.e. $\mathbf{x}$ is a minimizer of the functional Φ.

Notice that the previous relation is equivalent to

$$\frac{1}{2}\|\mathbf{y} - \mathbf{x}\|_{\mathrm{A}}^2 = \Phi(\mathbf{y}) - \Phi(\mathbf{x}), \tag{4.35}$$

where $\|\cdot\|_{\mathrm{A}}$ is the A-*norm* or *energy norm*, defined in (1.28).

The problem is thus to determine the minimizer $\mathbf{x}$ of Φ starting from a point $\mathbf{x}^{(0)} \in \mathbb{R}^n$ and, consequently, to select suitable directions along which moving to get as close as possible to the solution $\mathbf{x}$. The optimal direction, that joins the starting point $\mathbf{x}^{(0)}$ to the solution point $\mathbf{x}$, is obviously unknown a priori. Therefore, we must take a step from $\mathbf{x}^{(0)}$ along a given direction $\mathbf{p}^{(0)}$, and then fix along this latter a new point $\mathbf{x}^{(1)}$ from which to iterate the process until convergence.

Precisely, at the generic step k, $\mathbf{x}^{(k+1)}$ is computed as

$$\mathbf{x}^{(k+1)} = \mathbf{x}^{(k)} + \alpha_k \mathbf{p}^{(k)}, \tag{4.36}$$

where α_k is the value which fixes the length of the step along the direction $\mathbf{p}^{(k)}$. The most natural idea is to take as $\mathbf{p}^{(k)}$ the direction of maximum descent along the functional Φ in $\mathbf{x}^{(k)}$, which is given by $-\nabla\Phi(\mathbf{x}^{(k)})$. This yields the *gradient method*, also called *steepest descent method.*

Due to (4.34), $\nabla\Phi(\mathbf{x}^{(k)}) = \mathrm{A}\mathbf{x}^{(k)} - \mathbf{b} = -\mathbf{r}^{(k)}$, so that the direction of the gradient of Φ coincides with that of residual and can be immediately computed using the current iterate. This shows that the gradient method, as well as the Richardson method (4.24) with $\mathrm{P} = \mathrm{I}$, moves at each step k along the direction $\mathbf{p}^{(k)} = \mathbf{r}^{(k)} = -\nabla\Phi(\mathbf{x}^{(k)})$.

To compute the parameter α_k let us write explicitly $\Phi(\mathbf{x}^{(k+1)})$ as a function of a parameter α

$$\Phi(\mathbf{x}^{(k+1)}) = \frac{1}{2}(\mathbf{x}^{(k)} + \alpha\mathbf{r}^{(k)})^T \mathrm{A}(\mathbf{x}^{(k)} + \alpha\mathbf{r}^{(k)}) - (\mathbf{x}^{(k)} + \alpha\mathbf{r}^{(k)})^T\mathbf{b}.$$

Differentiating with respect to α and setting it equal to zero yields the desired value of α_k

$$\alpha_k = \frac{\mathbf{r}^{(k)T}\mathbf{r}^{(k)}}{\mathbf{r}^{(k)T}\mathrm{A}\mathbf{r}^{(k)}}, \tag{4.37}$$

which depends only on the residual at the k-th step. For this reason, the nonstationary nonpreconditioned Richardson method employing (4.37) to evaluate the acceleration parameter is also called the gradient method.

Summarizing, the gradient (or steepest descent) method can be described as follows:

given $\mathbf{x}^{(0)} \in \mathbb{R}^n$, set $\mathbf{r}^{(0)} = \mathbf{b} - \mathrm{A}\mathbf{x}^{(0)}$, and, for $k = 0, 1, \ldots$ until convergence, compute

$$\begin{aligned}
\alpha_k &= \frac{\mathbf{r}^{(k)T}\mathbf{r}^{(k)}}{\mathbf{r}^{(k)T}\mathrm{A}\mathbf{r}^{(k)}}, \\
\mathbf{x}^{(k+1)} &= \mathbf{x}^{(k)} + \alpha_k\mathbf{r}^{(k)}, \\
\mathbf{r}^{(k+1)} &= \mathbf{r}^{(k)} - \alpha_k\mathrm{A}\mathbf{r}^{(k)}.
\end{aligned}$$

Theorem 4.10 *Let* A *be a symmetric and positive definite matrix; then the gradient method is convergent for any choice of the initial datum* $\mathbf{x}^{(0)}$. *Moreover*

$$\|\mathbf{e}^{(k+1)}\|_\mathrm{A} \leq \frac{K_2(\mathrm{A}) - 1}{K_2(\mathrm{A}) + 1}\|\mathbf{e}^{(k)}\|_\mathrm{A}, \qquad k = 0, 1, \ldots, \tag{4.38}$$

where $\|\cdot\|_\mathrm{A}$ *is the energy norm defined in* (1.28).

Proof. Let $\mathbf{x}^{(k)}$ be the solution generated by the gradient method at the k-th step. Then, let $\mathbf{x}_R^{(k+1)}$ be the vector generated by taking one step of the nonpreconditioned Richardson method with optimal parameter starting from $\mathbf{x}^{(k)}$, i.e., $\mathbf{x}_R^{(k+1)} = \mathbf{x}^{(k)} + \alpha_{opt}\mathbf{r}^{(k)}$.

Due to Corollary 1 and to (4.28), we have

$$\|\mathbf{e}_R^{(k+1)}\|_{\mathrm{A}} \leq \frac{K_2(\mathrm{A})-1}{K_2(\mathrm{A})+1}\|\mathbf{e}^{(k)}\|_{\mathrm{A}},$$

where $\mathbf{e}_R^{(k+1)} = \mathbf{x}_R^{(k+1)} - \mathbf{x}$. Moreover, from (4.35) we have that the vector $\mathbf{x}^{(k+1)}$, generated by the gradient method, is the one that minimizes the A-norm of the error among all vectors of the form $\mathbf{x}^{(k)} + \theta\mathbf{r}^{(k)}$, with $\theta \in \mathbb{R}$. Therefore, $\|\mathbf{e}^{(k+1)}\|_{\mathrm{A}} \leq \|\mathbf{e}_R^{(k+1)}\|_{\mathrm{A}}$ which is the desired result. ◇

Let us now consider the preconditioned gradient method and assume that the matrix P is symmetric positive definite. In such a case the optimal value of α_k in algorithm (4.25) is

$$\alpha_k = \frac{\mathbf{z}^{(k)^T}\mathbf{r}^{(k)}}{\mathbf{z}^{(k)^T}\mathrm{A}\mathbf{z}^{(k)}}$$

and we have

$$\|\mathbf{e}^{(k+1)}\|_{\mathrm{A}} \leq \frac{K_2(\mathrm{P}^{-1}\mathrm{A})-1}{K_2(\mathrm{P}^{-1}\mathrm{A})+1}\|\mathbf{e}^{(k)}\|_{\mathrm{A}}.$$

For the proof of this convergence result see, e.g., [QV94], Section 2.4.1.
We notice that the line through $\mathbf{x}^{(k)}$ and $\mathbf{x}^{(k+1)}$ is tangent at the point $\mathbf{x}^{(k+1)}$ to the ellipsoidal level surface $\{\mathbf{x} \in \mathbb{R}^n : \Phi(\mathbf{x}) = \Phi(\mathbf{x}^{(k+1)})\}$ (see also Figure 4.5).

Relation (4.38) shows that convergence of the gradient method can be quite slow if $K_2(\mathrm{A}) = \lambda_1/\lambda_n$ is large. A simple geometric interpretation of this result can be given in the case $n = 2$. Suppose that $\mathrm{A}=\mathrm{diag}(\lambda_1, \lambda_2)$, with $0 < \lambda_2 \leq \lambda_1$ and $\mathbf{b} = (b_1, b_2)^T$.

In such a case, the curves corresponding to $\Phi(x_1, x_2) = c$, as c varies in $\mathbb{R}^+$, form a sequence of concentric ellipses whose semi-axes have length

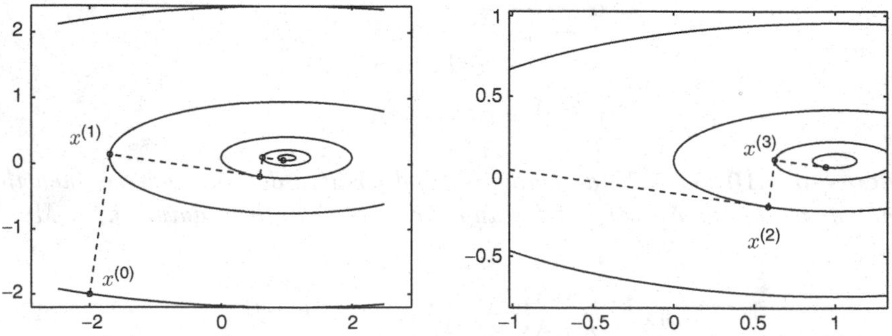

Fig. 4.5. The first iterates of the gradient method on the level curves of Φ

inversely proportional to the values λ_1 and λ_2. If $\lambda_1 = \lambda_2$, the ellipses degenerate into circles and the direction of the gradient crosses the center directly, in such a way that the gradient method converges in one iteration. Conversely, if $\lambda_1 \gg \lambda_2$, the ellipses become strongly eccentric and the method converges quite slowly, as shown in Figure 4.5, moving along a "zig-zag" trajectory.

Program 19 provides an implementation of the preconditioned gradient method. Here and in the programs reported in the remainder of the section, the input parameters `A`, `b`, `x`, `P`, `nmax` and `tol` respectively represent the coefficient matrix of the linear system, the right-hand side, the initial datum $\mathbf{x}^{(0)}$, a possible preconditioner, the maximum number of admissible iterations and a tolerance for the stopping test. This stopping test checks if the ratio $\|\mathbf{r}^{(k)}\|_2/\|\mathbf{b}\|_2$ is less than `tol`. The output parameters of the code are the the number of iterations `iter` required to fulfill the stopping test, the vector `x` with the solution computed after `iter` iterations and the normalized residual `relres` $= \|\mathbf{r}^{(\mathtt{niter})}\|_2/\|\mathbf{b}\|_2$. A null value of the parameter `flag` warns the user that the algorithm has actually satisfied the stopping test and it has not terminated due to reaching the maximum admissible number of iterations.

Program 19 - gradient : Preconditioned gradient method

```
function [x,relres,iter,flag]=gradient(A,b,x,P,nmax,tol)
%GRADIENT Gradient method
% [X,RELRES,ITER,FLAG]=GRADIENT(A,B,X0,NMAX,TOL,OMEGA) attempts
% to solve the system A*X=B with the gradient method. TOL specifies the
% tolerance of the method. NMAX specifies the maximum number of iterations.
% X0 specifies the initial guess. P is a preconditioner. RELRES is the relative
% residual. If FLAG is 1, then RELRES > TOL. ITER is the iteration number
% at which X is computed.
[n,m]=size(A);
if n ~= m, error('Only square systems'); end
flag = 0; iter = 0; bnrm2 = norm( b );
if bnrm2==0, bnrm2 = 1; end
r=b-A*x; relres=norm(r)/bnrm2;
if relres<tol, return, end
for iter=1:nmax
  z=P\r;
  rho=r'*z;
  q=A*z;
  alpha=rho/(z'*q);
  x=x+alpha*z;
  r=r-alpha*q;
  relres=norm(r)/bnrm2;
  if relres<=tol, break, end
end
if relres>tol, flag = 1; end
return
```

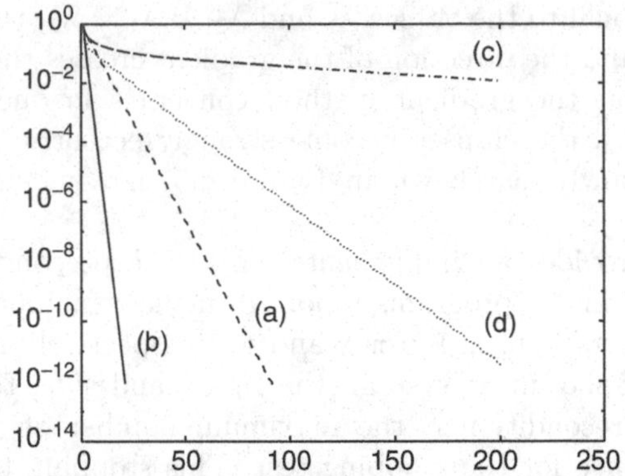

Fig. 4.6. The residual normalized to the starting one, as a function of the number of iterations, for the gradient method applied to the systems in Example 4.6. The curves labelled (a) and (b) refer to the case $m = 16$ with the nonpreconditioned and preconditioned method, respectively, while the curves labelled (c) and (d) refer to the case $m = 400$ with the nonpreconditioned and preconditioned method, respectively

Example 4.6 Let us solve with the gradient method the linear system with matrix $\mathrm{A}_m \in \mathbb{R}^{m\times m}$ generated with the MATLAB commands `G=numgrid('S',n);` `A=delsq(G)` where $m = (n-2)^2$. This matrix is associated with the discretization of the differential Laplace operator on the domain $[-1,1]^2$. The right-hand side $\mathbf{b}_m$ is selected in such a way that the exact solution is the vector $\mathbf{1} \in \mathbb{R}^m$. The matrix A_m is symmetric and positive definite for any m and becomes ill-conditioned for large values of m. We run Program 19 in the cases $m = 16$ and $m = 400$, with $\mathbf{x}^{(0)} = \mathbf{0}$, `tol`$=10^{-10}$ and `nmax`$=200$. If $m = 400$, the method fails to satisfy the stopping test within the admissible maximum number of iterations and exhibits an extremely slow reduction of the residual (see Figure 4.6). Actually, $K_2(\mathrm{A}_{400}) \simeq 258$. If, however, we precondition the system with the matrix $\mathrm{P} = \mathrm{R}_{in}^T\mathrm{R}_{in}$, where R_{in} is the lower triangular matrix in the Cholesky incomplete factorization of A, the algorithm fulfills the convergence within the maximum admissible number of iterations (indeed, now $K_2(\mathrm{P}^{-1}\mathrm{A}_{400}) \simeq 38$). •

4.3.4 The Conjugate Gradient Method

The gradient method consists essentially of two phases: choosing a direction $\mathbf{p}^{(k)}$ (which turns out to coincide with the one of the residual) and picking up a point of local minimum for Φ along that direction. The latter request can be accommodated by choosing α_k as the value of the parameter α such that $\Phi(\mathbf{x}^{(k)} + \alpha\mathbf{p}^{(k)})$ is minimized. Differentiating with respect to α and setting to zero the derivative at the minimizer, yields

$$\alpha_k = \frac{\mathbf{p}^{(k)T}\mathbf{r}^{(k)}}{\mathbf{p}^{(k)T}\mathrm{A}\mathbf{p}^{(k)}}. \tag{4.39}$$

(This reduces to (4.37) when $\mathbf{p}^{(k)} = \mathbf{r}^{(k)}$.) The question is whether a different choice of the search direction $\mathbf{p}^{(k)}$ exists, that might provide a faster convergence of the Richardson method in the case where $K_2(\mathrm{A})$ is large. Since, by (4.36), we have

$$\mathbf{r}^{(k+1)} = \mathbf{r}^{(k)} - \alpha_k \mathrm{A}\mathbf{p}^{(k)}, \tag{4.40}$$

using (4.39) shows that

$$(\mathbf{p}^{(k)})^T \mathbf{r}^{(k+1)} = 0,$$

that is, the new residual becomes orthogonal to the search direction. For the next iteration step, the strategy is thus to find a new search direction $\mathbf{p}^{(k+1)}$ in such a way that

$$(\mathrm{A}\mathbf{p}^{(j)})^T \mathbf{p}^{(k+1)} = 0, \qquad j = 0, \ldots, k. \tag{4.41}$$

To see how the $k+1$ relations (4.41) can be obtained in a practical way, we proceed as follows.

Assume that for $k \geq 1$, $\mathbf{p}^{(0)}, \mathbf{p}^{(1)}, \ldots, \mathbf{p}^{(k)}$ are *mutually conjugate orthogonal* (or *A-orthogonal*). This means that

$$(\mathrm{A}\mathbf{p}^{(i)})^T \mathbf{p}^{(j)} = 0, \qquad \forall i, j = 0, \ldots, k, \; i \neq j. \tag{4.42}$$

This makes sense (in exact arithmetic) provided $k < n$. Assume also, without loss of generality, that

$$(\mathbf{p}^{(j)})^T \mathbf{r}^{(k)} = 0, \qquad j = 0, 1, \ldots, k-1. \tag{4.43}$$

We claim that for every $k \geq 0$, the new residual $\mathbf{r}^{(k+1)}$ is orthogonal to the directions $\mathbf{p}^{(j)}$, $j = 0, \ldots, k$, that is

$$(\mathbf{p}^{(j)})^T \mathbf{r}^{(k+1)} = 0, \qquad j = 0, \ldots, k. \tag{4.44}$$

This can be proven by induction on k. For $k = 0$, $\mathbf{r}^{(1)} = \mathbf{r}^{(0)} - \alpha_0 \mathrm{A}\mathbf{r}^{(0)}$, thus $(\mathbf{p}^{(0)})^T \mathbf{r}^{(1)} = 0$ since $\alpha_0 = (\mathbf{p}^{(0)})^T \mathbf{r}^{(0)} / ((\mathbf{p}^{(0)})^T \mathrm{A}\mathbf{p}^{(0)})$, and (4.43) therefore holds. Equation (4.40) yields (since A is symmetric)

$$(\mathbf{p}^{(j)})^T \mathbf{r}^{(k+1)} = (\mathbf{p}^{(j)})^T \mathbf{r}^{(k)} - \alpha_k (\mathrm{A}\mathbf{p}^{(j)})^T \mathbf{p}^{(k)}.$$

Unless for $j = k$, $(\mathrm{A}\mathbf{p}^{(j)})^T \mathbf{p}^{(k)}$ vanishes owing to (4.42), whereas $(\mathbf{p}^{(j)})^T \mathbf{r}^{(k)}$ is zero due to the induction assumption. On the other hand, when $j = k$ the right hand side is zero due to the choice (4.39) of α_k.

It remains only to compute the sequence of search directions $\mathbf{p}^{(0)}, \mathbf{p}^{(1)}, \ldots,$ $\mathbf{p}^{(k)}$ in an efficient way to make them mutually A-orthogonal. To this end, let

$$\mathbf{p}^{(k+1)} = \mathbf{r}^{(k+1)} - \beta_k \mathbf{p}^{(k)}, \qquad k = 0, 1, \ldots, \tag{4.45}$$

where initially we let $\mathbf{p}^{(0)} = \mathbf{r}^{(0)}$ and $\beta_0, \beta_1, \ldots$ are still to be determined. Using (4.45) in (4.41) for $j = k$ yields

$$\beta_k = \frac{(\mathrm{A}\mathbf{p}^{(k)})^T \mathbf{r}^{(k+1)}}{(\mathrm{A}\mathbf{p}^{(k)})^T \mathbf{p}^{(k)}}, \qquad k = 0, 1, \ldots. \tag{4.46}$$

We also notice that, for every $j = 0, \ldots, k$, relation (4.45) implies

$$(\mathrm{A}\mathbf{p}^{(j)})^T \mathbf{p}^{(k+1)} = (\mathrm{A}\mathbf{p}^{(j)})^T \mathbf{r}^{(k+1)} - \beta_k (\mathrm{A}\mathbf{p}^{(j)})^T \mathbf{p}^{(k)}.$$

Now, by the induction assumption for $j \leq k-1$, the last scalar product is zero. To prove that also the first scalar product on the right hand side is zero, we proceed as follows. Let $V_k = \mathrm{span}(\mathbf{p}^{(0)}, \ldots, \mathbf{p}^{(k)})$. Then, if we choose $\mathbf{p}^{(0)} = \mathbf{r}^{(0)}$, using (4.45) we see that V_k has the alternative representation $V_k = \mathrm{span}(\mathbf{r}^{(0)}, \ldots, \mathbf{r}^{(k)})$. Hence, $\mathrm{A}\mathbf{p}^{(k)} \in V_{k+1}$ for all $k \geq 0$ owing to (4.40). Since $\mathbf{r}^{(k+1)}$ is orthogonal to any vector in V_k (see (4.44)), then

$$(\mathrm{A}\mathbf{p}^{(j)})^T \mathbf{r}^{(k+1)} = 0, \qquad j = 0, 1, \ldots, k-1.$$

We have therefore proven (4.41) by induction on k, provided the A-orthogonal directions are chosen as in (4.45) and (4.46).

The method obtained by choosing the search directions $\mathbf{p}^{(k)}$ as in (4.45) and the acceleration parameter α_k as in (4.39) is called *conjugate gradient method* (CG). The CG method reads as follows: given $\mathbf{x}^{(0)} \in \mathbb{R}^n$, set $\mathbf{r}^{(0)} = \mathbf{b} - \mathrm{A}\mathbf{x}^{(0)}$ and $\mathbf{p}^{(0)} = \mathbf{r}^{(0)}$ then, for $k = 0, 1, \ldots$, until convergence, compute

$$\begin{aligned}
\alpha_k &= \frac{\mathbf{p}^{(k)^T} \mathbf{r}^{(k)}}{\mathbf{p}^{(k)^T} \mathrm{A}\mathbf{p}^{(k)}}, \\
\mathbf{x}^{(k+1)} &= \mathbf{x}^{(k)} + \alpha_k \mathbf{p}^{(k)}, \\
\mathbf{r}^{(k+1)} &= \mathbf{r}^{(k)} - \alpha_k \mathrm{A}\mathbf{p}^{(k)}, \\
\beta_k &= \frac{(\mathrm{A}\mathbf{p}^{(k)})^T \mathbf{r}^{(k+1)}}{(\mathrm{A}\mathbf{p}^{(k)})^T \mathbf{p}^{(k)}}, \\
\mathbf{p}^{(k+1)} &= \mathbf{r}^{(k+1)} - \beta_k \mathbf{p}^{(k)}.
\end{aligned}$$

It can also be shown (see Exercise 12) that the two parameters α_k and β_k may be alternatively expressed as

$$\alpha_k = \frac{\|\mathbf{r}^{(k)}\|_2^2}{\mathbf{p}^{(k)^T} \mathrm{A}\mathbf{p}^{(k)}}, \; \beta_k = -\frac{\|\mathbf{r}^{(k+1)}\|_2^2}{\|\mathbf{r}^{(k)}\|_2^2}. \tag{4.47}$$

We finally notice that, eliminating the search directions from $\mathbf{r}^{(k+1)} = \mathbf{r}^{(k)} - \alpha_k \mathrm{A}\mathbf{p}^{(k)}$, the following recursive three-terms relation is obtained for the residuals (see Exercise 13)

$$\mathrm{A}\mathbf{r}^{(k)} = -\frac{1}{\alpha_k}\mathbf{r}^{(k+1)} + \left(\frac{1}{\alpha_k} - \frac{\beta_{k-1}}{\alpha_{k-1}}\right)\mathbf{r}^{(k)} + \frac{\beta_k}{\alpha_{k-1}}\mathbf{r}^{(k-1)}. \tag{4.48}$$

As for the convergence of the CG method, we have the following results.

Theorem 4.11 *Let* A *be a symmetric and positive definite matrix. Any method which employs conjugate directions to solve* (3.2) *terminates after at most n steps, yielding the exact solution.*

Proof. The directions $\mathbf{p}^{(0)}, \mathbf{p}^{(1)}, \ldots, \mathbf{p}^{(n-1)}$ form an A-orthogonal basis in $\mathbb{R}^n$. Moreover, from (4.43) it follows that $\mathbf{r}^{(k)}$ is orthogonal to the space $V_{k-1} = \mathrm{span}(\mathbf{p}^{(0)}, \mathbf{p}^{(1)}, \ldots, \mathbf{p}^{(k-1)})$. As a consequence, $\mathbf{r}^{(n)} \perp V_{n-1} = \mathbb{R}^n$ and thus $\mathbf{r}^{(n)} = \mathbf{0}$ which implies $\mathbf{x}^{(n)} = \mathbf{x}$. ◇

Going back to the example discussed in Section 4.3.3, Figure 4.7 shows the performance of the conjugate gradient (CG) method, compared to the gradient (G) method. In the present case ($n = 2$), the CG scheme converges in two iterations due to the property of A-orthogonality, while the gradient method converges very slowly, due to the above described "zig-zag" trajectory of the search directions.

Theorem 4.12 *Let* A *be a symmetric and positive definite matrix. The conjugate gradient method for solving* (3.2) *converges after at most n steps. Moreover, the error $\mathbf{e}^{(k)}$ at the k-th iteration (with $k < n$) is orthogonal to $\mathbf{p}^{(j)}$, for $j = 0, \ldots, k-1$ and*

$$\|\mathbf{e}^{(k)}\|_{\mathrm{A}} \leq \frac{2c^k}{1+c^{2k}}\|\mathbf{e}^{(0)}\|_{\mathrm{A}}, \text{ with } c = \frac{\sqrt{K_2(\mathrm{A})}-1}{\sqrt{K_2(\mathrm{A})}+1}. \tag{4.49}$$

Proof. The convergence of the CG method in n steps is a consequence of Theorem 4.11.

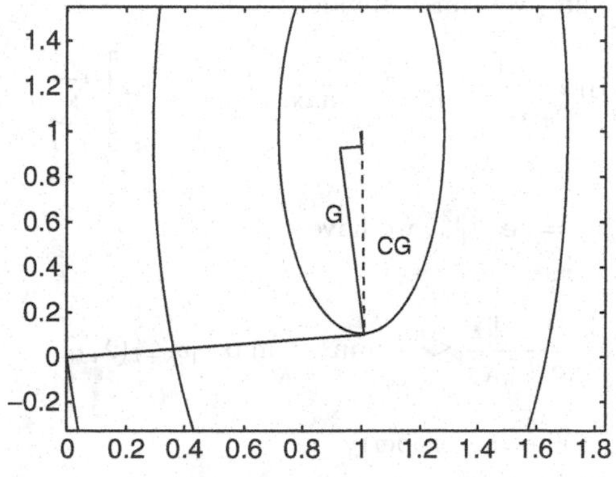

Fig. 4.7. Directions for the conjugate gradient method (denoted by CG, *dashed line*) and the gradient method (denoted by G, *solid line*). Notice that the CG method reaches the solution after two iterations

Let us prove the error estimate, assuming for simplicity that $\mathbf{x}^{(0)} = \mathbf{0}$. Notice first that, for fixed k

$$\mathbf{x}^{(k+1)} = \sum_{j=0}^{k} \gamma_j \mathrm{A}^j \mathbf{b},$$

for suitable $\gamma_j \in \mathbb{R}$. Moreover, by construction, $\mathbf{x}^{(k+1)}$ is the vector which minimizes the A-norm of the error at step $k+1$, among all vectors of the form $\mathbf{z} = \sum_{j=0}^{k} \delta_j \mathrm{A}^j \mathbf{b} = p_k(\mathrm{A})\mathbf{b}$, where $p_k(\xi) = \sum_{j=0}^{k} \delta_j \xi^j$ is a polynomial of degree k and $p_k(\mathrm{A})$ denotes the corresponding matrix polynomial. As a consequence

$$\|\mathbf{e}^{(k+1)}\|_{\mathrm{A}}^2 \leq (\mathbf{x}-\mathbf{z})^T \mathrm{A} (\mathbf{x}-\mathbf{z}) = \mathbf{x}^T q_{k+1}(\mathrm{A}) \mathrm{A} q_{k+1}(\mathrm{A}) \mathbf{x}, \qquad (4.50)$$

where $q_{k+1}(\xi) = 1 - p_k(\xi)\xi \in \mathbb{P}_{k+1}^{0,1}$, being $\mathbb{P}_{k+1}^{0,1} = \{q \in \mathbb{P}_{k+1} : q(0) = 1\}$ and $q_{k+1}(\mathrm{A})$ the associated matrix polynomial. From (4.50) we get

$$\|\mathbf{e}^{(k+1)}\|_{\mathrm{A}}^2 = \min_{q_{k+1} \in \mathbb{P}_{k+1}^{0,1}} \mathbf{x}^T q_{k+1}(\mathrm{A}) \mathrm{A} q_{k+1}(\mathrm{A}) \mathbf{x}. \qquad (4.51)$$

Since A is symmetric positive definite, there exists an orthogonal matrix Q such that $\mathrm{A} = \mathrm{Q}\Lambda\mathrm{Q}^T$ with $\Lambda = \mathrm{diag}(\lambda_1, \ldots, \lambda_n)$, with λ_1 and λ_n the largest and smallest eigenvalues of A, respectively. Noticing that $q_{k+1}(\mathrm{A}) = \mathrm{Q} q_{k+1}(\Lambda) \mathrm{Q}^T$, we get from (4.51)

$$\begin{aligned}
\|\mathbf{e}^{(k+1)}\|_{\mathrm{A}}^2 &= \min_{q_{k+1} \in \mathbb{P}_{k+1}^{0,1}} \mathbf{x}^T \mathrm{Q} q_{k+1}(\Lambda) \mathrm{Q}^T \mathrm{Q}\Lambda\mathrm{Q}^T \mathrm{Q} q_{k+1}(\Lambda) \mathrm{Q}^T \mathbf{x} \\
&= \min_{q_{k+1} \in \mathbb{P}_{k+1}^{0,1}} \mathbf{x}^T \mathrm{Q} q_{k+1}(\Lambda) \Lambda q_{k+1}(\Lambda) \mathrm{Q}^T \mathbf{x} \\
&= \min_{q_{k+1} \in \mathbb{P}_{k+1}^{0,1}} \mathbf{y}^T \mathrm{diag}(q_{k+1}(\lambda_i) \lambda_i q_{k+1}(\lambda_i)) \mathbf{y} \\
&= \min_{q_{k+1} \in \mathbb{P}_{k+1}^{0,1}} \sum_{i=1}^{n} y_i^2 \lambda_i (q_{k+1}(\lambda_i))^2,
\end{aligned}$$

having set $\mathbf{y} = \mathrm{Q}\mathbf{x}$. Thus, we can conclude that

$$\|\mathbf{e}^{(k+1)}\|_{\mathrm{A}}^2 \leq \left[\min_{q_{k+1} \in \mathbb{P}_{k+1}^{0,1}} \max_{\lambda_i \in \sigma(\mathrm{A})} (q_{k+1}(\lambda_i))^2 \right] \sum_{i=1}^{n} y_i^2 \lambda_i.$$

Recalling that $\sum_{i=1}^{n} y_i^2 \lambda_i = \|\mathbf{e}^{(0)}\|_{\mathrm{A}}^2$, we have

$$\frac{\|\mathbf{e}^{(k+1)}\|_{\mathrm{A}}}{\|\mathbf{e}^{(0)}\|_{\mathrm{A}}} \leq \min_{q_{k+1} \in \mathbb{P}_{k+1}^{0,1}} \max_{\lambda_i \in \sigma(\mathrm{A})} |q_{k+1}(\lambda_i)|.$$

Let us now recall the following property

Property 4.6 *The problem of minimizing* $\max_{\lambda_n \leq z \leq \lambda_1} |q(z)|$ *over the space* $\mathbb{P}_{k+1}^{0,1}([\lambda_n, \lambda_1])$ *admits a unique solution, given by the polynomial*

$$p_{k+1}(\xi) = T_{k+1}\left(\frac{\lambda_1 + \lambda_n - 2\xi}{\lambda_1 - \lambda_n}\right)/C_{k+1}, \qquad \xi \in [\lambda_n, \lambda_1],$$

where $C_{k+1} = T_{k+1}(\frac{\lambda_1+\lambda_n}{\lambda_1-\lambda_n})$ *and* T_{k+1} *is the Chebyshev polynomial of degree* $k+1$ *(see Section* 10.10*). The value of the minimum is* $1/C_{k+1}$.

Using this property we get

$$\frac{\|\mathbf{e}^{(k+1)}\|_A}{\|\mathbf{e}^{(0)}\|_A} \leq \frac{1}{T_{k+1}\left(\dfrac{\lambda_1 + \lambda_n}{\lambda_1 - \lambda_n}\right)},$$

from which the thesis follows since in the case of a symmetric positive definite matrix

$$\frac{1}{C_{k+1}} = \frac{2c^{k+1}}{1 + c^{2(k+1)}}.$$

◇

The generic k-th iteration of the conjugate gradient method is well defined only if the search direction $\mathbf{p}^{(k)}$ is nonnull. Besides, if $\mathbf{p}^{(k)} = \mathbf{0}$, then the iterate $\mathbf{x}^{(k)}$ must necessarily coincide with the solution $\mathbf{x}$ of the system. Moreover, irrespectively of the choice of the parameters β_k, one can show (see [Axe94], p. 463) that the sequence $\mathbf{x}^{(k)}$ generated by the CG method is such that either $\mathbf{x}^{(k)} \neq \mathbf{x}$, $\mathbf{p}^{(k)} \neq \mathbf{0}$, $\alpha_k \neq 0$ for any k, or there must exist an integer m such that $\mathbf{x}^{(m)} = \mathbf{x}$, where $\mathbf{x}^{(k)} \neq \mathbf{x}$, $\mathbf{p}^{(k)} \neq \mathbf{0}$ and $\alpha_k \neq 0$ for $k = 0, 1, \ldots, m-1$.

The particular choice made for β_k in (4.47) ensures that $m \leq n$. In absence of rounding errors, the CG method can thus be regarded as being a direct method, since it terminates after a finite number of steps. However, for matrices of large size, it is usually employed as an iterative scheme, where the iterations are stopped when the error gets below a fixed tolerance. In this respect, the dependence of the error reduction factor on the condition number of the matrix is more favorable than for the gradient method. We also notice that estimate (4.49) is often overly pessimistic and does not account for the fact that in this method, unlike what happens for the gradient method, the convergence is influenced by the *whole* spectrum of A, and not only by its extremal eigenvalues.

Remark 4.3 (Effect of rounding errors) The termination property of the CG method is rigorously valid only in exact arithmetic. The cumulating rounding errors prevent the search directions from being A-conjugate and can even generate null denominators in the computation of coefficients α_k and β_k. This latter phenomenon, known as *breakdown*, can be avoided by introducing suitable stabilization procedures; in such an event, we speak about stabilized gradient methods.

Despite the use of these strategies, it may happen that the CG method fails to converge (in finite arithmetic) after n iterations. In such a case, the only reasonable possibility is to restart the iterative process, taking as residual the last computed one. By so doing, the *cyclic CG method* or *CG method with restart* is obtained, for which, however, the convergence properties of the original CG method are no longer valid. ■

4.3.5 The Preconditioned Conjugate Gradient Method

If P is a symmetric and positive definite matrix, the preconditioned conjugate gradient method (PCG) consists of applying the CG method to the preconditioned system

$$\mathrm{P}^{-1/2}\mathrm{A}\mathrm{P}^{-1/2}\mathbf{y} = \mathrm{P}^{-1/2}\mathbf{b}, \qquad \text{with } \mathbf{y} = \mathrm{P}^{1/2}\mathbf{x}.$$

In practice, the method is implemented without explicitly requiring the computation of $\mathrm{P}^{1/2}$ or $\mathrm{P}^{-1/2}$. After some algebra, the following scheme is obtained:
given $\mathbf{x}^{(0)} \in \mathbb{R}^n$, set $\mathbf{r}^{(0)} = \mathbf{b} - \mathrm{A}\mathbf{x}^{(0)}$, $\mathbf{z}^{(0)} = \mathrm{P}^{-1}\mathbf{r}^{(0)}$ and $\mathbf{p}^{(0)} = \mathbf{z}^{(0)}$, for $k = 0, 1, \ldots$ until convergence, compute

$$\begin{aligned}
\alpha_k &= \frac{\mathbf{z}^{(k)^T}\mathbf{r}^{(k)}}{\mathbf{p}^{(k)^T}\mathrm{A}\mathbf{p}^{(k)}},\\
\mathbf{x}^{(k+1)} &= \mathbf{x}^{(k)} + \alpha_k \mathbf{p}^{(k)},\\
\mathbf{r}^{(k+1)} &= \mathbf{r}^{(k)} - \alpha_k \mathrm{A}\mathbf{p}^{(k)},\\
\mathrm{P}\mathbf{z}^{(k+1)} &= \mathbf{r}^{(k+1)},\\
\beta_k &= \frac{\mathbf{z}^{(k+1)^T}\mathbf{r}^{(k+1)}}{\mathbf{z}^{(k)^T}\mathbf{r}^{(k)}},\\
\mathbf{p}^{(k+1)} &= \mathbf{z}^{(k+1)} + \beta_k \mathbf{p}^{(k)}.
\end{aligned}$$

The computational cost is increased with respect to the CG method, as one needs to solve at each step the linear system $\mathrm{P}\mathbf{z}^{(k+1)} = \mathbf{r}^{(k+1)}$. For this system the symmetric preconditioners examined in Section 4.3.2 can be used. The error estimate is the same as for the nonpreconditioned method, provided the matrix A is replaced by $\mathrm{P}^{-1}\mathrm{A}$.

In Program 20 an implementation of the PCG method is reported. For a description of the input/output parameters, see Program 19.

Program 20 - conjgrad : Preconditioned conjugate gradient method

```
function [x,relres,iter,flag]=conjgrad(A,b,x,P,nmax,tol)
%CONJGRAD Conjugate gradient method
% [X,RELRES,ITER,FLAG]=CONJGRAD(A,B,X0,NMAX,TOL,OMEGA) attempts
% to solve the system A*X=B with the conjugate gradient method. TOL specifies
% the tolerance of the method. NMAX specifies the maximum number of iterations.
% X0 specifies the initial guess. P is a preconditioner. RELRES is the relative
```

```
% residual. If FLAG is 1, then RELRES > TOL. ITER is the iteration number at which
% X is computed.
flag=0; iter=0; bnrm2=norm(b);
if bnrm2==0, bnrm2=1; end
r=b-A*x; relres=norm(r)/bnrm2;
if relres<tol, return, end
for iter = 1:nmax
   z=P\r; rho=r'*z;
   if iter>1
      beta=rho/rho1;
      p=z+beta*p;
   else
      p=z;
   end
   q=A*p;
   alpha=rho/(p'*q);
   x=x+alpha*p;
   r=r-alpha*q;
   relres=norm(r)/bnrm2;
   if relres<=tol, break, end
   rho1 = rho;
end
if relres>tol, flag = 1; end
return
```

Example 4.7 Let us consider again the linear system of Example 4.6. The CG method has been run with the same input data as in the previous example. It converges in 3 iterations for $m = 16$ and in 45 iterations for $m = 400$. Using the

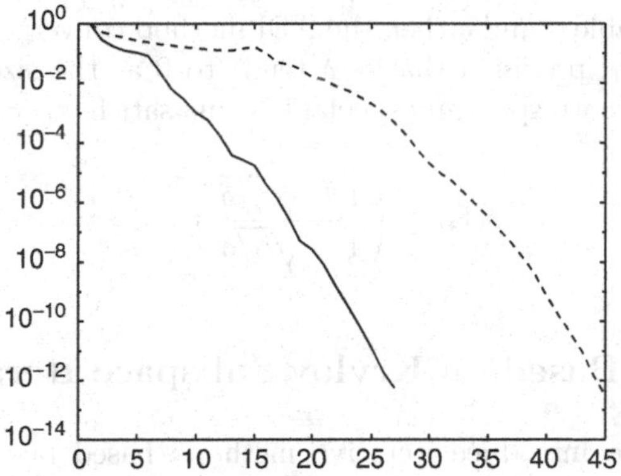

Fig. 4.8. Behavior of the residual, normalized to the right-hand side, as a function of the number of iterations for the conjugate gradient method applied to the systems of Example 4.6 in the case $m = 400$. The curve in dashed line refers to the nonpreconditioned method, while the curve in solid line refers to the preconditioned one

same preconditioner as in Example 4.6, the number of iterations decreases from 45 to 26, in the case $m = 400$. •

4.3.6 The Alternating-Direction Method

Assume that $A = A_1 + A_2$, with A_1 and A_2 symmetric and positive definite. The *alternating direction* method (ADI), as introduced by Peaceman and Rachford [PJ55], is an iterative scheme for (3.2) which consists of solving the following systems $\forall k \geq 0$

$$\begin{aligned} (I + \alpha_1 A_1)\mathbf{x}^{(k+1/2)} &= (I - \alpha_1 A_2)\mathbf{x}^{(k)} + \alpha_1 \mathbf{b}, \\ (I + \alpha_2 A_2)\mathbf{x}^{(k+1)} &= (I - \alpha_2 A_1)\mathbf{x}^{(k+1/2)} + \alpha_2 \mathbf{b}, \end{aligned} \tag{4.52}$$

where α_1 and α_2 are two real parameters. The ADI method can be cast in the form (4.2) setting

$$\begin{aligned} B &= (I + \alpha_2 A_2)^{-1}(I - \alpha_2 A_1)(I + \alpha_1 A_1)^{-1}(I - \alpha_1 A_2), \\ \mathbf{f} &= (I + \alpha_2 A_2)^{-1}\left[\alpha_1 (I - \alpha_2 A_1)(I + \alpha_1 A_1)^{-1} + \alpha_2 I\right]\mathbf{b}. \end{aligned}$$

Both B and $\mathbf{f}$ depend on α_1 and α_2. The following estimate holds

$$\rho(B) \leq \max_{i=1,\ldots,n} \left| \frac{1 - \alpha_2 \lambda_i^{(1)}}{1 + \alpha_1 \lambda_i^{(1)}} \right| \max_{i=1,\ldots,n} \left| \frac{1 - \alpha_1 \lambda_i^{(2)}}{1 + \alpha_2 \lambda_i^{(2)}} \right|,$$

where $\lambda_i^{(1)}$ and $\lambda_i^{(2)}$, for $i = 1, \ldots, n$, are the eigenvalues of A_1 and A_2, respectively. The method converges if $\rho(B) < 1$, which is always verified if $\alpha_1 = \alpha_2 = \alpha > 0$. Moreover (see [Axe94]) if $\gamma \leq \lambda_i^{(j)} \leq \delta$ $\forall i = 1, \ldots, n$, $\forall j = 1, 2$, for suitable γ and δ then the ADI method converges with the choice $\alpha_1 = \alpha_2 = 1/\sqrt{\delta\gamma}$, provided that γ/δ tends to 0 as the size of A grows. In such an event the corresponding spectral radius satisfies

$$\rho(B) \leq \left(\frac{1 - \sqrt{\gamma/\delta}}{1 + \sqrt{\gamma/\delta}} \right)^2.$$

4.4 Methods Based on Krylov Subspace Iterations

In this section we introduce iterative methods based on Krylov subspace iterations. For the proofs and further analysis, we refer to [Saa96], [Axe94], [Hac94] and [vdV03].

Consider the Richardson method (4.24) with $P = I$; the residual at the k-th step can be related to the initial residual as

$$\mathbf{r}^{(k)} = \prod_{j=0}^{k-1} (\mathrm{I} - \alpha_j \mathrm{A})\mathbf{r}^{(0)}, \tag{4.53}$$

so that $\mathbf{r}^{(k)} = p_k(\mathrm{A})\mathbf{r}^{(0)}$, where $p_k(\mathrm{A})$ is a polynomial in A of degree k. If we introduce the space

$$K_m(\mathrm{A}; \mathbf{v}) = \mathrm{span}\left\{\mathbf{v}, \mathrm{A}\mathbf{v}, \ldots, \mathrm{A}^{m-1}\mathbf{v}\right\}, \tag{4.54}$$

it immediately appears from (4.53) that $\mathbf{r}^{(k)} \in K_{k+1}(\mathrm{A}; \mathbf{r}^{(0)})$. The space defined in (4.54) is called the *Krylov subspace* of order m. It is a subspace of the set spanned by all the vectors $\mathbf{u} \in \mathbb{R}^n$ that can be written as $\mathbf{u} = p_{m-1}(\mathrm{A})\mathbf{v}$, where p_{m-1} is a polynomial in A of degree $\leq m-1$.

In an analogous manner as for (4.53), it is seen that the iterate $\mathbf{x}^{(k)}$ of the Richardson method is given by

$$\mathbf{x}^{(k)} = \mathbf{x}^{(0)} + \sum_{j=0}^{k-1} \alpha_j \mathbf{r}^{(j)},$$

so that $\mathbf{x}^{(k)}$ belongs to the following space

$$W_k = \left\{\mathbf{v} = \mathbf{x}^{(0)} + \mathbf{y},\ \mathbf{y} \in K_k(\mathrm{A}; \mathbf{r}^{(0)})\right\}. \tag{4.55}$$

Notice also that $\sum_{j=0}^{k-1} \alpha_j \mathbf{r}^{(j)}$ is a polynomial in A of degree less than $k-1$. In the nonpreconditioned Richardson method we are thus looking for an approximate solution to $\mathbf{x}$ in the space W_k. More generally, we can think of devising methods that search for approximate solutions of the form

$$\mathbf{x}^{(k)} = \mathbf{x}^{(0)} + q_{k-1}(\mathrm{A})\mathbf{r}^{(0)}, \tag{4.56}$$

where q_{k-1} is a polynomial selected in such a way that $\mathbf{x}^{(k)}$ be, in a sense that must be made precise, the best approximation of $\mathbf{x}$ in W_k. A method that looks for a solution of the form (4.56) with W_k defined as in (4.55) is called a *Krylov method.*

A first question concerning Krylov subspace iterations is whether the dimension of $K_m(\mathrm{A}; \mathbf{v})$ increases as the order m grows. A partial answer is provided by the following result.

Property 4.7 *Let* $\mathrm{A} \in \mathbb{R}^{n\times n}$ *and* $\mathbf{v} \in \mathbb{R}^n$. *The Krylov subspace* $K_m(\mathrm{A}; \mathbf{v})$ *has dimension equal to* m *iff the degree of* $\mathbf{v}$ *with respect to* A, *denoted by* $\mathrm{deg}_\mathrm{A}(\mathbf{v})$, *is not less than* m, *where the degree of* $\mathbf{v}$ *is defined as the minimum degree of a monic nonnull polynomial* p *in* A, *for which* $p(\mathrm{A})\mathbf{v} = \mathbf{0}$.

The dimension of $K_m(\mathrm{A}; \mathbf{v})$ is thus equal to the minimum between m and the degree of $\mathbf{v}$ with respect to A and, as a consequence, the dimension of the Krylov subspaces is certainly a nondecreasing function of m. Notice that the degree of $\mathbf{v}$ cannot be greater than n due to the Cayley-Hamilton Theorem (see Section 1.7).

Example 4.8 Consider the matrix $\mathrm{A} = \mathrm{tridiag}_4(-1,2,-1)$. The vector $\mathbf{v} = [1,1,1,1]^T$ has degree 2 with respect to A since $p_2(\mathrm{A})\mathbf{v} = \mathbf{0}$ with $p_2(\mathrm{A}) = \mathrm{I}_4 - 3\mathrm{A} + \mathrm{A}^2$, while there is no monic polynomial p_1 of degree 1 for which $p_1(\mathrm{A})\mathbf{v} = \mathbf{0}$. As a consequence, all Krylov subspaces from $K_2(\mathrm{A};\mathbf{v})$ on, have dimension equal to 2. The vector $\mathbf{w} = [1,1,-1,1]^T$ has, instead, degree 4 with respect to A. •

For a fixed m, it is possible to compute an orthonormal basis for $K_m(\mathrm{A};\mathbf{v})$ using the so-called *Arnoldi algorithm*.

Setting $\mathbf{v}_1 = \mathbf{v}/\|\mathbf{v}\|_2$, this method generates an orthonormal basis $\{\mathbf{v}_i\}$ for $K_m(\mathrm{A};\mathbf{v}_1)$ using the Gram-Schmidt procedure (see Section 3.4.3). For $k = 1,\ldots,m$, the Arnoldi algorithm computes

$$\begin{aligned} h_{ik} &= \mathbf{v}_i^T \mathrm{A}\mathbf{v}_k, \qquad\qquad i = 1,2,\ldots,k, \\ \mathbf{w}_k &= \mathrm{A}\mathbf{v}_k - \sum_{i=1}^{k} h_{ik}\mathbf{v}_i,\ h_{k+1,k} = \|\mathbf{w}_k\|_2. \end{aligned} \tag{4.57}$$

If $\mathbf{w}_k = \mathbf{0}$ the process terminates and in such a case we say that a *breakdown* of the algorithm has occurred; otherwise, we set $\mathbf{v}_{k+1} = \mathbf{w}_k/\|\mathbf{w}_k\|_2$ and the algorithm restarts, incrementing k by 1.

It can be shown that if the method terminates at the step m then the vectors $\mathbf{v}_1,\ldots,\mathbf{v}_m$ form a basis for $K_m(\mathrm{A};\mathbf{v})$. In such a case, if we denote by $\mathrm{V}_m \in \mathbb{R}^{n\times m}$ the matrix whose columns are the vectors $\mathbf{v}_i$, we have

$$\mathrm{V}_m^T \mathrm{A}\mathrm{V}_m = \mathrm{H}_m,\ \mathrm{V}_{m+1}^T \mathrm{A}\mathrm{V}_m = \widehat{\mathrm{H}}_m, \tag{4.58}$$

where $\widehat{\mathrm{H}}_m \in \mathbb{R}^{(m+1)\times m}$ is the upper Hessenberg matrix whose entries h_{ij} are given by (4.57) and $\mathrm{H}_m \in \mathbb{R}^{m\times m}$ is the restriction of $\widehat{\mathrm{H}}_m$ to the first m rows and m columns.

The algorithm terminates at an intermediate step $k < m$ iff $\deg_{\mathrm{A}}(\mathbf{v}_1) = k$. As for the stability of the procedure, all the considerations valid for the Gram-Schmidt method hold. For more efficient and stable computational variants of (4.57), we refer to [Saa96].

The functions `arnoldialg` and `GSarnoldi`, invoked by Program 21, provide an implementation of the Arnoldi algorithm. In output, the columns of `V` contain the vectors of the generated basis, while the matrix `H` stores the coefficients h_{ik} computed by the algorithm. If m steps are carried out, $\mathtt{V} = \mathrm{V}_m$ and $\mathtt{H}(1:m,1:m) = \mathrm{H}_m$.

Program 21 - arnoldialg : The Arnoldi algorithm

```
function [V,H]=arnoldialg(A,v,m)
% ARNOLDIALG Arnoldi algorithm
% [B,H]=ARNOLDIALG(A,V,M) computes for a fixed M an orthonormal basis B for
% K^M(A,V) such that V^T*A*V=H.
v=v/norm(v,2); V=v; H=[]; k=0;
```

```
while k <= m-1
   [k,V,H] = GSarnoldi(A,m,k,V,H);
end
return

function [k,V,H]=GSarnoldi(A,m,k,V,H)
% GSARNOLDI Gram-Schmidt method for the Arnoldi algorithm
k=k+1; H=[H,V(:,1:k)'*A*V(:,k)];
s=0;
for i=1:k
   s=s+H(i,k)*V(:,i);
end
w=A*V(:,k)-s; H(k+1,k)=norm(w,2);
if H(k+1,k)>=eps & k¡m
   V=[V,w/H(k+1,k)];
else
   k=m+1;
end
return
```

Having introduced an algorithm for generating the basis for a Krylov subspace of any order, we can now solve the linear system (3.2) by a Krylov method. As already noticed, for all of these methods the iterate $\mathbf{x}^{(k)}$ is always of the form (4.56) and, for a given $\mathbf{r}^{(0)}$, the vector $\mathbf{x}^{(k)}$ is selected as being the unique element in W_k which satisfies a criterion of minimal distance from $\mathbf{x}$. Thus, the feature distinguishing two different Krylov methods is the criterion for selecting $\mathbf{x}^{(k)}$.

The most natural idea consists of searching for $\mathbf{x}^{(k)} \in W_k$ as the vector which minimizes the Euclidean norm of the error. This approach, however, does not work in practice since $\mathbf{x}^{(k)}$ would depend on the (unknown) solution $\mathbf{x}$.

Two alternative strategies can be pursued:

1. compute $\mathbf{x}^{(k)} \in W_k$ enforcing that the residual $\mathbf{r}^{(k)}$ is orthogonal to any vector in $K_k(\mathrm{A}; \mathbf{r}^{(0)})$, i.e., we look for $\mathbf{x}^{(k)} \in W_k$ such that

$$\mathbf{v}^T(\mathbf{b} - \mathrm{A}\mathbf{x}^{(k)}) = 0 \qquad \forall \mathbf{v} \in K_k(\mathrm{A}; \mathbf{r}^{(0)}); \tag{4.59}$$

2. compute $\mathbf{x}^{(k)} \in W_k$ minimizing the Euclidean norm of the residual $\|\mathbf{r}^{(k)}\|_2$, i.e.

$$\|\mathbf{b} - \mathrm{A}\mathbf{x}^{(k)}\|_2 = \min_{\mathbf{v} \in W_k} \|\mathbf{b} - \mathrm{A}\mathbf{v}\|_2. \tag{4.60}$$

Satisfying (4.59) leads to the Arnoldi method for linear systems (more commonly known as FOM, *full orthogonalization method*), while satisfying (4.60) yields the GMRES (*generalized minimum residual*) method.

In the two forthcoming sections we shall assume that k steps of the Arnoldi algorithm have been carried out, in such a way that an orthonormal basis for

$K_k(\mathrm{A}; \mathbf{r}^{(0)})$ has been generated and stored into the column vectors of the matrix V_k with $\mathbf{v}_1 = \mathbf{r}^{(0)}/\|\mathbf{r}^{(0)}\|_2$. In such a case the new iterate $\mathbf{x}^{(k)}$ can always be written as

$$\mathbf{x}^{(k)} = \mathbf{x}^{(0)} + \mathrm{V}_k \mathbf{z}^{(k)}, \tag{4.61}$$

where $\mathbf{z}^{(k)}$ must be selected according to a fixed criterion.

4.4.1 The Arnoldi Method for Linear Systems

Let us enforce that $\mathbf{r}^{(k)}$ be orthogonal to $K_k(\mathrm{A}; \mathbf{r}^{(0)})$ by requiring that (4.59) holds for all the basis vectors $\mathbf{v}_i$, i.e.

$$\mathrm{V}_k^T \mathbf{r}^{(k)} = 0. \tag{4.62}$$

Since $\mathbf{r}^{(k)} = \mathbf{b} - \mathrm{A}\mathbf{x}^{(k)}$ with $\mathbf{x}^{(k)}$ of the form (4.61), relation (4.62) becomes

$$\mathrm{V}_k^T(\mathbf{b} - \mathrm{A}\mathbf{x}^{(0)}) - \mathrm{V}_k^T \mathrm{A} \mathrm{V}_k \mathbf{z}^{(k)} = \mathrm{V}_k^T \mathbf{r}^{(0)} - \mathrm{V}_k^T \mathrm{A} \mathrm{V}_k \mathbf{z}^{(k)} = 0. \tag{4.63}$$

Due to the orthonormality of the basis and the choice of $\mathbf{v}_1$, $\mathrm{V}_k^T \mathbf{r}^{(0)} = \|\mathbf{r}^{(0)}\|_2 \mathbf{e}_1$, $\mathbf{e}_1$ being the first unit vector of $\mathbb{R}^k$. Recalling (4.58), from (4.63) it turns out that $\mathbf{z}^{(k)}$ is the solution to the linear system

$$\mathrm{H}_k \mathbf{z}^{(k)} = \|\mathbf{r}^{(0)}\|_2 \mathbf{e}_1. \tag{4.64}$$

Once $\mathbf{z}^{(k)}$ is known, we can compute $\mathbf{x}^{(k)}$ from (4.61). Since H_k is an upper Hessenberg matrix, the linear system in (4.64) can be easily solved, for instance, resorting to the LU factorization of H_k.

We notice that the method, if working in exact arithmetic, cannot execute more than n steps and that it terminates after $m < n$ steps only if a breakdown in the Arnoldi algorithm occurs. As for the convergence of the method, the following result holds.

Theorem 4.13 *In exact arithmetic the Arnoldi method yields the solution of* (3.2) *after at most n iterations.*

Proof. If the method terminates at the n-th iteration, then it must necessarily be $\mathbf{x}^{(n)} = \mathbf{x}$ since $K_n(\mathrm{A}; \mathbf{r}^{(0)}) = \mathbb{R}^n$. Conversely, if a breakdown occurs after m iterations, for a suitable $m < n$, then $\mathbf{x}^{(m)} = \mathbf{x}$. Indeed, inverting the first relation in (4.58), we get

$$\mathbf{x}^{(m)} = \mathbf{x}^{(0)} + \mathrm{V}_m \mathbf{z}^{(m)} = \mathbf{x}^{(0)} + \mathrm{V}_m \mathrm{H}_m^{-1} \mathrm{V}_m^T \mathbf{r}^{(0)} = \mathrm{A}^{-1}\mathbf{b}.$$

◇

In its naive form, FOM does not require an explicit computation of the solution or the residual, unless a breakdown occurs. Therefore, monitoring its convergence (by computing, for instance, the residual at each step) might be computationally expensive. The residual, however, is available without explicitly requiring to compute the solution since at the k-th step we have

$$\|\mathbf{b} - \mathrm{A}\mathbf{x}^{(k)}\|_2 = h_{k+1,k}|\mathbf{e}_k^T \mathbf{z}_k|$$

and, as a consequence, one can decide to stop the method if

$$h_{k+1,k}|\mathbf{e}_k^T \mathbf{z}_k|/\|\mathbf{r}^{(0)}\|_2 \leq \varepsilon \qquad (4.65)$$

$\varepsilon > 0$ being a fixed tolerance.

The most relevant consequence of Theorem 4.13 is that FOM can be regarded as a direct method, since it yields the exact solution after a finite number of steps. However, this fails to hold when working in floating point arithmetic due to the cumulating rounding errors. Moreover, if we also account for the high computational effort, which, for a number of m steps and a sparse matrix of order n with n_z nonzero entries, is of the order of $2(n_z+mn)$ flops, and the large memory occupation needed to store the matrix V_m, we conclude that the Arnoldi method cannot be used in the practice, except for small values of m.

Several remedies to this drawback are available, one of which consisting of preconditioning the system (using, for instance, one of the preconditioners proposed in Section 4.3.2). Alternatively, we can also introduce some modified versions of the Arnoldi method following two approaches:

1. no more than m consecutive steps of FOM are taken, m being a small fixed number (usually, $m \simeq 10$). If the method fails to converge, we set $\mathbf{x}^{(0)} = \mathbf{x}^{(m)}$ and FOM is repeated for other m steps. This procedure is carried out until convergence is achieved. This method, known as FOM(m) or FOM with *restart*, reduces the memory occupation, only requiring to store matrices with m columns at most;
2. a limitation is set on the number of directions involved in the orthogonalization procedure in the Arnoldi algorithm, yielding the incomplete orthogonalization method or IOM. In the practice, the k-th step of the Arnoldi algorithm generates a vector $\mathbf{v}_{k+1}$ which is orthonormal, at most, to the q preceding vectors, where q is fixed according to the amount of available memory.

It is worth noticing that Theorem 4.13 does no longer hold for the methods stemming from the two strategies above.

Program 22 provides an implementation of the FOM algorithm with a stopping criterion based on the residual (4.65). The input parameter `m` is the maximum admissible size of the Krylov subspace that is being generated and represents, as a consequence, the maximum admissible number of iterations.

Program 22 - arnoldimet : The Arnoldi method for linear systems

```
function [x,iter]=arnoldimet(A,b,x0,m,tol)
%ARNOLDIMET Arnoldi method.
% [X,ITER]=ARNOLDIMET(A,B,X0,M,TOL) attempts to solve the system A*X=B
% with the Arnoldi method. TOL specifies the tolerance of the method.
```

```
% M specifies the maximum size of the Krylov subspace. X0 specifies
% the initial guess. ITER is the iteration number at which X is computed.
r0=b-A*x0; nr0=norm(r0,2);
if nr0 ~= 0
  v1=r0/nr0; V=[v1]; H=[]; iter=0; istop=0;
  while (iter <= m-1) & (istop == 0)
    [iter,V,H] = GSarnoldi(A,m,iter,V,H);
    [nr,nc]=size(H); e1=eye(nc);
    y=(e1(:,1)'*nr0)/H(1:nc,:);
    residual = H(nr,nc)*abs(y*e1(:,nc));
    if residual <= tol
      istop = 1; y=y';
    end
  end
  if istop==0
    [nr,nc]=size(H); e1=eye(nc);
    y=(e1(:,1)'*nr0)/H(1:nc,:); y=y';
  end
  x=x0+V(:,1:nc)*y;
else
  x=x0;
end
```

Example 4.9 Let us employ Program 22 to solve the linear system A$\mathbf{x} = \mathbf{b}$ with A $= \text{tridiag}_{100}(-1, 2, -1)$ and $\mathbf{b}$ such that the solution is $\mathbf{x} = \mathbf{1}$. The initial vector is $\mathbf{x}^{(0)} = \mathbf{0}$ and `tol`$=10^{-10}$. The method converges in 50 iterations and Figure 4.9 reports its convergence history. Notice the sudden, dramatic, reduction of the residual, which is a typical warning that the last generated subspace W_k is sufficiently rich to contain the exact solution of the system. •

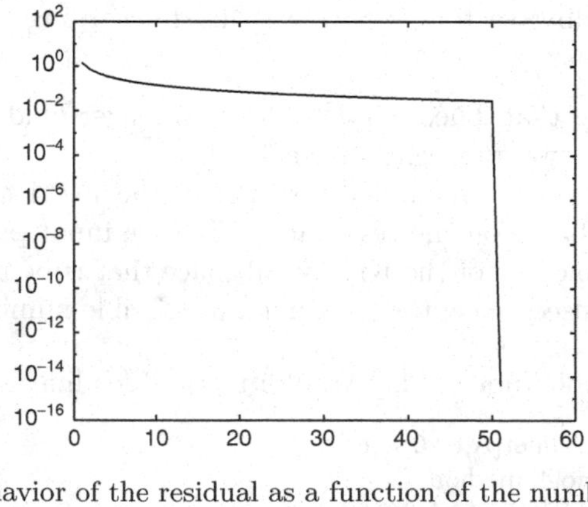

Fig. 4.9. The behavior of the residual as a function of the number of iterations for the Arnoldi method applied to the linear system in Example 4.9

4.4.2 The GMRES Method

This method is characterized by selecting $\mathbf{x}^{(k)}$ in such a way to minimize the Euclidean norm of the residual at each k-th step. Recalling (4.61) we have

$$\mathbf{r}^{(k)} = \mathbf{r}^{(0)} - \mathrm{A}\mathrm{V}_k\mathbf{z}^{(k)}, \tag{4.66}$$

but, since $\mathbf{r}^{(0)} = \mathbf{v}_1\|\mathbf{r}^{(0)}\|_2$ and (4.58) holds, relation (4.66) becomes

$$\mathbf{r}^{(k)} = \mathrm{V}_{k+1}(\|\mathbf{r}^{(0)}\|_2\mathbf{e}_1 - \widehat{\mathrm{H}}_k\mathbf{z}^{(k)}), \tag{4.67}$$

where $\mathbf{e}_1$ is the first unit vector of $\mathbb{R}^{k+1}$. Therefore, in the GMRES method the solution at step k can be computed through (4.61) as

$$\mathbf{z}^{(k)} \text{ chosen in such a way to minimize } \| \ \|\mathbf{r}^{(0)}\|_2\mathbf{e}_1 - \widehat{\mathrm{H}}_k\mathbf{z}^{(k)}\|_2 \tag{4.68}$$

(the matrix V_{k+1} appearing in (4.67) does not change the value of $\|\cdot\|_2$ since it is orthogonal). Having to solve at each step a least-squares problem of size k, the GMRES method will be the more effective the smaller is the number of iterations. Exactly as for the Arnoldi method, the GMRES method terminates at most after n iterations, yielding the exact solution. Premature stops are due to a breakdown in the orthonormalization Arnoldi algorithm. More precisely, we have the following result.

Property 4.8 *A breakdown occurs for the GMRES method at a step m (with $m < n$) iff the computed solution $\mathbf{x}^{(m)}$ coincides with the exact solution to the system.*

A basic implementation of the GMRES method is provided in Program 23. This latter requires in input the maximum admissible size `m` for the Krylov subspace and the tolerance `tol` on the Euclidean norm of the residual normalized to the initial residual. This implementation of the method computes the solution $\mathbf{x}^{(k)}$ at each step in order to evaluate the residual, with a consequent increase of the computational effort.

Program 23 - gmres : The GMRES method for linear systems

```
function [x,iter]=gmres(A,b,x0,m,tol)
%GMRES GMRES method.
% [X,ITER]=GMRES(A,B,X0,M,TOL) attempts to solve the system A*X=B
% with the GMRES method. TOL specifies the tolerance of the method.
% M specifies the maximum size of the Krylov subspace. X0 specifies
% the initial guess. ITER is the iteration number at which X is computed.
r0=b-A*x0; nr0=norm(r0,2);
if nr0 ~= 0
  v1=r0/nr0; V=[v1]; H=[]; iter=0; residual=1;
```

```
  while iter <= m-1 & residual > tol,
    [iter,V,H] = GSarnoldi(A,m,iter,V,H);
    [nr,nc]=size(H);   y=(H'*H) \ (H'*nr0*[1;zeros(nr-1,1)]);
    x=x0+V(:,1:nc)*y;   residual = norm(b-A*x,2)/nr0;
  end
else
  x=x0;
end
```

To improve the efficiency of the GMRES algorithm it is necessary to devise a stopping criterion which does not require the explicit evaluation of the residual at each step. This is possible, provided that the linear system with upper Hessenberg matrix $\widehat{\mathrm{H}}_k$ is appropriately solved.

In practice, $\widehat{\mathrm{H}}_k$ is transformed into an upper triangular matrix $\mathrm{R}_k \in \mathbb{R}^{(k+1)\times k}$ with $r_{k+1,k} = 0$ such that $\mathrm{Q}_k^T \mathrm{R}_k = \widehat{\mathrm{H}}_k$, where Q_k is a matrix obtained as the product of k Givens rotations (see Section 5.6.3). Then, since Q_k is orthogonal, it can be seen that minimizing $\| \|\mathbf{r}^{(0)}\|_2 \mathbf{e}_1 - \widehat{\mathrm{H}}_k \mathbf{z}^{(k)}\|_2$ is equivalent to minimize $\|\mathbf{f}_k - \mathrm{R}_k \mathbf{z}^{(k)}\|_2$, with $\mathbf{f}_k = \mathrm{Q}_k \|\mathbf{r}^{(0)}\|_2 \mathbf{e}_1$. It can also be shown that the $k+1$-th component of $\mathbf{f}_k$ is, in absolute value, the Euclidean norm of the residual at the k-th step.

As FOM, the GMRES method entails a high computational effort and a large amount of memory, unless convergence occurs after few iterations. For this reason, two variants of the algorithm are available, one named GMRES(m) and based on the *restart* after m steps, the other named Quasi-GMRES or QGMRES and based on stopping the Arnoldi orthogonalization process. It is worth noting that these two methods do not enjoy Property 4.8.

Remark 4.4 (Projection methods) Denoting by Y_k and L_k two generic m-dimensional subspaces of $\mathbb{R}^n$, we call *projection method* a process which generates an approximate solution $\mathbf{x}^{(k)}$ at step k, enforcing that $\mathbf{x}^{(k)} \in Y_k$ and that the residual $\mathbf{r}^{(k)} = \mathbf{b} - \mathrm{A}\mathbf{x}^{(k)}$ be orthogonal to L_k. If $Y_k = L_k$, the projection process is said to be *orthogonal*, *oblique* otherwise (see [Saa96]).

The Krylov subspace iterations can be regarded as being projection methods. For instance, the Arnoldi method is an orthogonal projection method where $L_k = Y_k = K_k(\mathrm{A}; \mathbf{r}^{(0)})$, while the GMRES method is an oblique projection method with $Y_k = K_k(\mathrm{A}; \mathbf{r}^{(0)})$ and $L_k = \mathrm{A}Y_k$. It is worth noticing that some classical methods introduced in previous sections fall into this category. For example, the Gauss-Seidel method is an orthogonal projection method where at the k-th step $K_k(\mathrm{A}; \mathbf{r}^{(0)}) = \mathrm{span}(\mathbf{e}_k)$, with $k = 1, \ldots, n$. The projection steps are carried out cyclically from 1 to n until convergence. ■

4.4.3 The Lanczos Method for Symmetric Systems

The Arnoldi algorithm simplifies considerably if A is symmetric since the matrix H_m is tridiagonal and symmetric (indeed, from (4.58) it turns out that

H_m must be symmetric, so that, being upper Hessenberg by construction, it must necessarily be tridiagonal). In such an event the method is more commonly known as the *Lanczos algorithm*. For ease of notation, we henceforth let $\alpha_i = h_{ii}$ and $\beta_i = h_{i-1,i}$.

An implementation of the Lanczos algorithm is provided in Program 24. Vectors `alpha` and `beta` contain the coefficients α_i and β_i computed by the scheme.

Program 24 - lanczos : The Lanczos algorithm

```
function [V,alpha,beta]=lanczos(A,m)
%LANCZOS Lanczos algorithm.
% [V,ALPHA,BETA]=LANCZOS(A,M) computes matrices V and H of dimension
% equal to M in (4.58).
n=size(A); V=[0*[1:n]',[1,0*[1:n-1]]'];
beta(1)=0; normb=1; k=1;
while k <= m & normb >= eps
  vk = V(:,k+1);          w = A*vk-beta(k)*V(:,k);
  alpha(k)= w'*vk;        w = w - alpha(k)*vk
  normb = norm(w,2);
  if normb ~= 0
    beta(k+1)=normb;    V=[V,w/normb];    k=k+1;
  end
end
[n,m]=size(V); V=V(:,2:m-1);
alpha=alpha(1:n); beta=beta(2:n);
```

The algorithm, which is far superior to Arnoldi's one as far as memory saving is concerned, is not numerically stable since only the first generated vectors are actually orthogonal. For this reason, several stable variants have been devised.

As in previous cases, also the Lanczos algorithm can be employed as a solver for linear systems, yielding a symmetric form of the FOM method. It can be shown that $\mathbf{r}^{(k)} = \gamma_k \mathbf{v}_{k+1}$, for a suitable γ_k (analogously to (4.65)) so that the residuals are all mutually orthogonal.

Remark 4.5 (The conjugate gradient method) If A is symmetric and positive definite, starting from the Lanczos method for linear systems it is possible to derive the conjugate gradient method already introduced in Section 4.3.4 (see [Saa96]). The conjugate gradient method is a variant of the Lanczos method where the orthonormalization process remains incomplete.

As a matter of fact, the A-conjugate directions of the CG method can be characterized as follows. If we carry out at the generic k-th step the LU factorization $H_k = L_k U_k$, with L_k (resp., U_k) lower (resp., upper) bidiagonal, the iterate $\mathbf{x}^{(k)}$ of the Lanczos method for systems reads

$$\mathbf{x}^{(k)} = \mathbf{x}^{(0)} + P_k L_k^{-1} \|\mathbf{r}^{(0)}\|_2 \mathbf{e}_1,$$

with $\mathrm{P}_k = \mathrm{V}_k\mathrm{U}_k^{-1}$. The column vectors of P_k are mutually A-conjugate. Indeed, $\mathrm{P}_k^T\mathrm{A}\mathrm{P}_k$ is symmetric and bidiagonal since

$$\mathrm{P}_k^T\mathrm{A}\mathrm{P}_k = \mathrm{U}_k^{-T}\mathrm{H}_k\mathrm{U}_k^{-1} = \mathrm{U}_k^{-T}\mathrm{L}_k,$$

so that it must necessarily be diagonal. As a result, $\left(\mathbf{p}^{(j)}\right)^T\mathrm{A}\mathbf{p}^{(i)} = 0$ if $i \neq j$, having denoted by $\mathbf{p}^{(i)}$ the i-th column vector of matrix P_k. ∎

As happens for the FOM method, also the GMRES method simplifies if A is symmetric. The resulting scheme is called *conjugate residuals* or CR method since it enjoys the property that the residuals are mutually A-conjugate. Variants of this method are the generalized conjugate residuals method (GCR) and the method commonly known as ORTHOMIN (obtained by truncation of the orthonormalization process as done for the IOM method).

4.5 The Lanczos Method for Unsymmetric Systems

The Lanczos orthogonalization process can be extended to deal with unsymmetric matrices through a *bi-orthogonalization* procedure as follows. Two bases, $\{\mathbf{v}_i\}_{i=1}^m$ and $\{\mathbf{z}_i\}_{i=1}^m$, are generated for the subspaces $\mathrm{K}_m(\mathrm{A};\mathbf{v}_1)$ and $\mathrm{K}_m(\mathrm{A}^T;\mathbf{z}_1)$, respectively, with $\mathbf{z}_1^T\mathbf{v}_1 = 1$, such that

$$\mathbf{z}_i^T\mathbf{v}_j = \delta_{ij}, \qquad i,j = 1,\ldots,m. \tag{4.69}$$

Two sets of vectors satisfying (4.69) are said to be *bi-orthogonal* and can be obtained through the following algorithm: setting $\beta_1 = \gamma_1 = 0$ and $\mathbf{z}_0 = \mathbf{v}_0 = \mathbf{0}^T$, at the generic k-th step, with $k = 1,\ldots,m$, we set $\alpha_k = \mathbf{z}_k^T\mathrm{A}\mathbf{v}_k$, then we compute

$$\tilde{\mathbf{v}}_{k+1} = \mathrm{A}\mathbf{v}_k - \alpha_k\mathbf{v}_k - \beta_k\mathbf{v}_{k-1},\ \tilde{\mathbf{z}}_{k+1} = \mathrm{A}^T\mathbf{z}_k - \alpha_k\mathbf{z}_k - \gamma_k\mathbf{z}_{k-1}.$$

If $\gamma_{k+1} = \sqrt{|\tilde{\mathbf{z}}_{k+1}^T\tilde{\mathbf{v}}_{k+1}|} = 0$ the algorithm is stopped, otherwise we set $\beta_{k+1} = \tilde{\mathbf{z}}_{k+1}^T\tilde{\mathbf{v}}_{k+1}/\gamma_{k+1}$ and generate two new vectors in the basis as

$$\mathbf{v}_{k+1} = \tilde{\mathbf{v}}_{k+1}/\gamma_{k+1},\ \mathbf{z}_{k+1} = \tilde{\mathbf{z}}_{k+1}/\beta_{k+1}.$$

If the process terminates after m steps, denoting by V_m and Z_m the matrices whose columns are the vectors of the basis that has been generated, we have

$$\mathrm{Z}_m^T\mathrm{A}\mathrm{V}_m = \mathrm{T}_m,$$

T_m being the following tridiagonal matrix

$$\mathrm{T}_m = \begin{bmatrix} \alpha_1 & \beta_2 & & 0 \\ \gamma_2 & \alpha_2 & \ddots & \\ & \ddots & \ddots & \beta_m \\ 0 & & \gamma_m & \alpha_m \end{bmatrix}.$$

As in the symmetric case, the bi-orthogonalization Lanczos algorithm can be utilized to solve the linear system (3.2). For this purpose, for m fixed, once the bases $\{\mathbf{v}_i\}_{i=1}^m$ and $\{\mathbf{z}_i\}_{i=1}^m$ have been constructed, it suffices to set

$$\mathbf{x}^{(m)} = \mathbf{x}^{(0)} + \mathrm{V}_m \mathbf{y}^{(m)},$$

where $\mathbf{y}^{(m)}$ is the solution to the linear system $\mathrm{T}_m \mathbf{y}^{(m)} = \|\mathbf{r}^{(0)}\|_2 \mathbf{e}_1$. It is also possible to introduce a stopping criterion based on the residual, without computing it explicitly, since

$$\|\mathbf{r}^{(m)}\|_2 = |\gamma_{m+1} \mathbf{e}_m^T \mathbf{y}^{(m)}| \; \|\mathbf{v}_{m+1}\|_2.$$

An implementation of the Lanczos method for unsymmetric systems is given in Program 25. If a breakdown of the algorithm occurs, i.e., if $\gamma_{k+1} = 0$, the method stops returning in output a negative value of the variable `niter` which denotes the number of iterations necessary to reduce the initial residual by a factor `tol`.

Program 25 - lanczosnosym : The Lanczos method for unsymmetric systems

```
function [xk,relres,iter]=lanczosnosym(A,b,x0,m,tol)
%LANCZOSNOSYM Lanczos method
% [X,RELRES,ITER]=LANCZOSNOSYM(A,B,X0,M,TOL) attempts to solve the
% system A*X=B with the Lanczos method. TOL specifies the tolerance of the
% method. M specifies the maximum number of iterations. X0 specifies the initial
guess. ITER is the iteration number at which X is computed.
r0=b-A*x0; relres0=norm(r0,2);
if relres0 ~= 0
   V=r0/relres0; Z=V; gamma(1)=0; beta(1)=0; k=1; relres=1;
   while k <= m & relres > tol
      vk=V(:,k); zk=Z(:,k);
      if k==1
         vk1=0*vk; zk1=0*zk;
      else
         vk1=V(:,k-1); zk1=Z(:,k-1);
      end
      alpha(k)=zk'*A*vk;
      tildev=A*vk-alpha(k)*vk-beta(k)*vk1;
      tildez=A'*zk-alpha(k)*zk-gamma(k)*zk1;
      gamma(k+1)=sqrt(abs(tildez'*tildev));
      if gamma(k+1) == 0
         k=m+2;
      else
         beta(k+1)=tildez'*tildev/gamma(k+1);
         Z=[Z,tildez/beta(k+1)];      V=[V,tildev/gamma(k+1)];
      end
```

```
        if k~=m+2
           if k==1
              Tk = alpha;
           else
              Tk=diag(alpha)+diag(beta(2:k),1)+diag(gamma(2:k),-1);
           end
           yk=Tk\(relres0*[1,0*[1:k-1]]');
           xk=x0+V(:,1:k)*yk;
           relres=abs(gamma(k+1)*[0*[1:k-1],1]*yk)*norm(V(:,k+1),2)/relres0;
           k=k+1;
        end
     end
else
   x=x0;
end
if k==m+2, iter=-k; else, iter=k-1; end
return
```

Example 4.10 Let us solve the linear system with matrix A $= \text{tridiag}_{100}(-0.5, 2, -1)$ and right-side $\mathbf{b}$ selected in such a way that the exact solution is $\mathbf{x} = \mathbf{1}$. Using Program 25 with `tol`$= 10^{-13}$ and a randomly generated `x0`, the algorithm converges in 59 iterations. Figure 4.10 shows the convergence history reporting the graph of $\|\mathbf{r}^{(k)}\|_2/\|\mathbf{r}^{(0)}\|_2$ as a function of the number of iterations. •

We conclude recalling that some variants of the unsymmetric Lanczos method have been devised, that are characterized by a reduced computational cost. We refer the interested reader to the bibliography below for a complete description of the algorithms and to the programs included in the MATLAB version of the public domain library `templates` for their efficient implementation [BBC$^+$94].

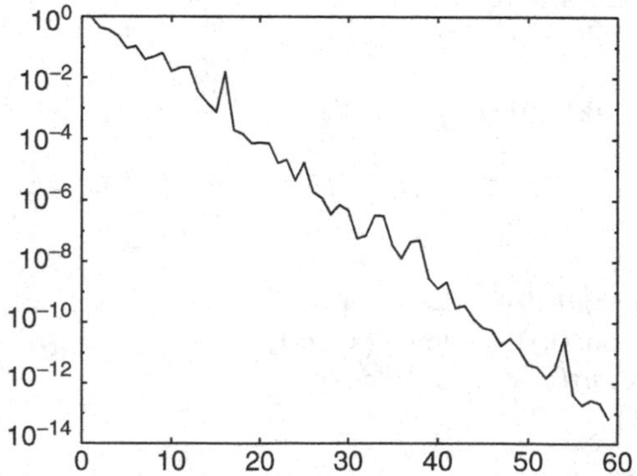

Fig. 4.10. Graph of the residual normalized to the initial residual as a function of the number of iterations for the Lanczos method applied to the system in Example 4.10

1. The *bi-conjugate gradient method* (BiCG): it can be derived by the unsymmetric Lanczos method in the same way as the conjugate gradient method is obtained from the FOM method [Fle75];
2. the *Quasi-Minimal Residual method* (QMR): it is analogous to the GMRES method, the only difference being the fact that the Arnoldi orthonormalization process is replaced by the Lanczos bi-orthogonalization;
3. the *conjugate gradient squared method* (CGS): the matrix-vector products involving the transposed matrix A^T are removed. A variant of this method, known as BiCGStab, is characterized by a more regular convergence than provided by the CGS method (see [Son89], [vdV92], [vdV03]).

4.6 Stopping Criteria

In this section we address the problem of how to estimate the error introduced by an iterative method and the number k_{min} of iterations needed to reduce the initial error by a factor ε.
In practice, k_{min} can be obtained by estimating the convergence rate of (4.2), i.e. the rate at which $\|\mathbf{e}^{(k)}\| \to 0$ as k tends to infinity. From (4.4), we get

$$\frac{\|\mathbf{e}^{(k)}\|}{\|\mathbf{e}^{(0)}\|} \leq \|\mathrm{B}^k\|,$$

so that $\|\mathrm{B}^k\|$ is an estimate of the reducing factor of the norm of the error after k steps. Typically, the iterative process is continued until $\|\mathbf{e}^{(k)}\|$ has reduced with respect to $\|\mathbf{e}^{(0)}\|$ by a certain factor $\varepsilon < 1$, that is

$$\|\mathbf{e}^{(k)}\| \leq \varepsilon \|\mathbf{e}^{(0)}\|. \tag{4.70}$$

If we assume that $\rho(\mathrm{B}) < 1$, then Property 1.13 implies that there exists a suitable matrix norm $\|\cdot\|$ such that $\|\mathrm{B}\| < 1$. As a consequence, $\|\mathrm{B}^k\|$ tends to zero as k tends to infinity, so that (4.70) can be satisfied for a sufficiently large k such that $\|\mathrm{B}^k\| \leq \varepsilon$ holds. However, since $\|\mathrm{B}^k\| < 1$, the previous inequality amounts to requiring that

$$k \geq \log(\varepsilon) / \left(\frac{1}{k} \log \|\mathrm{B}^k\| \right) = -\log(\varepsilon)/R_k(\mathrm{B}), \tag{4.71}$$

where $R_k(\mathrm{B})$ is the average convergence rate introduced in Definition 4.2. From a practical standpoint, (4.71) is useless, being nonlinear in k; if, however, the asymptotic convergence rate is adopted, instead of the average one, the following estimate for k_{min} is obtained

$$k_{min} \simeq -\log(\varepsilon)/R(\mathrm{B}). \tag{4.72}$$

This latter estimate is usually rather optimistic, as confirmed by Example 4.11.

Example 4.11 For the matrix A_3 of Example 4.2, in the case of Jacobi method, letting $\varepsilon = 10^{-5}$, condition (4.71) is satisfied with $k_{min} = 16$, while (4.72) yields $k_{min} = 15$, with a good agreement between the two estimates. Instead, on the matrix A_4 of Example 4.2, we find that (4.71) is satisfied with $k_{min} = 30$, while (4.72) yields $k_{min} = 26$. •

4.6.1 A Stopping Test Based on the Increment

From the recursive error relation $\mathbf{e}^{(k+1)} = B\mathbf{e}^{(k)}$, we get

$$\|\mathbf{e}^{(k+1)}\| \leq \|B\|\|\mathbf{e}^{(k)}\|. \tag{4.73}$$

Using the triangular inequality we get

$$\|\mathbf{e}^{(k+1)}\| \leq \|B\|(\|\mathbf{e}^{(k+1)}\| + \|\mathbf{x}^{(k+1)} - \mathbf{x}^{(k)}\|),$$

from which it follows that

$$\|\mathbf{x} - \mathbf{x}^{(k+1)}\| \leq \frac{\|B\|}{1 - \|B\|}\|\mathbf{x}^{(k+1)} - \mathbf{x}^{(k)}\|. \tag{4.74}$$

In particular, taking $k = 0$ in (4.74) and applying recursively (4.73) we also get

$$\|\mathbf{x} - \mathbf{x}^{(k+1)}\| \leq \frac{\|B\|^{k+1}}{1 - \|B\|}\|\mathbf{x}^{(1)} - \mathbf{x}^{(0)}\|,$$

which can be used to estimate the number of iterations necessary to fulfill the condition $\|\mathbf{e}^{(k+1)}\| \leq \varepsilon$, for a given tolerance ε.

In the practice, $\|B\|$ can be estimated as follows: since

$$\mathbf{x}^{(k+1)} - \mathbf{x}^{(k)} = -(\mathbf{x} - \mathbf{x}^{(k+1)}) + (\mathbf{x} - \mathbf{x}^{(k)}) = B(\mathbf{x}^{(k)} - \mathbf{x}^{(k-1)}),$$

a lower bound of $\|B\|$ is provided by $c = \delta_{k+1}/\delta_k$, where $\delta_{j+1} = \|\mathbf{x}^{(j+1)} - \mathbf{x}^{(j)}\|$, with $j = k-1, k$. Replacing $\|B\|$ by c, the right-hand side of (4.74) suggests using the following indicator for $\|\mathbf{e}^{(k+1)}\|$

$$\epsilon^{(k+1)} = \frac{\delta_{k+1}^2}{\delta_k - \delta_{k+1}}. \tag{4.75}$$

Due to the kind of approximation of $\|B\|$ that has been used, the reader is warned that $\epsilon^{(k+1)}$ should not be regarded as an upper bound for $\|\mathbf{e}^{(k+1)}\|$. However, often $\epsilon^{(k+1)}$ provides a reasonable indication about the true error behavior, as we can see in the following example.

Example 4.12 Consider the linear system $A\mathbf{x} = \mathbf{b}$ with

$$A = \begin{bmatrix} 4 & 1 & 1 \\ 2 & -9 & 0 \\ 0 & -8 & -6 \end{bmatrix}, \mathbf{b} = \begin{bmatrix} 6 \\ -7 \\ -14 \end{bmatrix},$$

which admits the unit vector as exact solution. Let us apply the Jacobi method and estimate the error at each step by using (4.75). Figure 4.11 shows an acceptable agreement between the behavior of the error $\|\mathbf{e}^{(k+1)}\|_\infty$ and that of its estimate $\epsilon^{(k+1)}$. •

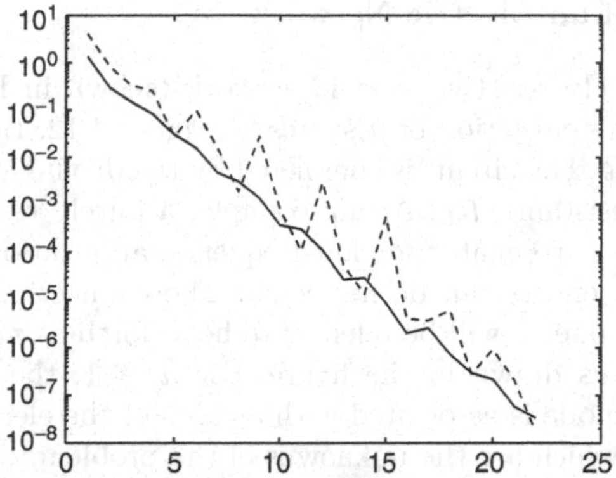

Fig. 4.11. Absolute error (*solid line*) versus the error estimated by (4.75) (*dashed line*). The number of iterations is indicated on the x-axis

4.6.2 A Stopping Test Based on the Residual

A different stopping criterion consists of continuing the iteration until $\|\mathbf{r}^{(k)}\| \leq \varepsilon$, ε being a fixed tolerance. Note that

$$\|\mathbf{x}-\mathbf{x}^{(k)}\| = \|\mathrm{A}^{-1}\mathbf{b}-\mathbf{x}^{(k)}\| = \|\mathrm{A}^{-1}\mathbf{r}^{(k)}\| \leq \|\mathrm{A}^{-1}\|\ \varepsilon.$$

Considering instead a normalized residual, i.e. stopping the iteration as soon as $\|\mathbf{r}^{(k)}\|/\|\mathbf{b}\| \leq \varepsilon$, we obtain the following control on the relative error

$$\frac{\|\mathbf{x}-\mathbf{x}^{(k)}\|}{\|\mathbf{x}\|} \leq \frac{\|\mathrm{A}^{-1}\|\,\|\mathbf{r}^{(k)}\|}{\|\mathbf{x}\|} \leq K(\mathrm{A})\frac{\|\mathbf{r}^{(k)}\|}{\|\mathbf{b}\|} \leq \varepsilon K(\mathrm{A}).$$

In the case of preconditioned methods, the residual is replaced by the preconditioned residual, so that the previous criterion becomes

$$\frac{\|\mathrm{P}^{-1}\mathbf{r}^{(k)}\|}{\|\mathrm{P}^{-1}\mathbf{r}^{(0)}\|} \leq \varepsilon,$$

where P is the preconditioning matrix.

4.7 Applications

In this section we consider two examples arising in electrical network analysis and structural mechanics which lead to the solution of large sparse linear systems.

4.7.1 Analysis of an Electric Network

We consider a purely resistive electric network (shown in Figure 4.12, left) which consists of a connection of n stages S (Figure 4.12, right) through the series resistances R. The circuit is completed by the driving current generator I_0 and the load resistance R_L. As an example, a purely resistive network is a model of a signal attenuator for low-frequency applications where capacitive and inductive effects can be neglected. The connecting points between the electrical components will be referred to henceforth as *nodes* and are progressively labeled as drawn in the figure. For $n \geq 1$, the total number of nodes is $4n$. Each node is associated with a value of the electric potential V_i, $i = 0, \ldots, 4n-1$, which are the unknowns of the problem.
The *nodal analysis* method is employed to solve the problem. Precisely, the *Kirchhoff current law* is written at any node of the network leading to the linear system $\tilde{\mathrm{Y}}\tilde{\mathbf{V}} = \tilde{\mathbf{I}}$, where $\tilde{\mathbf{V}} \in \mathbb{R}^{N+1}$ is the vector of nodal potentials, $\tilde{\mathbf{I}} \in \mathbb{R}^{N+1}$ is the load vector and the entries of the matrix $\tilde{\mathrm{Y}} \in \mathbb{R}^{(N+1)\times(N+1)}$, for $i, j = 0, \ldots, 4n-1$, are given by

$$\tilde{Y}_{ij} = \begin{cases} \sum\limits_{k \in \mathrm{adj}(i)} G_{ik}, & \text{for } i = j, \\ -G_{ij}, & \text{for } i \neq j, \end{cases}$$

where $\mathrm{adj}(i)$ is the index set of the neighboring nodes of node i and $G_{ij} = 1/R_{ij}$ is the admittance between node i and node j, provided R_{ij} denotes the resistance between the two nodes i and j. Since the potential is defined up to an additive constant, we arbitrarily set $V_0 = 0$ (*ground potential*). As a consequence, the number of independent nodes for potential difference computations is $N = 4n - 1$ and the linear system to be solved becomes $\mathrm{Y}\mathbf{V} = \mathbf{I}$, where $\mathrm{Y} \in \mathbb{R}^{N\times N}$, $\mathbf{V} \in \mathbb{R}^N$ and $\mathbf{I} \in \mathbb{R}^N$ are obtained eliminating the first row and column in $\tilde{\mathrm{Y}}$ and the first entry in $\tilde{\mathbf{V}}$ and $\tilde{\mathbf{I}}$, respectively.
The matrix Y is symmetric, diagonally dominant and positive definite. This last property follows by noting that

$$\tilde{\mathbf{V}}^T\tilde{\mathrm{Y}}\tilde{\mathbf{V}} = \sum_{i=1}^{N} \tilde{Y}_{ii} V_i^2 + \sum_{i,j=1}^{N} G_{ij}(V_i - V_j)^2,$$

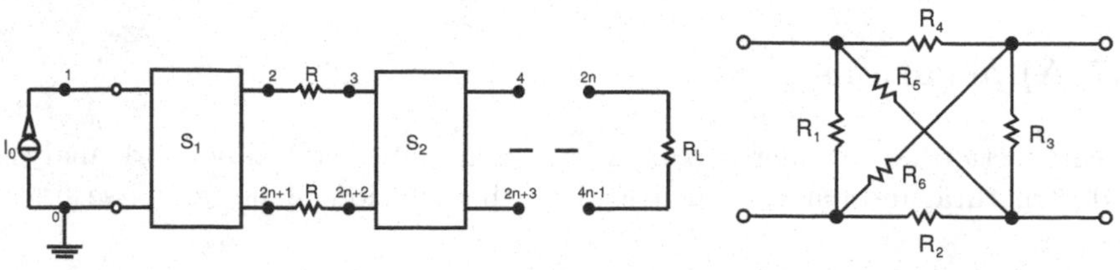

Fig. 4.12. Resistive electric network (*left*) and resistive stage S (*right*)

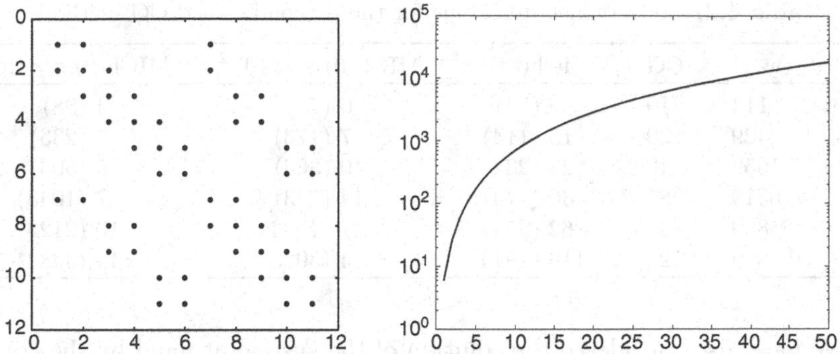

Fig. 4.13. Sparsity pattern of Y for $n = 3$ (*left*) and spectral condition number of Y as a function of n (*right*)

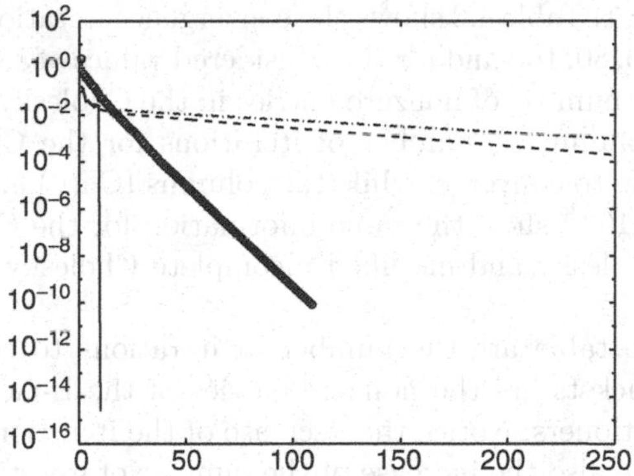

Fig. 4.14. Convergence history of several nonpreconditioned iterative methods

which is always a positive quantity, being equal to zero only if $\tilde{\mathbf{V}} = \mathbf{0}$. The sparsity pattern of Y in the case $n = 3$ is shown in Figure 4.13 (*left*) while the spectral condition number of Y as a function of the number of blocks n is reported in Figure 4.13 (*right*). Our numerical computations have been carried out setting the resistance values equal to 1 Ohm, while $I_0 = 1$ Ampère.

In Figure 4.14 we report the convergence history of several non preconditioned iterative methods in the case $n = 5$ corresponding to a matrix size of 19×19. The plots show the Euclidean norms of the residual normalized to the initial residual. The dashed curve refers to the Gauss-Seidel method, the dash-dotted line refers to the gradient method, while the solid and circled lines refer respectively to the conjugate gradient (CG) and SOR method (with an optimal value of the relaxation parameter $\omega \simeq 1.76$ computed according to (4.19) since Y is block tridiagonal symmetric positive definite). The SOR method converges in 109 iterations, while the CG method converges in 10 iterations.

Table 4.2. Convergence iterations for the preconditioned CG method

n	nz	CG	ICh(0)	MICh(0) $\varepsilon = 10^{-2}$	MICh(0) $\varepsilon = 10^{-3}$
5	114	10	9 (54)	6 (78)	4 (98)
10	429	20	15 (114)	7 (173)	5 (233)
20	1659	40	23 (234)	10 (363)	6 (503)
40	6519	80	36 (474)	14 (743)	7 (1043)
80	25839	160	62 (954)	21 (1503)	10 (2123)
160	102879	320	110 (1914)	34 (3023)	14 (4283)

We have also considered the solution of the system at hand by the conjugate gradient (CG) method using the Cholesky version of the ILU(0) and MILU(0) preconditioners, where drop tolerances equal to $\varepsilon = 10^{-2}, 10^{-3}$ have been chosen for the MILU(0) preconditioner (see Section 4.3.2). Calculations with both preconditioners have been done using the MATLAB functions `cholinc` and `michol`. Table 4.2 shows the convergence iterations of the method for $n = 5, 10, 20, 40, 80, 160$ and for the considered values of ε. We report in the second column the number of nonzero entries in the Cholesky factor of matrix Y, in the third column the number of iterations for the CG method without preconditioning to converge, while the columns ICh(0) and MICh(0) with $\varepsilon = 10^{-2}$ and $\varepsilon = 10^{-3}$ show the same information for the CG method using the incomplete Cholesky and modified incomplete Cholesky preconditioners, respectively.
The entries in the table are the number of iterations to converge and the number in the brackets are the nonzero entries of the L-factor of the corresponding preconditioners. Notice the decrease of the iterations as ε decreases, as expected. Notice also the increase of the number of iterations with respect to the increase of the size of the problem.

4.7.2 Finite Difference Analysis of Beam Bending

Consider the beam clamped at the endpoints that is drawn in Figure 4.15 (*left*). The structure, of length L, is subject to a distributed load P, varying along the free coordinate x and expressed in $[Kgm^{-1}]$. We assume henceforth that the beam has uniform rectangular section, of width r and depth s, momentum of inertia $J = rs^3/12$ and Young's module E, expressed in $[m^4]$ and $[Kg\,m^{-2}]$, respectively.
The transverse bending of the beam, under the assumption of small displacements, is governed by the following fourth-order differential equation

$$(EJu'')''(x) = P(x), \qquad 0 < x < L, \tag{4.76}$$

where $u = u(x)$ denotes the vertical displacement. The following boundary conditions (at the endpoints $x = 0$ and $x = L$)

$$u(0) = u(L) = 0, \qquad u'(0) = u'(L) = 0, \tag{4.77}$$

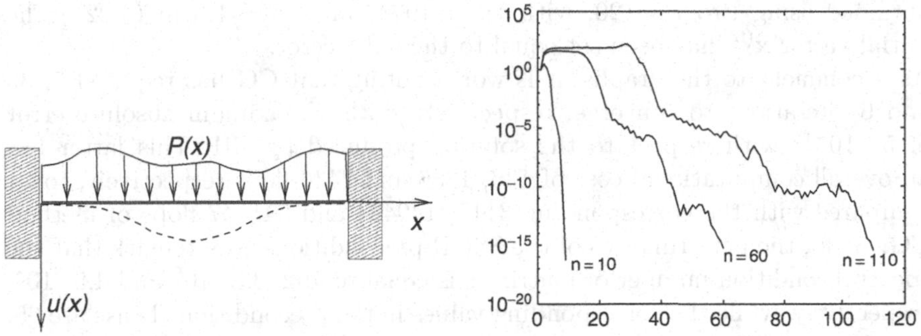

Fig. 4.15. Clamped beam (*left*); convergence histories for the preconditioned conjugate gradient method in the solution of system (4.78) (*right*)

model the effect of the two clampings (vanishing displacements and rotations). To solve numerically the boundary-value problem (4.76)-(4.77), we use the finite difference method (see Section 10.10.1 and Exercise 11 of Chapter 12).

With this aim, let us introduce the discretization nodes $x_j = jh$, with $h = L/N_h$ and $j = 0, \ldots, N_h$, and substitute at each node x_j the fourth-order derivative with an approximation through centered finite differences. Letting $f(x) = P(x)/(EJ)$, $f_j = f(x_j)$ and denoting by η_j the (approximate) *nodal displacement* of the beam at node x_j, the finite difference discretization of (4.76)-(4.77) is

$$\begin{cases} \eta_{j-2} - 4\eta_{j-1} + 6\eta_j - 4\eta_{j+1} + \eta_{j+2} = h^4 f_j, \ \forall j = 2, \ldots, N_h - 2, \\ \eta_0 = \eta_1 = \eta_{N_h-1} = \eta_{N_h} = 0. \end{cases} \tag{4.78}$$

The null displacement boundary conditions in (4.78) that have been imposed at the first and the last two nodes of the grid, require that $N_h \geq 4$. Notice that a fourth-order scheme has been used to approximate the fourth-order derivative, while, for sake of simplicity, a first-order approximation has been employed to deal with the boundary conditions (see Section 10.10.1).

The $N_h - 3$ discrete equations (4.78) yield a linear system of the form $\mathrm{A}\mathbf{x} = \mathbf{b}$ where the unknown vector $\mathbf{x} \in \mathbb{R}^{N_h-3}$ and the load vector $\mathbf{b} \in \mathbb{R}^{N_h-3}$ are given respectively by $\mathbf{x} = [\eta_2, \eta_3, \ldots, \eta_{N_h-2}]^T$ and $\mathbf{b} = [f_2, f_3, \ldots, f_{N_h-2}]^T$, while the coefficient matrix $\mathrm{A} \in \mathbb{R}^{(N_h-3)\times(N_h-3)}$ is pentadiagonal and symmetric, given by $\mathrm{A} = \mathrm{pentadiag}_{N_h-3}(1, -4, 6, -4, 1)$.

The matrix A is symmetric and positive definite. Therefore, to solve system $\mathrm{A}\mathbf{x} = \mathbf{b}$, the SSOR preconditioned conjugated gradient method (see Section 4.3.5) and the Cholesky factorization method have been employed. In the remainder of the section, the two methods are identified by the symbols (CG) and (CH).

The convergence histories of CG are reported in Figure 4.15 (right), where the sequences $\|\mathbf{r}^{(k)}\|_2/\|\mathbf{b}^{(k)}\|_2$, for the values $n = 10, 60, 110$, are plotted, $\mathbf{r}^{(k)} = \mathbf{b} - \mathrm{A}\mathbf{x}^{(k)}$ being the residual at the k-th step. The results have been

obtained using Program 20, with `tol`$=10^{-15}$ and $\omega = 1.8$ in (4.22). The initial vector $\mathbf{x}^{(0)}$ has been set equal to the null vector.
As a comment to the graphs, it is worth noting that CG has required 7, 33 and 64 iterations to converge, respectively, with a maximum absolute error of $5 \cdot 10^{-15}$ with respect to the solution produced by CH. This latter has an overall computational cost of 136, 1286 and 2436 flops respectively, to be compared with the corresponding 3117, 149424 and 541647 flops of method CG. As for the performances of the SSOR preconditioner, we remark that the spectral condition number of matrix A is equal to 192, $3.8 \cdot 10^5$ and $4.5 \cdot 10^6$, respectively, while the corresponding values in the preconditioned case are 65, $1.2 \cdot 10^4$ and $1.3 \cdot 10^5$.

4.8 Exercises

1. The spectral radius of the matrix

$$\mathrm{B} = \begin{bmatrix} a & 4 \\ 0 & a \end{bmatrix}$$

is $\rho(\mathrm{B}) = |a|$. Check that if $0 < a < 1$, then $\rho(\mathrm{B}) < 1$, while $\|\mathrm{B}^m\|_2^{1/m}$ can be greater than 1.
2. Let $\mathrm{A} \in \mathbb{R}^{n \times n}$ be a strictly diagonally dominant matrix by rows. Show that the Gauss-Seidel method for the solution of the linear system (3.2) is convergent.
3. Check that the matrix $\mathrm{A} = \mathrm{tridiag}(-1, \alpha, -1)$, with $\alpha \in \mathbb{R}$, has eigenvalues given by

$$\lambda_j = \alpha - 2\cos(j\theta),\ j = 1, \ldots, n,$$

where $\theta = \pi/(n+1)$ and the corresponding eigenvectors are

$$\mathbf{q}_j = [\sin(j\theta),\ \sin(2j\theta), \ldots, \sin(nj\theta)]^T.$$

Under which conditions on α is the matrix positive definite?
[*Solution* : $\alpha \geq 2$.]
4. Consider the pentadiagonal matrix $\mathrm{A} = \mathrm{pentadiag}_n(-1,-1,10,-1,-1)$. Assume $n = 10$ and $\mathrm{A} = \mathrm{M} + \mathrm{N} + \mathrm{D}$, with $\mathrm{D} = \mathrm{diag}(8, \ldots, 8) \in \mathbb{R}^{10\times 10}$, $\mathrm{M} = \mathrm{pentadiag}_{10}(-1,-1,1,0,0)$ and $\mathrm{N} = \mathrm{M}^T$. To solve $\mathrm{A}\mathbf{x} = \mathbf{b}$, analyze the convergence of the following iterative methods

$$(a)\ (\mathrm{M}+\mathrm{D})\mathbf{x}^{(k+1)} = -\mathrm{N}\mathbf{x}^{(k)} + \mathbf{b},$$

$$(b)\ \mathrm{D}\mathbf{x}^{(k+1)} = -(\mathrm{M}+\mathrm{N})\mathbf{x}^{(k)} + \mathbf{b},$$

$$(c)\ (\mathrm{M}+\mathrm{N})\mathbf{x}^{(k+1)} = -\mathrm{D}\mathbf{x}^{(k)} + \mathbf{b}.$$

[*Solution* : denoting respectively by ρ_a, ρ_b and ρ_c the spectral radii of the iteration matrices of the three methods, we have $\rho_a = 0.1450$, $\rho_b = 0.5$ and $\rho_c = 12.2870$ which implies convergence for methods (a) and (b) and divergence for method (c).]

5. For the solution of the linear system $\mathbf{A}\mathbf{x} = \mathbf{b}$ with

$$\mathrm{A} = \begin{bmatrix} 1 & 2 \\ 2 & 3 \end{bmatrix}, \quad \mathbf{b} = \begin{bmatrix} 3 \\ 5 \end{bmatrix},$$

consider the following iterative method:

$$\text{given } \mathbf{x}^{(0)} \in \mathbb{R}^2, \ \mathbf{x}^{(k+1)} = \mathrm{B}(\theta)\mathbf{x}^{(k)} + \mathbf{g}(\theta), \qquad k \geq 0,$$

where θ is a real parameter and

$$\mathrm{B}(\theta) = \frac{1}{4}\begin{bmatrix} 2\theta^2 + 2\theta + 1 & -2\theta^2 + 2\theta + 1 \\ -2\theta^2 + 2\theta + 1 & 2\theta^2 + 2\theta + 1 \end{bmatrix}, \qquad \mathbf{g}(\theta) = \begin{bmatrix} \frac{1}{2} - \theta \\ \frac{1}{2} - \theta \end{bmatrix}.$$

Check that the method is consistent $\forall\, \theta \in \mathbb{R}$. Then, determine the values of θ for which the method is convergent and compute the optimal value of θ (i.e., the value of the parameter for which the convergence rate is maximum).
[*Solution* : the method is convergent iff $-1 < \theta < 1/2$ and the convergence rate is maximum if $\theta = (1 - \sqrt{3})/2$.]

6. To solve the following block linear system

$$\begin{bmatrix} \mathrm{A}_1 & \mathrm{B} \\ \mathrm{B} & \mathrm{A}_2 \end{bmatrix}\begin{bmatrix} \mathbf{x} \\ \mathbf{y} \end{bmatrix} = \begin{bmatrix} \mathbf{b}_1 \\ \mathbf{b}_2 \end{bmatrix},$$

consider the two methods

$$(1)\ \mathrm{A}_1\mathbf{x}^{(k+1)} + \mathrm{B}\mathbf{y}^{(k)} = \mathbf{b}_1,\ \mathrm{B}\mathbf{x}^{(k)} + \mathrm{A}_2\mathbf{y}^{(k+1)} = \mathbf{b}_2;$$
$$(2)\ \mathrm{A}_1\mathbf{x}^{(k+1)} + \mathrm{B}\mathbf{y}^{(k)} = \mathbf{b}_1,\ \mathrm{B}\mathbf{x}^{(k+1)} + \mathrm{A}_2\mathbf{y}^{(k+1)} = \mathbf{b}_2.$$

Find sufficient conditions in order for the two schemes to be convergent for any choice of the initial data $\mathbf{x}^{(0)}$, $\mathbf{y}^{(0)}$.
[*Solution* : method (1) is a decoupled system in the unknowns $\mathbf{x}^{(k+1)}$ and $\mathbf{y}^{(k+1)}$. Assuming that A_1 and A_2 are invertible, method (1) converges if $\rho(\mathrm{A}_1^{-1}\mathrm{B}) < 1$ and $\rho(\mathrm{A}_2^{-1}\mathrm{B}) < 1$. In the case of method (2) we have a coupled system to solve at each step in the unknowns $\mathbf{x}^{(k+1)}$ and $\mathbf{y}^{(k+1)}$. Solving formally the first equation with respect to $\mathbf{x}^{(k+1)}$ (which requires A_1 to be invertible) and substituting into the second one we see that method (2) is convergent if $\rho(\mathrm{A}_2^{-1}\mathrm{B}\mathrm{A}_1^{-1}\mathrm{B}) < 1$ (again A_2 must be invertible).]

7. Consider the linear system $\mathbf{A}\mathbf{x} = \mathbf{b}$ with

$$\mathrm{A} = \begin{bmatrix} 62 & 24 & 1 & 8 & 15 \\ 23 & 50 & 7 & 14 & 16 \\ 4 & 6 & 58 & 20 & 22 \\ 10 & 12 & 19 & 66 & 3 \\ 11 & 18 & 25 & 2 & 54 \end{bmatrix}, \qquad \mathbf{b} = \begin{bmatrix} 110 \\ 110 \\ 110 \\ 110 \\ 110 \end{bmatrix}.$$

(1) Check if the Jacobi and Gauss-Seidel methods can be applied to solve this system. (2) Check if the stationary Richardson method with optimal parameter can be applied with $\mathrm{P} = \mathrm{I}$ and $\mathrm{P} = \mathrm{D}$, where D is the diagonal part of A, and compute the corresponding values of α_{opt} and ρ_{opt}.
[*Solution* : (1): matrix A is neither diagonally dominant nor symmetric positive definite, so that we must compute the spectral radii of the iteration matrices of

the Jacobi and Gauss-Seidel methods to verify if they are convergent. It turns out that $\rho_J = 0.9280$ and $\rho_{GS} = 0.3066$ which implies convergence for both methods. (2): in the case P = I all the eigenvalues of A are positive so that the Richardson method can be applied yielding $\alpha_{opt} = 0.015$ and $\rho_{opt} = 0.6452$. If P = D the method is still applicable and $\alpha_{opt} = 0.8510$, $\rho_{opt} = 0.6407$.]

8. Consider the linear system $\mathrm{A}\mathbf{x} = \mathbf{b}$ with

$$\mathrm{A} = \begin{bmatrix} 5 & 7 & 6 & 5 \\ 7 & 10 & 8 & 7 \\ 6 & 8 & 10 & 9 \\ 5 & 7 & 9 & 10 \end{bmatrix}, \quad \mathbf{b} = \begin{bmatrix} 23 \\ 32 \\ 33 \\ 31 \end{bmatrix}.$$

Analyze the convergence properties of the Jacobi and Gauss-Seidel methods applied to the system above in their point and block forms (for a 2×2 block partition of A).
[*Solution* : both methods are convergent, the block form being the faster one. Moreover, $\rho^2(\mathrm{B}_J) = \rho(\mathrm{B}_{GS})$.]

9. To solve the linear system $\mathrm{A}\mathbf{x} = \mathbf{b}$, consider the iterative method (4.6), with $\mathrm{P} = \mathrm{D} + \omega\mathrm{F}$ and $\mathrm{N} = -\beta\mathrm{F} - \mathrm{E}$, ω and β being real numbers. Check that the method is consistent only if $\beta = 1 - \omega$. In such a case, express the eigenvalues of the iteration matrix as a function of ω and determine for which values of ω the method is convergent, as well as the value of ω_{opt}, assuming that $\mathrm{A} = \mathrm{tridiag}_{10}(-1, 2, -1)$.
[*Hint* : Take advantage of the result in Exercise 3.]

10. Let $\mathrm{A} \in \mathbb{R}^{n\times n}$ be such that $\mathrm{A} = (1+\omega)\mathrm{P} - (\mathrm{N} + \omega\mathrm{P})$, with $\mathrm{P}^{-1}\mathrm{N}$ nonsingular and with real eigenvalues $1 > \lambda_1 \geq \lambda_2 \geq \ldots \geq \lambda_n$. Find the values of $\omega \in \mathbb{R}$ for which the following iterative method

$$(1+\omega)\mathrm{P}\mathbf{x}^{(k+1)} = (\mathrm{N} + \omega\mathrm{P})\mathbf{x}^{(k)} + \mathbf{b}, \qquad k \geq 0,$$

converges $\forall \mathbf{x}^{(0)}$ to the solution of the linear system (3.2). Determine also the value of ω for which the convergence rate is maximum.
[*Solution* : $\omega > -(1 + \lambda_n)/2$; $\omega_{opt} = -(\lambda_1 + \lambda_n)/2$.]

11. Consider the linear system

$$\mathrm{A}\mathbf{x} = \mathbf{b} \quad \text{with } \mathrm{A} = \begin{bmatrix} 3 & 2 \\ 2 & 6 \end{bmatrix}, \quad \mathbf{b} = \begin{bmatrix} 2 \\ -8 \end{bmatrix}.$$

Write the associated functional $\Phi(\mathbf{x})$ and give a graphical interpretation of the solution of the linear system. Perform some iterations of the gradient method, after proving convergence for it.

12. Show that the coefficients α_k and β_k in the conjugate gradient method can be written in the alternative form (4.47).
[*Solution*: notice that $\mathrm{A}\mathbf{p}^{(k)} = (\mathbf{r}^{(k)} - \mathbf{r}^{(k+1)})/\alpha_k$ and thus $(\mathrm{A}\mathbf{p}^{(k)})^T\mathbf{r}^{(k+1)} = -\|\mathbf{r}^{(k+1)}\|_2^2/\alpha_k$. Moreover, $\alpha_k(\mathrm{A}\mathbf{p}^{(k)})^T\mathbf{p}^{(k)} = -\|\mathbf{r}^{(k)}\|_2^2$.]

13. Prove the three-terms recursive relation (4.48) for the residual in the conjugate gradient method.
[*Solution*: subtract from both sides of $\mathrm{A}\mathbf{p}^{(k)} = (\mathbf{r}^{(k)} - \mathbf{r}^{(k+1)})/\alpha_k$ the quantity $\beta_{k-1}/\alpha_k\mathbf{r}^{(k)}$ and recall that $\mathrm{A}\mathbf{p}^{(k)} = \mathrm{A}\mathbf{r}^{(k)} - \beta_{k-1}\mathrm{A}\mathbf{p}^{(k-1)}$. Then, expressing the residual $\mathbf{r}^{(k)}$ as a function of $\mathbf{r}^{(k-1)}$ one immediately gets the desired relation.]

5

Approximation of Eigenvalues and Eigenvectors

In this chapter we deal with approximations of the eigenvalues and eigenvectors of a matrix $A \in \mathbb{C}^{n\times n}$. Two main classes of numerical methods exist to this purpose, *partial* methods, which compute the *extremal* eigenvalues of A (that is, those having maximum and minimum module), or *global* methods, which approximate the whole spectrum of A.
It is worth noting that methods which are introduced to solve the matrix eigenvalue problem are not necessarily suitable for calculating the matrix eigenvectors. For example, the *power method* (a partial method, see Section 5.3) provides an approximation to a *particular* eigenvalue/eigenvector pair.

The *QR method* (a global method, see Section 5.5) instead computes the real Schur form of A, a canonical form that displays *all* the eigenvalues of A but *not* its eigenvectors. These eigenvectors can be computed, starting from the real Schur form of A, with an extra amount of work, as described in Section 5.8.2.

Finally, some *ad hoc* methods for dealing effectively with the special case where A is a symmetric $(n \times n)$ matrix are considered in Section 5.10.

5.1 Geometrical Location of the Eigenvalues

Since the eigenvalues of A are the roots of the characteristic polynomial $p_A(\lambda)$ (see Section 1.7), iterative methods must be used for their approximation when $n \geq 5$. Knowledge of eigenvalue location in the complex plane can thus be helpful in accelerating the convergence of the process.

A first estimate is provided by Theorem 1.4,

$$|\lambda| \leq \|A\|, \qquad \forall \lambda \in \sigma(A), \tag{5.1}$$

for any consistent matrix norm $\|\cdot\|$. Inequality (5.1), which is often quite rough, states that *all* the eigenvalues of A are contained in a circle of radius $R_{\|A\|} = \|A\|$ centered at the origin of the Gauss plane.

Another result is obtained by extending Definition 1.23 to complex-valued matrices.

Theorem 5.1 *If* $A \in \mathbb{C}^{n\times n}$, *let*

$$H = (A + A^H)/2 \quad \textit{and} \quad iS = (A - A^H)/2$$

be the hermitian and skew-hermitian parts of A, *respectively, i being the imaginary unit. For any* $\lambda \in \sigma(A)$

$$\lambda_{min}(H) \leq \mathrm{Re}(\lambda) \leq \lambda_{max}(H), \quad \lambda_{min}(S) \leq \mathrm{Im}(\lambda) \leq \lambda_{max}(S). \tag{5.2}$$

Proof. From the definition of H and S it follows that $A = H + iS$. Let $\mathbf{u} \in \mathbb{C}^n$, $\|\mathbf{u}\|_2 = 1$, be the eigenvector associated with the eigenvalue λ; the Rayleigh quotient (introduced in Section 1.7) reads

$$\lambda = \mathbf{u}^H A\mathbf{u} = \mathbf{u}^H H\mathbf{u} + i\mathbf{u}^H S\mathbf{u}. \tag{5.3}$$

Notice that both H and S are hermitian matrices, whilst iS is skew-hermitian. Matrices H and S are thus unitarily similar to a real diagonal matrix (see Section 1.7), and therefore their eigenvalues are real. In such a case, (5.3) yields

$$\mathrm{Re}(\lambda) = \mathbf{u}^H H\mathbf{u}, \qquad \mathrm{Im}(\lambda) = \mathbf{u}^H S\mathbf{u},$$

from which (5.2) follows. ◇

An a priori bound for the eigenvalues of A is given by the following result.

Theorem 5.2 (of the Gershgorin circles) *Let* $A \in \mathbb{C}^{n\times n}$. *Then*

$$\sigma(A) \subseteq \mathcal{S}_{\mathcal{R}} = \bigcup_{i=1}^{n} \mathcal{R}_i, \qquad \mathcal{R}_i = \{z \in \mathbb{C} : |z - a_{ii}| \leq \sum_{\substack{j=1\\ j\neq i}}^{n} |a_{ij}|\}. \tag{5.4}$$

The sets $\mathcal{R}_i$ *are called Gershgorin circles.*

Proof. Let us decompose A as $A = D + E$, where D is the diagonal part of A, whilst $e_{ii} = 0$ for $i = 1, \ldots, n$. For $\lambda \in \sigma(A)$ (with $\lambda \neq a_{ii}$, $i = 1, \ldots, n$), let us introduce the matrix $B_\lambda = A - \lambda I = (D - \lambda I) + E$. Since B_λ is singular, there exists a non-null vector $\mathbf{x} \in \mathbb{C}^n$ such that $B_\lambda \mathbf{x} = \mathbf{0}$. This means that $((D - \lambda I) + E)\,\mathbf{x} = \mathbf{0}$, that is, passing to the $\|\cdot\|_\infty$ norm,

$$\mathbf{x} = -(D - \lambda I)^{-1}E\mathbf{x}, \qquad \|\mathbf{x}\|_\infty \leq \|(D - \lambda I)^{-1}E\|_\infty \|\mathbf{x}\|_\infty,$$

and thus

$$1 \leq \|(D - \lambda I)^{-1}E\|_\infty = \sum_{j=1}^{n} \frac{|e_{kj}|}{|a_{kk} - \lambda|} = \sum_{\substack{j=1\\ j\neq k}}^{n} \frac{|a_{kj}|}{|a_{kk} - \lambda|}, \tag{5.5}$$

for a certain k, $1 \leq k \leq n$. Inequality (5.5) implies $\lambda \in \mathcal{R}_k$ and thus (5.4). ◇

The bounds (5.4) ensure that any eigenvalue of A lies within the union of the circles $\mathcal{R}_i$. Moreover, since A and A^T share the same spectrum, Theorem 5.2 also holds in the form

$$\sigma(A) \subseteq \mathcal{S}_{\mathcal{C}} = \bigcup_{j=1}^{n} \mathcal{C}_j, \qquad \mathcal{C}_j = \{z \in \mathbb{C} : |z - a_{jj}| \le \sum_{\substack{i=1 \\ i \ne j}}^{n} |a_{ij}|\}. \tag{5.6}$$

The circles $\mathcal{R}_i$ in the complex plane are called row circles, and $\mathcal{C}_j$ column circles. The immediate consequence of (5.4) and (5.6) is the following.

Property 5.1 (First Gershgorin theorem) *For a given matrix* $A \in \mathbb{C}^{n\times n}$,

$$\forall \lambda \in \sigma(A), \qquad \lambda \in \mathcal{S}_{\mathcal{R}} \bigcap \mathcal{S}_{\mathcal{C}}. \tag{5.7}$$

The following two location theorems can also be proved (see [Atk89], pp. 588-590 and [Hou75], pp. 66-67).

Property 5.2 (Second Gershgorin theorem) *Let*

$$\mathcal{S}_1 = \bigcup_{i=1}^{m} \mathcal{R}_i, \quad \mathcal{S}_2 = \bigcup_{i=m+1}^{n} \mathcal{R}_i.$$

If $\mathcal{S}_1 \cap \mathcal{S}_2 = \emptyset$*, then* $\mathcal{S}_1$ *contains exactly* m *eigenvalues of* A*, each one being accounted for with its algebraic multiplicity, while the remaining eigenvalues are contained in* $\mathcal{S}_2$*.*

Remark 5.1 Properties 5.1 and 5.2 do not exclude the possibility that there exist circles containing no eigenvalues, as happens for the matrix in Exercise 1. ■

Definition 5.1 A matrix $A \in \mathbb{C}^{n\times n}$ is called *reducible* if there exists a permutation matrix P such that

$$PAP^T = \begin{bmatrix} B_{11} & B_{12} \\ 0 & B_{22} \end{bmatrix},$$

where B_{11} and B_{22} are square matrices; A is *irreducible* if it is not reducible. ■

To check if a matrix is reducible, the *oriented graph* of the matrix can be conveniently employed. Recall from Section 3.9 that the oriented graph of a real matrix A is obtained by joining n points (called vertices of the graph) $P_1, \ldots, P_n$ through a line oriented from P_i to P_j if the corresponding matrix entry $a_{ij} \neq 0$. An oriented graph is *strongly connected* if for any pair of distinct vertices P_i and P_j there exists an oriented path from P_i to P_j. The following result holds (see [Var62] for the proof).

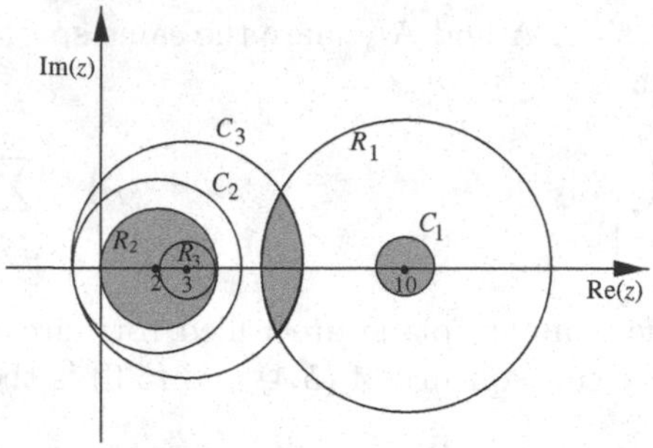

Fig. 5.1. Row and column circles for matrix A in Example 5.1

Property 5.3 *A matrix* $A \in \mathbb{C}^{n\times n}$ *is irreducible iff its oriented graph is strongly connected.*

Property 5.4 (Third Gershgorin theorem) *Let* $A \in \mathbb{C}^{n\times n}$ *be an irreducible matrix. An eigenvalue* $\lambda \in \sigma(A)$ *cannot lie on the boundary of* $\mathcal{S}_{\mathcal{R}}$ *unless it belongs to the boundary of every circle* $\mathcal{R}_i$, *for* $i = 1, \ldots, n$.

Example 5.1 Let us consider the matrix

$$A = \begin{bmatrix} 10 & 2 & 3 \\ -1 & 2 & -1 \\ 0 & 1 & 3 \end{bmatrix},$$

whose spectrum is (to four significant figures) $\sigma(A) = \{9.687, 2.656 \pm i0.693\}$. The following values of the norm of A: $\|A\|_1 = 11$, $\|A\|_2 = 10.72$, $\|A\|_\infty = 15$ and $\|A\|_F = 11.36$ can be used in the estimate (5.1). Estimate (5.2) provides instead $1.96 \le \mathrm{Re}(\lambda(A)) \le 10.34$, $-2.34 \le \mathrm{Im}(\lambda(A)) \le 2.34$, while the row and column circles are given respectively by $\mathcal{R}_1 = \{|z| : |z-10| \le 5\}$, $\mathcal{R}_2 = \{|z| : |z-2| \le 2\}$, $\mathcal{R}_3 = \{|z| : |z-3| \le 1\}$ and $\mathcal{C}_1 = \{|z| : |z-10| \le 1\}$, $\mathcal{C}_2 = \{|z| : |z-2| \le 3\}$, $\mathcal{C}_3 = \{|z| : |z-3| \le 4\}$.
In Figure 5.1, for $i = 1, 2, 3$ the $\mathcal{R}_i$ and $\mathcal{C}_i$ circles and the intersection $\mathcal{S}_{\mathcal{R}} \cap \mathcal{S}_{\mathcal{C}}$ (shaded areas) are drawn. In agreement with Property 5.2, we notice that an eigenvalue is contained in $\mathcal{C}_1$, which is disjoint from $\mathcal{C}_2$ and $\mathcal{C}_3$, while the remaining eigenvalues, thanks to Property 5.1, lie within the set $\mathcal{R}_2 \cup \{\mathcal{C}_3 \cap \mathcal{R}_1\}$. •

5.2 Stability and Conditioning Analysis

In this section we introduce some a priori and a posteriori estimates that are relevant in the stability analysis of the matrix eigenvalue and eigenvector problem. The presentation follows the guidelines that have been traced in Chapter 2.

5.2.1 A priori Estimates

Assume that $A \in \mathbb{C}^{n\times n}$ is a diagonalizable matrix and denote by $X = (\mathbf{x}_1, \ldots, \mathbf{x}_n) \in \mathbb{C}^{n\times n}$ the matrix of its right eigenvectors, where $\mathbf{x}_k \in \mathbb{C}^n$ for $k = 1, \ldots, n$, such that $D = X^{-1}AX = \mathrm{diag}(\lambda_1, \ldots, \lambda_n)$, λ_i being the eigenvalues of A, $i = 1, \ldots, n$. Moreover, let $E \in \mathbb{C}^{n\times n}$ be a perturbation of A. The following theorem holds.

Theorem 5.3 (Bauer-Fike) *Let μ be an eigenvalue of the matrix $A + E \in \mathbb{C}^{n\times n}$; then*

$$\min_{\lambda\in\sigma(A)} |\lambda - \mu| \leq K_p(X)\|E\|_p, \tag{5.8}$$

where $\|\cdot\|_p$ is any matrix p-norm and $K_p(X) = \|X\|_p\|X^{-1}\|_p$ is called the condition number of the eigenvalue problem for matrix A.

Proof. We first notice that if $\mu \in \sigma(A)$ then (5.8) is trivially verified, since $\|X\|_p\|X^{-1}\|_p\|E\|_p \geq 0$. Let us thus assume henceforth that $\mu \notin \sigma(A)$. From the definition of eigenvalue it follows that matrix $(A + E - \mu I)$ is singular, which means that, since X is invertible, the matrix $X^{-1}(A + E - \mu I)X = D + X^{-1}EX - \mu I$ is singular. Therefore, there exists a non-null vector $\mathbf{x} \in \mathbb{C}^n$ such that

$$\left((D - \mu I) + X^{-1}EX\right)\mathbf{x} = \mathbf{0}.$$

Since $\mu \notin \sigma(A)$, the diagonal matrix $(D - \mu I)$ is invertible and the previous equation can be written in the form

$$\left(I + (D - \mu I)^{-1}(X^{-1}EX)\right)\mathbf{x} = \mathbf{0}.$$

Passing to the $\|\cdot\|_p$ norm and proceeding as in the proof of Theorem 5.2, we get

$$1 \leq \|(D - \mu I)^{-1}\|_p K_p(X)\|E\|_p,$$

from which the estimate (5.8) follows, since

$$\|(D - \mu I)^{-1}\|_p = (\min_{\lambda\in\sigma(A)} |\lambda - \mu|)^{-1}.$$

◇

If A is a *normal* matrix, from the Schur decomposition theorem (see Section 1.8) it follows that the similarity transformation matrix X is unitary so that $K_2(X) = 1$. This implies that

$$\forall \mu \in \sigma(A + E), \quad \min_{\lambda\in\sigma(A)} |\lambda - \mu| \leq \|E\|_2, \tag{5.9}$$

hence the eigenvalue problem is *well-conditioned* with respect to the absolute error. This, however, does not prevent the matrix eigenvalue problem from being affected by significant *relative* errors, especially when A has a widely spread spectrum.

Table 5.1. Relative and absolute errors in the calculation of the eigenvalues of the Hilbert matrix (using the MATLAB intrinsic function `eig`). "Abs. Err." and "Rel. Err." denote respectively the absolute and relative errors (with respect to λ)

n	Abs. Err.	Rel. Err.	$\Vert \mathrm{E}_n \Vert_2$	$K_2(\mathrm{H}_n)$	$K_2(\mathrm{H}_n + \mathrm{E}_n)$
1	$1 \cdot 10^{-3}$	$1 \cdot 10^{-3}$	$1 \cdot 10^{-3}$	$1 \cdot 10^{-3}$	1
2	$1.677 \cdot 10^{-4}$	$1.446 \cdot 10^{-3}$	$2 \cdot 10^{-3}$	19.28	19.26
4	$5.080 \cdot 10^{-7}$	$2.207 \cdot 10^{-3}$	$4 \cdot 10^{-3}$	$1.551 \cdot 10^{4}$	$1.547 \cdot 10^{4}$
8	$1.156 \cdot 10^{-12}$	$3.496 \cdot 10^{-3}$	$8 \cdot 10^{-3}$	$1.526 \cdot 10^{10}$	$1.515 \cdot 10^{10}$
10	$1.355 \cdot 10^{-15}$	$4.078 \cdot 10^{-3}$	$1 \cdot 10^{-2}$	$1.603 \cdot 10^{13}$	$1.589 \cdot 10^{13}$

Example 5.2 Let us consider, for $1 \leq n \leq 10$, the calculation of the eigenvalues of the Hilbert matrix $\mathrm{H}_n \in \mathbb{R}^{n \times n}$ (see Example 3.2, Chapter 3). It is symmetric (thus, in particular, normal) and exhibits, for $n \geq 4$, a very large condition number. Let $\mathrm{E}_n \in \mathbb{R}^{n \times n}$ be a matrix having constant entries equal to $\eta = 10^{-3}$. We show in Table 5.1 the results of the computation of the minimum in (5.9). Notice how the absolute error is decreasing, since the eigenvalue of minimum module tends to zero, whilst the relative error is increasing as the size n of the matrix increases, due to the higher sensitivity of "small" eigenvalues with respect to rounding errors. •

The Bauer-Fike theorem states that the matrix eigenvalue problem is well-conditioned if A is a normal matrix. Failure to fulfil this property, however, does not necessarily imply that A must exhibit a "strong" numerical sensitivity to the computation of *every one* of its eigenvalues. In this respect, the following result holds, which can be regarded as an a priori estimate of the conditioning of the calculation of a particular eigenvalue of a matrix.

Theorem 5.4 *Let* $\mathrm{A} \in \mathbb{C}^{n \times n}$ *be a diagonalizable matrix; let* λ, $\mathbf{x}$ *and* $\mathbf{y}$ *be a simple eigenvalue of* A *and its associated right and left eigenvectors, respectively, with* $\Vert \mathbf{x} \Vert_2 = \Vert \mathbf{y} \Vert_2 = 1$. *Moreover, for* $\varepsilon > 0$, *let* $\mathrm{A}(\varepsilon) = \mathrm{A} + \varepsilon \mathrm{E}$, *with* $\mathrm{E} \in \mathbb{C}^{n \times n}$ *such that* $\Vert \mathrm{E} \Vert_2 = 1$. *Denoting by* $\lambda(\varepsilon)$ *and* $\mathbf{x}(\varepsilon)$ *the eigenvalue and the corresponding eigenvector of* $\mathrm{A}(\varepsilon)$, *such that* $\lambda(0) = \lambda$ *and* $\mathbf{x}(0) = \mathbf{x}$, *we have*

$$\left| \frac{\partial \lambda}{\partial \varepsilon}(0) \right| \leq \frac{1}{|\mathbf{y}^H \mathbf{x}|}. \tag{5.10}$$

Proof. Let us first prove that $\mathbf{y}^H \mathbf{x} \neq 0$. Setting $\mathrm{Y} = (\mathbf{y}_1, \ldots, \mathbf{y}_n) = (\mathrm{X}^H)^{-1}$, with $\mathbf{y}_k \in \mathbb{C}^n$ for $k = 1, \ldots, n$, it follows that $\mathbf{y}_k^H \mathrm{A} = \lambda_k \mathbf{y}_k^H$, i.e., the rows of $\mathrm{X}^{-1} = \mathrm{Y}^H$ are left eigenvectors of A. Then, since $\mathrm{Y}^H \mathrm{X} = \mathrm{I}$, $\mathbf{y}_i^H \mathbf{x}_j = \delta_{ij}$ for $i, j = 1, \ldots, n$, δ_{ij} being the Kronecker symbol. This result is equivalent to saying that the eigenvectors $\{\mathbf{x}\}$ of A and the eigenvectors $\{\mathbf{y}\}$ of A^H form a *bi-orthogonal* set (see (4.69)).

Let us now prove (5.10). Since the roots of the characteristic equation are continuous functions of the coefficients of the characteristic polynomial associated with $\mathrm{A}(\varepsilon)$, it follows that the eigenvalues of $\mathrm{A}(\varepsilon)$ are continuous functions of ε (see, for instance, [Hen74], p. 281). Therefore, in a neighborhood of $\varepsilon = 0$,

$$(\mathrm{A} + \varepsilon \mathrm{E}) \mathbf{x}(\varepsilon) = \lambda(\varepsilon) \mathbf{x}(\varepsilon).$$

Differentiating the previous equation with respect to ε and setting $\varepsilon = 0$ yields

$$\mathrm{A}\frac{\partial \mathbf{x}}{\partial \varepsilon}(0) + \mathrm{E}\mathbf{x} = \frac{\partial \lambda}{\partial \varepsilon}(0)\mathbf{x} + \lambda\frac{\partial \mathbf{x}}{\partial \varepsilon}(0),$$

from which, left-multiplying both sides by $\mathbf{y}^H$ and recalling that $\mathbf{y}^H$ is a left eigenvector of A,

$$\frac{\partial \lambda}{\partial \varepsilon}(0) = \frac{\mathbf{y}^H \mathrm{E}\mathbf{x}}{\mathbf{y}^H \mathbf{x}}.$$

Using the Cauchy-Schwarz inequality gives the desired estimate (5.10). ◇

Notice that $|\mathbf{y}^H\mathbf{x}| = |\cos(\theta_\lambda)|$, where θ_λ is the angle between the eigenvectors $\mathbf{y}^H$ and $\mathbf{x}$ (both having unit Euclidean norm). Therefore, if these two vectors are almost orthogonal the computation of the eigenvalue λ turns out to be ill-conditioned. The quantity

$$\kappa(\lambda) = \frac{1}{|\mathbf{y}^H\mathbf{x}|} = \frac{1}{|\cos(\theta_\lambda)|} \tag{5.11}$$

can thus be taken as the *condition number of the eigenvalue* λ. Obviously, $\kappa(\lambda) \geq 1$; when A is a normal matrix, since it is unitarily similar to a diagonal matrix, the left and right eigenvectors $\mathbf{y}$ and $\mathbf{x}$ coincide, yielding $\kappa(\lambda) = 1/\|\mathbf{x}\|_2^2 = 1$.

Inequality (5.10) can be roughly interpreted as stating that perturbations of the order of $\delta\varepsilon$ in the entries of matrix A induce changes of the order of $\delta\lambda = \delta\varepsilon/|\cos(\theta_\lambda)|$ in the eigenvalue λ. If normal matrices are considered, the calculation of λ is a well-conditioned problem; the case of a generic non-symmetric matrix A can be conveniently dealt with using methods based on *similarity transformations*, as will be seen in later sections.

It is interesting to check that the conditioning of the matrix eigenvalue problem remains *unchanged* if the transformation matrices are *unitary*. To this end, let $\mathrm{U} \in \mathbb{C}^{n\times n}$ be a unitary matrix and let $\widetilde{\mathrm{A}} = \mathrm{U}^H\mathrm{A}\mathrm{U}$. Also let λ_j be an eigenvalue of A and denote by κ_j the condition number (5.11). Moreover, let $\widetilde{\kappa}_j$ be the condition number of λ_j when it is regarded as an eigenvalue of $\widetilde{\mathrm{A}}$. Finally, let $\{\mathbf{x}_k\}$, $\{\mathbf{y}_k\}$ be the right and left eigenvectors of A respectively. Clearly, $\{\mathrm{U}^H\mathbf{x}_k\}$, $\{\mathrm{U}^H\mathbf{y}_k\}$ are the right and left eigenvectors of $\widetilde{\mathrm{A}}$. Thus, for any $j = 1, \ldots, n$,

$$\widetilde{\kappa}_j = \left|\mathbf{y}_j^H \mathrm{U}\mathrm{U}^H \mathbf{x}_j\right|^{-1} = \kappa_j,$$

from which it follows that the stability of the computation of λ_j is *not* affected by performing similarity transformations using unitary matrices. It can also be checked that unitary transformation matrices do not change the Euclidean length and the angles between vectors in $\mathbb{C}^n$. Moreover, the following a priori estimate holds (see [GL89], p. 317)

$$fl\left(\mathrm{X}^{-1}\mathrm{A}\mathrm{X}\right) = \mathrm{X}^{-1}\mathrm{A}\mathrm{X} + \mathrm{E}, \text{ with } \|\mathrm{E}\|_2 \simeq \mathrm{u}K_2(\mathrm{X})\|\mathrm{A}\|_2, \tag{5.12}$$

where $fl(\mathrm{M})$ is the machine representation of matrix M and $\mathbf{u}$ is the *roundoff* unit (see Section 2.5). From (5.12) it follows that using *nonunitary* transformation matrices in the eigenvalue computation can lead to an unstable process with respect to rounding errors.

We conclude this section with a stability result for the approximation of the eigenvector associated with a simple eigenvalue. Under the same assumptions of Theorem 5.4, the following result holds (see for the proof, [Atk89], Problem 6, pp. 649-650).

Property 5.5 *The eigenvectors* $\mathbf{x}_k$ *and* $\mathbf{x}_k(\varepsilon)$ *of the matrices* A *and* $\mathrm{A}(\varepsilon) = \mathrm{A} + \varepsilon\mathrm{E}$, *with* $\|\mathbf{x}_k(\varepsilon)\|_2 = \|\mathbf{x}_k\|_2 = 1$ *for* $k = 1, \ldots, n$, *satisfy*

$$\|\mathbf{x}_k(\varepsilon) - \mathbf{x}_k\|_2 \leq \frac{\varepsilon}{\min_{j \neq k} |\lambda_k - \lambda_j|} + \mathcal{O}(\varepsilon^2), \qquad \forall k = 1, \ldots, n.$$

Analogous to (5.11), the quantity

$$\kappa(\mathbf{x}_k) = \frac{1}{\min_{j \neq k} |\lambda_k - \lambda_j|}$$

can be regarded as being the *condition number of the eigenvector* $\mathbf{x}_k$. Computing $\mathbf{x}_k$ might be an ill-conditioned operation if some eigenvalues λ_j are "very close" to the eigenvalue λ_k associated with $\mathbf{x}_k$.

5.2.2 A posteriori Estimates

The a priori estimates examined in the previous section characterize the stability properties of the matrix eigenvalue and eigenvector problem. From the implementation standpoint, it is also important to dispose of a posteriori estimates that allow for a run-time control of the quality of the approximation that is being constructed. Since the methods that will be considered later are iterative processes, the results of this section can be usefully employed to devise reliable stopping criteria for these latter.

Theorem 5.5 *Let* $\mathrm{A} \in \mathbb{C}^{n \times n}$ *be an hermitian matrix and let* $(\widehat{\lambda}, \widehat{\mathbf{x}})$ *be the computed approximations of an eigenvalue/eigenvector pair* $(\lambda, \mathbf{x})$ *of* A. *Defining the residual as*

$$\widehat{\mathbf{r}} = \mathrm{A}\widehat{\mathbf{x}} - \widehat{\lambda}\widehat{\mathbf{x}}, \qquad \widehat{\mathbf{x}} \neq \mathbf{0},$$

it then follows that

$$\min_{\lambda_i \in \sigma(\mathrm{A})} |\widehat{\lambda} - \lambda_i| \leq \frac{\|\widehat{\mathbf{r}}\|_2}{\|\widehat{\mathbf{x}}\|_2}. \qquad (5.13)$$

Proof. Since A is hermitian, it admits a system of orthonormal eigenvectors $\{\mathbf{u}_k\}$ which can be taken as a basis of $\mathbb{C}^n$. In particular, $\widehat{\mathbf{x}} = \sum_{i=1}^{n} \alpha_i \mathbf{u}_i$ with $\alpha_i = \mathbf{u}_i^H \widehat{\mathbf{x}}$, and thus $\widehat{\mathbf{r}} = \sum_{i=1}^{n} \alpha_i (\lambda_i - \widehat{\lambda}) \mathbf{u}_i$. As a consequence

$$\left(\frac{\|\widehat{\mathbf{r}}\|_2}{\|\widehat{\mathbf{x}}\|_2}\right)^2 = \sum_{i=1}^{n} \beta_i(\lambda_i - \widehat{\lambda})^2, \quad \text{with } \beta_i = |\alpha_i|^2/(\sum_{j=1}^{n}|\alpha_j|^2). \tag{5.14}$$

Since $\sum_{i=1}^{n} \beta_i = 1$, the inequality (5.13) immediately follows from (5.14). ◇

The estimate (5.13) ensures that a small *absolute error* corresponds to a small *relative residual* in the computation of the eigenvalue of the matrix A which is closest to $\widehat{\lambda}$.
Let us now consider the following a posteriori estimate for the eigenvector $\widehat{\mathbf{x}}$ (for the proof, see [IK66], pp. 142-143).

Property 5.6 *Under the same assumptions of Theorem* 5.5, *suppose that* $|\lambda_i - \widehat{\lambda}| \le \|\widehat{\mathbf{r}}\|_2$ *for* $i = 1, \ldots, m$ *and that* $|\lambda_i - \widehat{\lambda}| \ge \delta > 0$ *for* $i = m+1, \ldots, n$. *Then*

$$d(\widehat{\mathbf{x}}, \mathbf{U}_m) \le \frac{\|\widehat{\mathbf{r}}\|_2}{\delta}, \tag{5.15}$$

where $d(\widehat{\mathbf{x}}, \mathbf{U}_m)$ *is the Euclidean distance between* $\widehat{\mathbf{x}}$ *and the space* $\mathbf{U}_m$ *generated by the eigenvectors* $\mathbf{u}_i$, $i = 1, \ldots, m$, *associated with the eigenvalues* λ_i *of* A.

Notice that the a posteriori estimate (5.15) ensures that a small *absolute error* corresponds to a small *residual* in the approximation of the eigenvector associated with the eigenvalue of A that is closest to $\widehat{\lambda}$, provided that the eigenvalues of A are well-separated (that is, if δ is sufficiently large).
In the general case of a nonhermitian matrix A, an a posteriori estimate can be given for the eigenvalue $\widehat{\lambda}$ only when the matrix of the eigenvectors of A is available. We have the following result (for the proof, we refer to [IK66], p. 146).

Property 5.7 *Let* $\mathrm{A} \in \mathbb{C}^{n\times n}$ *be a diagonalizable matrix, with matrix of eigenvectors* $\mathrm{X} = [\mathbf{x}_1, \ldots, \mathbf{x}_n]$. *If, for some* $\varepsilon > 0$,

$$\|\widehat{\mathbf{r}}\|_2 \le \varepsilon \|\widehat{\mathbf{x}}\|_2,$$

then

$$\min_{\lambda_i \in \sigma(\mathrm{A})} |\widehat{\lambda} - \lambda_i| \le \varepsilon \|\mathrm{X}^{-1}\|_2 \|\mathrm{X}\|_2.$$

This estimate is of little practical use, since it requires the knowledge of all the eigenvectors of A. Examples of a posteriori estimates that can actually be implemented in a numerical algorithm will be provided in Sections 5.3.1 and 5.3.2.

5.3 The Power Method

The *power method* is very good at approximating the *extremal* eigenvalues of the matrix, that is, the eigenvalues having largest and smallest module, denoted by λ_1 and λ_n respectively, as well as their associated eigenvectors.

Solving such a problem is of great interest in several real-life applications (geosysmic, machine and structural vibrations, electric network analysis, quantum mechanics,...) where the computation of λ_n (and its associated eigenvector $\mathbf{x}_n$) arises in the determination of the *proper frequency* (and the corresponding *fundamental mode*) of a given physical system. We shall come back to this point in Section 5.12.

Having approximations of λ_1 and λ_n can also be useful in the analysis of numerical methods. For instance, if A is symmetric and positive definite, one can compute the optimal value of the acceleration parameter of the Richardson method and estimate its error reducing factor (see Chapter 4), as well as perform the stability analysis of discretization methods for systems of ordinary differential equations (see Chapter 11).

5.3.1 Approximation of the Eigenvalue of Largest Module

Let $\mathrm{A} \in \mathbb{C}^{n\times n}$ be a diagonalizable matrix and let $\mathrm{X} \in \mathbb{C}^{n\times n}$ be the matrix of its right eigenvectors $\mathbf{x}_i$, for $i = 1, \ldots, n$. Let us also suppose that the eigenvalues of A are ordered as

$$|\lambda_1| > |\lambda_2| \geq |\lambda_3| \ldots \geq |\lambda_n|, \tag{5.16}$$

where λ_1 has algebraic multiplicity equal to 1. Under these assumptions, λ_1 is called the *dominant* eigenvalue of matrix A.

Given an arbitrary initial vector $\mathbf{q}^{(0)} \in \mathbb{C}^n$ of unit Euclidean norm, consider for $k = 1, 2, \ldots$ the following iteration based on the computation of powers of matrices, commonly known as the *power method*:

$$\begin{aligned} \mathbf{z}^{(k)} &= \mathrm{A}\mathbf{q}^{(k-1)}, \\ \mathbf{q}^{(k)} &= \mathbf{z}^{(k)}/\|\mathbf{z}^{(k)}\|_2, \\ \nu^{(k)} &= (\mathbf{q}^{(k)})^H \mathrm{A}\mathbf{q}^{(k)}. \end{aligned} \tag{5.17}$$

Let us analyze the convergence properties of method (5.17). By induction on k one can check that

$$\mathbf{q}^{(k)} = \frac{\mathrm{A}^k \mathbf{q}^{(0)}}{\|\mathrm{A}^k \mathbf{q}^{(0)}\|_2}, \qquad k \geq 1. \tag{5.18}$$

This relation explains the role played by the powers of A in the method. Because A is diagonalizable, its eigenvectors $\mathbf{x}_i$ form a basis of $\mathbb{C}^n$; it is thus possible to represent $\mathbf{q}^{(0)}$ as

$$\mathbf{q}^{(0)} = \sum_{i=1}^{n} \alpha_i \mathbf{x}_i, \qquad \alpha_i \in \mathbb{C}, \qquad i = 1, \ldots, n. \tag{5.19}$$

Moreover, since $\mathrm{A}\mathbf{x}_i = \lambda_i \mathbf{x}_i$, we have

$$\mathrm{A}^k \mathbf{q}^{(0)} = \alpha_1 \lambda_1^k \left(\mathbf{x}_1 + \sum_{i=2}^{n} \frac{\alpha_i}{\alpha_1} \left(\frac{\lambda_i}{\lambda_1} \right)^k \mathbf{x}_i \right), \; k = 1, 2, \ldots \tag{5.20}$$

Since $|\lambda_i/\lambda_1| < 1$ for $i = 2, \ldots, n$, as k increases the vector $\mathrm{A}^k \mathbf{q}^{(0)}$ (and thus also $\mathbf{q}^{(k)}$, due to (5.18)), tends to assume an increasingly significant component in the direction of the eigenvector $\mathbf{x}_1$, while its components in the other directions $\mathbf{x}_j$ decrease. Using (5.18) and (5.20), we get

$$\mathbf{q}^{(k)} = \frac{\alpha_1 \lambda_1^k (\mathbf{x}_1 + \mathbf{y}^{(k)})}{\|\alpha_1 \lambda_1^k (\mathbf{x}_1 + \mathbf{y}^{(k)})\|_2} = \mu_k \frac{\mathbf{x}_1 + \mathbf{y}^{(k)}}{\|\mathbf{x}_1 + \mathbf{y}^{(k)}\|_2},$$

where μ_k is the sign of $\alpha_1 \lambda_1^k$ and $\mathbf{y}^{(k)}$ denotes a vector that vanishes as $k \to \infty$.

As $k \to \infty$, the vector $\mathbf{q}^{(k)}$ thus aligns itself along the direction of eigenvector $\mathbf{x}_1$, and the following error estimate holds at each step k.

Theorem 5.6 *Let* $\mathrm{A} \in \mathbb{C}^{n \times n}$ *be a diagonalizable matrix whose eigenvalues satisfy* (5.16). *Assuming that* $\alpha_1 \neq 0$, *there exists a constant* $C > 0$ *such that*

$$\|\tilde{\mathbf{q}}^{(k)} - \mathbf{x}_1\|_2 \leq C \left| \frac{\lambda_2}{\lambda_1} \right|^k, \qquad k \geq 1, \tag{5.21}$$

where

$$\tilde{\mathbf{q}}^{(k)} = \frac{\mathbf{q}^{(k)} \|\mathrm{A}^k \mathbf{q}^{(0)}\|_2}{\alpha_1 \lambda_1^k} = \mathbf{x}_1 + \sum_{i=2}^{n} \frac{\alpha_i}{\alpha_1} \left(\frac{\lambda_i}{\lambda_1} \right)^k \mathbf{x}_i, \qquad k = 1, 2, \ldots \tag{5.22}$$

Proof. Since A is diagonalizable, without losing generality, we can pick up the nonsingular matrix X in such a way that its columns have unit Euclidean length, that is $\|\mathbf{x}_i\|_2 = 1$ for $i = 1, \ldots, n$. From (5.20) it thus follows that

$$\begin{aligned} \|\mathbf{x}_1 + \sum_{i=2}^{n} \left[\frac{\alpha_i}{\alpha_1} \left(\frac{\lambda_i}{\lambda_1} \right)^k \mathbf{x}_i \right] - \mathbf{x}_1\|_2 &= \|\sum_{i=2}^{n} \frac{\alpha_i}{\alpha_1} \left(\frac{\lambda_i}{\lambda_1} \right)^k \mathbf{x}_i\|_2 \\ \leq \left(\sum_{i=2}^{n} \left[\frac{\alpha_i}{\alpha_1} \right]^2 \left[\frac{\lambda_i}{\lambda_1} \right]^{2k} \right)^{1/2} &\leq \left| \frac{\lambda_2}{\lambda_1} \right|^k \left(\sum_{i=2}^{n} \left[\frac{\alpha_i}{\alpha_1} \right]^2 \right)^{1/2}, \end{aligned}$$

that is (5.21) with $C = \left(\sum_{i=2}^{n} (\alpha_i/\alpha_1)^2 \right)^{1/2}$. ◇

Estimate (5.21) expresses the convergence of the sequence $\tilde{\mathbf{q}}^{(k)}$ towards $\mathbf{x}_1$. Therefore the sequence of Rayleigh quotients

$$((\tilde{\mathbf{q}}^{(k)})^H \mathrm{A}\tilde{\mathbf{q}}^{(k)})/\|\tilde{\mathbf{q}}^{(k)}\|_2^2 = \left(\mathbf{q}^{(k)}\right)^H \mathrm{A}\mathbf{q}^{(k)} = \nu^{(k)}$$

will converge to λ_1. As a consequence, $\lim_{k\to\infty} \nu^{(k)} = \lambda_1$, and the convergence will be faster when the ratio $|\lambda_2/\lambda_1|$ is smaller.

If the matrix A is *real* and *symmetric* it can be proved, always assuming that $\alpha_1 \neq 0$, that (see [GL89], pp. 406-407)

$$|\lambda_1 - \nu^{(k)}| \leq |\lambda_1 - \lambda_n| \tan^2(\theta_0) \left|\frac{\lambda_2}{\lambda_1}\right|^{2k}, \qquad (5.23)$$

where $\cos(\theta_0) = |\mathbf{x}_1^T \mathbf{q}^{(0)}| \neq 0$. Inequality (5.23) outlines that the convergence of the sequence $\nu^{(k)}$ to λ_1 is *quadratic* with respect to the ratio $|\lambda_2/\lambda_1|$ (we refer to Section 5.3.3 for numerical results).

We conclude the section by providing a stopping criterion for the iteration (5.17). For this purpose, let us introduce the residual at step k

$$\mathbf{r}^{(k)} = \mathrm{A}\mathbf{q}^{(k)} - \nu^{(k)}\mathbf{q}^{(k)}, \qquad k \geq 1,$$

and, for $\varepsilon > 0$, the matrix $\varepsilon \mathrm{E}^{(k)} = -\mathbf{r}^{(k)} \left[\mathbf{q}^{(k)}\right]^H \in \mathbb{C}^{n\times n}$ with $\|\mathrm{E}^{(k)}\|_2 = 1$. Since

$$\varepsilon \mathrm{E}^{(k)} \mathbf{q}^{(k)} = -\mathbf{r}^{(k)}, \qquad k \geq 1, \qquad (5.24)$$

we obtain $\left(\mathrm{A} + \varepsilon \mathrm{E}^{(k)}\right) \mathbf{q}^{(k)} = \nu^{(k)}\mathbf{q}^{(k)}$. As a result, at each step of the power method $\nu^{(k)}$ is an *eigenvalue of the perturbed matrix* $\mathrm{A}+\varepsilon\mathrm{E}^{(k)}$. From (5.24) and from definition (1.20) it also follows that $\varepsilon = \|\mathbf{r}^{(k)}\|_2$ for $k = 1, 2, \ldots$. Plugging this identity back into (5.10) and approximating the partial derivative in (5.10) by the incremental ratio $|\lambda_1 - \nu^{(k)}|/\varepsilon$, we get

$$|\lambda_1 - \nu^{(k)}| \simeq \frac{\|\mathbf{r}^{(k)}\|_2}{|\cos(\theta_\lambda)|}, \qquad k \geq 1, \qquad (5.25)$$

where θ_λ is the angle between the right and the left eigenvectors, $\mathbf{x}_1$ and $\mathbf{y}_1$, associated with λ_1. Notice that, if A is an hermitian matrix, then $\cos(\theta_\lambda) = 1$, so that (5.25) yields an estimate which is analogue to (5.13).

In practice, in order to employ the estimate (5.25) it is necessary at each step k to replace $|\cos(\theta_\lambda)|$ with the module of the scalar product between two approximations $\mathbf{q}^{(k)}$ and $\mathbf{w}^{(k)}$ of $\mathbf{x}_1$ and $\mathbf{y}_1$, computed by the power method. The following a posteriori estimate is thus obtained

$$|\lambda_1 - \nu^{(k)}| \simeq \frac{\|\mathbf{r}^{(k)}\|_2}{|(\mathbf{w}^{(k)})^H \mathbf{q}^{(k)}|}, \qquad k \geq 1. \qquad (5.26)$$

Examples of applications of (5.26) will be provided in Section 5.3.3.

5.3.2 Inverse Iteration

In this section we look for an approximation of the eigenvalue of a matrix $A \in \mathbb{C}^{n\times n}$ which is *closest* to a given number $\mu \in \mathbb{C}$, where $\mu \notin \sigma(A)$. For this, the power iteration (5.17) can be applied to the matrix $(M_\mu)^{-1} = (A - \mu I)^{-1}$, yielding the so-called *inverse iteration* or *inverse power method.* The number μ is called a *shift.*

The eigenvalues of M_μ^{-1} are $\xi_i = (\lambda_i - \mu)^{-1}$; let us assume that there exists an integer m such that

$$|\lambda_m - \mu| < |\lambda_i - \mu|, \qquad \forall i = 1,\dots,n \qquad \text{and } i \neq m. \tag{5.27}$$

This amounts to requiring that the eigenvalue λ_m which is closest to μ has multiplicity equal to 1. Moreover, (5.27) shows that ξ_m is the eigenvalue of M_μ^{-1} with largest module; in particular, if $\mu = 0$, λ_m turns out to be the eigenvalue of A with smallest module.

Given an arbitrary initial vector $\mathbf{q}^{(0)} \in \mathbb{C}^n$ of unit Euclidean norm, for $k = 1, 2, \dots$ the following sequence is constructed:

$$\begin{aligned}(A - \mu I)\,\mathbf{z}^{(k)} &= \mathbf{q}^{(k-1)},\\ \mathbf{q}^{(k)} &= \mathbf{z}^{(k)}/\|\mathbf{z}^{(k)}\|_2,\\ \sigma^{(k)} &= (\mathbf{q}^{(k)})^H A\mathbf{q}^{(k)}.\end{aligned} \tag{5.28}$$

Notice that the eigenvectors of M_μ are the same as those of A since $M_\mu = X(\Lambda - \mu I_n)X^{-1}$, where $\Lambda = \mathrm{diag}(\lambda_1,\dots,\lambda_n)$. For this reason, the Rayleigh quotient in (5.28) is computed directly on the matrix A (and not on M_μ^{-1}). The main difference with respect to (5.17) is that at each step k a linear system with coefficient matrix $M_\mu = A - \mu I$ *must be solved.* For numerical convenience, the LU factorization of M_μ is computed once for all at $k = 1$, so that at each step only two triangular systems are to be solved, with a cost of the order of n^2 flops.

Although being more computationally expensive than the power method (5.17), the inverse iteration has the advantage that it can converge to any desired eigenvalue of A (namely, the one closest to the shift μ). Inverse iteration is thus ideally suited for refining an initial estimate μ of an eigenvalue of A, which can be obtained, for instance, by applying the localization techniques introduced in Section 5.1. Inverse iteration can be also effectively employed to compute the eigenvector associated with a given (approximate) eigenvalue, as described in Section 5.8.1.

In view of the convergence analysis of the iteration (5.28) we assume that A is diagonalizable, so that $\mathbf{q}^{(0)}$ can be represented in the form (5.19). Proceeding in the same way as in the power method, we let

$$\tilde{\mathbf{q}}^{(k)} = \mathbf{x}_m + \sum_{i=1, i\neq m}^{n} \frac{\alpha_i}{\alpha_m}\left(\frac{\xi_i}{\xi_m}\right)^k \mathbf{x}_i,$$

where $\mathbf{x}_i$ are the eigenvectors of M_μ^{-1} (and thus also of A), while α_i are as in (5.19). As a consequence, recalling the definition of ξ_i and using (5.27), we get

$$\lim_{k\to\infty} \tilde{\mathbf{q}}^{(k)} = \mathbf{x}_m, \qquad \lim_{k\to\infty} \sigma^{(k)} = \lambda_m.$$

Convergence will be faster when μ is closer to λ_m. Under the same assumptions made for proving (5.26), the following a posteriori estimate can be obtained for the approximation error on λ_m

$$|\lambda_m - \sigma^{(k)}| \simeq \frac{\|\widehat{\mathbf{r}}^{(k)}\|_2}{|(\widehat{\mathbf{w}}^{(k)})^H \mathbf{q}^{(k)}|}, \qquad k \geq 1, \tag{5.29}$$

where $\widehat{\mathbf{r}}^{(k)} = \mathrm{A}\mathbf{q}^{(k)} - \sigma^{(k)}\mathbf{q}^{(k)}$ and $\widehat{\mathbf{w}}^{(k)}$ is the k-th iterate of the inverse power method to approximate the left eigenvector associated with λ_m.

5.3.3 Implementation Issues

The convergence analysis of Section 5.3.1 shows that the effectiveness of the power method strongly depends on the dominant eigenvalues being *well-separated* (that is, $|\lambda_2|/|\lambda_1| \ll 1$). Let us now analyze the behavior of iteration (5.17) when *two* dominant eigenvalues of *equal* module exist (that is, $|\lambda_2| = |\lambda_1|$). Three cases must be distinguished:

1. $\lambda_2 = \lambda_1$: the two dominant eigenvalues are coincident. The method is still convergent, since for k sufficiently large (5.20) yields
$$\mathrm{A}^k \mathbf{q}^{(0)} \simeq \lambda_1^k (\alpha_1 \mathbf{x}_1 + \alpha_2 \mathbf{x}_2)$$
which is an eigenvector of A. For $k \to \infty$, the sequence $\tilde{\mathbf{q}}^{(k)}$ (after a suitable redefinition) converges to a vector lying in the subspace spanned by the eigenvectors $\mathbf{x}_1$ and $\mathbf{x}_2$, while the sequence $\nu^{(k)}$ still converges to λ_1.
2. $\lambda_2 = -\lambda_1$: the two dominant eigenvalues are opposite. In this case the eigenvalue of largest module can be approximated by applying the power method to the matrix A^2. Indeed, for $i = 1, \ldots, n$, $\lambda_i(\mathrm{A}^2) = [\lambda_i(\mathrm{A})]^2$, so that $\lambda_1^2 = \lambda_2^2$ and the analysis falls into the previous case, where the matrix is now A^2.
3. $\lambda_2 = \overline{\lambda}_1$: the two dominant eigenvalues are complex conjugate. Here, undamped oscillations arise in the sequence of vectors $\mathbf{q}^{(k)}$ and the power method is not convergent (see [Wil65], Chapter 9, Section 12).

As for the computer implementation of (5.17), it is worth noting that normalizing the vector $\mathbf{q}^{(k)}$ to 1 keeps away from *overflow* (when $|\lambda_1| > 1$) or *underflow* (when $|\lambda_1| < 1$) in (5.20). We also point out that the requirement $\alpha_1 \neq 0$ (which is a priori impossible to fulfil when no information about the eigenvector $\mathbf{x}_1$ is available) is not essential for the actual convergence of the algorithm.

Indeed, although it can be proved that, working in exact arithmetic, the sequence (5.17) converges to the pair $(\lambda_2, \mathbf{x}_2)$ if $\alpha_1 = 0$ (see Exercise 10), the arising of (unavoidable) rounding errors ensures that in practice the vector $\mathbf{q}^{(k)}$ contains a *non-null* component also in the direction of $\mathbf{x}_1$. This allows for the eigenvalue λ_1 to "show-up" and the power method to quickly converge to it.
An implementation of the power method is given in Program 26. Here and in the following algorithm, the convergence check is based on the a posteriori estimate (5.26).

Here and in the remainder of the chapter, the input data `z0`, `tol` and `nmax` are the initial vector, the tolerance for the stopping test and the maximum admissible number of iterations, respectively. In output, `lambda` is the approximate eigenvalue, `relres` is the vector contain the sequence $\{\|\mathbf{r}^{(k)}\|_2/|\cos(\theta_\lambda)|\}$ (see (5.26)), whilst `x` and `iter` are the approximation of the eigenvector $\mathbf{x}_1$ and the number of iterations taken by the algorithm to converge, respectively.

Program 26 - powerm : Power method

```
function [lambda,x,iter,relres]=powerm(A,z0,tol,nmax)
%POWERM Power method
% [LAMBDA,X,ITER,RELRES]=POWERM(A,Z0,TOL,NMAX) computes the
% eigenvalue LAMBDA of largest module of the matrix A and the corresponding
% eigenvector X of unit norm. TOL specifies the tolerance of the method.
% NMAX specifies the maximum number of iterations. Z0 specifies the initial
% guess. ITER is the iteration number at which X is computed.
q=z0/norm(z0); q2=q;
relres=tol+1; iter=0; z=A*q;
while relres(end)>=tol & iter<=nmax
 q=z/norm(z); z=A*q;
 lambda=q'*z; x=q;
 z2=q2'*A; q2=z2/norm(z2); q2=q2';
 y1=q2; costheta=abs(y1'*x);
 if costheta >= 5e-2
   iter=iter+1;
   temp=norm(z-lambda*q)/costheta;
   relres=[relres; temp];
 else
   fprintf('Multiple eigenvalue'); break;
 end
end
return
```

A coding of the inverse power method is provided in Program 27. The input parameter `mu` is the initial approximation of the eigenvalue. In output, `sigma` is the approximation of the computed eigenvalue and `relres` is a vector that contain the sequence $\{\|\widehat{\mathbf{r}}^{(k)}\|_2/|(\widehat{\mathbf{w}}^{(k)})^H\ \mathbf{q}^{(k)}|\}$ (see (5.29)). The LU factorization (with partial pivoting) of the matrix M_μ is carried out using the MATLAB intrinsic function `lu`.

Program 27 - invpower : Inverse power method

```
function [sigma,x,iter,relres]=invpower(A,z0,mu,tol,nmax)
%INVPOWER Inverse power method
% [SIGMA,X,ITER,RELRES]=INVPOWER(A,Z0,MU,TOL,NMAX) computes the
% eigenvalue LAMBDA of smallest module of the matrix A and the
% corresponding eigenvector X of unit norm. TOL specifies the tolerance of the
% method. NMAX specifies the maximum number of iterations. X0 specifies
% the initial guess. MU is the shift. ITER is the iteration number at which
% X is computed.
M=A-mu*eye(size(A)); [L,U,P]=lu(M);
q=z0/norm(z0); q2=q'; sigma=[];
relres=tol+1; iter=0;
while relres(end)>=tol & iter<=nmax
  iter=iter+1;
  b=P*q;
  y=L\b; z=U\y;
  q=z/norm(z); z=A*q; sigma=q'*z;
  b=q2'; y=U'\b; w=L'\y;
  q2=w'*P; q2=q2/norm(q2); costheta=abs(q2*q);
  if costheta>=5e-2
    temp=norm(z-sigma*q)/costheta; relres=[relres,temp];
  else
    fprintf('Multiple eigenvalue'); break;
  end
  x=q;
end
return
```

Example 5.3 Let us consider the following matrices

$$A = \begin{bmatrix} 15 & -2 & 2 \\ 1 & 10 & -3 \\ -2 & 1 & 0 \end{bmatrix}, \quad V = \begin{bmatrix} -0.944 & 0.393 & -0.088 \\ -0.312 & 0.919 & 0.309 \\ 0.112 & 0.013 & 0.947 \end{bmatrix}. \tag{5.30}$$

Matrix A has the following eigenvalues (to five significant figures): $\lambda_1 = 14.103$, $\lambda_2 = 10.385$ and $\lambda_3 = 0.512$, while the corresponding eigenvectors are the vector columns of matrix V.

To approximate the pair $(\lambda_1, \mathbf{x}_1)$, we have run the Program 26 with initial datum $\mathbf{z}^{(0)} = [1, 1, 1]^T$. After 71 iterations of the power method the absolute errors are $|\lambda_1 - \nu^{(71)}| = 2.2341 \cdot 10^{-10}$ and $\|\mathbf{x}_1 - \mathbf{x}_1^{(71)}\|_\infty = 1.42 \cdot 10^{-11}$.

In a second run, we have used $\mathbf{z}^{(0)} = \mathbf{x}_2 + \mathbf{x}_3$ (notice that with this choice $\alpha_1 = 0$). After 215 iterations the absolute errors are $|\lambda_1 - \nu^{(215)}| = 4.26 \cdot 10^{-14}$ and $\|\mathbf{x}_1 - \mathbf{x}_1^{(215)}\|_\infty = 1.38 \cdot 10^{-14}$.

Figure 5.2 (*left*) shows the reliability of the a posteriori estimate (5.26). The sequences $|\lambda_1 - \nu^{(k)}|$ (*solid line*) and the corresponding a posteriori estimates (5.26)

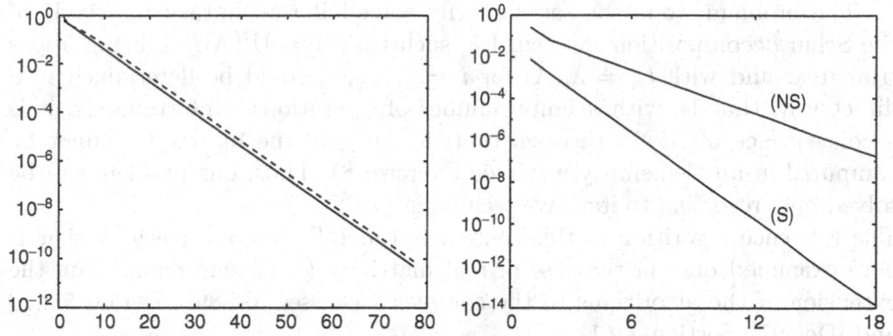

Fig. 5.2. Comparison between the a posteriori error estimate and the actual absolute error for matrix A in (5.30) (*left*); convergence curves for the power method applied to matrix A in (5.31) in its symmetric (S) and nonsymmetric (NS) forms (*right*)

(*dashed line*) are plotted as a function of the number of iterations (in abscissae). Notice the excellent agreement between the two curves.
Let us now consider the matrices

$$A = \begin{bmatrix} 1 & 3 & 4 \\ 3 & 1 & 2 \\ 4 & 2 & 1 \end{bmatrix}, \quad T = \begin{bmatrix} 8 & 1 & 6 \\ 3 & 5 & 7 \\ 4 & 9 & 2 \end{bmatrix} \tag{5.31}$$

where A has the following spectrum: $\lambda_1 = 7.047$, $\lambda_2 = -3.1879$ and $\lambda_3 = -0.8868$ (to five significant figures).
It is interesting to compare the behaviour of the power method when computing λ_1 for the symmetric matrix A and for its similar matrix $M = T^{-1}AT$, where T is the nonsingular (and nonorthogonal) matrix in (5.31).
Running Program 26 with $\mathbf{z}^{(0)} = [1, 1, 1]^T$, the power method converges to the eigenvalue λ_1 in 18 and 30 iterations, for matrices A and M, respectively. The sequence of absolute errors $|\lambda_1 - \nu^{(k)}|$ is plotted in Figure 5.2 (right) where (S) and (NS) refer to the computations on A and M, respectively. Notice the rapid error reduction in the symmetric case, according to the quadratic convergence properties of the power method (see Section 5.3.1).
We finally employ the inverse power method (5.28) to compute the eigenvalue of smallest module $\lambda_3 = 0.512$ of matrix A in (5.30). Running Program 27 with $\mathbf{q}^{(0)} = [1, 1, 1]^T/\sqrt{3}$, the method converges in 9 iterations, with absolute errors $|\lambda_3 - \sigma^{(9)}| = 1.194 \cdot 10^{-12}$ and $\|\mathbf{x}_3 - \mathbf{x}_3^{(9)}\|_\infty = 4.59 \cdot 10^{-13}$. •

5.4 The QR Iteration

In this section we present some iterative techniques for *simultaneously* approximating *all* the eigenvalues of a given matrix A. The basic idea consists of reducing A, by means of suitable similarity transformations, into a form for which the calculation of the eigenvalues is easier than on the starting matrix.

The problem would be satisfactorily solved if the unitary matrix U of the Schur decomposition theorem 1.5, such that $T = U^H AU$, T being upper triangular and with $t_{ii} = \lambda_i(A)$ for $i = 1, \ldots, n$, could be determined in a direct way, that is, with a finite number of operations. Unfortunately, it is a consequence of Abel's theorem that, for $n \geq 5$, the matrix U cannot be computed in an elementary way (see Exercise 8). Thus, our problem can be solved only resorting to iterative techniques.
The reference algorithm in this context is the *QR iteration* method, that is here examined only in the case of real matrices. (For some remarks on the extension of the algorithms to the complex case, see [GL89], Section 5.2.10 and [Dem97], Section 4.2.1).

Let $A \in \mathbb{R}^{n\times n}$; given an orthogonal matrix $Q^{(0)} \in \mathbb{R}^{n\times n}$ and letting $T^{(0)} = (Q^{(0)})^T AQ^{(0)}$, for $k = 1, 2, \ldots$, until convergence, the QR iteration consists of:

$$\begin{aligned}
&\text{determine } Q^{(k)}, R^{(k)} \text{ such that}\\
&Q^{(k)}R^{(k)} = T^{(k-1)} \qquad \text{(QR factorization)};\\
&\text{then, let}\\
&T^{(k)} = R^{(k)}Q^{(k)}.
\end{aligned} \tag{5.32}$$

At each step $k \geq 1$, the first phase of the iteration is the factorization of the matrix $T^{(k-1)}$ into the product of an orthogonal matrix $Q^{(k)}$ with an upper triangular matrix $R^{(k)}$ (see Section 5.6.3). The second phase is a simple matrix product. Notice that

$$\begin{aligned}
T^{(k)} &= R^{(k)}Q^{(k)} = (Q^{(k)})^T(Q^{(k)}R^{(k)})Q^{(k)} = (Q^{(k)})^T T^{(k-1)}Q^{(k)}\\
&= (Q^{(0)}Q^{(1)}\cdots Q^{(k)})^T A(Q^{(0)}Q^{(1)}\cdots Q^{(k)}), \qquad k \geq 0,
\end{aligned} \tag{5.33}$$

i.e., every matrix $T^{(k)}$ is *orthogonally similar* to A. This is particularly relevant for the *stability* of the method, since, as shown in Section 5.2, the conditioning of the matrix eigenvalue problem for $T^{(k)}$ is not worse than it is for A (see also [GL89], p. 360).

A basic implementation of the QR iteration (5.32), assuming $Q^{(0)} = I_n$, is examined in Section 5.5, while a more computationally efficient version, starting from $T^{(0)}$ in upper Hessenberg form, is described in detail in Section 5.6. If A has real eigenvalues, distinct in module, it will be seen in Section 5.5 that the limit of $T^{(k)}$ is an upper triangular matrix (with the eigenvalues of A on the main diagonal). However, if A has complex eigenvalues the limit of $T^{(k)}$ *cannot* be an upper triangular matrix T. Indeed if it were T would necessarily have real eigenvalues, although it is similar to A.

Failure to converge to a triangular matrix may also happen in more general situations, as addressed in Example 5.9.

For this, it is necessary to introduce variants of the QR iteration (5.32), based on deflation and *shift* techniques (see Section 5.7 and, for a more detailed

discussion of the subject, [GL89], Chapter 7, [Dat95], Chapter 8 and [Dem97], Chapter 4).
These techniques allow for $T^{(k)}$ to converge to an upper *quasi-triangular* matrix, known as the *real Schur decomposition* of A, for which the following result holds (for the proof we refer to [GL89], pp. 341-342).

Property 5.8 *Given a matrix* $A \in \mathbb{R}^{n\times n}$, *there exists an orthogonal matrix* $Q \in \mathbb{R}^{n\times n}$ *such that*

$$Q^T A Q = \begin{bmatrix} R_{11} & R_{12} & \dots & R_{1m} \\ 0 & R_{22} & \dots & R_{2m} \\ \vdots & \vdots & \ddots & \vdots \\ 0 & 0 & \dots & R_{mm} \end{bmatrix}, \tag{5.34}$$

where each block R_{ii} *is either a real number or a matrix of order 2 having complex conjugate eigenvalues, and*

$$Q = \lim_{k\to\infty} \left[Q^{(0)} Q^{(1)} \cdots Q^{(k)} \right] \tag{5.35}$$

$Q^{(k)}$ *being the orthogonal matrix generated by the k-th factorization step of the QR iteration* (5.32).

The QR iteration can be also employed to compute all the eigenvectors of a given matrix. For this purpose, we describe in Section 5.8 two possible approaches, one based on the coupling between (5.32) and the inverse iteration (5.28), the other working on the real Schur form (5.34).

5.5 The Basic QR Iteration

In the basic version of the QR method, one sets $Q^{(0)} = I_n$ in such a way that $T^{(0)} = A$. At each step $k \geq 1$ the QR factorization of the matrix $T^{(k-1)}$ can be carried out using the modified Gram-Schmidt procedure introduced in Section 3.4.3, with a cost of the order of $2n^3$ flops (for a full matrix A). The following convergence result holds (for the proof, see [GL89], Theorem 7.3.1, or [Wil65], pp. 517-519).

Property 5.9 (Convergence of QR method) *Let* $A \in \mathbb{R}^{n\times n}$ *be a matrix with real eigenvalues such that*

$$|\lambda_1| > |\lambda_2| > \ldots > |\lambda_n|.$$

Then

$$\lim_{k \to +\infty} \mathrm{T}^{(k)} = \begin{bmatrix} \lambda_1 & t_{12} & \ldots & t_{1n} \\ 0 & \lambda_2 & t_{23} & \ldots \\ \vdots & \vdots & \ddots & \vdots \\ 0 & 0 & \ldots & \lambda_n \end{bmatrix}. \tag{5.36}$$

As for the convergence rate, we have

$$|t_{i,i-1}^{(k)}| = \mathcal{O}\left(\left|\frac{\lambda_i}{\lambda_{i-1}}\right|^k\right), \qquad i = 2, \ldots, n, \qquad \text{for } k \to +\infty. \tag{5.37}$$

Under the additional assumption that A *is symmetric, the sequence* $\{\mathrm{T}^{(k)}\}$ *tends to a diagonal matrix.*

If the eigenvalues of A, although being distinct, are *not well-separated*, it follows from (5.37) that the convergence of $\mathrm{T}^{(k)}$ towards a triangular matrix can be quite slow. With the aim of accelerating it, one can resort to the so-called *shift* technique, which will be addressed in Section 5.7.

Remark 5.2 It is always possible to reduce the matrix A into a triangular form by means of an iterative algorithm employing *nonorthogonal* similarity transformations. In such a case, the so-called *LR iteration* (known also as *Rutishauser method*, [Rut58]) can be used, from which the QR method has actually been derived (see also [Fra61], [Wil65]). The LR iteration is based on the factorization of the matrix A into the product of two matrices L and R, respectively unit lower triangular and upper triangular, and on the (nonorthogonal) similarity transformation

$$\mathrm{L}^{-1}\mathrm{A}\mathrm{L} = \mathrm{L}^{-1}(\mathrm{L}\mathrm{R})\mathrm{L} = \mathrm{R}\mathrm{L}.$$

The rare use of the LR method in practical computations is due to the loss of accuracy that can arise in the LR factorization because of the increase in module of the upper diagonal entries of R. This aspect, together with the details of the implementation of the algorithm and some comparisons with the QR method, is examined in [Wil65], Chapter 8. ■

Example 5.4 We apply the QR method to the symmetric matrix A$\in \mathbb{R}^{4\times 4}$ such that $a_{ii} = 4$, for $i = 1, \ldots, 4$, and $a_{ij} = 4 + i - j$ for $i < j \le 4$, whose eigenvalues are (to three significant figures) $\lambda_1 = 11.09$, $\lambda_2 = 3.41$, $\lambda_3 = 0.90$ and $\lambda_4 = 0.59$. After 20 iterations, we get

$$\mathrm{T}^{(20)} = \begin{bmatrix} \boxed{11.09} & 6.44 \cdot 10^{-10} & -3.62 \cdot 10^{-15} & 9.49 \cdot 10^{-15} \\ 6.47 \cdot 10^{-10} & \boxed{3.41} & 1.43 \cdot 10^{-11} & 4.60 \cdot 10^{-16} \\ 1.74 \cdot 10^{-21} & 1.43 \cdot 10^{-11} & \boxed{0.90} & 1.16 \cdot 10^{-4} \\ 2.32 \cdot 10^{-25} & 2.68 \cdot 10^{-15} & 1.16 \cdot 10^{-4} & \boxed{0.58} \end{bmatrix}.$$

Notice the "almost-diagonal" structure of the matrix $T^{(20)}$ and, at the same time, the effect of rounding errors which slightly alter its expected symmetry. Good agreement can also be found between the under-diagonal entries and the estimate (5.37). •

A computer implementation of the basic QR iteration is given in Program 28. The QR factorization is executed using the modified Gram-Schmidt method (Program 8). The input parameter `nmax` denotes the maximum admissible number of iterations, while the output parameters `T`, `Q` and `R` are the matrices T, Q and R in (5.32) after `nmax` iterations of the QR procedure.

Program 28 - basicqr : Basic QR iteration

```
function [T,Q,R]=basicqr(A,nmax)
%BASICQR Basic QR iteration
% [T,Q,R]=BASICQR(A,NMAX) performs NMAX iterations of the basic QR
% algorithm.
T=A;
for i=1:nmax
   [Q,R]=modgrams(T);
   T=R*Q;
end
return
```

5.6 The QR Method for Matrices in Hessenberg Form

The naive implementation of the QR method discussed in the previous section requires (for a full matrix) a computational effort of the order of n^3 flops per iteration. In this section we illustrate a variant for the QR iteration, known as *Hessenberg-QR iteration*, with a greatly reduced computational cost. The idea consists of starting the iteration from a matrix $T^{(0)}$ in *upper Hessenberg* form, that is, $t_{ij}^{(0)} = 0$ for $i > j + 1$. Indeed, it can be checked that with this choice the computation of $T^{(k)}$ in (5.32) requires only an order of n^2 flops per iteration.

To achieve maximum efficiency and stability of the algorithm, suitable *transformation matrices* are employed. Precisely, the preliminary reduction of matrix A into upper Hessenberg form is realized with Householder matrices, whilst the QR factorization of $T^{(k)}$ is carried out using Givens matrices, instead of the modified Gram-Schmidt procedure introduced in Section 3.4.3. We briefly describe Householder and Givens matrices in the next section, referring to Section 5.6.5 for their implementation. The algorithm and examples of computations of the real Schur form of A starting from its upper Hessenberg form are then discussed in Section 5.6.4.

5.6.1 Householder and Givens Transformation Matrices

For any vector $\mathbf{v} \in \mathbb{R}^n$, let us introduce the orthogonal and symmetric matrix

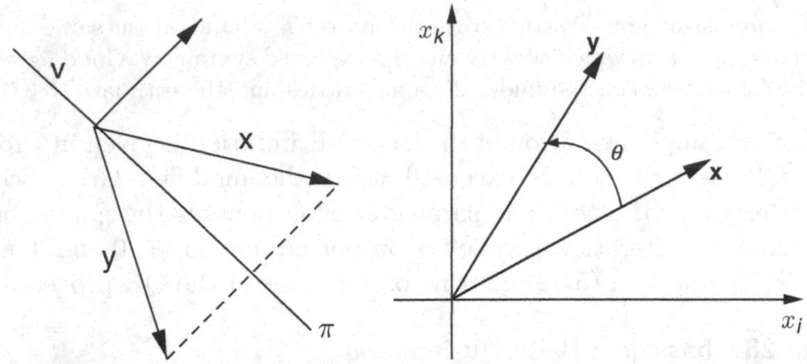

Fig. 5.3. Reflection across the hyperplane orthogonal to **v** (*left*); rotation by an angle θ in the plane (x_i, x_k) (*right*)

$$\mathrm{P} = \mathrm{I} - 2\mathbf{v}\mathbf{v}^T/\|\mathbf{v}\|_2^2. \tag{5.38}$$

Given a vector $\mathbf{x} \in \mathbb{R}^n$, the vector $\mathbf{y} = \mathrm{P}\mathbf{x}$ is the reflection of $\mathbf{x}$ with respect to the hyperplane $\pi = \mathrm{span}\{\mathbf{v}\}^\perp$ formed by the set of the vectors that are orthogonal to **v** (see Figure 5.3, *left*). Matrix P and the vector **v** are called the *Householder reflection matrix* and the *Householder vector*, respectively.
Householder matrices can be used to *set to zero* a block of components of a given vector $\mathbf{x} \in \mathbb{R}^n$. If, in particular, one would like to set to zero all the components of **x**, except the m-th one, the Householder vector ought to be chosen as

$$\mathbf{v} = \mathbf{x} \pm \|\mathbf{x}\|_2 \mathbf{e}_m, \tag{5.39}$$

$\mathbf{e}_m$ being the m-th unit vector of $\mathbb{R}^n$. The matrix P computed by (5.38) depends on the vector **x** itself, and it can be checked that

$$\mathrm{P}\mathbf{x} = \left[0, 0, \ldots, \underbrace{\pm\|\mathbf{x}\|_2}_{m}, 0, \ldots, 0\right]^T. \tag{5.40}$$

Example 5.5 Let $\mathbf{x} = [1, 1, 1, 1]^T$ and $m = 3$; then

$$\mathbf{v} = \begin{bmatrix} 1 \\ 1 \\ 3 \\ 1 \end{bmatrix}, \ \mathrm{P} = \frac{1}{6}\begin{bmatrix} 5 & -1 & -3 & -1 \\ -1 & 5 & -3 & -1 \\ -3 & -3 & -3 & -3 \\ -1 & -1 & -3 & 5 \end{bmatrix}, \ \mathrm{P}\mathbf{x} = \begin{bmatrix} 0 \\ 0 \\ -2 \\ 0 \end{bmatrix}.$$

•

If, for some $k \geq 1$, the first k components of **x** must remain unaltered, while the components from $k+2$ on are to be set to zero, the Householder matrix $\mathrm{P} = \mathrm{P}_{(k)}$ takes the following form

$$\mathrm{P}_{(k)} = \begin{bmatrix} \mathrm{I}_k & 0 \\ & \\ 0 & \mathrm{R}_{n-k} \end{bmatrix}, \ \mathrm{R}_{n-k} = \mathrm{I}_{n-k} - 2\frac{\mathbf{w}^{(k)}(\mathbf{w}^{(k)})^T}{\|\mathbf{w}^{(k)}\|_2^2}. \tag{5.41}$$

As usual, I_k is the identity matrix of order k, while R_{n-k} is the elementary Householder matrix of order $n-k$ associated with the reflection across the hyperplane orthogonal to the vector $\mathbf{w}^{(k)} \in \mathbb{R}^{n-k}$. According to (5.39), the Householder vector is given by

$$\mathbf{w}^{(k)} = \mathbf{x}^{(n-k)} \pm \|\mathbf{x}^{(n-k)}\|_2 \mathbf{e}_1^{(n-k)}, \tag{5.42}$$

where $\mathbf{x}^{(n-k)} \in \mathbb{R}^{n-k}$ is the vector formed by the last $n-k$ components of $\mathbf{x}$ and $\mathbf{e}_1^{(n-k)}$ is the first unit vector of the canonical basis of $\mathbb{R}^{n-k}$. We notice that $\mathrm{P}_{(k)}$ is a function of $\mathbf{x}$ through $\mathbf{w}^{(k)}$. The criterion for fixing the sign in the definition of $\mathbf{w}^{(k)}$ will be discussed in Section 5.6.5.

The components of the transformed vector $\mathbf{y} = \mathrm{P}_{(k)}\,\mathbf{x}$ read

$$\begin{cases} y_j = x_j & j = 1, \ldots, k, \\ y_j = 0 & j = k+2, \ldots, n, \\ y_{k+1} = \pm\|\mathbf{x}^{(n-k)}\|_2. \end{cases}$$

The Householder matrices will be employed in Section 5.6.2 to carry out the reduction of a given matrix A to a matrix $\mathrm{H}^{(0)}$ in upper Hessenberg form. This is the first step for an efficient implementation of the QR iteration (5.32) with $\mathrm{T}^{(0)} = \mathrm{H}^{(0)}$ (see Section 5.6).

Example 5.6 Let $\mathbf{x}=[1,2,3,4,5]^T$ and $k = 1$ (this means that we want to set to zero the components x_j, with $j = 3,4,5$). The matrix $\mathrm{P}_{(1)}$ and the transformed vector $\mathbf{y}=\mathrm{P}_{(1)}\,\mathbf{x}$ are given by

$$\mathrm{P}_{(1)} = \begin{bmatrix} 1 & 0 & 0 & 0 & 0 \\ 0 & 0.2722 & 0.4082 & 0.5443 & 0.6804 \\ 0 & 0.4082 & 0.7710 & -0.3053 & -0.3816 \\ 0 & 0.5443 & -0.3053 & 0.5929 & -0.5089 \\ 0 & 0.6804 & -0.3816 & -0.5089 & 0.3639 \end{bmatrix}, \ \mathbf{y} = \begin{bmatrix} 1 \\ 7.3485 \\ 0 \\ 0 \\ 0 \end{bmatrix}.$$

•

The *Givens elementary matrices* are orthogonal rotation matrices that allow for setting to zero in a selective way the entries of a vector or matrix. For a given pair of indices i and k, and a given angle θ, these matrices are defined as

$$\mathrm{G}(i,k,\theta) = \mathrm{I}_n - \mathrm{Y}, \tag{5.43}$$

where $\mathrm{Y} \in \mathbb{R}^{n\times n}$ is a null matrix except for the following entries: $y_{ii} = y_{kk} = 1 - \cos(\theta)$, $y_{ik} = -\sin(\theta) = -y_{ki}$. A Givens matrix is of the form

$$
\overset{\displaystyle i \qquad\qquad k}{G(i,k,\theta) = \begin{bmatrix} 1 & & & & & & 0 \\ & 1 & & & & & \\ & & \ddots & & & & \\ & & & \cos(\theta) & & \sin(\theta) & \\ & & & & \ddots & & \\ & & & -\sin(\theta) & & \cos(\theta) & \\ & & & & & & \ddots \\ & & & & & & 1 \\ 0 & & & & & & & 1 \end{bmatrix} \begin{matrix} \\ \\ \\ i \\ \\ k \\ \\ \\ \end{matrix}}
$$

For a given vector $\mathbf{x} \in \mathbb{R}^n$, the product $\mathbf{y} = (G(i,k,\theta))^T \mathbf{x}$ is equivalent to rotating $\mathbf{x}$ counterclockwise by an angle θ in the coordinate plane (x_i, x_k) (see Figure 5.3, *right*). After letting $c = \cos\theta$, $s = \sin\theta$, it follows that

$$
y_j = \begin{cases} x_j & j \neq i, k, \\ cx_i - sx_k & j = i, \\ sx_i + cx_k & j = k. \end{cases} \tag{5.44}
$$

Let $\alpha_{ik} = \sqrt{x_i^2 + x_k^2}$ and notice that if c and s satisfy $c = x_i/\alpha_{ik}$, $s = -x_k/\alpha_{ik}$ (in such a case, $\theta = \arctan(-x_k/x_i)$), we get $y_k = 0$, $y_i = \alpha_{ik}$ and $y_j = x_j$ for $j \neq i, k$. Similarly, if $c = x_k/\alpha_{ik}$, $s = x_i/\alpha_{ik}$ (that is, $\theta = \arctan(x_i/x_k)$), then $y_i = 0$, $y_k = \alpha_{ik}$ and $y_j = x_j$ for $j \neq i, k$.

The Givens rotation matrices will be employed in Section 5.6.3 to carry out the QR factorization step in the algorithm (5.32) and in Section 5.10.1 where the Jacobi method for symmetric matrices is considered.

Remark 5.3 (Householder deflation for power iterations) The elementary Householder tranformations can be conveniently employed to compute the first (largest or smallest) eigenvalues of a given matrix $A \in \mathbb{R}^{n \times n}$. Assume that the eigenvalues of A are ordered as in (5.16) and suppose that the eigenvalue/eigenvector pair $(\lambda_1, \mathbf{x}_1)$ has been computed using the power method. Then the matrix A can be transformed into the following block form (see for the proof [Dat95], Theorem 8.5.4, p. 418)

$$
A_1 = HAH = \begin{bmatrix} \lambda_1 & \mathbf{b}^T \\ 0 & A_2 \end{bmatrix},
$$

where $\mathbf{b} \in \mathbb{R}^{n-1}$, H is the Householder matrix such that $H\mathbf{x}_1 = \alpha\mathbf{x}_1$ for some $\alpha \in \mathbb{R}$, the matrix $A_2 \in \mathbb{R}^{(n-1)\times(n-1)}$ and the eigenvalues of A_2 are the same as those of A except for λ_1. The matrix H can be computed using (5.38) with $\mathbf{v} = \mathbf{x}_1 \pm \|\mathbf{x}_1\|_2 \mathbf{e}_1$.

The *deflation* procedure consists of computing the second dominant (subdominant) eigenvalue of A by applying the power method to A_2 provided that $|\lambda_2| \neq |\lambda_3|$. Once λ_2 is available, the corresponding eigenvector $\mathbf{x}_2$ can be computed by applying the inverse power iteration to the matrix A taking $\mu = \lambda_2$ (see Section 5.3.2) and proceeding in the same manner with the remaining eigenvalue/eigenvector pairs. An example of deflation will be presented in Section 5.12.2. ■

5.6.2 Reducing a Matrix in Hessenberg Form

A given matrix $A \in \mathbb{R}^{n\times n}$ can be transformed by similarity transformations into *upper Hessenberg form* with a cost of the order of n^3 flops. The algorithm takes $n-2$ steps and the similarity transformation Q can be computed as the product of Householder matrices $P_{(1)} \cdots P_{(n-2)}$. For this, the reduction procedure is commonly known as the *Householder method.*

Precisely, the k-th step consists of a similarity transformation of A through the Householder matrix $P_{(k)}$ which aims at setting to zero the elements in positions $k+2, \ldots, n$ of the k-th column of A, for $k = 1, \ldots, n-2$ (see Section 5.6.1). For example, in the case $n = 4$ the reduction process yields

$$\begin{bmatrix} \bullet\bullet\bullet\bullet \\ \bullet\bullet\bullet\bullet \\ \bullet\bullet\bullet\bullet \\ \bullet\bullet\bullet\bullet \end{bmatrix} \underset{P_{(1)}}{\longrightarrow} \begin{bmatrix} \bullet\bullet\bullet\bullet \\ \bullet\bullet\bullet\bullet \\ 0\,\bullet\bullet\bullet \\ 0\,\bullet\bullet\bullet \end{bmatrix} \underset{P_{(2)}}{\longrightarrow} \begin{bmatrix} \bullet\bullet\bullet\bullet \\ \bullet\bullet\bullet\bullet \\ 0\,\bullet\bullet\bullet \\ 0\,0\,\bullet\bullet \end{bmatrix},$$

having denoted by • the entries of the matrices that are a priori nonzero. Given $A^{(0)} = A$, the method generates a sequence of matrices $A^{(k)}$ that are orthogonally similar to A

$$\begin{aligned} A^{(k)} &= P_{(k)}^T A^{(k-1)} P_{(k)} = (P_{(1)} \cdots P_{(k)})^T A (P_{(1)} \cdots P_{(k)}) \\ &= Q_{(k)}^T A Q_{(k)}, \qquad k \geq 1. \end{aligned} \tag{5.45}$$

For any $k \geq 1$ the matrix $P_{(k)}$ is given by (5.41), where $\mathbf{x}$ is substituted by the k-th column vector in matrix $A^{(k-1)}$. From the definition (5.41) it is easy to check that the operation $P_{(k)}^T\ A^{(k-1)}$ leaves the first k rows of $A^{(k-1)}$ unchanged, whilst $P_{(k)}^T\ A^{(k-1)}\ P_{(k)} = A^{(k)}$ does the same on the first k columns. After $n-2$ steps of the Householder reduction, we obtain a matrix $H = A^{(n-2)}$ in upper Hessenberg form.

Remark 5.4 (The symmetric case) If A is symmetric, the transformation (5.45) maintains such a property. Indeed

$$(A^{(k)})^T = (Q_{(k)}^T A Q_{(k)})^T = A^{(k)}, \qquad \forall k \geq 1,$$

so that H must be *tridiagonal*. Its eigenvalues can be efficiently computed using the *method of Sturm sequences* with a cost of the order of n flops, as will be addressed in Section 5.10.2. ■

A coding of the Householder reduction method is provided in Program 29. To compute the Householder vector, Program 32 is employed. In output, the two matrices H and Q, respectively in Hessenberg form and orthogonal, are such that $\mathrm{H} = \mathrm{Q}^T\mathrm{AQ}$.

Program 29 - houshess : Hessenberg-Householder method

```
function [H,Q]=houshess(A)
%HOUSHESS Hessenberg-Householder method.
%  [H,Q]=HOUSHESS(A) computes the matrices H and Q such that H=Q'AQ.
[n,m]=size(A);
if n~=m; error('Only square matrices'); end
Q=eye(n); H=A;
for k=1:n-2
   [v,beta]=vhouse(H(k+1:n,k)); I=eye(k); N=zeros(k,n-k);
   m=length(v);
   R=eye(m)-beta*v*v';
   H(k+1:n,k:n)=R*H(k+1:n,k:n);
   H(1:n,k+1:n)=H(1:n,k+1:n)*R; P=[I, N; N', R]; Q=Q*P;
end
return
```

The algorithm coded in Program 29 requires a cost of $10n^3/3$ flops and is well-conditioned with respect to rounding errors. Indeed, the following estimate holds (see [Wil65], p. 351)

$$\widehat{\mathrm{H}} = \mathrm{Q}^T(\mathrm{A}+\mathrm{E})\,\mathrm{Q}, \qquad \|\mathrm{E}\|_F \leq cn^2\mathrm{u}\|\mathrm{A}\|_F, \tag{5.46}$$

where $\widehat{\mathrm{H}}$ is the Hessenberg matrix computed by Program 29, Q is an orthogonal matrix, c is a constant, u is the *roundoff* unit and $\|\cdot\|_F$ is the Frobenius norm (see (1.18)).

Example 5.7 Consider the reduction in upper Hessenberg form of the Hilbert matrix $\mathrm{H}_4 \in \mathbb{R}^{4\times 4}$. Since H_4 is symmetric, its Hessenberg form should be a tridiagonal symmetric matrix. Program 29 yields the following results

$$\mathrm{Q} = \begin{bmatrix} 1.00 & 0 & 0 & 0 \\ 0 & 0.77 & -0.61 & 0.20 \\ 0 & 0.51 & 0.40 & -0.76 \\ 0 & 0.38 & 0.69 & 0.61 \end{bmatrix}, \quad \mathrm{H} = \begin{bmatrix} 1.00 & 0.65 & 0 & 0 \\ 0.65 & 0.65 & 0.06 & 0 \\ 0 & 0.06 & 0.02 & 0.001 \\ 0 & 0 & 0.001 & 0.0003 \end{bmatrix}.$$

The accuracy of the transformation procedure (5.45) can be measured by computing the Frobenius norm (1.18) of the difference between H and $\mathrm{Q}^T\mathrm{H}_4\mathrm{Q}$. This yields $\|\mathrm{H}-\mathrm{Q}^T\mathrm{H}_4\mathrm{Q}\|_F = 3.38\cdot 10^{-17}$, which confirms the stability estimate (5.46). •

5.6.3 QR Factorization of a Matrix in Hessenberg Form

In this section we explain how to efficiently implement the generic step of the QR iteration, starting from a matrix $\mathrm{T}^{(0)} = \mathrm{H}^{(0)}$ in upper Hessenberg form.

For any $k \geq 1$, the first phase consists of computing the QR factorization of $\mathrm{H}^{(k-1)}$ by means of $n-1$ Givens rotations

$$\left(\mathrm{Q}^{(k)}\right)^T \mathrm{H}^{(k-1)} = \left(\mathrm{G}_1^{(k)} \cdots \mathrm{G}_{n-1}^{(k)}\right)^T \mathrm{H}^{(k-1)} = \mathrm{R}^{(k)}, \qquad (5.47)$$

where, for any $j = 1, \ldots, n-1$, $\mathrm{G}_j^{(k)} = \mathrm{G}(j, j+1, \theta_j)^{(k)}$ is, for any $k \geq 1$, the j-th Givens rotation matrix (5.43) in which θ_j is chosen according to (5.44) in such a way that the entry of indices $(j+1, j)$ of the matrix $\left(\mathrm{G}_1^{(k)} \cdots \mathrm{G}_j^{(k)}\right)^T \mathrm{H}^{(k-1)}$ is set equal to zero. The product (5.47) requires a computational cost of the order of $3n^2$ flops.

The next step consists of completing the orthogonal similarity transformation

$$\mathrm{H}^{(k)} = \mathrm{R}^{(k)}\mathrm{Q}^{(k)} = \mathrm{R}^{(k)} \left(\mathrm{G}_1^{(k)} \cdots \mathrm{G}_{n-1}^{(k)}\right). \qquad (5.48)$$

The orthogonal matrix $\mathrm{Q}^{(k)} = \left(\mathrm{G}_1^{(k)} \cdots \mathrm{G}_{n-1}^{(k)}\right)$ is in upper Hessenberg form. Indeed, taking for instance $n = 3$, and recalling Section 5.6.1, we get

$$\mathrm{Q}^{(k)} = \mathrm{G}_1^{(k)}\mathrm{G}_2^{(k)} = \begin{bmatrix} \bullet & \bullet & 0 \\ \bullet & \bullet & 0 \\ 0 & 0 & 1 \end{bmatrix} \begin{bmatrix} 1 & 0 & 0 \\ 0 & \bullet & \bullet \\ 0 & \bullet & \bullet \end{bmatrix} = \begin{bmatrix} \bullet & \bullet & \bullet \\ \bullet & \bullet & \bullet \\ 0 & \bullet & \bullet \end{bmatrix}.$$

Also (5.48) requires a cost of the order of $3n^2$ operations, for an overall effort of the order of $6n^2$ flops. In conclusion, performing the QR factorization with elementary Givens rotations on a starting matrix in upper Hessenberg form yields a reduction of the operation count of *one order of magnitude* with respect to the corresponding factorization with the modified Gram-Schmidt procedure of Section 5.5.

5.6.4 The Basic QR Iteration Starting from Upper Hessenberg Form

A basic implementation of the QR iteration to generate the real Schur decomposition of a matrix A is given in Program 30.

This program uses Program 29 to reduce A in upper Hessenberg form; then each QR factorization step in (5.32) is carried out with Program 31 which utilizes Givens rotations. The overall efficiency of the algorithm is ensured by pre- and post-multiplying with Givens matrices as explained in Section 5.6.5, and by constructing the matrix $\mathrm{Q}^{(k)} = \mathrm{G}_1^{(k)} \ldots \mathrm{G}_{n-1}^{(k)}$ in the function `prodgiv`, with a cost of $n^2 - 2$ flops and *without* explicitly forming the Givens matrices $\mathrm{G}_j^{(k)}$, for $j = 1, \ldots, n-1$.

As for the stability of the QR iteration with respect to rounding error propagation, it can be shown that the computed real Schur form $\widehat{T}$ is orthogonally similar to a matrix "close" to A, i.e.

$$\widehat{T} = Q^T(A + E)Q,$$

where Q is orthogonal and $\|E\|_2 \simeq u\|A\|_2$, u being the machine *roundoff* unit.

Program 30 returns in output, after `nmax` iterations of the QR procedure, the matrices T, Q and R in (5.32).

Program 30 - hessqr : Hessenberg-QR method

```
function [T,Q,R]=hessqr(A,nmax)
%HESSQR Hessenberg-QR method.
% [T,Q,R]=QR(A,NMAX) computes the real Schur decomposition of the matrix
% A in the Hessenberg form after NMAX iterations.
[n,m]=size(A);
if n~=m, error('Only square matrices'); end
[T,Qhess]=houshess(A);
for j=1:nmax
   [Q,R,c,s]= qrgivens(T);
   T=R;
   for k=1:n-1,
      T=gacol(T,c(k),s(k),1,k+1,k,k+1);
   end
end
return
```

Program 31 - qrgivens : QR factorization with Givens rotations

```
function [Q,R,c,s]= qrgivens(H)
%QRGIVENS QR factorization with Givens rotations.
[m,n]=size(H);
for k=1:n-1
   [c(k),s(k)]=givcos(H(k,k),H(k+1,k));
   H=garow(H,c(k),s(k),k,k+1,k,n);
end
R=H; Q=prodgiv(c,s,n);
return

function Q=prodgiv(c,s,n)
n1=n-1; n2=n-2;
Q=eye(n); Q(n1,n1)=c(n1); Q(n,n)=c(n1);
Q(n1,n)=s(n1); Q(n,n1)=-s(n1);
for k=n2:-1:1,
  k1=k+1; Q(k,k)=c(k); Q(k1,k)=-s(k);
  q=Q(k1,k1:n); Q(k,k1:n)=s(k)*q;
  Q(k1,k1:n)=c(k)*q;
end
return
```

Example 5.8 Consider the matrix A (already in Hessenberg form)

$$\mathrm{A} = \begin{bmatrix} 3 & 17 & -37 & 18 & -40 \\ 1 & 0 & 0 & 0 & 0 \\ 0 & 1 & 0 & 0 & 0 \\ 0 & 0 & 1 & 0 & 0 \\ 0 & 0 & 0 & 1 & 0 \end{bmatrix}.$$

To compute its eigenvalues, given by -4, $\pm i$, 2 and 5, we apply the QR method and we compute the matrix $\mathrm{T}^{(40)}$ after 40 iterations of Program 30. Notice that the algorithm converges to the real Schur decomposition of A (5.34), with three blocks R_{ii} of order 1 ($i = 1, 2, 3$) and with the block $\mathrm{R}_{44} = \mathrm{T}^{(40)}(4:5, 4:5)$ having eigenvalues equal to $\pm i$

$$\mathrm{T}^{(40)} = \begin{bmatrix} 4.9997 & 18.9739 & -34.2570 & 32.8760 & -28.4604 \\ 0 & -3.9997 & 6.7693 & -6.4968 & 5.6216 \\ 0 & 0 & 2 & -1.4557 & 1.1562 \\ 0 & 0 & 0 & 0.3129 & -0.8709 \\ 0 & 0 & 0 & 1.2607 & -0.3129 \end{bmatrix}.$$

•

Example 5.9 Let us now employ the QR method to generate the Schur real decomposition of the matrix A below, after reducing it to upper Hessenberg form

$$\mathrm{A} = \begin{bmatrix} 17 & 24 & 1 & 8 & 15 \\ 23 & 5 & 7 & 14 & 16 \\ 4 & 6 & 13 & 20 & 22 \\ 10 & 12 & 19 & 21 & 3 \\ 11 & 18 & 25 & 2 & 9 \end{bmatrix}.$$

The eigenvalues of A are real and given (to four significant figures) by $\lambda_1 = 65$, $\lambda_{2,3} = \pm 21.28$ and $\lambda_{4,5} = \pm 13.13$. After 40 iterations of Program 30, the computed matrix reads

$$\mathrm{T}^{(40)} = \begin{bmatrix} 65 & 0 & 0 & 0 & 0 \\ 0 & 14.6701 & 14.2435 & 4.4848 & -3.4375 \\ 0 & 16.6735 & -14.6701 & -1.2159 & 2.0416 \\ 0 & 0 & 0 & -13.0293 & -0.7643 \\ 0 & 0 & 0 & -3.3173 & 13.0293 \end{bmatrix}.$$

It is *not* upper triangular, but block upper triangular, with a diagonal block $R_{11} = 65$ and the two blocks

$$R_{22} = \begin{bmatrix} 14.6701 & 14.2435 \\ 16.6735 & -14.6701 \end{bmatrix}, \quad R_{33} = \begin{bmatrix} -13.0293 & -0.7643 \\ -3.3173 & 13.0293 \end{bmatrix},$$

having spectrums given by $\sigma(R_{22}) = \lambda_{2,3}$ and $\sigma(R_{33}) = \lambda_{4,5}$ respectively.

It is important to recognize that matrix $T^{(40)}$ is *not* the real Schur decomposition of A, but only a "cheating" version of it. In fact, in order for the QR method to converge to the real Schur decomposition of A, it is mandatory to resort to the *shift* techniques introduced in Section 5.7. •

5.6.5 Implementation of Transformation Matrices

In the definition (5.42) it is convenient to choose the minus sign, obtaining $\mathbf{w}^{(k)} = \mathbf{x}^{(n-k)} - \|\mathbf{x}^{(n-k)}\|_2 \mathbf{e}_1^{(n-k)}$, in such a way that the vector $R_{n-k}\mathbf{x}^{(n-k)}$ is a positive multiple of $\mathbf{e}_1^{(n-k)}$. If x_{k+1} is positive, in order to avoid numerical cancellations, the computation can be rationalized as follows

$$w_1^{(k)} = \frac{x_{k+1}^2 - \|\mathbf{x}^{(n-k)}\|_2^2}{x_{k+1} + \|\mathbf{x}^{(n-k)}\|_2} = \frac{-\sum_{j=k+2}^{n} x_j^2}{x_{k+1} + \|\mathbf{x}^{(n-k)}\|_2}.$$

The construction of the Householder vector is performed by Program 32, which takes as input a vector $\mathbf{p} \in \mathbb{R}^{n-k}$ (formerly, the vector $\mathbf{x}^{(n-k)}$) and returns a vector $\mathbf{q} \in \mathbb{R}^{n-k}$ (the Householder vector $\mathbf{w}^{(k)}$), with a cost of the order of n flops.

If $M \in \mathbb{R}^{m\times m}$ is the generic matrix to which the Householder matrix P (5.38) is applied (where I is the identity matrix of order m and $\mathbf{v} \in \mathbb{R}^m$), letting $\mathbf{w} = M^T\mathbf{v}$, then

$$PM = M - \beta\mathbf{v}\mathbf{w}^T, \qquad \beta = 2/\|\mathbf{v}\|_2^2. \tag{5.49}$$

Therefore, performing the product PM amounts to a matrix-vector product ($\mathbf{w} = M^T\mathbf{v}$) plus an external product vector-vector ($\mathbf{v}\mathbf{w}^T$). The overall computational cost of the product PM is thus equal to $2(m^2 + m)$ flops. Similar considerations hold in the case where the product MP is to be computed; defining $\mathbf{w} = M\mathbf{v}$, we get

$$MP = M - \beta\mathbf{w}\mathbf{v}^T. \tag{5.50}$$

Notice that (5.49) and (5.50) do *not* require the explicit construction of the matrix P. This reduces the computational cost to an order of m^2 flops, whilst executing the product PM *without* taking advantage of the special structure of P would increase the operation count to an order of m^3 flops.

Program 32 - vhouse : Construction of the Householder vector

```
function [v,beta]=vhouse(x)
%VHOUSE Householder vector
n=length(x); x=x/norm(x); s=x(2:n)'*x(2:n); v=[1; x(2:n)];
if s==0
   beta=0;
else
   mu=sqrt(x(1)^2+s);
   if x(1)¡=0
      v(1)=x(1)-mu;
   else
      v(1)=-s/(x(1)+mu);
   end
   beta=2*v(1)^2/(s+v(1)^2); v=v/v(1);
end
return
```

Concerning the Givens rotation matrices, the computation of c and s is carried out as follows. Let i and k be two fixed indices and assume that the k-th component of a given vector $\mathbf{x} \in \mathbb{R}^n$ must be set to zero. Letting $r = \sqrt{x_i^2 + x_k^2}$, relation (5.44) yields

$$\begin{bmatrix} c & -s \\ s & c \end{bmatrix} \begin{bmatrix} x_i \\ x_k \end{bmatrix} = \begin{bmatrix} r \\ 0 \end{bmatrix}, \qquad (5.51)$$

from which it turns out that there is no need of explicitly computing θ, nor evaluating any trigonometric function.

Executing Program 33 to solve system (5.51), requires 5 flops, plus the evaluation of a square root. As already noticed in the case of Householder matrices, even for Givens rotations we don't have to explicitly compute the matrix $\mathrm{G}(i,k,\theta)$ to perform its product with a given matrix $\mathrm{M} \in \mathbb{R}^{m\times m}$. For that purpose Programs 34 and 35 are used, both at the cost of $6m$ flops. Looking at the structure (5.43) of matrix $\mathrm{G}(i,k,\theta)$, it is clear that the first algorithm only modifies rows i and k of M, whilst the second one only changes columns i and k of M.

We conclude by noticing that the computation of the Householder vector $\mathbf{v}$ and of the Givens sine and cosine (c, s), are *well-conditioned* operations with respect to rounding errors (see [GL89], pp. 212-217 and the references therein).

The solution of system (5.51) is implemented in Program 33. The input parameters are the vector components x_i and x_k, whilst the output data are the Givens cosine and sine c and s.

Program 33 - givcos : Computation of Givens cosine and sine

```
function [c,s]=givcos(xi, xk)
%GIVCOS Computes the Givens cosine and sine.
if xk==0
```

```
    c=1; s=0;
else
    if abs(xk)>abs(xi)
        t=-xi/xk; s=1/sqrt(1+t^2); c=s*t;
    else
        t=-xk/xi; c=1/sqrt(1+t^2); s=c*t;
    end
end
return
```

Programs 34 and 35 compute $G(i,k,\theta)^T M$ and $MG(i,k,\theta)$ respectively. The input parameters c and s are the Givens cosine and sine. In Program 34, the indices i and k identify the rows of the matrix M that are being affected by the update $M \leftarrow G(i,k,\theta)^T M$, while j1 and j2 are the indices of the columns involved in the computation. Similarly, in Program 35 i and k identify the columns effected by the update $M \leftarrow MG(i,k,\theta)$, while j1 and j2 are the indices of the rows involved in the computation.

Program 34 - garow : Product $G(i,k,\theta)^T M$

```
function [M]=garow(M,c,s,i,k,j1,j2)
%GAROW Product of the transpose of a Givens rotation matrix with M.
for j=j1:j2
    t1=M(i,j);
    t2=M(k,j);
    M(i,j)=c*t1-s*t2;
    M(k,j)=s*t1+c*t2;
end
return
```

Program 35 - gacol : Product $MG(i,k,\theta)$

```
function [M]=gacol(M,c,s,j1,j2,i,k)
%GACOL Product of M with a Givens rotation matrix.
for j=j1:j2
    t1=M(j,i);
    t2=M(j,k);
    M(j,i)=c*t1-s*t2;
    M(j,k)=s*t1+c*t2;
end
return
```

5.7 The QR Iteration with Shifting Techniques

Example 5.9 reveals that the QR iteration does not always converge to the real Schur form of a given matrix A. To make this happen, an effective approach consists of incorporating in the QR iteration (5.32) a shifting technique similar to that introduced for inverse iteration in Section 5.3.2.

This leads to the *QR method with single shift* described in Section 5.7.1, which is used to accelerate the convergence of the QR iteration when A has eigenvalues with moduli very close to each other.

In Section 5.7.2, a more sophisticated shifting technique is considered, which guarantees the convergence of the QR iteration to the (approximate) *Schur form* of matrix A (see Property 5.8). The resulting method (known as *QR iteration with double shift*) is the most popular version of the QR iteration (5.32) for solving the matrix eigenvalue problem, and is implemented in the MATLAB intrinsic function `eig`.

5.7.1 The QR Method with Single Shift

Given $\mu \in \mathbb{R}$, the *shifted* QR iteration is defined as follows. For $k = 1, 2, \ldots$, until convergence:

$$\begin{aligned}
&\text{determine } \mathrm{Q}^{(k)}, \mathrm{R}^{(k)} \text{ such that} \\
&\mathrm{Q}^{(k)}\mathrm{R}^{(k)} = \mathrm{T}^{(k-1)} - \mu \mathrm{I} \qquad \text{(QR factorization);} \\
&\text{then, let} \\
&\mathrm{T}^{(k)} = \mathrm{R}^{(k)}\mathrm{Q}^{(k)} + \mu \mathrm{I},
\end{aligned} \tag{5.52}$$

where $\mathrm{T}^{(0)} = \left(\mathrm{Q}^{(0)}\right)^T \mathrm{A}\mathrm{Q}^{(0)}$ is in upper Hessenberg form. Since the QR factorization in (5.52) is performed on the shifted matrix $\mathrm{T}^{(k-1)} - \mu\mathrm{I}$, the scalar μ is called *shift*. The sequence of matrices $\mathrm{T}^{(k)}$ generated by (5.52) is still similar to the initial matrix A, since for any $k \geq 1$

$$\begin{aligned}
\mathrm{R}^{(k)}\mathrm{Q}^{(k)} + \mu\mathrm{I} &= \left(\mathrm{Q}^{(k)}\right)^T \left(\mathrm{Q}^{(k)}\mathrm{R}^{(k)}\mathrm{Q}^{(k)} + \mu\mathrm{Q}^{(k)}\right) \\
&= \left(\mathrm{Q}^{(k)}\right)^T \left(\mathrm{Q}^{(k)}\mathrm{R}^{(k)} + \mu\mathrm{I}\right)\mathrm{Q}^{(k)} = \left(\mathrm{Q}^{(k)}\right)^T \mathrm{T}^{(k-1)}\mathrm{Q}^{(k)} \\
&= (\mathrm{Q}^{(0)}\mathrm{Q}^{(1)}\cdots\mathrm{Q}^{(k)})^T\mathrm{A}(\mathrm{Q}^{(0)}\mathrm{Q}^{(1)}\cdots\mathrm{Q}^{(k)}), \qquad k \geq 0.
\end{aligned}$$

Assume μ is fixed and that the eigenvalues of A are ordered in such a way that

$$|\lambda_1 - \mu| \geq |\lambda_2 - \mu| \geq \ldots \geq |\lambda_n - \mu|.$$

Then it can be shown that, for $1 < j \leq n$, the subdiagonal entry $t_{j,j-1}^{(k)}$ tends to zero with a rate that is proportional to the ratio

$$|(\lambda_j - \mu)/(\lambda_{j-1} - \mu)|^k.$$

This extends the convergence result (5.37) to the shifted QR method (see [GL89], Sections 7.5.2 and 7.3).

The result above suggests that if μ is chosen in such a way that

$$|\lambda_n - \mu| < |\lambda_i - \mu|, \qquad i = 1, \ldots, n-1,$$

then the matrix entry $t_{n,n-1}^{(k)}$ in the iteration (5.52) tends rapidly to zero as k increases. (In the limit, if μ were equal to an eigenvalue of $\mathrm{T}^{(k)}$, that is of A, then $t_{n,n-1}^{(k)} = 0$ and $t_{n,n}^{(k)} = \mu$). In practice one takes

$$\mu = t_{n,n}^{(k)}, \tag{5.53}$$

yielding the so called *QR iteration with single shift*. Correspondingly, the convergence to zero of the sequence $\left\{t_{n,n-1}^{(k)}\right\}$ is *quadratic* in the sense that if $|t_{n,n-1}^{(k)}|/\|\mathrm{T}^{(0)}\|_2 = \eta_k < 1$, for some $k \geq 0$, then $|t_{n,n-1}^{(k+1)}|/\|\mathrm{T}^{(0)}\|_2 = \mathcal{O}(\eta_k^2)$ (see [Dem97], pp. 161-163 and [GL89], pp. 354-355).

This can be profitably taken into account when programming the QR iteration with single shift by monitoring the size of the subdiagonal entry $|t_{n,n-1}^{(k)}|$. In practice, $t_{n,n-1}^{(k)}$ is set equal to zero if

$$|t_{n,n-1}^{(k)}| \leq \varepsilon(|t_{n-1,n-1}^{(k)}| + |t_{n,n}^{(k)}|), \qquad k \geq 0, \tag{5.54}$$

for a prescribed ε, in general of the order of the *roundoff* unit. (This convergence test is adopted in the library EISPACK). If A is an Hessenberg matrix, when for a certain k $a_{n,n-1}^{(k)}$ is set to zero, $t_{n,n}^{(k)}$ provides the desired approximation of λ_n. Then the QR iteration with shift can continue on the matrix $\mathrm{T}^{(k)}(1:n-1, 1:n-1)$, and so on. This is a *deflation* algorithm (for another example see Remark 5.3).

Example 5.10 We consider again the matrix A as in Example 5.9. Program 36, with `tol` equal to the *roundoff* unit, converges in 14 iterations to the following approximate real Schur form of A, which displays the correct eigenvalues of matrix A on its diagonal (to six significant figures)

$$\mathrm{T}^{(40)} = \begin{bmatrix} 65 & 0 & 0 & 0 & 0 \\ 0 & -21.2768 & 2.5888 & -0.0445 & -4.2959 \\ 0 & 0 & -13.1263 & -4.0294 & -13.079 \\ 0 & 0 & 0 & 21.2768 & -2.6197 \\ 0 & 0 & 0 & 0 & 13.1263 \end{bmatrix}.$$

We also report in Table 5.2 the convergence rate $p^{(k)}$ of the sequence $\left\{t_{n,n-1}^{(k)}\right\}$ $(n = 5)$ computed as

$$p^{(k)} = 1 + \frac{1}{\log(\eta_k)} \log \frac{|t_{n,n-1}^{(k)}|}{|t_{n,n-1}^{(k-1)}|}, \qquad k \geq 1.$$

The results show good agreement with the expected quadratic rate. •

The coding of the QR iteration with single shift (5.52) is given in Program 36. The code utilizes Program 29 to reduce the matrix A in upper Hessenberg form and Program 31 to perform the QR factorization step. The input

Table 5.2. Convergence rate of the sequence $\left\{t_{n,n-1}^{(k)}\right\}$ in the QR iteration with single shift

k	$\|t_{n,n-1}^{(k)}\|/\|\mathrm{T}^{(0)}\|_2$	$p^{(k)}$
0	0.13865	
1	$1.5401 \cdot 10^{-2}$	2.1122
2	$1.2213 \cdot 10^{-4}$	2.1591
3	$1.8268 \cdot 10^{-8}$	1.9775
4	$8.9036 \cdot 10^{-16}$	1.9449

parameters `tol` and `itmax` are the tolerance ε in (5.54) and the maximum admissible number of iterations, respectively. In output, the program returns the (approximate) real Schur form of A and the number of iterations needed for its computation.

Program 36 - qrshift : QR iteration with single shift

```
function [T,iter]=qrshift(A,tol,nmax)
%QRSHIFT QR iteration with single shifting technique.
% [T,ITER]=QRSHIFT(A,TOL,NMAX) computes after ITER iterations the real
% Schur form T of the matrix A with a tolerance TOL. NMAX specifies the
% maximum number of iterations.
[n,m]=size(A);
if n~=m, error('Only square matrices'); end
iter=0; [T,Q]=houshess(A);
for k=n:-1:2
   I=eye(k);
   while abs(T(k,k-1))>tol*(abs(T(k,k))+abs(T(k-1,k-1)))
      iter=iter+1;
      if iter > nmax
         return
      end
      mu=T(k,k); [Q,R,c,s]=qrgivens(T(1:k,1:k)-mu*I);
      T(1:k,1:k)=R*Q+mu*I;
   end
   T(k,k-1)=0;
end
return
```

5.7.2 The QR Method with Double Shift

The single-shift QR iteration (5.52) with the choice (5.53) for μ is effective if the eigenvalues of A are real, but not necessarily when complex conjugate eigenvalues are present, as happens in the following example.

Example 5.11 The matrix $A \in \mathbb{R}^{4\times4}$ (reported below to five significant figures)

$$A = \begin{bmatrix} 1.5726 & -0.6392 & 3.7696 & -1.3143 \\ 0.2166 & -0.0420 & 0.4006 & -1.2054 \\ 0.0226 & 0.3592 & 0.2045 & -0.1411 \\ -0.1814 & 1.1146 & -3.2330 & 1.2648 \end{bmatrix}$$

has eigenvalues $\{\pm i, 1, 2\}$, i being the imaginary unit. Running Program 36 with `tol` equal to the *roundoff* unit yields after 100 iterations

$$T^{(101)} = \begin{bmatrix} 2 & 1.1999 & 0.5148 & 4.9004 \\ 0 & -0.0001 & -0.8575 & 0.7182 \\ 0 & 1.1662 & 0.0001 & -0.8186 \\ 0 & 0 & 0 & 1 \end{bmatrix}.$$

The obtained matrix is the real Schur form of A, where the 2×2 block $T^{(101)}(2{:}3, 2{:}3)$ has complex conjugate eigenvalues $\pm i$. These eigenvalues cannot be computed by the algorithm (5.52)-(5.53) since μ is real. •

The problem with this example is that working with real matrices necessarily yields a real shift, whereas a *complex* one would be needed. The *QR iteration with double shift* is set up to account for complex eigenvalues and allows for removing the 2×2 diagonal blocks of the real Schur form of A.

Precisely, suppose that the QR iteration with single shift (5.52) detects at some step k a 2×2 diagonal block $R_{kk}^{(k)}$ that cannot be reduced into upper triangular form. Since the iteration is converging to the real Schur form of the matrix A the two eigenvalues of $R_{kk}^{(k)}$ are complex conjugate and will be denoted by $\lambda^{(k)}$ and $\bar{\lambda}^{(k)}$. The *double shift* strategy consists of the following steps:

$$\begin{aligned} &\text{determine } Q^{(k)}, R^{(k)} \text{ such that} \\ &Q^{(k)}R^{(k)} = T^{(k-1)} - \lambda^{(k)}I \qquad \text{(first QR factorization);} \\ &\text{then, let} \\ &T^{(k)} = R^{(k)}Q^{(k)} + \lambda^{(k)}I; \\ &\text{determine } Q^{(k+1)}, R^{(k+1)} \text{ such that} \\ &Q^{(k+1)}R^{(k+1)} = T^{(k)} - \bar{\lambda}^{(k)}I \qquad \text{(second QR factorization);} \\ &\text{then, let} \\ &T^{(k+1)} = R^{(k+1)}Q^{(k+1)} + \bar{\lambda}^{(k)}I. \end{aligned} \tag{5.55}$$

Once the double shift has been carried out the QR iteration with single shift is continued until a situation analogous to the one above is encountered.

The QR iteration incorporating the double shift strategy is the most effective algorithm for computing eigenvalues and yields the approximate Schur form of a given matrix A. Its actual implementation is far more sophisticated than the outline above and is called QR iteration with *Francis shift* (see [Fra61], and, also, [GL89], Section 7.5 and [Dem97], Section 4.4.5). As for the case of the QR iteration with single shift, quadratic convergence can also be proven for the QR method with Francis shift. However, special matrices have recently been found for which the method fails to converge (see for an example Exercise 14 and Remark 5.13). We refer for some analysis and remedies to [Bat90], [Day96], although the finding of a shift strategy that guarantees convergence of the QR iteration for all matrices is still an open problem.

Example 5.12 Let us apply the QR iteration with double shift to the matrix A in Example 5.11. After 97 iterations of Program 37, with `tol` equal to the *roundoff* unit, we get the following (approximate) Schur form of A, which displays on its diagonal the four eigenvalues of A

$$\mathrm{T}^{(97)} = \begin{bmatrix} 2 & 1+2i & -2.33+0.86i & 4.90 \\ 0 & 5.02\cdot 10^{-14}+i & -2.02+6.91\cdot 10^{-14}i & 0.72 \\ t_{31}^{(97)} & 0 & -1.78\cdot 10^{-14}-i & -0.82 \\ t_{41}^{(97)} & t_{42}^{(97)} & 0 & 1 \end{bmatrix},$$

where $t_{31}^{(97)} = 2.06\cdot 10^{-17}+7.15\cdot 10^{-49}i$, $t_{41}^{(97)} = -5.59\cdot 10^{-17}$ and $t_{42}^{(97)} = -4.26\cdot 10^{-18}$, respectively. •

Example 5.13 Consider the pseudo-spectral differentiation matrix (10.73) of order 5. This matrix is singular, with a unique eigenvalue $\lambda = 0$ of algebraic multiplicity equal to 5 (see [CHQZ06], p. 44). In this case the QR method with double shift provides an inaccurate approximation of the spectrum of the matrix. Indeed, using Program 37, with `tol=eps`, the method converges after 59 iterations to an upper triangular matrix with diagonal entries given by 0.0020, $0.0006 \pm 0.0019i$ and $-0.0017 \pm 0.0012i$, respectively. Using the MATLAB intrinsic function `eig` yields instead the eigenvalues -0.0024, $-0.0007 \pm 0.0023i$ and $0.0019 \pm 0.0014i$. •

A basic implementation of the QR iteration with double shift is provided in Program 37. The input/output parameters are the same as those of Program 36. The output matrix `T` is the approximate Schur form of matrix A.

Program 37 - qr2shift : QR iteration with double shift

```
function [T,iter]=qr2shift(A,tol,nmax)
%QR2SHIFT QR iteration with double shifting technique.
% [T,ITER]=QR2SHIFT(A,TOL,NMAX) computes after ITER iterations the real
% Schur form T of the matrix A with a tolerance TOL. NMAX specifies the
% maximum number of iterations.
[n,m]=size(A);
if n~=m, error('Only square matrices'); end
```

```
iter=0; [T,Q]=houshess(A);
for k=n:-1:2
  I=eye(k);
  while abs(T(k,k-1))>tol*(abs(T(k,k))+abs(T(k-1,k-1)))
    iter=iter+1;
    if iter > nmax, return, end
    mu=T(k,k); [Q,R,c,s]=qrgivens(T(1:k,1:k)-mu*I);
    T(1:k,1:k)=R*Q+mu*I;
    if k > 2
      Tdiag2=abs(T(k-1,k-1))+abs(T(k-2,k-2));
      if abs(T(k-1,k-2))<=tol*Tdiag2;
        [lambda]=eig(T(k-1:k,k-1:k));
        [Q,R,c,s]=qrgivens(T(1:k,1:k)-lambda(1)*I);
        T(1:k,1:k)=R*Q+lambda(1)*I;
        [Q,R,c,s]=qrgivens(T(1:k,1:k)-lambda(2)*I);
        T(1:k,1:k)=R*Q+lambda(2)*I;
      end
    end
  end
  T(k,k-1)=0;
end
I=eye(2);
while (abs(T(2,1))>tol*(abs(T(2,2))+abs(T(1,1)))) & (iter<=nmax)
  iter=iter+1;
  mu=T(2,2);
  [Q,R,c,s]=qrgivens(T(1:2,1:2)-mu*I);
  T(1:2,1:2)=R*Q+mu*I;
end
return
```

5.8 Computing the Eigenvectors and the SVD of a Matrix

The power and inverse iterations described in Section 5.3.2 can be used to compute a selected number of eigenvalue/eigenvector pairs. If all the eigenvalues and eigenvectors of a matrix are needed, the QR iteration can be profitably employed to compute the eigenvectors as shown in Sections 5.8.1 and 5.8.2. In Section 5.8.3 we deal with the computation of the singular value decomposition (SVD) of a given matrix.

5.8.1 The Hessenberg Inverse Iteration

For any approximate eigenvalue λ computed by the QR iteration as described in Section 5.7.2, the inverse iteration (5.28) can be applied to the matrix $\mathrm{H} = \mathrm{Q}^T\mathrm{AQ}$ in Hessenberg form, yielding an approximate eigenvector $\mathbf{q}$. Then, the eigenvector $\mathbf{x}$ associated with λ is computed as $\mathbf{x} = \mathrm{Q}\mathbf{q}$. Clearly, one can take advantage of the structure of the Hessenberg matrix for an efficient solution of the linear system at each step of (5.28). Typically, only one iteration is required to produce an adequate approximation of the desired eigenvector $\mathbf{x}$ (see [GL89], Section 7.6.1 and [PW79] for more details).

5.8.2 Computing the Eigenvectors from the Schur Form of a Matrix

Suppose that the (approximate) Schur form $\mathrm{Q}^H\mathrm{AQ}=\mathrm{T}$ of a given matrix $\mathrm{A}\in\mathbb{R}^{n\times n}$ has been computed by the QR iteration with double shift, Q being a unitary matrix and T being upper triangular.

Then, if $\mathrm{A}\mathbf{x}=\lambda\mathbf{x}$, we have $\mathrm{Q}^H\mathrm{AQQ}^H\mathbf{x}= \mathrm{Q}^H\lambda\mathbf{x}$, i.e., letting $\mathbf{y}=\mathrm{Q}^H\mathbf{x}$, we have that $\mathrm{T}\,\mathbf{y}=\lambda\mathbf{y}$ holds. Therefore $\mathbf{y}$ is an eigenvector of T, so that to compute the eigenvectors of A we can work directly on the Schur form T.

Assume for simplicity that $\lambda = t_{kk} \in \mathbb{C}$ is a simple eigenvalue of A. Then the upper triangular matrix T can be decomposed as

$$\mathrm{T} = \begin{bmatrix} \mathrm{T}_{11} & \mathbf{v} & \mathrm{T}_{13} \\ 0 & \lambda & \mathbf{w}^T \\ 0 & 0 & \mathrm{T}_{33} \end{bmatrix},$$

where $\mathrm{T}_{11} \in \mathbb{C}^{(k-1)\times(k-1)}$ and $\mathrm{T}_{33} \in \mathbb{C}^{(n-k)\times(n-k)}$ are upper triangular matrices, $\mathbf{v}\in \mathbb{C}^{k-1}$, $\mathbf{w}\in \mathbb{C}^{n-k}$ and $\lambda \notin \sigma(\mathrm{T}_{11}) \cup \sigma(\mathrm{T}_{33})$.

Thus, letting $\mathbf{y} = [\mathbf{y}_{k-1}^T, y, \mathbf{y}_{n-k}^T]^T$, with $\mathbf{y}_{k-1} \in \mathbb{C}^{k-1}$, $y \in \mathbb{C}$ and $\mathbf{y}_{n-k} \in \mathbb{C}^{n-k}$, the matrix eigenvector problem (T - λI) $\mathbf{y}=\mathbf{0}$ can be written as

$$\begin{cases} (\mathrm{T}_{11} - \lambda\mathrm{I}_{k-1})\mathbf{y}_{k-1} + \mathbf{v}y + \mathrm{T}_{13}\mathbf{y}_{n-k} = \mathbf{0}, \\ \mathbf{w}^T\mathbf{y}_{n-k} = 0, \\ (\mathrm{T}_{33} - \lambda\mathrm{I}_{n-k})\mathbf{y}_{n-k} = \mathbf{0}. \end{cases} \tag{5.56}$$

Since λ is simple, both matrices $\mathrm{T}_{11} - \lambda\mathrm{I}_{k-1}$ and $\mathrm{T}_{33} - \lambda\mathrm{I}_{n-k}$ are nonsingular, so that the third equation in (5.56) yields $\mathbf{y}_{n-k} = \mathbf{0}$ and the first equation becomes

$$(\mathrm{T}_{11} - \lambda\mathrm{I}_{k-1})\mathbf{y}_{k-1} = -\mathbf{v}y.$$

Setting arbitrarily $y = 1$ and solving the triangular system above for $\mathbf{y}_{k-1}$ yields (formally)

$$\mathbf{y} = \begin{pmatrix} -(\mathrm{T}_{11} - \lambda\mathrm{I}_{k-1})^{-1}\mathbf{v} \\ 1 \\ 0 \end{pmatrix}.$$

The desired eigenvector $\mathbf{x}$ can then be computed as $\mathbf{x}=\mathrm{Q}\mathbf{y}$.

An efficient implementation of the above procedure is carried out in the intrinsic MATLAB function `eig`. Invoking this function with the format `[V, D]= eig(A)` yields the matrix `V` whose columns are the right eigenvectors of A and the diagonal matrix `D` contains its eigenvalues. Further details can be found in the `strvec` subroutine in the LAPACK library, while for the computation of eigenvectors in the case where A is symmetric, we refer to [GL89], Chapter 8 and [Dem97], Section 5.3.

5.8.3 Approximate Computation of the SVD of a Matrix

In this section we describe the Golub-Kahan-Reinsch algorithm for the computation of the SVD of a matrix $A \in \mathbb{R}^{m\times n}$ with $m \geq n$ (see [GL89], Section 5.4). The method consists of two phases, a direct one and an iterative one.

In the first phase A is transformed into an upper trapezoidal matrix of the form

$$\mathcal{U}^T A \mathcal{V} = \begin{bmatrix} B \\ 0 \end{bmatrix}, \tag{5.57}$$

where $\mathcal{U}$ and $\mathcal{V}$ are two orthogonal matrices and $B \in \mathbb{R}^{n\times n}$ is upper bidiagonal. The matrices $\mathcal{U}$ and $\mathcal{V}$ are generated using $n+m-3$ Householder matrices $\mathcal{U}_1, \ldots, \mathcal{U}_n, \mathcal{V}_1, \ldots, \mathcal{V}_{n-2}$ as follows.

The algorithm initially generates $\mathcal{U}_1$ in such a way that the matrix $A^{(1)} = \mathcal{U}_1 A$ has $a_{i1}^{(1)} = 0$ if $i > 1$. Then, $\mathcal{V}_1$ is determined so that $A^{(2)} = A^{(1)}\mathcal{V}_1$ has $a_{1j}^{(2)} = 0$ for $j > 2$, preserving at the same time the null entries of the previous step. The procedure is repeated starting from $A^{(2)}$, and taking $\mathcal{U}_2$ such that $A^{(3)} = \mathcal{U}_2 A^{(2)}$ has $a_{i2}^{(3)} = 0$ for $i > 2$ and $\mathcal{V}_2$ in such a way that $A^{(4)} = A^{(3)}\mathcal{V}_2$ has $a_{2j}^{(4)} = 0$ for $j > 3$, yet preserving the null entries already generated. For example, in the case $m = 5$, $n = 4$ the first two steps of the reduction process yield

$$A^{(1)} = \mathcal{U}_1 A = \begin{bmatrix} \bullet & \bullet & \bullet & \bullet \\ 0 & \bullet & \bullet & \bullet \\ 0 & \bullet & \bullet & \bullet \\ 0 & \bullet & \bullet & \bullet \\ 0 & \bullet & \bullet & \bullet \end{bmatrix} \longrightarrow A^{(2)} = A^{(1)}\mathcal{V}_1 = \begin{bmatrix} \bullet & \bullet & 0 & 0 \\ 0 & \bullet & \bullet & \bullet \\ 0 & \bullet & \bullet & \bullet \\ 0 & \bullet & \bullet & \bullet \\ 0 & \bullet & \bullet & \bullet \end{bmatrix},$$

having denoted by $\bullet$ the entries of the matrices that in principle are different than zero. After at most $m-1$ steps, we find (5.57) with

$$\mathcal{U} = \mathcal{U}_1\mathcal{U}_2 \cdots \mathcal{U}_{m-1}, \; \mathcal{V} = \mathcal{V}_1\mathcal{V}_2 \cdots \mathcal{V}_{n-2}.$$

In the second phase, the obtained matrix B is reduced into a diagonal matrix Σ using the QR iteration. Precisely, a sequence of upper bidiagonal matrices $B^{(k)}$ are constructed such that, as $k \to \infty$, their off-diagonal entries tend to zero quadratically and the diagonal entries tend to the singular values σ_i of A. In the limit, the process generates two orthogonal matrices $\mathcal{W}$ and $\mathcal{Z}$ such that

$$\mathcal{W}^T \mathrm{B} \mathcal{Z} = \Sigma = \mathrm{diag}(\sigma_1, \ldots, \sigma_n).$$

The SVD of A is then given by

$$\mathrm{U}^T \mathrm{AV} = \begin{bmatrix} \Sigma \\ 0 \end{bmatrix},$$

with $\mathrm{U} = \mathcal{U}\mathrm{diag}(\mathcal{W}, \mathrm{I}_{m-n})$ and $\mathrm{V} = \mathcal{V}\mathcal{Z}$.

The computational cost of this procedure is $2m^2n + 4mn^2 + \frac{9}{2}n^3$ flops, which reduces to $2mn^2 - \frac{2}{3}n^3$ flops if only the singular values are computed. In this case, recalling what was stated in Section 3.13 about $\mathrm{A}^T\mathrm{A}$, the method described in the present section is preferable to computing directly the eigenvalues of $\mathrm{A}^T\mathrm{A}$ and then taking their square roots.

As for the stability of this procedure, it can be shown that the computed σ_i turn out to be the singular values of the matrix $\mathrm{A} + \delta\mathrm{A}$ with

$$\|\delta\mathrm{A}\|_2 \le C_{mn}\mathrm{u}\|\mathrm{A}\|_2,$$

C_{mn} being a constant dependent on n, m and the *roundoff* unit u. For other approaches to the computation of the SVD of a matrix, see [Dat95] and [GL89].

5.9 The Generalized Eigenvalue Problem

Let $\mathrm{A}, \mathrm{B} \in \mathbb{C}^{n \times n}$ be two given matrices; for any $z \in \mathbb{C}$, we call $\mathrm{A} - z\mathrm{B}$ a *matrix pencil* and denote it by (A,B). The set σ(A,B) of the eigenvalues of (A,B) is defined as

$$\sigma(\mathrm{A}, \mathrm{B}) = \{\mu \in \mathbb{C} : \det(\mathrm{A} - \mu\mathrm{B}) = 0\}.$$

The *generalized matrix eigenvalue problem* can be formulated as: find $\lambda \in \sigma$(A,B) and a nonnull vector $\mathbf{x} \in \mathbb{C}^n$ such that

$$\mathrm{A}\mathbf{x} = \lambda\mathrm{B}\mathbf{x}. \tag{5.58}$$

The pair $(\lambda, \mathbf{x})$ satisfying (5.58) is an eigenvalue/eigenvector pair of the pencil (A,B). Note that by setting $\mathrm{B}=\mathrm{I}_n$ in (5.58) we recover the standard matrix eigenvalue problem considered thus far.

Problems like (5.58) arise frequently in engineering applications, e.g., in the study of vibrations of structures (buildings, aircrafts and bridges) or in the mode analysis for waveguides (see [Inm94] and [Bos93]). Another example is the computation of the extremal eigenvalues of a preconditioned matrix $\mathrm{P}^{-1}\mathrm{A}$ (in which case $\mathrm{B} = \mathrm{P}$ in (5.58)) when solving a linear system with an iterative method (see Remark 4.2).

Let us introduce some definitions. We say that the pencil (A,B) is *regular* if $\det(A-zB)$ is not identically zero, otherwise the pencil is *singular*. When (A,B) is regular, $p(z) = \det(A - zB)$ is the *characteristic polynomial* of the pencil; denoting by k the degree of p, the eigenvalues of (A,B) are defined as:

1. the roots of $p(z) = 0$, if $k = n$;
2. ∞ if $k < n$ (with multiplicity equal to $n - k$).

Example 5.14 (Taken from [Par80], [Saa92] and [GL89])

$$A = \begin{bmatrix} -1 & 0 \\ 0 & 1 \end{bmatrix}, B = \begin{bmatrix} 0 & 1 \\ 1 & 0 \end{bmatrix} \quad p(z) = z^2 + 1 \Longrightarrow \sigma(A, B) = \pm i;$$

$$A = \begin{bmatrix} -1 & 0 \\ 0 & 0 \end{bmatrix}, B = \begin{bmatrix} 0 & 0 \\ 0 & 1 \end{bmatrix} \quad p(z) = z \qquad \Longrightarrow \sigma(A, B) = \{0, \infty\};$$

$$A = \begin{bmatrix} 1 & 2 \\ 0 & 0 \end{bmatrix}, B = \begin{bmatrix} 1 & 0 \\ 0 & 0 \end{bmatrix} \quad p(z) = 0 \qquad \Longrightarrow \sigma(A, B) = \mathbb{C}.$$

The first pair of matrices shows that symmetric pencils, unlike symmetric matrices, may exhibit *complex conjugate* eigenvalues. The second pair is a regular pencil displaying an eigenvalue equal to infinity, while the third pair is an example of singular pencil. •

5.9.1 Computing the Generalized Real Schur Form

The definitions and examples above imply that the pencil (A,B) has n finite eigenvalues iff B is nonsingular.

In such a case, a possible approach to the solution of problem (5.58) is to transform it into the equivalent eigenvalue problem $C\mathbf{x} = \lambda\mathbf{x}$, where the matrix C is the solution of the system BC = A, then apply the QR iteration to C. For actually computing the matrix C, one can use Gauss elimination with pivoting or the techniques shown in Section 3.6. This procedure can yield inaccurate results if B is ill-conditioned, since computing C is affected by rounding errors of the order of $\mathbf{u}\ \|A\|_2\|B^{-1}\|_2$ (see [GL89], p. 376).

A more attractive approach is based on the following result, which generalizes the Schur decomposition theorem 1.5 to the case of regular pencils (for a proof, see [Dat95], p. 497).

Property 5.10 (Generalized Schur decomposition) *Let* (A,B) *be a regular pencil. Then, there exist two unitary matrices* U *and* Z *such that* $U^H AZ = T$, $U^H BZ = S$, *where* T *and* S *are upper triangular. For* $i = 1, \ldots, n$ *the eigenvalues of* (A,B) *are given by*

$$\lambda_i = t_{ii}/s_{ii}, \ \textit{if}\ s_{ii} \neq 0,$$

$$\lambda_i = \infty, \qquad \textit{if}\ t_{ii} \neq 0,\ s_{ii} = 0.$$

Exactly as in the matrix eigenvalue problem, the generalized Schur form cannot be explicitly computed, so the counterpart of the real Schur form (5.34) has to be computed. Assuming that the matrices A and B are real, it can be shown that there exist two orthogonal matrices $\tilde{\mathrm{U}}$ and $\tilde{\mathrm{Z}}$ such that $\tilde{\mathrm{T}} = \tilde{\mathrm{U}}^T\mathrm{A}\tilde{\mathrm{Z}}$ is upper quasi-triangular and $\tilde{\mathrm{S}} = \tilde{\mathrm{U}}^T\mathrm{B}\tilde{\mathrm{Z}}$ is upper triangular. This decomposition is known as the *generalized real Schur decomposition* of a pair (A,B) and can be computed by a suitably modified version of the QR algorithm, known as *QZ iteration*, which consists of the following steps (for a more detailed description, see [GL89], Section 7.7, [Dat95], Section 9.3):

1. reduce A and B into upper Hessenberg form and upper triangular form, respectively, i.e., find two orthogonal matrices Q and Z such that $\mathcal{A} = \mathrm{Q}^T\mathrm{AZ}$ is upper Hessenberg and $\mathcal{B} = \mathrm{Q}^T\mathrm{BZ}$ is upper triangular;
2. the QR iteration is applied to the matrix $\mathcal{A}\mathcal{B}^{-1}$ to reduce it to real Schur form.

To save computational resources, the QZ algorithm overwrites the matrices A and B on their upper Hessenberg and triangular forms and requires $30n^3$ flops; an additional cost of $36n^3$ operations is required if Q and Z are also needed. The method is implemented in the LAPACK library in the subroutine `sgges` and can be invoked in the MATLAB environment with the command `eig(A,B)`.

5.9.2 Generalized Real Schur Form of Symmetric-Definite Pencils

A remarkable situation occurs when both A and B are symmetric, and one of them, say B, is also positive definite. In such a case, the pair (A,B) forms a *symmetric-definite* pencil for which the following result holds.

Theorem 5.7 *The symmetric-definite pencil* (A,B) *has real eigenvalues and linearly independent eigenvectors. Moreover, the matrices* A *and* B *can be simultaneously diagonalized. Precisely, there exists a nonsingular matrix* $\mathrm{X} \in \mathbb{R}^{n\times n}$ *such that*

$$\mathrm{X}^T\mathrm{AX} = \Lambda = \mathrm{diag}(\lambda_1, \lambda_2, \ldots, \lambda_n),\ \mathrm{X}^T\mathrm{BX} = \mathrm{I}_n,$$

where for $i = 1, \ldots, n$, λ_i *are the eigenvalues of the pencil* (A, B).

Proof. Since B is symmetric positive definite, it admits a unique Cholesky factorization $\mathrm{B} = \mathrm{H}^T\mathrm{H}$, where H is upper triangular (see Section 3.4.2). From (5.58) we deduce that $\mathrm{C}\mathbf{z} = \lambda\mathbf{z}$ with $\mathrm{C} = \mathrm{H}^{-T}\mathrm{AH}^{-1}$, $\mathbf{z} = \mathrm{H}\mathbf{x}$, where $(\lambda, \mathbf{x})$ is an eigenvalue/eigenvector pair of (A,B).

The matrix C is symmetric; therefore, its eigenvalues are real and a set of orthonormal eigenvectors $[\mathbf{y}_1, \ldots, \mathbf{y}_n] = \mathrm{Y}$ exists. As a consequence, letting $\mathrm{X} = \mathrm{H}^{-1}\mathrm{Y}$ allows for simultaneously diagonalizing both A and B since

$$\mathrm{X}^T\mathrm{AX} = \mathrm{Y}^T\mathrm{H}^{-T}\mathrm{AH}^{-1}\mathrm{Y} = \mathrm{Y}^T\mathrm{CY} = \Lambda = \mathrm{diag}(\lambda_1, \ldots, \lambda_n),$$

$$\mathrm{X}^T\mathrm{BX} = \mathrm{Y}^T\mathrm{H}^{-T}\mathrm{BH}^{-1}\mathrm{Y} = \mathrm{Y}^T\mathrm{Y} = \mathrm{I}_n.$$

◇

The following QR-Cholesky algorithm computes the eigenvalues λ_i and the corresponding eigenvectors $\mathbf{x}_i$ of a symmetric-definite pencil (A,B), for $i = 1, \ldots, n$ (see for more details [GL89], Section 8.7, [Dat95], Section 9.5):

1. compute the Cholesky factorization $\mathrm{B} = \mathrm{H}^T\mathrm{H}$;
2. compute $\mathrm{C} = \mathrm{H}^{-T}\mathrm{A}\mathrm{H}^{-1}$;
3. for $i = 1, \ldots, n$, compute the eigenvalues λ_i and eigenvectors $\mathbf{z}_i$ of the symmetric matrix C using the QR iteration. Then construct from the set $\{\mathbf{z}_i\}$ an orthonormal set of eigenvectors $\{\mathbf{y}_i\}$ (using, for instance, the modified Gram-Schmidt procedure of Section 3.4.3);
4. for $i = 1, \ldots, n$, compute the eigenvectors $\mathbf{x}_i$ of the pencil (A,B) by solving the systems $\mathrm{H}\mathbf{x}_i = \mathbf{y}_i$.

This algorithm requires an order of $14n^3$ flops and it can be shown (see [GL89], p. 464) that, if $\hat{\lambda}$ is a computed eigenvalue, then

$$\hat{\lambda} \in \sigma(\mathrm{H}^{-T}\mathrm{A}\mathrm{H}^{-1} + \mathrm{E}), \qquad \text{with } \|\mathrm{E}\|_2 \simeq \mathrm{u}\|\mathrm{A}\|_2\|\mathrm{B}^{-1}\|_2.$$

Thus, the generalized eigenvalue problem in the symmetric-definite case may become unstable with respect to rounding errors propagation if B is ill-conditioned. For a stabilized version of the QR-Cholesky method, see [GL89], p. 464 and the references cited therein.

5.10 Methods for Eigenvalues of Symmetric Matrices

In this section we deal with the computation of the eigenvalues of a symmetric matrix $\mathrm{A} \in \mathbb{R}^{n\times n}$. Besides the QR method previously examined, specific algorithms which take advantage of the symmetry of A are available.

Among these, we first consider the Jacobi method, which generates a sequence of matrices orthogonally similar to A and converging to the diagonal Schur form of A. Then, the Sturm sequence and Lanczos procedures are presented, for handling the case of tridiagonal matrices and large sparse matrices respectively.

5.10.1 The Jacobi Method

The Jacobi method generates a sequence of matrices $\mathrm{A}^{(k)}$ that are orthogonally similar to matrix A and converge to a diagonal matrix whose entries are the eigenvalues of A. This is done using the Givens similarity transformations (5.43) as follows.

Given $\mathrm{A}^{(0)} = \mathrm{A}$, for any $k = 1, 2, \ldots$, a pair of indices p and q is fixed, with $1 \leq p < q \leq n$. Next, letting $\mathrm{G}_{pq} = \mathrm{G}(p, q, \theta)$, the matrix $\mathrm{A}^{(k)} = (\mathrm{G}_{pq})^T\mathrm{A}^{(k-1)}\mathrm{G}_{pq}$, orthogonally similar to A, is constructed in such a way that

$$a_{ij}^{(k)} = 0 \quad \text{if} \quad (i, j) = (p, q). \tag{5.59}$$

Letting $c = \cos\theta$ and $s = \sin\theta$, the procedure for computing the entries of $A^{(k)}$ that are changed with respect to those of $A^{(k-1)}$, can be written as

$$\begin{bmatrix} a_{pp}^{(k)} & a_{pq}^{(k)} \\ a_{pq}^{(k)} & a_{qq}^{(k)} \end{bmatrix} = \begin{bmatrix} c & s \\ -s & c \end{bmatrix}^T \begin{bmatrix} a_{pp}^{(k-1)} & a_{pq}^{(k-1)} \\ a_{pq}^{(k-1)} & a_{qq}^{(k-1)} \end{bmatrix} \begin{bmatrix} c & s \\ -s & c \end{bmatrix}. \qquad (5.60)$$

If $a_{pq}^{(k-1)} = 0$, we can satisfy (5.59) by taking $c = 1$ and $s = 0$. If $a_{pq}^{(k-1)} \neq 0$, letting $t = s/c$, (5.60) requires the solution of the following algebraic equation

$$t^2 + 2\eta t - 1 = 0, \qquad \eta = \frac{a_{qq}^{(k-1)} - a_{pp}^{(k-1)}}{2a_{pq}^{(k-1)}}. \qquad (5.61)$$

The root $t = 1/(\eta + \sqrt{1+\eta^2})$ is chosen in (5.61) if $\eta \geq 0$, otherwise we take $t = -1/(-\eta + \sqrt{1+\eta^2})$; next, we let

$$c = \frac{1}{\sqrt{1+t^2}}, \qquad s = ct. \qquad (5.62)$$

To examine the rate at which the off-diagonal entries of $A^{(k)}$ tend to zero, it is convenient to introduce, for any matrix $M \in \mathbb{R}^{n\times n}$, the nonnegative quantity

$$\Psi(M) = \left(\sum_{\substack{i,j=1 \\ i\neq j}}^{n} m_{ij}^2 \right)^{1/2} = \left(\|M\|_F^2 - \sum_{i=1}^{n} m_{ii}^2 \right)^{1/2}. \qquad (5.63)$$

The Jacobi method ensures that $\Psi(A^{(k)}) \leq \Psi(A^{(k-1)})$ for any $k \geq 1$. Indeed, the computation of (5.63) for matrix $A^{(k)}$ yields

$$(\Psi(A^{(k)}))^2 = (\Psi(A^{(k-1)}))^2 - 2\left(a_{pq}^{(k-1)}\right)^2 \leq (\Psi(A^{(k-1)}))^2. \qquad (5.64)$$

The estimate (5.64) suggests that, at each step k, the optimal choice of the indices p and q is that corresponding to the entry in $A^{(k-1)}$ such that

$$|a_{pq}^{(k-1)}| = \max_{i\neq j} |a_{ij}^{(k-1)}|.$$

The computational cost of this strategy is of the order of n^2 flops for the search of the maximum module entry, while the updating step $A^{(k)} = (G_{pq})^T A^{(k-1)} G_{pq}$ requires only a cost of the order of n flops, as already noticed in Section 5.6.5. It is thus convenient to resort to the so called *row cyclic Jacobi method*, in which the choice of the indices p and q is done by a row-sweeping of the matrix $A^{(k)}$ according to the following algorithm: for any $k = 1, 2, \ldots$ and for any i-th row of $A^{(k)}$ ($i = 1, \ldots, n-1$), we set $p = i$ and $q = i+1, \ldots, n$. Each complete sweep requires $N = n(n-1)/2$ Jacobi transformations. Assuming that $|\lambda_i - \lambda_j| \geq \delta$ for $i \neq j$, it can be shown that the cyclic Jacobi method converges quadratically, that is (see [Wil65], [Wil62])

Table 5.3. Convergence of the cyclic Jacobi algorithm

Sweep	$\Psi(\mathrm{H}_4^{(k)})$	*Sweep*	$\Psi(\mathrm{H}_4^{(k)})$	*Sweep*	$\Psi(\mathrm{H}_4^{(k)})$
1	$5.262 \cdot 10^{-2}$	2	$3.824 \cdot 10^{-5}$	3	$5.313 \cdot 10^{-16}$

$$\Psi(\mathrm{A}^{(k+N)}) \leq \frac{1}{\delta\sqrt{2}}(\Psi(\mathrm{A}^{(k)}))^2, \qquad k = 1, 2, \ldots$$

For further details of the algorithm, we refer to [GL89], Section 8.4.

Example 5.15 Let us apply the cyclic Jacobi method to the Hilbert matrix H_4, whose eigenvalues read (to five significant figures) $\lambda_1 = 1.5002$, $\lambda_2 = 1.6914 \cdot 10^{-1}$, $\lambda_3 = 6.7383 \cdot 10^{-3}$ and $\lambda_4 = 9.6702 \cdot 10^{-5}$. Running Program 40 with $\texttt{tol} = 10^{-15}$, the method converges in 3 sweeps to a matrix whose diagonal entries coincide with the eigenvalues of H_4 unless $4.4409 \cdot 10^{-16}$. As for the off-diagonal entries, the values attained by $\Psi(\mathrm{H}_4^{(k)})$ are reported in Table 5.3. •

Formulae (5.63) and (5.62) are implemented in Programs 38 and 39.

Program 38 - psinorm : Evaluation of $\Psi(\mathrm{A})$

```
function [psi]=psinorm(A)
%PSINORM Evaluation of Psi(A).
[n,m]=size(A);
if n~=m, error('Only square matrices'); end
psi=0;
for i=1:n-1
   j=[i+1:n];
   psi = psi + sum(A(i,j).^2+A(j,i).^2');
end
psi=sqrt(psi);
return
```

Program 39 - symschur : Evaluation of c and s

```
function [c,s]=symschur(A,p,q)
%SYMSCHUR Evaluation of parameters c and s in (5.62).
if A(p,q)==0
   c=1; s=0;
else
   eta=(A(q,q)-A(p,p))/(2*A(p,q));
   if eta>=0
      t=1/(eta+sqrt(1+eta^2));
   else
      t=-1/(-eta+sqrt(1+eta^2));
   end
   c=1/sqrt(1+t^2); s=c*t;
end
return
```

A coding of the cyclic Jacobi method is implemented in Program 40. This program gets as input parameters the symmetric matrix $\mathrm{A} \in \mathbb{R}^{n\times n}$ and a tolerance `tol`. The program returns a matrix $\mathrm{D} = \mathrm{G}^T\mathrm{AG}$, G being orthogonal, such that $\Psi(\mathrm{D}) \leq \mathtt{tol}\|\mathrm{A}\|_F$, the value of $\Psi(\mathrm{D})$ and the number of sweeps to achieve convergence.

Program 40 - cycjacobi : Cyclic Jacobi method for symmetric matrices

```
function [D,sweep,psi]=cycjacobi(A,tol,nmax)
%CYCJACOBI Cyclic Jacobi method.
% [D,SWEEP,PSI]=CYCJACOBI(A,TOL) computes the eigenvalues D of the symmetric
% matrix A. TOL specifies the tolerance of the method. PSI=PSINORM(D) and
% SWEEP is the number of sweeps. NMAX specifies the maximum number of iterations.
[n,m]=size(A);
if n~=m, error('Only square matrices'); end
D=A;
psi=norm(A,'fro');
epsi=tol*psi;
psi=psinorm(D);
sweep=0;
iter=0;
while psi>epsi&iter<=nmax
   iter = iter + 1;
   sweep=sweep+1;
   for p=1:n-1
      for q=p+1:n
         [c,s]=symschur(D,p,q);
         [D]=gacol(D,c,s,1,n,p,q);
         [D]=garow(D,c,s,p,q,1,n);
      end
   end
   psi=psinorm(D);
end
return
```

5.10.2 The Method of Sturm Sequences

In this section we deal with the calculation of the eigenvalues of a real, tridiagonal and symmetric matrix T. Typical instances of such a problem arise when applying the Householder transformation to a given symmetric matrix A (see Section 5.6.2) or when solving boundary value problems in one spatial dimension (see for an example Section 5.12.1).

We analyze the *method of Sturm sequences*, or *Givens method*, introduced in [Giv54]. For $i = 1, \ldots, n$, we denote by d_i the diagonal entries of T and by b_i, $i = 1, \ldots, n-1$, the elements of the upper and lower subdiagonals of T. We shall assume that $b_i \neq 0$ for any i. Otherwise, indeed, the computation reduces to problems of less complexity.

Letting T_i be the principal submatrix of order i of matrix T and $p_0(x) = 1$, we define for $i = 1, \ldots, n$ the following sequence of polynomials $p_i(x) = \det(\mathrm{T}_i - x\mathrm{I}_i)$

$$\begin{aligned} p_1(x) &= d_1 - x, \\ p_i(x) &= (d_i - x)p_{i-1}(x) - b_{i-1}^2 p_{i-2}(x), \; i = 2, \ldots, n. \end{aligned} \tag{5.65}$$

It can be checked that p_n is the characteristic polynomial of T; the computational cost of its evaluation at point x is of the order of $2n$ flops. The sequence (5.65) is called the *Sturm sequence* owing to the following result, for whose proof we refer to [Wil65], Chapter 2, Section 47 and Chapter 5, Section 37.

Property 5.11 (of Sturm sequence) *For $i = 2, \ldots, n$ the eigenvalues of T_{i-1} strictly separate those of T_i, that is*

$$\lambda_i(\mathrm{T}_i) < \lambda_{i-1}(\mathrm{T}_{i-1}) < \lambda_{i-1}(\mathrm{T}_i) < \ldots < \lambda_2(\mathrm{T}_i) < \lambda_1(\mathrm{T}_{i-1}) < \lambda_1(\mathrm{T}_i).$$

Moreover, letting for any real number μ

$$\mathcal{S}_\mu = \{p_0(\mu), p_1(\mu), \ldots, p_n(\mu)\},$$

the number $s(\mu)$ of sign changes in $\mathcal{S}_\mu$ yields the number of eigenvalues of T that are strictly less than μ, with the convention that $p_i(\mu)$ has opposite sign to $p_{i-1}(\mu)$ if $p_i(\mu) = 0$ (two consecutive elements in the sequence cannot vanish at the same value of μ).

Example 5.16 Let T be the tridiagonal part of the Hilbert matrix $\mathrm{H}_4 \in \mathbb{R}^{4\times 4}$, having entries $h_{ij} = 1/(i + j - 1)$. The eigenvalues of T are (to five significant figures) $\lambda_1 = 1.2813$, $\lambda_2 = 0.4205$, $\lambda_3 = -0.1417$ and $\lambda_4 = 0.1161$. Taking $\mu = 0$, Program 41 computes the following Sturm sequence

$$\mathcal{S}_0 = \{p_0(0), p_1(0), p_2(0), p_3(0), p_4(0)\} = \{1, 1, 0.0833, -0.0458, -0.0089\},$$

from which, applying Property 5.11, one concludes that matrix T has one eigenvalue less than 0. In the case of matrix $\mathrm{T} = \mathrm{tridiag}_4(-1, 2, -1)$, with eigenvalues $\{0.38, 1.38, 2.62, 3.62\}$ (to three significant figures), we get, taking $\mu = 3$

$$\{p_0(3), p_1(3), p_2(3), p_3(3), p_4(3)\} = \{1, -1, 0, 1, -1\},$$

which shows that matrix T has three eigenvalues less than 3, since three sign changes occur. •

The Givens method for the calculation of the eigenvalues of T proceeds as follows. Letting $b_0 = b_n = 0$, Theorem 5.2 yields the interval $\mathcal{J} = [\alpha, \beta]$ which contains the spectrum of T, where

$$\alpha = \min_{1\le i\le n} \left[d_i - (|b_{i-1}| + |b_i|)\right], \qquad \beta = \max_{1\le i\le n} \left[d_i + (|b_{i-1}| + |b_i|)\right].$$

Table 5.4. Convergence of the Givens method for the calculation of the eigenvalue λ_2 of the matrix T in Example 5.16

k	$a^{(k)}$	$b^{(k)}$	$c^{(k)}$	$s^{(k)}$	k	$a^{(k)}$	$b^{(k)}$	$c^{(k)}$	$s^{(k)}$
0	0	4.000	2.0000	2	7	2.5938	2.625	2.6094	2
1	2.0000	4.000	3.0000	3	8	2.6094	2.625	2.6172	2
2	2.0000	3.000	2.5000	2	9	2.6094	2.625	2.6172	2
3	2.5000	3.000	2.7500	3	10	2.6172	2.625	2.6211	3
4	2.5000	2.750	2.6250	3	11	2.6172	2.621	2.6191	3
5	2.5000	2.625	2.5625	2	12	2.6172	2.619	2.6182	3
6	2.5625	2.625	2.5938	2	13	2.6172	2.618	2.6177	2

The set $\mathcal{J}$ is used as an initial guess in the search for generic eigenvalues λ_i of matrix T, for $i = 1, \ldots, n$, using the bisection method (see Chapter 6).

Precisely, given $a^{(0)} = \alpha$ and $b^{(0)} = \beta$, we let $c^{(0)} = (\alpha+\beta)/2$ and compute $s(c^{(0)})$; then, recalling Property 5.11, we let $b^{(1)} = c^{(0)}$ if $s(c^{(0)}) > (n-i)$, otherwise we set $a^{(1)} = c^{(0)}$. After r iterations, the value $c^{(r)} = (a^{(r)} + b^{(r)})/2$ provides an approximation of λ_i within $(|\alpha| + |\beta|) \cdot 2^{-(r+1)}$, as is shown in (6.9).

A systematic procedure can be set up to store any information about the position within the interval $\mathcal{J}$ of the eigenvalues of T that are being computed by the Givens method. The resulting algorithm generates a sequence of neighboring subintervals $a_j^{(r)}, b_j^{(r)}$, for $j = 1, \ldots, n$, each one of arbitrarily small length and containing one eigenvalue λ_j of T (for further details, see [BMW67]).

Example 5.17 Let us employ the Givens method to compute the eigenvalue $\lambda_2 \simeq 2.62$ of matrix T considered in Example 5.16. Letting `tol`$=10^{-4}$ in Program 42 we obtain the results reported in Table 5.4, which demonstrate the convergence of the sequence $c^{(k)}$ to the desired eigenvalue in 13 iterations. We have denoted for brevity, $s^{(k)} = s(c^{(k)})$. Similar results are obtained by running Program 42 to compute the remaining eigenvalues of T. •

An implementation of the polynomial evaluation (5.65) is given in Program 41. This program receives in input the vectors `dd` and `bb` containing the main and the upper diagonals of T. The output values $p_i(x)$ are stored, for $i = 0, \ldots, n$, in the vector `p`.

Program 41 - sturm : Sturm sequence evaluation

```
function [p]=sturm(dd,bb,x)
%STURM Sturm sequence
% P=STURM(D,B,X) evaluates the Sturm sequence (5.65) at X.
n=length(dd);
p(1)=1;
p(2)=dd(1)-x;
for i=2:n
```

```
    p(i+1)=(dd(i)-x)*p(i)-bb(i-1)^2*p(i-1);
end
return
```

A basic implementation of the Givens method is provided in Program 42. In input, `ind` is the pointer to the searched eigenvalue, while the other parameters are similar to those in Program 41. In output the values of the elements of sequences $a^{(k)}$, $b^{(k)}$ and $c^{(k)}$ are returned, together with the required number of iterations `niter` and the sequence of sign changes $s(c^{(k)})$.

Program 42 - givsturm : Givens method using the Sturm sequence

```
function [ak,bk,ck,nch,niter]=givsturm(dd,bb,ind,tol)
%GIVSTURM Givens method with Sturm sequence
[a, b]=bound(dd,bb); dist=abs(b-a); s=abs(b)+abs(a);
n=length(dd); niter=0; nch=[];
while dist>tol*s
    niter=niter+1;
    c=(b+a)/2;
    ak(niter)=a;
    bk(niter)=b;
    ck(niter)=c;
    nch(niter)=chcksign(dd,bb,c);
    if nch(niter)>n-ind
        b=c;
    else
        a=c;
    end
    dist=abs(b-a); s=abs(b)+abs(a);
end
return
```

Program 43 - chcksign : Sign changes in the Sturm sequence

```
function nch=chcksign(dd,bb,x)
%CHCKSIGN Determines the sign changes in the Sturm sequence.
[p]=sturm(dd,bb,x);
n=length(dd);
nch=0;
s=0;
for i=2:n+1
    if p(i)*p(i-1)<=0
        nch=nch+1;
    end
    if p(i)==0
        s=s+1;
    end
end
nch=nch-s;
return
```

Program 44 - bound : Calculation of the interval $\mathcal{J} = [\alpha, \beta]$

```
function [alfa,beta]=bound(dd,bb)
%BOUND Calculation of the interval [ALPHA,BETA] for the Givens method.
n=length(dd);
alfa=dd(1)-abs(bb(1));
temp=dd(n)-abs(bb(n-1));
if temp<alfa
   alfa=temp;
end
for i=2:n-1
   temp=dd(i)-abs(bb(i-1))-abs(bb(i));
   if temp<alfa
      alfa=temp;
   end
end
beta=dd(1)+abs(bb(1)); temp=dd(n)+abs(bb(n-1));
if temp>beta
   beta=temp;
end
for i=2:n-1
   temp=dd(i)+abs(bb(i-1))+abs(bb(i));
   if temp>beta
      beta=temp;
   end
end
return
```

5.11 The Lanczos Method

Let $\mathrm{A} \in \mathbb{R}^{n\times n}$ be a symmetric sparse matrix, whose (real) eigenvalues are ordered as

$$\lambda_1 \geq \lambda_2 \geq \ldots \geq \lambda_{n-1} \geq \lambda_n. \tag{5.66}$$

When n is very large, the Lanczos method [Lan50] described in Section 4.4.3 can be applied to approximate the extremal eigenvalues λ_n and λ_1. It generates a sequence of tridiagonal matrices H_m whose extremal eigenvalues rapidly converge to the extremal eigenvalues of A.

To estimate the convergence of the tridiagonalization process, we introduce the Rayleigh quotient $r(\mathbf{x}) = (\mathbf{x}^T\mathrm{A}\mathbf{x})/(\mathbf{x}^T\mathbf{x})$ associated with a nonnull vector $\mathbf{x} \in \mathbb{R}^n$. The following result, known as Courant-Fisher Theorem, holds (for the proof see [GL89], p. 394)

$$\lambda_1(\mathrm{A}) = \max_{\substack{\mathbf{x}\in\mathbb{R}^n\\ \mathbf{x}\neq\mathbf{0}}} r(\mathbf{x}), \qquad \lambda_n(\mathrm{A}) = \min_{\substack{\mathbf{x}\in\mathbb{R}^n\\ \mathbf{x}\neq\mathbf{0}}} r(\mathbf{x}).$$

Its application to the matrix $\mathrm{H}_m = \mathrm{V}_m^T \mathrm{A} \mathrm{V}_m$, yields

$$\begin{aligned}\lambda_1(\mathrm{H}_m) &= \max_{\substack{\mathbf{x}\in\mathbb{R}^n\\ \mathbf{x}\neq\mathbf{0}}} \frac{(\mathrm{V}_m\mathbf{x})^T \mathrm{A}(\mathrm{V}_m\mathbf{x})}{\mathbf{x}^T\mathbf{x}} = \max_{\|\mathbf{x}\|_2=1} r(\mathrm{H}_m\mathbf{x}) \leq \lambda_1(\mathrm{A}),\\ \lambda_m(\mathrm{H}_m) &= \min_{\substack{\mathbf{x}\in\mathbb{R}^n\\ \mathbf{x}\neq\mathbf{0}}} \frac{(\mathrm{V}_m\mathbf{x})^T \mathrm{A}(\mathrm{V}_m\mathbf{x})}{\mathbf{x}^T\mathbf{x}} = \min_{\|\mathbf{x}\|_2=1} r(\mathrm{H}_m\mathbf{x}) \geq \lambda_n(\mathrm{A}).\end{aligned} \tag{5.67}$$

At each step of the Lanczos method, the estimates (5.67) provide a lower and upper bound for the extremal eigenvalues of A. The convergence of the sequences $\{\lambda_1(\mathrm{H}_m)\}$ and $\{\lambda_m(\mathrm{H}_m)\}$ to λ_1 and λ_n, respectively, is governed by the following property, for whose proof we refer to [GL89], pp. 475-477.

Property 5.12 *Let* $\mathrm{A} \in \mathbb{R}^{n\times n}$ *be a symmetric matrix with eigenvalues ordered as in* (5.66) *and let* $\mathbf{u}_1, \ldots, \mathbf{u}_n$ *be the corresponding orthonormal eigenvectors. If* $\eta_1, \ldots, \eta_m$ *denote the eigenvalues of* H_m*, with* $\eta_1 \geq \eta_2 \geq \ldots \geq \eta_m$*, then*

$$\lambda_1 \geq \eta_1 \geq \lambda_1 - \frac{(\lambda_1 - \lambda_n)(\tan\phi_1)^2}{(T_{m-1}(1+2\rho_1))^2},$$

where $\cos\phi_1 = |(\mathbf{q}^{(1)})^T\mathbf{u}_1|$, $\rho_1 = (\lambda_1 - \lambda_2)/(\lambda_2 - \lambda_n)$ *and* $T_{m-1}(x)$ *is the Chebyshev polynomial of degree* $m-1$ *(see Section* 10.1.1*).*

A similar result holds of course for the convergence estimate of the eigenvalues η_m to λ_n

$$\lambda_n \leq \eta_m \leq \lambda_n + \frac{(\lambda_1 - \lambda_n)(\tan\phi_n)^2}{(T_{m-1}(1+2\rho_n))^2},$$

where $\rho_n = (\lambda_{n-1} - \lambda_n)/(\lambda_1 - \lambda_{n-1})$ and $\cos\phi_n = |(\mathbf{q}^{(n)})^T\mathbf{u}_n|$.

A naive implementation of the Lanczos algorithm can be affected by numerical instability due to propagation of rounding errors. In particular, the Lanczos vectors will not verify the mutual orthogonality relation, making the extremal properties (5.67) false. This requires careful programming of the Lanczos iteration by incorporating suitable reorthogonalization procedures as described in [GL89], Sections 9.2.3-9.2.4.

Despite this limitation, the Lanczos method has two relevant features: it *preserves* the sparsity pattern of the matrix (unlike Householder tridiagonalization), and such a property makes it quite attractive when dealing with large size matrices; furthermore, it converges to the extremal eigenvalues of A much more rapidly than the power method does (see [Kan66], [GL89], p. 477).

The Lanczos method can be generalized to compute the extremal eigenvalues of an unsymmetric matrix along the same lines as in Section 4.5 in the case of the solution of a linear system. Details on the practical implementation of the algorithm and a theoretical convergence analysis can be found in [LS96] and [Jia95], while some documentation of the latest software can be found in www.caam.rice.edu/software/ARPACK (see also the MATLAB command `eigs`).

An implementation of the Lanczos algorithm is provided in Program 45. The input parameter m is the size of the Krylov subspace in the tridiagonalization procedure, while tol is a tolerance monitoring the size of the increment of the computed eigenvalues between two successive iterations. The output vectors lmin, lmax and deltaeig contain the sequences of the approximate extremal eigenvalues and of their increments between successive iterations. Program 42 is invoked for computing the eigenvalues of the tridiagonal matrix H_m.

Program 45 - eiglancz : Extremal eigenvalues of a symmetric matrix

```
function [lmin,lmax,deltaeig,k]=eiglancz(A,m,tol)
%EIGLANCZ Lanczos method.
% [LMIN,LMAX,DELTAEIG,ITER]=EIGLANCZ(A,M,TOL) computes the extremal
% eigenvalues LMIN and LMAX for the symmetric matrix A after ITER iterations.
% TOL specifies the tolerance of the method and M is the dimension of the Krylov
% subspace.
[n,dim]=size(A);
if n~=dim, error('Only square matrices'); end
V=[0*[1:n]',[1,0*[1:n-1]]'];
beta(1)=0; normb=1; k=1; deltaeig(1)=1;
while k<=m & normb>=eps & deltaeig(k)>tol
  vk = V(:,k+1);   w = A*vk-beta(k)*V(:,k);
  alpha(k)= w'*vk; w = w - alpha(k)*vk;
  normb = norm(w,2); beta(k+1)=normb;
  if normb ~= 0
    V=[V,w/normb];
    if k==1
      lmin(1)=alpha;
      lmax(1)=alpha;
      k=k+1;
      deltaeig(k)=1;
    else
      d=alpha;
      b=beta(2:length(beta)-1);
      [ak,bk,ck,nch,niter]=givsturm(d,b,1,tol);
      lmax(k)=(ak(niter)+bk(niter))/2;
      [ak,bk,ck,nch,niter]=givsturm(d,b,k,tol);
      lmin(k)=(ak(niter)+bk(niter))/2;
      deltaeig(k+1)=max(abs(lmin(k)-lmin(k-1)),abs(lmax(k)-lmax(k-1)));
      k=k+1;
    end
  else
    fprintf('Breakdown');
    d=alpha; b=beta(2:length(beta)-1);
    [ak,bk,ck,nch,niter]=givsturm(d,b,1,tol);
    lmax(k)=(ak(niter)+bk(niter))/2;
    [ak,bk,ck,nch,niter]=givsturm(d,b,k,tol);
    lmin(k)=(ak(niter)+bk(niter))/2;
    deltaeig(k+1)=max(abs(lmin(k)-lmin(k-1)),abs(lmax(k)-lmax(k-1)));
```

```
    k=k+1;
  end
end
k=k-1;
return
```

Example 5.18 Consider the eigenvalue problem for the matrix A$\in \mathbb{R}^{n\times n}$ with $n = 100$, having diagonal entries equal to 2 and off-diagonal entries equal to -1 on the upper and lower tenth diagonal. Program 45, with `m=100` and `tol=eps`, takes 10 iterations to approximate the extremal eigenvalues of A with an absolute error of the order of the machine precision. •

5.12 Applications

A classical problem in engineering is to determine the *proper or natural frequencies* of a system (mechanical, structural or electric). Typically, this leads to solving a matrix eigenvalue problem. Two examples coming from structural applications are presented in the forthcoming sections where the buckling problem of a beam and the study of the free vibrations of a bridge are considered.

5.12.1 Analysis of the Buckling of a Beam

Consider the homogeneous and thin beam of length L shown in Figure 5.4. The beam is simply supported at the end and is subject to a normal compression load P at $x = L$. Denote by $y(x)$ the vertical displacement of the beam; the structure constraints demand that $y(0) = y(L) = 0$. Let us consider the problem of the buckling of the beam. This amounts to determining the *critical load* P_{cr}, i.e. the smallest value of P such that an equilibrium configuration of the beam exists which is *different* from being rectilinear. Reaching the condition of critical load is a warning of structure *instability*, so that it is quite important to determine its value accurately.

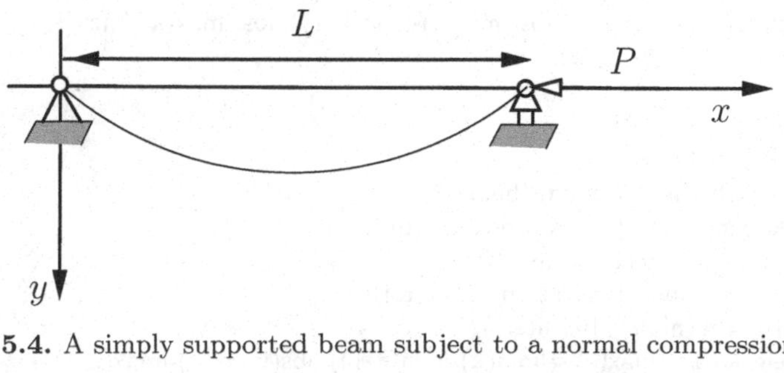

Fig. 5.4. A simply supported beam subject to a normal compression load

The explicit computation of the critical load can be worked out under the assumption of small displacements, writing the equilibrium equation for the structure in its deformed configuration (see Figure 5.4)

$$\begin{cases} -E\,(J(x)y'(x))' = M_e(x), & 0 < x < L, \\ y(0) = y(L) = 0, \end{cases} \tag{5.68}$$

where E is the constant Young's modulus of the beam and $M_e(x) = Py(x)$ is the momentum of the load P with respect to a generic point of the beam of abscissa x. In (5.68) we are assuming that the momentum of inertia J can be varying along the beam, which indeed happens if the beam has nonuniform cross-section.

Equation (5.68) expresses the equilibrium between the external momentum M_e and the internal momentum $M_i = -E(Jy')'$ which tends to restore the rectilinear equilibrium configuration of the beam. If the stabilizing reaction M_i prevails on the unstabilizing action M_e, the equilibrium of the initial rectilinear configuration is stable. The critical situation (buckling of the beam) clearly arises when $M_i = M_e$.
Assume that J is constant and let $\alpha^2 = P/(EJ)$; solving the boundary value problem (5.68), we get the equation $C \sin\alpha L = 0$, which admits nontrivial solutions $\alpha = (k\pi)/L, k = 1, 2, \ldots$. Taking $k = 1$ yields the value of the critical load $P_{cr} = \frac{\pi^2 EJ}{L^2}$.

To solve numerically the boundary value problem (5.68) it is convenient to introduce for $n \geq 1$, the discretization nodes $x_j = jh$, with $h = L/(n+1)$ and $j = 1, \ldots, n$, thus defining the vector of *nodal approximate displacements* u_j at the internal nodes x_j (where $u_0 = y(0) = 0$, $u_{n+1} = y(L) = 0$). Then, using the finite difference method (see Section 12.2), the calculation of the critical load amounts to determining the *smallest* eigenvalue of the tridiagonal symmetric and positive definite matrix $\mathrm{A} = \mathrm{tridiag}_n(-1, 2, -1) \in \mathbb{R}^{n\times n}$.
It can indeed be checked that the finite difference discretization of problem (5.68) by centered differences leads to the following matrix eigenvalue problem

$$\mathrm{A}\mathbf{u} = \alpha^2 h^2 \mathbf{u},$$

where $\mathbf{u} \in \mathbb{R}^n$ is the vector of nodal displacements u_j. The discrete counterpart of condition $C\sin(\alpha) = 0$ requires that $Ph^2/(EJ)$ coincides with the eigenvalues of A as P varies.

Denoting by λ_{min} and P_{cr}^h, the smallest eigenvalue of A and the (approximate) value of the critical load, respectively, then $P_{cr}^h = (\lambda_{min} EJ)/h^2$. Letting $\theta = \pi/(n+1)$, it can be checked (see Exercise 3, Chapter 4) that the eigenvalues of matrix A are

$$\lambda_j = 2(1 - \cos(j\theta)), \qquad j = 1, \ldots, n. \tag{5.69}$$

The numerical calculation of λ_{min} has been carried out using the Givens algorithm described in Section 5.10.2 and assuming $n = 10$. Running the

Program 42 with an absolute tolerance equal to the *roundoff* unit, the solution $\lambda_{min} \simeq 0.081$ has been obtained after 57 iterations.

It is also interesting to analyze the case where the beam has nonuniform cross-section, since the value of the critical load, unlike the previous situation, is not exactly known *a priori*. We assume that, for each $x \in [0, L]$, the section of the beam is rectangular, with depth a fixed and height σ that varies according to the rule

$$\sigma(x) = s\left[1 + \left(\frac{S}{s} - 1\right)\left(\frac{x}{L} - 1\right)^2\right], \qquad 0 \le x \le L,$$

where S and s are the values at the ends, with $S \ge s > 0$. The momentum of inertia, as a function of x, is given by $J(x) = (1/12)a\sigma^3(x)$; proceeding similarly as before, we end up with a system of linear algebraic equations of the form

$$\tilde{\mathrm{A}}\mathbf{u} = (P/E)h^2\mathbf{u},$$

where this time $\tilde{\mathrm{A}} = \mathrm{tridiag}_n(\mathbf{b}, \mathbf{d}, \mathbf{b})$ is a tridiagonal, symmetric and positive definite matrix having diagonal entries $d_i = J(x_{i-1/2}) + J(x_{i+1/2})$, for $i = 1, \ldots, n$, and off-diagonal entries $b_i = -J(x_{i+1/2})$, for $i = 1, \ldots, n-1$.

Assume the following values of the parameters: $a = 0.4\,[m]$, $s = a$, $S = 0.5\,[m]$ and $L = 10\,[m]$. To ensure a correct dimensional comparison, we have multiplied by $\bar{J} = a^4/12$ the smallest eigenvalue of the matrix A in the uniform case (corresponding to $S = s = a$), obtaining $\lambda_{min} = 1.7283 \cdot 10^{-4}$. Running Program 42, with $n = 10$, yields in the nonuniform case the value $\lambda_{min} = 2.243 \cdot 10^{-4}$. This result confirms that the critical load increases for a beam having a wider section at $x = 0$, that is, the structure enters the instability regime for higher values of the load than in the uniform cross-section case.

5.12.2 Free Dynamic Vibration of a Bridge

We are concerned with the analysis of the free response of a bridge whose schematic structure is shown in Figure 5.5. The number of the nodes of the structure is equal to $2n$ while the number of the beams is $5n$. Each horizontal and vertical beam has a mass equal to m while the diagonal beams have mass equal to $m\sqrt{2}$. The stiffness of each beam is represented by the spring constant κ. The nodes labeled by "0" and "$2n+1$" are constrained to ground.

Denoting by $\mathbf{x}$ and $\mathbf{y}$ the vectors of the $2n$ nodal horizontal and vertical displacements the free response of the bridge can be studied by solving the generalized eigenvalue problems

$$\mathrm{M}\mathbf{x} = \lambda\mathrm{K}\mathbf{x}, \qquad \mathrm{M}\mathbf{y} = \lambda\mathrm{K}\mathbf{y}, \tag{5.70}$$

where $\mathrm{M} = m\mathrm{diag}_{2n}(\alpha, \mathbf{b}, \alpha, \gamma, \mathbf{b}, \gamma)$, where $\alpha = 3 + \sqrt{2}$, $\mathbf{b} = [\beta, \ldots, \beta]^T \in \mathbb{R}^{n-2}$ with $\beta = 3/2 + \sqrt{2}$ and $\gamma = 1 + \sqrt{2}$,

$$\mathrm{K} = \kappa \begin{bmatrix} K_{11} & K_{12} \\ K_{12} & K_{11} \end{bmatrix}$$

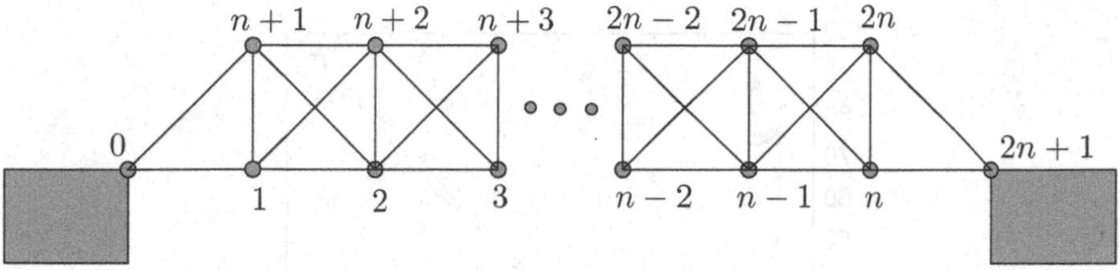

Fig. 5.5. Schematic structure of a bridge

for a positive constant κ and where $K_{12} = \text{tridiag}_n(-1, -1, -1)$, $K_{11} = \text{tridiag}_n(-1, \mathbf{d}, -1)$ with $\mathbf{d} = [4, 5, \ldots, 5, 4]^T \in \mathbb{R}^n$. The diagonal matrix M is the *mass matrix* while the symmetric and positive definite matrix K is the *stiffness* matrix.
For $k = 1, \ldots, 2n$ we denote by $(\lambda_k, \mathbf{z}_k)$ any eigenvalue/eigenvector pair of (5.70) and call $\omega_k = \sqrt{\lambda_k}$ the *natural frequencies* and $\mathbf{z}_k$ the *modes* of vibration of the bridge. The study of the free vibrations is of primary importance in the design of a structure like a bridge or a multi-story building. Indeed, if the excitation frequency of an external force (vehicles, wind or, even worse, an earthquake) coincides with one of the natural frequencies of the structure then a condition of *resonance* occurs and, as a result, large oscillations may dangerously arise.

Let us now deal with the numerical solution of the matrix eigenvalue problem (5.70). For this purpose we introduce the change of variable $\mathbf{z} = \mathrm{M}^{1/2}\mathbf{x}$ (or $\mathbf{z} = \mathrm{M}^{1/2}\mathbf{y}$) so that each generalized eigenvalue problem in (5.70) can be conveniently reformulated as

$$\mathrm{C}\mathbf{z} = \tilde{\lambda}\mathbf{z},$$

where $\tilde{\lambda} = 1/\lambda$ and the matrix $\mathrm{C} = \mathrm{M}^{-1/2}\mathrm{K}\mathrm{M}^{-1/2}$ is symmetric positive definite. This property allows us to use the Lanczos method described in Section 5.11 and also ensures quadratic convergence of the power iterations (see Section 5.11).

We approximate the first two subdominant eigenvalues $\tilde{\lambda}_{2n}$ and $\tilde{\lambda}_{2n-1}$ of the matrix C (i.e., its smallest and second smallest eigenvalues) in the case $m = \kappa = 1$ using the deflation procedure considered in Remark 5.3. The inverse power iteration and the Lanczos method are compared in the computation of $\tilde{\lambda}_{2n}$ and $\tilde{\lambda}_{2n-1}$ in Figure 5.6.

The results show the superiority of the Lanczos method over the inverse iterations only when the matrix C is of small size. This is to be ascribed to the fact that, as n grows, the progressive influence of the rounding errors causes a loss of mutual orthogonality of the Lanczos vectors and, in turn, an increase in the number of iterations for the method to converge. Suitable reorthogonalization procedures are thus needed to improve the performances of the Lanczos iteration as pointed out in Section 5.11.

We conclude the free response analysis of the bridge showing in Figure 5.7 (in the case $n = 5$, $m = 10$ and $\kappa = 1$) the modes of vibration $\mathbf{z}_8$ and $\mathbf{z}_{10}$

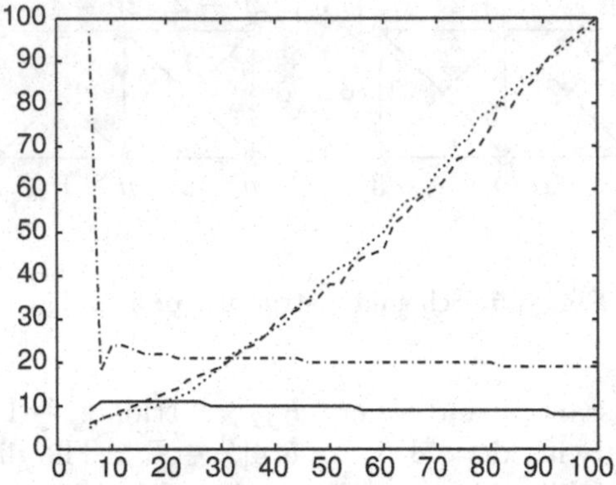

Fig. 5.6. Iterations number of the Lanczos method and of the inverse power method versus the size $2n$ of matrix C. The solid and the dash-dotted curves refer to the inverse power method (for $\tilde{\lambda}_{2n}$ and $\tilde{\lambda}_{2n-1}$ respectively), while the dashed and the dotted curves refer to the Lanczos method (still for $\tilde{\lambda}_{2n}$ and $\tilde{\lambda}_{2n-1}$, respectively)

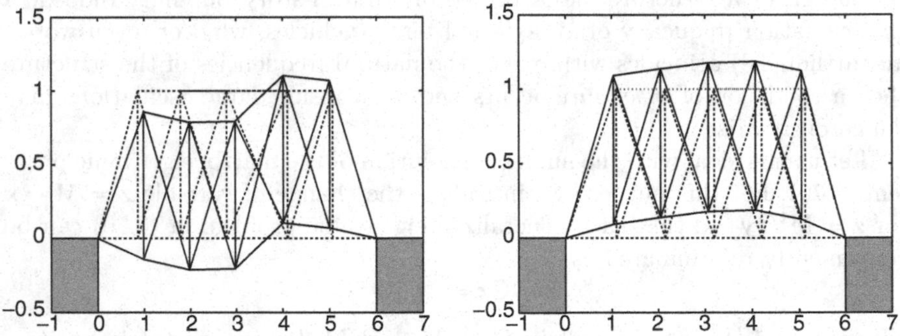

Fig. 5.7. Modes of vibration corresponding to the natural frequencies ω_8 (*left*) and ω_{10} (*right*). The undeformed configuration of the bridge is drawn in dotted line

corresponding to the natural frequencies $\omega_8 = 990.42$ and $\omega_{10} = 2904.59$. The MATLAB built-in function `eig` has been employed to solve the generalized eigenvalue problems (5.70) as explained in Section 5.9.1.

5.13 Exercises

1. Using the Gershgorin theorems, localize the eigenvalues of the matrix A which is obtained setting $A = (P^{-1}DP)^T$ and then $a_{13} = 0$, $a_{23} = 0$, where $D = \text{diag}_3(1, 50, 100)$ and

$$P = \begin{bmatrix} 1 & 1 & 1 \\ 10 & 20 & 30 \\ 100 & 50 & 60 \end{bmatrix}.$$

[*Solution* : $\sigma(A) = \{-151.84, 80.34, 222.5\}$.]

2. Localize the spectrum of the matrix

$$A = \begin{bmatrix} 1 & 2 & -1 \\ 2 & 7 & 0 \\ -1 & 0 & 5 \end{bmatrix}.$$

[*Solution* : $\sigma(A) \subset [-2, 9]$.]

3. Draw the oriented graph of the matrix

$$A = \begin{bmatrix} 1 & 3 & 0 \\ 0 & 2 & -1 \\ -1 & 0 & 2 \end{bmatrix}.$$

4. Check if the following matrices are reducible.

$$A_1 = \begin{bmatrix} 1 & 0 & -1 & 0 \\ 2 & 3 & -2 & 1 \\ -1 & 0 & -2 & 0 \\ 1 & -1 & 1 & 4 \end{bmatrix}, \quad A_2 = \begin{bmatrix} 0 & 0 & 1 & 0 \\ 0 & 0 & 0 & 1 \\ 0 & 1 & 0 & 0 \\ 1 & 0 & 0 & 0 \end{bmatrix}.$$

[*Solution* : A_1, reducible; A_2, irreducible.]

5. Provide an estimate of the number of complex eigenvalues of the matrix

$$A = \begin{bmatrix} -4 & 0 & 0 & 0.5 & 0 \\ 2 & 2 & 4 & -3 & 1 \\ 0.5 & 0 & -1 & 0 & 0 \\ 0.5 & 0 & 0.2 & 3 & 0 \\ 2 & 0.5 & -1 & 3 & 4 \end{bmatrix}.$$

[*Hint* : Check that A can be reduced to the form

$$A = \begin{bmatrix} M_1 & M_2 \\ 0 & M_3 \end{bmatrix},$$

where $M_1 \in \mathbb{R}^{2\times 2}$ and $M_2 \in \mathbb{R}^{3\times 3}$. Then, study the eigenvalues of blocks M_1 and M_2 using the Gershgorin theorems and check that A has no complex eigenvalues.]

6. Let $A \in \mathbb{C}^{n\times n}$ be a diagonal matrix and let $\widetilde{A} = A + E$ be a perturbation of A with $e_{ii} = 0$ for $i = 1, \ldots, n$. Show that

$$|\lambda_i(\widetilde{A}) - \lambda_i(A)| \le \sum_{j=1}^{n} |e_{ij}|, \qquad i = 1, \ldots, n. \tag{5.71}$$

7. Apply estimate (5.71) to the case in which A and E are, for $\varepsilon \ge 0$, the matrices

$$A = \begin{bmatrix} 1 & 0 \\ 0 & 2 \end{bmatrix}, \qquad E = \begin{bmatrix} 0 & \varepsilon \\ \varepsilon & 0 \end{bmatrix}.$$

[*Solution* : $\sigma(A) = \{1, 2\}$ and $\sigma(\widetilde{A}) = \{(3 \mp \sqrt{1+4\varepsilon^2})/2\}$.]
8. Check that finding the zeros of a polynomial of degree $\le n$ with real coefficients

$$p_n(x) = \sum_{k=0}^{n} a_k x^k = a_0 + a_1 x + \ldots + a_n x^n, \quad a_n \ne 0, \quad a_k \in \mathbb{R}, \ k = 0, \ldots, n$$

is equivalent to determining the spectrum of the Frobenius matrix $C \in \mathbb{R}^{n\times n}$ associated with p_n (known as the *companion matrix*)

$$C = \begin{bmatrix} -(a_{n-1}/a_n) & -(a_{n-2}/a_n) & \ldots & -(a_1/a_n) & -(a_0/a_n) \\ 1 & 0 & \ldots & 0 & 0 \\ 0 & 1 & \ldots & 0 & 0 \\ \vdots & \vdots & \ddots & \vdots & \vdots \\ 0 & 0 & \ldots & 1 & 0 \end{bmatrix}. \tag{5.72}$$

An important consequence of the result above is that, due to Abel's theorem, there exist in general no direct methods for computing the eigenvalues of a given matrix, for $n \ge 5$.
9. Show that if matrix $A \in \mathbb{C}^{n\times n}$ admits eigenvalue/eigenvector pairs $(\lambda, \mathbf{x})$, then the matrix $U^H A U$, with U unitary, admits eigenvalue/eigenvector pairs $(\lambda, U^H\mathbf{x})$. (Similarity transformation using an orthogonal matrix).
10. Suppose that all the assumptions needed to apply the power method are satisfied except for the requirement $\alpha_1 \ne 0$ (see Section 5.3.1). Show that in such a case the sequence (5.17) converges to the eigenvalue/eigenvector pair $(\lambda_2, \mathbf{x}_2)$. Then, study experimentally the behaviour of the method, computing the pair $(\lambda_1, \mathbf{x}_1)$ for the matrix

$$A = \begin{bmatrix} 1 & -1 & 2 \\ -2 & 0 & 5 \\ 6 & -3 & 6 \end{bmatrix}.$$

For this, use Program 26, taking $\mathbf{q}^{(0)} = \mathbf{1}/\sqrt{3}$ and $\mathbf{q}^{(0)} = \mathbf{w}^{(0)}/\|\mathbf{w}^{(0)}\|_2$, respectively, where $\mathbf{w}^{(0)} = (1/3)\mathbf{x}_2 - (2/3)\mathbf{x}_3$.
[*Solution* : $\lambda_1 = 5$, $\lambda_2 = 3$, $\lambda_3 = -1$ and $\mathbf{x}_1 = [5, 16, 18]^T$, $\mathbf{x}_2 = [1, 6, 4]^T$, $\mathbf{x}_3 = [5, 16, 18]^T$.]

11. Show that the *companion matrix* associated with the polynomial $p_n(x) = x^n + a_n x^{n-1} + \ldots + a_1$, can be written in the alternative form (5.72)

$$A = \begin{bmatrix} 0 & a_1 & & & & 0 \\ -1 & 0 & a_2 & & & \\ & \ddots & \ddots & \ddots & & \\ & & -1 & 0 & a_{n-1} \\ 0 & & & -1 & a_n \end{bmatrix}.$$

12. (From [FF63]) Suppose that a real matrix $A \in \mathbb{R}^{n\times n}$ has two maximum module complex eigenvalues given by $\lambda_1 = \rho e^{i\theta}$ and $\lambda_2 = \rho e^{-i\theta}$, with $\theta \neq 0$. Assume, moreover, that the remaining eigenvalues have modules less than ρ. The power method can then be modified as follows:
let $\mathbf{q}^{(0)}$ be a real vector and $\mathbf{q}^{(k)}$ be the vector provided by the power method without normalization. Then, set $x_k = \mathbf{q}_{n_0}^{(k)}$ for some n_0, with $1 \le n_0 \le n$. Prove that

$$\rho^2 = \frac{x_k x_{k+2} - x_{k+1}^2}{x_{k-1}x_{k+1} - x_k^2} + \mathcal{O}\left(\left|\frac{\lambda_3}{\rho}\right|^k\right),$$

$$\cos(\theta) = \frac{\rho x_{k-1} + r^{-1} x_{k+1}}{2x_k} + \mathcal{O}\left(\left|\frac{\lambda_3}{\rho}\right|^k\right).$$

[*Hint* : first, show that

$$x_k = C(\rho^k \cos(k\theta + \alpha)) + \mathcal{O}\left(\left|\frac{\lambda_3}{\rho}\right|^k\right),$$

where α depends on the components of the initial vector along the directions of the eigenvectors associated with λ_1 and λ_2.]

13. Apply the modified power method of Exercise 12 to the matrix

$$A = \begin{bmatrix} 1 & -\frac{1}{4} & \frac{1}{4} \\ 1 & 0 & 0 \\ 0 & 1 & 0 \end{bmatrix},$$

and compare the obtained results with those yielded by the standard power method.

14. (Taken from [Dem97]). Apply the QR iteration with double shift to compute the eigenvalues of the matrix

$$A = \begin{bmatrix} 0 & 0 & 1 \\ 1 & 0 & 0 \\ 0 & 1 & 0 \end{bmatrix}.$$

Run Program 37 setting `tol=eps`, `itmax=100` and comment about the form of the obtained matrix $\mathtt{T}^{(\mathtt{iter})}$ after `iter` iterations of the algorithm.
[*Solution* : the eigenvalues of A are the solution of $\lambda^3 - 1 = 0$, i.e., $\sigma(A) = \{1, -1/2 \pm \sqrt{3}/2i\}$. After `iter=100` iterations, Program 37 yields the matrix

$$\mathrm{T}^{(100)} = \begin{bmatrix} 0 & 0 & -1 \\ 1 & 0 & 0 \\ 0 & -1 & 0 \end{bmatrix},$$

which means that the QR iteration leaves A unchanged (except for sign changes that are nonrelevant for eigenvalues computation). This is a simple but glaring example of matrix for which the QR method with double shift fails to converge.]

Part III

Around Functions and Functionals

6

Rootfinding for Nonlinear Equations

This chapter deals with the numerical approximation of the zeros of a real-valued function of one variable, that is

$$\text{given } f : \mathcal{I} = (a,b) \subseteq \mathbb{R} \to \mathbb{R}, \text{ find } \alpha \in \mathbb{C} \text{ such that } f(\alpha) = 0. \tag{6.1}$$

The analysis of problem (6.1) in the case of systems of nonlinear equations will be addressed in Chapter 7. It is important to notice that, although f is assumed to be real-valued, its zeros can be complex. This is, e.g., the case when f is an algebraic polynomial of degree $\leq n$, which will be addressed in Section 6.4.

Methods for the numerical approximation of a zero of f are usually iterative. The aim is to generate a sequence of values $x^{(k)}$ such that

$$\lim_{k\to\infty} x^{(k)} = \alpha.$$

The convergence of the iteration is characterized by the following definition.

Definition 6.1 A sequence $\{x^{(k)}\}$ generated by a numerical method is said to *converge to α with order $p \geq 1$* if

$$\exists C > 0 : \frac{|x^{(k+1)} - \alpha|}{|x^{(k)} - \alpha|^p} \leq C, \ \forall k \geq k_0, \tag{6.2}$$

where $k_0 \geq 0$ is a suitable integer. In such a case, the method is said to be of *order p*. Notice that if p is equal to 1, in order for $x^{(k)}$ to converge to α it is necessary that $C < 1$ in (6.2). In such an event, the constant C is called the *convergence factor* of the method. ■

Unlike the case of linear systems, convergence of iterative methods for rootfinding of nonlinear equations depends in general on the choice of the initial datum $x^{(0)}$. This allows for establishing only *local* convergence results, that is, holding for any $x^{(0)}$ which belongs to a suitable neighborhood of the root α. Methods for which convergence to α holds *for any* choice of $x^{(0)}$ in the interval $\mathcal{I}$, are said to be *globally convergent* to α.

6.1 Conditioning of a Nonlinear Equation

Consider the nonlinear equation $f(x) = \varphi(x) - d = 0$ and assume that f is a continuously differentiable function. Let us analyze the sensitivity of finding the roots of f with respect to changes in the datum d.

The problem is well posed only if the function φ is invertible. In such a case, indeed, one gets $\alpha = \varphi^{-1}(d)$ from which, using the notation of Chapter 2, the resolvent G is φ^{-1}. On the other hand, $(\varphi^{-1})'(d) = 1/\varphi'(\alpha)$, so that formula (2.7) for the approximate condition number (relative and absolute) yields

$$K(d) \simeq \frac{|d|}{|\alpha||f'(\alpha)|}, \qquad K_{abs}(d) \simeq \frac{1}{|f'(\alpha)|}. \tag{6.3}$$

The problem is thus ill-conditioned when $f'(\alpha)$ is "small" and well-conditioned if $f'(\alpha)$ is "large".

The analysis which leads to (6.3) can be generalized to the case in which α is a root of f with multiplicity $m > 1$ as follows. Expanding φ in a Taylor series around α up to the m-th order term, we get

$$d + \delta d = \varphi(\alpha + \delta\alpha) = \varphi(\alpha) + \sum_{k=1}^{m} \frac{\varphi^{(k)}(\alpha)}{k!}(\delta\alpha)^k + o((\delta\alpha)^m).$$

Since $\varphi^{(k)}(\alpha) = 0$ for $k = 1, \ldots, m-1$, we obtain

$$\delta d = f^{(m)}(\alpha)(\delta\alpha)^m/m!,$$

so that an approximation to the absolute condition number is

$$K_{abs}(d) \simeq \left|\frac{m!\delta d}{f^{(m)}(\alpha)}\right|^{1/m} \frac{1}{|\delta d|}. \tag{6.4}$$

Notice that (6.3) is the special case of (6.4) where $m = 1$. From this it also follows that, even if δd is sufficiently small to make $|m!\delta d/f^{(m)}(\alpha)| < 1$, $K_{abs}(d)$ could nevertheless be a large number. We therefore conclude that the problem of rootfinding of a nonlinear equation is well-conditioned if α is a simple root and $|f'(\alpha)|$ is definitely different from zero, ill-conditioned otherwise.

Let us now consider the following problem, which is closely connected with the previous analysis. Assume $d = 0$ and let α be a simple root of f; moreover, for $\hat{\alpha} \neq \alpha$, let $f(\hat{\alpha}) = \hat{r} \neq 0$. We seek a bound for the difference $\hat{\alpha} - \alpha$ as a function of the *residual* $\hat{r}$. Applying (6.3) yields

$$K_{abs}(0) \simeq \frac{1}{|f'(\alpha)|}.$$

Therefore, letting $\delta x = \hat{\alpha} - \alpha$ and $\delta d = \hat{r}$ in the definition of K_{abs} (see (2.5)), we get

$$\frac{|\hat{\alpha}-\alpha|}{|\alpha|} \lesssim \frac{|\hat{r}|}{|f'(\alpha)||\alpha|}, \tag{6.5}$$

where the following convention has been adopted: if $a \leq b$ and $a \simeq c$, then we write $a \lesssim c$. If α has multiplicity $m > 1$, using (6.4) instead of (6.3) and proceeding as above, we get

$$\frac{|\hat{\alpha}-\alpha|}{|\alpha|} \lesssim \left(\frac{m!}{|f^{(m)}(\alpha)||\alpha|^m}\right)^{1/m} |\hat{r}|^{1/m}. \tag{6.6}$$

These estimates will be useful in the analysis of stopping criteria for iterative methods (see Section 6.5).

A remarkable example of a nonlinear problem is when f is a polynomial p_n of degree n, in which case it admits exactly n roots α_i, real or complex, each one counted with its multiplicity. We want to investigate the sensitivity of the roots of p_n with respect to the changes of its coefficients.

To this end, let $\hat{p}_n = p_n + q_n$, where q_n is a perturbation polynomial of degree n, and let $\hat{\alpha}_i$ be the corresponding roots of $\hat{p}_n$. A direct use of (6.6) yields for any root α_i the following estimate

$$E^i_{rel} = \frac{|\hat{\alpha}_i-\alpha_i|}{|\alpha_i|} \lesssim \left(\frac{m!}{|p_n^{(m)}(\alpha_i)||\alpha_i|^m}\right)^{1/m} |q_n(\hat{\alpha}_i)|^{1/m} = S^i, \tag{6.7}$$

where m is the multiplicity of the root at hand and $q_n(\hat{\alpha}_i) = -p_n(\hat{\alpha}_i)$ is the "residual" of the polynomial p_n evaluated at the perturbed root.

Remark 6.1 A formal analogy exists between the a priori estimates so far obtained for the nonlinear problem $\varphi(\alpha) = d$ and those developed in Section 3.1.2 for linear systems, provided that A corresponds to φ and $\mathbf{b}$ to d. More precisely, (6.5) is the analogue of (3.9) if δA=0, and the same holds for (6.7) (for $m = 1$) if $\boldsymbol{\delta}\mathbf{b} = \mathbf{0}$. ■

Example 6.1 Let $p_4(x) = (x-1)^4$, and let $\hat{p}_4(x) = (x-1)^4 - \varepsilon$, with $0 < \varepsilon \ll 1$. The roots of the perturbed polynomial are simple and equal to $\hat{\alpha}_i = \alpha_i + \sqrt[4]{\varepsilon}$, where $\alpha_i = 1$ are the (coincident) zeros of p_4. They lie with intervals of $\pi/2$ on the circle of radius $\sqrt[4]{\varepsilon}$ and center $z = (1,0)$ in the complex plane.

The problem is stable (that is $\lim_{\varepsilon\to 0} \hat{\alpha}_i = 1$), but is *ill-conditioned* since

$$\frac{|\hat{\alpha}_i-\alpha_i|}{|\alpha_i|} = \sqrt[4]{\varepsilon}, \qquad i = 1, \ldots 4,$$

For example, if $\varepsilon = 10^{-4}$ the relative change is 10^{-1}. Notice that the right-side of (6.7) is just $\sqrt[4]{\varepsilon}$, so that, in this case, (6.7) becomes an equality. •

Example 6.2 (Wilkinson). Consider the following polynomial

$$p_{10}(x) = \Pi_{k=1}^{10}(x+k) = x^{10} + 55x^9 + \ldots + 10!.$$

Table 6.1. Relative error and estimated error using (6.7) for the Wilkinson polynomial of degree 10

i	E^i_{rel}	S^i	i	E^i_{rel}	S^i
1	$3.039 \cdot 10^{-13}$	$3.285 \cdot 10^{-13}$	6	$6.956 \cdot 10^{-5}$	$6.956 \cdot 10^{-5}$
2	$7.562 \cdot 10^{-10}$	$7.568 \cdot 10^{-10}$	7	$1.589 \cdot 10^{-4}$	$1.588 \cdot 10^{-4}$
3	$7.758 \cdot 10^{-8}$	$7.759 \cdot 10^{-8}$	8	$1.984 \cdot 10^{-4}$	$1.987 \cdot 10^{-4}$
4	$1.808 \cdot 10^{-6}$	$1.808 \cdot 10^{-6}$	9	$1.273 \cdot 10^{-4}$	$1.271 \cdot 10^{-4}$
5	$1.616 \cdot 10^{-5}$	$1.616 \cdot 10^{-5}$	10	$3.283 \cdot 10^{-5}$	$3.286 \cdot 10^{-5}$

Let $\hat{p}_{10} = p_{10} + \varepsilon x^9$, with $\varepsilon = 2^{-23} \simeq 1.2 \cdot 10^{-7}$. Let us study the conditioning of finding the roots of p_{10}. Using (6.7) with $m = 1$, we report for $i = 1, \ldots, 10$ in Table 6.1 the relative errors E^i_{rel} and the corresponding estimates S^i.

These results show that the problem is ill-conditioned, since the maximum relative error for the root $\alpha_8 = -8$ is three orders of magnitude larger than the corresponding absolute perturbation. Moreover, excellent agreement can be observed between the a priori estimate and the actual relative error. •

6.2 A Geometric Approach to Rootfinding

In this section we introduce the following methods for finding roots: the bisection method, the chord method, the secant method, the false position (or *Regula Falsi*) method and Newton's method. The order of the presentation reflects the growing complexity of the algorithms. In the case of the bisection method, indeed, the only information that is being used is the *sign* of the function f at the end points of any bisection (sub)interval, whilst the remaining algorithms also take into account the *values* of the function and/or its derivative.

6.2.1 The Bisection Method

The bisection method is based on the following property.

Property 6.1 (theorem of zeros for continuous functions) *Given a continuous function* $f : [a, b] \to \mathbb{R}$*, such that* $f(a)f(b) < 0$*, then* $\exists\, \alpha \in (a, b)$ *such that* $f(\alpha) = 0$.

Starting from $\mathcal{I}_0 = [a, b]$, the bisection method generates a sequence of subintervals $\mathcal{I}_k = [a^{(k)}, b^{(k)}]$, $k \geq 0$, with $\mathcal{I}_k \subset \mathcal{I}_{k-1}$, $k \geq 1$, and enjoys the property that $f(a^{(k)})f(b^{(k)}) < 0$. Precisely, we set $a^{(0)} = a$, $b^{(0)} = b$ and $x^{(0)} = (a^{(0)} + b^{(0)})/2$; then, for $k \geq 0$:

$$\text{set } a^{(k+1)} = a^{(k)},\ b^{(k+1)} = x^{(k)} \qquad \text{if } f(x^{(k)})f(a^{(k)}) < 0;$$

$$\text{set } a^{(k+1)} = x^{(k)},\ b^{(k+1)} = b^{(k)} \qquad \text{if } f(x^{(k)})f(b^{(k)}) < 0;$$

$$\text{finally, set } x^{(k+1)} = (a^{(k+1)} + b^{(k+1)})/2.$$

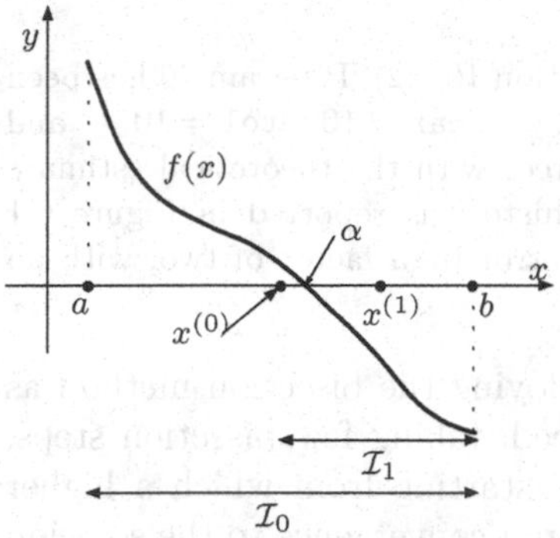

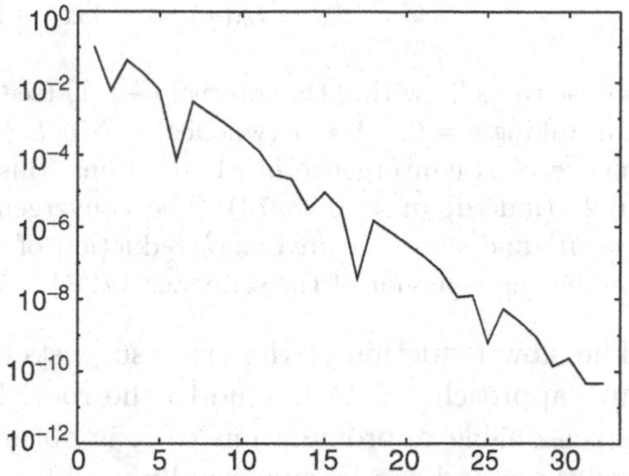

Fig. 6.1. The bisection method. The first two steps (*left*); convergence history for the Example 6.3 (*right*). The number of iterations and the absolute error as a function of k are reported on the x- and y-axis, respectively

The bisection iteration terminates at the m-th step for which $|x^{(m)} - \alpha| \leq |\mathcal{I}_m| \leq \varepsilon$, where ε is a fixed tolerance and $|\mathcal{I}_m|$ is the length of $\mathcal{I}_m$. As for the *speed of convergence* of the bisection method, notice that $|\mathcal{I}_0| = b - a$, while

$$|\mathcal{I}_k| = |\mathcal{I}_0|/2^k = (b-a)/2^k, \qquad k \geq 0. \tag{6.8}$$

Denoting by $e^{(k)} = x^{(k)} - \alpha$ the *absolute error* at step k, from (6.8) it follows that $|e^{(k)}| < |\mathcal{I}_k|/2 = (b-a)/2^{k+1}$, $k \geq 0$, which implies $\lim_{k\to\infty} |e^{(k)}| = 0$.

The bisection method is therefore *globally convergent.* Moreover, to get $|x^{(m)} - \alpha| \leq \varepsilon$ we must take

$$m \geq \log_2\left(\frac{b-a}{\varepsilon}\right) - 1 = \frac{\log((b-a)/\varepsilon)}{\log(2)} - 1 \simeq \frac{\log((b-a)/\varepsilon)}{0.6931} - 1. \tag{6.9}$$

In particular, to gain a significant figure in the accuracy of the approximation of the root (that is, to have $|x^{(k)} - \alpha| = |x^{(j)} - \alpha|/10$), one needs $k - j = \log_2(10) - 1 \simeq 2.32$ bisections. This singles out the bisection method as an algorithm of certain, but slow, convergence. We must also point out that the bisection method does not generally guarantee a *monotone* reduction of the absolute error between two successive iterations, that is, we cannot ensure a priori that

$$|e^{(k+1)}| \leq \mathcal{M}_k |e^{(k)}|, \qquad \text{for any } k \geq 0, \tag{6.10}$$

with $\mathcal{M}_k < 1$. For this purpose, consider the situation depicted in Figure 6.1 (*left*), where clearly $|e^{(1)}| > |e^{(0)}|$. Failure to satisfy (6.10) does not allow for qualifying the bisection method as a method of order 1, in the sense of Definition 6.1.

Example 6.3 Let us check the convergence properties of the bisection method in the approximation of the root $\alpha \simeq 0.9062$ of the Legendre polynomial of degree 5

$$L_5(x) = \frac{x}{8}(63x^4 - 70x^2 + 15),$$

whose roots lie within the interval $(-1, 1)$ (see Section 10.1.2). Program 46 has been run taking `a` $= 0.6$, `b` $= 1$ (whence, $L_5(a) \cdot L_5(b) < 0$), **`nmax`** $= 100$, `tol` $= 10^{-10}$ and has reached convergence in 31 iterations, this agrees with the theoretical estimate (6.9) (indeed, $m \geq 30.8974$). The convergence history is reported in Figure 6.1 (*right*) and shows an (average) reduction of the error by a factor of two, with an oscillating behavior of the sequence $\{x^{(k)}\}$. •

The slow reduction of the error suggests employing the bisection method as an "approaching" technique to the root. Indeed, taking few bisection steps, a reasonable approximation to α is obtained, starting from which a higher order method can be successfully used for a rapid convergence to the solution within the fixed tolerance. An example of such a procedure will be addressed in Section 6.7.1.

The bisection algorithm is implemented in Program 46. The input parameters, here and in the remainder of this chapter, have the following meaning: `a` and `b` denote the end points of the search interval, `fun` is the variable containing the expression of the function f, `tol` is a fixed tolerance and **`nmax`** is the maximum admissible number of steps for the iterative process.

In the output vectors `xvect`, `xdif` and `fx` the sequences $\{x^{(k)}\}$, $\{|x^{(k+1)} - x^{(k)}|\}$ and $\{f(x^{(k)})\}$, for $k \geq 0$, are respectively stored, while `nit` denotes the number of iterations needed to satisfy the stopping criteria. In the case of the bisection method, the code returns as soon as the half-length of the search interval is less than `tol`.

Program 46 - bisect : BISECT method

```
function [xvect,xdif,fx,nit]=bisect(a,b,tol,nmax,fun)
%BISECT Bisection method
% [XVECT,XDIF,FX,NIT]=BISECT(A,B,TOL,NMAX,FUN) tries to find a zero
% of the continuous function FUN in the interval [A,B] using the bisection
% method. FUN accepts real scalar input x and returns a real scalar value.
% XVECT is the vector of iterates, XDIF the vector of the differences between
% consecutive iterates, FX the residual. TOL specifies the tolerance of the
% method.
err=tol+1;
nit=0;
xvect=[]; fx=[]; xdif=[];
while nit<nmax & err>tol
   nit=nit+1;
   c=(a+b)/2; x=c; fc=eval(fun); xvect=[xvect;x];
   fx=[fx;fc]; x=a;
   if fc*eval(fun)>0
      a=c;
   else
      b=c;
   end
```

```
  err=0.5*abs(b-a); xdif=[xdif;err];
end
return
```

6.2.2 The Methods of Chord, Secant and Regula Falsi and Newton's Method

In order to devise algorithms with better convergence properties than the bisection method, it is necessary to include information from the values attained by f and, possibly, also by its derivative f' (if f is differentiable) or by a suitable approximation.

For this purpose, let us expand f in a Taylor series around α and truncate the expansion at the first order. The following *linearized* version of problem (6.1) is obtained

$$f(\alpha) = 0 = f(x) + (\alpha - x)f'(\xi), \tag{6.11}$$

for a suitable ξ between α and x. Equation (6.11) prompts the following iterative method: for any $k \geq 0$, given $x^{(k)}$, determine $x^{(k+1)}$ by solving equation $f(x^{(k)})+(x^{(k+1)}-x^{(k)})q_k = 0$, where q_k is a suitable approximation of $f'(x^{(k)})$.

The method described here amounts to finding the intersection between the x-axis and the straight line of slope q_k passing through the point $(x^{(k)}, f(x^{(k)}))$, and thus can be more conveniently set up in the form

$$x^{(k+1)} = x^{(k)} - q_k^{-1} f(x^{(k)}), \qquad \forall k \geq 0.$$

We consider below four particular choices of q_k.

The chord method. We let

$$q_k = q = \frac{f(b) - f(a)}{b - a}, \qquad \forall k \geq 0,$$

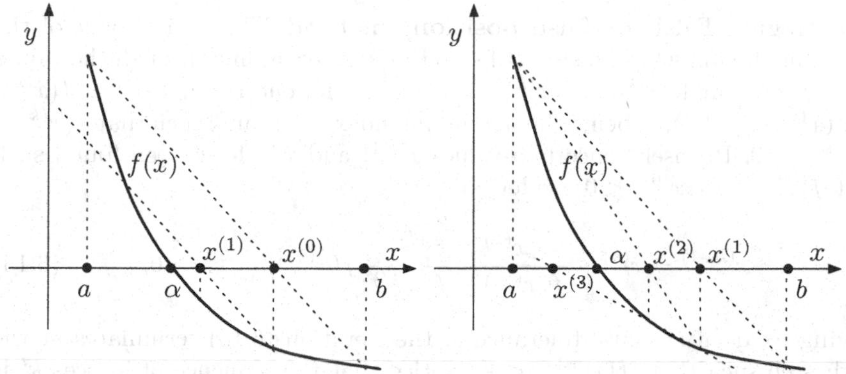

Fig. 6.2. The first step of the chord method (*left*) and the first three steps of the secant method (*right*). For this method we set $x^{(-1)} = b$ and $x^{(0)} = a$

from which, given an initial value $x^{(0)}$, the following recursive relation is obtained

$$x^{(k+1)} = x^{(k)} - \frac{b-a}{f(b)-f(a)} f(x^{(k)}), \qquad k \geq 0. \tag{6.12}$$

In Section 6.3.1, we shall see that the sequence $\{x^{(k)}\}$ generated by (6.12) converges to the root α with order of convergence $p = 1$.

The secant method. We let

$$q_k = \frac{f(x^{(k)}) - f(x^{(k-1)})}{x^{(k)} - x^{(k-1)}}, \qquad \forall k \geq 0, \tag{6.13}$$

from which, giving *two initial values* $x^{(-1)}$ and $x^{(0)}$, we obtain the following relation

$$x^{(k+1)} = x^{(k)} - \frac{x^{(k)} - x^{(k-1)}}{f(x^{(k)}) - f(x^{(k-1)})} f(x^{(k)}), \qquad k \geq 0. \tag{6.14}$$

If compared with the chord method, the iterative process (6.14) requires an extra initial point $x^{(-1)}$ and the corresponding function value $f(x^{(-1)})$, as well as, for any k, computing the incremental ratio (6.13). The benefit due to the increase in the computational cost is the higher speed of convergence of the secant method, as stated in the following property which can be regarded as a first example of the *local convergence* theorem (for the proof see [IK66], pp. 99-101).

Property 6.2 *Let $f \in C^2(\mathcal{J})$, $\mathcal{J}$ being a suitable neighborhood of the root α and assume that $f'(\alpha) \neq 0$. Then, if the initial data $x^{(-1)}$ and $x^{(0)}$ are chosen in $\mathcal{J}$ sufficiently close to α, the sequence* (6.14) *converges to α with order $p = (1+\sqrt{5})/2 \simeq 1.63$.*

The Regula Falsi (or false position) method. This is a variant of the secant method in which, instead of selecting the secant line through the values $(x^{(k)}, f(x^{(k)})$ and $(x^{(k-1)}, f(x^{(k-1)})$, we take the one through $(x^{(k)}, f(x^{(k)})$ and $(x^{(k')}, f(x^{(k')})$, k' being the maximum index less than k such that $f(x^{(k')}) \cdot f(x^{(k)}) < 0$. Precisely, once two values $x^{(-1)}$ and $x^{(0)}$ have been found such that $f(x^{(-1)}) \cdot f(x^{(0)}) < 0$, we let

$$x^{(k+1)} = x^{(k)} - \frac{x^{(k)} - x^{(k')}}{f(x^{(k)}) - f(x^{(k')})} f(x^{(k)}), \qquad k \geq 0. \tag{6.15}$$

Having fixed an absolute tolerance ε, the iteration (6.15) terminates at the m-th step such that $|f(x^{(m)})| < \varepsilon$. Notice that the sequence of indices k' is nondecreasing; therefore, in order to find at step k the *new* value of k', it is not necessary to sweep all the sequence back, but it suffices to stop at the value

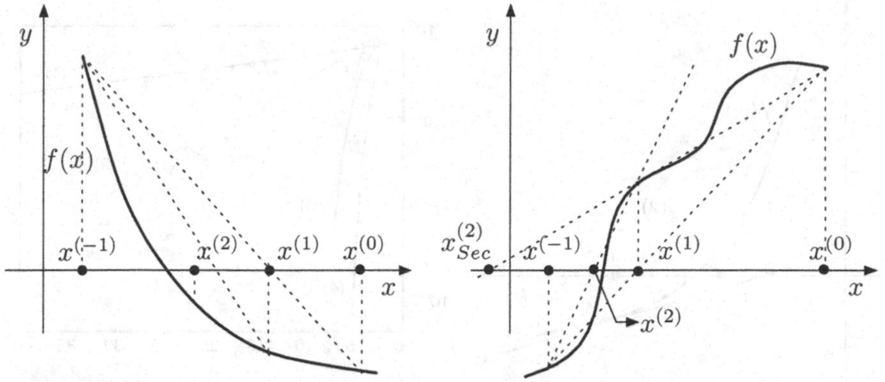

Fig. 6.3. The first two steps of the *Regula Falsi* method for two different functions

of k' that has been determined at the previous step. We show in Figure 6.3 (*left*) the first two steps of (6.15) in the special case in which $x^{(k')}$ coincides with $x^{(-1)}$ for any $k \geq 0$.

The *Regula Falsi* method, though of the same complexity as the secant method, has linear convergence order (see, for example, [RR78], pp. 339-340). However, unlike the secant method, the iterates generated by (6.15) are all contained within the starting interval $[x^{(-1)}, x^{(0)}]$.

In Figure 6.3 (*right*), the first two iterations of both the secant and *Regula Falsi* methods are shown, starting from the same initial data $x^{(-1)}$ and $x^{(0)}$. Notice that the iterate $x^{(1)}$ computed by the secant method coincides with that computed by the *Regula Falsi* method, while the value $x^{(2)}$ computed by the former method (and denoted in the figure by $x^{(2)}_{Sec}$) falls outside the searching interval $[x^{(-1)}, x^{(0)}]$.

In this respect, the *Regula Falsi* method, as well as the bisection method, can be regarded as a *globally convergent* method.

Newton's method. Assuming that $f \in C^1(\mathcal{I})$ and that $f'(\alpha) \neq 0$ (i.e., α is a simple root of f), if we let

$$q_k = f'(x^{(k)}), \qquad \forall k \geq 0$$

and assign the initial value $x^{(0)}$, we obtain the so called *Newton's method*

$$x^{(k+1)} = x^{(k)} - \frac{f(x^{(k)})}{f'(x^{(k)})}, \qquad k \geq 0. \tag{6.16}$$

At the k-th iteration, Newton's method requires the *two* functional evaluations $f(x^{(k)})$ and $f'(x^{(k)})$. The increasing computational cost with respect to the methods previously considered is more than compensated for by a higher order of convergence, Newton's method being of order 2 (see Section 6.3.1).

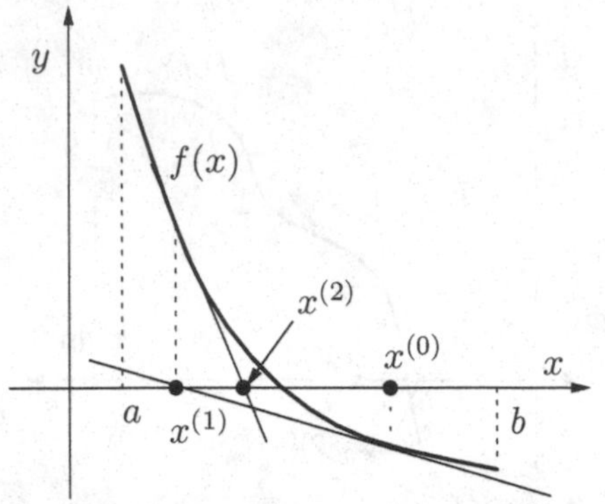

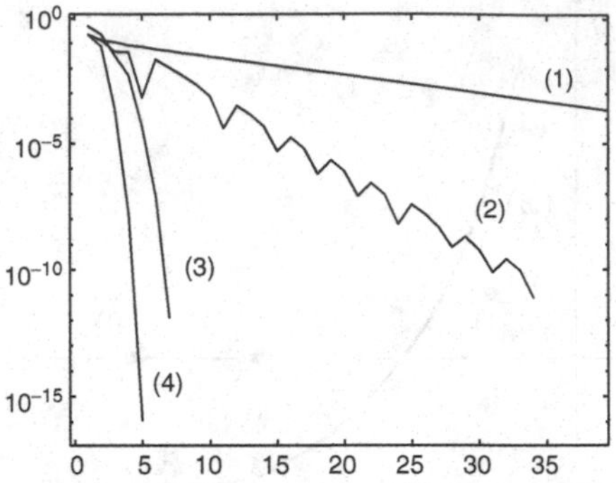

Fig. 6.4. The first two steps of Newton's method (*left*); convergence histories in Example 6.4 for the chord method (1), bisection method (2), secant method (3) and Newton's method (4) (*right*). The number of iterations and the absolute error as a function of k are shown on the x-axis and y-axis, respectively

Example 6.4 Let us compare the methods introduced so far for the approximation of the root $\alpha \simeq 0.5149$ of the function $f(x) = \cos^2(2x) - x^2$ in the interval $(0, 1.5)$. The tolerance ε on the absolute error has been taken equal to 10^{-10} and the convergence histories are drawn in Figure 6.4 (*right*). For all methods, the initial guess $x^{(0)}$ has been set equal to 0.75. For the secant method we chose $x^{(-1)} = 0$.

The analysis of the results singles out the slow convergence of the chord method. The error curve for the *Regula Falsi* method is similar to that of secant method, thus it was not reported in Figure 6.4.

It is interesting to compare the performances of Newton's and secant methods (both having order $p > 1$), in terms of their computational effort. It can indeed be proven that it is more convenient to employ the secant method whenever the number of floating point operations to evaluate f' are about twice those needed for evaluating f (see [Atk89], pp. 71-73). In the example at hand, Newton's method converges to α in 6 iterations, instead of 7, but the secant method takes 94 flops instead of 177 flops required by Newton's method. •

The chord, secant, *Regula Falsi* and Newton's methods are implemented in Programs 47, 48, 49 and 50, respectively. Here and in the rest of the chapter, `x0` and `xm1` denote the initial data $x^{(0)}$ and $x^{(-1)}$. In the case of the *Regula Falsi* method the stopping test checks is $|f(x^{(k)})| <$ `tol`, while for the other methods the test is $|x^{(k+1)} - x^{(k)}| <$ `tol`. The string `dfun` contains the expression of f' to be used in the Newton method.

Program 47 - chord : The chord method

```
function [xvect,xdif,fx,nit]=chord(a,b,x0,tol,nmax,fun)
%CHORD Chord method
% [XVECT,XDIF,FX,NIT]=CHORD(A,B,X0,TOL,NMAX,FUN) tries to find a zero
% of the continuous function FUN in the interval [A,B] using the chord method.
% FUN accepts real scalar input x and returns a real scalar value. XVECT is the
```

```
% vector of iterates, XDIF the vector of the differences between consecutive
% iterates, FX the residual. TOL specifies the tolerance of the method.
x=a; fa=eval(fun);
x=b; fb=eval(fun);
r=(fb-fa)/(b-a);
err=tol+1; nit=0; xvect=x0; x=x0; fx=eval(fun); xdif=[];
while nit<nmax & err>tol
    nit=nit+1;
    x=xvect(nit);
    xn=x-fx(nit)/r;
    err=abs(xn-x);
    xdif=[xdif; err];
    x=xn;
    xvect=[xvect;x];    fx=[fx;eval(fun)];
end
return
```

Program 48 - secant : The secant method

```
function [xvect,xdif,fx,nit]=secant(xm1,x0,tol,nmax,fun)
%SECANT Secant method
% [XVECT,XDIF,FX,NIT]=SECANT(XM1,X0,TOL,NMAX,FUN) tries to find a zero
% of the continuous function FUN  using the secant method. FUN accepts real
% scalar input x and returns a real scalar value. XVECT is the vector of iterates,
% XDIF the vector of the differences between consecutive iterates, FX the residual.
% TOL specifies the tolerance of the method.
x=xm1; fxm1=eval(fun);
xvect=[x]; fx=[fxm1];
x=x0; fx0=eval(fun);
xvect=[xvect;x]; fx=[fx;fx0];
err=tol+1; nit=0; xdif=[];
while nit<nmax & err>tol
    nit=nit+1;
    x=x0-fx0*(x0-xm1)/(fx0-fxm1);
    xvect=[xvect;x];
    fnew=eval(fun);    fx=[fx;fnew];
    err=abs(x0-x);
    xdif=[xdif;err];
    xm1=x0; fxm1=fx0;
    x0=x; fx0=fnew;
end
return
```

Program 49 - regfalsi : The *Regula Falsi* method

```
function [xvect,xdif,fx,nit]=regfalsi(xm1,x0,tol,nmax,fun)
%REGFALSI Regula Falsi method
% [XVECT,XDIF,FX,NIT]=REGFALSI(XM1,X0,TOL,NMAX,FUN) tries to find a zero
% of the continuous function FUN in the interval [XM1,X0] using the Regula Falsi
```

```
% method. FUN accepts real scalar input x and returns a real scalar value. XVECT
% is the vector of iterates, XDIF the vector of the differences between consecutive
% iterates, FX the residual. TOL specifies the tolerance of the method.
nit=0;
x=xm1; f=eval(fun); fx=[f];
x=x0; f=eval(fun); fx=[fx, f];
xvect=[xm1,x0]; xdif=[]; f=tol+1; kprime=1;
while nit<nmax & abs(f)>tol
   nit=nit+1;
   dim=length(xvect);
   x=xvect(dim);
   fxk=eval(fun);
   xk=x; i=dim;
   while i>=kprime
      i=i-1; x=xvect(i); fxkpr=eval(fun);
      if fxkpr*fxk<0
         xkpr=x; kprime=i; break;
      end
   end
   x=xk-fxk*(xk-xkpr)/(fxk-fxkpr);
   xvect=[xvect, x]; f=eval(fun);
   fx=[fx, f]; err=abs(x-xkpr); xdif=[xdif, err];
end
return
```

Program 50 - newton : Newton's method

```
function [xvect,xdif,fx,nit]=newton(x0,tol,nmax,fun,dfun)
%NEWTON Newton method
% [XVECT,XDIF,FX,NIT]=NEWTON(XM1,X0,TOL,NMAX,FUN,DFUN) tries
% to find a zero of the continuous function FUN using the Newton method
% starting from the initial guess X0. FUN and DFUN accept real scalar
% input x and return a real scalar value. XVECT is the vector of iterates,
% XDIF the vector of the differences between consecutive iterates, FX the
% residual. TOL specifies the tolerance of the method.
err=tol+1; nit=0; xvect=x0; x=x0; fx=eval(fun); xdif=[];
while nit<nmax & err>tol
   nit=nit+1;
   x=xvect(nit);
   dfx=eval(dfun);
   if dfx==0
      err=tol*1.e-10;
      fprintf('Stop for vanishing dfun');
   else
      xn=x-fx(nit)/dfx; err=abs(xn-x); xdif=[xdif; err];
      x=xn; xvect=[xvect;x]; fx=[fx;eval(fun)];
   end
end
return
```

6.2.3 The Dekker-Brent Method

The Dekker-Brent method combines the bisection and secant methods, providing a synthesis of the advantages of both. This algorithm carries out an iteration in which three abscissas a, b and c are present at each stage. Normally, b is the latest iterate and closest approximation to the zero, a is the previous iterate and c is the previous or an older iterate so that $f(b)$ and $f(c)$ have opposite signs. At all times b and c bracket the zero and $|f(b)| \leq |f(c)|$.

Once an interval $[a, b]$ containing at least one root α of the function $y = f(x)$ is found with $f(a)f(b) < 0$, the algorithm generates a sequence of values a, b and c such that α always lies between b and c and, at convergence, the half-length $|c - b|/2$ is less than a fixed tolerance. If the function f is sufficiently smooth around the desired root, then the order of convergence of the algorithm is more than linear (see [Dek69], [Bre73] Chapter 4 and [Atk89], pp. 91-93).

In the following we describe the main lines of the algorithm as implemented in the MATLAB function `fzero`. Throughout the parameter d will be a correction to the point b since it is best to arrange formulae so that they express the desired quantity as a small correction to a good approximation. For example, if the new value of b were computed as $(b+c)/2$ (bisection step) a numerical cancellation might occur, while computing b as $b+(c-b)/2$ gives a more stable formula.

Denote by ε a suitable tolerance (usually the machine precision) and let $c = b$; then, the Dekker-Brent method proceeds as follows:

First, check if $f(b) = 0$. Should this be the case, the algorithm terminates and returns b as the approximate zero of f. Otherwise, the following steps are executed:

1. if $f(b)f(c) > 0$, set $c = a$, $d = b - a$ and $e = d$.
2. If $|f(c)| < |f(b)|$, perform the exchanges $a \leftarrow b$, $b \leftarrow c$ and $c \leftarrow a$.
3. Set $\delta = 2\varepsilon \max\{|b|, 1\}$ and $m = (c - b)/2$. If $|m| \leq \delta$ or $f(b) = 0$ then the algorithm terminates and returns b as the approximate zero of f.
4. Choose bisection or interpolation.
   a) If $|e| < \delta$ or $|f(a)| \leq |f(b)|$ then a bisection step is taken, i.e., set $d = m$ and $e = m$; otherwise, the interpolation step is executed.
   b) if $a = c$ execute *linear interpolation*, i.e., compute the zero of the straight line passing through the points $(b, f(b))$ and $(c, f(c))$ as a correction δb to the point b. This amounts to taking a step of the *secant method* on the interval having b and c as end points.
   If $a \neq c$ execute *inverse quadratic interpolation*, i.e., construct the second-degree polynomial *with respect to* y, that interpolates at the points $(f(a), a)$, $(f(b), b)$ and $(f(c), c)$ and its value at $y = 0$ is computed as a correction δb to the point b. Notice that at this stage the values $f(a)$, $f(b)$ and $f(c)$ are different one from the others, being $|f(a)| > |f(b)|$, $f(b)f(c) < 0$ and $a \neq c$.
   Then the algorithm checks whether the point $b + \delta b$ can be accepted.

Table 6.2. Solution of the equation $\cos^2(2x) - x^2 = 0$ using the Dekker-Brent algorithm. The integer k denotes the current iteration

k	a	b	c	$f(b)$
0	2.1	0.3	2.1	0·5912
1	0.3	0.5235	0.3	$-2.39\cdot10^{-2}$
2	0.5235	0.5148	0.5235	$3.11\cdot10^{-4}$
3	0.5148	0.5149	0.5148	$-8.8\cdot10^{-7}$
4	0.5149	0.5149	0.5148	$-3.07\cdot10^{-11}$

This is a rather technical issue but essentially it amounts to ascertaining if the point is inside the current interval and not too close to the end points. This guarantees that the length of the interval decreases by a large factor when the function is well behaved. If the point is accepted then $e = d$ and $d = \delta b$, i.e., the interpolation is actually carried out, else a bisection step is executed by setting $d = m$ and $e = m$.

5. The algorithm now updates the current iterate. Set $a = b$ and if $|d| > \delta$ then $b = b + d$, else $b = b + \delta\text{sign}(m)$ and go back to step 1.

Example 6.5 Let us consider the finding of roots of the function f considered in Example 6.4, taking ε equal to the *roundoff* unit. The MATLAB function `fzero` has been employed. It automatically determines the values a and b, starting from a given initial guess ξ provided by the user. Starting from $\xi = 1.5$, the algorithm finds the values $a = 0.3$ and $b = 2.1$; convergence is achieved in 5 iterations and the sequences of the values a, b, c and $f(b)$ are reported in Table 6.2.

Notice that the tabulated values refer to the state of the algorithm before step 3., and thus, in particular, after possible exchanges between a and b. •

6.3 Fixed-point Iterations for Nonlinear Equations

In this section a completely general framework for finding the roots of a nonlinear function is provided. The method is based on the fact that, for a given $f : [a,b] \to \mathbb{R}$, it is always possible to transform the problem $f(x) = 0$ into an equivalent problem $x - \phi(x) = 0$, where the auxiliary function $\phi : [a,b] \to \mathbb{R}$ has to be chosen in such a way that $\phi(\alpha) = \alpha$ whenever $f(\alpha) = 0$. Approximating the zeros of a function has thus become the problem of finding the *fixed points* of the mapping ϕ, which is done by the following iterative algorithm:

$$\text{given } x^{(0)}, \quad x^{(k+1)} = \phi(x^{(k)}), \qquad k \geq 0. \tag{6.17}$$

We say that (6.17) is a *fixed-point iteration* and ϕ is its associated *iteration function.* Sometimes, (6.17) is also referred to as *Picard iteration* or *functional iteration* for the solution of $f(x) = 0$. Notice that by construction the methods of the form (6.17) are *strongly consistent* in the sense of the definition given in Section 2.2.

The choice of ϕ is not unique. For instance, any function of the form $\phi(x) = x + F(f(x))$, where F is a continuous function such that $F(0) = 0$, is an admissible iteration function.

The next two results provide *sufficient* conditions in order for the fixed-point method (6.17) to converge to the root α of problem (6.1). These conditions are stated precisely in the following theorem.

Theorem 6.1 (Convergence of fixed-point iterations) *Consider the sequence $x^{(k+1)} = \phi(x^{(k)})$, for $k \geq 0$, being $x^{(0)}$ given. Assume that:*

1. $\phi : [a, b] \to [a, b]$;
2. $\phi \in C^1([a, b])$;
3. $\exists K < 1 : |\phi'(x)| \leq K \ \forall x \in [a, b]$.

Then, ϕ has a unique fixed point α in $[a, b]$ and the sequence $\{x^{(k)}\}$ converges to α for any choice of $x^{(0)} \in [a, b]$. Moreover, we have

$$\lim_{k\to\infty} \frac{x^{(k+1)} - \alpha}{x^{(k)} - \alpha} = \phi'(\alpha). \tag{6.18}$$

Proof. The assumption *1.* and the continuity of ϕ ensure that the iteration function ϕ has at least one fixed point in $[a, b]$. Assumption *3.* states that ϕ is a *contraction mapping* and ensures the uniqueness of the fixed point. Indeed, suppose that there exist two distinct values $\alpha_1, \alpha_2 \in [a, b]$ such that $\phi(\alpha_1) = \alpha_1$ and $\phi(\alpha_2) = \alpha_2$. Expanding ϕ in a Taylor series around α_1 and truncating it at first order, it follows that

$$|\alpha_2 - \alpha_1| = |\phi(\alpha_2) - \phi(\alpha_1)| = |\phi'(\eta)(\alpha_2 - \alpha_1)| \leq K|\alpha_2 - \alpha_1| < |\alpha_2 - \alpha_1|,$$

for $\eta \in (\alpha_1, \alpha_2)$, from which it must necessarily be that $\alpha_2 = \alpha_1$.

The convergence analysis for the sequence $\{x^{(k)}\}$ is again based on a Taylor series expansion. Indeed, for any $k \geq 0$ there exists a value $\eta^{(k)}$ between α and $x^{(k)}$ such that

$$x^{(k+1)} - \alpha = \phi(x^{(k)}) - \phi(\alpha) = \phi'(\eta^{(k)})(x^{(k)} - \alpha) \tag{6.19}$$

from which $|x^{(k+1)} - \alpha| \leq K|x^{(k)} - \alpha| \leq K^{k+1}|x^{(0)} - \alpha| \to 0$ for $k \to \infty$. Thus, $x^{(k)}$ converges to α and (6.19) implies that

$$\lim_{k\to\infty} \frac{x^{(k+1)} - \alpha}{x^{(k)} - \alpha} = \lim_{k\to\infty} \phi'(\eta^{(k)}) = \phi'(\alpha),$$

that is (6.18). ◇

The quantity $|\phi'(\alpha)|$ is called the asymptotic convergence factor and, in analogy with the case of iterative methods for linear systems, the asymptotic convergence rate can be defined as

$$R = -\log(|\phi'(\alpha)|). \tag{6.20}$$

Theorem 6.1 ensures linear convergence of the sequence $\{x^{(k)}\}$ to the root α for *any choice* of the initial value $x^{(0)} \in [a,b]$. As such, it represents an example of a *global* convergence result.

In practice, however, it is often quite difficult to determine a priori the width of the interval $[a,b]$; in such a case the following convergence result can be useful (see for the proof, [OR70]).

Property 6.3 (Ostrowski theorem) *Let α be a fixed point of a function ϕ, which is continuous and differentiable in a neighborhood $\mathcal{J}$ of α. If $|\phi'(\alpha)| < 1$ then there exists $\delta > 0$ such that the sequence $\{x^{(k)}\}$ converges to α, for any $x^{(0)}$ such that $|x^{(0)} - \alpha| < \delta$.*

Remark 6.2 If $|\phi'(\alpha)| > 1$ it follows from (6.19) that if $x^{(n)}$ is sufficiently close to α, so that $|\phi'(x^{(n)})| > 1$, then $|\alpha - x^{(n+1)}| > |\alpha - x^{(n)}|$, thus no convergence is possible. In the case $|\phi'(\alpha)| = 1$ no general conclusion can be stated since both convergence and nonconvergence may be possible, depending on the problem at hand. ■

Example 6.6 Let $\phi(x) = x - x^3$, which admits $\alpha = 0$ as fixed point. Although $\phi'(\alpha) = 1$, if $x^{(0)} \in [-1,1]$ then $x^{(k)} \in (-1,1)$ for $k \geq 1$ and it converges (very slowly) to α (if $x^{(0)} = \pm 1$, we even have $x^{(k)} = \alpha$ for any $k \geq 1$). Starting from $x^{(0)} = 1/2$ the absolute error after 2000 iterations is 0.0158. Let now $\phi(x) = x + x^3$ having also $\alpha = 0$ as fixed point. Again, $\phi'(\alpha) = 1$ but in this case the sequence $x^{(k)}$ diverges for any choice $x^{(0)} \neq 0$. •

We say that a fixed-point method has *order* p (p non necessarily being an integer) if the sequence that is generated by the method converges to the fixed point α with order p according to Definition 6.1.

Property 6.4 *If $\phi \in C^{p+1}(\mathcal{J})$ for a suitable neighborhood $\mathcal{J}$ of α and an integer $p \geq 1$, and if $\phi^{(i)}(\alpha) = 0$ for $1 \leq i \leq p$ and $\phi^{(p+1)}(\alpha) \neq 0$, then the fixed-point method with iteration function ϕ has order $p+1$ and*

$$\lim_{k\to\infty} \frac{x^{(k+1)} - \alpha}{(x^{(k)} - \alpha)^{p+1}} = \frac{\phi^{(p+1)}(\alpha)}{(p+1)!}. \tag{6.21}$$

Proof. Let us expand ϕ in a Taylor series around $x = \alpha$ obtaining

$$x^{(k+1)} - \alpha = \sum_{i=0}^{p} \frac{\phi^{(i)}(\alpha)}{i!}(x^{(k)} - \alpha)^i + \frac{\phi^{(p+1)}(\eta)}{(p+1)!}(x^{(k)} - \alpha)^{p+1} - \phi(\alpha),$$

for a certain η between $x^{(k)}$ and α. Thus, we have

$$\lim_{k\to\infty} \frac{x^{(k+1)} - \alpha}{(x^{(k)} - \alpha)^{p+1}} = \lim_{k\to\infty} \frac{\phi^{(p+1)}(\eta)}{(p+1)!} = \frac{\phi^{(p+1)}(\alpha)}{(p+1)!}.$$

◇

The convergence of the sequence to the root α will be faster, for a fixed order p, when the quantity at right-side in (6.21) is smaller.
The fixed-point method (6.17) is implemented in Program 51. The variable `phi` contains the expression of the iteration function ϕ.

Program 51 - fixpoint : Fixed-point method

```
function [xvect,xdif,fx,nit]=fixpoint(x0,tol,nmax,fun,phi)
%FIXPOINT Fixed-point iteration
% [XVECT,XDIF,FX,NIT]=FIXPOINT(X0,TOL,NMAX,FUN,PHI) tries to find a zero
% of the continuous function FUN  using the fixed-point iteration X=PHI(X), starting
% from the initial guess X0. XVECT is the vector of iterates, XDIF the vector of the
% differences between consecutive iterates, FX the residual. TOL specifies the
% tolerance of the  method.
err=tol+1; nit=0;
xvect=x0; x=x0; fx=eval(fun); xdif=[];
while nit<nmax & err>tol
  nit=nit+1;
  x=xvect(nit);
  xn=eval(phi);
  err=abs(xn-x);
  xdif=[xdif; err];
  x=xn; xvect=[xvect;x]; fx=[fx;eval(fun)];
end
return
```

6.3.1 Convergence Results for Some Fixed-point Methods

Theorem 6.1 provides a theoretical tool for analyzing some of the iterative methods introduced in Section 6.2.2.
The chord method. Equation (6.12) is a special instance of (6.17), in which we let $\phi(x) = \phi_{chord}(x) = x - q^{-1}f(x) = x - (b-a)/(f(b) - f(a))f(x)$. If $f'(\alpha) = 0$, $\phi'_{chord}(\alpha) = 1$ and the method is not guaranteed to converge. Otherwise, the condition $|\phi'_{chord}(\alpha)| < 1$ is equivalent to requiring that $0 < q^{-1}f'(\alpha) < 2$.

Therefore, the slope q of the chord must have the same sign as $f'(\alpha)$, and the search interval $[a, b]$ has to satisfy the constraint

$$b - a < 2\frac{f(b) - f(a)}{f'(\alpha)}.$$

The chord method converges in one iteration if f is a straight line, otherwise it converges linearly, apart the (lucky) case when $f'(\alpha) = (f(b) - f(a))/(b - a)$, for which $\phi'_{chord}(\alpha) = 0$.

Newton's method. Equation (6.16) can be cast in the general framework (6.17) letting

$$\phi_{Newt}(x) = x - \frac{f(x)}{f'(x)}.$$

Assuming $f'(\alpha) \neq 0$ (that is, α is a simple root)

$$\phi'_{Newt}(\alpha) = 0, \qquad \phi''_{Newt}(\alpha) = \frac{f''(\alpha)}{f'(\alpha)}.$$

If the root α has multiplicity $m > 1$, then the method (6.16) is no longer second-order convergent. Indeed we have (see Exercise 2)

$$\phi'_{Newt}(\alpha) = 1 - \frac{1}{m}. \tag{6.22}$$

If the value of m is known a priori, then the quadratic convergence of Newton's method can be recovered by resorting to the so-called *modified Newton's method*

$$x^{(k+1)} = x^{(k)} - m\frac{f(x^{(k)})}{f'(x^{(k)})}, \qquad k \geq 0. \tag{6.23}$$

To check the convergence order of the iteration (6.23), see Exercise 2.

6.4 Zeros of Algebraic Equations

In this section we address the special case in which f is a polynomial of degree $n \geq 0$, i.e., a function of the form

$$p_n(x) = \sum_{k=0}^{n} a_k x^k, \tag{6.24}$$

where $a_k \in \mathbb{R}$ are given coefficients.

The above representation of p_n is not the only one possible. Actually, one can also write

$$p_n(x) = a_n(x - \alpha_1)^{m_1} \cdots (x - \alpha_k)^{m_k}, \qquad \sum_{l=1}^{k} m_l = n,$$

where α_i and m_i denote the i-th root of p_n and its multiplicity, respectively. Other representations are available as well, see Section 6.4.1.

Notice that, since the coefficients a_k are real, if α is a zero of p_n, then its complex conjugate $\bar{\alpha}$ is a zero of p_n too.

Abel's theorem states that for $n \geq 5$ there does not exist an explicit formula for the zeros of p_n (see, for instance, [MM71], Theorem 10.1). This, in turn, motivates numerical solutions of the nonlinear equation $p_n(x) = 0$. Since the methods introduced so far must be provided by a suitable search interval $[a, b]$ or an initial guess $x^{(0)}$, we recall two results that can be useful to *localize* the zeros of a polynomial.

Property 6.5 (Descartes' rule of signs) *Let $p_n \in \mathbb{P}_n$. Denote by ν the number of sign changes in the set of coefficients $\{a_j\}$ and by k the number of real positive roots of p_n (each counted with its multiplicity). Then, $k \leq \nu$ and $\nu - k$ is an even number.*

Property 6.6 (Cauchy's Theorem) *All zeros of p_n are contained in the circle Γ in the complex plane*

$$\Gamma = \{z \in \mathbb{C}: \ |z| \leq 1 + \eta_k\}, \qquad \text{where } \eta_k = \max_{0\leq k\leq n-1} |a_k/a_n|.$$

This second property is of little use if $\eta_k \gg 1$. In such an event, it is convenient to perform a *translation* through a suitable change of coordinates.

6.4.1 The Horner Method and Deflation

In this section we describe the Horner method for efficiently evaluating a polynomial (and its derivative) at a given point z. The algorithm allows for generating automatically a procedure, called *deflation*, for the sequential approximation of *all* the roots of a polynomial.

Horner's method is based on the observation that any polynomial $p_n \in \mathbb{P}_n$ can be written as

$$p_n(x) = a_0 + x(a_1 + x(a_2 + \ldots + x(a_{n-1} + a_n x)\ldots)). \qquad (6.25)$$

Formulae (6.24) and (6.25) are completely equivalent from an algebraic standpoint; nevertheless, (6.24) requires n sums and $2n-1$ multiplications to evaluate $p_n(x)$, while (6.25) requires n sums and n multiplications. The second expression, known as *nested multiplications* algorithm, is the basic ingredient of Horner's method. This method efficiently evaluates the polynomial p_n at a point z through the following *synthetic division* algorithm

$$b_n = a_n, \ b_k = a_k + b_{k+1}z, \ k = n-1, n-2, ..., 0, \qquad (6.26)$$

which is implemented in Program 52. The coefficients a_j of the polynomial are stored in vector a ordered from a_n back to a_0.

Program 52 - horner : Synthetic division algorithm

```
function [pnz,b] = horner(a,n,z)
%HORNER Polynomial synthetic division algorithm.
% [PNZ,B]=HORNER(A,N,Z) evaluates with the Horner method a polynomial
% of degree N having coefficients A(1),...,A(N) at a point Z.
b(1)=a(1);
for j=2:n+1
   b(j)=a(j)+b(j-1)*z;
end
pnz=b(n+1);
return
```

All the coefficients b_k in (6.26) depend on z and $b_0 = p_n(z)$. The polynomial

$$q_{n-1}(x;z) = b_1 + b_2 x + ... + b_n x^{n-1} = \sum_{k=1}^{n} b_k x^{k-1} \tag{6.27}$$

has degree $n-1$ in the variable x and depends on the parameter z through the coefficients b_k; it is called the *associated polynomial* of p_n.
Let us now recall the following property of *polynomial division*:

given two polynomials $h_n \in \mathbb{P}_n$ and $g_m \in \mathbb{P}_m$ with $m \leq n$, there exists a unique polynomial $\delta \in \mathbb{P}_{n-m}$ and an unique polynomial $\rho \in \mathbb{P}_{m-1}$ such that

$$h_n(x) = g_m(x)\delta(x) + \rho(x). \tag{6.28}$$

Then, dividing p_n by $x - z$, from (6.28) it follows that

$$p_n(x) = b_0 + (x-z)q_{n-1}(x;z),$$

having denoted by q_{n-1} the quotient and by b_0 the remainder of the division. If z is a zero of p_n, then $b_0 = p_n(z) = 0$ and thus $p_n(x) = (x-z)q_{n-1}(x;z)$. In such a case, the algebraic equation $q_{n-1}(x;z) = 0$ yields the $n-1$ remaining roots of $p_n(x)$. This observation suggests adopting the following *deflation* procedure for finding the roots of p_n. For $m = n, n-1, \ldots, 1$:

1. find a root r of p_m using a suitable approximation method;
2. evaluate $q_{m-1}(x;r)$ by (6.26);
3. let $p_{m-1} = q_{m-1}$.

In the two forthcoming sections some deflation methods will be addressed, making a precise choice for the scheme at point 1.

6.4.2 The Newton-Horner Method

A first example of deflation employs Newton's method for computing the root r at step 1. of the procedure in the previous section. Implementing Newton's method fully benefits from Horner's algorithm (6.26). Indeed, if q_{n-1} is the associated polynomial of p_n defined in (6.27), since $p_n'(x) = q_{n-1}(x;z) + (x - z)q_{n-1}'(x;z)$ then $p_n'(z) = q_{n-1}(z;z)$. Thanks to this identity, the Newton-Horner method for the approximation of a root (real or complex) r_j of p_n ($j = 1, \ldots, n$) takes the following form:
given an initial estimate $r_j^{(0)}$ of the root, solve for any $k \geq 0$

$$r_j^{(k+1)} = r_j^{(k)} - \frac{p_n(r_j^{(k)})}{p_n'(r_j^{(k)})} = r_j^{(k)} - \frac{p_n(r_j^{(k)})}{q_{n-1}(r_j^{(k)}; r_j^{(k)})}. \tag{6.29}$$

Once convergence has been achieved for the iteration (6.29), polynomial deflation is performed, this deflation being helped by the fact that $p_n(x) =$

$(x - r_j)p_{n-1}(x)$. Then, the approximation of a root of $p_{n-1}(x)$ is carried out until all the roots of p_n have been computed.

Denoting by $n_k = n - k$ the degree of the polynomial that is obtained at each step of the deflation process, for $k = 0, \ldots, n-1$, the computational cost of each Newton-Horner iteration (6.29) is equal to $4n_k$. If $r_j \in \mathbb{C}$, it is necessary to work in complex arithmetic and take $r_j^{(0)} \in \mathbb{C}$; otherwise, indeed, the Newton-Horner method (6.29) would yield a sequence $\{r_j^{(k)}\}$ of *real* numbers.

The deflation procedure might be affected by rounding error propagation and, as a consequence, can lead to inaccurate results. For the sake of stability, it is therefore convenient to approximate first the root r_1 of minimum module, which is the most sensitive to ill-conditioning of the problem (see Example 2.7, Chapter 2) and then to continue with the successive roots $r_2, \ldots$, until the root of maximum module is computed. To localize r_1, the techniques described in Section 5.1 or the method of *Sturm sequences* can be used (see [IK66], p. 126).

A further increase in accuracy can be obtained, once an approximation $\widetilde{r}_j$ of the root r_j is available, by going back to the *original* polynomial p_n and generating through the Newton-Horner method (6.29) a new approximation to r_j, taking as initial guess $r_j^{(0)} = \widetilde{r}_j$. This combination of deflation and successive correction of the root is called the Newton-Horner method *with refinement.*

Example 6.7 Let us examine the performance of the Newton-Horner method in two cases: in the first one, the polynomial admits real roots, while in the second one there are two pairs of complex conjugate roots. To single out the importance of refinement, we have implemented (6.29) both switching it on and off (methods `NwtRef` and `Nwt`, respectively). The approximate roots obtained using method `Nwt` are denoted by r_j, while s_j are those computed by method `NwtRef`. As for the numerical experiments, the computations have been done in complex arithmetic, with $x^{(0)} = 0 + i\,0$, i being the imaginary unit, `nmax` $= 100$ and `tol` $= 10^{-5}$. The tolerance for the stopping test in the refinement cycle has been set to 10^{-3}`tol`.

1) $p_5(x) = x^5 + x^4 - 9x^3 - x^2 + 20x - 12 = (x-1)^2(x-2)(x+2)(x+3)$.

We report in Tables 6.3(a) and 6.3(b) the approximate roots r_j $(j = 1, \ldots, 5)$ and the number of Newton iterations (`Nit`) needed to get each of them; in the case of method `NwtRef` we also show the number of extra Newton iterations for the refinement (`Extra`).

Notice a neat increase in the accuracy of rootfinding due to refinement, even with few extra iterations.

2) $p_6(x) = x^6 - 2x^5 + 5x^4 - 6x^3 + 2x^2 + 8x - 8$.

The zeros of p_6 are the complex numbers $\{1, -1, 1 \pm i, \pm 2i\}$. We report below, denoting them by r_j, $(j = 1, \ldots, 6)$, the approximations to the roots of p_6 obtained using method `Nwt`, with a number of iterations equal to 2, 1, 1, 7, 7 and 1, respectively.

Table 6.3. Roots of the polynomial p_5. Roots computed by the Newton-Horner method without refinement (*left*), and with refinement (*right*)

(a)

r_j	Nit
0.99999348047830	17
$1 - i3.56 \cdot 10^{-25}$	6
$2 - i2.24 \cdot 10^{-13}$	9
$-2 - i1.70 \cdot 10^{-10}$	7
$-3 + i5.62 \cdot 10^{-6}$	1

(b)

s_j	Nit	Extra
0.9999999899210124	17	10
$1 - i2.40 \cdot 10^{-28}$	6	10
$2 + i1.12 \cdot 10^{-22}$	9	1
$-2 + i8.18 \cdot 10^{-22}$	7	1
$-3 - i7.06 \cdot 10^{-21}$	1	2

Table 6.4. Roots of the polynomial p_6 obtained using the Newton-Horner method without (*left*) and with (*right*) refinement

r_j	Nwt	s_j	NwtRef
r_1	1	s_1	1
r_2	$-0.99 - i9.54 \cdot 10^{-17}$	s_2	$-1 + i1.23 \cdot 10^{-32}$
r_3	1+i	s_3	1+i
r_4	1−i	s_4	1−i
r_5	-1.31 $\cdot 10^{-8} + i2$	s_5	$-5.66 \cdot 10^{-17} + i2$
r_6	$-i2$	s_6	$-i2$

Beside, we also show the corresponding approximations s_j computed by method `NwtRef` and obtained with a maximum number of 2 extra iterations. •

A coding of the Newton-Horner algorithm is provided in Program 53. The input parameters are `A` (a vector containing the polynomial coefficients), `n` (the degree of the polynomial), `tol` (tolerance on the maximum variation between successive iterates in Newton's method), `x0` (initial value, with $x^{(0)} \in \mathbb{R}$), `nmax` (maximum number of admissible iterations for Newton's method) and `iref` (if `iref` = 1, then the refinement procedure is activated). For dealing with the general case of complex roots, the initial datum is automatically converted into the complex number $z = x^{(0)} + ix^{(0)}$, where $i = \sqrt{-1}$.

The program returns as output the variables `xn` (a vector containing the sequence of iterates for each zero of $p_n(x)$), `iter` (a vector containing the number of iterations needed to approximate each root), `itrefin` (a vector containing the Newton iterations required to refine each estimate of the computed root) and `root` (vector containing the computed roots).

Program 53 - newthorn : Newton-Horner method with refinement

```
function [xn,iter,root,itrefin]=newthorn(A,n,tol,x0,nmax,iref)
%NEWTHORN Newton-Horner method with refinement.
%   [XN,ITER,ROOT,ITREFIN]=NEWTHORN(A,N,X0,TOL,NMAX,IREF) tries
%   to compute all the roots of a polynomial of degree N having coefficients
%   A(1),...,A(N). TOL specifies the tolerance of the method. X0 is an initial
```

```
%   guess. NMAX specifies the maximum number of iterations. If the flag IREF
%   is equal 1, then the refinement procedure is activated.
apoly=A;
for i=1:n, it=1; xn(it,i)=x0+sqrt(-1)*x0; err=tol+1; Ndeg=n-i+1;
   if Ndeg == 1
      it=it+1; xn(it,i)=-A(2)/A(1);
   else
      while it<nmax & err>tol
         [px,B]=horner(A,Ndeg,xn(it,i)); [pdx,C]=horner(B,Ndeg-1,xn(it,i));
         it=it+1;
         if pdx ~=0
            xn(it,i)=xn(it-1,i)-px/pdx;
            err=max(abs(xn(it,i)-xn(it-1,i)),abs(px));
         else
            fprintf(' Stop due to a vanishing p'' ');
            err=0; xn(it,i)=xn(it-1,i);
         end
      end
   end
   A=B;
   if iref==1
      alfa=xn(it,i); itr=1; err=tol+1;
      while err>tol*1e-3 & itr<nmax
         [px,B]=horner(apoly,n,alfa); [pdx,C]=horner(B,n-1,alfa);
         itr=itr+1;
         if pdx~=0
            alfa2=alfa-px/pdx;
            err=max(abs(alfa2-alfa),abs(px));
            alfa=alfa2;
         else
            fprintf(' Stop due to a vanishing p'' ');
            err=0;
         end
      end
      itrefin(i)=itr-1; xn(it,i)=alfa;
   end
   iter(i)=it-1; root(i)=xn(it,i); x0=root(i);
end
return
```

6.4.3 The Muller Method

A second example of deflation employs Muller's method for finding an approximation to the root r at step 1. of the procedure described in Section 6.4.1 (see [Mul56]). Unlike Newton's or secant methods, Muller's method is able to compute complex zeros of a given function f, even starting from a real initial datum; moreover, its order of convergence is almost quadratic.

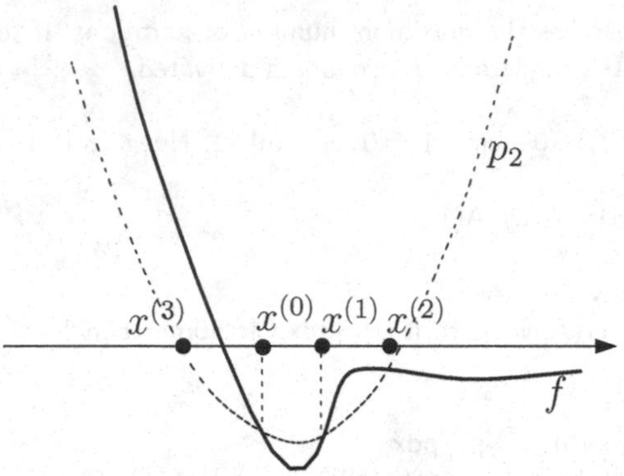

Fig. 6.5. The first step of Muller's method

The action of Muller's method is drawn in Figure 6.5. The scheme extends the secant method, substituting the linear polynomial introduced in (6.13) with a second-degree polynomial as follows. Given three distinct values $x^{(0)}$, $x^{(1)}$ and $x^{(2)}$, the new point $x^{(3)}$ is determined by setting $p_2(x^{(3)}) = 0$, where $p_2 \in \mathbb{P}_2$ is the unique polynomial that interpolates f at the points $x^{(i)}$, $i = 0, 1, 2$, that is, $p_2(x^{(i)}) = f(x^{(i)})$ for $i = 0, 1, 2$. Therefore,

$$p_2(x) = f(x^{(2)}) + (x - x^{(2)})f[x^{(2)}, x^{(1)}] + (x - x^{(2)})(x - x^{(1)})f[x^{(2)}, x^{(1)}, x^{(0)}],$$

where

$$f[\xi, \eta] = \frac{f(\eta) - f(\xi)}{\eta - \xi}, \; f[\xi, \eta, \tau] = \frac{f[\eta, \tau] - f[\xi, \eta]}{\tau - \xi}$$

are the *divided differences* of order 1 and 2 associated with the points ξ, η and τ (see Section 8.2.1). Noticing that $x - x^{(1)} = (x - x^{(2)}) + (x^{(2)} - x^{(1)})$, we get

$$p_2(x) = f(x^{(2)}) + w(x - x^{(2)}) + f[x^{(2)}, x^{(1)}, x^{(0)}](x - x^{(2)})^2,$$

having defined

$$\begin{aligned} w &= f[x^{(2)}, x^{(1)}] + (x^{(2)} - x^{(1)})f[x^{(2)}, x^{(1)}, x^{(0)}] \\ &= f[x^{(2)}, x^{(1)}] + f[x^{(2)}, x^{(0)}] - f[x^{(0)}, x^{(1)}]. \end{aligned}$$

Requiring that $p_2(x^{(3)}) = 0$ it follows that

$$x^{(3)} = x^{(2)} + \frac{-w \pm \left\{w^2 - 4f(x^{(2)})f[x^{(2)}, x^{(1)}, x^{(0)}]\right\}^{1/2}}{2f[x^{(2)}, x^{(1)}, x^{(0)}]}.$$

Similar computations must be done for getting $x^{(4)}$ starting from $x^{(1)}$, $x^{(2)}$ and $x^{(3)}$ and, more generally, to find $x^{(k+1)}$ starting from $x^{(k-2)}$, $x^{(k-1)}$ and $x^{(k)}$,

Table 6.5. Roots of polynomial p_6 with Muller's method without (r_j) and with (s_j) refinement

r_j		s_j	
r_1	$1 + i2.2 \cdot 10^{-15}$	s_1	$1 + i9.9 \cdot 10^{-18}$
r_2	$-1 - i8.4 \cdot 10^{-16}$	s_2	-1
r_3	$0.99 + i$	s_3	$1 + i$
r_4	$0.99 - i$	s_4	$1 - i$
r_5	$-1.1 \cdot 10^{-15} + i1.99$	s_5	$i2$
r_6	$-1.0 \cdot 10^{-15} - i2$	s_6	-$i2$

with $k \geq 2$, according with the following formula (notice that the numerator has been rationalized)

$$x^{(k+1)} = x^{(k)} - \frac{2f(x^{(k)})}{w \mp \{w^2 - 4f(x^{(k)})f[x^{(k)}, x^{(k-1)}, x^{(k-2)}]\}^{1/2}}. \qquad (6.30)$$

The sign in (6.30) is chosen in such a way that the module of the denominator is maximized. Assuming that $f \in C^3(\mathcal{J})$ in a suitable neighborhood $\mathcal{J}$ of the root α, with $f'(\alpha) \neq 0$, the order of convergence is almost quadratic. Precisely, the error $e^{(k)} = \alpha - x^{(k)}$ obeys the following relation (see for the proof [Hil87])

$$\lim_{k \to \infty} \frac{|e^{(k+1)}|}{|e^{(k)}|^p} = \frac{1}{6}\left|\frac{f'''(\alpha)}{f'(\alpha)}\right|, \qquad p \simeq 1.84.$$

Example 6.8 Let us employ Muller's method to approximate the roots of the polynomial p_6 examined in Example 6.7. The tolerance on the stopping test is `tol` $= 10^{-6}$, while $x^{(0)} = -5$, $x^{(1)} = 0$ and $x^{(2)} = 5$ are the inputs to (6.30). We report in Table 6.5 the approximate roots of p_6, denoted by s_j and r_j ($j = 1, \ldots, 6$), where, as in Example 6.7, s_j and r_j have been obtained by switching the refinement procedure on and off, respectively. To compute the roots r_j, 12, 11, 9, 9, 2 and 1 iterations are needed, respectively, while only one extra iteration is taken to refine all the roots.
Even in this example, one can notice the effectiveness of the refinement procedure, based on Newton's method, on the accuracy of the solution yielded by (6.30). •

The Muller method is implemented in Program 54, in the special case where f is a polynomial of degree n. The deflation process also includes a refinement phase; the evaluation of $f(x^{(k-2)})$, $f(x^{(k-1)})$ and $f(x^{(k)})$, with $k \geq 2$, is carried out using Program 52. The input/output parameters are analogous to those described in Program 53.

Program 54 - mulldefl : Muller's method with refinement

```
function [xn,iter,root,itrefin]=mulldefl(A,n,tol,x0,x1,x2,nmax,iref)
%MULLDEFL Muller method with refinement.
%   [XN,ITER,ROOT,ITREFIN]=MULLDEFL(A,N,TOL,X0,X1,X2,NMAX,IREF) tries
```

```
%  to compute all the roots of a polynomial of degree N having coefficients
%  A(1),...,A(N). TOL specifies the tolerance of the method. X0 is an initial
%  guess. NMAX specifies the maximum number of iterations. If the flag IREF
%  is equal 1, then the refinement procedure is activated.
apoly=A;
for i=1:n
   xn(1,i)=x0; xn(2,i)=x1; xn(3,i)=x2;
   it=0; err=tol+1; k=2; Ndeg=n-i+1;
   if Ndeg==1
      it=it+1; k=0; xn(it,i)=-A(2)/A(1);
   else
      while err>tol & it<nmax
         k=k+1; it=it+1;
         [f0,B]=horner(A,Ndeg,xn(k-2,i)); [f1,B]=horner(A,Ndeg,xn(k-1,i));
         [f2,B]=horner(A,Ndeg,xn(k,i));
         f01=(f1-f0)/(xn(k-1,i)-xn(k-2,i));  f12=(f2-f1)/(xn(k,i)-xn(k-1,i));
         f012=(f12-f01)/(xn(k,i)-xn(k-2,i));
         w=f12+(xn(k,i)-xn(k-1,i))*f012;
         arg=w^2-4*f2*f012; d1=w-sqrt(arg);
         d2=w+sqrt(arg); den=max(d1,d2);
         if den~=0
            xn(k+1,i)=xn(k,i)-(2*f2)/den;
            err=abs(xn(k+1,i)-xn(k,i));
         else
            fprintf(' Vanishing denominator ');
            return
         end
      end
   end
   radix=xn(k+1,i);
   if iref==1
      alfa=radix; itr=1; err=tol+1;
      while err>tol*1e-3 & itr<nmax
         [px,B]=horner(apoly,n,alfa); [pdx,C]=horner(B,n-1,alfa);
         if pdx == 0
            fprintf(' Vanishing derivative '); err=0;
         end
         itr=itr+1;
         if pdx~=0
            alfa2=alfa-px/pdx; err=abs(alfa2-alfa); alfa=alfa2;
         end
      end
      itrefin(i)=itr-1; xn(k+1,i)=alfa; radix=alfa;
   end
   iter(i)=it; root(i)=radix; [px,B]=horner(A,Ndeg-1,xn(k+1,i)); A=B;
end
return
```

6.5 Stopping Criteria

Suppose that $\{x^{(k)}\}$ is a sequence converging to a zero α of the function f. In this section we provide some stopping criteria for terminating the iterative process that approximates α. Analogous to Section 4.6, where the case of iterative methods for linear systems has been examined, there are two possible criteria: a stopping test based on the residual and on the increment. Below, ε is a fixed tolerance on the approximate calculation of α and $e^{(k)} = \alpha - x^{(k)}$ denotes the absolute error. We shall moreover assume that f is continuously differentiable in a suitable neighborhood of the root.

1. **Control of the residual**: *the iterative process terminates at the first step* k *such that* $|f(x^{(k)})| < \varepsilon$.
Situations can arise where the test turns out to be either too restrictive or excessively optimistic (see Figure 6.6). Applying the estimate (6.6) to the case at hand yields

$$\frac{|e^{(k)}|}{|\alpha|} \lesssim \left(\frac{m!}{|f^{(m)}(\alpha)||\alpha|^m}\right)^{1/m} |f(x^{(k)})|^{1/m}.$$

In particular, in the case of simple roots, the error is bound to the residual by the factor $1/|f'(\alpha)|$ so that the following conclusions can be drawn:

1. if $|f'(\alpha)| \simeq 1$, then $|e^{(k)}| \simeq \varepsilon$; therefore, the test provides a satisfactory indication of the error;
2. if $|f'(\alpha)| \ll 1$, the test is not reliable since $|e^{(k)}|$ could be quite large with respect to ε;
3. if, finally, $|f'(\alpha)| \gg 1$, we get $|e^{(k)}| \ll \varepsilon$ and the test is too restrictive.

We refer to Figure 6.6 for an illustration of the last two cases.
The conclusions that we have drawn agree with those in Example 2.4. Indeed, when $f'(\alpha) \simeq 0$, the condition number of the problem $f(x) = 0$ is very high

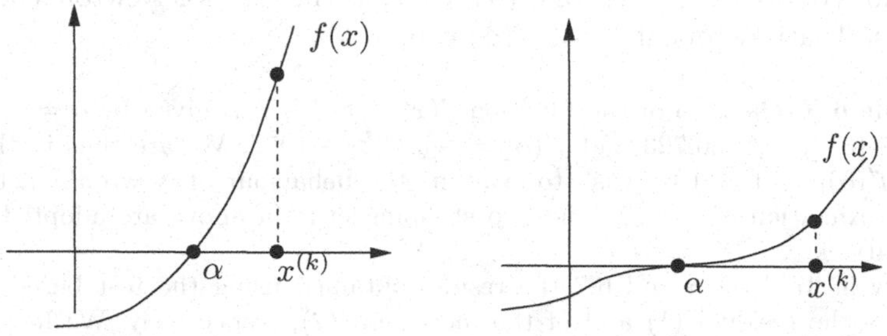

Fig. 6.6. Two situations where the stopping test based on the residual is either too restrictive (when $|e^{(k)}| \ll |f(x^{(k)})|$, left) or too optimistic (when $|e^{(k)}| \gg |f(x^{(k)})|$, right)

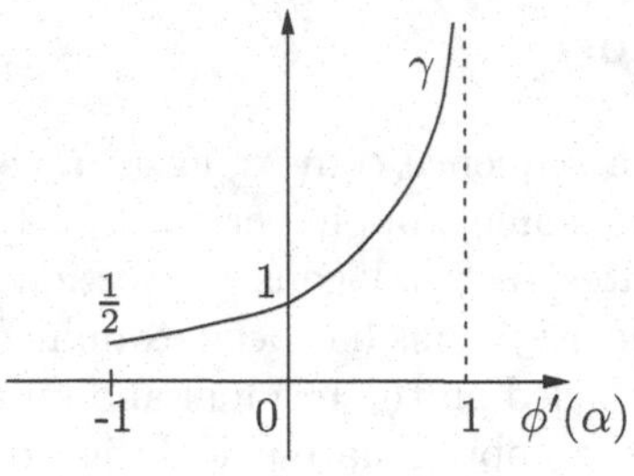

Fig. 6.7. Behavior of $\gamma = 1/(1-\phi'(\alpha))$ as a function of $\phi'(\alpha)$

and, as a consequence, the residual does not provide a significant indication of the error.

2. **Control of the increment**: *the iterative process terminates as soon as* $|x^{(k+1)} - x^{(k)}| < \varepsilon$.
Let $\{x^{(k)}\}$ be generated by the fixed-point method $x^{(k+1)} = \phi(x^{(k)})$. Using the mean value theorem, we get

$$e^{(k+1)} = \phi(\alpha) - \phi(x^{(k)}) = \phi'(\xi^{(k)})e^{(k)},$$

where $\xi^{(k)}$ lies between $x^{(k)}$ and α. Then,

$$x^{(k+1)} - x^{(k)} = e^{(k)} - e^{(k+1)} = \left(1 - \phi'(\xi^{(k)})\right) e^{(k)}$$

so that, assuming that we can replace $\phi'(\xi^{(k)})$ with $\phi'(\alpha)$, it follows that

$$e^{(k)} \simeq \frac{1}{1-\phi'(\alpha)}(x^{(k+1)} - x^{(k)}). \tag{6.31}$$

As shown in Figure 6.7, we can conclude that the test:
- is unsatisfactory if $\phi'(\alpha)$ is close to 1;
- provides an optimal balancing between increment and error in the case of methods of order 2 for which $\phi'(\alpha) = 0$ as is the case for Newton's method;
- is still satisfactory if $-1 < \phi'(\alpha) < 0$.

Example 6.9 The zero of the function $f(x) = e^{-x} - \eta$ is given by $\alpha = -\log(\eta)$. For $\eta = 10^{-9}$, $\alpha \simeq 20.723$ and $f'(\alpha) = -e^{-\alpha} \simeq -10^{-9}$. We are thus in the case where $|f'(\alpha)| \ll 1$ and we wish to examine the behaviour of Newton's method in the approximation of α when the two stopping criteria above are adopted in the computations.
We show in Tables 6.6 and 6.7 the results obtained using the test based on the control of the residual (1) and of the increment (2), respectively. We have taken $x^{(0)} = 0$ and used two different values of the tolerance. The number of iterations required by the method is denoted by `nit`.

According to (6.31), since $\phi'(\alpha) = 0$, the stopping test based on the increment reveals to be reliable for both the values (which are quite differing) of the stop

Table 6.6. Newton's method for the approximation of the root of $f(x) = e^{-x} - \eta = 0$. The stopping test is based on the control of the residual

ε	nit	$\vert f(x^{(\mathrm{nit})})\vert$	$\vert\alpha - x^{(\mathrm{nit})}\vert$	$\vert\alpha - x^{(\mathrm{nit})}\vert/\alpha$
10^{-10}	22	$5.9 \cdot 10^{-11}$	$5.7 \cdot 10^{-2}$	0.27
10^{-3}	7	$9.1 \cdot 10^{-4}$	13.7	66.2

Table 6.7. Newton's method for the approximation of the root of $f(x) = e^{-x} - \eta = 0$. The stopping test is based on the control of the increment

ε	nit	$\vert x^{(\mathrm{nit})} - x^{(\mathrm{nit}-1)}\vert$	$\vert\alpha - x^{(\mathrm{nit})}\vert$	$\vert\alpha - x^{(\mathrm{nit})}\vert/\alpha$
10^{-10}	26	$8.4 \cdot 10^{-13}$	$\simeq 0$	$\simeq 0$
10^{-3}	25	$1.3 \cdot 10^{-6}$	$8.4 \cdot 10^{-13}$	$4 \cdot 10^{-12}$

tolerance ε. The test based on the residual, instead, yields an acceptable estimate of the root only for very small tolerances, while it is completely wrong for large values of ε. •

6.6 Post-processing Techniques for Iterative Methods

We conclude this chapter by introducing two algorithms that aim at accelerating the convergence of iterative methods for finding the roots of a function.

6.6.1 Aitken's Acceleration

We describe this technique in the case of linearly convergent fixed-point methods, referring to [IK66], pp. 104–108, for the case of methods of higher order.

Consider a fixed-point iteration that is linearly converging to a zero α of a given function f. Denoting by λ an approximation of $\phi'(\alpha)$ to be suitably determined and recalling (6.18) we have, for $k \geq 1$

$$\begin{aligned}\alpha &\simeq \frac{x^{(k)} - \lambda x^{(k-1)}}{1-\lambda} = \frac{x^{(k)} - \lambda x^{(k)} + \lambda x^{(k)} - \lambda x^{(k-1)}}{1-\lambda} \\ &= x^{(k)} + \frac{\lambda}{1-\lambda}(x^{(k)} - x^{(k-1)}).\end{aligned} \tag{6.32}$$

Aitken's method provides a simple way of computing λ that is able to accelerate the convergence of the sequence $\{x^{(k)}\}$ to the root α. With this aim, let us consider for $k \geq 2$ the following ratio

$$\lambda^{(k)} = \frac{x^{(k)} - x^{(k-1)}}{x^{(k-1)} - x^{(k-2)}}, \tag{6.33}$$

and check that

$$\lim_{k\to\infty} \lambda^{(k)} = \phi'(\alpha). \tag{6.34}$$

Indeed, for k sufficiently large

$$x^{(k+2)} - \alpha \simeq \phi'(\alpha)(x^{(k+1)} - \alpha)$$

and thus, elaborating (6.33), we get

$$\lim_{k\to\infty} \lambda^{(k)} = \lim_{k\to\infty} \frac{x^{(k)} - x^{(k-1)}}{x^{(k-1)} - x^{(k-2)}} = \lim_{k\to\infty} \frac{(x^{(k)} - \alpha) - (x^{(k-1)} - \alpha)}{(x^{(k-1)} - \alpha) - (x^{(k-2)} - \alpha)}$$

$$= \lim_{k\to\infty} \frac{\dfrac{x^{(k)} - \alpha}{x^{(k-1)} - \alpha} - 1}{1 - \dfrac{x^{(k-2)} - \alpha}{x^{(k-1)} - \alpha}} = \frac{\phi'(\alpha) - 1}{1 - \dfrac{1}{\phi'(\alpha)}} = \phi'(\alpha)$$

which is (6.34). Substituting in (6.32) λ with its approximation $\lambda^{(k)}$ given by (6.33), yields the updated estimate of α

$$\alpha \simeq x^{(k)} + \frac{\lambda^{(k)}}{1 - \lambda^{(k)}}(x^{(k)} - x^{(k-1)}) \tag{6.35}$$

which, rigorously speaking, is significant only for a sufficiently large k. However, assuming that (6.35) holds for any $k \geq 2$, we denote by $\widehat{x}^{(k)}$ the new approximation of α that is obtained by plugging (6.33) back into (6.35)

$$\widehat{x}^{(k)} = x^{(k)} - \frac{(x^{(k)} - x^{(k-1)})^2}{(x^{(k)} - x^{(k-1)}) - (x^{(k-1)} - x^{(k-2)})}, \qquad k \geq 2. \tag{6.36}$$

This relation is known as *Aitken's extrapolation formula.*
Letting, for $k \geq 2$,

$$\triangle x^{(k)} = x^{(k)} - x^{(k-1)}, \qquad \triangle^2 x^{(k)} = \triangle(\triangle x^{(k)}) = \triangle x^{(k)} - \triangle x^{(k-1)},$$

formula (6.36) can be written as

$$\widehat{x}^{(k)} = x^{(k)} - \frac{(\triangle x^{(k)})^2}{\triangle^2 x^{(k)}}, \qquad k \geq 2. \tag{6.37}$$

Form (6.37) explains the reason why method (6.36) is more commonly known as *Aitken's $\triangle^2$ method.*
For the convergence analysis of Aitken's method, it is useful to write (6.36) as a fixed-point method in the form (6.17), by introducing the iteration function

$$\phi_\triangle(x) = \frac{x\phi(\phi(x)) - \phi^2(x)}{\phi(\phi(x)) - 2\phi(x) + x}. \tag{6.38}$$

This function is indeterminate at $x = \alpha$ since $\phi(\alpha) = \alpha$; however, by applying L'Hospital's rule one can easily check that $\lim_{x\to\alpha} \phi_\Delta(x) = \alpha$ under the assumption that ϕ is differentiable at α and $\phi'(\alpha) \neq 1$. Thus, ϕ_Δ is consistent and has a continuos extension at α, the same being also true if α is a multiple root of f. Moreover, it can be shown that the fixed points of (6.38) coincide with those of ϕ even in the case where α is a multiple root of f (see [IK66], pp. 104–106).

From (6.38) we conclude that Aitken's method can be applied to a fixed-point method $x = \phi(x)$ of arbitrary order. Actually, the following convergence result holds.

Property 6.7 (Convergence of Aitken's method) *Let $x^{(k+1)} = \phi(x^{(k)})$ be a fixed-point iteration of order $p \geq 1$ for the approximation of a simple zero α of a function f. If $p = 1$, Aitken's method converges to α with order 2, while if $p \geq 2$ the convergence order is $2p - 1$. In particular, if $p = 1$, Aitken's method is convergent even if the fixed-point method is not. If α has multiplicity $m \geq 2$ and the method $x^{(k+1)} = \phi(x^{(k)})$ is first-order convergent, then Aitken's method converges linearly, with convergence factor $C = 1 - 1/m$.*

Example 6.10 Consider the computation of the simple zero $\alpha = 1$ for the function $f(x) = (x-1)e^x$. For this, we use three fixed-point methods whose iteration functions are, respectively, $\phi_0(x) = \log(xe^x)$, $\phi_1(x) = (e^x + x)/(e^x + 1)$ and $\phi_2(x) = (x^2 - x + 1)/x$ (for $x \neq 0$). Notice that, since $|\phi_0'(1)| = 2$, the corresponding fixed-point method is not convergent, while in the other two cases the methods have order 1 and 2, respectively.

Let us check the performance of Aitken's method, running Program 55 with $x^{(0)} = 2$, `tol` $= 10^{-10}$ and working in complex arithmetic. Notice that in the case of ϕ_0 this produces complex numbers if $x^{(k)}$ happens to be negative. According to Property 6.7, Aitken's method applied to the iteration function ϕ_0 converges in 8 steps to the value $x^{(8)} = 1.000002 + i\,0.000002$. In the other two cases, the method of order 1 converges to α in 18 iterations, to be compared with the 4 iterations required by Aitken's method, while in the case of the iteration function ϕ_2 convergence holds in 7 iterations against 5 iterations required by Aitken's method. •

Aitken's method is implemented in Program 55. The input/output parameters are the same as those of previous programs in this chapter.

Program 55 - aitken : Aitken's extrapolation

```
function [xvect,xdif,fx,nit]=aitken(x0,nmax,tol,phi,fun)
%AITKEN Aitken's extrapolation
% [XVECT,XDIF,FX,NIT]=AITKEN(X0,TOL,NMAX,FUN,PHI) tries to find a
% zero of the continuous function FUN  using the Aitken's extrapolation on the
% fixed-point iteration X=PHI(X), starting from the initial guess X0. XVECT is
% the vector of iterates, XDIF the vector of the differences between consecutive
% iterates, FX the residual. TOL specifies the tolerance of the method.
nit=0; xvect=[x0]; x=x0; fxn=eval(fun);
```

```
fx=[fxn]; xdif=[]; err=tol+1;
while err>=tol & nit<=nmax
   nit=nit+1; xv=xvect(nit); x=xv; phix=eval(phi);
   x=phix; phixx=eval(phi); den=phixx-2*phix+xv;
   if den == 0
      err=tol*1.e-01;
   else
      xn=(xv*phixx-phix^2)/den;
      xvect=[xvect; xn];
      xdif=[xdif; abs(xn-xv)];
      x=xn; fxn=abs(eval(fun));
      fx=[fx; fxn]; err=fxn;
   end
end
return
```

6.6.2 Techniques for Multiple Roots

As previously noticed in deriving Aitken's acceleration, taking the incremental ratios of successive iterates $\lambda^{(k)}$ in (6.33) provides a way to estimate the asymptotic convergence factor $\phi'(\alpha)$.

This information can be employed also to estimate the multiplicity of the root of a nonlinear equation and, as a consequence, it provides a tool for modifying Newton's method in order to recover its quadratic convergence (see (6.23)). Indeed, define the sequence $m^{(k)}$ through the relation $\lambda^{(k)} = 1 - 1/m^{(k)}$, and recalling (6.22), it follows that $m^{(k)}$ tends to m as $k \to \infty$. If the multiplicity m is known a priori, it is clearly convenient to use the modified Newton method (6.23). In other cases, the following *adaptive Newton algorithm* can be used

$$x^{(k+1)} = x^{(k)} - m^{(k)} \frac{f(x^{(k)})}{f'(x^{(k)})}, \qquad k \geq 2, \tag{6.39}$$

where we have set

$$m^{(k)} = \frac{1}{1 - \lambda^{(k)}} = \frac{x^{(k-1)} - x^{(k-2)}}{2x^{(k-1)} - x^{(k)} - x^{(k-2)}}. \tag{6.40}$$

Example 6.11 Let us check the performances of Newton's method in its three versions proposed so far (standard (6.16), modified (6.23) and adaptive (6.39)), to approximate the multiple zero $\alpha = 1$ of the function $f(x) = (x^2-1)^p \log x$ (for $p \geq 1$ and $x > 0$). The desired root has multiplicity $m = p + 1$. The values $p = 2, 4, 6$ have been considered and $x^{(0)} = 0.8$, `tol`$=10^{-10}$ have always been taken in numerical computations.

The obtained results are summarized in Table 6.8, where for each method the number of iterations n_{it} required to converge are reported. In the case of the adaptive method, beside the value of n_{it} we have also shown in braces the estimate $m^{(n_{it})}$ of the multiplicity m that is yielded by Program 56. •

Table 6.8. Solution of problem $(x^2 - 1)^p \log x = 0$ in the interval $[0.5, 1.5]$, with $p = 2, 4, 6$

m	standard	adaptive	modified
3	51	13 (2.9860)	4
5	90	16 (4.9143)	5
7	127	18 (6.7792)	5

In Example 6.11, the adaptive Newton method converges more rapidly than the standard method, but less rapidly than the modified Newton method. It must be noticed, however, that the adaptive method yields as a useful by-product a good estimate of the multiplicity of the root, which can be profitably employed in a deflation procedure for the approximation of the roots of a polynomial.

The algorithm 6.39, with the adaptive estimate (6.40) of the multiplicity of the root, is implemented in Program 56. To avoid the onset of numerical instabilities, the updating of $m^{(k)}$ is performed only when the variation between two consecutive iterates is sufficiently diminished. The input/output parameters are the same as those of previous programs in this chapter.

Program 56 - adptnewt : Adaptive Newton's method

```
function [xvect,xdif,fx,nit]=adptnewt(x0,tol,nmax,fun,dfun)
%ADPTNEWT Adaptive Newton's method
% [XVECT,XDIF,FX,NIT]=ADPTNEWT(X0,TOL,NMAX,FUN,DFUN) tries to find a
% zero of the continuous function FUN using the adaptive Newton method starting
% from the initial guess X0. FUN and DFUN accept real scalar input x and
% return a real scalar value. XVECT is the vector of iterates, XDIF the vector
% of the differences between consecutive iterates, FX the residual. TOL specifies
% the tolerance of the method.
xvect=x0;
nit=0; r=[1]; err=tol+1; m=[1]; xdif=[];
while nit<nmax & err>tol
   nit=nit+1;
   x=xvect(nit); fx(nit)=eval(fun); f1x=eval(dfun);
   if f1x == 0
      fprintf(' Stop due to vanishing derivative ');
      return
   end;
   x=x-m(nit)*fx(nit)/f1x;
   xvect=[xvect;x]; fx=[fx;eval(fun)];
   rd=err; err=abs(xvect(nit+1)-xvect(nit)); xdif=[xdif;err];
   ra=err/rd; r=[r;ra]; diff=abs(r(nit+1)-r(nit));
   if diff<1.e-3 & r(nit+1)>1.e-2
      m(nit+1)=max(m(nit),1/abs(1-r(nit+1)));
```

Table 6.9. Convergence of Newton's method to the root of equation (6.41)

$v^{(0)}$	N_{it}	$v^{(0)}$	N_{it}	$v^{(0)}$	N_{it}	$v^{(0)}$	N_{it}
10^{-4}	47	10^{-2}	7	10^{-3}	21	10^{-1}	5

```
    else
        m(nit+1)=m(nit);
    end
end
return
```

6.7 Applications

We apply iterative methods for nonlinear equations considered so far in the solution of two problems arising in the study of the thermal properties of gases and electronics, respectively.

6.7.1 Analysis of the State Equation for a Real Gas

For a mole of a perfect gas, the state equation $Pv = RT$ establishes a relation between the pressure P of the gas (in Pascals $[Pa]$), the specific volume v (in cubic meters per kilogram $[m^3 Kg^{-1}]$) and its temperature T (in Kelvin $[K]$), R being the universal gas constant, expressed in $[JKg^{-1}K^{-1}]$ (joules per kilogram per Kelvin).

For a real gas, the deviation from the state equation of perfect gases is due to van der Waals and takes into account the intermolecular interaction and the space occupied by molecules of finite size (see [Sla63]).

Denoting by α and β the gas constants according to the van der Waals model, in order to determine the specific volume v of the gas, once P and T are known, we must solve the nonlinear equation

$$f(v) = (P + \alpha/v^2)(v - \beta) - RT = 0. \tag{6.41}$$

With this aim, let us consider Newton's method (6.16) in the case of carbon dioxide (CO_2), at the pressure of $P = 10[atm]$ (equal to $1013250[Pa]$) and at the temperature of $T = 300[K]$. In such a case, $\alpha = 188.33[Pa\, m^6 Kg^{-2}]$ and $\beta = 9.77 \cdot 10^{-4}[m^3 Kg^{-1}]$; as a comparison, the solution computed by assuming that the gas is perfect is $\tilde{v} \simeq 0.056[m^3 Kg^{-1}]$.

We report in Table 6.9 the results obtained by running Program 50 for different choices of the initial guess $v^{(0)}$. We have denoted by N_{it} the number of iterations needed by Newton's method to converge to the root v^* of $f(v) = 0$ using an absolute tolerance equal to the *roundoff* unit.

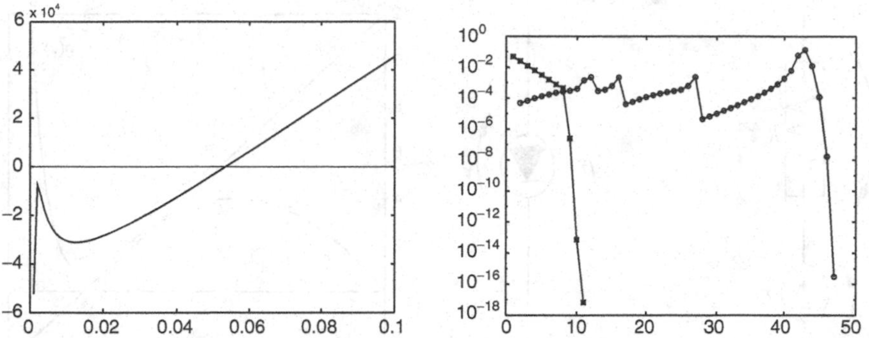

Fig. 6.8. Graph of the function f in (6.41) (*left*); increments $|v^{(k+1)} - v^{(k)}|$ computed by the Newton's method (*circled curve*) and bisection-Newton's method (*starred curve*)

The computed approximation of v^* is $v^{(N_{it})} \simeq 0.0535$. To analyze the causes of the strong dependence of N_{it} on the value of $v^{(0)}$, let us examine the derivative $f'(v) = P - \alpha v^{-2} + 2\alpha\beta v^{-3}$. For $v > 0$, $f'(v) = 0$ at $v_M \simeq 1.99 \cdot 10^{-3}[m^3Kg^{-1}]$ (relative maximum) and at $v_m \simeq 1.25 \cdot 10^{-2}[m^3Kg^{-1}]$ (relative minimum), as can be seen in the graph of Figure 6.8 (*left*).
A choice of $v^{(0)}$ in the interval $(0, v_m)$ (with $v^{(0)} \neq v_M$) thus necessarily leads to a slow convergence of Newton's method, as demonstrated in Figure 6.8 (*right*), where, in solid circled line, the sequence $\{|v^{(k+1)} - v^{(k)}|\}$ is shown, for $k \geq 0$.

A possible remedy consists of resorting to a polyalgorithmic approach, based on the sequential use of the bisection method and Newton's method (see Section 6.2.1). Running the bisection-Newton's method with the endpoints of the search interval equal to $a = 10^{-4}[m^3Kg^{-1}]$ and $b = 0.1[m^3Kg^{-1}]$ and an absolute tolerance of $10^{-3}[m^3Kg^{-1}]$, yields an overall convergence of the algorithm to the root v^* in 11 iterations, with an accuracy of the order of the *roundoff* unit. The plot of the sequence $\{|v^{(k+1)} - v^{(k)}|\}$, for $k \geq 0$, is shown in solid and starred lines in Figure 6.8 (*right*).

6.7.2 Analysis of a Nonlinear Electrical Circuit

Let us consider the electrical circuit in Figure 6.9 (*left*), where v and j denote respectively the voltage drop across the device D (called a *tunneling diode*) and the current flowing through D, while R and E are a resistor and a voltage generator of given values.

The circuit is commonly employed as a biasing circuit for electronic devices working at high frequency (see [Col66]). In such applications the parameters R and E are designed in such a way that v attains a value internal to the interval for which $g'(v) < 0$, where g is the function which describes

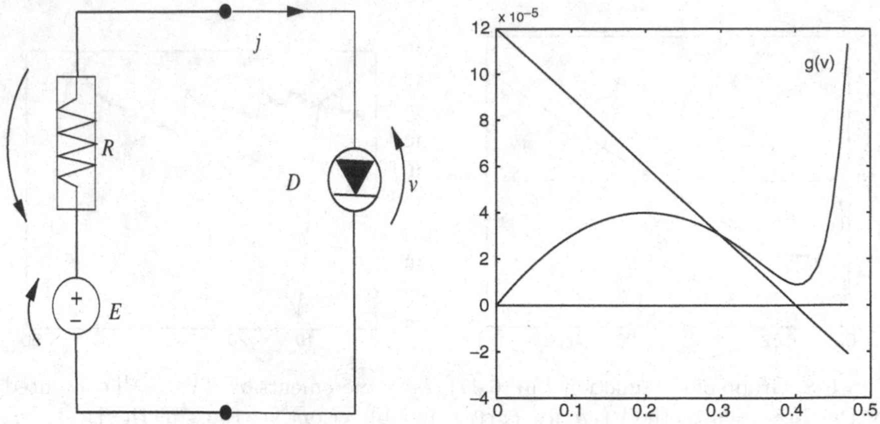

Fig. 6.9. Tunneling diode circuit (*left*) and working point computation (*right*)

the bound between current and voltage for D and is drawn in Figure 6.9 (*right*). Explicitly, $g = \alpha(e^{v/\beta} - 1) - \mu v(v - \gamma)$, for suitable constants α, β, γ and μ.

Our aim is to determine the *working point* of the circuit at hand, that is, the values attained by v and j for given parameters R and E. For that, we write Kirchhoff's law for the voltages across the loop, obtaining the following nonlinear equation

$$f(v) = v\left(\frac{1}{R} + \mu\gamma\right) - \mu v^2 + \alpha(e^{v/\beta} - 1) - \frac{E}{R} = 0. \qquad (6.42)$$

From a graphical standpoint, finding out the working point of the circuit amounts to determining the intersection between the function g and the straight line of equation $j = (E - v)/R$, as shown in Figure 6.9 (*right*). Assume the following (real-life) values for the parameters of the problem: $E/R = 1.2 \cdot 10^{-4}$ $[A]$, $\alpha = 10^{-12}$ $[A]$, $\beta^{-1} = 40$ $[V^{-1}]$, $\mu = 10^{-3}$ $[AV^{-2}]$ and $\gamma = 0.4$ $[V]$. The solution of (6.42), which is also unique for the considered values of the parameters, is $v^* \simeq 0.3$ $[V]$.

To approximate v^*, we compare the main iterative methods introduced in this chapter. We have taken $v^{(0)} = 0\,[V]$ for Newton's method, $\xi = 0$ for the Dekker-Brent algorithm (for the meaning of ξ, see Example 6.5), while for all the other schemes the search interval has been taken equal to $[0, 0.5]$. The stopping tolerance `tol` has been set to 10^{-10}. The obtained results are reported in Table 6.10 where `nit` and $f^{(\mathtt{nit})}$ denote respectively the number of iterations needed by the method to converge and the value of f at the computed solution.

Notice the extremely slow convergence of the *Regula Falsi* method, due to the fact that the value $v^{(k')}$ always coincides with the right end-point $v = 0.5$

Table 6.10. Convergence of the methods for the approximation of the root of equation (6.42)

Method	nit	$f^{(\texttt{nit})}$	Method	nit	$f^{(\texttt{nit})}$
bisection	32	$-1.12 \cdot 10^{-15}$	Dekker-Brent	11	$1.09 \cdot 10^{-14}$
Regula Falsi	225	$-9.77 \cdot 10^{-11}$	secant	11	$2.7 \cdot 10^{-20}$
chord	186	$-9.80 \cdot 10^{-14}$	Newton's	8	$-1.35 \cdot 10^{-20}$

and the function f around v^* has derivative very close to zero. An analogous interpretation holds for the chord method.

6.8 Exercises

1. Derive geometrically the sequence of the first iterates computed by bisection, *Regula Falsi*, secant and Newton's methods in the approximation of the zero of the function $f(x) = x^2 - 2$ in the interval $[1, 3]$.
2. Let f be a continuous function that is m-times differentiable ($m \geq 1$), such that $f(\alpha) = \ldots = f^{(m-1)}(\alpha) = 0$ and $f^{(m)}(\alpha) \neq 0$. Prove (6.22) and check that the modified Newton method (6.23) has order of convergence equal to 2.
[*Hint*: let $f(x) = (x-\alpha)^m h(x)$, h being a function such that $h(\alpha) \neq 0$].
3. Let $f(x) = \cos^2(2x) - x^2$ be the function in the interval $0 \leq x \leq 1.5$ examined in Example 6.4. Having fixed a tolerance $\varepsilon = 10^{-10}$ on the absolute error, determine experimentally the subintervals for which Newton's method is convergent to the zero $\alpha \simeq 0.5149$.
[*Solution*: for $0 < x^{(0)} \leq 0.02$, $0.94 \leq x^{(0)} \leq 1.13$ and $1.476 \leq x^{(0)} \leq 1.5$, the method converges to the solution $-\alpha$. For any other value of $x^{(0)}$ in $[0, 1.5]$, the method converges to α].
4. Check the following properties:
   a) $0 < \phi'(\alpha) < 1$: *monotone* convergence, that is, the error $x^{(k)} - \alpha$ maintains a constant sign as k varies;
   b) $-1 < \phi'(\alpha) < 0$: oscillatory convergence that is, $x^{(k)} - \alpha$ changes sign as k varies;
   c) $|\phi'(\alpha)| > 1$: divergence. More precisely, if $\phi'(\alpha) > 1$, the sequence is monotonically diverging, while for $\phi'(\alpha) < -1$ it diverges with oscillatory sign.
5. Consider for $k \geq 0$ the fixed-point method, known as *Steffensen's method*

$$x^{(k+1)} = x^{(k)} - \frac{f(x^{(k)})}{\varphi(x^{(k)})}, \; \varphi(x^{(k)}) = \frac{f(x^{(k)} + f(x^{(k)})) - f(x^{(k)})}{f(x^{(k)})},$$

   and prove that it is a second-order method. Implement the Steffensen method in a MATLAB code and employ it to approximate the root of the nonlinear equation $e^{-x} - \sin(x) = 0$.
6. Analyze the convergence of the fixed-point method $x^{(k+1)} = \phi_j(x^{(k)})$ for computing the zeros $\alpha_1 = -1$ and $\alpha_2 = 2$ of the function $f(x) = x^2 - x - 2$, when the following iteration functions are used: $\phi_1(x) = x^2 - 2$, $\phi_2(x) = \sqrt{2+x}$ $\phi_3(x) = -\sqrt{2+x}$ and $\phi_4(x) = 1 + 2/x$, $x \neq 0$.

[*Solution*: the method is non convergent with ϕ_1, it converges only to α_2, with ϕ_2 and ϕ_4, while it converges only to α_1 with ϕ_3].

7. For the approximation of the zeros of the function $f(x) = (2x^2 - 3x - 2)/(x - 1)$, consider the following fixed-point methods:
(1) $x^{(k+1)} = g(x^{(k)})$, where $g(x) = (3x^2 - 4x - 2)/(x - 1)$;
(2) $x^{(k+1)} = h(x^{(k)})$, where $h(x) = x - 2 + x/(x - 1)$.
Analyze the convergence properties of the two methods and determine in particular their order. Check the behavior of the two schemes using Program 51 and provide, for the second method, an experimental estimate of the interval such that if $x^{(0)}$ is chosen in the interval then the method converges to $\alpha = 2$.
[*Solution*: zeros: $\alpha_1 = -1/2$ and $\alpha_2 = 2$. Method (1) is not convergent, while (2) can approximate only α_2 and is second-order. Convergence holds for any $x^{(0)} > 1$].
8. Propose at least two fixed-point methods for approximating the root $\alpha \simeq 0.5885$ of equation $e^{-x} - \sin(x) = 0$ and analyze their convergence.
9. Using Descartes's rule of signs, determine the number of real roots of the polynomials $p_6(x) = x^6 - x - 1$ and $p_4(x) = x^4 - x^3 - x^2 + x - 1$.
[*Solution*: both p_6 and p_4 have one negative and one positive real root].
10. Let $g : \mathbb{R} \to \mathbb{R}$ be defined as $g(x) = \sqrt{1 + x^2}$. Show that the iterates of Newton's method for the equation $g'(x) = 0$ satisfy the following properties:

$$(a) \quad |x^{(0)}| < 1 \Rightarrow g(x^{(k+1)}) < g(x^{(k)}),\ k \geq 0,\ \lim_{k\to\infty} x^{(k)} = 0,$$

$$(b) \quad |x^{(0)}| > 1 \Rightarrow g(x^{(k+1)}) > g(x^{(k)}),\ k \geq 0,\ \lim_{k\to\infty} |x^{(k)}| = +\infty.$$

7

Nonlinear Systems and Numerical Optimization

In this chapter we address the numerical solution of systems of nonlinear equations and the minimization of a function of several variables.

The first problem generalizes to the n-dimensional case the search for the zeros of a function, which was considered in Chapter 6, and can be formulated as follows: given $\mathbf{F} : \mathbb{R}^n \rightarrow \mathbb{R}^n$,

$$\text{find } \mathbf{x}^* \in \mathbb{R}^n \text{ such that } \mathbf{F}(\mathbf{x}^*) = \mathbf{0}. \tag{7.1}$$

Problem (7.1) will be solved by extending to several dimensions some of the schemes that have been proposed in Chapter 6.
The basic formulation of the second problem reads: given $f : \mathbb{R}^n \rightarrow \mathbb{R}$, called an *objective function*,

$$\text{minimize } f(\mathbf{x}) \text{ in } \mathbb{R}^n, \tag{7.2}$$

and is called an *unconstrained optimization problem.*

A typical example consists of determining the optimal allocation of n resources, $x_1, x_2, \ldots, x_n$, in competition with each other and ruled by a specific law. Generally, such resources are not unlimited; this circumstance, from a mathematical standpoint, amounts to requiring that the minimizer of the objective function lies within a subset $\Omega \subset \mathbb{R}^n$, and, possibly, that some equality or inequality constraints must be satisfied.

When these constraints exist the optimization problem is called *constrained* and can be formulated as follows: given the objective function f,

$$\text{minimize } f(\mathbf{x}) \text{ in } \Omega \subset \mathbb{R}^n. \tag{7.3}$$

Remarkable instances of (7.3) are those in which Ω is characterized by conditions like $\mathbf{h}(\mathbf{x}) = \mathbf{0}$ (equality constraints) or $\mathbf{h}(\mathbf{x}) \leq \mathbf{0}$ (inequality constraints), where $\mathbf{h} : \mathbb{R}^n \rightarrow \mathbb{R}^m$, with $m \leq n$, is a given function, called *cost function*, and the condition $\mathbf{h}(\mathbf{x}) \leq \mathbf{0}$ means $h_i(\mathbf{x}) \leq 0$, for $i = 1, \ldots, m$.

If the function $\mathbf{h}$ is continuous and Ω is connected, problem (7.3) is usually referred to as a *nonlinear programming* problem. Notable examples in this area are:

1. *convex programming* if f is a convex function and $\mathbf{h}$ has convex components (see (7.21));
2. *linear programming* if f and $\mathbf{h}$ are linear;
3. *quadratic programming* if f is quadratic and $\mathbf{h}$ is linear.

Problems (7.1) and (7.2) are strictly related to one another. Indeed, if we denote by F_i the components of $\mathbf{F}$, then a point $\mathbf{x}^*$, a solution of (7.1), is a minimizer of the function $f(\mathbf{x}) = \sum_{i=1}^n F_i^2(\mathbf{x})$. Conversely, assuming that f is differentiable and setting the partial derivatives of f equal to zero at a point $\mathbf{x}^*$ at which f is minimum leads to a system of nonlinear equations. Thus, any system of nonlinear equations can be associated with a suitable minimization problem, and vice versa. We shall take advantage of this observation when devising efficient numerical methods.

7.1 Solution of Systems of Nonlinear Equations

Before considering problem (7.1), let us set some notation which will be used throughout the chapter.

For $k \geq 0$, we denote by $C^k(D)$ the set of k-continuously differentiable functions from D to $\mathbb{R}^n$, where $D \subseteq \mathbb{R}^n$ is a set that will be made precise from time to time. We shall always assume that $\mathbf{F} \in C^1(D)$, i.e., $\mathbf{F} : \mathbb{R}^n \to \mathbb{R}^n$ is a continuously differentiable function on D.

We denote also by $\mathrm{J}_\mathbf{F}(\mathbf{x})$ the Jacobian matrix associated with $\mathbf{F}$ and evaluated at the point $\mathbf{x} = [x_1, \ldots, x_n]^T$ of $\mathbb{R}^n$, defined as

$$(\mathrm{J}_\mathbf{F}(\mathbf{x}))_{ij} = \left(\frac{\partial F_i}{\partial x_j}\right)(\mathbf{x}), \qquad i, j = 1, \ldots, n.$$

Given any vector norm $\|\cdot\|$, we shall henceforth denote the sphere of radius R with center $\mathbf{x}^*$ by

$$B(\mathbf{x}^*; R) = \{\mathbf{y} \in \mathbb{R}^n : \ \|\mathbf{y} - \mathbf{x}^*\| < R\}.$$

7.1.1 Newton's Method and Its Variants

An immediate extension to the vector case of Newton's method (6.16) for scalar equations can be formulated as follows: given $\mathbf{x}^{(0)} \in \mathbb{R}^n$, for $k = 0, 1, \ldots$, until convergence

$$\begin{aligned} &\text{solve} \quad \mathrm{J}_\mathbf{F}(\mathbf{x}^{(k)})\boldsymbol{\delta}\mathbf{x}^{(k)} = -\mathbf{F}(\mathbf{x}^{(k)}); \\ &\text{set} \quad\ \ \mathbf{x}^{(k+1)} = \mathbf{x}^{(k)} + \boldsymbol{\delta}\mathbf{x}^{(k)}. \end{aligned} \tag{7.4}$$

Thus, at each step k the solution of a linear system with matrix $\mathrm{J}_\mathbf{F}(\mathbf{x}^{(k)})$ is required.

Example 7.1 Consider the nonlinear system

$$\begin{cases} e^{x_1^2+x_2^2} - 1 = 0, \\ e^{x_1^2-x_2^2} - 1 = 0, \end{cases}$$

which admits the unique solution $\mathbf{x}^* = \mathbf{0}$. In this case, $\mathbf{F}(\mathbf{x}) = [e^{x_1^2+x_2^2} - 1, e^{x_1^2-x_2^2} - 1]^T$. Running Program 57, leads to convergence in 15 iterations to the pair $[0.61 \cdot 10^{-5}, 0.61 \cdot 10^{-5}]^T$, starting from the initial datum $\mathbf{x}^{(0)} = [0.1, 0.1]^T$, thus demonstrating a fairly rapid convergence rate. The results, however, dramatically change as the choice of the initial guess is varied. For instance, picking up $\mathbf{x}^{(0)} = [10, 10]^T$, 220 iterations are needed to obtain a solution comparable to the previous one, while, starting from $\mathbf{x}^{(0)} = [20, 20]^T$, Newton's method fails to converge. •

The previous example points out the high sensitivity of Newton's method on the choice of the initial datum $\mathbf{x}^{(0)}$, as confirmed by the following local convergence result.

Theorem 7.1 *Let $\mathbf{F} : \mathbb{R}^n \rightarrow \mathbb{R}^n$ be a C^1 function in a convex open set D of $\mathbb{R}^n$ that contains $\mathbf{x}^*$. Suppose that $\mathrm{J}_{\mathbf{F}}^{-1}(\mathbf{x}^*)$ exists and that there exist positive constants R, C and L, such that $\|\mathrm{J}_{\mathbf{F}}^{-1}(\mathbf{x}^*)\| \leq C$ and*

$$\|\mathrm{J}_{\mathbf{F}}(\mathbf{x}) - \mathrm{J}_{\mathbf{F}}(\mathbf{y})\| \leq L\|\mathbf{x} - \mathbf{y}\| \quad \forall \mathbf{x}, \mathbf{y} \in B(\mathbf{x}^*; R),$$

having denoted by the same symbol $\|\cdot\|$ two consistent vector and matrix norms. Then, there exists $r > 0$ such that, for any $\mathbf{x}^{(0)} \in B(\mathbf{x}^*; r)$, the sequence (7.4) is uniquely defined and converges to $\mathbf{x}^*$ with

$$\|\mathbf{x}^{(k+1)} - \mathbf{x}^*\| \leq CL\|\mathbf{x}^{(k)} - \mathbf{x}^*\|^2. \tag{7.5}$$

Proof. Proceeding by induction on k, let us check (7.5) and, moreover, that $\mathbf{x}^{(k+1)} \in B(\mathbf{x}^*; r)$, where $r = \min(R, 1/(2CL))$. First, we prove that for any $\mathbf{x}^{(0)} \in B(\mathbf{x}^*; r)$, the inverse matrix $\mathrm{J}_{\mathbf{F}}^{-1}(\mathbf{x}^{(0)})$ exists. Indeed

$$\|\mathrm{J}_{\mathbf{F}}^{-1}(\mathbf{x}^*)[\mathrm{J}_{\mathbf{F}}(\mathbf{x}^{(0)}) - \mathrm{J}_{\mathbf{F}}(\mathbf{x}^*)]\| \leq \|\mathrm{J}_{\mathbf{F}}^{-1}(\mathbf{x}^*)\| \, \|\mathrm{J}_{\mathbf{F}}(\mathbf{x}^{(0)}) - \mathrm{J}_{\mathbf{F}}(\mathbf{x}^*)\| \leq CLr \leq \frac{1}{2},$$

and thus, thanks to Theorem 1.5, we can conclude that $\mathrm{J}_{\mathbf{F}}^{-1}(\mathbf{x}^{(0)})$ exists, since

$$\|\mathrm{J}_{\mathbf{F}}^{-1}(\mathbf{x}^{(0)})\| \leq \frac{\|\mathrm{J}_{\mathbf{F}}^{-1}(\mathbf{x}^*)\|}{1 - \|\mathrm{J}_{\mathbf{F}}^{-1}(\mathbf{x}^*)[\mathrm{J}_{\mathbf{F}}(\mathbf{x}^{(0)}) - \mathrm{J}_{\mathbf{F}}(\mathbf{x}^*)]\|} \leq 2\|\mathrm{J}_{\mathbf{F}}^{-1}(\mathbf{x}^*)\| \leq 2C.$$

As a consequence, $\mathbf{x}^{(1)}$ is well defined and

$$\mathbf{x}^{(1)} - \mathbf{x}^* = \mathbf{x}^{(0)} - \mathbf{x}^* - \mathrm{J}_{\mathbf{F}}^{-1}(\mathbf{x}^{(0)})[\mathbf{F}(\mathbf{x}^{(0)}) - \mathbf{F}(\mathbf{x}^*)].$$

Factoring out $\mathrm{J}_{\mathbf{F}}^{-1}(\mathbf{x}^{(0)})$ on the right-hand side and passing to the norms, we get

$$\begin{aligned} \|\mathbf{x}^{(1)} - \mathbf{x}^*\| &\leq \|\mathrm{J}_{\mathbf{F}}^{-1}(\mathbf{x}^{(0)})\| \, \|\mathbf{F}(\mathbf{x}^*) - \mathbf{F}(\mathbf{x}^{(0)}) - \mathrm{J}_{\mathbf{F}}(\mathbf{x}^{(0)})[\mathbf{x}^* - \mathbf{x}^{(0)}]\| \\ &\leq 2C\frac{L}{2}\|\mathbf{x}^* - \mathbf{x}^{(0)}\|^2, \end{aligned}$$

where the remainder of Taylor's series of $\mathbf{F}$ has been used. The previous relation proves (7.5) in the case $k = 0$; moreover, since $\mathbf{x}^{(0)} \in B(\mathbf{x}^*; r)$, we have $\|\mathbf{x}^* - \mathbf{x}^{(0)}\| \leq 1/(2CL)$, from which $\|\mathbf{x}^{(1)} - \mathbf{x}^*\| \leq \dfrac{1}{2}\|\mathbf{x}^* - \mathbf{x}^{(0)}\|$.
This ensures that $\mathbf{x}^{(1)} \in B(\mathbf{x}^*; r)$.

By a similar proof, one can check that, should (7.5) be true for a certain k, then the same inequality would follow also for $k+1$ in place of k. This proves the theorem. ◇

Theorem 7.1 thus confirms that Newton's method is quadratically convergent only if $\mathbf{x}^{(0)}$ is sufficiently close to the solution $\mathbf{x}^*$ and if the Jacobian matrix is nonsingular. Moreover, it is worth noting that the computational effort needed to solve the linear system (7.4) can be excessively high as n gets large. Also, $\mathrm{J}_{\mathbf{F}}(\mathbf{x}^{(k)})$ could be ill-conditioned, which makes it quite difficult to obtain an accurate solution. For these reasons, several modifications to Newton's method have been proposed, which will be briefly considered in the later sections, referring to the specialized literature for further details (see [OR70], [DS83], [Erh97], [BS90], [SM03], [Deu04] and the references therein).

Remark 7.1 Let $\mathbf{G}(\mathbf{x}) = \mathbf{x} - \mathbf{F}(\mathbf{x})$ and denote by $\mathbf{r}^{(k)} = \mathbf{F}(\mathbf{x}^{(k)})$ the residual at step k. Then, from (7.4) it turns out that Newton's method can be alternatively formulated as

$$\left(\mathrm{I} - \mathrm{J}_{\mathbf{G}}(\mathbf{x}^{(k)})\right)\left(\mathbf{x}^{(k+1)} - \mathbf{x}^{(k)}\right) = -\mathbf{r}^{(k)},$$

where $\mathrm{J}_{\mathbf{G}}$ denotes the Jacobian matrix associated with $\mathbf{G}$. This equation allows us to interpret Newton's method as a preconditioned stationary Richardson method. This prompts introducing a parameter α_k in order to accelerate the convergence of the iteration

$$\left(\mathrm{I} - \mathrm{J}_{\mathbf{G}}(\mathbf{x}^{(k)})\right)\left(\mathbf{x}^{(k+1)} - \mathbf{x}^{(k)}\right) = -\alpha_k \mathbf{r}^{(k)}.$$

The problem of how to select α_k will be addressed in Section 7.2.6. ■

7.1.2 Modified Newton's Methods

Several modifications of Newton's method have been proposed in order to reduce its cost when the computed solution is sufficiently close to $\mathbf{x}^*$. Further variants, that are globally convergent, will be introduced for the solution of the minimization problem (7.2).

1. Cyclic updating of the Jacobian matrix

An efficient alternative to method (7.4) consists of keeping the Jacobian matrix (more precisely, its factorization) unchanged for a certain number, say $p \geq 2$, of steps. Generally, a deterioration of convergence rate is accompanied by a gain in computational efficiency.

Program 57 implements Newton's method in the case in which the LU factorization of the Jacobian matrix is updated once every `p` steps. The programs used to solve the triangular systems have been described in Chapter 3.

Here and in later codings in this chapter, we denote by `x0` the initial vector, by `F` and `J` the variables containing the functional expressions of $\mathbf{F}$ and of its Jacobian matrix $J_{\mathbf{F}}$, respectively. The parameters `tol` and `nmax` represent the stopping tolerance in the convergence of the iterative process and the maximum admissible number of iterations, respectively. In output, the vector `x` contains the approximation to the searched zero of $\mathbf{F}$, while `nit` denotes the number of iterations necessary to converge.

Program 57 - newtonsys : Newton's method for nonlinear systems

```
function [x,iter]=newtonsys(F,J,x0,tol,nmax,p)
%NEWTONSYS Newton method for nonlinear systems
% [X, ITER] = NEWTONSYS(F, J, X0, TOL, NMAX, P) attempts to solve the
% nonlinear system F(X)=0 with the Newton method. F and J are strings
% containing the functional expressions of the nonlinear equations and of
% the Jacobian matrix. X0 specifies the initial guess. TOL specifies the
% tolerance of the method. NMAX specifies the maximum number of iterations.
% P specifies the number of consecutive steps during which the Jacobian is
% mantained fixed. ITER is the iteration number at which X is computed.
[n,m]=size(F);
if n ~= m, error('Only square systems'); end
iter=0; Fxn=zeros(n,1); x=x0; err=tol+1;
for i=1:n
   for j=1:n
      Jxn(i,j)=eval(J((i-1)*n+j,:));
   end
end
[L,U,P]=lu(Jxn);
step=0;
while err>tol
   if step == p
      step = 0;
      for i=1:n
         Fxn(i)=eval(F(i,:));
         for j=1:n; Jxn(i,j)=eval(J((i-1)*n+j,:)); end
      end
      [L,U,P]=lu(Jxn);
   else
      for i=1:n, Fxn(i)=eval(F(i,:)); end
   end
   iter=iter+1; step=step+1; Fxn=-P*Fxn;
   y=forwardcol(L,Fxn);
   deltax=backwardcol(U,y);
   x = x + deltax;
```

```
    err=norm(deltax);
  if iter > nmax
    error(' Fails to converge within maximum number of iterations ');
  end
end
return
```

2. Inexact solution of the linear systems

Another possibility consists of solving the linear system (7.4) by an iterative method, where the maximum number of admissible iterations is fixed a priori. The resulting schemes are identified as Newton-Jacobi, Newton-SOR or Newton-Krylov methods, according to the iterative process that is used for the linear system (see [BS90], [Kel99]). Here, we limit ourselves to describing the Newton-SOR method.

In analogy with what was done in Section 4.2.1, let us decompose the Jacobian matrix at step k as

$$J_{\mathbf{F}}(\mathbf{x}^{(k)}) = D_k - E_k - F_k, \tag{7.6}$$

where $D_k = D(\mathbf{x}^{(k)})$, $-E_k = -E(\mathbf{x}^{(k)})$ and $-F_k = -F(\mathbf{x}^{(k)})$, the diagonal part and the lower and upper triangular portions of the matrix $J_{\mathbf{F}}(\mathbf{x}^{(k)})$, respectively. We suppose also that D_k is nonsingular. The SOR method for solving the linear system in (7.4) is organized as follows: setting $\boldsymbol{\delta}\mathbf{x}_0^{(k)} = \mathbf{0}$, solve

$$\boldsymbol{\delta}\mathbf{x}_r^{(k)} = M_k \boldsymbol{\delta}\mathbf{x}_{r-1}^{(k)} - \omega_k (D_k - \omega_k E_k)^{-1} \mathbf{F}(\mathbf{x}^{(k)}), \quad r = 1, 2, \ldots, \tag{7.7}$$

where M_k is the iteration matrix of SOR method

$$M_k = [D_k - \omega_k E_k]^{-1} \ [(1 - \omega_k) D_k + \omega_k F_k],$$

and ω_k is a positive relaxation parameter whose optimal value can rarely be determined a priori. Assume that only $r = m$ steps of the method are carried out. Recalling that $\boldsymbol{\delta}\mathbf{x}_r^{(k)} = \mathbf{x}_r^{(k)} - \mathbf{x}^{(k)}$ and still denoting by $\mathbf{x}^{(k+1)}$ the approximate solution computed after m steps, we find that this latter can be written as (see Exercise 1)

$$\mathbf{x}^{(k+1)} = \mathbf{x}^{(k)} - \omega_k \left(M_k^{m-1} + \ldots + I \right) (D_k - \omega_k E_k)^{-1} \mathbf{F}(\mathbf{x}^{(k)}). \tag{7.8}$$

This method is thus a composite iteration, in which at each step k, starting from $\mathbf{x}^{(k)}$, m steps of the SOR method are carried out to solve approximately system (7.4).

The integer m, as well as ω_k, can depend on the iteration index k; the simplest choice amounts to performing, at each Newton's step, only one iteration of the SOR method, thus obtaining for $r = 1$ from (7.7) the one-step Newton-SOR method

$$\mathbf{x}^{(k+1)} = \mathbf{x}^{(k)} - \omega_k \left(\mathrm{D}_k - \omega_k \mathrm{E}_k\right)^{-1} \mathbf{F}(\mathbf{x}^{(k)}).$$

In a similar way, the preconditioned Newton-Richardson method with matrix P_k, if truncated at the m-th iteration, is

$$\mathbf{x}^{(k+1)} = \mathbf{x}^{(k)} - \left[\mathrm{I} + \mathrm{M}_k + \ldots + \mathrm{M}_k^{m-1}\right] \mathrm{P}_k^{-1}\mathbf{F}(\mathbf{x}^{(k)}),$$

where P_k is the preconditioner of $\mathrm{J}_\mathbf{F}$ and

$$\mathrm{M}_k = \mathrm{P}_k^{-1}\mathrm{N}_k, \, \mathrm{N}_k = \mathrm{P}_k - \mathrm{J}_\mathbf{F}(\mathbf{x}^{(k)}).$$

For an efficient implementation of these techniques we refer to the MATLAB software package developed in [Kel99].

3. Difference approximations of the Jacobian matrix

Another possibility consists of replacing $\mathrm{J}_\mathbf{F}(\mathbf{x}^{(k)})$ (whose explicit computation is often very expensive) with an approximation through n-dimensional differences of the form

$$(\mathrm{J}_h^{(k)})_j = \frac{\mathbf{F}(\mathbf{x}^{(k)} + h_j^{(k)}\mathbf{e}_j) - \mathbf{F}(\mathbf{x}^{(k)})}{h_j^{(k)}}, \qquad \forall k \geq 0, \tag{7.9}$$

where $\mathbf{e}_j$ is the j-th vector of the canonical basis of $\mathbb{R}^n$ and $h_j^{(k)} > 0$ are increments to be suitably chosen at each step k of the iteration (7.4). The following result can be shown.

Property 7.1 *Let* $\mathbf{F}$ *and* $\mathbf{x}^*$ *be such that the hypotheses of Theorem 7.1 are fulfilled, where* $\|\cdot\|$ *denotes the* $\|\cdot\|_1$ *vector norm and the corresponding induced matrix norm. If there exist two positive constants* ε *and* h *such that* $\mathbf{x}^{(0)} \in B(\mathbf{x}^*, \varepsilon)$ *and* $0 < |h_j^{(k)}| \leq h$ *for* $j = 1, \ldots, n$ *then the sequence defined by*

$$\mathbf{x}^{(k+1)} = \mathbf{x}^{(k)} - \left[\mathrm{J}_h^{(k)}\right]^{-1} \mathbf{F}(\mathbf{x}^{(k)}), \tag{7.10}$$

is well defined and converges linearly to $\mathbf{x}^*$. *Moreover, if there exists a positive constant* C *such that* $\max_j |h_j^{(k)}| \leq C\|\mathbf{x}^{(k)} - \mathbf{x}^*\|$ *or, equivalently, there exists a positive constant* c *such that* $\max_j |h_j^{(k)}| \leq c\|\mathbf{F}(\mathbf{x}^{(k)})\|$, *then the sequence* (7.10) *is convergent quadratically.*

This result does not provide any constructive indication as to how to compute the increments $h_j^{(k)}$. In this regard, the following remarks can be made. The first-order truncation error with respect to $h_j^{(k)}$, which arises from the divided difference (7.9), can be reduced by reducing the sizes of $h_j^{(k)}$. On the other hand, a too small value for $h_j^{(k)}$ can lead to large rounding errors. A trade-off must therefore be made between the need of limiting the truncation errors and ensuring a certain accuracy in the computations.

A possible choice is to take

$$h_j^{(k)} = \sqrt{\epsilon_M} \max\left\{|x_j^{(k)}|, M_j\right\} \text{sign}(x_j),$$

where M_j is a parameter that characterizes the typical size of the component x_j of the solution. Further improvements can be achieved using higher-order divided differences to approximate derivatives, like

$$(\mathrm{J}_h^{(k)})_j = \frac{\mathbf{F}(\mathbf{x}^{(k)} + h_j^{(k)}\mathbf{e}_j) - \mathbf{F}(\mathbf{x}^{(k)} - h_j^{(k)}\mathbf{e}_j)}{2h_j^{(k)}}, \qquad \forall k \geq 0.$$

For further details on this subject, see, for instance, [BS90].

7.1.3 Quasi-Newton Methods

By this term, we denote all those schemes in which globally convergent methods are coupled with Newton-like methods that are only locally convergent, but with an order greater than one.

In a quasi-Newton method, given a continuously differentiable function $\mathbf{F} : \mathbb{R}^n \to \mathbb{R}^n$, and an initial value $\mathbf{x}^{(0)} \in \mathbb{R}^n$, at each step k one has to accomplish the following operations:

1. compute $\mathbf{F}(\mathbf{x}^{(k)})$;
2. choose $\tilde{\mathrm{J}}_{\mathbf{F}}(\mathbf{x}^{(k)})$ as being either the exact $\mathrm{J}_{\mathbf{F}}(\mathbf{x}^{(k)})$ or an approximation of it;
3. solve the linear system $\tilde{\mathrm{J}}_{\mathbf{F}}(\mathbf{x}^{(k)})\boldsymbol{\delta}\mathbf{x}^{(k)} = -\mathbf{F}(\mathbf{x}^{(k)})$;
4. set $\mathbf{x}^{(k+1)} = \mathbf{x}^{(k)} + \alpha_k \boldsymbol{\delta}\mathbf{x}^{(k)}$, where α_k are suitable *damping parameters.*

Step 4. is thus the characterizing element of this family of methods. It will be addressed in Section 7.2.6, where a criterion for selecting the "direction" $\boldsymbol{\delta}\mathbf{x}^{(k)}$ will be provided.

7.1.4 Secant-like Methods

These methods are constructed starting from the secant method introduced in Section 6.2 for scalar functions. Precisely, given two vectors $\mathbf{x}^{(0)}$ and $\mathbf{x}^{(1)}$, at the generic step $k \geq 1$ we solve the linear system

$$\mathrm{Q}_k \boldsymbol{\delta}\mathbf{x}^{(k+1)} = -\mathbf{F}(\mathbf{x}^{(k)}) \tag{7.11}$$

and we set $\mathbf{x}^{(k+1)} = \mathbf{x}^{(k)} + \boldsymbol{\delta}\mathbf{x}^{(k+1)}$. Q_k is an $n \times n$ matrix such that

$$\mathrm{Q}_k \boldsymbol{\delta}\mathbf{x}^{(k)} = \mathbf{F}(\mathbf{x}^{(k)}) - \mathbf{F}(\mathbf{x}^{(k-1)}) = \mathbf{b}^{(k)}, \qquad k \geq 1,$$

and is obtained by a formal generalization of (6.13). However, the algebraic relation above does not suffice to uniquely determine Q_k. For this purpose we require Q_k for $k \geq n$ to be a solution to the following set of n systems

$$Q_k \left(\mathbf{x}^{(k)} - \mathbf{x}^{(k-j)}\right) = \mathbf{F}(\mathbf{x}^{(k)}) - \mathbf{F}(\mathbf{x}^{(k-j)}), \qquad j = 1, \ldots, n. \qquad (7.12)$$

If the vectors $\mathbf{x}^{(k-j)}$, ..., $\mathbf{x}^{(k)}$ are linearly independent, system (7.12) allows for calculating all the unknown coefficients $\{(Q_k)_{lm},\ l, m = 1, \ldots, n\}$ of Q_k. Unfortunately, in practice the above vectors tend to become linearly dependent and the resulting scheme is unstable, not to mention the need for storing all the previous n iterates.

For these reasons, an alternative approach is pursued which aims at preserving the information already provided by the method at step k. Precisely, Q_k is looked for in such a way that the difference between the following linear approximants to $\mathbf{F}(\mathbf{x}^{(k-1)})$ and $\mathbf{F}(\mathbf{x}^{(k)})$, respectively

$$\mathbf{F}(\mathbf{x}^{(k)}) + Q_k(\mathbf{x} - \mathbf{x}^{(k)}),\ \mathbf{F}(\mathbf{x}^{(k-1)}) + Q_{k-1}(\mathbf{x} - \mathbf{x}^{(k-1)}),$$

is minimized jointly with the constraint that Q_k satisfies system (7.12). Using (7.12) with $j = 1$, the difference between the two approximants is found to be

$$\mathbf{d}_k = (Q_k - Q_{k-1}) \left(\mathbf{x} - \mathbf{x}^{(k-1)}\right). \qquad (7.13)$$

Let us decompose the vector $\mathbf{x} - \mathbf{x}^{(k-1)}$ as

$$\mathbf{x} - \mathbf{x}^{(k-1)} = \alpha \boldsymbol{\delta}\mathbf{x}^{(k)} + \mathbf{s},$$

where $\alpha \in \mathbb{R}$ and $\mathbf{s}^T \boldsymbol{\delta}\mathbf{x}^{(k)} = 0$. Therefore, (7.13) becomes

$$\mathbf{d}_k = \alpha\,(Q_k - Q_{k-1})\,\boldsymbol{\delta}\mathbf{x}^{(k)} + (Q_k - Q_{k-1})\,\mathbf{s}.$$

Only the second term in the relation above can be minimized since the first one is independent of Q_k, being

$$(Q_k - Q_{k-1})\boldsymbol{\delta}\mathbf{x}^{(k)} = \mathbf{b}^{(k)} - Q_{k-1}\boldsymbol{\delta}\mathbf{x}^{(k)}.$$

The problem has thus become: find the matrix Q_k such that $(Q_k - Q_{k-1})\,\mathbf{s}$ is minimized $\forall \mathbf{s}$ orthogonal to $\boldsymbol{\delta}\mathbf{x}^{(k)}$ with the constraint that (7.12) holds. It can be shown that such a matrix exists and can be recursively computed as follows

$$Q_k = Q_{k-1} + \frac{(\mathbf{b}^{(k)} - Q_{k-1}\boldsymbol{\delta}\mathbf{x}^{(k)})\boldsymbol{\delta}\mathbf{x}^{(k)^T}}{\boldsymbol{\delta}\mathbf{x}^{(k)^T}\boldsymbol{\delta}\mathbf{x}^{(k)}}. \qquad (7.14)$$

The method (7.11), with the choice (7.14) of matrix Q_k is known as the *Broyden method.* To initialize (7.14), we set Q_0 equal to the matrix $J_{\mathbf{F}}(\mathbf{x}^{(0)})$ or to any approximation of it, for instance, the one yielded by (7.9). As for the convergence of Broyden's method, the following result holds.

Property 7.2 *If the assumptions of Theorem* 7.1 *are satisfied and there exist two positive constants* ε *and* γ *such that*

$$\|\mathbf{x}^{(0)} - \mathbf{x}^*\| \leq \varepsilon, \ \|\mathrm{Q}_0 - \mathrm{J}_{\mathbf{F}}(\mathbf{x}^*)\| \leq \gamma,$$

then the sequence of vectors $\mathbf{x}^{(k)}$ *generated by Broyden's method is well defined and converges superlinearly to* $\mathbf{x}^*$, *that is*

$$\|\mathbf{x}^{(k)} - \mathbf{x}^*\| \leq c_k \|\mathbf{x}^{(k-1)} - \mathbf{x}^*\| \tag{7.15}$$

where the constants c_k *are such that* $\lim_{k\to\infty} c_k = 0$.

Under further assumptions, it is also possible to prove that the sequence Q_k converges to $\mathrm{J}_{\mathbf{F}}(\mathbf{x}^*)$, a property that does not necessarily hold for the above method as demonstrated in Example 7.3.

There exist several variants to Broyden's method which aim at reducing its computational cost, but are usually less stable (see [DS83], Chapter 8). Program 58 implements Broyden's method (7.11)-(7.14). We have denoted by `Q` the initial approximation Q_0 in (7.14).

Program 58 - broyden : Broyden's method for nonlinear systems

```
function [x,iter]=broyden(F,Q,x0,tol,nmax)
%BROYDEN Broyden method for nonlinear systems
% [X, ITER] = BROYDEN(F, Q, X0, TOL, NMAX) attempts to solve the
% nonlinear system F(X)=0 with the Broyden method. F is a string variable
% containing the functional expressions of the nonlinear equations. Q is a
% starting approximation of the Jacobian. X0 specifies the initial guess.
% TOL specifies the tolerance of the method. NMAX specifies the maximum
% number of iterations. ITER is the iteration number at which X is computed.
[n,m]=size(F);
if n ~= m, error('Only square systems'); end
iter=0; err=1+tol; fk=zeros(n,1); fk1=fk; x=x0;
for i=1:n
   fk(i)=eval(F(i,:)); end
   while iter < nmax & err > tol
      s=-Q \ fk;
      x=s+x;
      err=norm(s,inf);
      if err > tol
         for i=1:n,  fk1(i)=eval(F(i,:));  end
         Q=Q+1/(s'*s)*fk1*s';
      end
      iter=iter+1;
      fk=fk1;
   end
end
return
```

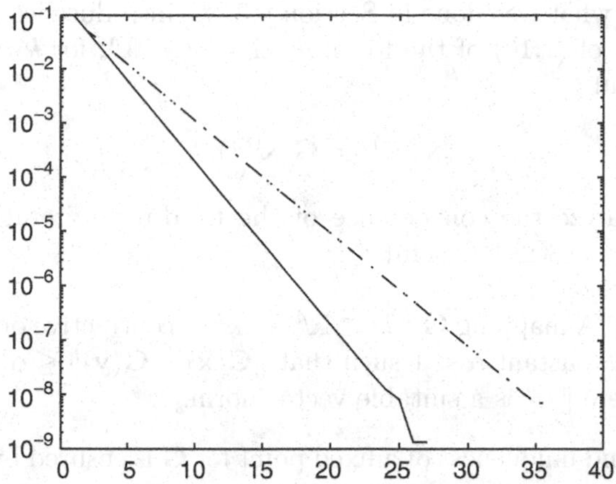

Fig. 7.1. Euclidean norm of the error for the Newton method (*solid line*) and the Broyden method (*dashed line*) in the case of the nonlinear system of Example 7.1

Example 7.2 Let us solve using Broyden's method the nonlinear system of Example 7.1. The method converges in 35 iterations to the value $(0.7 \cdot 10^{-8}, 0.7 \cdot 10^{-8})^T$ compared with the 26 iterations required by Newton's method starting from the same initial guess ($\mathbf{x}^{(0)} = [0.1, 0.1]^T$). The matrix Q_0 has been set equal to the Jacobian matrix evaluated at $\mathbf{x}^{(0)}$. Figure 7.1 shows the behavior of the Euclidean norm of the error for both methods. •

Example 7.3 Suppose we wish to solve using the Broyden method the nonlinear system $\mathbf{F}(\mathbf{x}) = [x_1 + x_2 - 3, x_1^2 + x_2^2 - 9]^T = \mathbf{0}$. This system admits the two solutions $[0,3]^T$ and $[3,0]^T$. Broyden's method converges in 8 iterations to the solution $[0,3]^T$ starting from $\mathbf{x}^{(0)} = [2,4]^T$. However, the sequence of Q_k, stored in the variable `Q` of Program 58, does not converge to the Jacobian matrix, since

$$\lim_{k\to\infty} \mathrm{Q}^{(k)} = \begin{bmatrix} 1 & 1 \\ 1.5 & 1.75 \end{bmatrix} \neq \mathrm{J}_{\mathbf{F}}([0,3]^T) = \begin{bmatrix} 1 & 1 \\ 0 & 6 \end{bmatrix}.$$

•

7.1.5 Fixed-point Methods

We conclude the analysis of methods for solving systems of nonlinear equations by extending to n-dimensions the fixed-point techniques introduced in the scalar case. For this, we reformulate problem (7.1) as

$$\text{given } \mathbf{G} : \mathbb{R}^n \to \mathbb{R}^n, \text{ find } \mathbf{x}^* \in \mathbb{R}^n \text{ such that } \mathbf{G}(\mathbf{x}^*) = \mathbf{x}^* \qquad (7.16)$$

where $\mathbf{G}$ is related to $\mathbf{F}$ through the following property: if $\mathbf{x}^*$ is a fixed point of $\mathbf{G}$, then $\mathbf{F}(\mathbf{x}^*) = \mathbf{0}$.

Analogously to what was done in Section 6.3, we introduce iterative methods for the solution of (7.16) of the form: given $\mathbf{x}^{(0)} \in \mathbb{R}^n$, for $k = 0, 1, \ldots$ until convergence, find

$$\mathbf{x}^{(k+1)} = \mathbf{G}(\mathbf{x}^{(k)}). \tag{7.17}$$

In order to analyze the convergence of the fixed-point iteration (7.17) the following definition will be useful.

Definition 7.1 A mapping $\mathbf{G} : D \subset \mathbb{R}^n \rightarrow \mathbb{R}^n$ is contractive on a set $D_0 \subset D$ if there exists a constant $\alpha < 1$ such that $\|\mathbf{G}(\mathbf{x}) - \mathbf{G}(\mathbf{y})\| \leq \alpha \|\mathbf{x} - \mathbf{y}\|$ for all $\mathbf{x}$, $\mathbf{y}$ in D_0, where $\|\cdot\|$ is a suitable vector norm. ■

The existence and uniqueness of a fixed point for $\mathbf{G}$ is ensured by the following theorem.

Theorem 7.2 (Contraction-mapping theorem) *Suppose that* $\mathbf{G} : D \subset \mathbb{R}^n \rightarrow \mathbb{R}^n$ *is contractive on a closed set* $D_0 \subset D$ *and that* $\mathbf{G}(\mathbf{x}) \subset D_0$ *for all* $\mathbf{x} \in D_0$. *Then* $\mathbf{G}$ *has a unique fixed point in* D_0.

Proof. Let us first prove the uniqueness of the fixed point. For this, assume that there exist two distinct fixed points, $\mathbf{x}^*$, $\mathbf{y}^*$. Then

$$\|\mathbf{x}^* - \mathbf{y}^*\| = \|\mathbf{G}(\mathbf{x}^*) - \mathbf{G}(\mathbf{y}^*)\| \leq \alpha \|\mathbf{x}^* - \mathbf{y}^*\|$$

from which $(1-\alpha)\|\mathbf{x}^* - \mathbf{y}^*\| \leq 0$. Since $(1-\alpha) > 0$, it must necessarily be that $\|\mathbf{x}^* - \mathbf{y}^*\| = 0$, i.e., $\mathbf{x}^* = \mathbf{y}^*$.

To prove the existence we show that $\mathbf{x}^{(k)}$ given by (7.17) is a Cauchy sequence. This in turn implies that $\mathbf{x}^{(k)}$ is convergent to a point $\mathbf{x}^{(*)} \in D_0$. Take $\mathbf{x}^{(0)}$ arbitrarily in D_0. Then, since the image of $\mathbf{G}$ is included in D_0, the sequence $\mathbf{x}^{(k)}$ is well defined and

$$\|\mathbf{x}^{(k+1)} - \mathbf{x}^{(k)}\| = \|\mathbf{G}(\mathbf{x}^{(k)}) - \mathbf{G}(\mathbf{x}^{(k-1)})\| \leq \alpha \|\mathbf{x}^{(k)} - \mathbf{x}^{(k-1)}\|.$$

After p steps, $p \geq 1$, we obtain

$$\begin{aligned}\|\mathbf{x}^{(k+p)} - \mathbf{x}^{(k)}\| &\leq \sum_{i=1}^{p} \|\mathbf{x}^{(k+i)} - \mathbf{x}^{(k+i-1)}\| \leq \left(\alpha^{p-1} + \ldots + 1\right) \|\mathbf{x}^{(k+1)} - \mathbf{x}^{(k)}\| \\ &\leq \frac{\alpha^k}{1-\alpha} \|\mathbf{x}^{(1)} - \mathbf{x}^{(0)}\|.\end{aligned}$$

Owing to the continuity of $\mathbf{G}$ it follows that $\lim_{k \rightarrow \infty} \mathbf{G}(\mathbf{x}^{(k)}) = \mathbf{G}(\mathbf{x}^{(*)})$ which proves that $\mathbf{x}^{(*)}$ is a fixed point for $\mathbf{G}$. ◇

The following result provides a sufficient condition for the iteration (7.17) to converge (for the proof see [OR70], pp. 299-301), and extends the analogous Theorem 6.3 in the scalar case.

Property 7.3 *Suppose that* $\mathbf{G} : D \subset \mathbb{R}^n \to \mathbb{R}^n$ *has a fixed point* $\mathbf{x}^*$ *in the interior of* D *and that* $\mathbf{G}$ *is continuously differentiable in a neighborhood of* $\mathbf{x}^*$*. Denote by* $\mathrm{J}_\mathbf{G}$ *the Jacobian matrix of* $\mathbf{G}$ *and assume that* $\rho(\mathrm{J}_\mathbf{G}(\mathbf{x}^{(*)})) < 1$*. Then there exists a neighborhood* S *of* $\mathbf{x}^*$ *such that* $S \subset D$ *and, for any* $\mathbf{x}^{(0)} \in S$*, the iterates defined by (7.17) all lie in* D *and converge to* $\mathbf{x}^*$*.*

As usual, since the spectral radius is the infimum of the induced matrix norms, in order for convergence to hold it suffices to check that $\|\mathrm{J}_\mathbf{G}(\mathbf{x})\| < 1$ for some matrix norm.

Example 7.4 Consider the nonlinear system

$$\mathbf{F}(\mathbf{x}) = [x_1^2 + x_2^2 - 1, 2x_1 + x_2 - 1]^T = \mathbf{0},$$

whose solutions are $\mathbf{x}_1^* = [0, 1]^T$ and $\mathbf{x}_2^* = [4/5, -3/5]^T$. To solve it, let us use two fixed-point schemes, respectively defined by the following iteration functions

$$\mathbf{G}_1(\mathbf{x}) = \begin{bmatrix} \dfrac{1-x_2}{2} \\ \sqrt{1-x_1^2} \end{bmatrix}, \quad \mathbf{G}_2(\mathbf{x}) = \begin{bmatrix} \dfrac{1-x_2}{2} \\ -\sqrt{1-x_1^2} \end{bmatrix}. \tag{7.18}$$

It can be checked that $\mathbf{G}_i(\mathbf{x}_i^*) = \mathbf{x}_i^*$ for $i = 1, 2$ and that the Jacobian matrices of $\mathbf{G}_1$ and $\mathbf{G}_2$, evaluated at $\mathbf{x}_1^*$ and $\mathbf{x}_2^*$ respectively, are

$$\mathrm{J}_{\mathbf{G}_1}(\mathbf{x}_1^*) = \begin{bmatrix} 0 & -\frac{1}{2} \\ 0 & 0 \end{bmatrix}, \quad \mathrm{J}_{\mathbf{G}_2}(\mathbf{x}_2^*) = \begin{bmatrix} 0 & -\frac{1}{2} \\ \frac{4}{3} & 0 \end{bmatrix}.$$

The spectral radii are $\rho(\mathrm{J}_{\mathbf{G}_1}(\mathbf{x}_1^*)) = 0$ and $\rho(\mathrm{J}_{\mathbf{G}_2}(\mathbf{x}_2^*)) = \sqrt{2/3} \simeq 0.817 < 1$ so that both methods are convergent in a suitable neighborhood of their respective fixed points.

Running Program 59, with a tolerance of 10^{-10} on the maximum absolute difference between two successive iterates, the first scheme converges to $\mathbf{x}_1^*$ in 9 iterations, starting from $\mathbf{x}^{(0)} = [-0.9, 0.9]^T$, while the second one converges to $\mathbf{x}_2^*$ in 115 iterations, starting from $\mathbf{x}^{(0)} = [0.9, 0.9]^T$. The dramatic change in the convergence behavior of the two methods can be explained in view of the difference between the spectral radii of the corresponding iteration matrices. •

Remark 7.2 Newton's method can be regarded as a fixed-point method with iteration function

$$\mathbf{G}_N(\mathbf{x}) = \mathbf{x} - \mathrm{J}_\mathbf{F}^{-1}(\mathbf{x})\mathbf{F}(\mathbf{x}). \tag{7.19}$$

■

An implementation of the fixed-point method (7.17) is provided in Program 59. We have denoted by `dim` the size of the nonlinear system and by `Phi` the variables containing the functional expressions of the iteration function **G**. In output, the vector `alpha` contains the approximation of the sought zero of **F** and the vector `res` contains the sequence of the maximum norms of the residuals of $\mathbf{F}(\mathbf{x}^{(k)})$.

Program 59 - fixposys : Fixed-point method for nonlinear systems

```
function [alpha,res,iter]=fixposys(F,Phi,x0,tol,nmax,dim)
%FIXPOSYS Fixed-point method for nonlinear systems
% [ALPHA, RES, ITER] = FIXPOSYS(F, PHI, X0, TOL, NMAX, DIM) attempts
% to solve the nonlinear system F(X)=0 with the Fixed Point method. F and PHI are
% string variables containing the functional expressions of the nonlinear equations
% and of the iteration function. X0 specifies the initial guess. TOL specifies the
% tolerance of the method. NMAX specifies the maximum number of iterations. DIM is
% the size of the nonlinear system. ITER is the iteration number at which ALPHA is
% computed. RES is the system residual computed at ALPHA.
x = x0; alpha=[x']; res = 0;
for k=1:dim
   r=abs(eval(F(k,:))); if (r > res), res = r; end
end;
iter = 0;
residual(1)=res;
while ((iter <= nmax) & (res >= tol)),
  iter = iter + 1;
  for k = 1:dim
     xnew(k) = eval(Phi(k,:));
  end
  x = xnew; res = 0; alpha=[alpha;x]; x=x';
  for k = 1:dim
     r = abs(eval(F(k,:)));
     if (r > res), res=r; end,
  end
  residual(iter+1)=res;
end
res=residual';
return
```

7.2 Unconstrained Optimization

We turn now to minimization problems. The point $\mathbf{x}^*$, the solution of (7.2), is called a *global minimizer* of f, while $\mathbf{x}^*$ is a *local minimizer* of f if $\exists R > 0$ such that

$$f(\mathbf{x}^*) \leq f(\mathbf{x}), \qquad \forall \mathbf{x} \in B(\mathbf{x}^*; R).$$

Throughout this section we shall always assume that $f \in C^1(\mathbb{R}^n)$, and we refer to [Lem89] for the case in which f is nondifferentiable. We shall denote by

$$\nabla f(\mathbf{x}) = \left(\frac{\partial f}{\partial x_1}(\mathbf{x}), \ldots, \frac{\partial f}{\partial x_n}(\mathbf{x}) \right)^T,$$

the *gradient* of f at a point $\mathbf{x}$. If $\mathbf{d}$ is a nonnull vector in $\mathbb{R}^n$, then the directional derivative of f with respect to $\mathbf{d}$ is

$$\frac{\partial f}{\partial \mathbf{d}}(\mathbf{x}) = \lim_{\alpha \to 0} \frac{f(\mathbf{x} + \alpha \mathbf{d}) - f(\mathbf{x})}{\alpha}$$

and satisfies $\partial f(\mathbf{x})/\partial \mathbf{d} = \nabla f(\mathbf{x})^T \mathbf{d}$. Moreover, denoting by $(\mathbf{x}, \mathbf{x} + \alpha \mathbf{d})$ the segment in $\mathbb{R}^n$ joining the points $\mathbf{x}$ and $\mathbf{x}+\alpha\mathbf{d}$, with $\alpha \in \mathbb{R}$, Taylor's expansion ensures that $\exists \boldsymbol{\xi} \in (\mathbf{x}, \mathbf{x} + \alpha \mathbf{d})$ such that

$$f(\mathbf{x} + \alpha \mathbf{d}) - f(\mathbf{x}) = \alpha \nabla f(\boldsymbol{\xi})^T \mathbf{d}. \tag{7.20}$$

If $f \in C^2(\mathbb{R}^n)$, we shall denote by $\mathrm{H}(\mathbf{x})$ (or $\nabla^2 f(\mathbf{x})$) the *Hessian matrix* of f evaluated at a point $\mathbf{x}$, whose entries are

$$h_{ij}(\mathbf{x}) = \frac{\partial^2 f(\mathbf{x})}{\partial x_i \partial x_j}, \; i, j = 1, \ldots, n.$$

In such a case it can be shown that, if $\mathbf{d} \neq \mathbf{0}$, the second-order directional derivative exists and we have

$$\frac{\partial^2 f}{\partial \mathbf{d}^2}(\mathbf{x}) = \mathbf{d}^T \mathrm{H}(\mathbf{x}) \mathbf{d}.$$

For a suitable $\boldsymbol{\xi} \in (\mathbf{x}, \mathbf{x} + \mathbf{d})$ we also have

$$f(\mathbf{x} + \mathbf{d}) - f(\mathbf{x}) = \nabla f(\mathbf{x})^T \mathbf{d} + \frac{1}{2} \mathbf{d}^T \mathrm{H}(\boldsymbol{\xi}) \mathbf{d}.$$

Existence and uniqueness of solutions for (7.2) are not guaranteed in $\mathbb{R}^n$. Nevertheless, the following optimality conditions can be proved.

Property 7.4 *Let* $\mathbf{x}^* \in \mathbb{R}^n$ *be a local minimizer of* f *and assume that* $f \in C^1(B(\mathbf{x}^*; R))$ *for a suitable* $R > 0$. *Then* $\nabla f(\mathbf{x}^*) = \mathbf{0}$. *Moreover, if* $f \in C^2(B(\mathbf{x}^*; R))$ *then* $\mathrm{H}(\mathbf{x}^*)$ *is positive semidefinite. Conversely, if* $\nabla f(\mathbf{x}^*) = \mathbf{0}$ *and* $\mathrm{H}(\mathbf{x}^*)$ *is positive definite, then* $\mathbf{x}^*$ *is a local minimizer of* f *in* $B(\mathbf{x}^*; R)$.

A point $\mathbf{x}^*$ such that $\nabla f(\mathbf{x}^*) = \mathbf{0}$, is said to be a *critical point* for f. This condition is necessary for optimality to hold. However, this condition also becomes sufficient if f is a convex function on $\mathbb{R}^n$, i.e., such that $\forall \mathbf{x}, \mathbf{y} \in \mathbb{R}^n$ and for any $\alpha \in [0, 1]$

$$f[\alpha \mathbf{x} + (1 - \alpha) \mathbf{y}] \leq \alpha f(\mathbf{x}) + (1 - \alpha) f(\mathbf{y}). \tag{7.21}$$

For further and more general existence results, see [Ber82].

7.2.1 Direct Search Methods

In this section we deal with *direct* methods for solving problem (7.2), which only require f to be continuous. In later sections, we shall introduce the so-called *descent* methods, which also involve values of the derivatives of f and have, in general, better convergence properties.

Direct methods are employed when f is not differentiable or if the computation of its derivatives is a nontrivial task. They can also be used to provide an approximate solution to employ as an initial guess for a descent method. For further details, we refer to [Wal75] and [Wol78].

The Hooke and Jeeves Method

Assume we are searching for the minimizer of f starting from a given initial point $\mathbf{x}^{(0)}$ and requiring that the error on the residual is less than a certain fixed tolerance ϵ. The Hooke and Jeeves method computes a new point $\mathbf{x}^{(1)}$ using the values of f at suitable points along the orthogonal coordinate directions around $\mathbf{x}^{(0)}$. The method consists of two steps: an *exploration* step and an *advancing* step.

The exploration step starts by evaluating $f(\mathbf{x}^{(0)} + h_1\mathbf{e}_1)$, where $\mathbf{e}_1$ is the first vector of the canonical basis of $\mathbb{R}^n$ and h_1 is a positive real number to be suitably chosen.

If $f(\mathbf{x}^{(0)} + h_1\mathbf{e}_1) < f(\mathbf{x}^{(0)})$, then a success is recorded and the starting point is moved in $\mathbf{x}^{(0)} + h_1\mathbf{e}_1$, from which an analogous check is carried out at point $\mathbf{x}^{(0)} + h_1\mathbf{e}_1 + h_2\mathbf{e}_2$ with $h_2 \in \mathbb{R}^+$.

If, instead, $f(\mathbf{x}^{(0)} + h_1\mathbf{e}_1) \geq f(\mathbf{x}^{(0)})$, then a failure is recorded and a similar check is performed at $\mathbf{x}^{(0)} - h_1\mathbf{e}_1$. If a success is registered, the method explores, as previously, the behavior of f in the direction $\mathbf{e}_2$ starting from this new point, while, in case of a failure, the method passes directly to examining direction $\mathbf{e}_2$, keeping $\mathbf{x}^{(0)}$ as starting point for the exploration step.

To achieve a certain accuracy, the step lengths h_i must be selected in such a way that the quantities

$$|f(\mathbf{x}^{(0)} \pm h_j\mathbf{e}_j) - f(\mathbf{x}^{(0)}|, \quad j = 1, \ldots, n \tag{7.22}$$

have comparable sizes.

The exploration step terminates as soon as all the n Cartesian directions have been examined. Therefore, the method generates a new point, $\mathbf{y}^{(0)}$, after at most $2n + 1$ functional evaluations. Only two possibilities may arise:

1. $\mathbf{y}^{(0)} = \mathbf{x}^{(0)}$. In such a case, if $\max_{i=1,\ldots,n} h_i \leq \epsilon$ the method terminates and yields the approximate solution $\mathbf{x}^{(0)}$. Otherwise, the step lengths h_i are halved and another exploration step is performed starting from $\mathbf{x}^{(0)}$;
2. $\mathbf{y}^{(0)} \neq \mathbf{x}^{(0)}$. If $\max_{i=1,\ldots,n} |h_i| < \epsilon$, then the method terminates yielding $\mathbf{y}^{(0)}$ as an approximate solution, otherwise the advancing step starts. The advancing step consists of moving further from $\mathbf{y}^{(0)}$ along the direction $\mathbf{y}^{(0)} - \mathbf{x}^{(0)}$

(which is the direction that recorded the maximum decrease of f during the exploration step), rather then simply setting $\mathbf{y}^{(0)}$ as a new starting point $\mathbf{x}^{(1)}$.

This new starting point is instead set equal to $2\mathbf{y}^{(0)} - \mathbf{x}^{(0)}$. From this point a new series of exploration moves is started. If this exploration leads to a point $\mathbf{y}^{(1)}$ such that $f(\mathbf{y}^{(1)}) < f(\mathbf{y}^{(0)} - \mathbf{x}^{(0)})$, then a new starting point for the next exploration step has been found, otherwise the initial guess for further explorations is set equal to $\mathbf{y}^{(1)} = \mathbf{y}^{(0)} - \mathbf{x}^{(0)}$.

The method is now ready to restart from the point $\mathbf{x}^{(1)}$ just computed.

Program 60 provides an implementation of the Hooke and Jeeves method. The input parameters are the size `n` of the problem, the vector `h` of the initial steps along the Cartesian directions, the variable `f` containing the functional expression of f in terms of the components `x(1)`, ..., `x(n)`, the initial point `x0` and the stopping tolerance `tol` equal to ϵ. In output, the code returns the approximate minimizer of f, `x`, the value `minf` attained by f at `x` and the number of iterations needed to compute `x` up to the desired accuracy. The exploration step is performed by Program 61.

Program 60 - hookejeeves : The method of Hooke and Jeeves (HJ)

```
function [x,minf,iter]=hookejeeves(f,n,h,x0,tol)
%HOOKEJEEVES HOOKE and JEEVES method for function minimization.
% [X, MINF, ITER] = HOOKEJEEVES(F, N, H, X0, TOL) attempts to compute the
% minimizer of a function of N variables with the Hooke and Jeeves method. F is
% a string variable containing the functional expression of f. H is an initial
% step. X0 specifies the initial guess. TOL specifies the tolerance of the method.
% ITER is the iteration number at which X is computed. MINF is the value of F at
% the mimimizer X.
x = x0; minf = eval(f); iter = 0;
while h > tol
   [y] = explore(f,n,h,x);
   if y == x
      h = h/2;
   else
      x = 2*y-x;
      [z] = explore(f,n,h,x);
      if z == x
         x = y;
      else
         x = z;
      end
   end
   iter = iter +1;
end
minf = eval(f);
return
```

Program 61 - explore : Exploration step in the HJ method

```
function [x]=explore(f,n,h,x0)
%EXPLORE Exploration step for function minimization.
% [X] = EXPLORE(F, N, H, X0) executes one exploration step of size H in the Hooke
% and Jeeves method for function minimization.
x = x0; f0 = eval(f);
for i=1:n
   x(i) = x(i) + h(i); ff = eval(f);
   if ff < f0
      f0 = ff;
   else
      x(i) = x0(i) - h(i);
      ff = eval(f);
      if ff < f0
         f0 = ff;
      else
         x(i) = x0(i);
      end
   end
end
return
```

The Method of Nelder and Mead

This method, proposed in [NM65], employs local linear approximants of f to generate a sequence of points $\mathbf{x}^{(k)}$, approximations of $\mathbf{x}^*$, starting from simple geometrical considerations. To explain the details of the algorithm, we begin by noticing that a plane in $\mathbb{R}^n$ is uniquely determined by fixing $n+1$ points that must not be lying on a hyperplane.
Denote such points by $\mathbf{x}^{(k)}$, for $k = 0, \ldots, n$. They could be generated as

$$\mathbf{x}^{(k)} = \mathbf{x}^{(0)} + h_k \mathbf{e}_k, \ k = 1, \ldots, n, \tag{7.23}$$

having selected the steplengths $h_k \in \mathbb{R}^+$ in such a way that the variations (7.22) are of comparable size.

Let us now denote by $\mathbf{x}^{(M)}$, $\mathbf{x}^{(m)}$ and $\mathbf{x}^{(\mu)}$ those points of the set $\{\mathbf{x}^{(k)}\}$ at which f respectively attains its maximum and minimum value and the value immediately preceding the maximum. Moreover, denote by $\mathbf{x}_c^{(k)}$ the *centroid* of point $\mathbf{x}^{(k)}$ defined as

$$\mathbf{x}_c^{(k)} = \frac{1}{n} \sum_{j=0, j \neq k}^{n} \mathbf{x}^{(j)}.$$

The method generates a sequence of approximations of $\mathbf{x}^*$, starting from $\mathbf{x}^{(k)}$, by employing only three possible transformations: *reflections* with respect to centroids, *dilations* and *contractions*. Let us examine the details of the algorithm assuming that $n+1$ initial points are available.

1. Determine the points $\mathbf{x}^{(M)}$, $\mathbf{x}^{(m)}$ and $\mathbf{x}^{(\mu)}$.
2. Compute as an approximation of $\mathbf{x}^*$ the point

$$\bar{\mathbf{x}} = \frac{1}{n+1}\sum_{i=0}^{n}\mathbf{x}^{(i)}$$

and check if $\bar{\mathbf{x}}$ is sufficiently close (in a sense to be made precise) to $\mathbf{x}^*$. Typically, one requires that the standard deviation of the values $f(\mathbf{x}^{(0)}),\ldots,f(\mathbf{x}^{(n)})$ from

$$\bar{f} = \frac{1}{n+1}\sum_{i=0}^{n} f(\mathbf{x}^{(i)})$$

are less than a fixed tolerance ε, that is

$$\frac{1}{n}\sum_{i=0}^{n}\left(f(\mathbf{x}^{(i)}) - \bar{f}\right)^2 < \varepsilon.$$

Otherwise, $\mathbf{x}^{(M)}$ is reflected with respect to $\mathbf{x}_c^{(M)}$, that is, the following new point $\mathbf{x}^{(r)}$ is computed

$$\mathbf{x}^{(r)} = (1+\alpha)\mathbf{x}_c^{(M)} - \alpha\mathbf{x}^{(M)},$$

where $\alpha \geq 0$ is a suitable reflection factor. Notice that the method has moved along the "opposite" direction to $\mathbf{x}^{(M)}$. This statement has a geometrical interpretation in the case $n = 2$, since the points $\mathbf{x}^{(k)}$ coincide with $\mathbf{x}^{(M)}$, $\mathbf{x}^{(m)}$ and $\mathbf{x}^{(\mu)}$. They thus define a plane whose slope points from $\mathbf{x}^{(M)}$ towards $\mathbf{x}^{(m)}$ and the method provides a step along this direction.
3. If $f(\mathbf{x}^{(m)}) \leq f(\mathbf{x}^{(r)}) \leq f(\mathbf{x}^{(\mu)})$, the point $\mathbf{x}^{(M)}$ is replaced by $\mathbf{x}^{(r)}$ and the algorithm returns to step 2.
4. If $f(\mathbf{x}^{(r)}) < f(\mathbf{x}^{(m)})$ then the reflection step has produced a new minimizer. This means that the minimizer could lie outside the set defined by the convex hull of the considered points. Therefore, this set must be expanded by computing the new vertex

$$\mathbf{x}^{(e)} = \beta\mathbf{x}^{(r)} + (1-\beta)\mathbf{x}_c^{(M)},$$

where $\beta > 1$ is an expansion factor. Then, before coming back to step 2., two possibilities arise:

4a. if $f(\mathbf{x}^{(e)}) < f(\mathbf{x}^{(m)})$ then $\mathbf{x}^{(M)}$ is replaced by $\mathbf{x}^{(e)}$;

4b. $f(\mathbf{x}^{(e)}) \geq f(\mathbf{x}^{(m)})$ then $\mathbf{x}^{(M)}$ is replaced by $\mathbf{x}^{(r)}$ since $f(\mathbf{x}^{(r)}) < f(\mathbf{x}^{(m)})$.

5. If $f(\mathbf{x}^{(r)}) > f(\mathbf{x}^{(\mu)})$ then the minimizer probably lies within a subset of the convex hull of points $\{\mathbf{x}^{(k)}\}$ and, therefore, two different approaches can be pursued to contract this set. If $f(\mathbf{x}^{(r)}) < f(\mathbf{x}^{(M)})$, the contraction generates a new point of the form

$$\mathbf{x}^{(co)} = \gamma \mathbf{x}^{(r)} + (1-\gamma)\mathbf{x}_c^{(M)}, \quad \gamma \in (0,1),$$

otherwise,

$$\mathbf{x}^{(co)} = \gamma \mathbf{x}^{(M)} + (1-\gamma)\mathbf{x}_c^{(M)}, \quad \gamma \in (0,1),$$

Finally, before returning to step 2., if $f(\mathbf{x}^{(co)}) < f(\mathbf{x}^{(M)})$ and $f(\mathbf{x}^{(co)}) < f(\mathbf{x}^{(r)})$, the point $\mathbf{x}^{(M)}$ is replaced by $\mathbf{x}^{(co)}$, while if $f(\mathbf{x}^{(co)}) \geq f(\mathbf{x}^{(M)})$ or if $f(\mathbf{x}^{(co)}) > f(\mathbf{x}^{(r)})$, then n new points $\mathbf{x}^{(k)}$ are generated, with $k = 1, \ldots, n$, by halving the distances between the original points and $\mathbf{x}^{(0)}$.

As far as the choice of the parameters α, β and γ is concerned, the following values are empirically suggested in [NM65]: $\alpha = 1$, $\beta = 2$ and $\gamma = 1/2$. The resulting scheme is known as the *Simplex method* (that must not be confused with a method sharing the same name used in linear programming), since the set of the points $\mathbf{x}^{(k)}$, together with their convex combinations, form a simplex in $\mathbb{R}^n$.

The convergence rate of the method is strongly affected by the orientation of the starting simplex. To address this concern, in absence of information about the behavior of f, the initial choice (7.23) turns out to be satisfactory in most cases.

We finally mention that the Simplex method is the basic ingredient of the MATLAB function `fmins` for function minimization in n dimensions.

Example 7.5 Let us compare the performances of the Simplex method with the Hooke and Jeeves method, in the minimization of the Rosembrock function

$$f(\mathbf{x}) = 100(x_2 - x_1^2)^2 + (1 - x_1)^2. \tag{7.24}$$

This function has a minimizer at $[1,1]^T$ and represents a severe benchmark for testing numerical methods in minimization problems. The starting point for both methods is set equal to $\mathbf{x}^{(0)} = [-1.2, 1]^T$, while the step sizes are taken equal to $h_1 = 0.6$ and $h_2 = 0.5$, in such a way that (7.23) is satisfied. The stopping tolerance on the residual is set equal to 10^{-4}. For the implementation of Simplex method, we have used the MATLAB function `fmins`.

Figure 7.2 shows the iterates computed by the Hooke and Jeeves method (of which one in every ten iterates have been reported, for the sake of clarity) and by the Simplex method, superposed to the level curves of the Rosembrock function. The graph demonstrates the difficulty of this benchmark: actually, the function is like a curved, narrow valley, which attains its minimum along the parabola of equation $x_1^2 - x_2 = 0$.

The Simplex method converges in only 165 iterations, while 935 are needed for the Hooke and Jeeves method to converge. The former scheme yields a solution equal to $[0.999987, 0.999978]^T$, while the latter gives the vector $[0.9655, 0.9322]^T$. •

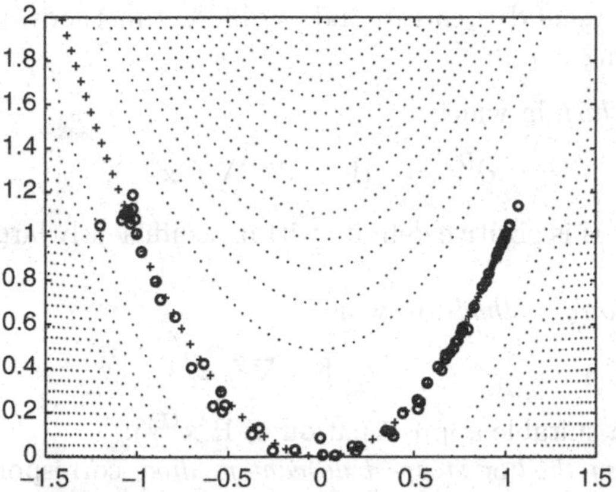

Fig. 7.2. Convergence histories of the Hooke and Jeeves method (*crossed-line*) and the Simplex method (*circled-line*). The level curves of the minimized function (7.24) are reported in dashed line

7.2.2 Descent Methods

In this section we introduce iterative methods that are more sophisticated than those examined in Section 7.2.1. They can be formulated as follows:

given an initial vector $\mathbf{x}^{(0)} \in \mathbb{R}^n$, compute for $k \geq 0$ until convergence

$$\mathbf{x}^{(k+1)} = \mathbf{x}^{(k)} + \alpha_k \mathbf{d}^{(k)}, \tag{7.25}$$

where $\mathbf{d}^{(k)}$ is a suitably chosen direction and α_k is a positive parameter (called *stepsize*) that measures the step along the direction $\mathbf{d}^{(k)}$. This direction $\mathbf{d}^{(k)}$ is a *descent direction* if

$$\begin{aligned} &\mathbf{d}^{(k)T} \nabla f(\mathbf{x}^{(k)}) < 0 \quad &\text{if } \nabla f(\mathbf{x}^{(k)}) \neq \mathbf{0}, \\ &\mathbf{d}^{(k)} = \mathbf{0} &\text{if } \nabla f(\mathbf{x}^{(k)}) = \mathbf{0}. \end{aligned} \tag{7.26}$$

A *descent method* is a method like (7.25), in which the vectors $\mathbf{d}^{(k)}$ are descent directions.

Property (7.20) ensures that there exists $\alpha_k > 0$, sufficiently small, such that

$$f(\mathbf{x}^{(k)} + \alpha_k \mathbf{d}^{(k)}) < f(\mathbf{x}^{(k)}), \tag{7.27}$$

provided that f is continuously differentiable. Actually, taking in (7.20) $\xi = \mathbf{x}^{(k)} + \vartheta \alpha_k \mathbf{d}^{(k)}$ with $\vartheta \in (0,1)$, and employing the continuity of ∇f, we get

$$f(\mathbf{x}^{(k)} + \alpha_k \mathbf{d}^{(k)}) - f(\mathbf{x}^{(k)}) = \alpha_k \nabla f(\mathbf{x}^{(k)})^T \mathbf{d}^{(k)} + \varepsilon, \tag{7.28}$$

where ε tends to zero as α_k tends to zero. As a consequence, if $\alpha_k > 0$ is sufficiently small, the sign of the left-side of (7.28) coincides with the sign of $\nabla f(\mathbf{x}^{(k)})^T \mathbf{d}^{(k)}$, so that (7.27) is satisfied if $\mathbf{d}^{(k)}$ is a descent direction.

Different choices of $\mathbf{d}^{(k)}$ correspond to different methods. In particular, we recall the following ones:

- *Newton's method*, in which
$$\mathbf{d}^{(k)} = -\mathrm{H}^{-1}(\mathbf{x}^{(k)})\nabla f(\mathbf{x}^{(k)}),$$
provided that H is positive definite within a sufficiently large neighborhood of point $\mathbf{x}^*$;
- *inexact Newton's methods*, in which
$$\mathbf{d}^{(k)} = -\mathrm{B}_k^{-1}\nabla f(\mathbf{x}^{(k)}),$$
where B_k is a suitable approximation of $\mathrm{H}(\mathbf{x}^{(k)})$;
- the *gradient method* or *steepest descent method*, corresponding to setting $\mathbf{d}^{(k)} = -\nabla f(\mathbf{x}^{(k)})$. This method is thus an inexact Newton's method, in which $\mathrm{B}_k = \mathrm{I}$. It can also be regarded as a gradient-like method, since ${\mathbf{d}^{(k)}}^T\nabla f(\mathbf{x}^{(k)}) = -\|\nabla f(\mathbf{x}^{(k)})\|_2^2$;
- the *conjugate gradient method*, for which
$$\mathbf{d}^{(k)} = -\nabla f(\mathbf{x}^{(k)}) + \beta_k \mathbf{d}^{(k-1)},$$
where β_k is a scalar to be suitably selected in such a way that the directions $\{\mathbf{d}^{(k)}\}$ turn out to be mutually orthogonal with respect to a suitable scalar product.

Selecting $\mathbf{d}^{(k)}$ is not enough to completely identify a descent method, since it remains an open problem how to determine α_k in such a way that (7.27) is fulfilled without resorting to excessively small stepsizes α_k (and, thus, to methods with a slow convergence).

A method for computing α_k consists of solving the following minimization problem in one dimension:

$$\text{find } \alpha \text{ such that } \phi(\alpha) = f(\mathbf{x}^{(k)} + \alpha\mathbf{d}^{(k)}) \text{ is minimized.} \qquad (7.29)$$

In such a case we have the following result.

Theorem 7.3 *Consider the descent method* (7.25). *If at the generic step k, the parameter* α_k *is set equal to the exact solution of* (7.29), *then the following orthogonality property holds*

$$\nabla f(\mathbf{x}^{(k+1)})^T\mathbf{d}^{(k)} = 0.$$

Proof. Let α_k be a solution to (7.29). Then, the first derivative of ϕ, given by

$$\phi'(\alpha) = \sum_{i=1}^{n} \frac{\partial f}{\partial x_i}(\mathbf{x}^{(k)} + \alpha_k\mathbf{d}^{(k)})\frac{\partial}{\partial\alpha}(x_i^{(k)} + \alpha d_i^{(k)}) = \nabla f(\mathbf{x}^{(k)} + \alpha_k\mathbf{d}^{(k)})^T\mathbf{d}^{(k)},$$

vanishes at $\alpha = \alpha_k$. The thesis then follows, recalling the definition of $\mathbf{x}^{(k+1)}$. ◇

Unfortunately, except for in special cases (which are nevetherless quite relevant, see Section 7.2.4), providing an exact solution of (7.29) is not feasible, since this is a nonlinear problem. One possible strategy consists of approximating f along the straight line $\mathbf{x}^{(k)} + \alpha\mathbf{d}^{(k)}$ through an interpolating polynomial and then minimizing this polynomial (see the quadratic interpolation Powell methods and cubic interpolation Davidon methods in [Wal75]).

Generally speaking, a process that leads to an approximate solution to (7.29) is said to be a *line search technique* and is addressed in the next section.

7.2.3 Line Search Techniques

The methods that we are going to deal with in this section, are iterative techniques that terminate as soon as some accuracy stopping criterion on α_k is satisfied. We shall assume that (7.26) holds.

Practical experience reveals that it is not necessary to solve accurately for (7.29) in order to devise efficient methods, rather, it is crucial to enforce some limitation on the step lengths (and, thus, on the admissible values for α_k). Actually, without introducing any limitation, a reasonable request on α_k would seem be that the new iterate $\mathbf{x}^{(k+1)}$ satisfies the inequality

$$f(\mathbf{x}^{(k+1)}) < f(\mathbf{x}^{(k)}), \tag{7.30}$$

where $\mathbf{x}^{(k)}$ and $\mathbf{d}^{(k)}$ have been fixed. For this purpose, the procedure based on starting from a (sufficiently large) value of the step length α_k and halve this value until (7.30) is fulfilled, can yield completely wrong results (see, [DS83]).

More stringent criteria than (7.30) should be adopted in the choice of possible values for α_k. To this end, we notice that two kinds of difficulties arise with the above examples: a slow descent rate of the sequence and the use of small stepsizes.

The first difficulty can be overcome by requiring that

$$\begin{aligned} 0 \geq v_M(\mathbf{x}^{(k+1)}) &= \frac{1}{\alpha_k}\left[f(\mathbf{x}^{(k)}) - f(\mathbf{x}^{(k)} + \alpha_k\mathbf{d}^{(k)})\right] \\ &\geq -\sigma\nabla f(\mathbf{x}^{(k)})^T\mathbf{d}^{(k)}, \end{aligned} \tag{7.31}$$

with $\sigma \in (0, 1/2)$. This amounts to requiring that the average descent rate v_M of f along $\mathbf{d}^{(k)}$, evaluated at $\mathbf{x}^{(k+1)}$, be at least equal to a given fraction of the initial descent rate at $\mathbf{x}^{(k)}$. To avoid the generation of too small stepsizes, we require that the descent rate in the direction $\mathbf{d}^{(k)}$ at $\mathbf{x}^{(k+1)}$ is not less than a given fraction of the descent rate at $\mathbf{x}^{(k)}$

$$|\nabla f(\mathbf{x}^{(k)} + \alpha_k\mathbf{d}^{(k)})^T\mathbf{d}^{(k)}| \leq \beta|\nabla f(\mathbf{x}^{(k)})^T\mathbf{d}^{(k)}|, \tag{7.32}$$

with $\beta \in (\sigma, 1)$ in such a way as to also satisfy (7.31). In computational practice, $\sigma \in [10^{-5}, 10^{-1}]$ and $\beta \in [10^{-1}, \frac{1}{2}]$ are usual choices. Sometimes, (7.32) is replaced by the milder condition

$$\nabla f(\mathbf{x}^{(k)} + \alpha_k\mathbf{d}^{(k)})^T\mathbf{d}^{(k)} \geq \beta\nabla f(\mathbf{x}^{(k)})^T\mathbf{d}^{(k)} \tag{7.33}$$

(recall that $\nabla f(\mathbf{x}^{(k)})^T\mathbf{d}^{(k)}$ is negative, since $\mathbf{d}^{(k)}$ is a descent direction).

The following property ensures that, under suitable assumptions, it is possible to find out values of α_k which satisfy (7.31)-(7.32) or (7.31)-(7.33).

Property 7.5 *Assume that $f(x) \geq M$ for any $x \in \mathbb{R}^n$. Then there exists an interval $I = [c, C]$ for the descent method, with $0 < c < C$, such that $\forall \alpha_k \in I$,* (7.31), (7.32) *(or* (7.31)-(7.33)*) are satisfied, with $\sigma \in (0, 1/2)$ and $\beta \in (\sigma, 1)$.*

Under the constraint of fulfilling conditions (7.31) and (7.32), several choices for α_k are available. Among the most *up-to-date* strategies, we recall here the *backtracking techniques*: having fixed $\sigma \in (0, 1/2)$, then start with $\alpha_k = 1$ and then keep on reducing its value by a suitable scale factor $\rho \in (0, 1)$ (*backtrack* step) until (7.31) is satisfied. This procedure is implemented in Program 62, which requires as input parameters the vector x containing $\mathbf{x}^{(k)}$, the macros f and J of the functional expressions of f and its Jacobian, the vector d of the direction $\mathbf{d}^{(k)}$, and a value for σ (usually of the order of 10^{-4}) and the scale factor ρ. In output, the code returns the vector $\mathbf{x}^{(k+1)}$, computed using a suitable value of α_k.

Program 62 - backtrackr : Backtraking for line search

```
function [xnew]= backtrackr(f,J,x,d,sigma,rho)
%BACKTRACKR Backtraking method for line search.
% [XNEW] = BACKTRACKR(F, J, X, D, SIGMA, RHO) attempts to compute the new
% minimizer XNEW with the line search method. F and J are string variables
% containing the functional expressions of f and of its Jacobian. X is the present
% minimizer. D is a given direction. SIGMA and RHO are given parameters.
alphak = 1; fk = eval(f); Jfk = eval (J);
xx = x; x = x + alphak * d; fk1 = eval (f);
while fk1 > fk + sigma * alphak * Jfk'*d
   alphak = alphak*rho;
   x = xx + alphak*d;
   fk1 = eval(f);
end
xnew = x;
return
```

Other commonly used strategies are those developed by *Armijo* and *Goldstein* (see [Arm66], [GP67]). Both use $\sigma \in (0, 1/2)$. In the Armijo formula, one takes $\alpha_k = \beta^{m_k}\bar{\alpha}$, where $\beta \in (0, 1)$, $\bar{\alpha} > 0$ and m_k is the first nonnegative integer such that (7.31) is satisfied. In the Goldstein formula, the parameter α_k is determined in such a way that

$$\sigma \leq \frac{f(\mathbf{x}^{(k)} + \alpha_k \mathbf{d}^{(k)}) - f(\mathbf{x}^{(k)})}{\alpha_k \nabla f(\mathbf{x}^{(k)})^T \mathbf{d}^{(k)}} \leq 1 - \sigma. \tag{7.34}$$

A procedure for computing α_k that satisfies (7.34) is provided in [Ber82], Chapter 1. Of course, one can even choose $\alpha_k = \bar{\alpha}$ for any k, which is clearly convenient when evaluating f is a costly task.

In any case, a good choice of the value $\bar{\alpha}$ is mandatory. In this respect, one can proceed as follows. For a given value $\bar{\alpha}$, the second degree polynomial Π_2 along the direction $\mathbf{d}^{(k)}$ is constructed, subject to the following interpolation constraints

$$\begin{aligned}
\Pi_2(\mathbf{x}^{(k)}) &= f(\mathbf{x}^{(k)}),\\
\Pi_2(\mathbf{x}^{(k)} + \bar{\alpha}\mathbf{d}^{(k)}) &= (\mathbf{x}^{(k)} + \bar{\alpha}\mathbf{d}^{(k)}),\\
\Pi_2'(\mathbf{x}^{(k)}) &= \nabla f(\mathbf{x}^{(k)})^T \mathbf{d}^{(k)}.
\end{aligned}$$

Next, the value $\tilde{\alpha}$ is computed such that Π_2 is minimized, then, we let $\bar{\alpha} = \tilde{\alpha}$.

7.2.4 Descent Methods for Quadratic Functions

A case of remarkable interest, where the parameter α_k can be exactly computed, is the problem of minimizing the quadratic function

$$f(\mathbf{x}) = \frac{1}{2}\mathbf{x}^T \mathrm{A}\mathbf{x} - \mathbf{b}^T\mathbf{x}, \tag{7.35}$$

where $\mathrm{A} \in \mathbb{R}^{n\times n}$ is a symmetric and positive definite matrix and $\mathbf{b} \in \mathbb{R}^n$. In such a case, as already seen in Section 4.3.3, a necessary condition for $\mathbf{x}^*$ to be a minimizer for f is that $\mathbf{x}^*$ is the solution of the linear system (3.2). Actually, it can be checked that if f is a quadratic function

$$\nabla f(\mathbf{x}) = \mathrm{A}\mathbf{x} - \mathbf{b} = -\mathbf{r},\ \mathrm{H}(\mathbf{x}) = \mathrm{A}.$$

As a consequence, all gradient-like iterative methods developed in Section 4.3.3 for linear systems, can be extended *tout-court* to solve minimization problems.

In particular, having fixed a descent direction $\mathbf{d}^{(k)}$, we can determine the optimal value of the acceleration parameter α_k that appears in (7.25), in such a way as to find the point where the function f, restricted to the direction $\mathbf{d}^{(k)}$, is minimized. Setting to zero the directional derivative, we get

$$\frac{\mathrm{d}}{\mathrm{d}\alpha_k} f(\mathbf{x}^{(k)} + \alpha_k \mathbf{d}^{(k)}) = -\mathbf{d}^{(k)^T}\mathbf{r}^{(k)} + \alpha_k \mathbf{d}^{(k)^T}\mathrm{A}\mathbf{d}^{(k)} = 0$$

from which the following expression for α_k is obtained

$$\alpha_k = \frac{\mathbf{d}^{(k)^T}\mathbf{r}^{(k)}}{\mathbf{d}^{(k)^T}\mathrm{A}\mathbf{d}^{(k)}}. \tag{7.36}$$

The error introduced by the iterative process (7.25) at the k-th step is

$$\begin{aligned}
\|\mathbf{x}^{(k+1)} - \mathbf{x}^*\|_{\mathrm{A}}^2 &= \left(\mathbf{x}^{(k+1)} - \mathbf{x}^*\right)^T \mathrm{A}\left(\mathbf{x}^{(k+1)} - \mathbf{x}^*\right)\\
&= \|\mathbf{x}^{(k)} - \mathbf{x}^*\|_{\mathrm{A}}^2 + 2\alpha_k \mathbf{d}^{(k)^T}\mathrm{A}\left(\mathbf{x}^{(k)} - \mathbf{x}^*\right)\\
&\quad + \alpha_k^2 \mathbf{d}^{(k)^T}\mathrm{A}\mathbf{d}^{(k)}.
\end{aligned} \tag{7.37}$$

On the other hand $\|\mathbf{x}^{(k)} - \mathbf{x}^*\|_{\mathrm{A}}^2 = \mathbf{r}^{(k)^T}\mathrm{A}^{-1}\mathbf{r}^{(k)}$, so that from (7.37) it follows that

$$\|\mathbf{x}^{(k+1)} - \mathbf{x}^*\|_{\mathrm{A}}^2 = \rho_k \|\mathbf{x}^{(k)} - \mathbf{x}^*\|_{\mathrm{A}}^2, \qquad (7.38)$$

having denoted by $\rho_k = 1 - \sigma_k$, with

$$\sigma_k = (\mathbf{d}^{(k)^T}\mathbf{r}^{(k)})^2 / \left(\left(\mathbf{d}^{(k)}\right)^T \mathrm{A}\mathbf{d}^{(k)} \left(\mathbf{r}^{(k)}\right)^T \mathrm{A}^{-1}\mathbf{r}^{(k)} \right).$$

Since A is symmetric and positive definite, σ_k is always positive. Moreover, it can be directly checked that ρ_k is strictly less than 1, except when $\mathbf{d}^{(k)}$ is orthogonal to $\mathbf{r}^{(k)}$, in which case $\rho_k = 1$.

The choice $\mathbf{d}^{(k)} = \mathbf{r}^{(k)}$, which leads to the *steepest descent method*, prevents this last circumstance from arising. In such a case, from (7.38) we get

$$\|\mathbf{x}^{(k+1)} - \mathbf{x}^*\|_{\mathrm{A}} \leq \frac{\lambda_{max} - \lambda_{min}}{\lambda_{max} + \lambda_{min}} \|\mathbf{x}^{(k)} - \mathbf{x}^*\|_{\mathrm{A}} \qquad (7.39)$$

having employed the following result.

Lemma 7.1 (Kantorovich inequality) *Let* $\mathrm{A} \in \mathbb{R}^{n\times n}$ *be a symmetric positive definite matrix whose eigenvalues with largest and smallest module are given by* λ_{max} *and* λ_{min}*, respectively. Then,* $\forall \mathbf{y} \in \mathbb{R}^n$, $\mathbf{y} \neq \mathbf{0}$,

$$\frac{(\mathbf{y}^T\mathbf{y})^2}{(\mathbf{y}^T\mathrm{A}\mathbf{y})(\mathbf{y}^T\mathrm{A}^{-1}\mathbf{y})} \geq \frac{4\lambda_{max}\lambda_{min}}{(\lambda_{max} + \lambda_{min})^2}.$$

It follows from (7.39) that, if A is ill-conditioned, the error reducing factor for the *steepest descent* method is close to 1, yielding a slow convergence to the minimizer $\mathbf{x}^*$. As done in Chapter 4, this drawback can be overcome by introducing directions $\mathbf{d}^{(k)}$ that are mutually A-conjugate, i.e.

$$\mathbf{d}^{(k)^T}\mathrm{A}\mathbf{d}^{(m)} = 0 \ \text{ if } k \neq m.$$

The corresponding methods enjoy the following finite termination property.

Property 7.6 *A method for computing the minimizer* $\mathbf{x}^*$ *of the quadratic function* (7.35) *which employs* A*-conjugate directions terminates after at most* n *steps if the acceleration parameter* α_k *is selected as in* (7.36)*. Moreover, for any* k, $\mathbf{x}^{(k+1)}$ *is the minimizer of* f *over the subspace generated by the vectors* $\mathbf{x}^{(0)}, \mathbf{d}^{(0)}, \ldots, \mathbf{d}^{(k)}$ *and*

$$\mathbf{r}^{(k+1)^T}\mathbf{d}^{(m)} = 0 \ \forall m \leq k.$$

The A-conjugate directions can be determined by following the procedure described in Section 4.3.4. Given $\mathbf{x}^{(0)} \in \mathbb{R}^n$ and letting $\mathbf{d}^{(0)} = \mathbf{r}^{(0)}$, the *conjugate gradient method* for function minimization is

$$\mathbf{d}^{(k+1)} = \mathbf{r}^{(k)} + \beta_k \mathbf{d}^{(k)},$$

$$\beta_k = -\frac{\mathbf{r}^{(k+1)^T} \mathrm{A}\mathbf{d}^{(k)}}{\mathbf{d}^{(k)^T} \mathrm{A}\mathbf{d}^{(k)}} = \frac{\mathbf{r}^{(k+1)^T} \mathbf{r}^{(k+1)}}{\mathbf{r}^{(k)^T} \mathbf{r}^{(k)}},$$

$$\mathbf{x}^{(k+1)} = \mathbf{x}^{(k)} + \alpha_k \mathbf{d}^{(k)}.$$

It satisfies the following error estimate

$$\|\mathbf{x}^{(k)} - \mathbf{x}^*\|_{\mathrm{A}} \leq 2 \left(\frac{\sqrt{K_2(\mathrm{A})} - 1}{\sqrt{K_2(\mathrm{A})} + 1} \right)^k \|\mathbf{x}^{(0)} - \mathbf{x}^*\|_{\mathrm{A}},$$

which can be improved by lowering the condition number of A, i.e., resorting to the preconditioning techniques that have been dealt with in Section 4.3.2.

Remark 7.3 (The nonquadratic case) The conjugate gradient method can be extended to the case in which f is a nonquadratic function. However, in such an event, the acceleration parameter α_k cannot be exactly determined a priori, but requires the solution of a local minimization problem. Moreover, the parameters β_k can no longer be uniquely found. Among the most reliable formulae, we recall the one due to Fletcher-Reeves,

$$\beta_1 = 0, \ \beta_k = \frac{\|\nabla f(\mathbf{x}^{(k)})\|_2^2}{\|\nabla f(\mathbf{x}^{(k-1)})\|_2^2}, \quad \text{for } k > 1$$

and the one due to Polak-Ribiére

$$\beta_1 = 0, \ \beta_k = \frac{\nabla f(\mathbf{x}^{(k)})^T (\nabla f(\mathbf{x}^{(k)}) - \nabla f(\mathbf{x}^{(k-1)}))}{\|\nabla f(\mathbf{x}^{(k-1)})\|_2^2}, \qquad \text{for } k > 1.$$

■

7.2.5 Newton-like Methods for Function Minimization

Another example of descent method is provided by Newton's method, which differs from its version for nonlinear systems in that now it is no longer applied to f, but to its gradient.

Using the notation of Section 7.2.2, Newton's method for function minimization amounts to computing, given $\mathbf{x}^{(0)} \in \mathbb{R}^n$, for $k = 0, 1, \ldots$, until convergence

$$\begin{aligned} \mathbf{d}^{(k)} &= -\mathrm{H}_k^{-1} \nabla f(\mathbf{x}^{(k)}), \\ \mathbf{x}^{(k+1)} &= \mathbf{x}^{(k)} + \mathbf{d}^{(k)}, \end{aligned} \qquad (7.40)$$

having set $\mathrm{H}_k = \mathrm{H}(\mathbf{x}^{(k)})$. The method can be derived by truncating Taylor's expansion of $f(\mathbf{x}^{(k)})$ at the second-order,

$$f(\mathbf{x}^{(k)} + \mathbf{p}) \simeq f(\mathbf{x}^{(k)}) + \nabla f(\mathbf{x}^{(k)})^T \mathbf{p} + \frac{1}{2} \mathbf{p}^T \mathrm{H}_k \mathbf{p}. \quad (7.41)$$

Selecting $\mathbf{p}$ in (7.41) in such a way that the new vector $\mathbf{x}^{(k+1)} = \mathbf{x}^{(k)} + \mathbf{p}$ satisfies $\nabla f(\mathbf{x}^{k+1}) = \mathbf{0}$, we end up with method (7.40), which thus converges in one step if f is quadratic.

In the general case, a result analogous to Theorem 7.1 also holds for function minimization. Method (7.40) is therefore locally quadratically convergent to the minimizer $\mathbf{x}^*$. However, it is not convenient to use Newton's method from the beginning of the computation, unless $\mathbf{x}^{(0)}$ is sufficiently close to $\mathbf{x}^*$. Otherwise, indeed, H_k could not be invertible and the directions $\mathbf{d}^{(k)}$ could fail to be descent directions. Moreover, if H_k is not positive definite, nothing prevents the scheme (7.40) from converging to a saddle point or a maximizer, which are points where ∇f is equal to zero. All these drawbacks, together with the high computational cost (recall that a linear system with matrix H_k must be solved at each iteration), prompt suitably modifying method (7.40), which leads to the so-called *quasi-Newton* methods.

A first modification, which applies to the case where H_k is not positive definite, yields the so-called *Newton's method with shift.* The idea is to prevent Newton's method from converging to non-minimizers of f, by applying the scheme to a new Hessian matrix $\tilde{\mathrm{H}}_k = \mathrm{H}_k + \mu_k \mathrm{I}_n$, where, as usual, I_n denotes the identity matrix of order n and μ_k is selected in such a way that $\tilde{\mathrm{H}}_k$ is positive definite. The problem is to determine the *shift* μ_k with a reduced effort. This can be done, for instance, by applying the Gershgorin theorem to the matrix $\tilde{\mathrm{H}}_k$ (see Section 5.1). For further details on the subject, see [DS83] and [GMW81].

7.2.6 Quasi-Newton Methods

At the generic k-th iteration, a *quasi-Newton method* for function minimization performs the following steps:

1. compute the Hessian matrix H_k, or a suitable approximation B_k;
2. find a descent direction $\mathbf{d}^{(k)}$ (not necessarily coinciding with the direction provided by Newton's method), using H_k or B_k;
3. compute the acceleration parameter α_k;
4. update the solution, setting $\mathbf{x}^{(k+1)} = \mathbf{x}^{(k)} + \alpha_k \mathbf{d}^{(k)}$, according to a global convergence criterion.

In the particular case where $\mathbf{d}^{(k)} = -\mathrm{H}_k^{-1} \nabla f(\mathbf{x}^{(k)})$, the resulting scheme is called the *damped Newton's method.* To compute H_k or B_k, one can resort to either Newton's method or secant-like methods, which will be considered in Section 7.2.7.

The criteria for selecting the parameter α_k, that have been discussed in Section 7.2.3, can now be usefully employed to devise globally convergent methods. Property 7.5 ensures that there exist values of α_k satisfying (7.31), (7.33) or (7.31), (7.32).

Let us then assume that a sequence of iterates $\mathbf{x}^{(k)}$, generated by a descent method for a given $\mathbf{x}^{(0)}$, converge to a vector $\mathbf{x}^*$. This vector will not be, in general, a critical point for f. The following result gives some conditions on the directions $\mathbf{d}^{(k)}$ which ensure that the limit $\mathbf{x}^*$ of the sequence is also a critical point of f.

Property 7.7 (Convergence) *Let $f : \mathbb{R}^n \to \mathbb{R}$ be a continuously differentiable function, and assume that there exists $L > 0$ such that*

$$\|\nabla f(\mathbf{x}) - \nabla f(\mathbf{y})\|_2 \leq L\|\mathbf{x} - \mathbf{y}\|_2.$$

Then, if $\{\mathbf{x}^{(k)}\}$ is a sequence generated by a gradient-like method which fulfills (7.31) and (7.33), then, one (and only one) of the following events can occur:

1. $\nabla f(\mathbf{x}^{(k)}) = \mathbf{0}$ *for some* k;
2. $\lim_{k\to\infty} f(\mathbf{x}^{(k)}) = -\infty$;
3. $\lim_{k\to\infty} \dfrac{\nabla f(\mathbf{x}^{(k)})^T \mathbf{d}^{(k)}}{\|\mathbf{d}^{(k)}\|_2} = 0.$

Thus, unless the pathological cases where the directions $\mathbf{d}^{(k)}$ become too large or too small with respect to $\nabla f(\mathbf{x}^{(k)})$ or, even, are orthogonal to $\nabla f(\mathbf{x}^{(k)})$, any limit of the sequence $\{\mathbf{x}^{(k)}\}$ is a critical point of f.

The convergence result for the sequence $\mathbf{x}^{(k)}$ can also be extended to the sequence $f(\mathbf{x}^{(k)})$. Indeed, the following result holds.

Property 7.8 *Let $\{\mathbf{x}^{(k)}\}$ be a convergent sequence generated by a gradient-like method, i.e., such that any limit of the sequence is also a critical point of f. If the sequence $\{\mathbf{x}^{(k)}\}$ is bounded, then $\nabla f(\mathbf{x}^{(k)})$ tends to zero as $k \to \infty$.*

For the proofs of the above results, see [Wol69] and [Wol71].

7.2.7 Secant-like methods

In quasi-Newton methods the Hessian matrix H is replaced by a suitable approximation. Precisely, the generic iterate is

$$\mathbf{x}^{(k+1)} = \mathbf{x}^{(k)} - \mathrm{B}_k^{-1}\nabla f(\mathbf{x}^{(k)}) = \mathbf{x}^{(k)} + \mathbf{s}^{(k)}.$$

Assume that $f : \mathbb{R}^n \to \mathbb{R}$ is of class C^2 on an open convex set $D \subset \mathbb{R}^n$. In such a case, H is symmetric and, as a consequence, approximants B_k of H ought to be symmetric. Moreover, if B_k were symmetric at a point $\mathbf{x}^{(k)}$, we would also like the next approximant B_{k+1} to be symmetric at $\mathbf{x}^{(k+1)} = \mathbf{x}^{(k)} + \mathbf{s}^{(k)}$.
To generate B_{k+1} starting from B_k, consider the Taylor expansion

$$\nabla f(\mathbf{x}^{(k)}) = \nabla f(\mathbf{x}^{(k+1)}) + \mathrm{B}_{k+1}(\mathbf{x}^{(k)} - \mathbf{x}^{(k+1)}),$$

from which we get

$$\mathrm{B}_{k+1}\mathbf{s}^{(k)} = \mathbf{y}^{(k)}, \text{ with } \mathbf{y}^{(k)} = \nabla f(\mathbf{x}^{(k+1)}) - \nabla f(\mathbf{x}^{(k)}).$$

Using again a series expansion of B, we end up with the following first-order approximation of H

$$\mathrm{B}_{k+1} = \mathrm{B}_k + \frac{(\mathbf{y}^{(k)} - \mathrm{B}_k\mathbf{s}^{(k)})\mathbf{c}^T}{\mathbf{c}^T\mathbf{s}^{(k)}}, \tag{7.42}$$

where $\mathbf{c} \in \mathbb{R}^n$ and having assumed that $\mathbf{c}^T\mathbf{s}^{(k)} \neq 0$. We notice that taking $\mathbf{c} = \mathbf{s}^{(k)}$ yields Broyden's method, already discussed in Section 7.1.4 in the case of systems of nonlinear equations.

Since (7.42) does not guarantee that B_{k+1} is symmetric, it must be suitably modified. A way for constructing a symmetric approximant B_{k+1} consists of choosing $\mathbf{c} = \mathbf{y}^{(k)} - \mathrm{B}_k\mathbf{s}^{(k)}$ in (7.42), assuming that $(\mathbf{y}^{(k)} - \mathrm{B}_k\mathbf{s}^{(k)})^T\mathbf{s}^{(k)} \neq 0$. By so doing, the following symmetric first-order approximation is obtained

$$\mathrm{B}_{k+1} = \mathrm{B}_k + \frac{(\mathbf{y}^{(k)} - \mathrm{B}_k\mathbf{s}^{(k)})(\mathbf{y}^{(k)} - \mathrm{B}_k\mathbf{s}^{(k)})^T}{(\mathbf{y}^{(k)} - \mathrm{B}_k\mathbf{s}^{(k)})^T\mathbf{s}^{(k)}}. \tag{7.43}$$

From a computational standpoint, disposing of an approximation for H is not completely satisfactory, since the inverse of the approximation of H appears in the iterative methods that we are dealing with. Using the Sherman-Morrison formula (3.57), with $\mathrm{C}_k = \mathrm{B}_k^{-1}$, yields the following recursive formula for the computation of the inverse

$$\mathrm{C}_{k+1} = \mathrm{C}_k + \frac{(\mathbf{s}^{(k)} - \mathrm{C}_k\mathbf{y}^{(k)})(\mathbf{s}^{(k)} - \mathrm{C}_k\mathbf{y}^{(k)})^T}{(\mathbf{s}^{(k)} - \mathrm{C}_k\mathbf{y}^{(k)})^T\mathbf{y}^{(k)}}, \quad k = 0, 1, \ldots, \tag{7.44}$$

having assumed that $\mathbf{y}^{(k)} = \mathrm{B}\mathbf{s}^{(k)}$, where B is a symmetric nonsingular matrix, and that $(\mathbf{s}^{(k)} - \mathrm{C}_k\mathbf{y}^{(k)})^T\mathbf{y}^{(k)} \neq 0$.

An algorithm that employs the approximations (7.43) or (7.44), is potentially unstable when $(\mathbf{s}^{(k)} - \mathrm{C}_k\mathbf{y}^{(k)})^T\mathbf{y}^{(k)} \simeq 0$, due to rounding errors. For this reason, it is convenient to set up the previous scheme in a more stable form. To this end, instead of (7.42), we introduce the approximation

$$\mathrm{B}_{k+1}^{(1)} = \mathrm{B}_k + \frac{(\mathbf{y}^{(k)} - \mathrm{B}_k\mathbf{s}^{(k)})\mathbf{c}^T}{\mathbf{c}^T\mathbf{s}^{(k)}},$$

then, we define $\mathrm{B}_{k+1}^{(2)}$ as being the symmetric part

$$\mathrm{B}_{k+1}^{(2)} = \frac{\mathrm{B}_{k+1}^{(1)} + (\mathrm{B}_{k+1}^{(1)})^T}{2}.$$

The procedure can be iterated as follows

$$\begin{aligned} \mathrm{B}_{k+1}^{(2j+1)} &= \mathrm{B}_{k+1}^{(2j)} + \frac{(\mathbf{y}^{(k)} - \mathrm{B}_{k+1}^{(2j)}\mathbf{s}^{(k)})\mathbf{c}^T}{\mathbf{c}^T\mathbf{s}^{(k)}}, \\ \mathrm{B}_{k+1}^{(2j+2)} &= \frac{\mathrm{B}_{k+1}^{(2j+1)} + (\mathrm{B}_{k+1}^{(2j+1)})^T}{2}, \end{aligned} \tag{7.45}$$

with $k = 0, 1, \ldots$ and having set $\mathrm{B}_{k+1}^{(0)} = \mathrm{B}_k$. It can be shown that the limit as j tends to infinity of (7.45) is

$$\begin{aligned} \lim_{j\to\infty} \mathrm{B}_{k+1}^{(j)} = \mathrm{B}_{k+1} = \mathrm{B}_k &+ \frac{(\mathbf{y}^{(k)} - \mathrm{B}_k\mathbf{s}^{(k)})\mathbf{c}^T + \mathbf{c}(\mathbf{y}^{(k)} - \mathrm{B}_k\mathbf{s}^{(k)})^T}{\mathbf{c}^T\mathbf{s}^{(k)}} \\ &- \frac{(\mathbf{y}^{(k)} - \mathrm{B}_k\mathbf{s}^{(k)})^T\mathbf{s}^{(k)}}{(\mathbf{c}^T\mathbf{s}^{(k)})^2}\mathbf{c}\mathbf{c}^T, \end{aligned} \tag{7.46}$$

having assumed that $\mathbf{c}^T\mathbf{s}^{(k)} \neq 0$. If $\mathbf{c} = \mathbf{s}^{(k)}$, the method employing (7.46) is known as the *symmetric Powell-Broyden method.* Denoting by B_{SPB} the corresponding matrix B_{k+1}, it can be shown that B_{SPB} is the unique solution to the problem:

$$\text{find } \bar{\mathrm{B}} \text{ such that } \|\bar{\mathrm{B}} - \mathrm{B}\|_F \text{ is minimized,}$$

where $\bar{\mathrm{B}}\mathbf{s}^{(k)} = \mathbf{y}^{(k)}$ and $\|\cdot\|_F$ is the Frobenius norm.
As for the error made approximating $\mathrm{H}(\mathbf{x}^{(k+1)})$ with B_{SPB}, it can be proved that

$$\|\mathrm{B}_{SPB} - \mathrm{H}(\mathbf{x}^{(k+1)})\|_F \leq \|\mathrm{B}_k - \mathrm{H}(\mathbf{x}^{(k)})\|_F + 3L\|\mathbf{s}^{(k)}\|,$$

where it is assumed that H is Lipschitz continuous, with Lipschitz constant L, and that the iterates $\mathbf{x}^{(k+1)}$ and $\mathbf{x}^{(k)}$ belong to D.
To deal with the particular case in which the Hessian matrix is not only symmetric but also positive definite, we refer to [DS83], Section 9.2.

7.3 Constrained Optimization

The simplest case of constrained optimization can be formulated as follows. Given $f : \mathbb{R}^n \to \mathbb{R}$,

$$\text{minimize } f(\mathbf{x}), \text{ with } \mathbf{x} \in \Omega \subset \mathbb{R}^n. \tag{7.47}$$

More precisely, the point $\mathbf{x}^*$ is said to be a *global minimizer* in Ω if it satisfies (7.47), while it is a *local minimizer* if $\exists R > 0$ such that

$$f(\mathbf{x}^*) \leq f(\mathbf{x}), \forall \mathbf{x} \in B(\mathbf{x}^*; R) \subset \Omega.$$

Existence of solutions to problem (7.47) is, for instance, ensured by the Weierstrass theorem, in the case in which f is continuous and Ω is a closed and bounded set. Under the assumption that Ω is a convex set, the following optimality conditions hold.

Property 7.9 *Let $\Omega \subset \mathbb{R}^n$ be a convex set, $\mathbf{x}^* \in \Omega$ and $f \in C^1(B(\mathbf{x}^*; R))$, for a suitable $R > 0$. Then:*

1. *if $\mathbf{x}^*$ is a local minimizer of f then*
$$\nabla f(\mathbf{x}^*)^T(\mathbf{x} - \mathbf{x}^*) \geq 0, \forall \mathbf{x} \in \Omega; \tag{7.48}$$
2. *moreover, if f is convex on Ω (see (7.21)) and (7.48) is satisfied, then $\mathbf{x}^*$ is a global minimizer of f.*

We recall that $f : \Omega \rightarrow \mathbb{R}$ is a strongly convex function if $\exists \rho > 0$ such that

$$f[\alpha \mathbf{x} + (1-\alpha)\mathbf{y}] \leq \alpha f(\mathbf{x}) + (1-\alpha) f(\mathbf{y}) - \alpha(1-\alpha)\rho\|\mathbf{x} - \mathbf{y}\|_2^2, \tag{7.49}$$

$\forall \mathbf{x}, \mathbf{y} \in \Omega$ and $\forall \alpha \in [0, 1]$. The following result holds.

Property 7.10 *Let $\Omega \subset \mathbb{R}^n$ be a closed and convex set and f be a strongly convex function in Ω. Then there exists a unique local minimizer $\mathbf{x}^* \in \Omega$.*

Throughout this section, we refer to [Avr76], [Ber82], [CCP70], [Lue73] and [Man69], for the proofs of the quoted results and further details.

A remarkable instance of (7.47) is the following problem: given $f : \mathbb{R}^n \rightarrow \mathbb{R}$,

$$\text{minimize } f(\mathbf{x}), \text{ under the constraint that } \mathbf{h}(\mathbf{x}) = \mathbf{0}, \tag{7.50}$$

where $\mathbf{h} : \mathbb{R}^n \rightarrow \mathbb{R}^m$, with $m \leq n$, is a given function of components $h_1, \ldots, h_m$. The analogues of critical points in problem (7.50) are called the *regular points.*

Definition 7.2 A point $\mathbf{x}^* \in \mathbb{R}^n$, such that $\mathbf{h}(\mathbf{x}^*) = \mathbf{0}$, is said to be *regular* if the column vectors of the Jacobian matrix $\mathrm{J}_{\mathbf{h}}(\mathbf{x}^*)$ are linearly independent, having assumed that $h_i \in C^1(B(\mathbf{x}^*; R))$, for a suitable $R > 0$ and $i = 1, \ldots, m$. ■

Our aim now is to convert problem (7.50) into an unconstrained minimization problem of the form (7.2), to which the methods introduced in Section 7.2 can be applied.

For this purpose, we introduce the *Lagrangian function* $\mathcal{L} : \mathbb{R}^n \times \mathbb{R}^m \rightarrow \mathbb{R}$

$$\mathcal{L}(\mathbf{x}, \boldsymbol{\lambda}) = f(\mathbf{x}) + \boldsymbol{\lambda}^T \mathbf{h}(\mathbf{x}),$$

where the vector $\boldsymbol{\lambda}$ is called the *Lagrange multiplier.* Moreover, let us denote by $\mathrm{J}_{\mathcal{L}}$ the Jacobian matrix associated with $\mathcal{L}$, but where the partial derivatives are only taken with respect to the variables $x_1, \ldots, x_n$. The link between (7.2) and (7.50) is then expressed by the following result.

Property 7.11 *Let $\mathbf{x}^*$ be a local minimizer for* (7.50) *and suppose that, for a suitable $R > 0$, $f, h_i \in C^1(B(\mathbf{x}^*; R))$, for $i = 1, \ldots, m$. Then there exists a unique vector $\boldsymbol{\lambda}^* \in \mathbb{R}^m$ such that $\mathrm{J}_{\mathcal{L}}(\mathbf{x}^*, \boldsymbol{\lambda}^*) = \mathbf{0}$.*

Conversely, assume that $\mathbf{x}^* \in \mathbb{R}^n$ satisfies $\mathbf{h}(\mathbf{x}^*) = \mathbf{0}$ and that, for a suitable $R > 0$ and $i = 1, \ldots, m$, $f, h_i \in C^2(B(\mathbf{x}^*; R))$. Let $\mathrm{H}_{\mathcal{L}}$ be the matrix of entries $\partial^2 \mathcal{L}/\partial x_i \partial x_j$ for $i, j = 1, \ldots, n$. If there exists a vector $\boldsymbol{\lambda}^* \in \mathbb{R}^m$ such that $\mathrm{J}_{\mathcal{L}}(\mathbf{x}^*, \boldsymbol{\lambda}^*) = \mathbf{0}$ and

$$\mathbf{z}^T \mathrm{H}_{\mathcal{L}}(\mathbf{x}^*, \boldsymbol{\lambda}^*)\mathbf{z} > 0 \ \forall \mathbf{z} \neq \mathbf{0}, \quad \textit{with} \ \ \nabla \mathbf{h}(\mathbf{x}^*)^T \mathbf{z} = 0,$$

then $\mathbf{x}^*$ is a strict local minimizer of (7.50).

The last class of problems that we are going to deal with includes the case where inequality constraints are also present, i.e.: given $f : \mathbb{R}^n \to \mathbb{R}$,

$$\text{minimize } f(\mathbf{x}), \ \ \text{under the constraint that } \mathbf{h}(\mathbf{x}) = \mathbf{0} \text{ and } \mathbf{g}(\mathbf{x}) \leq \mathbf{0}, \tag{7.51}$$

where $\mathbf{h} : \mathbb{R}^n \to \mathbb{R}^m$, with $m \leq n$, and $\mathbf{g} : \mathbb{R}^n \to \mathbb{R}^r$ are two given functions. It is understood that $\mathbf{g}(\mathbf{x}) \leq \mathbf{0}$ means $g_i(\mathbf{x}) \leq 0$ for $i = 1, \ldots, r$. Inequality constraints give rise to some extra formal complication with respect to the case previously examined, but do not prevent converting the solution of (7.51) into the minimization of a suitable Lagrangian function.

In particular, Definition 7.2 becomes

Definition 7.3 Assume that $h_i, g_j \in C^1(B(\mathbf{x}^*; R))$ for a suitable $R > 0$ with $i = 1, \ldots, m$ and $j = 1, \ldots, r$, and denote by $\mathcal{J}(\mathbf{x}^*)$ the set of indices j such that $g_j(\mathbf{x}^*) = 0$. A point $\mathbf{x}^* \in \mathbb{R}^n$ such that $\mathbf{h}(\mathbf{x}^*) = \mathbf{0}$ and $\mathbf{g}(\mathbf{x}^*) \leq \mathbf{0}$ is said to be *regular* if the column vectors of the Jacobian matrix $\mathrm{J}_{\mathbf{h}}(\mathbf{x}^*)$ together with the vectors $\nabla g_j(\mathbf{x}^*)$, $j \in \mathcal{J}(\mathbf{x}^*)$ form a set of linearly independent vectors. ■

Finally, an analogue of Property 7.11 holds, provided that the following Lagrangian function is used

$$\mathcal{M}(\mathbf{x}, \boldsymbol{\lambda}, \boldsymbol{\mu}) = f(\mathbf{x}) + \boldsymbol{\lambda}^T \mathbf{h}(\mathbf{x}) + \boldsymbol{\mu}^T \mathbf{g}(\mathbf{x})$$

instead of $\mathcal{L}$ and that further assumptions on the constraints are made.

For the sake of simplicity, we report in this case only the following necessary condition for optimality of problem (7.51) to hold.

Property 7.12 *Let $\mathbf{x}^*$ be a regular local minimizer for* (7.51) *and suppose that, for a suitable $R > 0$, $f, h_i, g_j \in C^1(B(\mathbf{x}^*; R))$ with $i = 1, \ldots, m$, $j = 1, \ldots, r$. Then, there exist only two vectors $\boldsymbol{\lambda}^* \in \mathbb{R}^m$ and $\boldsymbol{\mu}^* \in \mathbb{R}^r$, such that $\mathrm{J}_{\mathcal{M}}(\mathbf{x}^*, \boldsymbol{\lambda}^*, \boldsymbol{\mu}^*) = \mathbf{0}$ with $\mu_j^* \geq 0$ and $\mu_j^* g_j(\mathbf{x}^*) = 0 \ \forall j = 1, \ldots, r$.*

7.3.1 Kuhn-Tucker Necessary Conditions for Nonlinear Programming

In this section we recall some results, known as *Kuhn-Tucker conditions* [KT51], that ensure in general the existence of a local solution for the nonlinear programming problem. Under suitable assumptions they also guarantee the existence of a global solution. Throughout this section we suppose that a minimization problem can always be reformulated as a maximization one.

Let us consider the general nonlinear programming problem:

$$\begin{aligned} &\text{given } f : \mathbb{R}^n \to \mathbb{R}, \\ &\text{maximize } f(\mathbf{x}), \text{ subject to} \\ &g_i(\mathbf{x}) \leq b_i, \quad i = 1, \ldots, l, \\ &g_i(\mathbf{x}) \geq b_i, \quad i = l+1, \ldots, k, \\ &g_i(\mathbf{x}) = b_i, \quad i = k+1, \ldots, m, \\ &\mathbf{x} \geq \mathbf{0}. \end{aligned} \tag{7.52}$$

A vector $\mathbf{x}$ that satisfies the constraints above is called a *feasible solution* of (7.52) and the set of the feasible solutions is called the *feasible region*. We assume henceforth that $f, g_i \in C^1(\mathbb{R}^n)$, $i = 1, \ldots, m$, and define the sets $I_= = \{i : g_i(\mathbf{x}^*) = b_i\}$, $I_{\neq} = \{i : g_i(\mathbf{x}^*) \neq b_i\}$, $J_= = \{i : x_i^* = 0\}$, $J_> = \{i : x_i^* > 0\}$, having denoted by $\mathbf{x}^*$ a local maximizer of f. We associate with (7.52) the following Lagrangian

$$\mathcal{L}(\mathbf{x}, \boldsymbol{\lambda}) = f(\mathbf{x}) + \sum_{i=1}^{m} \lambda_i \left[b_i - g_i(\mathbf{x})\right] - \sum_{i=m+1}^{m+n} \lambda_i x_{i-m}.$$

The following result can be proved.

Property 7.13 (Kuhn-Tucker conditions I and II) *If f has a constrained local maximum at the point $\mathbf{x} = \mathbf{x}^*$, it is necessary that a vector $\boldsymbol{\lambda}^* \in \mathbb{R}^{m+n}$ exists such that (first Kuhn-Tucker condition)*

$$\nabla_{\mathbf{x}} \mathcal{L}(\mathbf{x}^*, \boldsymbol{\lambda}^*) \leq 0,$$

where strict equality holds for every component $i \in J_>$. Moreover (second Kuhn-Tucker condition)

$$\nabla_{\mathbf{x}} \mathcal{L}(\mathbf{x}^*, \boldsymbol{\lambda}^*)^T \mathbf{x}^* = 0.$$

The other two necessary Kuhn-Tucker conditions are as follows.

Property 7.14 *Under the same hypothesis as in Property 7.13, the third Kuhn-Tucker condition requires that:*

$$\nabla_{\boldsymbol{\lambda}}\mathcal{L}(\mathbf{x}^*, \boldsymbol{\lambda}^*) \geq 0 \quad i = 1, \ldots, l,$$
$$\nabla_{\boldsymbol{\lambda}}\mathcal{L}(\mathbf{x}^*, \boldsymbol{\lambda}^*) \leq 0 \quad i = l+1, \ldots, k,$$
$$\nabla_{\boldsymbol{\lambda}}\mathcal{L}(\mathbf{x}^*, \boldsymbol{\lambda}^*) = 0 \quad i = k+1, \ldots, m.$$

Moreover (fourth Kuhn-Tucker condition)

$$\nabla_{\boldsymbol{\lambda}}\mathcal{L}(\mathbf{x}^*, \boldsymbol{\lambda}^*)^T \mathbf{x}^* = 0.$$

It is worth noticing that the Kuhn-Tucker conditions hold provided that the vector $\boldsymbol{\lambda}^*$ exists. To ensure this, it is necessary to introduce a further geometric condition that is known as *constraint qualification* (see [Wal75], p. 48).

We conclude this section by the following fundamental theorem which establishes when the Kuhn-Tucker conditions become also sufficient for the existence of a global maximizer for f.

Property 7.15 *Assume that the function f in* (7.52) *is a concave function (i.e., $-f$ is convex) in the feasible region. Suppose also that the point* $(\mathbf{x}^*, \boldsymbol{\lambda}^*)$ *satisfies all the Kuhn-Tucker necessary conditions and that the functions g_i for which $\lambda_i^* > 0$ are convex while those for which $\lambda_i^* < 0$ are concave. Then $f(\mathbf{x}^*)$ is the constrained global maximizer of f for problem* (7.52).

7.3.2 The Penalty Method

The basic idea of this method is to eliminate, partly or completely, the constraints in order to transform the constrained problem into an unconstrained one. This new problem is characterized by the presence of a parameter that yields a measure of the accuracy at which the constraint is actually imposed.

Let us consider the constrained problem (7.50), assuming we are searching for the solution $\mathbf{x}^*$ only in $\Omega \subset \mathbb{R}^n$. Suppose that such a problem admits at least one solution in Ω and write it in the following penalized form

$$\text{minimize } \mathcal{L}_\alpha(\mathbf{x}) \quad \text{for } \mathbf{x} \in \Omega, \tag{7.53}$$

where

$$\mathcal{L}_\alpha(\mathbf{x}) = f(\mathbf{x}) + \frac{1}{2}\alpha\|\mathbf{h}(\mathbf{x})\|_2^2.$$

The function $\mathcal{L}_\alpha : \mathbb{R}^n \to \mathbb{R}$ is called the *penalized Lagrangian*, and α is called the *penalty parameter*. It is clear that if the constraint was exactly satisfied then minimizing f would be equivalent to minimizing $\mathcal{L}_\alpha$.
The penalty method is an iterative technique for solving (7.53).

For $k = 0, 1, \ldots,$ until convergence, one must solve the sequence of problems:

$$\text{minimize } \mathcal{L}_{\alpha_k}(\mathbf{x}) \quad \text{with } \mathbf{x} \in \Omega, \tag{7.54}$$

where $\{\alpha_k\}$ is an increasing monotonically sequence of positive penalty parameters, such that $\alpha_k \to \infty$ as $k \to \infty$. As a consequence, after choosing α_k, at each step of the penalty process we have to solve a minimization problem with respect to the variable $\mathbf{x}$, leading to a sequence of values $\mathbf{x}_k^*$, solutions to (7.54). By doing so, the objective function $\mathcal{L}_{\alpha_k}(\mathbf{x})$ tends to infinity, unless $\mathbf{h}(\mathbf{x})$ is equal to zero.

The minimization problems can then be solved by one of the methods introduced in Section 7.2. The following property ensures the convergence of the penalty method in the form (7.53).

Property 7.16 *Assume that $f : \mathbb{R}^n \to \mathbb{R}$ and $\mathbf{h} : \mathbb{R}^n \to \mathbb{R}^m$, with $m \leq n$, are continuous functions on a closed set $\Omega \subset \mathbb{R}^n$ and suppose that the sequence of penalty parameters $\alpha_k > 0$ is monotonically divergent. Finally, let $\mathbf{x}_k^*$ be the global minimizer of problem (7.54) at step k. Then, taking the limit as $k \to \infty$, the sequence $\mathbf{x}_k^*$ converges to $\mathbf{x}^*$, which is a global minimizer of f in Ω and satisfies the constraint $\mathbf{h}(\mathbf{x}^*) = \mathbf{0}$.*

Regarding the selection of the parameters α_k, it can be shown that large values of α_k make the minimization problem in (7.54) ill-conditioned, thus making its solution quite prohibitive unless the initial guess is particularly close to $\mathbf{x}^*$. On the other hand, the sequence α_k must not grow too slowly, since this would negatively affect the overall convergence of the method.

A choice that is commonly made in practice is to pick up a not too large value of α_0 and then set $\alpha_k = \beta\alpha_{k-1}$ for $k > 0$, where β is an integer number between 4 and 10 (see [Ber82]). Finally, the starting point for the numerical method used to solve the minimization problem (7.54) can be set equal to the last computed iterate.

The penalty method is implemented in Program 63. This requires as input parameters the functions `f`, `h`, an initial value `alpha0` for the penalty parameter and the number `beta`.

Program 63 - lagrpen : Penalty method

```
function [x,vinc,iter]=lagrpen(f,h,x0,h,tol,alpha0,beta)
%LAGRPEN Penalty method for constrained function optimization
% [X,VINC,ITER]=LAGRPEN(F,H,X0,TOL,ALPHA0,BETA) attempts to compute
% the minimizer X of a function F with the Penalty method. F is a string containing
% the functional expressions of the function. X0 specifies the initial guess. H is a
% string variable containing the constraint. TOL specifies the tolerance of the method.
% ALPHA0 and BETA are given parameters. ITER is the iteration number at which X is
% computed. VINC is the accuracy at which the constraint is satisfied.
x = x0; [r,c]=size(h); vinc = 0;
for i=1:r
   vinc = max(vinc,eval(h(i,1:c)));
end
norm2h=['(',h(1,1:c),')^2'];
for i=2:r
   norm2h=[norm2h,'+(',h(i,1:c),')^2'];
end
```

```
alpha = alpha0;
options(1)=0; options(2)=tol*0.1;
iter = 0;
while vinc > tol
   g=[f,'+0.5*',num2str(alpha,16),'*',norm2h];
   [x]=fmins(g,x,options);
   vinc=0;
   iter = iter + 1;
   for i=1:r
      vinc = max(vinc,eval(h(i,1:c)));
   end
   alpha=alpha*beta;
end
return
```

Example 7.6 Let us employ the penalty method to compute the minimizer of $f(\mathbf{x}) = 100(x_2 - x_1^2)^2 + (1 - x_1)^2$ under the constraint $h(\mathbf{x}) = (x_1 + 0.5)^2 + (x_2 + 0.5)^2 - 0.25 = 0$. The crosses in Figure 7.3 denote the sequence of iterates computed by Program 63 starting from $\mathbf{x}^{(0)} = [1, 1]^T$ and choosing $\alpha_0 = 0.1$, $\beta = 6$. The method converges in 12 iterations to the value $\mathbf{x} = [-0.2463, -0.0691]^T$, satisfying the constraint up to a tolerance of 10^{-4}. •

7.3.3 The Method of Lagrange Multipliers

A variant of the penalty method makes use of (instead of $\mathcal{L}_\alpha(\mathbf{x})$ in (7.53)) the *augmented Lagrangian* function $\mathcal{G}_\alpha : \mathbb{R}^n \times \mathbb{R}^m \to \mathbb{R}$ given by

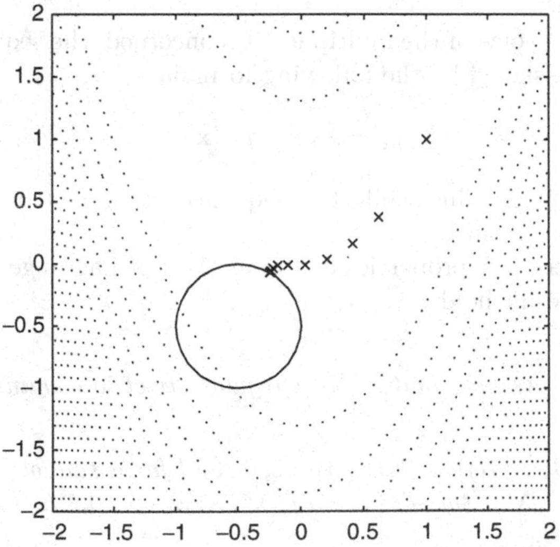

Fig. 7.3. Convergence history of the penalty method in Example 7.6

$$\mathcal{G}_\alpha(\mathbf{x}, \boldsymbol{\lambda}) = f(\mathbf{x}) + \boldsymbol{\lambda}^T \mathbf{h}(\mathbf{x}) + \frac{1}{2}\alpha \|\mathbf{h}(\mathbf{x})\|_2^2, \tag{7.55}$$

where $\boldsymbol{\lambda} \in \mathbb{R}^m$ is a *Lagrange multiplier.* Clearly, if $\mathbf{x}^*$ is a solution to problem (7.50), then it will also be a solution to (7.55), but with the advantage, with respect to (7.53), of disposing of the further degree of freedom $\boldsymbol{\lambda}$. The penalty method applied to (7.55) reads: for $k = 0, 1, \ldots$, until convergence, solve the sequence of problems

$$\text{minimize } \mathcal{G}_{\alpha_k}(\mathbf{x}, \boldsymbol{\lambda}_k) \quad \text{for } \mathbf{x} \in \Omega, \tag{7.56}$$

where $\{\boldsymbol{\lambda}_k\}$ is a bounded sequence of unknown vectors in $\mathbb{R}^m$, and the parameters α_k are defined as above (notice that if $\boldsymbol{\lambda}_k$ were zero, then we would recover method (7.54)).

Property 7.16 also holds for method (7.56), provided that the multipliers are assumed to be bounded. Notice that the existence of the minimizer of (7.56) is not guaranteed, even in the case where f has a unique global minimizer (see Example 7.7). This circumstance can be overcome by adding further non quadratic terms to the augmented Lagrangian function (e.g., of the form $\|\mathbf{h}\|_2^p$, with p large).

Example 7.7 Let us find the minimizer of $f(x) = -x^4$ under the constraint $x = 0$. Such problem clearly admits the solution $x^* = 0$. If, instead, one considers the augmented Lagrangian function

$$\mathcal{L}_{\alpha_k}(x, \lambda_k) = -x^4 + \lambda_k x + \frac{1}{2}\alpha_k x^2,$$

one finds that it no longer admits a minimum at $x = 0$, though vanishing there, for any α_k different from zero. •

As far as the choice of the multipliers is concerned, the sequence of vectors $\boldsymbol{\lambda}_k$ is typically assigned by the following formula

$$\boldsymbol{\lambda}_{k+1} = \boldsymbol{\lambda}_k + \alpha_k \mathbf{h}(\mathbf{x}^{(k)}),$$

where $\boldsymbol{\lambda}_0$ is a given value while the sequence of α_k can be set a priori or modified during run-time.
As for the convergence properties of the method of Lagrange multipliers, the following local result holds.

Property 7.17 *Assume that* $\mathbf{x}^*$ *is a regular strict local minimizer of* (7.50) *and that:*

1. $f, h_i \in C^2(B(\mathbf{x}^*; R))$ *with* $i = 1, \ldots, m$ *and for a suitable* $R > 0$;
2. *the pair* $(\mathbf{x}^*, \boldsymbol{\lambda}^*)$ *satisfies* $\mathbf{z}^T \mathrm{H}_{\mathcal{G}_0}(\mathbf{x}^*, \boldsymbol{\lambda}^*)\mathbf{z} > 0$, $\forall \mathbf{z} \neq \mathbf{0}$ *such that* $\mathrm{J}_{\mathbf{h}}(\mathbf{x}^*)^T \mathbf{z} = 0$;
3. $\exists \bar{\alpha} > 0$ *such that* $\mathrm{H}_{\mathcal{G}_{\bar{\alpha}}}(\mathbf{x}^*, \boldsymbol{\lambda}^*) > 0$.

Then, there exist three positive scalars δ, γ and M such that, for any pair $(\boldsymbol{\lambda}, \alpha) \in V = \{(\boldsymbol{\lambda}, \alpha) \in \mathbb{R}^{m+1} \ : \ \|\boldsymbol{\lambda} - \boldsymbol{\lambda}^*\|_2 < \delta\alpha, \ \alpha \geq \bar{\alpha}\}$, the problem

$$\textit{minimize } \mathcal{G}_\alpha(\mathbf{x}, \boldsymbol{\lambda}), \textit{ with } \mathbf{x} \in B(\mathbf{x}^*; \gamma),$$

admits a unique solution $\mathbf{x}(\boldsymbol{\lambda}, \alpha)$, differentiable with respect to its arguments. Moreover, $\forall(\boldsymbol{\lambda}, \alpha) \in V$

$$\|\mathbf{x}(\boldsymbol{\lambda}, \alpha) - \mathbf{x}^*\|_2 \leq M\|\boldsymbol{\lambda} - \boldsymbol{\lambda}^*\|_2.$$

Under further assumptions (see [Ber82], Proposition 2.7), it can be proved that the Lagrange multipliers method converges. Moreover, if $\alpha_k \to \infty$, as $k \to \infty$, then

$$\lim_{k \to \infty} \frac{\|\boldsymbol{\lambda}_{k+1} - \boldsymbol{\lambda}^*\|_2}{\|\boldsymbol{\lambda}_k - \boldsymbol{\lambda}^*\|_2} = 0$$

and the convergence of the method is more than linear.

In the case where the sequence α_k has an upper bound, the method converges linearly.

Finally, we notice that, unlike the penalty method, it is no longer necessary that the sequence of α_k tends to infinity. This, in turn, limits the ill-conditioning of problem (7.56) as α_k is growing. Another advantage concerns the convergence rate of the method, which turns out to be independent of the growth rate of the penalty parameter, in the case of the Lagrange multipliers technique. This of course implies a considerable reduction of the computational cost.

The method of Lagrange multipliers is implemented in Program 64. Compared with Program 63, this further requires in input the initial value `lambda0` of the multiplier.

Program 64 - lagrmult : Method of Lagrange multipliers

```
function [x,vinc,iter]=lagrmult(f,h,x0,lambda0,tol,alpha0,beta)
%LAGRMULT Method of Lagrange multipliers for constrained function optimization
% [X,VINC,ITER]=LAGRMULT(F,H,X0,LAMBDA0,TOL,ALPHA0,BETA) attempts
% to compute the minimizer X of a function F with the method of Lagrange
% multipliers. F ia a string containing the functional expressions of the function.
% X0 and LAMBDA0 specify the initial guesses. H is a string variable containing the
% constraint. TOL specifies the tolerance of the method. ALPHA0 and BETA are given
% parameters. ITER is the iteration number at which X is computed. VINC is the
% accuracy at which the constraint is satisfied.
x = x0; [r,c]=size(h); vinc = 0; lambda = lambda0;
for i=1:r
   vinc = max(vinc,eval(h(i,1:c)));
end
norm2h=['(',h(1,1:c),')^2'];
```

```
for i=2:r
   norm2h=[norm2h,'+(',h(i,1:c),')^2'];
end
alpha = alpha0;
options(1)=0; options(2)=tol*0.1;
iter = 0;
while vinc > tol
   lh=['(',h(1,1:c),')*',num2str(lambda(1))];
   for i=2:r
      lh=[lh,'+(',h(i,1:c),')*',num2str(lambda(i))];
   end
   g=[f,'+0.5*',num2str(alpha,16),'*',norm2h,'+',lh];
   [x]=fmins(g,x,options);
   vinc=0;
   iter = iter + 1;
   for i=1:r
      vinc = max(vinc,eval(h(i,1:c)));
   end
   alpha=alpha*beta;
   for i=1:r
      lambda(i)=lambda(i)+alpha*eval(h(i,1:c));
   end
end
return
```

Example 7.8 We use the method of Lagrange multipliers to solve the problem presented in Example 7.6. Set $\boldsymbol{\lambda} = 10$ and leave the remaining parameters unchanged. The method converges in 6 iterations and the crosses in Figure 7.4 show the iterates computed by Program 64. The constraint is here satisfied up to machine precision. •

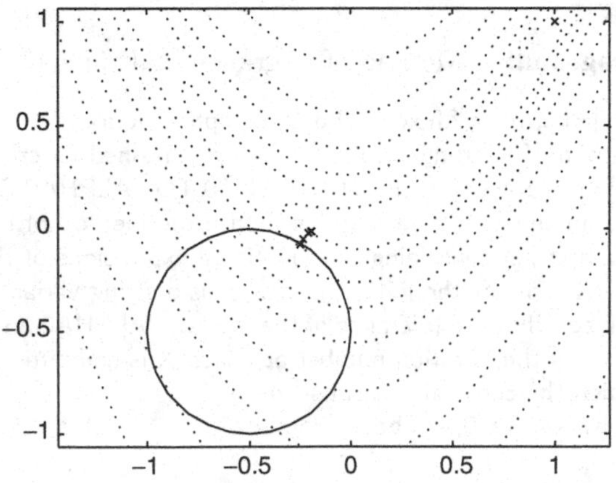

Fig. 7.4. Convergence history for the method of Lagrange multipliers in Example 7.8

7.4 Applications

The two applications of this section are concerned with nonlinear systems arising in the simulation of the electric potential in a semiconductor device and in the triangulation of a two-dimensional polygon.

7.4.1 Solution of a Nonlinear System Arising from Semiconductor Device Simulation

Let us consider the nonlinear system in the unknown $\mathbf{u} \in \mathbb{R}^n$

$$\mathbf{F}(\mathbf{u}) = \mathrm{A}\mathbf{u} + \boldsymbol{\phi}(\mathbf{u}) - \mathbf{b} = \mathbf{0}, \tag{7.57}$$

where $\mathrm{A} = (\lambda/h)^2\mathrm{tridiag}_n(-1, 2-1)$, for $h = 1/(n+1)$, $\phi_i(\mathbf{u}) = 2K\sinh(u_i)$ for $i = 1, \ldots, n$, where λ and K are two positive constants and $\mathbf{b} \in \mathbb{R}^n$ is a given vector. Problem (7.57) arises in the numerical simulation of semiconductor devices in microelectronics, where $\mathbf{u}$ and $\mathbf{b}$ represent electric potential and doping profile, respectively.

In Figure 7.5 (*left*) we show schematically the particular device considered in the numerical example, a $p-n$ junction diode of unit normalized length, subject to an external bias $\Delta V = V_b - V_a$, together with the doping profile of the device, normalized to 1 (*right*). Notice that $b_i = b(x_i)$, for $i = 1, \ldots, n$, where $x_i = ih$. The mathematical model of the problem at hand comprises a nonlinear Poisson equation for the electric potential and two continuity equations of advection-diffusion type, as those addressed in Chapter 12, for the current densities. For the complete derivation of the model and its analysis see, for instance, [Mar86] and [Jer96].
Solving system (7.57) corresponds to finding the minimizer in $\mathbb{R}^n$ of the function $f : \mathbb{R}^n \to \mathbb{R}$ defined as

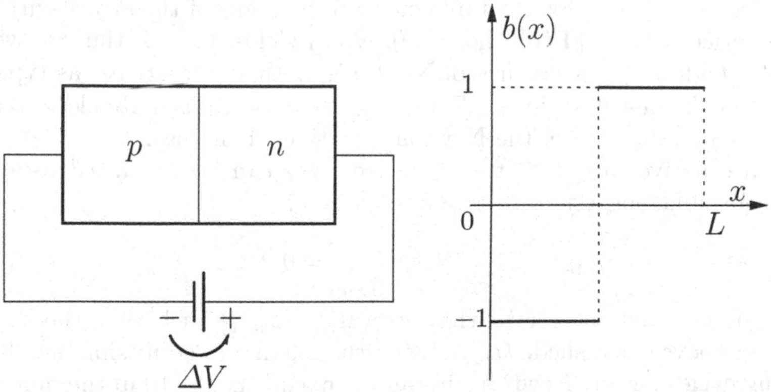

Fig. 7.5. Scheme of a semiconductor device (*left*); doping profile (*right*)

$$f(\mathbf{u}) = \frac{1}{2}\mathbf{u}^T \mathrm{A}\mathbf{u} + 2\sum_{i=1}^{n} \cosh(u_i) - \mathbf{b}^T\mathbf{u}. \tag{7.58}$$

It can be checked (see Exercise 5) that for any $\mathbf{u}, \mathbf{v} \in \mathbb{R}^n$, with $\mathbf{u} \neq \mathbf{v}$, and for any $\lambda \in (0, 1)$

$$\lambda f(\mathbf{u}) + (1-\lambda) f(\mathbf{v}) - f(\lambda\mathbf{u} + (1-\lambda)\mathbf{v}) > (1/2)\lambda(1-\lambda)\|\mathbf{u}-\mathbf{v}\|_{\mathrm{A}}^2,$$

where $\|\cdot\|_{\mathrm{A}}$ denotes the energy norm introduced in (1.28). This implies that $f(\mathbf{u})$ is an uniformly convex function in $\mathbb{R}^n$, that is, it strictly satisfies (7.49) with $\rho = 1/2$.

Property 7.10 ensures, in turn, that the function in (7.58) admits a unique minimizer $\mathbf{u}^* \in \mathbb{R}^n$ and it can be shown (see Theorem 14.4.3, p. 503 [OR70]) that there exists a sequence $\{\alpha_k\}$ such that the iterates of the damped Newton method introduced in Section 7.2.6 converge to $\mathbf{u}^* \in \mathbb{R}^n$ (at least) superlinearly.

Thus, using the damped Newton method for solving system (7.57) leads to the following sequence of linearized problems: given $\mathbf{u}^{(0)} \in \mathbb{R}^n$, for $k = 0, 1, \ldots$ until convergence solve

$$\left[\mathrm{A} + 2K\, \mathrm{diag}_n(\cosh(u_i^{(k)}))\right] \boldsymbol{\delta}\mathbf{u}^{(k)} = \mathbf{b} - \left(\mathrm{A}\mathbf{u}^{(k)} + \boldsymbol{\phi}(\mathbf{u}^{(k)})\right), \tag{7.59}$$

then set $\mathbf{u}^{(k+1)} = \mathbf{u}^{(k)} + \alpha_k \boldsymbol{\delta}\mathbf{u}^{(k)}$.

Let us now address two possible choices of the acceleration parameters α_k. The first one has been proposed in [BR81] and is

$$\alpha_k = \frac{1}{1 + \rho_k \|\mathbf{F}(\mathbf{u}^{(k)})\|_\infty}, \qquad k = 0, 1, \ldots, \tag{7.60}$$

where the coefficients $\rho_k \geq 0$ are suitable acceleration parameters picked in such a way that the descent condition $\|\mathbf{F}(\mathbf{u}^{(k)} + \alpha_k \delta\mathbf{u}^{(k)})\|_\infty < \|\mathbf{F}(\mathbf{u}^{(k)})\|_\infty$ is satisfied (see [BR81] for the implementation details of the algorithm).

We notice that, as $\|\mathbf{F}(\mathbf{u}^{(k)})\|_\infty \to 0$, (7.60) yields $\alpha_k \to 1$, thus recovering the full (quadratic) convergence of Newton's method. Otherwise, as typically happens in the first iterations, $\|\mathbf{F}(\mathbf{u}^{(k)})\|_\infty \gg 1$ and α_k is quite close to zero, with a strong reduction of the Newton variation (damping).

As an alternative to (7.60), the sequence $\{\alpha_k\}$ can be generated using the simpler formula, suggested in [Sel84], Chapter 7

$$\alpha_k = 2^{-i(i-1)/2}, \qquad k = 0, 1, \ldots, \tag{7.61}$$

where i is the first integer in the interval $[1, It_{max}]$ such that the descent condition above is satisfied, It_{max} being the maximum admissible number of damping cycles for any Newton's iteration (fixed equal to 10 in the numerical experiments).

As a comparison, both damped and standard Newton's methods have been implemented, the former one with both choices (7.60) and (7.61) for the coefficients α_k. In the case of Newton's method, we have set in (7.59) $\alpha_k = 1$ for any $k \geq 0$.

The numerical examples have been performed with $n = 49$, $b_i = -1$ for $i \leq n/2$ and the remaining values b_i equal to 1. Moreover, we have taken $\lambda^2 = 1.67 \cdot 10^{-4}$, $K = 6.77 \cdot 10^{-6}$ and fixed the first $n/2$ components of the initial vector $\mathbf{u}^{(0)}$ equal to V_a and the remaining ones equal to V_b, where $V_a = 0$ and $V_b = 10$.

The tolerance on the maximum change between two successive iterates, which monitors the convergence of damped Newton's method (7.59), has been set equal to 10^{-4}.
Figure 7.6 (*left*) shows the log-scale absolute error for the three algorithms as functions of the iteration number. Notice the rapid convergence of the damped Newton's method (8 and 10 iterations in the case of (7.60) and (7.61), respectively), compared with the extremely slow convergence of the standard Newton's method (192 iterations). Moreover, it is interesting to analyze in Figure 7.6 (*right*) the plot of the sequences of parameters α_k as functions of the iteration number.

The starred and the circled curves refer to the choices (7.60) and (7.61) for the coefficients α_k, respectively. As previously observed, the α_k's start from very small values, to converge quickly to 1 as the damped Newton method (7.59) enters the attraction region of the minimizer $\mathbf{x}^*$.

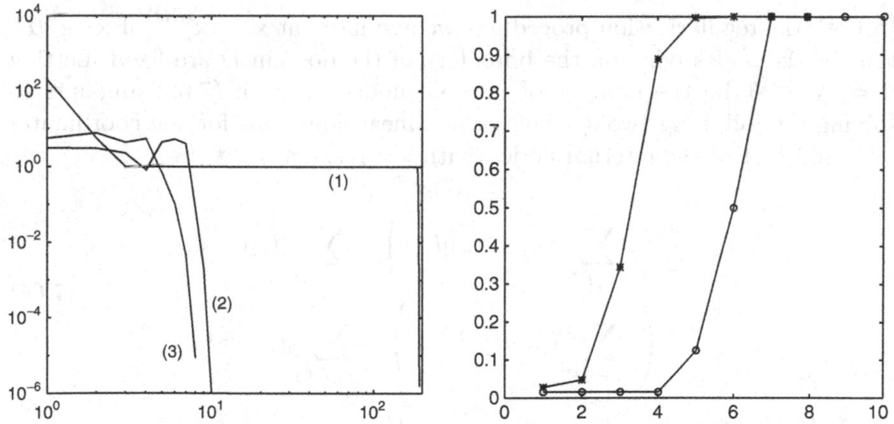

Fig. 7.6. Absolute error (*left*) and damping parameters α_k (*right*). The error curve for standard Newton's method is denoted by (1), while (2) and (3) refer to damped Newton's method with the choices (7.61) and (7.60) for the coefficients α_k, respectively

7.4.2 Nonlinear Regularization of a Discretization Grid

In this section we go back to the problem of regularizing a discretization grid that has been introduced in Section 3.14.2. There, we considered the technique of barycentric regularization, which leads to solving a linear system, typically of large size and featuring a sparse coefficient matrix.

In this section we address two alternative techniques, denoted as regularization *by edges* and *by areas*. The main difference with respect to the method described in Section 3.14.2 lies in the fact that these new approaches lead to systems of *nonlinear* equations.

Using the notation of Section 3.14.2, for each pair of nodes $\mathbf{x}_j, \mathbf{x}_k \in \mathcal{Z}_i$, denote by l_{jk} the edge on the boundary $\partial\mathcal{P}_i$ of $\mathcal{P}_i$ which connects them and by $\mathbf{x}_{jk}$ the midpoint of l_{jk}, while for each triangle $T \in \mathcal{P}_i$ we denote by $\mathbf{x}_{b,T}$ the centroid of T. Moreover, let $n_i = \dim(\mathcal{Z}_i)$ and denote for any geometric entity (side or triangle) by $|\cdot|$ its measure in $\mathbb{R}^1$ or $\mathbb{R}^2$.

In the case of regularization by edges, we let

$$\mathbf{x}_i = \left(\sum_{l_{jk}\in\partial\mathcal{P}_i} \mathbf{x}_{jk}|l_{jk}| \right) / |\partial\mathcal{P}_i|, \qquad \forall \mathbf{x}_i \in \mathcal{N}_h, \tag{7.62}$$

while in the case of regularization by areas, we let

$$\mathbf{x}_i = \left(\sum_{T\in\mathcal{P}_i} \mathbf{x}_{b,T}|T| \right) / |\mathcal{P}_i|, \qquad \forall \mathbf{x}_i \in \mathcal{N}_h. \tag{7.63}$$

In both the regularization procedures we assume that $\mathbf{x}_i = \mathbf{x}_i^{(\partial D)}$ if $\mathbf{x}_i \in \partial D$, that is, the nodes lying on the boundary of the domain D are fixed. Letting $n = N - N_b$ be the number of internal nodes, relation (7.62) amounts to solving the following two systems of nonlinear equations for the coordinates $\{x_i\}$ and $\{y_i\}$ of the internal nodes, with $i = 1, \ldots, n$

$$\begin{aligned} x_i - \frac{1}{2}\left(\sum_{l_{jk}\in\partial\mathcal{P}_i} (x_j + x_k)|l_{jk}| \right) / \sum_{l_{jk}\in\partial\mathcal{P}_i} |l_{jk}| = 0, \\ y_i - \frac{1}{2}\left(\sum_{l_{jk}\in\partial\mathcal{P}_i} (y_j + y_k)|l_{jk}| \right) / \sum_{l_{jk}\in\partial\mathcal{P}_i} |l_{jk}| = 0. \end{aligned} \tag{7.64}$$

Similarly, (7.63) leads to the following nonlinear systems, for $i = 1, \ldots, n$

$$\begin{aligned} x_i - \frac{1}{3}\left(\sum_{T\in\mathcal{P}_i} (x_{1,T} + x_{2,T} + x_{3,T})|T| \right) / \sum_{T\in\mathcal{P}_i} |T| = 0, \\ y_i - \frac{1}{3}\left(\sum_{T\in\mathcal{P}_i} (y_{1,T} + y_{2,T} + y_{3,T})|T| \right) / \sum_{T\in\mathcal{P}_i} |T| = 0, \end{aligned} \tag{7.65}$$

where $\mathbf{x}_{s,T} = [x_{s,T}, y_{s,T}]^T$, for $s = 1, 2, 3$, are the coordinates of the vertices of each triangle $T \in \mathcal{P}_i$. Notice that the nonlinearity of systems (7.64) and (7.65) is due to the presence of terms $|l_{jk}|$ and $|T|$.

Both systems (7.64) and (7.65) can be cast in the form (7.1), denoting, as usual, by f_i the i-th nonlinear equation of the system, for $i = 1, \ldots, n$. The complex functional dependence of f_i on the unknowns makes it prohibitive to use Newton's method (7.4), which would require the explicit computation of the Jacobian matrix $\mathrm{J}_{\mathbf{F}}$.

A convenient alternative is provided by the *nonlinear Gauss-Seidel method* (see [OR70], Chapter 7), which generalizes the corresponding method proposed in Chapter 4 for linear systems and can be formulated as follows.

Denote by z_i, for $i = 1, \ldots, n$, either of the unknown x_i or y_i. Given the initial vector $\mathbf{z}^{(0)} = [z_1^{(0)}, \ldots, z_n^{(0)}]^T$, for $k = 0, 1, \ldots$ until convergence, solve

$$f_i(z_1^{(k+1)}, \ldots, z_{i-1}^{(k+1)}, \xi, z_{i+1}^{(k)}, \ldots, z_n^{(k)}) = 0, \qquad i = 1, \ldots, n, \tag{7.66}$$

then, set $z_i^{(k+1)} = \xi$. Thus, the nonlinear Gauss-Seidel method converts problem (7.1) into the successive solution of n scalar nonlinear equations. In the case of system (7.64), each of these equations is *linear* in the unknown $z_i^{(k+1)}$ (since ξ does not explicitly appear in the bracketed term at the right side of (7.64)). This allows for its exact solution in one step.

In the case of system (7.65), the equation (7.66) is genuinely nonlinear with respect to ξ, and is solved taking one step of a fixed-point iteration.

The nonlinear Gauss-Seidel (7.66) has been implemented in MATLAB to solve systems (7.64) and (7.65) in the case of the initial triangulation shown in Figure 7.7 (*left*). Such a triangulation covers the external region of a two dimensional wing section of type NACA 2316. The grid contains $N_T = 534$ triangles and $n = 198$ internal nodes.

The algorithm reached convergence in 42 iterations for both kinds of regularization, having used as stopping criterion the test $\|\mathbf{z}^{(k+1)} - \mathbf{z}^{(k)}\|_\infty \leq 10^{-4}$. In Figure 7.7 (*right*) the discretization grid obtained after the regularization

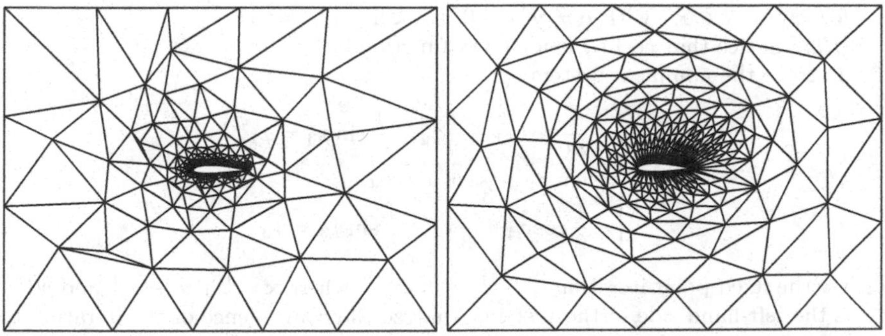

Fig. 7.7. Triangulation before (*left*) and after (*right*) the regularization

by areas is shown (a similar result has been provided by the regularization by edges). Notice the higher uniformity of the triangles with respect to those of the starting grid.

7.5 Exercises

1. Prove (7.8) for the m-step Newton-SOR method.
[*Hint*: use the SOR method for solving a linear system $\mathbf{A}\mathbf{x}=\mathbf{b}$ with A=D-E-F and express the k-th iterate as a function of the initial datum $\mathbf{x}^{(0)}$, obtaining
$$\mathbf{x}^{(k+1)} = \mathbf{x}^{(0)} + (\mathrm{M}^{k+1} - \mathrm{I})\mathbf{x}^{(0)} + (\mathrm{M}^k + \ldots + \mathrm{I})\mathrm{B}^{-1}\mathbf{b},$$
where $\mathrm{B}=\omega^{-1}(\mathrm{D}-\omega\mathrm{E})$ and $\mathrm{M} = \mathrm{B}^{-1}\omega^{-1}[(1-\omega)\mathrm{D} + \omega F]$. Since $\mathrm{B}^{-1}\mathrm{A} = \mathrm{I}-\mathrm{M}$ and
$$(\mathrm{I} + \ldots + \mathrm{M}^k)(\mathrm{I} - \mathrm{M}) = \mathrm{I} - \mathrm{M}^{k+1}$$
then (7.8) follows by suitably identifying the matrix and the right-side of the system.]
2. Prove that using the gradient method for minimizing $f(x) = x^2$ with the directions $p^{(k)} = -1$ and the parameters $\alpha_k = 2^{-k+1}$, does not yield the minimizer of f.
3. Show that for the *steepest descent* method applied to minimizing a quadratic functional f of the form (7.35) the following inequality holds
$$f(\mathbf{x}^{(k+1)}) \le \left(\frac{\lambda_{max} - \lambda_{min}}{\lambda_{max} + \lambda_{min}}\right)^2 f(\mathbf{x}^{(k)}),$$
where $\lambda_{max}, \lambda_{min}$ are the eigenvalues of maximum and minimum module, respectively, of the matrix A that appears in (7.35).
[*Hint*: proceed as done for (7.38).]
4. Check that the parameters α_k of Exercise 2 do not fulfill the conditions (7.31) and (7.32).
5. Consider the function $f : \mathbb{R}^n \to \mathbb{R}$ introduced in (7.58) and check that it is uniformly convex on $\mathbb{R}^n$, that is
$$\lambda f(\mathbf{u}) + (1-\lambda)f(\mathbf{v}) - f(\lambda\mathbf{u} + (1-\lambda)\mathbf{v}) > (1/2)\lambda(1-\lambda)\|\mathbf{u} - \mathbf{v}\|_{\mathrm{A}}^2$$
for any $\mathbf{u}$, $\mathbf{v} \in \mathbb{R}^n$ with $\mathbf{u} \neq \mathbf{v}$ and $0 < \lambda < 1$.
[*Hint*: notice that $\cosh(\cdot)$ is a convex function.]
6. To solve the nonlinear system
$$\begin{cases} -\dfrac{1}{81}\cos x_1 + \dfrac{1}{9}x_2^2 + \dfrac{1}{3}\sin x_3 = x_1, \\ \dfrac{1}{3}\sin x_1 + \dfrac{1}{3}\cos x_3 = x_2, \\ -\dfrac{1}{9}\cos x_1 + \dfrac{1}{3}x_2 + \dfrac{1}{6}\sin x_3 = x_3, \end{cases}$$
use the fixed-point iteration $\mathbf{x}^{(n+1)} = \Psi(\mathbf{x}^{(n)})$, where $\mathbf{x} = [x_1, x_2, x_3]^T$ and $\Psi(\mathbf{x})$ is the left-hand side of the system. Analyze the convergence of the iteration to compute the fixed point $\boldsymbol{\alpha} = [0, 1/3, 0]^T$.
[*Solution*: the fixed-point method is convergent since $\|\Psi(\boldsymbol{\alpha})\|_\infty = 1/2$.]

7. Using Program 50 implementing Newton's method, determine the global maximizer of the function

$$f(x) = e^{-\frac{x^2}{2}} - \frac{1}{4}\cos(2x)$$

and analyze the performance of the method (input data: `xv=1; tol=1e-6; nmax=500`). Solve the same problem using the following fixed-point iteration

$$x_{(k+1)} = g(x_k) \qquad \text{with } g(x) = \sin(2x)\left[\frac{e^{\frac{x^2}{2}}\left(x\sin(2x) + 2\cos(2x)\right) - 2}{2\left(x\sin(2x) + 2\cos(2x)\right)}\right].$$

Analyze the performance of this second scheme, both theoretically and experimentally, and compare the results obtained using the two methods.
[*Solution*: the function f has a global maximum at $x = 0$. This point is a double zero for f'. Thus, Newton's method is only linearly convergent. Conversely, the proposed fixed-point method is third-order convergent.]

8

Polynomial Interpolation

This chapter is addressed to the approximation of a function which is known through its values at a given number of points.

Precisely, given $m+1$ pairs (x_i, y_i), the problem consists of finding a function $\Phi = \Phi(x)$ such that $\Phi(x_i) = y_i$ for $i = 0, \ldots, m$, y_i being some given values, and say that Φ *interpolates* $\{y_i\}$ at the *nodes* $\{x_i\}$. We speak about *polynomial interpolation* if Φ is an algebraic polynomial, *trigonometric approximation* if Φ is a trigonometric polynomial or *piecewise polynomial interpolation* (or *spline interpolation*) if Φ is only locally a polynomial.

The numbers y_i may represent the values attained at the nodes x_i by a function f that is known in closed form, as well as experimental data. In the former case, the approximation process aims at replacing f with a simpler function to deal with, in particular in view of its numerical integration or derivation. In the latter case, the primary goal of approximation is to provide a compact representation of the available data, whose number is often quite large.

Polynomial interpolation is addressed in Sections 8.1 and 8.2, while piecewise polynomial interpolation is introduced in Sections 8.4, 8.5 and 8.6. Finally, univariate and parametric splines are addressed in Sections 8.7 and 8.8. Interpolation processes based on trigonometric or algebraic orthogonal polynomials will be considered in Chapter 10.

8.1 Polynomial Interpolation

Let us consider $n+1$ pairs (x_i, y_i). The problem is to find a polynomial $\Pi_m \in \mathbb{P}_m$, called *interpolating polynomial*, such that

$$\Pi_m(x_i) = a_m x_i^m + \ldots + a_1 x_i + a_0 = y_i, \; i = 0, \ldots, n. \tag{8.1}$$

The points x_i are called *interpolation nodes*. If $n \neq m$ the problem is over or under-determined and will be addressed in Section 10.7.1. If $n = m$, the following result holds.

Theorem 8.1 *Given $n+1$ distinct nodes $x_0, \ldots, x_n$ and $n+1$ corresponding values $y_0, \ldots, y_n$, there exists a unique polynomial $\Pi_n \in \mathbb{P}_n$ such that $\Pi_n(x_i) = y_i$ for $i = 0, \ldots, n$.*

Proof. To prove existence, let us use a constructive approach, providing an expression for Π_n. Denoting by $\{l_i\}_{i=0}^n$ a basis for $\mathbb{P}_n$, then Π_n admits a representation on such a basis of the form $\Pi_n(x) = \sum_{i=0}^n b_i l_i(x)$ with the property that

$$\Pi_n(x_i) = \sum_{j=0}^{n} b_j l_j(x_i) = y_i, \quad i = 0, \ldots, n. \tag{8.2}$$

If we define

$$l_i \in \mathbb{P}_n : l_i(x) = \prod_{\substack{j=0 \\ j \neq i}}^{n} \frac{x - x_j}{x_i - x_j}, \quad i = 0, \ldots, n, \tag{8.3}$$

then $l_i(x_j) = \delta_{ij}$ and we immediately get from (8.2) that $b_i = y_i$.
The polynomials $\{l_i,\, i = 0, \ldots, n\}$ form a basis for $\mathbb{P}_n$ (see Exercise 1). As a consequence, the interpolating polynomial exists and has the following form (called *Lagrange* form)

$$\Pi_n(x) = \sum_{i=0}^{n} y_i l_i(x). \tag{8.4}$$

To prove uniqueness, suppose that another interpolating polynomial Ψ_m of degree $m \leq n$ exists, such that $\Psi_m(x_i) = y_i$ for $i = 0, ..., n$. Then, the difference polynomial $\Pi_n - \Psi_m$ vanishes at $n+1$ distinct points x_i and thus coincides with the null polynomial. Therefore, $\Psi_m = \Pi_n$.

An alternative approach to prove existence and uniqueness of Π_n is provided in Exercise 2. ◇

It can be checked that (see Exercise 3)

$$\Pi_n(x) = \sum_{i=0}^{n} \frac{\omega_{n+1}(x)}{(x - x_i)\omega'_{n+1}(x_i)} y_i, \tag{8.5}$$

where ω_{n+1} is the *nodal polynomial* of degree $n+1$ defined as

$$\omega_{n+1}(x) = \prod_{i=0}^{n} (x - x_i). \tag{8.6}$$

Formula (8.4) is called the *Lagrange form* of the interpolating polynomial, while the polynomials $l_i(x)$ are the *characteristic polynomials.* In Figure 8.1 we show the characteristic polynomials $l_2(x)$, $l_3(x)$ and $l_4(x)$, in the case of degree $n = 6$, on the interval $[-1, 1]$ where equally spaced nodes are taken, including the end points.
Notice that $|l_i(x)|$ can be greater than 1 within the interpolation interval.

If $y_i = f(x_i)$ for $i = 0, \ldots, n$, f being a given function, the interpolating polynomial $\Pi_n(x)$ will be denoted by $\Pi_n f(x)$.

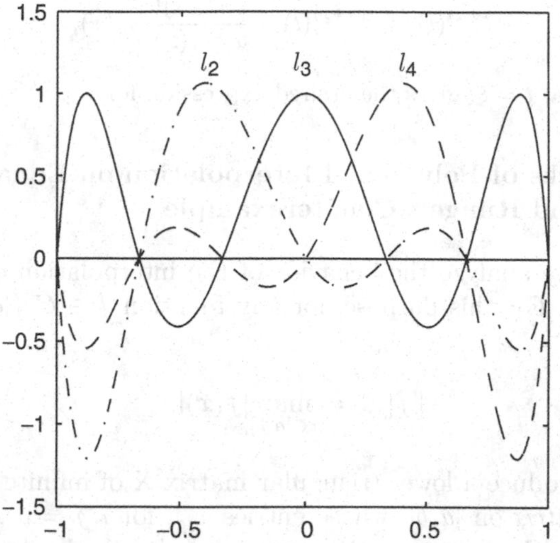

Fig. 8.1. Lagrange characteristic polynomials

8.1.1 The Interpolation Error

In this section we estimate the interpolation error that is made when replacing a given function f with its interpolating polynomial $\Pi_n f$ at the nodes $x_0, x_1, \ldots, x_n$ (for further results, we refer the reader to [Wen66], [Dav63]).

Theorem 8.2 *Let $x_0, x_1, \ldots, x_n$ be $n+1$ distinct nodes and let x be a point belonging to the domain of a given function f. Assume that $f \in C^{n+1}(I_x)$, where I_x is the smallest interval containing the nodes $x_0, x_1, \ldots, x_n$ and x. Then the interpolation error at the point x is given by*

$$E_n(x) = f(x) - \Pi_n f(x) = \frac{f^{(n+1)}(\xi)}{(n+1)!}\omega_{n+1}(x), \tag{8.7}$$

where $\xi \in I_x$ and ω_{n+1} is the nodal polynomial of degree $n+1$.

Proof. The result is obviously true if x coincides with any of the interpolation nodes. Otherwise, define, for any $t \in I_x$, the function $G(t) = E_n(t) - \omega_{n+1}(t)E_n(x)/\omega_{n+1}(x)$. Since $f \in C^{(n+1)}(I_x)$ and ω_{n+1} is a polynomial, then $G \in C^{(n+1)}(I_x)$ and it has at least $n+2$ distinct zeros in I_x, since

$$G(x_i) = E_n(x_i) - \omega_{n+1}(x_i)E_n(x)/\omega_{n+1}(x) = 0, \quad i = 0, \ldots, n,$$
$$G(x) = E_n(x) - \omega_{n+1}(x)E_n(x)/\omega_{n+1}(x) = 0.$$

Then, thanks to the mean value theorem, G' has at least $n+1$ distinct zeros and, by recursion, $G^{(j)}$ admits at least $n+2-j$ distinct zeros. As a consequence, $G^{(n+1)}$ has at least one zero, which we denote by ξ. On the other hand, since $E_n^{(n+1)}(t) = f^{(n+1)}(t)$ and $\omega_{n+1}^{(n+1)}(x) = (n+1)!$ we get

$$G^{(n+1)}(t) = f^{(n+1)}(t) - \frac{(n+1)!}{\omega_{n+1}(x)} E_n(x),$$

which, evaluated at $t = \xi$, gives the desired expression for $E_n(x)$. ◇

8.1.2 Drawbacks of Polynomial Interpolation on Equally Spaced Nodes and Runge's Counterexample

In this section we analyze the behavior of the interpolation error (8.7) as n tends to infinity. For this purpose, for any function $f \in C^0([a,b])$, define its *maximum norm*

$$\|f\|_\infty = \max_{x \in [a,b]} |f(x)|. \qquad (8.8)$$

Then, let us introduce a lower triangular matrix X of infinite size, called the *interpolation matrix* on $[a,b]$, whose entries x_{ij}, for $i,j = 0,1,\ldots$, represent points of $[a,b]$, with the assumption that on each row the entries are all distinct.

Thus, for any $n \geq 0$, the $n+1$-th row of X contains $n+1$ distinct values that we can identify as nodes, so that, for a given function f, we can uniquely define an interpolating polynomial $\Pi_n f$ of degree n at those nodes (any polynomial $\Pi_n f$ depends on X, as well as on f).

Having fixed f and an interpolation matrix X, let us define the interpolation error

$$E_{n,\infty}(\mathrm{X}) = \|f - \Pi_n f\|_\infty, \quad n = 0,1,\ldots. \qquad (8.9)$$

Next, denote by $p_n^* \in \mathbb{P}_n$ the *best approximation polynomial*, for which

$$E_n^* = \|f - p_n^*\|_\infty \leq \|f - q_n\|_\infty \qquad \forall q_n \in \mathbb{P}_n.$$

The following comparison result holds (for the proof, see [Riv74]).

Property 8.1 *Let $f \in C^0([a,b])$ and X be an interpolation matrix on $[a,b]$. Then*

$$E_{n,\infty}(\mathrm{X}) \leq E_n^* \left(1 + \Lambda_n(\mathrm{X})\right), \qquad n = 0,1,\ldots, \qquad (8.10)$$

where $\Lambda_n(\mathrm{X})$ denotes the Lebesgue constant of X, defined as

$$\Lambda_n(\mathrm{X}) = \left\| \sum_{j=0}^{n} |l_j^{(n)}| \right\|_\infty, \qquad (8.11)$$

and where $l_j^{(n)} \in \mathbb{P}_n$ is the j-th characteristic polynomial associated with the $n+1$-th row of X, that is, satisfying $l_j^{(n)}(x_{nk}) = \delta_{jk}$, $j,k = 0,1,\ldots$.

Since E_n^* does not depend on X, all the information concerning the effects of X on $E_{n,\infty}(X)$ must be looked for in $\Lambda_n(X)$. Although there exists an interpolation matrix X^* such that $\Lambda_n(X)$ is minimized, it is not in general a simple task to determine its entries explicitly. We shall see in Section 10.3, that the zeros of the Chebyshev polynomials provide on the interval $[-1, 1]$ an interpolation matrix with a very small value of the Lebesgue constant.

On the other hand, for any possible choice of X, there exists a constant $C > 0$ such that (see [Erd61])

$$\Lambda_n(X) > \frac{2}{\pi}\log(n+1) - C, \qquad n = 0, 1, \ldots .$$

This property shows that $\Lambda_n(X) \to \infty$ as $n \to \infty$. This fact has important consequences: in particular, it can be proved (see [Fab14]) that, given an interpolation matrix X on an interval $[a, b]$, there always exists a continuous function f in $[a, b]$, such that $\Pi_n f$ does not converge uniformly (that is, in the maximum norm) to f. Thus, polynomial interpolation does not allow for approximating *any* continuous function, as demonstrated by the following example.

Example 8.1 (Runge's counterexample) Suppose we approximate the following function

$$f(x) = \frac{1}{1+x^2}, \qquad -5 \le x \le 5, \tag{8.12}$$

using Lagrange interpolation on equally spaced nodes. It can be checked that some points x exist within the interpolation interval such that

$$\lim_{n\to\infty} |f(x) - \Pi_n f(x)| \neq 0.$$

In particular, Lagrange interpolation diverges for $|x| > 3.63\ldots$. This phenomenon is particularly evident in the neighborhood of the end points of the interpolation interval, as shown in Figure 8.2, and is due to the choice of equally spaced nodes. We shall in Chapter 10 that resorting to suitably chosen nodes will allow for uniform convergence of the interpolating polynomial to the function f to hold. •

8.1.3 Stability of Polynomial Interpolation

Let us consider a set of function values $\left\{\widetilde{f}(x_i)\right\}$ which is a perturbation of the data $f(x_i)$ relative to the nodes x_i, with $i = 0, \ldots, n$, in an interval $[a, b]$. The perturbation may be due, for instance, to the effect of rounding errors, or may be caused by an error in the experimental measure of the data.

Denoting by $\Pi_n \widetilde{f}$ the interpolating polynomial on the set of values $\widetilde{f}(x_i)$, we have

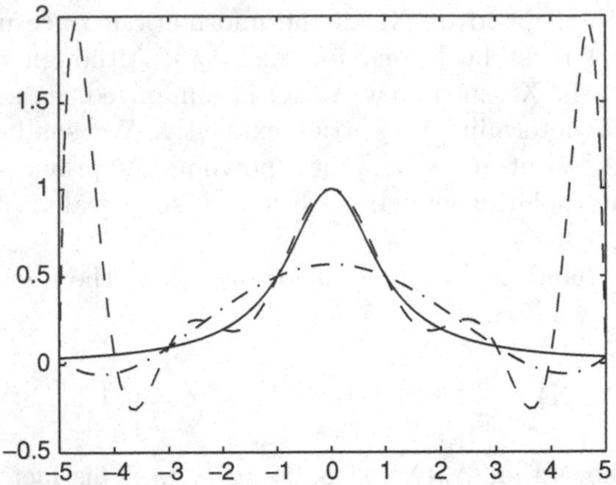

Fig. 8.2. Lagrange interpolation on equally spaced nodes for the function $f(x) = 1/(1+x^2)$: the interpolating polynomials $\Pi_5 f$ and $\Pi_{10} f$ are shown in dotted and dashed line, respectively

$$\begin{aligned}\|\Pi_n f - \Pi_n \widetilde{f}\|_\infty &= \max_{a \le x \le b} \left| \sum_{j=0}^{n} (f(x_j) - \widetilde{f}(x_j)) l_j(x) \right| \\ &\le \Lambda_n(\mathrm{X}) \max_{i=0,\ldots,n} |f(x_i) - \widetilde{f}(x_i)|.\end{aligned}$$

As a consequence, small changes on the data give rise to small changes on the interpolating polynomial only if the Lebesgue constant is small. This constant plays the role of the *condition number* for the interpolation problem.

As previously noticed, Λ_n grows as $n \to \infty$ and in particular, in the case of Lagrange interpolation on equally spaced nodes, it can be proved that (see [Nat65])

$$\Lambda_n(\mathrm{X}) \simeq \frac{2^{n+1}}{e n \log n},$$

where $e \simeq 2.7183$ is the naeperian number. This shows that, for n large, this form of interpolation can become unstable. Notice also that so far we have completely neglected the errors generated by the interpolation process in constructing $\Pi_n f$. However, it can be shown that the effect of such errors is generally negligible (see [Atk89]).

Example 8.2 On the interval $[-1, 1]$ let us interpolate the function $f(x) = \sin(2\pi x)$ at 22 equally spaced nodes x_i. Next, we generate a perturbed set of values $\widetilde{f}(x_i)$ of the function evaluations $f(x_i) = \sin(2\pi x_i)$ with $\max_{i=0,\ldots,21} |f(x_i) - \widetilde{f}(x_i)| \simeq 9.5 \cdot 10^{-4}$. In Figure 8.3 we compare the polynomials $\Pi_{21} f$ and $\Pi_{21} \widetilde{f}$: notice how the difference between the two interpolating polynomials, around the end points of the interpolation interval, is much larger than the impressed perturbation (actually, $\|\Pi_{21} f - \Pi_{21} \widetilde{f}\|_\infty \simeq 1.5926$ and $\Lambda_{21} \simeq 24000$). •

8.2 Newton Form of the Interpolating Polynomial

The Lagrange form (8.4) of the interpolating polynomial is not the most convenient from a practical standpoint. In this section we introduce an alternative form characterized by a cheaper computational cost. Our goal is the following:

given $n+1$ pairs $\{x_i, y_i\}$, $i = 0, \ldots, n$, we want to represent Π_n (with $\Pi_n(x_i) = y_i$ for $i = 0, \ldots, n$) as the sum of Π_{n-1} (with $\Pi_{n-1}(x_i) = y_i$ for $i = 0, \ldots, n-1$) and a polynomial of degree n which depends on the nodes x_i and on only one unknown coefficient. We thus set

$$\Pi_n(x) = \Pi_{n-1}(x) + q_n(x), \tag{8.13}$$

where $q_n \in \mathbb{P}_n$. Since $q_n(x_i) = \Pi_n(x_i) - \Pi_{n-1}(x_i) = 0$ for $i = 0, \ldots, n-1$, it must necessarily be that

$$q_n(x) = a_n(x - x_0) \cdots (x - x_{n-1}) = a_n \omega_n(x).$$

To determine the unknown coefficient a_n, suppose that $y_i = f(x_i)$, $i = 0, \ldots, n$, where f is a suitable function, not necessarily known in explicit form. Since $\Pi_n f(x_n) = f(x_n)$, from (8.13) it follows that

$$a_n = \frac{f(x_n) - \Pi_{n-1} f(x_n)}{\omega_n(x_n)}. \tag{8.14}$$

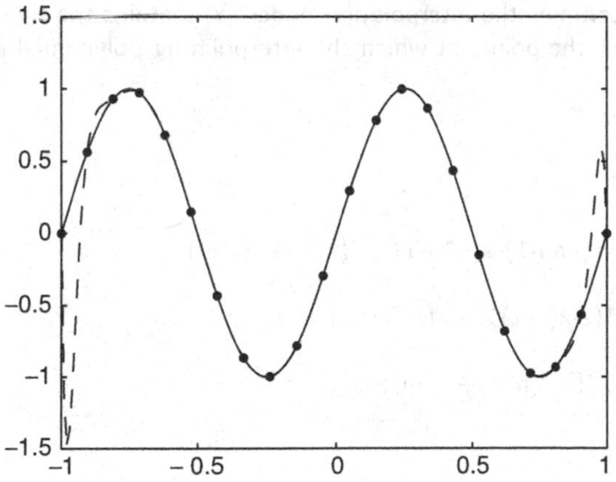

Fig. 8.3. Instability of Lagrange interpolation. In solid line $\Pi_{21} f$, on unperturbed data, in dashed line $\Pi_{21} \widetilde{f}$, on perturbed data, for Example 8.2

The coefficient a_n is called the n-th *Newton divided difference* and is generally denoted by

$$a_n = f[x_0, x_1, \ldots, x_n] \tag{8.15}$$

for $n \geq 1$. As a consequence, (8.13) becomes

$$\Pi_n f(x) = \Pi_{n-1} f(x) + \omega_n(x) f[x_0, x_1, \ldots, x_n]. \tag{8.16}$$

If we let $y_0 = f(x_0) = f[x_0]$ and $\omega_0 = 1$, by recursion on n we can obtain from (8.16) the following formula

$$\Pi_n f(x) = \sum_{k=0}^{n} \omega_k(x) f[x_0, \ldots, x_k]. \tag{8.17}$$

Uniqueness of the interpolating polynomial ensures that the above expression yields the same interpolating polynomial generated by the Lagrange form. Form (8.17) is commonly known as the *Newton divided difference formula* for the interpolating polynomial.
Program 65 provides an implementation of Newton's formula. The input vectors **x** and **y** contain the interpolation nodes and the corresponding functional evaluations of f, respectively, while vector **z** contains the abscissae where the polynomial $\Pi_n f$ is to be evaluated. This polynomial is stored in the output vector `f`.

Program 65 - interpol : Lagrange polynomial using Newton's formula

```
function [f]=interpol(x,y,z)
%INTERPOL Lagrange polynomial interpolation
% [F] = INTERPOL(X, Y, Z) computes the Lagrange interpolating polynomial of
% a function. X contains the interpolation nodes. Y contains the function values
% at X. Z contains the points at which the interpolating polynomial F must be
% evaluated.
[m n] = size(y);
for j = 1:m
  a (:,1) = y (j,:)';
  for i = 2:n
     a (i:n,i) = ( a(i:n,i-1)-a(i-1,i-1) )./(x(i:n)-x(i-1))';
  end
  f(j,:) = a(n,n).*(z-x(n-1)) + a(n-1,n-1);
  for i = 2:n-1
     f(j,:) = f(j,:).*(z-x(n-i))+a(n-i,n-i);
  end
end
return
```

8.2.1 Some Properties of Newton Divided Differences

The n-th divided difference $f[x_0, \ldots, x_n] = a_n$ can be further characterized by noticing that it is the coefficient of x^n in $\Pi_n f$. Isolating such a coefficient from (8.5) and equating it with the corresponding coefficient in the Newton formula (8.17), we end up with the following explicit representation

$$f[x_0, \ldots, x_n] = \sum_{i=0}^{n} \frac{f(x_i)}{\omega'_{n+1}(x_i)}. \qquad (8.18)$$

This formula has remarkable consequences:

1. the value attained by the divided difference is invariant with respect to permutations of the indexes of the nodes. This instance can be profitably employed when stability problems suggest exchanging the indexes (for example, if x is the point where the polynomial must be computed, it is convenient to introduce a permutation of the indexes such that $|x - x_k| \leq |x - x_{k-1}|$ with $k = 1, \ldots, n$);
2. if $f = \alpha g + \beta h$ for some $\alpha, \beta \in \mathbb{R}$, then
$$f[x_0, \ldots, x_n] = \alpha g[x_0, \ldots, x_n] + \beta h[x_0, \ldots, x_n];$$
3. if $f = gh$, the following formula (called the Leibniz formula) holds (see [Die93])
$$f[x_0, \ldots, x_n] = \sum_{j=0}^{n} g[x_0, \ldots, x_j] h[x_j, \ldots, x_n];$$
4. an algebraic manipulation of (8.18) (see Exercise 7) yields the following *recursive formula* for computing divided differences
$$f[x_0, \ldots, x_n] = \frac{f[x_1, \ldots, x_n] - f[x_0, \ldots, x_{n-1}]}{x_n - x_0}, \quad n \geq 1. \qquad (8.19)$$

Program 66 implements the recursive formula (8.19). The evaluations of f at the interpolation nodes **x** are stored in vector **y**, while the output matrix **d** (lower triangular) contains the divided differences, which are stored in the following form

$$\begin{array}{c|lllll}
x_0 & f[x_0] & & & & \\
x_1 & f[x_1] & f[x_0, x_1] & & & \\
x_2 & f[x_2] & f[x_1, x_2] & f[x_0, x_1, x_2] & & \\
\vdots & \vdots & & \vdots & \ddots & \\
x_n & f[x_n] & f[x_{n-1}, x_n] & f[x_{n-2}, x_{n-1}, x_n] & \ldots & f[x_0, \ldots, x_n]
\end{array}$$

The coefficients involved in the Newton formula are the diagonal entries of the matrix.

Program 66 - dividif : Newton divided differences

```
function [d]=dividif(x,y)
%DIVIDIF Newton divided differences
% [D] = DIVIDIF(X, Y) computes the divided difference of order n. X contains the
% interpolation nodes. Y contains the function values at X. D contains the divided
% difference of order n.
[n,m]=size(y);
if n == 1, n = m; end
n = n-1;
d = zeros (n+1,n+1);
d(:,1) = y';
for j = 2:n+1
   for i = j:n+1
      d (i,j) = ( d (i-1,j-1)-d (i,j-1))/(x (i-j+1)-x (i));
   end
end
return
```

Using (8.19), $n(n+1)$ sums and $n(n+1)/2$ divisions are needed to generate the whole matrix. If a new evaluation of f were available at a new node x_{n+1}, only the calculation of a new row of the matrix would be required ($f[x_n, x_{n+1}]$, ..., $f[x_0, x_1, \ldots, x_{n+1}]$). Thus, in order to construct $\Pi_{n+1}f$ from $\Pi_n f$, it suffices to add to $\Pi_n f$ the term $a_{n+1}\omega_{n+1}(x)$, with a computational cost of $(n+1)$ divisions and $2(n+1)$ sums. For the sake of notational simplicity, we write below $D^r f_i = f[x_i, x_{i+1}, \ldots, x_{i+r}]$.

Example 8.3 In Table 8.1 we show the divided differences on the interval (0,2) for the function $f(x) = 1 + \sin(3x)$. The values of f and the corresponding divided differences have been computed using 16 significant figures, although only at most 5 figures are reported. If the value of f were available at node $x = 0.2$, updating the divided difference table would require only to computing the entries denoted by italics in Table 8.1. •

Notice that $f[x_0, \ldots, x_n] = 0$ for any $f \in \mathbb{P}_{n-1}$. This property, however, is not always verified numerically, since the computation of divided differences might be highly affected by rounding errors.

Example 8.4 Consider again the divided differences for the function $f(x) = 1+\sin(3x)$ on the interval $(0, 0.0002)$. The function behaves like $1+3x$ in a sufficiently small neighbourhood of 0, so that we expect to find smaller numbers as the order of divided differences increases. However, the results obtained running Program 66, and shown in Table 8.2 in exponential notation up to at most 4 significant figures (although 16 digits have been employed in the calculations), exhibit a substantially different pattern. The small rounding errors introduced in the computation of divided differences of low order have dramatically propagated on the higher order divided differences. •

Table 8.1. Divided differences for the function $f(x) = 1 + \sin(3x)$ in the case in which the evaluation of f at $x = 0.2$ is also available. The newly computed values are denoted by italics

x_i	$f(x_i)$	$f[x_i, x_{i-1}]$	$D^2 f_i$	$D^3 f_i$	$D^4 f_i$	$D^5 f_i$	$D^6 f_i$
0	1.0000						
0.2	*1.5646*	*2.82*					
0.4	1.9320	*1.83*	*−2.46*				
0.8	1.6755	−0.64	*−4.13*	*−2.08*			
1.2	0.5575	−2.79	−2.69	*1.43*	*2.93*		
1.6	0.0038	−1.38	1.76	3.71	*1.62*	*−0.81*	
2.0	0.7206	1.79	3.97	1.83	−1.17	*−1.55*	*−0.36*

Table 8.2. Divided differences for the function $f(x) = 1 + \sin(3x)$ on the interval (0,0.0002). Notice the completely wrong value in the last column (it should be approximately equal to 0), due to the propagation of rounding errors throughout the algorithm

x_i	$f(x_i)$	$f[x_i, x_{i-1}]$	$D^2 f_i$	$D^3 f_i$	$D^4 f_i$	$D^5 f_i$
0	1.0000					
4.0e-5	1.0001	3.000				
8.0e-5	1.0002	3.000	−5.39e-4			
1.2e-4	1.0004	3.000	−1.08e-3	−4.50		
1.6e-4	1.0005	3.000	−1.62e-3	−4.49	1.80e+1	
2.0e-4	1.0006	3.000	−2.15e-3	−4.49	−7.23	−1.2e + 5

8.2.2 The Interpolation Error Using Divided Differences

Consider the nodes $x_0, \ldots, x_n$ and let $\Pi_n f$ be the interpolating polynomial of f on such nodes. Now let x be a node distinct from the previous ones; letting $x_{n+1} = x$, we denote by $\Pi_{n+1} f$ the interpolating polynomial of f at the nodes x_k, $k = 0, \ldots, n+1$. Using the Newton divided differences formula, we get

$$\Pi_{n+1} f(t) = \Pi_n f(t) + (t - x_0) \cdots (t - x_n) f[x_0, \ldots, x_n, t].$$

Since $\Pi_{n+1} f(x) = f(x)$, we obtain the following formula for the interpolation error at $t = x$

$$\begin{aligned} E_n(x) &= f(x) - \Pi_n f(x) = \Pi_{n+1} f(x) - \Pi_n f(x) \\ &= (x - x_0) \cdots (x - x_n) f[x_0, \ldots, x_n, x] \qquad (8.20) \\ &= \omega_{n+1}(x) f[x_0, \ldots, x_n, x]. \end{aligned}$$

Assuming $f \in C^{(n+1)}(I_x)$ and comparing (8.20) with (8.7), yields

$$f[x_0, \ldots, x_n, x] = \frac{f^{(n+1)}(\xi)}{(n+1)!} \qquad (8.21)$$

for a suitable $\xi \in I_x$. Since (8.21) resembles the remainder of the Taylor series expansion of f, the Newton formula (8.17) for the interpolating polynomial is often regarded as being a truncated expansion around x_0 provided that $|x_n - x_0|$ is not too big.

8.3 Barycentric Lagrange Interpolation

The main drawbacks of the Lagrange form (8.4) of the interpolation can be summarized as follows:

1. each evaluation of Π_n requires $\mathcal{O}(n^2)$ additions and multiplications;
2. adding a new data pair (x_{n+1}, y_{n+1}) requires a new computation from scratch (a drawback that is overcome by the Newton form);
3. the computations can be numerically unstable.

A representation of Π_n alternative to (8.4) has been advocated in ([Rut90]) and, more recently, investigated in ([BT04]) and can be obtained as follows.

The first point of this approach is to rewrite the Lagrange formula in such a way that it can be evaluated and updated in $\mathcal{O}(n)$ operations, just like its Newton counterpart. To this end, it suffices to note that the generic l_j in (8.4) can be written as

$$l_j(x) = \omega_{n+1}(x)\frac{w_j}{x - x_j},$$

where ω_{n+1} is the nodal polynomial (8.6). The coefficients

$$w_j = \frac{1}{\prod\limits_{k \neq j}(x_j - x_k)} \tag{8.22}$$

are called *barycentric weights*. Then (8.4) can be rewritten as

$$\Pi_n(x) = \omega_{n+1}(x)\sum_{j=0}^{n}\frac{w_j}{x - x_j}y_j. \tag{8.23}$$

This is called *first form of the barycentric interpolation formula*. The computation of the $n+1$ coefficients requires $\mathcal{O}(n^2)$ operations (off-line); then for every x only $\mathcal{O}(n)$ operations are necessary for evaluating Π_n.

If a new pair (x_{n+1}, y_{n+1}) is added, the following $\mathcal{O}(n)$ operations are required:

1. for $j = 0, \ldots, n$ divide each w_j by $x_j - x_{n+1}$ for a cost of $n+1$ operations;
2. computing w_{n+1} with formula (8.22) requires $n+1$ further operations.

Formula (8.23) can be modified in a way that is often used in practice. Suppose we interpolate the constant values $y_i = 1$ for $i = 0, \ldots, n$. From (8.23), we get for all x, the following expression

$$1 = \sum_{j=0}^{n} l_j(x) = \omega_{n+1}(x) \sum_{j=0}^{n} \frac{w_j}{x - x_j}.$$

Dividing the right-hand side of (8.23) by this expression and cancelling the common factor $\omega_{n+1}(x)$, we obtain the *second form of the barycentric interpolation formula*barycentric!interpolation formula for Π_n:

$$\Pi_n(x) = \frac{\displaystyle\sum_{j=0}^{n} \frac{w_j}{x - x_j} y_j}{\displaystyle\sum_{j=0}^{n} \frac{w_j}{x - x_j}}. \tag{8.24}$$

Formula (8.24) is a Lagrange formula with a special symmetry: the weights w_j appear in the denominator exactly as in the numerator, except without the data factors y_i. Then any possible common factor in all the weights w_j may be cancelled without affecting the value of $\Pi_n(x)$. Like (8.23), (8.24) can also take advantage of the updating of the weights w_j in $\mathcal{O}(n)$ operations to incorporate a new data pair (x_{n+1}, y_{n+1}). Another advantage is that the barycentric interpolation formula is more stable than the Newton formula if special care is introduced in the computation of weigths in order to avoid division by zero in the expression of weights w_j. Note that only the case when $x = x_k$ for some k will require a special treatment. In fact, when $x \simeq x_j$ the quantity $w_j/(x - x_j)$ will be very large and we would expect a risk of inaccuracy in this number associated with the substraction of two nearby quantities in the denominator. However, as pointed out in [Hen79], this is not a problem since the same inaccurate number appears in both numerator and denominator of (8.24) and these inaccuracies cancel out (see Section 10.3).

For special sets of nodes $\{x_j\}$, one can give explicit formulas for the barycentric weights w_j. In particular, for *equidistant nodes* with spacing $h = 2/n$ on the interval $[-1, 1]$,

$$w_j = (-1)^{n-j} \binom{n}{j} / (h^n n!) = (-1)^j \binom{n}{j}. \tag{8.25}$$

Note that for a generic interval $[a, b]$, the expression in (8.25) should be multiplied by $2^n (b - a)^{-n}$.

An application of the barycentric formula is connected with the estimation of the Lebesgue's constant $\Lambda_n(\mathrm{X})$ (see, Section 8.1.2). It is possible to prove that, for any interpolation matrix X, using the computed weights (8.22), the following lower-bound holds (see, [BM97])

$$\Lambda_n(\mathrm{X}) \geq \frac{1}{2n^2} \frac{\max\limits_{j=0,\ldots,n} |w_j|}{\min\limits_{j=0,\ldots,n} |w_j|}.$$

Then we conclude that if the barycentric weights vary widely (as in (8.25) the interpolation problem must be ill-conditioned. Other applications are for rational interpolation and differentiation of polynomial interpolants (see, for details, [BT04]).

8.4 Piecewise Lagrange Interpolation

In Section 8.1.2 we have outlined the fact that, for equally spaced interpolating nodes, uniform convergence of $\Pi_n f$ to f is not guaranteed as $n \to \infty$. On the other hand, using equally spaced nodes is clearly computationally convenient and, moreover, Lagrange interpolation of low degree is sufficiently accurate, provided sufficiently small interpolation intervals are considered.

Therefore, it is natural to introduce a partition $\mathcal{T}_h$ of $[a, b]$ into K subintervals $I_j = [x_j, x_{j+1}]$ of length h_j, with $h = \max_{0\leq j\leq K-1} h_j$, such that $[a, b] = \cup_{j=0}^{K-1} I_j$ and then to employ Lagrange interpolation on each I_j using $k+1$ equally spaced nodes $\left\{x_j^{(i)},\ 0 \leq i \leq k\right\}$ with a small k.

For $k \geq 1$, we introduce on $\mathcal{T}_h$ the piecewise polynomial space

$$X_h^k = \left\{v \in C^0([a, b]) : v|_{I_j} \in \mathbb{P}_k(I_j)\, \forall I_j \in \mathcal{T}_h\right\} \tag{8.26}$$

which is the space of the continuous functions over $[a, b]$ whose restrictions on each I_j are polynomials of degree $\leq k$. Then, for any continuous function f in $[a, b]$, the *piecewise interpolation polynomial* $\Pi_h^k f$ coincides on each I_j with the interpolating polynomial of $f_{|I_j}$ at the $k+1$ nodes $\left\{x_j^{(i)},\ 0 \leq i \leq k\right\}$. As a consequence, if $f \in C^{k+1}([a, b])$, using (8.7) within each interval we obtain the following error estimate

$$\|f - \Pi_h^k f\|_\infty \leq C h^{k+1}\, \|f^{(k+1)}\|_\infty. \tag{8.27}$$

Note that a small interpolation error can be obtained even for low k provided that h is sufficiently "small".

Example 8.5 Let us go back to the function of Runge's counterexample. Now, piecewise polynomials of degree $k = 1$ and $k = 2$ are employed. We check experimentally for the behavior of the error as h decreases. In Table 8.3 we show the absolute errors measured in the maximum norm over the interval $[-5, 5]$ and the corresponding estimates of the convergence order p with respect to h. Except when using an excessively small number of subintervals, the results confirm the theoretical estimate (8.27), that is $p = k + 1$. •

Table 8.3. Interpolation error for Lagrange piecewise interpolation of degree $k = 1$ and $k = 2$, in the case of Runge's function (8.12); p denotes the trend of the exponent of h. Notice that, as $h \to 0$, $p \to k + 1$, as predicted by (8.27)

h	$\|f - \Pi_h^1\|_\infty$	p	$\|f - \Pi_h^2\|_\infty$	p
5	0.4153		0.0835	
2.5	0.1787	1.216	0.0971	−0.217
1.25	0.0631	1.501	0.0477	1.024
0.625	0.0535	0.237	0.0082	2.537
0.3125	0.0206	1.374	0.0010	3.038
0.15625	0.0058	1.819	1.3828e-04	2.856
0.078125	0.0015	1.954	1.7715e-05	2.964

Besides estimate (8.27), convergence results in integral norms exist (see [QV94], [EEHJ96]). For this purpose, we introduce the following space

$$\mathrm{L}^2(a, b) = \{f : (a, b) \to \mathbb{R}, \int_a^b |f(x)|^2 dx < +\infty\}, \tag{8.28}$$

with

$$\|f\|_{\mathrm{L}^2(a,b)} = \left(\int_a^b |f(x)|^2 dx \right)^{1/2}. \tag{8.29}$$

Formula (8.29) defines a norm for $\mathrm{L}^2(a, b)$. (We recall that norms and seminorms of functions can be defined in a manner similar to what was done in Definition 1.17 in the case of vectors). We warn the reader that the integral of the function $|f|^2$ in (8.28) has to be intended in the Lebesgue sense (see, e.g., [Rud83]). In particular, f needs not be continuous everywhere.

Theorem 8.3 *Let $0 \leq m \leq k+1$, with $k \geq 1$ and assume that $f^{(m)} \in \mathrm{L}^2(a, b)$ for $0 \leq m \leq k + 1$; then there exists a positive constant C, independent of h, such that*

$$\|(f - \Pi_h^k f)^{(m)}\|_{\mathrm{L}^2(a,b)} \leq C h^{k+1-m} \|f^{(k+1)}\|_{\mathrm{L}^2(a,b)}. \tag{8.30}$$

In particular, for $k = 1$, and $m = 0$ or $m = 1$, we obtain

$$\begin{aligned} \|f - \Pi_h^1 f\|_{\mathrm{L}^2(a,b)} &\leq C_1 h^2 \|f''\|_{\mathrm{L}^2(a,b)}, \\ \|(f - \Pi_h^1 f)'\|_{\mathrm{L}^2(a,b)} &\leq C_2 h \|f''\|_{\mathrm{L}^2(a,b)}, \end{aligned} \tag{8.31}$$

for two suitable positive constants C_1 and C_2.

Proof. We only prove (8.31) and refer to [QV94], Chapter 3 for the proof of (8.30) in the general case.

Define $e = f - \Pi_h^1 f$. Since $e(x_j) = 0$ for all $j = 0, \ldots, K$, Rolle's theorem infers the existence of $\xi_j \in (x_j, x_{j+1})$, for $j = 0, \ldots, K - 1$ such that $e'(\xi_j) = 0$.

Since $\Pi_h^1 f$ is a linear function on each I_j, for $x \in I_j$ we obtain

$$e'(x) = \int_{\xi_j}^{x} e''(s)ds = \int_{\xi_j}^{x} f''(s)ds,$$

whence

$$|e'(x)| \leq \int_{x_j}^{x_{j+1}} |f''(s)|ds, \qquad \text{for } x \in [x_j, x_{j+1}]. \tag{8.32}$$

We recall the *Cauchy-Schwarz inequality*

$$\left| \int_\alpha^\beta u(x)v(x)dx \right| \leq \left(\int_\alpha^\beta u^2(x)dx \right)^{1/2} \left(\int_\alpha^\beta v^2(x)dx \right)^{1/2}, \tag{8.33}$$

which holds if $u, v \in \mathrm{L}^2(\alpha, \beta)$. If we apply this inequality to (8.32) we obtain

$$\begin{aligned} |e'(x)| &\leq \left(\int_{x_j}^{x_{j+1}} 1^2 dx \right)^{1/2} \left(\int_{x_j}^{x_{j+1}} |f''(s)|^2 ds \right)^{1/2} \\ &\leq h^{1/2} \left(\int_{x_j}^{x_{j+1}} |f''(s)|^2 ds \right)^{1/2}. \end{aligned} \tag{8.34}$$

To find a bound for $|e(x)|$, we notice that

$$e(x) = \int_{x_j}^{x} e'(s)ds,$$

so that, applying (8.34), we get

$$|e(x)| \leq \int_{x_j}^{x_{j+1}} |e'(s)|ds \leq h^{3/2} \left(\int_{x_j}^{x_{j+1}} |f''(s)|^2 ds \right)^{1/2}. \tag{8.35}$$

Then

$$\int_{x_j}^{x_{j+1}} |e'(x)|^2 dx \leq h^2 \int_{x_j}^{x_{j+1}} |f''(s)|^2 ds \text{ and } \int_{x_j}^{x_{j+1}} |e(x)|^2 dx \leq h^4 \int_{x_j}^{x_{j+1}} |f''(s)|^2 ds,$$

from which, summing over the index j from 0 to $K-1$ and taking the square root of both sides, we obtain

$$\left(\int_a^b |e'(x)|^2 dx \right)^{1/2} \leq h \left(\int_a^b |f''(x)|^2 dx \right)^{1/2},$$

and

$$\left(\int_a^b |e(x)|^2 dx \right)^{1/2} \leq h^2 \left(\int_a^b |f''(x)|^2 dx \right)^{1/2},$$

which is the desired estimate (8.31), with $C_1 = C_2 = 1$. ◇

8.5 Hermite-Birkoff Interpolation

Lagrange polynomial interpolation can be generalized to the case in which also the values of the derivatives of a function f are available at some (or all) of the nodes x_i.

Let us then suppose that $(x_i, f^{(k)}(x_i))$ are given data, with $i = 0, \ldots, n$, $k = 0, \ldots, m_i$ and $m_i \in \mathbb{N}$. Letting $N = \sum_{i=0}^{n}(m_i + 1)$, it can be proved (see [Dav63]) that, if the nodes $\{x_i\}$ are distinct, there exists a unique polynomial $H_{N-1} \in \mathbb{P}_{N-1}$, called the *Hermite interpolation polynomial*, such that

$$H_{N-1}^{(k)}(x_i) = y_i^{(k)}, \; i = 0, \ldots, n \quad k = 0, \ldots, m_i,$$

of the form

$$H_{N-1}(x) = \sum_{i=0}^{n}\sum_{k=0}^{m_i} y_i^{(k)} L_{ik}(x), \tag{8.36}$$

where $y_i^{(k)} = f^{(k)}(x_i)$, $i = 0, \ldots, n$, $k = 0, \ldots, m_i$.
The functions $L_{ik} \in \mathbb{P}_{N-1}$ are called the *Hermite characteristic polynomials* and are defined through the relations

$$\frac{d^p}{dx^p}(L_{ik})(x_j) = \begin{cases} 1 \text{ if } i = j \text{ and } k = p, \\ 0 \text{ otherwise.} \end{cases}$$

Defining the polynomials

$$l_{ij}(x) = \frac{(x - x_i)^j}{j!} \prod_{\substack{k=0 \\ k \neq i}}^{n} \left(\frac{x - x_k}{x_i - x_k} \right)^{m_k + 1}, \; i = 0, \ldots, n, \; j = 0, \ldots, m_i,$$

and letting $L_{im_i}(x) = l_{im_i}(x)$ for $i = 0, \ldots, n$, we have the following recursive formula for the polynomials L_{ij}

$$L_{ij}(x) = l_{ij}(x) - \sum_{k=j+1}^{m_i} l_{ij}^{(k)}(x_i) L_{ik}(x), \qquad j = m_i - 1, m_i - 2, \ldots, 0.$$

As for the interpolation error, the following estimate holds

$$f(x) - H_{N-1}(x) = \frac{f^{(N)}(\xi)}{N!} \Omega_N(x) \; \forall x \in \mathbb{R},$$

where $\xi \in \mathrm{I}(x; x_0, \ldots, x_n)$ and Ω_N is the polynomial of degree N defined by

$$\Omega_N(x) = (x - x_0)^{m_0+1}(x - x_1)^{m_1+1} \cdots (x - x_n)^{m_n+1}. \tag{8.37}$$

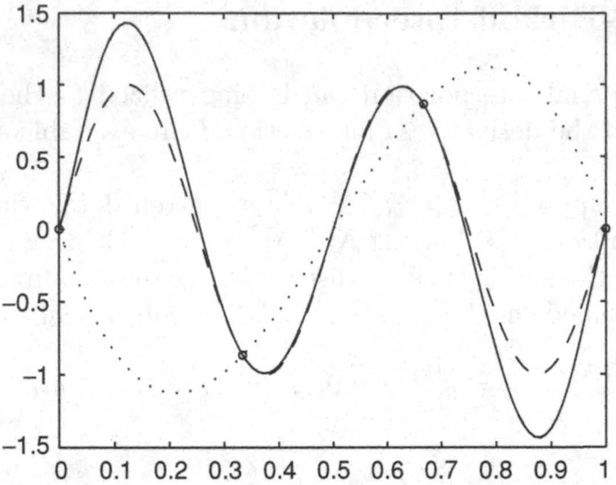

Fig. 8.4. Lagrange (*dotted line*) and Hermite interpolation (*solid line*) for the function $f(x) = \sin(4\pi x)$ (*dashed line*) on the interval $[0, 1]$

Example 8.6 (Osculatory interpolation) Let us set $m_i = 1$ for $i = 0, \ldots, n$. In this case $N = 2n + 2$ and the interpolating Hermite polynomial is called the *osculating polynomial*, and it is given by

$$H_{N-1}(x) = \sum_{i=0}^{n} \left(y_i A_i(x) + y_i^{(1)} B_i(x) \right),$$

where $A_i(x) = (1 - 2(x - x_i)l_i'(x_i))l_i(x)^2$ and $B_i(x) = (x - x_i)l_i(x)^2$, for $i = 0, \ldots, n$, with

$$l_i'(x_i) = \sum_{k=0, k \neq i}^{n} \frac{1}{x_i - x_k}, \qquad i = 0, \ldots, n.$$

As a comparison, we use Programs 65 and 67 to compute the Lagrange and Hermite interpolating polynomials of the function $f(x) = \sin(4\pi x)$ on the interval $[0, 1]$ taking four equally spaced nodes ($n = 3$). Figure 8.4 shows the superposed graphs of the function f and of the two polynomials $\Pi_n f$ and H_{N-1}. •

Program 67 computes the values of the osculating polynomial at the abscissae contained in the vector `z`. The input vectors `x`, `y` and `dy` contain the interpolation nodes and the corresponding function evaluations of f and f', respectively.

Program 67 - hermpol : Osculating polynomial

```
function [herm] = hermpol(x,y,dy,z)
%HERMPOL Hermite polynomial interpolation
% [HERM] = HERMPOL(X, Y, DY, Z) computes the Hermite interpolating polynomial
% of a function. X contains the interpolation nodes. Y and DY contain the values
% of the function and of its derivative at X. Z contains the points at which the
% interpolating polynomial HERM must be evaluated.
```

```
n = max(size(x));
m = max(size(z));
herm = [];
for j = 1:m
   xx = z(j); hxv = 0;
   for i = 1:n
      den = 1; num = 1; xn = x(i); derLi = 0;
      for k = 1:n
         if k ~= i
            num = num*(xx-x(k)); arg = xn-x(k);
            den = den*arg; derLi = derLi+1/arg;
         end
      end
      Lix2 = (num/den)^2; p = (1-2*(xx-xn)*derLi)*Lix2;
      q = (xx-xn)*Lix2; hxv = hxv+(y(i)*p+dy(i)*q);
   end
   herm = [herm, hxv];
end
return
```

8.6 Extension to the Two-Dimensional Case

In this section we briefly address the extension of the previous concepts to the two-dimensional case, referring to [SL89], [CHQZ06], [QV94] for more details. We denote by Ω a bounded domain in $\mathbb{R}^2$ and by $\mathbf{x} = (x, y)$ the coordinate vector of a point in Ω.

8.6.1 Polynomial Interpolation

A particularly simple situation occurs when $\Omega = [a, b] \times [c, d]$, i.e., the interpolation domain Ω is the tensor product of two intervals. In such a case, introducing the nodes $a = x_0 < x_1 < \ldots < x_n = b$ and $c = y_0 < y_1 < \ldots < y_m = d$, the interpolating polynomial $\Pi_{n,m}f$ can be written as $\Pi_{n,m}f(x, y) = \sum_{i=0}^{n} \sum_{j=0}^{m} \alpha_{ij} l_i(x) l_j(y)$, where $l_i \in \mathbb{P}_n$, $i = 0, \ldots, n$, and $l_j \in \mathbb{P}_m$, $j = 0, \ldots, m$, are the characteristic one-dimensional Lagrange polynomials with respect to the x and y variables respectively, and where $\alpha_{ij} = f(x_i, y_j)$.

The drawbacks of one-dimensional Lagrange interpolation are inherited by the two-dimensional case, as confirmed by the example in Figure 8.5.

Remark 8.1 (The general case) If Ω is not a rectangular domain or if the interpolation nodes are not uniformly distributed over a Cartesian grid, the interpolation problem is difficult to solve, and, generally speaking, it is preferable to resort to a least-squares solution (see Section 10.7). We also point out that in d dimensions (with $d \geq 2$) the problem of finding an interpolating

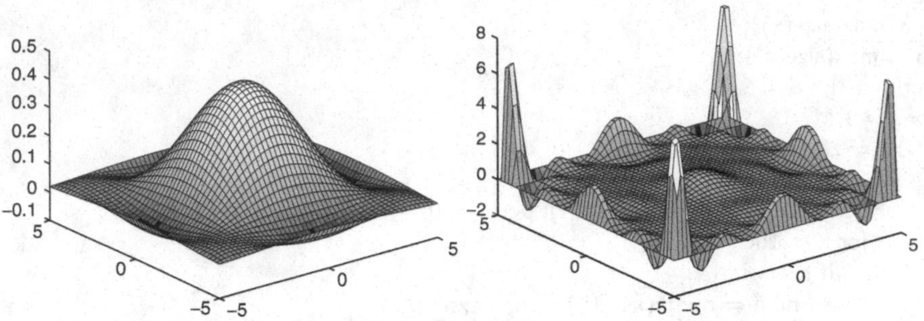

Fig. 8.5. Runge's counterexample extended to the two-dimensional case: interpolating polynomial on a 6×6 nodes grid (*left*) and on a 11×11 nodes grid (*right*). Notice the change in the vertical scale between the two plots

polynomial of degree n with respect to each space variable on $n+1$ distinct nodes might be ill-posed.

Consider, for example, a polynomial of degree 1 with respect to x and y of the form $p(x, y) = a_3xy + a_2x + a_1y + a_0$ to interpolate a function f at the nodes $(-1, 0)$, $(0, -1)$, $(1, 0)$ and $(0, 1)$. Although the nodes are distinct, the problem (which is nonlinear) does not in general admit a unique solution; actually, imposing the interpolation constraints, we end up with a system that is satisfied by any value of the coefficient a_3. ■

8.6.2 Piecewise Polynomial Interpolation

In the multidimensional case, the higher flexibility of piecewise interpolation allows for easy handling of domains of complex shape. Let us suppose that Ω is a polygon in $\mathbb{R}^2$. Then, Ω can be partitioned into K nonoverlapping triangles (or *elements*) T, which define the so called *triangulation* of the domain which will be denoted by $\mathcal{T}_h$. Clearly, $\overline{\Omega} = \bigcup_{T \in \mathcal{T}_h} T$. Suppose that the maximum length of the edges of the triangles is less than a positive number h. As shown in Figure 8.6 (*left*), not any arbitrary triangulation is allowed. Precisely, the admissible ones are those for which any pair of nondisjoint triangles may have a vertex or an edge in common.

Any element $T \in \mathcal{T}_h$, of area equal to $|T|$, is the image through the affine map $\mathbf{x} = F_T(\hat{\mathbf{x}}) = \mathrm{B}_T\hat{\mathbf{x}} + \mathbf{b}_T$ of the *reference triangle* $\widehat{T}$, of vertices (0,0), (1,0) and (0,1) in the $\hat{\mathbf{x}} = (\hat{x}, \hat{y})$ plane (see Figure 8.6, *right*), where the invertible matrix B_T and the right-hand side $\mathbf{b}_T$ are given respectively by

$$\mathrm{B}_T = \begin{bmatrix} x_2 - x_1 & x_3 - x_1 \\ y_2 - y_1 & y_3 - y_1 \end{bmatrix}, \ \mathbf{b}_T = (x_1, y_1)^T, \tag{8.38}$$

while the coordinates of the vertices of T are denoted by $\mathbf{a}_l^T = (x_l, y_l)^T$ for $l = 1, 2, 3$.

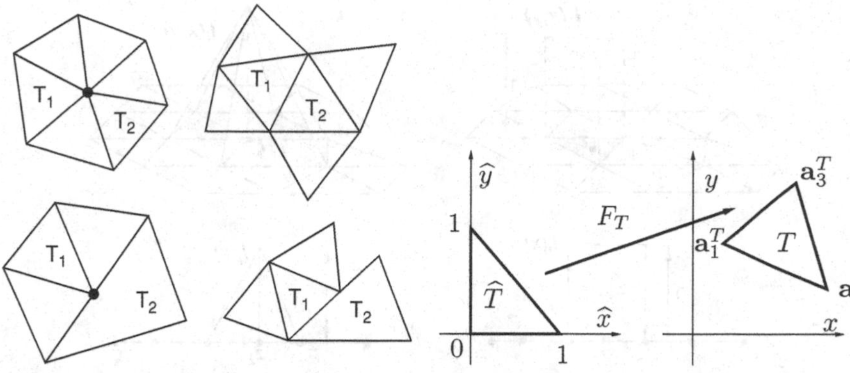

Fig. 8.6. The left side picture shows admissible (*above*) and nonadmissible (*below*) triangulations while the right side picture shows the affine map from the reference triangle $\hat{T}$ to the generic element $T \in \mathcal{T}_h$

The affine map (8.38) is of remarkable importance in practical computations, since, once a basis has been generated for representing the piecewise polynomial interpolant on $\hat{T}$, it is possible, by applying the change of coordinates $\mathbf{x} = F_T(\hat{\mathbf{x}})$, to reconstruct the polynomial on each element T of $\mathcal{T}_h$. We are thus interested in devising local basis functions, which can be fully described over each triangle without needing any information from adjacent triangles. For this purpose, let us introduce on $\mathcal{T}_h$ the set $\mathcal{Z}$ of the *piecewise interpolation nodes* $\mathbf{z}_i = (x_i, y_i)^T$, for $i = 1, \ldots, N$, and denote by $\mathbb{P}_k(\Omega)$, $k \geq 0$, the space of algebraic polynomials of degree $\leq k$ in the space variables x, y

$$\mathbb{P}_k(\Omega) = \left\{ p(x,y) = \sum_{\substack{i,j=0 \\ i+j \leq k}}^{k} a_{ij} x^i y^j,\ x, y \in \Omega \right\}. \tag{8.39}$$

Finally, for $k \geq 0$, let $\mathbb{P}_k^c(\Omega)$ be the space of piecewise polynomials of degree $\leq k$, such that, for any $p \in \mathbb{P}_k^c(\Omega)$, $p|_T \in \mathbb{P}_k(T)$ for any $T \in \mathcal{T}_h$. An elementary basis for $\mathbb{P}_k^c(\Omega)$ consists of the *Lagrange characteristic polynomials* $l_i = l_i(x, y)$, such that $l_i \in \mathbb{P}_k^c(\Omega)$ and

$$l_i(\mathbf{z}_j) = \delta_{ij}, \qquad i, j = 1, \ldots, N, \tag{8.40}$$

where δ_{ij} is the Kronecker symbol. We show in Figure 8.7 the functions l_i for $k = 0$, 1, together with their corresponding one-dimensional counterparts. In the case $k = 0$, the interpolation nodes are collocated at the *centers of gravity* of the triangles, while in the case $k = 1$ the nodes coincide with the *vertices* of the triangles. This choice, that we are going to maintain henceforth, is not the only one possible. The midpoints of the edges of the triangles could be used as well, giving rise to a discontinuous piecewise polynomial over Ω.

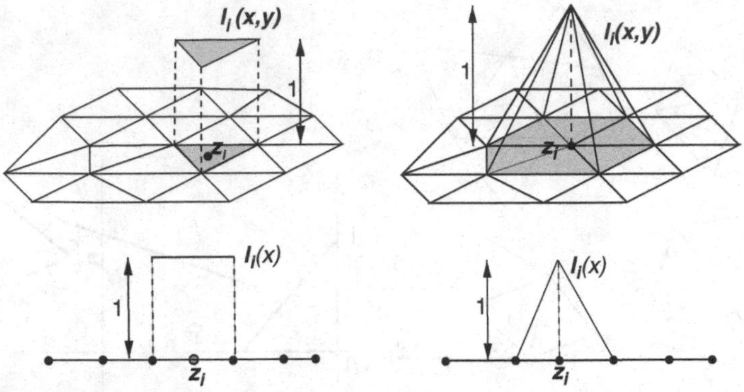

Fig. 8.7. Characteristic piecewise Lagrange polynomial, in two and one space dimensions. Left, $k = 0$; right, $k = 1$

For $k \geq 0$, the *Lagrange piecewise interpolating polynomial* of f, $\Pi_h^k f \in \mathbb{P}_k^c(\Omega)$, is defined as

$$\Pi_h^k f(x, y) = \sum_{i=1}^{N} f(\mathbf{z}_i) l_i(x, y). \tag{8.41}$$

Notice that $\Pi_h^0 f$ is a piecewise constant function, while $\Pi_h^1 f$ is a linear function over each triangle, continuous at the vertices, and thus globally continuous.

For any $T \in \mathcal{T}_h$, we shall denote by $\Pi_T^k f$ the restriction of the piecewise interpolating polynomial of f over the element T. By definition, $\Pi_T^k f \in \mathbb{P}_k(T)$; noticing that $d_k = \dim\mathbb{P}_k(T) = (k+1)(k+2)/2$, we can therefore write

$$\Pi_T^k f(x, y) = \sum_{m=0}^{d_k - 1} f(\tilde{\mathbf{z}}_T^{(m)}) l_{m,T}(x, y), \qquad \forall T \in \mathcal{T}_h. \tag{8.42}$$

In (8.42), we have denoted by $\tilde{\mathbf{z}}_T^{(m)}$, for $m = 0, \ldots, d_k - 1$, the piecewise interpolation nodes on T and by $l_{m,T}(x, y)$ the restriction to T of the Lagrange characteristic polynomial having index i in (8.41) which corresponds in the list of the "global" nodes $\mathbf{z}_i$ to that of the "local" node $\tilde{\mathbf{z}}_T^{(m)}$.

Keeping on with this notation, we have $l_{j,T}(\mathbf{x}) = \hat{l}_j \circ F_T^{-1}(\mathbf{x})$, where $\hat{l}_j = \hat{l}_j(\hat{\mathbf{x}})$ is, for $j = 0, \ldots, d_k - 1$, the j-th Lagrange basis function for $\mathbb{P}_k(\hat{T})$ generated on the reference element $\hat{T}$. We notice that if $k = 0$ then $d_0 = 1$, that is, only one local interpolation node exists (and it is the center of gravity of the triangle T), while if $k = 1$ then $d_1 = 3$, that is, three local interpolation nodes exist, coinciding with the vertices of T. In Figure 8.8 we draw the local interpolation nodes on $\hat{T}$ for $k = 0$, 1 and 2.

As for the interpolation error estimate, denoting for any $T \in \mathcal{T}_h$ by h_T the maximum length of the edges of T, the following result holds (see for the proof, [CL91], Theorem 16.1, pp. 125-126 and [QV94], Remark 3.4.2, pp. 89-90)

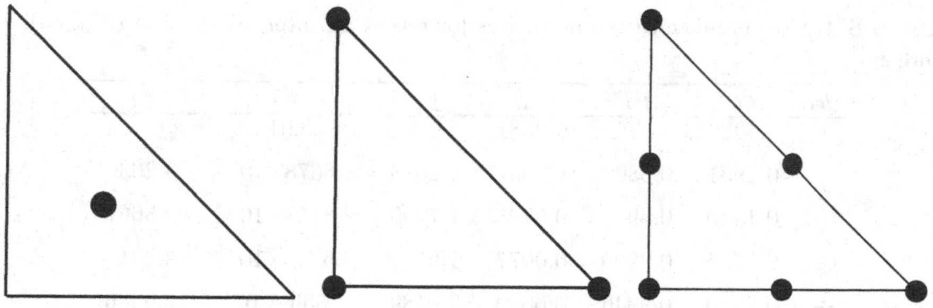

Fig. 8.8. Local interpolation nodes on $\hat{T}$ for $k = 0$ (*left*), $k = 1$ (*center*), $k = 2$ (*right*)

$$\|f - \Pi_T^k f\|_{\infty,T} \leq C h_T^{k+1} \|f^{(k+1)}\|_{\infty,T}, \qquad k \geq 0, \qquad (8.43)$$

where for every $g \in C^0(T)$, $\|g\|_{\infty,T} = \max_{\mathbf{x}\in T} |g(\mathbf{x})|$. In (8.43), C is a positive constant independent of h_T and f.

Let us assume that the triangulation $\mathcal{T}_h$ is *regular*, i.e., there exists a positive constant σ such that

$$\max_{T\in\mathcal{T}_h} \frac{h_T}{\rho_T} \leq \sigma,$$

where $\forall T \in \mathcal{T}_h$, ρ_T is the diameter of the inscribed circle to T. Then, it is possible to derive from (8.43) the following interpolation error estimate over the whole domain Ω

$$\|f - \Pi_h^k f\|_{\infty,\Omega} \leq C h^{k+1} \|f^{(k+1)}\|_{\infty,\Omega}, \qquad k \geq 0, \qquad \forall f \in C^{k+1}(\Omega). \quad (8.44)$$

The theory of piecewise interpolation is a basic tool of the *finite element method*, a computational technique that is widely used in the numerical approximation of partial differential equations (see Chapter 12 for the one-dimensional case and [QV94] for a complete presentation of the method).

Example 8.7 We compare the convergence of the piecewise polynomial interpolation of degree 0, 1 and 2, on the function $f(x, y) = e^{-(x^2+y^2)}$ on $\Omega = (-1, 1)^2$. We show in Table 8.4 the error $\mathcal{E}_k = \|f - \Pi_h^k f\|_{\infty,\Omega}$, for $k = 0, 1, 2$, and the order of convergence p_k as a function of the mesh size $h = 2/N$ for $N = 2, \ldots, 32$. Clearly, linear convergence is observed for interpolation of degree 0 while the order of convergence is quadratic with respect to h for interpolation of degree 1 and cubic for interpolation of degree 2. •

8.7 Approximation by Splines

In this section we address the matter of approximating a given function using *splines*, which allow for a piecewise interpolation with a global smoothness.

Table 8.4. Convergence rates and orders for piecewise interpolations of degree 0, 1 and 2

h	$\mathcal{E}_0$	p_0	$\mathcal{E}_1$	p_1	$\mathcal{E}_2$	p_2
1	0.4384		0.2387		0.016	
$\frac{1}{2}$	0.2931	0.5809	0.1037	1.2028	$1.6678 \cdot 10^{-3}$	3.2639
$\frac{1}{4}$	0.1579	0.8924	0.0298	1.7990	$2.8151 \cdot 10^{-4}$	2.5667
$\frac{1}{8}$	0.0795	0.9900	0.0077	1.9524	$3.5165 \cdot 10^{-5}$	3.001
$\frac{1}{16}$	0.0399	0.9946	0.0019	2.0189	$4.555 \cdot 10^{-6}$	2.9486

Definition 8.1 Let $x_0, \ldots, x_n$, be $n+1$ distinct nodes of $[a, b]$, with $a = x_0 < x_1 < \ldots < x_n = b$. The function $s_k(x)$ on the interval $[a, b]$ is a *spline* of degree k relative to the nodes x_j if

$$s_{k|[x_j, x_{j+1}]} \in \mathbb{P}_k, \; j = 0, 1, \ldots, n-1, \tag{8.45}$$

$$s_k \in C^{k-1}[a, b]. \tag{8.46}$$

■

Denoting by $\mathcal{S}_k$ the space of splines s_k on $[a, b]$ relative to $n+1$ distinct nodes, then $\dim \mathcal{S}_k = n+k$. Obviously, any polynomial of degree k on $[a, b]$ is a spline; however, in the practice a spline is represented by a different polynomial on each subinterval and for this reason there could be a discontinuity in its k-th derivative at the internal nodes $x_1, \ldots, x_{n-1}$. The nodes for which this actually happens are called *active* nodes.

It is simple to check that conditions (8.45) and (8.46) do not suffice to characterize a spline of degree k. Indeed, the restriction $s_{k,j} = s_{k|[x_j, x_{j+1}]}$ can be represented as

$$s_{k,j}(x) = \sum_{i=0}^{k} s_{ij}(x - x_j)^i, \text{ if } x \in [x_j, x_{j+1}], \tag{8.47}$$

so that $(k+1)n$ coefficients s_{ij} must be determined. On the other hand, from (8.46) it follows that

$$s_{k,j-1}^{(m)}(x_j) = s_{k,j}^{(m)}(x_j), \; j = 1, \ldots, n-1, \; m = 1, \ldots, k-1$$

which amounts to setting $k(n-1)$ conditions. As a consequence, the remaining degrees of freedom are $(k+1)n - k(n-1) = k + n$.
Even if the spline were *interpolatory*, that is, such that $s_k(x_j) = f_j$ for $j = 0, \ldots, n$, where $f_0, \ldots, f_n$ are given values, there would still be $k-1$ unsaturated degrees of freedom. For this reason further constraints are usually imposed, which lead to:

1. *periodic splines*, if

$$s_k^{(m)}(a) = s_k^{(m)}(b),\ m = 0, 1, \ldots, k-1; \tag{8.48}$$

2. *natural splines*, if for $k = 2l - 1$, with $l \geq 2$

$$s_k^{(l+j)}(a) = s_k^{(l+j)}(b) = 0,\ j = 0, 1, \ldots, l-2. \tag{8.49}$$

From (8.47) it turns out that a spline can be conveniently represented using $k + n$ spline basis functions, such that (8.46) is automatically satisfied. The simplest choice, which consists of employing a suitably enriched monomial basis (see Exercise 10), is not satisfactory from the numerical standpoint, since it is ill-conditioned. In Sections 8.7.1 and 8.7.2 possible examples of spline basis functions will be provided: cardinal splines for the specific case $k = 3$ and B-splines for a generic k.

8.7.1 Interpolatory Cubic Splines

Interpolatory cubic splines are particularly significant since: *i.* they are the splines of minimum degree that yield C^2 approximations; *ii.* they are sufficiently smooth in the presence of small curvatures.

Let us thus consider, in $[a, b]$, $n + 1$ ordered nodes $a = x_0 < x_1 < \ldots < x_n = b$ and the corresponding evaluations f_i, $i = 0, \ldots, n$. Our aim is to provide an efficient procedure for constructing the cubic spline interpolating those values. Since the spline is of degree 3, its second-order derivative must be continuous. Let us introduce the following notation

$$f_i = s_3(x_i),\ m_i = s_3'(x_i),\ M_i = s_3''(x_i),\ i = 0, \ldots, n.$$

Since $s_{3,i-1} \in \mathbb{P}_3$, $s_{3,i-1}''$ is linear and

$$s_{3,i-1}''(x) = M_{i-1}\frac{x_i - x}{h_i} + M_i\frac{x - x_{i-1}}{h_i} \quad \text{for } x \in [x_{i-1}, x_i] \tag{8.50}$$

where $h_i = x_i - x_{i-1}$, $i = 1, \ldots, n$. Integrating (8.50) twice we get

$$s_{3,i-1}(x) = M_{i-1}\frac{(x_i - x)^3}{6h_i} + M_i\frac{(x - x_{i-1})^3}{6h_i} + C_{i-1}(x - x_{i-1}) + \widetilde{C}_{i-1},$$

and the constants C_{i-1} and $\widetilde{C}_{i-1}$ are determined by imposing the end point values $s_3(x_{i-1}) = f_{i-1}$ and $s_3(x_i) = f_i$. This yields, for $i = 1, \ldots, n-1$

$$\widetilde{C}_{i-1} = f_{i-1} - M_{i-1}\frac{h_i^2}{6},\ C_{i-1} = \frac{f_i - f_{i-1}}{h_i} - \frac{h_i}{6}(M_i - M_{i-1}).$$

Let us now enforce the continuity of the first derivatives at x_i; we get

$$s_3'(x_i^-) = \frac{h_i}{6}M_{i-1} + \frac{h_i}{3}M_i + \frac{f_i - f_{i-1}}{h_i}$$
$$= -\frac{h_{i+1}}{3}M_i - \frac{h_{i+1}}{6}M_{i+1} + \frac{f_{i+1} - f_i}{h_{i+1}} = s_3'(x_i^+),$$

where $s_3'(x_i^{\pm}) = \lim\limits_{t\to 0} s_3'(x_i \pm t)$. This leads to the following linear system (called M-continuity system)

$$\mu_i M_{i-1} + 2M_i + \lambda_i M_{i+1} = d_i \quad i = 1, \ldots, n-1, \tag{8.51}$$

where we have set

$$\mu_i = \frac{h_i}{h_i + h_{i+1}}, \qquad \lambda_i = \frac{h_{i+1}}{h_i + h_{i+1}},$$
$$d_i = \frac{6}{h_i + h_{i+1}}\left(\frac{f_{i+1} - f_i}{h_{i+1}} - \frac{f_i - f_{i-1}}{h_i}\right), \quad i = 1, \ldots, n-1.$$

System (8.51) has $n+1$ unknowns and $n-1$ equations; thus, $2(= k-1)$ conditions are still lacking. In general, these conditions can be of the form

$$2M_0 + \lambda_0 M_1 = d_0, \; \mu_n M_{n-1} + 2M_n = d_n,$$

with $0 \leq \lambda_0, \mu_n \leq 1$ and d_0, d_n given values. For instance, in order to obtain the natural splines (satisfying $s_3''(a) = s_3''(b) = 0$), we must set the above coefficients equal to zero. A popular choice sets $\lambda_0 = \mu_n = 1$ and $d_0 = d_1$, $d_n = d_{n-1}$, which corresponds to prolongating the spline outside the end points of the interval $[a, b]$ and treating a and b as internal points. This strategy produces a spline with a "smooth" behavior. In general, the resulting linear system is tridiagonal of the form

$$\begin{bmatrix} 2 & \lambda_0 & 0 & \ldots & 0 \\ \mu_1 & 2 & \lambda_1 & & \vdots \\ 0 & \ddots & \ddots & \ddots & 0 \\ \vdots & & \mu_{n-1} & 2 & \lambda_{n-1} \\ 0 & \ldots & 0 & \mu_n & 2 \end{bmatrix} \begin{bmatrix} M_0 \\ M_1 \\ \vdots \\ M_{n-1} \\ M_n \end{bmatrix} = \begin{bmatrix} d_0 \\ d_1 \\ \vdots \\ d_{n-1} \\ d_n \end{bmatrix} \tag{8.52}$$

and it can be efficiently solved using the Thomas algorithm (3.53).

A closure condition for system (8.52), which can be useful when the derivatives $f'(a)$ and $f'(b)$ are not available, consists of enforcing the continuity of $s_3'''(x)$ at x_1 and x_{n-1}. Since the nodes x_1 and x_{n-1} do not actually contribute in constructing the cubic spline, it is called a *not-a-knot spline*, with "active" knots $\{x_0, x_2, \ldots, x_{n-2}, x_n\}$ and interpolating f at all the nodes $\{x_0, x_1, x_2, \ldots, x_{n-2}, x_{n-1}, x_n\}$.

Remark 8.2 (Specific software) Several packages exist for dealing with interpolating splines. In the case of cubic splines, we mention the command `spline`, which uses the *not-a-knot* condition introduced above, or, in general, the `spline toolbox` of MATLAB [dB90] and the library FITPACK [Die87a], [Die87b]. ■

A different approach for generating s_3 consists of providing a basis $\{\varphi_i\}$ for the space $\mathcal{S}_3$ of cubic splines, whose dimension is equal to $n+3$. We consider here the case in which the $n+3$ basis functions φ_i have global support in the interval $[a, b]$, referring to Section 8.7.2 for the case of a basis with local support.

Functions φ_i, for $i, j = 0, \ldots, n$, are defined through the following interpolation constraints

$$\varphi_i(x_j) = \delta_{ij}, \qquad \varphi_i'(x_0) = \varphi_i'(x_n) = 0,$$

and two suitable splines must be added, φ_{n+1} and φ_{n+2}. For instance, if the spline must satisfy some assigned conditions on the derivative at the end points, we ask that

$$\begin{array}{ll} \varphi_{n+1}(x_j) = 0, & j = 0, \ldots, n, \ \varphi_{n+1}'(x_0) = 1, \ \varphi_{n+1}'(x_n) = 0, \\ \varphi_{n+2}(x_j) = 0, & j = 0, \ldots, n, \ \varphi_{n+2}'(x_0) = 0, \ \varphi_{n+2}'(x_n) = 1. \end{array}$$

By doing so, the spline takes the form

$$s_3(x) = \sum_{i=0}^{n} f_i \varphi_i(x) + f_0' \varphi_{n+1}(x) + f_n' \varphi_{n+2}(x),$$

where f_0' and f_n' are two given values. The resulting basis $\{\varphi_i, \ i = 0, ..., n+2\}$ is called a *cardinal spline basis* and is frequently employed in the numerical solution of differential or integral equations. Figure 8.9 shows a generic cardinal spline, which is computed over a virtually unbounded interval where the interpolation nodes x_j are the integers. The spline changes sign in any adjacent intervals $[x_{j-1}, x_j]$ and $[x_j, x_{j+1}]$ and rapidly decays to zero.

Restricting ourselves to the positive axis, it can be shown (see [SL89]) that the extremant of the function on the interval $[x_j, x_{j+1}]$ is equal to the extremant on the interval $[x_{j+1}, x_{j+2}]$ multiplied by a decaying factor $\lambda \in (0, 1)$. In such a way, possible errors arising over an interval are rapidly damped on the next one, thus ensuring the stability of the algorithm.

Let us summarize the main properties of interpolating cubic splines, referring to [Sch81] and [dB83] for the proofs and more general results.

Property 8.2 *Let $f \in C^2([a, b])$, and let s_3 be the natural cubic spline interpolating f. Then*

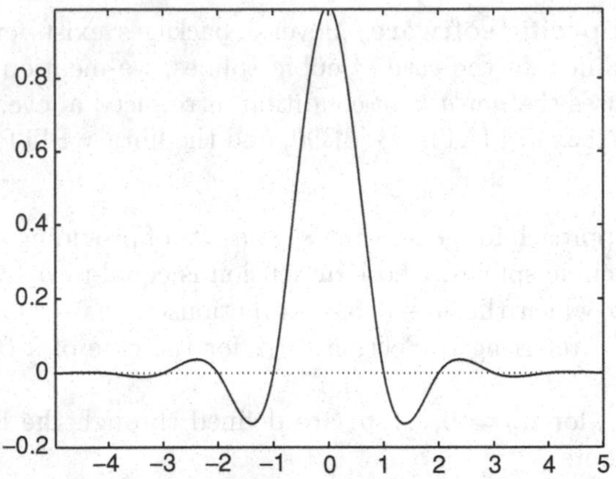

Fig. 8.9. Cardinal spline

$$\int_a^b [s_3''(x)]^2 dx \leq \int_a^b [f''(x)]^2 dx, \tag{8.53}$$

where equality holds if and only if $f = s_3$.

The above result is known as the *minimum norm property* and has the meaning of the minimum energy principle in mechanics. Property (8.53) still holds if conditions on the first derivative of the spline at the end points are assigned instead of natural conditions (in such a case, the spline is called *constrained*, see Exercise 11).

The cubic interpolating spline s_f of a function $f \in C^2([a,b])$, with $s_f'(a) = f'(a)$ and $s_f'(b) = f'(b)$, also satisfies the following property

$$\int_a^b [f''(x) - s_f''(x)]^2 dx \leq \int_a^b [f''(x) - s''(x)]^2 dx, \ \forall s \in S_3.$$

As far as the error estimate is concerned, the following result holds.

Property 8.3 *Let* $f \in C^4([a,b])$ *and fix a partition of* $[a,b]$ *into subintervals of width* h_i *such that* $h = \max_i h_i$ *and* $\beta = h/\min_i h_i$. *Let* s_3 *be the cubic spline interpolating* f. *Then*

$$\|f^{(r)} - s_3^{(r)}\|_\infty \leq C_r h^{4-r} \|f^{(4)}\|_\infty, \qquad r = 0,1,2,3, \tag{8.54}$$

with $C_0 = 5/384$, $C_1 = 1/24$, $C_2 = 3/8$ *and* $C_3 = (\beta + \beta^{-1})/2$.

As a consequence, spline s_3 and its first and second order derivatives uniformly converge to f and to its derivatives, as h tends to zero. The third order derivative converges as well, provided that β is uniformly bounded.

Example 8.8 Figure 8.10 shows the cubic spline approximating the function in the Runge's example, and its first, second and third order derivatives, on a grid of 11 equally spaced nodes, while in Table 8.5 the error $\|s_3 - f\|_\infty$ is reported as a function of h together with the computed order of convergence p. The results clearly demonstrate that p tends to 4 (the theoretical order) as h tends to zero. •

8.7.2 B-splines

Let us go back to splines of a generic degree k, and consider the B-spline (or *bell-spline*) basis, referring to divided differences introduced in Section 8.2.1.

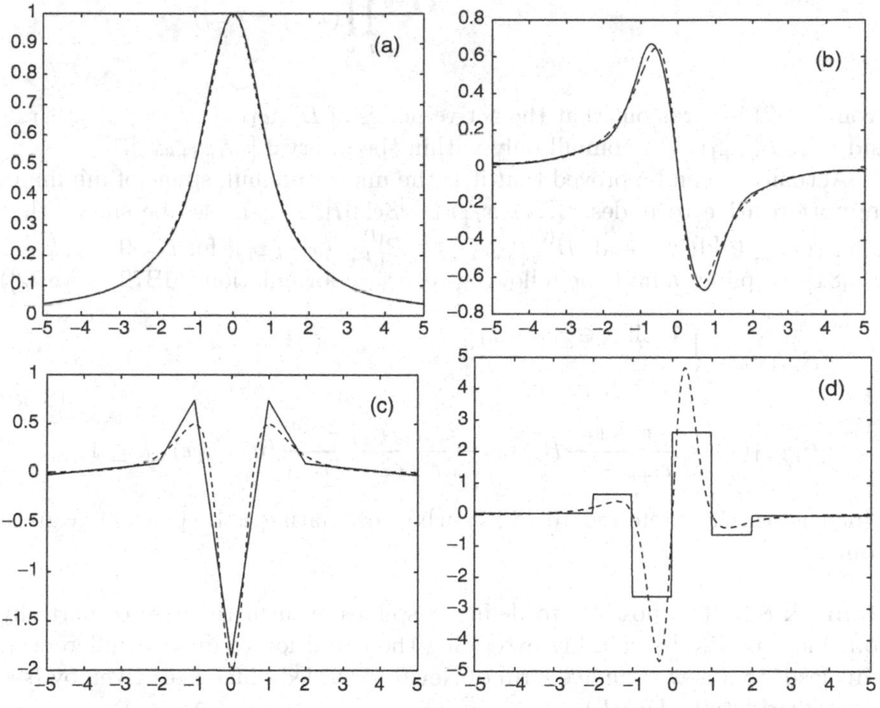

Fig. 8.10. Interpolating spline (a) and its first (b), second (c) and third (d) order derivatives (*solid line*) for the function of Runge's example (*dashed line*)

Table 8.5. Experimental interpolation error for Runge's function using cubic splines

h	1	0.5	0.25	0.125	0.0625
$\|s_3 - f\|_\infty$	0.022	0.0032	2.7741e-4	1.5983e-5	9.6343e-7
p	–	2.7881	3.5197	4.1175	4.0522

Definition 8.2 The *normalized B-spline* $B_{i,k+1}$ of degree k relative to the distinct nodes $x_i, \ldots, x_{i+k+1}$ is defined as

$$B_{i,k+1}(x) = (x_{i+k+1} - x_i)g[x_i, \ldots, x_{i+k+1}], \tag{8.55}$$

where

$$g(t) = (t-x)_+^k = \begin{cases} (t-x)^k & \text{if } x \le t, \\ 0 & \text{otherwise.} \end{cases} \tag{8.56}$$

■

Substituting (8.18) into (8.55) yields the following explicit representation

$$B_{i,k+1}(x) = (x_{i+k+1} - x_i)\sum_{j=0}^{k+1} \frac{(x_{j+i} - x)_+^k}{\prod\limits_{\substack{l=0\\ l\neq j}}^{k+1} (x_{i+j} - x_{i+l})}. \tag{8.57}$$

From (8.57) it turns out that the active nodes of $B_{i,k+1}(x)$ are $x_i, \ldots, x_{i+k+1}$ and that $B_{i,k+1}(x)$ is nonnull only within the interval $[x_i, x_{i+k+1}]$.

Actually, it can be proved that it is the unique nonnull spline of minimum support relative to nodes $x_i, \ldots, x_{i+k+1}$ [Sch67]. It can also be shown that $B_{i,k+1}(x) \ge 0$ [dB83] and $|B^{(l)}_{i,k+1}(x_i)| = |B^{(l)}_{i,k+1}(x_{i+k+1})|$ for $l = 0, \ldots, k-1$ [Sch81]. B-splines admit the following recursive formulation ([dB72], [Cox72])

$$\begin{aligned} B_{i,1}(x) &= \begin{cases} 1 & \text{if } x \in [x_i, x_{i+1}], \\ 0 & \text{otherwise,} \end{cases} \\ B_{i,k+1}(x) &= \frac{x - x_i}{x_{i+k} - x_i} B_{i,k}(x) + \frac{x_{i+k+1} - x}{x_{i+k+1} - x_{i+1}} B_{i+1,k}(x), \ k \ge 1, \end{aligned} \tag{8.58}$$

which is usually preferred to (8.57) when evaluating a B-spline at a given point.

Remark 8.3 It is possible to define B-splines even in the case of partially coincident nodes, by suitably extending the definition of divided differences. This leads to a new recursive form of Newton divided differences given by (see for further details [Die93])

$$f[x_0, \ldots, x_n] = \begin{cases} \dfrac{f[x_1, \ldots, x_n] - f[x_0, \ldots, x_{n-1}]}{x_n - x_0} & \text{if } x_0 < x_1 < \ldots < x_n, \\[2ex] \dfrac{f^{(n+1)}(x_0)}{(n+1)!} & \text{if } x_0 = x_1 = \ldots = x_n. \end{cases}$$

Assuming that m (with $1 < m < k+2$) of the $k+2$ nodes $x_i, \ldots, x_{i+k+1}$ are coincident and equal to λ, then (8.47) will contain a linear combination of the functions $(\lambda - x)_+^{k+1-j}$, for $j = 1, \ldots, m$. As a consequence, the B-spline can have continuous derivatives at λ only up to order $k - m$ and, therefore, it is discontinuous if $m = k + 1$. It can be checked [Die93] that, if $x_{i-1} < x_i = \ldots = x_{i+k} < x_{i+k+1}$, then

$$B_{i,k+1}(x) = \begin{cases} \left(\dfrac{x_{i+k+1} - x}{x_{i+k+1} - x_i}\right)^k & \text{if } x \in [x_i, x_{i+k+1}], \\ 0 & \text{otherwise,} \end{cases}$$

while for $x_i < x_{i+1} = \ldots = x_{i+k+1} < x_{i+k+2}$

$$B_{i,k+1}(x) = \begin{cases} \left(\dfrac{x - x_i}{x_{i+k+1} - x_i}\right)^k & \text{if } x \in [x_i, x_{i+k+1}], \\ 0 & \text{otherwise.} \end{cases}$$

Combining these formulae with the recursive relation (8.58) allows for constructing B-splines with coincident nodes. ■

Example 8.9 Let us examine the special case of cubic B-splines on equally spaced nodes $x_{i+1} = x_i + h$ for $i = 0, ..., n - 1$. Equation (8.57) becomes

$$6h^3 B_{i,4}(x) =$$

$$\begin{cases} (x - x_i)^3, & \text{if} \quad x \in [x_i, x_{i+1}], \\ h^3 + 3h^2(x - x_{i+1}) + 3h(x - x_{i+1})^2 - 3(x - x_{i+1})^3, & \text{if} \quad x \in [x_{i+1}, x_{i+2}], \\ h^3 + 3h^2(x_{i+3} - x) + 3h(x_{i+3} - x)^2 - 3(x_{i+3} - x)^3, & \text{if } x \in [x_{i+2}, x_{i+3}], \\ (x_{i+4} - x)^3, & \text{if} \quad x \in [x_{i+3}, x_{i+4}], \\ 0 & \text{otherwise.} \end{cases}$$

In Figure 8.11 the graph of $B_{i,4}$ is shown in the case of distinct nodes and of partially coincident nodes. •

Given $n + 1$ distinct nodes x_j, $j = 0, \ldots, n$, $n - k$ linearly independent B-splines of degree k can be constructed, though $2k$ degrees of freedom are still available to generate a basis for $\mathcal{S}_k$. One way of proceeding consists of introducing $2k$ fictitious nodes

$$\begin{aligned} &x_{-k} \le x_{-k+1} \le \ldots \le x_{-1} \le x_0 = a, \\ &b = x_n \le x_{n+1} \le \ldots \le x_{n+k}, \end{aligned} \tag{8.59}$$

which the B-splines $B_{i,k+1}$, with $i = -k, \ldots, -1$ and $i = n - k, \ldots, n - 1$, are associated with. By doing so, any spline $s_k \in \mathcal{S}_k$ can be uniquely written as

$$s_k(x) = \sum_{i=-k}^{n-1} c_i B_{i,k+1}(x). \tag{8.60}$$

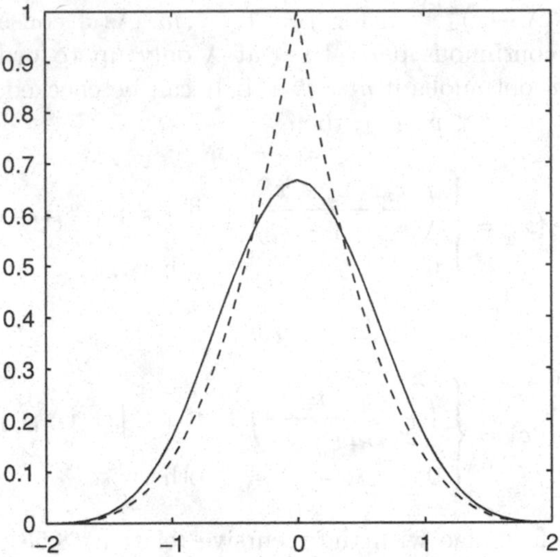

Fig. 8.11. B-spline with distinct nodes (*solid line*) and with three coincident nodes at the origin (*dashed line*). Notice the discontinuity of the first derivative

The real numbers c_i are the *B-spline coefficients* of s_k. Nodes (8.59) are usually chosen as coincident or periodic:

1. *coincident*: this choice is suitable for enforcing the values attained by a spline at the end points of its definition interval. In such a case, indeed, thanks to Remark 8.3 about B-splines with coincident nodes, we get
$$s_k(a) = c_{-k},\ s_k(b) = c_{n-1}; \tag{8.61}$$
2. *periodic*, that is
$$x_{-i} = x_{n-i} - b + a,\ x_{i+n} = x_i + b - a,\ i = 1, \ldots, k.$$
This choice is useful if the periodicity conditions (8.48) have to be imposed.

Using B-splines instead of cardinal splines is advantageous when handling, with a reduced computational effort, a given configuration of nodes for which a spline s_k is known. In particular, assume that the coefficients c_i of s_k (in form (8.60)) are available over the nodes $x_{-k}, x_{-k+1}, \ldots, x_{n+k}$, and that we wish to add to these a new node $\widetilde{x}$.

The spline $\widetilde{s}_k \in \mathcal{S}_k$, defined over the new set of nodes, admits the following representation with respect to a new B-spline basis $\left\{\tilde{B}_{i,k+1}\right\}$
$$\widetilde{s}_k(x) = \sum_{i=-k}^{n-1} d_i \widetilde{B}_{i,k+1}(x).$$

The new coefficients d_i can be computed starting from the known coefficients c_i using the following algorithm [Boe80]:

let $\widetilde{x} \in [x_j, x_{j+1})$; then, construct a new set of nodes $\{y_i\}$ such that

$$y_i = x_i \quad \text{for } i = -k, \ldots, j, \qquad y_{j+1} = \widetilde{x},$$
$$y_i = x_{i-1} \quad \text{for } i = j+2, \ldots, n+k+1;$$

define

$$\omega_i = \begin{cases} 1 & \text{for } i = -k, \ldots, j-k, \\ \dfrac{y_{j+1} - y_i}{y_{i+k+1} - y_i} & \text{for } i = j-k+1, \ldots, j, \\ 0 & \text{for } i = j+1, \ldots, n; \end{cases}$$

compute

$$d_i = \omega_i c_i + (1 - \omega_i) c_i \qquad \text{for } i = -k, ..., n-1.$$

This algorithm has good stability properties and can be generalized to the case where more than one node is inserted at the same time (see [Die93]).

8.8 Splines in Parametric Form

Using interpolating splines presents the following two drawbacks:

1. the resulting approximation is of good quality only if the function f does not exhibit large derivatives (in particular, we require that $|f'(x)| < 1$ for every x). Otherwise, oscillating behaviors may arise in the spline, as demonstrated by the example considered in Figure 8.12 which shows, in solid line, the cubic interpolating spline over the following set of data (from [SL89])

x_i	8.125	8.4	9	9.845	9.6	9.959	10.166	10.2
f_i	0.0774	0.099	0.28	0.6	0.708	1.3	1.8	2.177

2. s_k depends on the choice of the coordinate system. In fact, performing a clockwise rotation of 36 degrees of the coordinate system in the above example, would lead to the spline without spurious oscillations reported in the boxed frame in Figure 8.12.

All the interpolation procedures considered so far depend on the chosen Cartesian reference system, which is a negative feature if the spline is used for a graphical representation of a given figure (for instance, an ellipse). Indeed, we would like such a representation to be independent of the reference system, that is, to have a geometric invariance property.

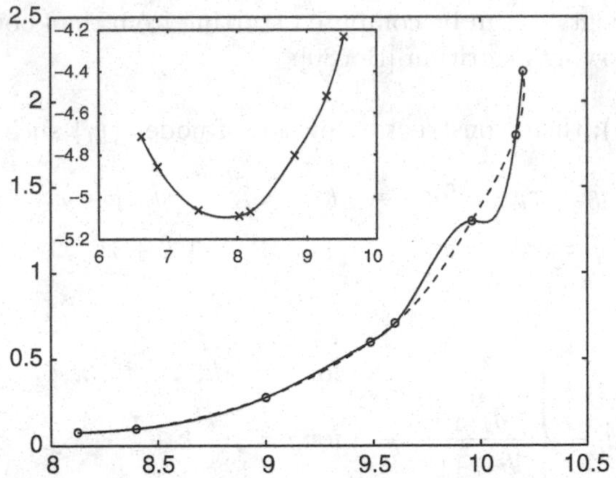

Fig. 8.12. Geometric noninvariance for an interpolating cubic spline s_3: the set of data for s_3 in the boxed frame is the same as in the main figure, rotated by 36 degrees. The rotation diminishes the slope of the interpolated curve and eliminates any oscillation from s_3. Notice that resorting to a parametric spline (*dashed line*) removes the oscillations in s_3 without any rotation of the reference system

A solution is provided by parametric splines, in which any component of the curve, written in parametric form, is approximated by a spline function. Consider a plane curve in parametric form $\mathbf{P}(t) = (x(t), y(t))$, with $t \in [0, T]$, then take the set of the points in the plane of coordinates $\mathbf{P}_i = (x_i, y_i)$, for $i = 0, \ldots, n$, and introduce a partition onto $[0, T]$: $0 = t_0 < t_1 < \ldots < t_n = T$.

Using the two sets of values $\{t_i, x_i\}$ and $\{t_i, y_i\}$ as interpolation data, we obtain the two splines $s_{k,x}$ and $s_{k,y}$, with respect to the independent variable t, that interpolate $x(t)$ and $y(t)$, respectively. The parametric curve $\mathbf{S}_k(t) = (s_{k,x}(t), s_{k,y}(t))$ is called the *parametric spline*. Obviously, different parameterizations of the interval $[0, T]$ yield different splines (see Figure 8.13).

A reasonable choice of the parameterization makes use of the length of each segment $\mathbf{P}_{i-1}\mathbf{P}_i$,

$$l_i = \sqrt{(x_i - x_{i-1})^2 + (y_i - y_{i-1})^2}, \; i = 1, \ldots, n.$$

Setting $t_0 = 0$ and $t_i = \sum_{k=1}^{i} l_k$ for $i = 1, \ldots, n$, every t_i represents the cumulative length of the piecewise line that joins the points from $\mathbf{P}_0$ to $\mathbf{P}_i$. This function is called the *cumulative length spline* and approximates satisfactorily even those curves with large curvature. Moreover, it can also be proved (see [SL89]) that it is geometrically invariant.

Program 68 implements the construction of cumulative parametric cubic splines in two dimensions (it can be easily generalized to the three-dimensional case). Composite parametric splines can be generated as well by enforcing suitable continuity conditions (see [SL89]).

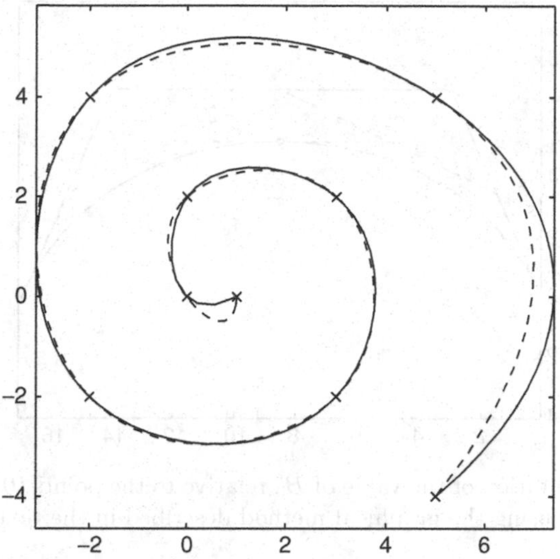

Fig. 8.13. Parametric splines for a spiral-like node distribution. The spline of cumulative length is drawn in solid line

Program 68 - parspline : Parametric splines

```
function [xi,yi] = parspline (x,y)
%PARSPLINE Parametric cubic spline interpolation
% [XI, YI] = PARSPLINE(X, Y) constructs a two-dimensional cumulative cubic
% spline. X and Y contain the interpolation data. XI and YI contain the
% parametric components of the cubic spline with respect to x and y axes.
t (1) = 0;
for i = 1:length (x)-1
   t (i+1) = t (i) + sqrt ( (x(i+1)-x(i))^2 + (y(i+1)-y(i))^2 );
end
z = [t(1):(t(length(t))-t(1))/100:t(length(t))];
xi = spline (t,x,z);
yi = spline (t,y,z);
```

8.8.1 Bézier Curves and Parametric B-splines

The Bézier curves and parametric B-splines are widely employed in graphical applications, where the nodes' locations might be affected by some uncertainty.

Let $\mathbf{P}_0, \mathbf{P}_1, \ldots, \mathbf{P}_n$ be $n+1$ points ordered in the plane. The oriented polygon formed by them is called the *characteristic polygon* or *Bézier polygon*. Let us introduce the Bernstein polynomials over the interval $[0,1]$ defined as

$$b_{n,k}(t) = \binom{n}{k} t^k (1-t)^{n-k} = \frac{n!}{k!(n-k)!} t^k (1-t)^{n-k},$$

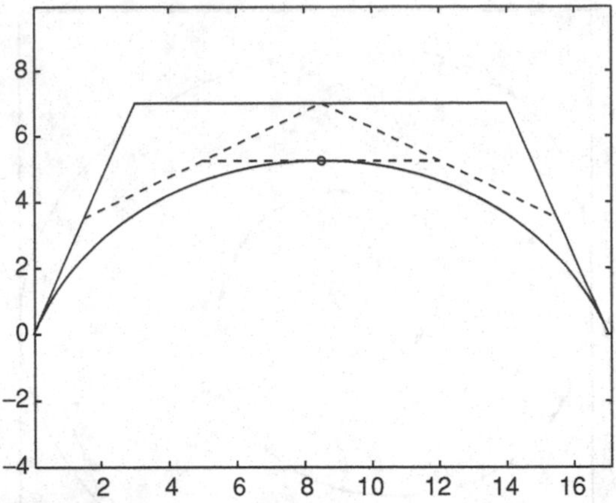

Fig. 8.14. Computation of the value of B_3 relative to the points (0,0), (3,7), (14,7), (17,0) for $t = 0.5$, using the graphical method described in the text

for $n = 0, 1, \ldots$ and $k = 0, \ldots, n$. They can be obtained by the following recursive formula

$$\begin{cases} b_{n,0}(t) = (1-t)^n, \\ b_{n,k}(t) = (1-t)b_{n-1,k}(t) + tb_{n-1,k-1}(t),\ k = 1, \ldots, n,\ t \in [0,1]. \end{cases}$$

It is easily seen that $b_{n,k} \in \mathbb{P}_n$, for $k = 0, \ldots, n$. Also, $\{b_{n,k},\ k = 0, \ldots, n\}$ provides a basis for $\mathbb{P}_n$. The Bézier curve is defined as follows

$$B_n(\mathbf{P}_0, \mathbf{P}_1, \ldots, \mathbf{P}_n, t) = \sum_{k=0}^{n} \mathbf{P}_k b_{n,k}(t), \qquad 0 \leq t \leq 1. \tag{8.62}$$

This expression can be regarded as a weighted average of the points $\mathbf{P}_k$, with weights $b_{n,k}(t)$.

The Bézier curves can also be obtained by a pure geometric approach starting from the characteristic polygon. Indeed, for any fixed $t \in [0,1]$, we define $\mathbf{P}_{i,1}(t) = (1-t)\mathbf{P}_i + t\mathbf{P}_{i+1}$ for $i = 0, \ldots, n-1$ and, for t fixed, the piecewise line that joins the new nodes $\mathbf{P}_{i,1}(t)$ forms a polygon of $n-1$ edges. We can now repeat the procedure by generating the new vertices $\mathbf{P}_{i,2}(t)$ ($i = 0, \ldots, n-2$), and terminating as soon as the polygon comprises only the vertices $\mathbf{P}_{0,n-1}(t)$ and $\mathbf{P}_{1,n-1}(t)$. It can be shown that

$$\mathbf{P}_{0,n}(t) = (1-t)\mathbf{P}_{0,n-1}(t) + t\mathbf{P}_{1,n-1}(t) = B_n(\mathbf{P}_0, \mathbf{P}_1, \ldots, \mathbf{P}_n, t),$$

that is, $\mathbf{P}_{0,n}(t)$ is equal to the value of the Bézier curve B_n at the points corresponding to the fixed value of t. Repeating the process for several values of the parameter t yields the construction of the curve in the considered region of the plane.

Notice that, for a given node configuration, several curves can be constructed according to the ordering of points $\mathbf{P}_i$. Moreover, the Bézier curve

$B_n(\mathbf{P}_0, \mathbf{P}_1, \ldots, \mathbf{P}_n, t)$ coincides with $B_n(\mathbf{P}_n, \mathbf{P}_{n-1}, \ldots, \mathbf{P}_0, t)$, apart from the orientation.

Program 69 computes $b_{n,k}$ at the point x for $x \in [0, 1]$.

Program 69 - bernstein : Bernstein polynomials

```
function [bnk]=bernstein (n,k,x)
%BERNSTEIN Bernstein polynomial of degree n
% [BNK] = BERNSTEIN(N, K, X) constructs the Bernstein polynomial b'n,k at X.
if k == 0,
  C = 1;
else,
  C = prod ([1:n])/( prod([1:k])*prod([1:n-k]));
end
bnk = C * x^k * (1-x)^(n-k);
```

Program 70 plots the Bézier curve relative to the set of points $(\mathbf{x}, \mathbf{y})$.

Program 70 - bezier : Bézier curves

```
function [bezx,bezy] = bezier (n, x, y)
%BEZIER Bezier curves
% [BEZX, BEZY] = BEZIER(N, X, Y) constructs the Bezier curve (BEZX, BEZY)
% associated with a given set of points (X,Y) in the plane.
i = 0; k = 0; for t = 0:0.01:1,
    i = i + 1; bnk = bernstein (n,k,t); ber(i) = bnk;
end
bezx = ber * x (1); bezy = ber * y (1);
for k = 1:n
  i = 0;
  for t = 0:0.01:1
    i = i + 1; bnk = bernstein (n,k,t); ber(i) = bnk;
  end
  bezx = bezx + ber * x (k+1); bezy = bezy + ber * y (k+1);
end
plot(bezx,bezy)
```

In practice, the Bézier curves are rarely used since they do not provide a sufficiently accurate approximation to the characteristic polygon. For this reason, in the 70's the *parametric B-splines* were introduced, and they are used in (8.62) instead of the Bernstein polynomials. Parametric B-splines are widely employed in packages for computer graphics since they enjoy the following properties:

1. perturbing a single vertex of the characteristic polygon yields a local perturbation of the curve only around the vertex itself;
2. the parametric B-spline better approximates the control polygon than the corresponding Bézier curve does, and it is always contained within the convex hull of the polygon.

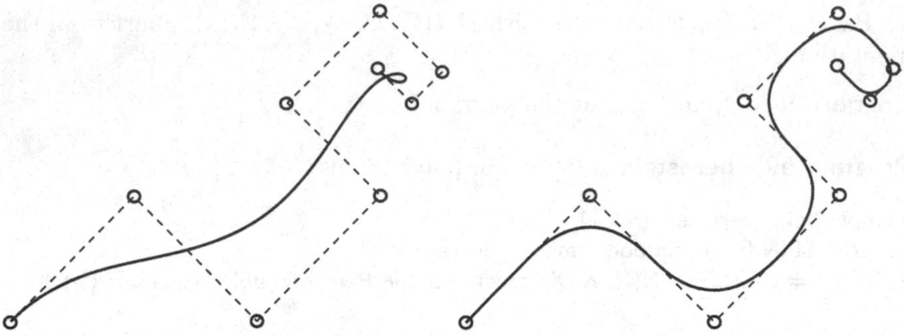

Fig. 8.15. Comparison of a Bézier curve (*left*) and a parametric B-spline (*right*). The vertices of the characteristic polygon are denoted by ∘

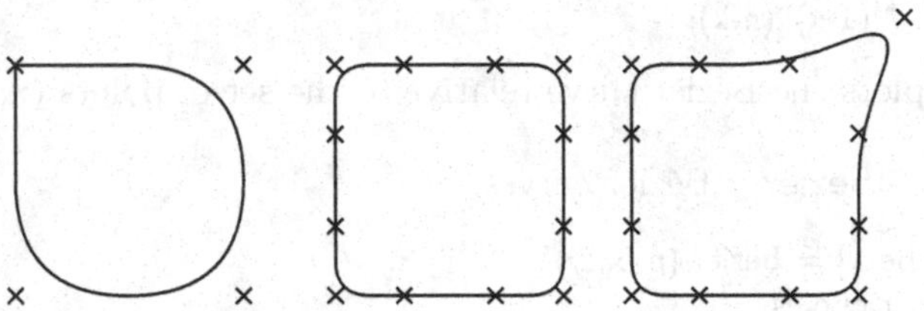

Fig. 8.16. Some parametric B-splines as functions of the number and positions of the vertices of the characteristic polygon. Notice in the third figure (*right*) the localization effects due to moving a single vertex

In Figure 8.15 a comparison is made between Bézier curves and parametric B-splines for the approximation of a given characteristic polygon.

We conclude this section by noticing that parametric cubic B-splines allow for obtaining locally straight lines by aligning four consecutive vertices (see Figure 8.16) and that a parametric B-spline can be constrained at a specific point of the characteristic polygon by simply making three consecutive points of the polygon coincide with the desired point.

8.9 Applications

In this section we consider two problems arising from the solution of fourth-order differential equations and from the reconstruction of images in axial tomographies.

8.9.1 Finite Element Analysis of a Clamped Beam

Let us employ piecewise Hermite polynomials (see Section 8.5) for the numerical approximation of the transversal bending of a clamped beam. This problem was already considered in Section 4.7.2 where centered finite differences were used.

The mathematical model is the fourth-order boundary value problem (4.76), here presented in the following general formulation

$$\begin{cases} (\alpha(x)u''(x))'' = f(x), & 0 < x < \mathcal{L}, \\ u(0) = u(\mathcal{L}) = 0, & u'(0) = u'(\mathcal{L}) = 0. \end{cases} \tag{8.63}$$

In the particular case of (4.76) we have $\alpha = EJ$ and $f = P$; we assume henceforth that α is a positive and bounded function over $(0, \mathcal{L})$ and that $f \in \mathrm{L}^2(0, \mathcal{L})$.

We multiply (8.63) by a sufficiently smooth arbitrary function v, then, we integrate by parts twice, to obtain

$$\int_0^{\mathcal{L}} \alpha u'' v'' dx - [\alpha u''' v]_0^{\mathcal{L}} + [\alpha u'' v']_0^{\mathcal{L}} = \int_0^{\mathcal{L}} f v dx.$$

Problem (8.63) is then replaced by the following problem in integral form

$$\text{find } u \in V \text{ such that } \int_0^{\mathcal{L}} \alpha u'' v'' dx = \int_0^{\mathcal{L}} f v dx, \qquad \forall v \in V, \tag{8.64}$$

where

$$V = \left\{ v : v^{(r)} \in \mathrm{L}^2(0, \mathcal{L}),\, r = 0, 1, 2,\, v^{(r)}(0) = v^{(r)}(\mathcal{L}) = 0,\, r = 0, 1 \right\}.$$

Problem (8.64) admits a unique solution, which represents the deformed configuration that minimizes the total potential energy of the beam over the space V (see, for instance, [Red86], p. 156)

$$J(u) = \int_0^{\mathcal{L}} \left(\frac{1}{2} \alpha (u'')^2 - f u \right) dx.$$

In view of the numerical solution of problem (8.64), we introduce a partition $\mathcal{T}_h$ of $[0, \mathcal{L}]$ into K subintervals $T_k = [x_{k-1}, x_k]$, $(k = 1, \ldots, K)$ of uniform length $h = \mathcal{L}/K$, with $x_k = kh$, and the finite dimensional space

$$\begin{aligned} V_h = \Big\{ & v_h \in C^1([0, \mathcal{L}]),\, v_h|_T \in \mathbb{P}_3(T) \\ & \forall T \in \mathcal{T}_h, v_h^{(r)}(0) = v_h^{(r)}(\mathcal{L}) = 0,\, r = 0, 1 \Big\}. \end{aligned} \tag{8.65}$$

Let us equip V_h with a basis. With this purpose, we associate with each internal node x_i $(i = 1, \ldots, K-1)$ a support $\sigma_i = T_i \cup T_{i+1}$ and *two* functions φ_i, ψ_i defined as follows: for any k, $\varphi_i|_{T_k} \in \mathbb{P}_3(T_k)$, $\psi_i|_{T_k} \in \mathbb{P}_3(T_k)$ and for any $j = 0, \ldots, K$,

$$\begin{cases} \varphi_i(x_j) = \delta_{ij}, \, \varphi_i'(x_j) = 0, \\ \psi_i(x_j) = 0, \quad \psi_i'(x_j) = \delta_{ij}. \end{cases} \tag{8.66}$$

Notice that the above functions belong to V_h and define a basis

$$B_h = \{\varphi_i, \, \psi_i, \, i = 1, \ldots, K-1\}. \tag{8.67}$$

These basis functions can be brought back to the reference interval $\hat{T} = [0,1]$ for $0 \leq \hat{x} \leq 1$, by the affine maps $x = h\hat{x} + x_{k-1}$ between $\hat{T}$ and T_k, for $k = 1, \ldots, K$.

Therefore, let us introduce on the interval $\hat{T}$ the basis functions $\hat{\varphi}_0^{(0)}$ and $\hat{\varphi}_0^{(1)}$, associated with the node $\hat{x} = 0$, and $\hat{\varphi}_1^{(0)}$ and $\hat{\varphi}_1^{(1)}$, associated with node $\hat{x} = 1$. Each of these is of the form $\hat{\varphi} = a_0 + a_1\hat{x} + a_2\hat{x}^2 + a_3\hat{x}^3$; in particular, the functions with superscript "0" must satisfy the first two conditions in (8.66), while those with superscript "1" must fulfill the remaining two conditions. Solving the (4×4) associated system, we get

$$\begin{aligned} &\hat{\varphi}_0^{(0)}(\hat{x}) = 1 - 3\hat{x}^2 + 2\hat{x}^3, \, \hat{\varphi}_0^{(1)}(\hat{x}) = \hat{x} - 2\hat{x}^2 + \hat{x}^3, \\ &\hat{\varphi}_1^{(0)}(\hat{x}) = 3\hat{x}^2 - 2\hat{x}^3, \qquad \hat{\varphi}_1^{(1)}(\hat{x}) = -\hat{x}^2 + \hat{x}^3. \end{aligned} \tag{8.68}$$

The graphs of the functions (8.68) are drawn in Figure 8.17 (*left*), where (0), (1), (2) and (3) denote $\hat{\varphi}_0^{(0)}$, $\hat{\varphi}_1^{(0)}$, $\hat{\varphi}_0^{(1)}$ and $\hat{\varphi}_1^{(1)}$, respectively.

The function $u_h \in V_h$ can be written as

$$u_h(x) = \sum_{i=1}^{K-1} u_i \varphi_i(x) + \sum_{i=1}^{K-1} u_i^{(1)} \psi_i(x). \tag{8.69}$$

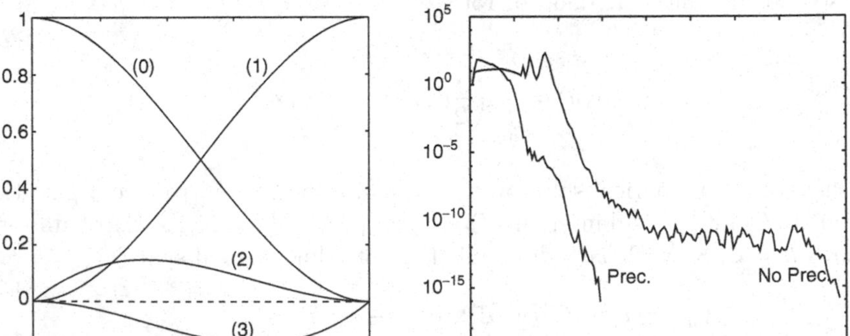

Fig. 8.17. Canonical Hermite basis on the reference interval $0 \leq \hat{x} \leq 1$ (*left*); convergence histories for the conjugate gradient method in the solution of system (8.73) (*right*). On the x-axis the number of iterations k is shown, while the y-axis represents the quantity $\|\mathbf{r}^{(k)}\|_2 / \|\mathbf{b}_1\|_2$, where $\mathbf{r}^{(k)}$ is the residual of system (8.73) at the k-th iteration

The coefficients and the *degrees of freedom* of u_h have the following meaning: $u_i = u_h(x_i)$, $u_i^{(1)}(x_i) = u_h'(x_i)$ for $i = 1, \ldots, K-1$. Notice that (8.69) is a special instance of (8.36), having set $m_i = 1$.

The discretization of problem (8.64) reads

$$\text{find } u_h \in V_h \text{ such that } \int_0^{\mathcal{L}} \alpha u_h'' v_h'' dx = \int_0^{\mathcal{L}} f v_h dx, \qquad \forall v_h \in B_h. \tag{8.70}$$

This is called the *Galerkin finite element* approximation of the differential problem (8.63). We refer to Chapter 12, Sections 12.4 and 12.4.5, for a more comprehensive discussion and analysis of the method.

Using the representation (8.69) we end up with the following system in the $2K-2$ unknowns $u_1, u_2, \ldots, u_{K-1}, u_1^{(1)}, u_2^{(1)}, \ldots u_{K-1}^{(1)}$

$$\begin{cases} \sum_{j=1}^{K-1} \left\{ u_j \int_0^{\mathcal{L}} \alpha \varphi_j'' \varphi_i'' dx + u_j^{(1)} \int_0^{\mathcal{L}} \alpha \psi_j'' \varphi_i'' dx \right\} = \int_0^{\mathcal{L}} f \varphi_i dx, \\ \sum_{j=1}^{K-1} \left\{ u_j \int_0^{\mathcal{L}} \alpha \varphi_j'' \psi_i'' dx + u_j^{(1)} \int_0^{\mathcal{L}} \alpha \psi_j'' \psi_i'' dx \right\} = \int_0^{\mathcal{L}} f \psi_i dx, \end{cases} \tag{8.71}$$

for $i = 1, \ldots, K-1$. Assuming, for the sake of simplicity, that the beam has unit length $\mathcal{L}$, that α and f are two constants and computing the integrals in (8.71), the final system reads in matrix form

$$\begin{cases} \mathrm{A}\mathbf{u} + \mathrm{B}\mathbf{p} = \mathbf{b}_1, \\ \mathrm{B}^T\mathbf{u} + \mathrm{C}\mathbf{p} = \mathbf{0}, \end{cases} \tag{8.72}$$

where the vectors $\mathbf{u}, \mathbf{p} \in \mathbb{R}^{K-1}$ contain the nodal unknowns u_i and $u_i^{(1)}$, $\mathbf{b}_1 \in \mathbb{R}^{K-1}$ is the vector of components equal to $h^4 f/\alpha$, while

$$\begin{aligned} \mathrm{A} &= \mathrm{tridiag}_{K-1}(-12, 24, -12), \\ \mathrm{B} &= \mathrm{tridiag}_{K-1}(-6, 0, 6), \\ \mathrm{C} &= \mathrm{tridiag}_{K-1}(2, 8, 2). \end{aligned}$$

System (8.72) has size equal to $2(K-1)$; eliminating the unknown $\mathbf{p}$ from the second equation, we get the reduced system (of size $K-1$)

$$\left(\mathrm{A} - \mathrm{B}\mathrm{C}^{-1}\mathrm{B}^T\right)\mathbf{u} = \mathbf{b}_1. \tag{8.73}$$

Since B is skew-symmetric and A is symmetric and positive definite (s.p.d.), the matrix $\mathrm{M} = \mathrm{A} - \mathrm{B}\mathrm{C}^{-1}\mathrm{B}^T$ is s.p.d. too. Using Cholesky factorization for solving system (8.73) is impractical as C^{-1} is full. An alternative is thus the conjugate gradient method (CG) supplied with a suitable preconditioner as the spectral condition number of M is of the order of $h^{-4} = K^4$.

We notice that computing the residual at each step $k \geq 0$ requires solving a linear system whose right side is the vector $\mathrm{B}^T\mathbf{u}^{(k)}$, $\mathbf{u}^{(k)}$ being the current iterate of CG method, and whose coefficient matrix is matrix C. This system can be solved using the Thomas algorithm (3.53) with a cost of the order of K flops.

The CG algorithm terminates in correspondence to the lowest value of k for which $\|\mathbf{r}^{(k)}\|_2 \leq \mathrm{u}\|\mathbf{b}_1\|_2$, where $\mathbf{r}^{(k)}$ is the residual of system (8.73) and u is the roundoff unit.

The results obtained running the CG method in the case of a uniform partition of $[0,1]$ with $K = 50$ elements and setting $\alpha = f = 1$ are summarized in Figure 8.17 (*right*), which shows the convergence histories of the method in both nonpreconditioned form (denoted by "Non Prec.") and with SSOR preconditioner (denoted by "Prec."), having set the relaxation parameter $\omega = 1.95$.

We notice that the CG method does not converge within $K - 1$ steps due to the effect of the rounding errors. Notice also the effectiveness of the SSOR preconditioner in terms of the reduction of the number of iterations. However, the high computational cost of this preconditioner prompts us to devise another choice. Looking at the structure of the matrix M a natural preconditioner is $\mathcal{M} = \mathrm{A} - \mathrm{B}\widetilde{\mathrm{C}}^{-1}\mathrm{B}^T$, where $\widetilde{\mathrm{C}}$ is the diagonal matrix whose entries are $\widetilde{c}_{ii} = \sum_{j=1}^{K-1} |c_{ij}|$. The matrix $\mathcal{M}$ is banded so that its inversion requires a strongly reduced cost than for the SSOR preconditioner. Moreover, as shown in Table 8.6, using $\mathcal{M}$ provides a dramatic decrease of the number of iterations to converge.

8.9.2 Geometric Reconstruction Based on Computer Tomographies

A typical application of the algorithms presented in Section 8.8 deals with the reconstruction of the three-dimensional structure of internal organs of human body based on *computer tomographies* (CT).

The CT usually provides a sequence of images which represent the sections of an organ at several horizontal planes; as a convention, we say that the CT produces sections of the x, y plane in correspondance of several values of z. The

Table 8.6. Number of iterations as a function of K

K	Without Precond.	SSOR	$\mathcal{M}$
25	51	27	12
50	178	61	25
100	685	118	33
200	2849	237	34

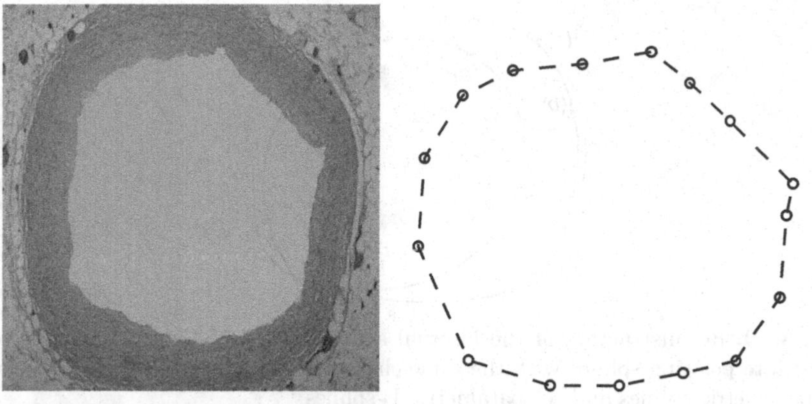

Fig. 8.18. Cross-section of a blood vessel (*left*) and an associated characteristic polygon using 16 points $\mathbf{P}_i$ (*right*)

result is analogous to what we would get by sectioning the organ at different values of z and taking the picture of the corresponding sections. Obviously, the great advantage in using the CT is that the organ under investigation can be visualized without being hidden by the neighboring ones, as happens in other kinds of medical images, e.g., angiographies.

The image that is obtained for each section is coded into a matrix of *pixels* (abbreviation of *pictures elements*) in the x, y plane; a certain value is associated with each pixel expressing the level of grey of the image at that point. This level is determined by the density of X rays which are collected by a detector after passing through the human body. In practice, the information contained in a CT at a given value of z is expressed by a set of points (x_i, y_i) which identify the boundary of the organ at z.

To improve the diagnostics it is often useful to reconstruct the three-dimensional structure of the organ under examination starting from the sections provided by the CT. With this aim, it is necessary to convert the information coded by pixels into a parametric representation which can be expressed by suitable functions interpolating the image at some significant points on its boundary. This reconstruction can be carried out by using the methods described in Section 8.8 as shown in Figure 8.19.
A set of curves like those shown in Figure 8.19 can be suitably stacked to provide an overall three-dimensional view of the organ under examination.

8.10 Exercises

1. Prove that the characteristic polynomials $l_i \in \mathbb{P}_n$ defined in (8.3) form a basis for $\mathbb{P}_n$.
2. An alternative approach to the method in Theorem 8.1, for constructing the interpolating polynomial, consists of directly enforcing the $n + 1$ interpolation

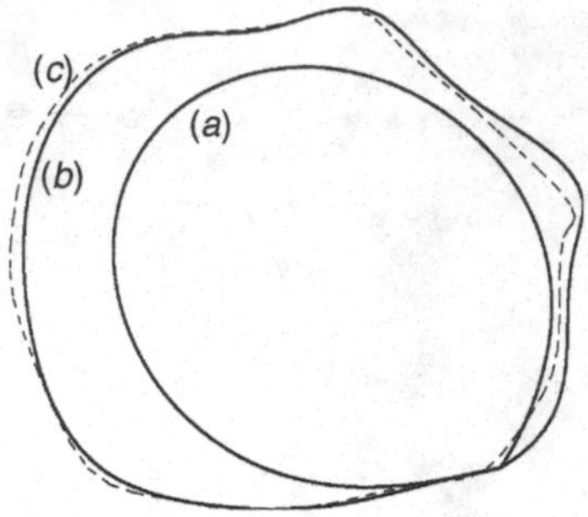

Fig. 8.19. Reconstruction of the internal arterial vessel of Figure 8.18 using different interpolating splines with the same characteristic polygon: (a) Bézier curves, (b) parametric splines and (c) parametric B-splines

constraints on Π_n and then computing the coefficients a_i. By doing so, we end up with a linear system X**a**= **y**, with $\mathbf{a} = (a_0, \ldots, a_n)^T$, $\mathbf{y} = (y_0, \ldots, y_n)^T$ and $\mathrm{X} = [x_i^{j-1}]$. X is called *Vandermonde matrix.* Prove that X is nonsingular if the nodes x_i are distinct.

[*Hint*: show that $\det(\mathrm{X}) = \prod_{0\le j<i\le n} (x_i - x_j)$ by recursion on n.]

3. Prove that $\omega'_{n+1}(x_i) = \prod_{\substack{j=0\\ j\neq i}}^{n} (x_i - x_j)$ where ω_{n+1} is the nodal polynomial (8.6).
   Then, check (8.5).
4. Provide an estimate of $\|\omega_{n+1}\|_\infty$, in the cases $n = 1$ and $n = 2$, for a distribution of equally spaced nodes.
5. Prove that

$$(n-1)!h^{n-1}|(x - x_{n-1})(x - x_n)| \le |\omega_{n+1}(x)| \le n!h^{n-1}|(x - x_{n-1})(x - x_n)|,$$

   where n is even, $-1 = x_0 < x_1 < \ldots < x_{n-1} < x_n = 1$, $x \in (x_{n-1}, x_n)$ and $h = 2/n$.
   [*Hint* : let $N = n/2$ and show first that

$$\begin{aligned}\omega_{n+1}(x) = (x + Nh)(x + (N-1)h)\ldots(x+h)x \\ (x-h)\ldots(x-(N-1)h)(x-Nh).\end{aligned} \tag{8.74}$$

   Then, take $x = rh$ with $N - 1 < r < N$.]
6. Under the assumptions of Exercise 5, show that $|\omega_{n+1}|$ is maximum if $x \in (x_{n-1}, x_n)$ (notice that $|\omega_{n+1}|$ is an even function).
   [*Hint* : use (8.74) to prove that $|\omega_{n+1}(x+h)/\omega_{n+1}(x)| > 1$ for any $x \in (0, x_{n-1})$ with x not coinciding with any interpolation node.]
7. Prove the recursive relation (8.19) for Newton divided differences.
8. Determine an interpolating polynomial $Hf \in \mathbb{P}_n$ such that

$$(Hf)^{(k)}(x_0) = f^{(k)}(x_0), \qquad k = 0, \ldots, n,$$

and check that

$$Hf(x) = \sum_{j=0}^{n} \frac{f^{(j)}(x_0)}{j!}(x - x_0)^j,$$

that is, the Hermite interpolating polynomial on one node coincides with the *Taylor polynomial.*

9. Given the following set of data

$$\left\{f_0 = f(-1) = 1,\ f_1 = f'(-1) = 1,\ f_2 = f'(1) = 2,\ f_3 = f(2) = 1\right\},$$

prove that the Hermite-Birkoff interpolating polynomial H_3 does not exist for them.
[*Solution* : letting $H_3(x) = a_3x^3 + a_2x^2 + a_1x + a_0$, one must check that the matrix of the linear system $H_3(x_i) = f_i$ for $i = 0, \ldots, 3$ is singular.]

10. Check that any $s_k \in \mathcal{S}_k[a, b]$ admits a representation of the form

$$s_k(x) = \sum_{i=0}^{k} b_i x^i + \sum_{i=1}^{g} c_i(x - x_i)_+^k,$$

that is, $1, x, x^2, \ldots, x^k, (x - x_1)_+^k, \ldots, (x - x_g)_+^k$ form a basis for $\mathcal{S}_k[a, b]$.

11. Prove Property 8.2 and check its validity even in the case where the spline s satisfies conditions of the form $s'(a) = f'(a)$, $s'(b) = f'(b)$.
[*Hint*: start from

$$\int_a^b \left[f''(x) - s''(x)\right] s''(x)dx = \sum_{i=1}^{n} \int_{x_{i-1}}^{x_i} \left[f''(x) - s''(x)\right] s''dx$$

and integrate by parts twice.]

12. Let $f(x) = \cos(x) = 1 - \frac{x^2}{2!} + \frac{x^4}{4!} - \frac{x^6}{6!} + \ldots$; then, consider the following rational approximation

$$r(x) = \frac{a_0 + a_2x^2 + a_4x^4}{1 + b_2x^2}, \tag{8.75}$$

called the *Padé approximation*. Determine the coefficients of r in such a way that

$$f(x) - r(x) = \gamma_8 x^8 + \gamma_{10} x^{10} + \ldots.$$

[*Solution*: $a_0 = 1$, $a_2 = -7/15$, $a_4 = 1/40$, $b_2 = 1/30$.]

13. Assume that the function f of the previous exercise is known at a set of n equally spaced points $x_i \in (-\pi/2, \pi/2)$ with $i = 0, \ldots, n$. Repeat Exercise 12, determining, by using MATLAB, the coefficients of r in such a way that the quantity $\sum_{i=0}^{n} |f(x_i) - r(x_i)|^2$ is minimized. Consider the cases $n = 5$ and $n = 10$.

9

Numerical Integration

In this chapter we present the most commonly used methods for numerical integration. We will mainly consider one-dimensional integrals over bounded intervals, although in Sections 9.8 and 9.9 an extension of the techniques to integration over unbounded intervals (or integration of functions with singularities) and to the multidimensional case will be considered.

9.1 Quadrature Formulae

Let f be a real integrable function over the interval $[a, b]$. Computing explicitly the definite integral $I(f) = \int_a^b f(x)dx$ may be difficult or even impossible. Any explicit formula that is suitable for providing an approximation of $I(f)$ is said to be a *quadrature formula* or *numerical integration formula.*

An example can be obtained by replacing f with an approximation f_n, depending on the integer $n \geq 0$, then computing $I(f_n)$ instead of $I(f)$. Letting $I_n(f) = I(f_n)$, we have

$$I_n(f) = \int_a^b f_n(x)dx, \qquad n \geq 0. \tag{9.1}$$

The dependence on the end points a, b is always understood, so we write $I_n(f)$ instead of $I_n(f; a, b)$.

If $f \in C^0([a, b])$, the *quadrature error* $E_n(f) = I(f) - I_n(f)$ satisfies

$$|E_n(f)| \leq \int_a^b |f(x) - f_n(x)|dx \leq (b-a)\|f - f_n\|_\infty.$$

Therefore, if for some n, $\|f - f_n\|_\infty < \varepsilon$, then $|E_n(f)| \leq \varepsilon(b - a)$.

The approximant f_n must be easily integrable, which is the case if, for example, $f_n \in \mathbb{P}_n$. In this respect, a natural approach consists of using

$f_n = \Pi_n f$, the interpolating Lagrange polynomial of f over a set of $n+1$ distinct nodes $\{x_i\}$, with $i = 0, \ldots, n$. By doing so, from (9.1) it follows that

$$I_n(f) = \sum_{i=0}^{n} f(x_i) \int_a^b l_i(x)dx, \tag{9.2}$$

where l_i is the characteristic Lagrange polynomial of degree n associated with node x_i (see Section 8.1). We notice that (9.2) is a special instance of the following quadrature formula

$$I_n(f) = \sum_{i=0}^{n} \alpha_i f(x_i), \tag{9.3}$$

where the coefficients α_i of the linear combination are given by $\int_a^b l_i(x)dx$.

Formula (9.3) is a weighted sum of the values of f at the points x_i, for $i = 0, \ldots, n$. These points are said to be the *nodes* of the quadrature formula, while the numbers $\alpha_i \in \mathbb{R}$ are its *coefficients* or *weights*. Both weights and nodes depend in general on n; again, for notational simplicity, this dependence is always understood.

Formula (9.2), called the *Lagrange* quadrature formula, can be generalized to the case where also the values of the derivative of f are available. This leads to the *Hermite* quadrature formula (see Section 9.5)

$$I_n(f) = \sum_{k=0}^{1} \sum_{i=0}^{n} \alpha_{ik} f^{(k)}(x_i), \tag{9.4}$$

where the weights are now denoted by α_{ik}.

Both (9.2) and (9.4) are *interpolatory quadrature formulae*, since the function f has been replaced by its interpolating polynomial (Lagrange and Hermite polynomials, respectively). We define the *degree of exactness* of a quadrature formula as the maximum integer $r \geq 0$ for which

$$I_n(f) = I(f), \qquad \forall f \in \mathbb{P}_r.$$

Any interpolatory quadrature formula that makes use of $n+1$ distinct nodes has degree of exactness equal to at least n. Indeed, if $f \in \mathbb{P}_n$, then $\Pi_n f = f$ and thus $I_n(\Pi_n f) = I(\Pi_n f)$. The converse statement is also true, that is, a quadrature formula using $n+1$ distinct nodes and having degree of exactness equal at least to n is necessarily of interpolatory type (for the proof see [IK66], p. 316).

As we will see in Section 10.2, the degree of exactness of a Lagrange quadrature formula can be as large as $2n+1$ in the case of the so-called Gaussian quadrature formulae.

9.2 Interpolatory Quadratures

We consider three remarkable instances of formula (9.2), corresponding to $n = 0$, 1 and 2.

9.2.1 The Midpoint or Rectangle Formula

This formula is obtained by replacing f over $[a, b]$ with the constant function equal to the value attained by f at the midpoint of $[a, b]$ (see Figure 9.1, left). This yields

$$I_0(f) = (b-a)f\left(\frac{a+b}{2}\right), \tag{9.5}$$

with weight $\alpha_0 = b - a$ and node $x_0 = (a + b)/2$. If $f \in C^2([a, b])$, the quadrature error is

$$E_0(f) = \frac{h^3}{3} f''(\xi),\ h = \frac{b-a}{2}, \tag{9.6}$$

where ξ lies within the interval (a, b).

Indeed, expanding f in a Taylor's series around $c = (a + b)/2$ and truncating at the second-order, we get

$$f(x) = f(c) + f'(c)(x-c) + f''(\eta(x))(x-c)^2/2,$$

from which, integrating on (a, b) and using the mean-value theorem, (9.6) follows. From this, it turns out that (9.5) is exact for constant and affine functions (since in both cases $f''(\xi) = 0$ for any $\xi \in (a, b)$), so that the midpoint rule has *degree of exactness* equal to 1.
It is worth noting that if the width of the integration interval $[a, b]$ is not sufficiently small, the quadrature error (9.6) can be quite large. This drawback is common to all the numerical integration formulae that will be described in the three forthcoming sections and can be overcome by resorting to their composite counterparts as discussed in Section 9.4.

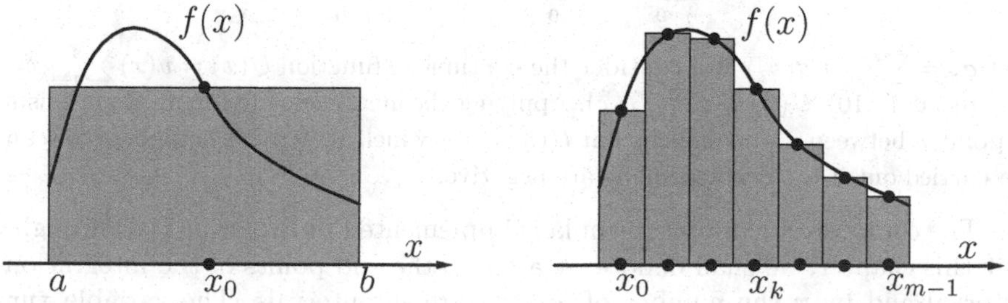

Fig. 9.1. The midpoint formula (*left*); the composite midpoint formula (*right*)

Suppose now that we approximate the integral $I(f)$ by replacing f over $[a,b]$ with its composite interpolating polynomial of degree zero, constructed on m subintervals of width $H=(b-a)/m$, for $m \geq 1$ (see Figure 9.1, right). Introducing the quadrature nodes $x_k = a+(2k+1)H/2$, for $k=0,\ldots,m-1$, we get the composite midpoint formula

$$I_{0,m}(f) = H\sum_{k=0}^{m-1} f(x_k), \qquad m \geq 1. \tag{9.7}$$

The quadrature error $E_{0,m}(f) = I(f) - I_{0,m}(f)$ is given by

$$E_{0,m}(f) = \frac{b-a}{24}H^2 f''(\xi),\ H = \frac{b-a}{m} \tag{9.8}$$

provided that $f \in C^2([a,b])$ and where $\xi \in (a,b)$. From (9.8) we conclude that (9.7) has degree of exactness equal to 1; (9.8) can be proved by recalling (9.6) and using the additivity of integrals. Indeed, for $k=0,\ldots,m-1$ and $\xi_k \in (a+kH, a+(k+1)H)$,

$$E_{0,m}(f) = \sum_{k=0}^{m-1} f''(\xi_k)(H/2)^3/3 = \sum_{k=0}^{m-1} f''(\xi_k)\frac{H^2}{24}\frac{b-a}{m} = \frac{b-a}{24}H^2 f''(\xi).$$

The last equality is a consequence of the following theorem, that is applied letting $u=f''$ and $\delta_j = 1$ for $j=0,\ldots,m-1$.

Theorem 9.1 (Discrete mean-value theorem) *Let $u \in C^0([a,b])$ and let x_j be $s+1$ points in $[a,b]$ and δ_j be $s+1$ constants, all having the same sign. Then there exists $\eta \in [a,b]$ such that*

$$\sum_{j=0}^{s} \delta_j u(x_j) = u(\eta)\sum_{j=0}^{s}\delta_j. \tag{9.9}$$

Proof. Let $u_m = \min_{x\in[a,b]} u(x) = u(\bar{x})$ and $u_M = \max_{x\in[a,b]} u(x) = u(\bar{\bar{x}})$, where $\bar{x}$ and $\bar{\bar{x}}$ are two points in (a,b). Then

$$u_m\sum_{j=0}^{s}\delta_j \leq \sum_{j=0}^{s}\delta_j u(x_j) \leq u_M\sum_{j=0}^{s}\delta_j. \tag{9.10}$$

Let $\sigma_s = \sum_{j=0}^{s}\delta_j u(x_j)$ and consider the continuous function $U(x) = u(x)\sum_{j=0}^{s}\delta_j$. Thanks to (9.10), $U(\bar{x}) \leq \sigma_s \leq U(\bar{\bar{x}})$. Applying the mean-value theorem, there exists a point η between a and b such that $U(\eta) = \sigma_s$, which is (9.9). A similar proof can be carried out if the coefficients δ_j are negative. ◇

The composite midpoint formula is implemented in Program 71. Throughout this chapter, we shall denote by `a` and `b` the end points of the integration interval and by `m` the number of quadrature subintervals. The variable `fun` contains the expression of the function f, while the output variable `int` contains the value of the approximate integral.

Program 71 - midpntc : Composite midpoint formula

```
function int = midpntc(a,b,m,fun)
%MIDPNTC Composite midpoint formula
% INT=MIDPNTC(A,B,M,FUN) computes an approximation of the integral of the
% function FUN over (A,B) via the midpoint method (with M equispaced intervals).
% FUN accepts a real vector input x and returns a real vector value.
h=(b-a)/m;
x=[a+h/2:h:b];
dim=length(x);
y=eval(fun);
if size(y)==1
   y=diag(ones(dim))*y;
end
int=h*sum(y);
return
```

9.2.2 The Trapezoidal Formula

This formula is obtained by replacing f with $\Pi_1 f$, its Lagrange interpolating polynomial of degree 1, relative to the nodes $x_0 = a$ and $x_1 = b$ (see Figure 9.2, left). The resulting quadrature, having nodes $x_0 = a$, $x_1 = b$ and weights $\alpha_0 = \alpha_1 = (b-a)/2$, is

$$I_1(f) = \frac{b-a}{2}\left[f(a) + f(b)\right]. \tag{9.11}$$

If $f \in C^2([a,b])$, the quadrature error is given by

$$E_1(f) = -\frac{h^3}{12} f''(\xi),\ h = b - a, \tag{9.12}$$

where ξ is a point within the integration interval.

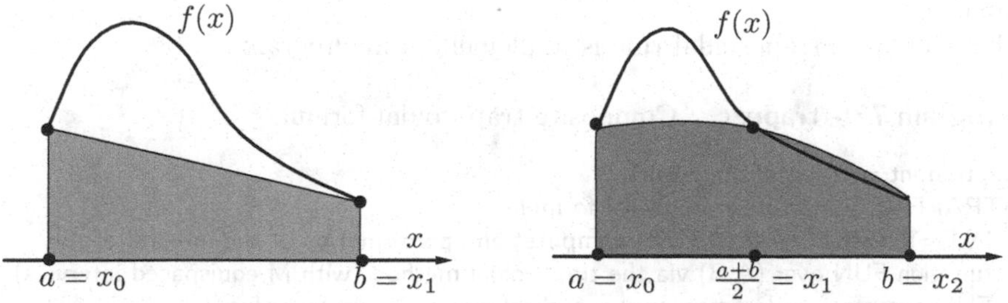

Fig. 9.2. Trapezoidal formula (*left*) and Cavalieri-Simpson formula (*right*)

Indeed, from the expression of the interpolation error (8.7) one gets

$$E_1(f) = \int_a^b (f(x) - \Pi_1 f(x))dx = -\frac{1}{2}\int_a^b f''(\xi(x))(x-a)(b-x)dx.$$

Since $\omega_2(x) = (x-a)(x-b) < 0$ in (a,b), the mean-value theorem yields

$$E_1(f) = (1/2)f''(\xi)\int_a^b \omega_2(x)dx = -f''(\xi)(b-a)^3/12,$$

for some $\xi \in (a,b)$, which is (9.12). The trapezoidal quadrature therefore has degree of exactness equal to 1, as is the case with the midpoint rule.

To obtain the composite trapezoidal formula, we proceed as in the case where $n = 0$, by replacing f over $[a,b]$ with its composite Lagrange polynomial of degree 1 on m subintervals, with $m \geq 1$. Introduce the quadrature nodes $x_k = a + kH$, for $k = 0, \ldots, m$ and $H = (b-a)/m$, getting

$$I_{1,m}(f) = \frac{H}{2}\sum_{k=0}^{m-1}(f(x_k) + f(x_{k+1})), \qquad m \geq 1. \tag{9.13}$$

Each term in (9.13) is counted twice, except the first and the last one, so that the formula can be written as

$$I_{1,m}(f) = H\left[\frac{1}{2}f(x_0) + f(x_1) + \ldots + f(x_{m-1}) + \frac{1}{2}f(x_m)\right]. \tag{9.14}$$

As was done for (9.8), it can be shown that the quadrature error associated with (9.14) is

$$E_{1,m}(f) = -\frac{b-a}{12}H^2 f''(\xi),$$

provided that $f \in C^2([a,b])$, where $\xi \in (a,b)$. The degree of exactness is again equal to 1.

The composite trapezoidal rule is implemented in Program 72.

Program 72 - trapezc : Composite trapezoidal formula

```
function int = trapezc(a,b,m,fun)
%TRAPEZC Composite trapezoidal formula
% INT=TRAPEZC(A,B,M,FUN) computes an approximation of the integral of the
% function FUN over (A,B) via the trapezoidal method (with M equispaced intervals).
% FUN accepts a real vector input x and returns a real vector value.
h=(b-a)/m;
x=[a:h:b];
dim=length(x);
```

```
y=eval(fun);
if size(y)==1
  y=diag(ones(dim))*y;
end
int=h*(0.5*y(1)+sum(y(2:m))+0.5*y(m+1));
return
```

9.2.3 The Cavalieri-Simpson Formula

The Cavalieri-Simpson formula can be obtained by replacing f over $[a, b]$ with its interpolating polynomial of degree 2 at the nodes $x_0 = a$, $x_1 = (a+b)/2$ and $x_2 = b$ (see Figure 9.2, right). The weights are given by $\alpha_0 = \alpha_2 = (b-a)/6$ and $\alpha_1 = 4(b-a)/6$, and the resulting formula reads

$$I_2(f) = \frac{b-a}{6}\left[f(a) + 4f\left(\frac{a+b}{2}\right) + f(b)\right]. \tag{9.15}$$

It can be shown that the quadrature error is

$$E_2(f) = -\frac{h^5}{90}f^{(4)}(\xi), \quad h = \frac{b-a}{2}, \tag{9.16}$$

provided that $f \in C^4([a, b])$, and where ξ lies within (a, b). From (9.16) it turns out that (9.15) has degree of exactness equal to 3.

Replacing f with its composite polynomial of degree 2 over $[a, b]$ yields the composite formula corresponding to (9.15). Introducing the quadrature nodes $x_k = a + kH/2$, for $k = 0, \ldots, 2m$ and letting $H = (b-a)/m$, with $m \geq 1$ gives

$$I_{2,m} = \frac{H}{6}\left[f(x_0) + 2\sum_{r=1}^{m-1} f(x_{2r}) + 4\sum_{s=0}^{m-1} f(x_{2s+1}) + f(x_{2m})\right]. \tag{9.17}$$

The quadrature error associated with (9.17) is

$$E_{2,m}(f) = -\frac{b-a}{180}(H/2)^4 f^{(4)}(\xi),$$

provided that $f \in C^4([a, b])$ and where $\xi \in (a, b)$; the degree of exactness of the formula is 3.

The composite Cavalieri-Simpson quadrature is implemented in Program 73.

Program 73 - simpsonc : Composite Cavalieri-Simpson formula

```
function int = simpsonc(a,b,m,fun)
%SIMPSONC Composite Simpson formula
% INT=SIMPSONC(A,B,M,FUN) computes an approximation of the integral of the
```

Table 9.1. Absolute error for midpoint, trapezoidal and Cavalieri-Simpson composite formulae in the approximate evaluation of integral (9.18)

m	$\|E_{0,m}\|$	$\mathcal{R}_m$	$\|E_{1,m}\|$	$\mathcal{R}_m$	$\|E_{2,m}\|$	$\mathcal{R}_m$
1	0.9751		1.589e-01		7.030e-01	
2	1.037	0.9406	0.5670	0.2804	0.5021	1.400
4	0.1221	8.489	0.2348	2.415	$3.139 \cdot 10^{-3}$	159.96
8	$2.980 \cdot 10^{-2}$	4.097	$5.635 \cdot 10^{-2}$	4.167	$1.085 \cdot 10^{-3}$	2.892
16	$6.748 \cdot 10^{-3}$	4.417	$1.327 \cdot 10^{-2}$	4.245	$7.381 \cdot 10^{-5}$	14.704
32	$1.639 \cdot 10^{-3}$	4.118	$3.263 \cdot 10^{-3}$	4.068	$4.682 \cdot 10^{-6}$	15.765
64	$4.066 \cdot 10^{-4}$	4.030	$8.123 \cdot 10^{-4}$	4.017	$2.936 \cdot 10^{-7}$	15.946
128	$1.014 \cdot 10^{-4}$	4.008	$2.028 \cdot 10^{-4}$	4.004	$1.836 \cdot 10^{-8}$	15.987
256	$2.535 \cdot 10^{-5}$	4.002	$5.070 \cdot 10^{-5}$	4.001	$1.148 \cdot 10^{-9}$	15.997

```
% function FUN over (A,B) via the Simpson method (with M equispaced intervals). FUN
% accepts a real vector input x and returns a real vector value.
h=(b-a)/m;
x=[a:h/2:b];
dim= length(x);
y=eval(fun);
if size(y)==1
  y=diag(ones(dim))*y;
end
int=(h/6)*(y(1)+2*sum(y(3:2:2*m-1))+4*sum(y(2:2:2*m))+y(2*m+1));
return
```

Example 9.1 Let us employ the midpoint, trapezoidal and Cavalieri-Simpson composite formulae to compute the integral

$$\int_0^{2\pi} xe^{-x}\cos(2x)dx = \frac{\left[3(e^{-2\pi}-1) - 10\pi e^{-2\pi}\right]}{25} \simeq -0.122122. \tag{9.18}$$

Table 9.1 shows in even columns the behavior of the absolute value of the error when halving H (thus, doubling m), while in odd columns the ratio $\mathcal{R}_m = |E_m|/|E_{2m}|$ between two consecutive errors is given. As predicted by the previous theoretical analysis, $\mathcal{R}_m$ tends to 4 for the midpoint and trapezoidal rules and to 16 for the Cavalieri-Simpson formula. •

9.3 Newton-Cotes Formulae

These formulae are based on Lagrange interpolation with *equally spaced* nodes in $[a, b]$. For a fixed $n \geq 0$, let us denote the quadrature nodes by $x_k = x_0 + kh$, $k = 0, \ldots, n$. The midpoint, trapezoidal and Simpson formulae are special instances of the Newton-Cotes formulae, taking $n = 0$, $n = 1$ and $n = 2$ respectively. In the general case, we define:

- *closed formulae*, those where $x_0 = a$, $x_n = b$ and $h = \dfrac{b-a}{n}$ $(n \geq 1)$;
- *open formulae*, those where $x_0 = a + h$, $x_n = b - h$ and $h = \dfrac{b-a}{n+2}$ $(n \geq 0)$.

A significant property of the Newton-Cotes formulae is that the quadrature weights α_i depend explicitly only on n and h, but not on the integration interval $[a, b]$. To check this property in the case of closed formulae, let us introduce the change of variable $x = \Psi(t) = x_0 + th$. Noting that $\Psi(0) = a$, $\Psi(n) = b$ and $x_k = a + kh$, we get

$$\frac{x - x_k}{x_i - x_k} = \frac{a + th - (a + kh)}{a + ih - (a + kh)} = \frac{t-k}{i-k}.$$

Therefore, if $n \geq 1$

$$l_i(x) = \prod_{k=0, k\neq i}^{n} \frac{t-k}{i-k} = \varphi_i(t), \qquad 0 \leq i \leq n.$$

The following expression for the quadrature weights is obtained

$$\alpha_i = \int_a^b l_i(x)dx = \int_0^n \varphi_i(t)hdt = h\int_0^n \varphi_i(t)dt,$$

from which we get the formula

$$I_n(f) = h\sum_{i=0}^{n} w_i f(x_i), \qquad w_i = \int_0^n \varphi_i(t)dt.$$

Open formulae can be interpreted in a similar manner. Actually, using again the mapping $x = \Psi(t)$, we get $x_0 = a + h$, $x_n = b - h$ and $x_k = a + h(k+1)$ for $k = 1, \ldots, n-1$. Letting, for sake of coherence, $x_{-1} = a$, $x_{n+1} = b$ and proceeding as in the case of closed formulae, we get $\alpha_i = h\int_{-1}^{n+1} \varphi_i(t)dt$, and thus

$$I_n(f) = h\sum_{i=0}^{n} w_i f(x_i), \qquad w_i = \int_{-1}^{n+1} \varphi_i(t)dt.$$

In the special case where $n = 0$, since $l_0(x) = \varphi_0(t) = 1$, we get $w_0 = 2$.

The coefficients w_i do not depend on a, b, h and f, but only depend on n, and can therefore be tabulated a priori. In the case of closed formulae, the polynomials φ_i and φ_{n-i}, for $i = 0, \ldots, n-1$, have by symmetry the same

Table 9.2. Weights of closed (*left*) and open Newton-Cotes formulae (*right*)

n	1	2	3	4	5	6
w_0	$\frac{1}{2}$	$\frac{1}{3}$	$\frac{3}{8}$	$\frac{14}{45}$	$\frac{95}{288}$	$\frac{41}{140}$
w_1	0	$\frac{4}{3}$	$\frac{9}{8}$	$\frac{64}{45}$	$\frac{375}{288}$	$\frac{216}{140}$
w_2	0	0	0	$\frac{24}{45}$	$\frac{250}{288}$	$\frac{27}{140}$
w_3	0	0	0	0	0	$\frac{272}{140}$

n	0	1	2	3	4	5
w_0	2	$\frac{3}{2}$	$\frac{8}{3}$	$\frac{55}{24}$	$\frac{66}{20}$	$\frac{4277}{1440}$
w_1	0	0	$-\frac{4}{3}$	$\frac{5}{24}$	$-\frac{84}{20}$	$-\frac{3171}{1440}$
w_2	0	0	0	0	$\frac{156}{20}$	$\frac{3934}{1440}$

integral, so that also the corresponding weights w_i and w_{n-i} are equal for $i = 0, \ldots, n-1$. In the case of open formulae, the weights w_i and w_{n-i} are equal for $i = 0, \ldots, n$. For this reason, we show in Table 9.2 only the first half of the weights.
Notice the presence of *negative weights* in open formulae for $n \geq 2$. This can be a source of numerical instability, in particular due to rounding errors.

Besides its degree of exactness, a quadrature formula can also be qualified by its *order of infinitesimal* with respect to the integration stepsize h, which is defined as the maximum integer p such that $|I(f) - I_n(f)| = \mathcal{O}(h^p)$. Regarding this, the following result holds

Theorem 9.2 *For any Newton-Cotes formula corresponding to an even value of n, the following error characterization holds*

$$E_n(f) = \frac{M_n}{(n+2)!} h^{n+3} f^{(n+2)}(\xi), \tag{9.19}$$

provided $f \in C^{n+2}([a,b])$, where $\xi \in (a,b)$ and

$$M_n = \begin{cases} \displaystyle\int_0^n t\ \pi_{n+1}(t)dt < 0 & \textit{for closed formulae,} \\ \displaystyle\int_{-1}^{n+1} t\ \pi_{n+1}(t)dt > 0 & \textit{for open formulae,} \end{cases}$$

having defined $\pi_{n+1}(t) = \prod_{i=0}^{n}(t-i)$. From (9.19), it turns out that the degree of exactness is equal to $n+1$ and the order of infinitesimal is $n+3$.

Similarly, for odd values of n, the following error characterization holds

$$E_n(f) = \frac{K_n}{(n+1)!} h^{n+2} f^{(n+1)}(\eta), \tag{9.20}$$

provided $f \in C^{n+1}([a,b])$, where $\eta \in (a,b)$ and

$$K_n = \begin{cases} \int_0^n \pi_{n+1}(t)dt < 0 \quad \textit{for closed formulae}, \\ \int_{-1}^{n+1} \pi_{n+1}(t)dt > 0 \ \textit{for open formulae}. \end{cases}$$

The degree of exactness is thus equal to n and the order of infinitesimal is n+2.

Proof. We give a proof in the particular case of closed formulae with n even, referring to [IK66], pp. 308-314, for a complete demonstration of the theorem.

Thanks to (8.20), we have

$$E_n(f) = I(f) - I_n(f) = \int_a^b f[x_0, \ldots, x_n, x]\omega_{n+1}(x)dx. \tag{9.21}$$

Set $W(x) = \int_a^x \omega_{n+1}(t)dt$. Clearly, $W(a) = 0$; moreover, $\omega_{n+1}(t)$ is an odd function with respect to the midpoint $(a+b)/2$ so that $W(b) = 0$. Integrating by parts (9.21) we get

$$E_n(f) = \int_a^b f[x_0, \ldots, x_n, x]W'(x)dx = -\int_a^b \frac{d}{dx} f[x_0, \ldots, x_n, x]W(x)dx$$

$$= -\int_a^b \frac{f^{(n+2)}(\xi(x))}{(n+2)!} W(x)dx.$$

In deriving the formula above we have used the following identity (see Exercise 4)

$$\frac{\mathrm{d}}{\mathrm{d}x} f[x_0, \ldots, x_n, x] = f[x_0, \ldots, x_n, x, x]. \tag{9.22}$$

Since $W(x) > 0$ for $a < x < b$ (see [IK66], p. 309), using the mean-value theorem we obtain

$$E_n(f) = -\frac{f^{(n+2)}(\xi)}{(n+2)!} \int_a^b W(x)dx = -\frac{f^{(n+2)}(\xi)}{(n+2)!} \int_a^b \int_a^x \omega_{n+1}(t)\ dt\ dx, \tag{9.23}$$

where ξ lies within (a, b). Exchanging the order of integration, letting $s = x_0 + \tau h$, for $0 \le \tau \le n$, and recalling that $a = x_0$, $b = x_n$, yields

$$\int_a^b W(x)dx = \int_a^b \int_s^b (s - x_0) \cdots (s - x_n)dxds$$

$$= \int_{x_0}^{x_n} (s - x_0) \cdots (s - x_{n-1})(s - x_n)(x_n - s)ds$$

$$= -h^{n+3} \int_0^n \tau(\tau - 1) \cdots (\tau - n + 1)(\tau - n)^2 d\tau.$$

Finally, letting $t = n - \tau$ and combining this result with (9.23), we get (9.19). ◇

Table 9.3. Degree of exactness and error constants for closed Newton-Cotes formulae

n	r_n	$\mathcal{M}_n$	$\mathcal{K}_n$	n	r_n	$\mathcal{M}_n$	$\mathcal{K}_n$	n	r_n	$\mathcal{M}_n$	$\mathcal{K}_n$
1	1		$\frac{1}{12}$	3	3		$\frac{3}{80}$	5	5		$\frac{275}{12096}$
2	3	$\frac{1}{90}$		4	5	$\frac{8}{945}$		6	7	$\frac{9}{1400}$	

Relations (9.19) and (9.20) are *a priori estimates* for the quadrature error (see Chapter 2, Section 2.3). Their use in generating *a posteriori estimates* of the error in the frame of adaptive algorithms will be examined in Section 9.7.

In the case of closed Newton-Cotes formulae, we show in Table 9.3, for $1 \leq n \leq 6$, the degree of exactness (that we denote henceforth by r_n) and the absolute value of the constant $\mathcal{M}_n = M_n/(n+2)!$ (if n is even) or $\mathcal{K}_n = K_n/(n+1)!$ (if n is odd).

Example 9.2 The purpose of this example is to assess the importance of the regularity assumption on f for the error estimates (9.19) and (9.20). Consider the closed Newton-Cotes formulae, for $1 \leq n \leq 6$, to approximate the integral $\int_0^1 x^{5/2}dx = 2/7 \simeq 0.2857$. Since f is only $C^2([0,1])$, we do not expect a substantial increase of the accuracy as n gets larger. Actually, this is confirmed by Table 9.4, where the results obtained by running Program 74 are reported.

For $n = 1, \ldots, 6$, we have denoted by $E_n^c(f)$ the module of the absolute error, by q_n^c the computed order of infinitesimal and by q_n^s the corresponding theoretical value predicted by (9.19) and (9.20) under optimal regularity assumptions for f. As is clearly seen, q_n^c is definitely less than the potential theoretical value q_n^s. •

Example 9.3 From a brief analysis of error estimates (9.19) and (9.20), we could be led to believe that only non-smooth functions can be a source of trouble when dealing with Newton-Cotes formulae. Thus, it is a little surprising to see results like those in Table 9.5, concerning the approximation of the integral

$$I(f) = \int_{-5}^{5} \frac{1}{1+x^2} dx = 2 \arctan 5 \simeq 2.747, \tag{9.24}$$

where $f(x) = 1/(1+x^2)$ is Runge's function (see Section 8.1.2), which belongs to $C^\infty(\mathbb{R})$. The results clearly demonstrate that the error remains almost unchanged

Table 9.4. Error in the approximation of $\int_0^1 x^{5/2}dx$

n	$E_n^c(f)$	q_n^c	q_n^s	n	$E_n^c(f)$	q_n^c	q_n^s
1	0.2143	3	3	4	$5.009 \cdot 10^{-5}$	4.7	7
2	$1.196 \cdot 10^{-3}$	3.2	5	5	$3.189 \cdot 10^{-5}$	2.6	7
3	$5.753 \cdot 10^{-4}$	3.8	5	6	$7.857 \cdot 10^{-6}$	3.7	9

Table 9.5. Relative error $E_n(f) = [I(f) - I_n(f)]/I(f)$ in the approximate evaluation of (9.24) using closed Newton-Cotes formulae

n	$E_n(f)$	n	$E_n(f)$	n	$E_n(f)$
1	0.8601	3	0.2422	5	0.1599
2	−1.474	4	0.1357	6	−0.4091

Table 9.6. Weights of the closed Newton-Cotes formula with 9 nodes

n	w_0	w_1	w_2	w_3	w_4	r_n	M_n
8	$\frac{3956}{14175}$	$\frac{23552}{14175}$	$-\frac{3712}{14175}$	$\frac{41984}{14175}$	$-\frac{18160}{14175}$	9	$\frac{2368}{467775}$

as n grows. This is due to the fact that singularities on the imaginary axis may also affect the convergence properties of a quadrature formula. This is indeed the case with the function at hand, which exhibits two singularities at $\pm\sqrt{-1}$ (see [DR75], pp. 64-66). •

To increase the accuracy of an interpolatory quadrature rule, it is by no means convenient to increase the value of n. By doing so, the same drawbacks of Lagrange interpolation on equally spaced nodes would arise. For example, the weights of the closed Newton-Cotes formula with $n = 8$ do not have the same sign (see Table 9.6 and recall that $w_i = w_{n-i}$ for $i = 0, \ldots, n-1$).

This can give rise to numerical instabilities, due to rounding errors (see Chapter 2), and makes this formula useless in the practice, as happens for all the Newton-Cotes formulae using more than 8 nodes. As an alternative, one can resort to composite formulae, whose error analysis is addressed in Section 9.4, or to Gaussian formulae, which will be dealt with in Chapter 10 and which yield maximum degree of exactness with a nonequally spaced nodes distribution.
The closed Newton-Cotes formulae, for $1 \le n \le 6$, are implemented in Program 74.

Program 74 - newtcot : Closed Newton-Cotes formulae

```
function int = newtcot(a,b,n,fun)
%NEWTCOT Newton-Cotes formulae.
% INT=NEWTCOT(A,B,N,FUN) computes an approximation of the integral of the
% function FUN over (A,B) via a closed Newton-Cotes formula with N nodes.
% FUN accepts a real vector input x and returns a real vector value.
h=(b-a)/n;
n2=fix(n/2);
if n > 6, error('Maximum value of n equal to 6'); end
a03=1/3; a08=1/8; a45=1/45; a288=1/288; a140=1/140;
alpha=[0.5    0      0       0; ...
       a03    4*a03  0       0; ...
       3*a08  9*a08  0       0; ...
```

```
      14*a45  64*a45   24*a45   0; ...
      95*a288 375*a288 250*a288 0; ...
      41*a140 216*a140 27*a140  272*a140];
x=a; y(1)=eval(fun);
for j=2:n+1
   x=x+h; y(j)=eval(fun);
end
int=0;
j=[1:n2+1];   int=sum(y(j).*alpha(n,j));
j=[n2+2:n+1]; int=int+sum(y(j).*alpha(n,n-j+2));
int=int*h;
return
```

9.4 Composite Newton-Cotes Formulae

The examples of Section 9.2 have already pointed out that composite Newton-Cotes formulae can be constructed by replacing f with its composite Lagrange interpolating polynomial, introduced in Section 8.4.

The general procedure consists of partitioning the integration interval $[a, b]$ into m subintervals $T_j = [y_j, y_{j+1}]$ such that $y_j = a + jH$ for $j = 0, \ldots, m$, where $H = (b - a)/m$. Then, over each subinterval, an interpolatory formula with nodes $\{x_k^{(j)}, 0 \leq k \leq n\}$ and weights $\{\alpha_k^{(j)}, 0 \leq k \leq n\}$ is used. Since

$$I(f) = \int_a^b f(x)dx = \sum_{j=0}^{m-1} \int_{T_j} f(x)dx,$$

a composite interpolatory quadrature formula is obtained by replacing $I(f)$ with

$$I_{n,m}(f) = \sum_{j=0}^{m-1} \sum_{k=0}^{n} \alpha_k^{(j)} f(x_k^{(j)}). \tag{9.25}$$

The quadrature error is defined as $E_{n,m}(f) = I(f) - I_{n,m}(f)$. In particular, over each subinterval T_j one can resort to a Newton-Cotes formula with $n + 1$ equally spaced nodes: in such a case, the weights $\alpha_k^{(j)} = hw_k$ are still independent of T_j.

Using the same notation as in Theorem 9.2, the following convergence result holds for composite formulae.

Theorem 9.3 *Let a composite Newton-Cotes formula, with n even, be used. If $f \in C^{n+2}([a, b])$, then*

$$E_{n,m}(f) = \frac{b-a}{(n+2)!} \frac{M_n}{\gamma_n^{n+3}} H^{n+2} f^{(n+2)}(\xi), \tag{9.26}$$

where $\xi \in (a,b)$. Therefore, the quadrature error is an infinitesimal in H of order $n+2$ and the formula has degree of exactness equal to $n+1$.

For a composite Newton-Cotes formula, with n odd, if $f \in C^{n+1}([a,b])$

$$E_{n,m}(f) = \frac{b-a}{(n+1)!}\frac{K_n}{\gamma_n^{n+2}}H^{n+1}f^{(n+1)}(\eta), \tag{9.27}$$

where $\eta \in (a,b)$. Thus, the quadrature error is an infinitesimal in H of order $n+1$ and the formula has degree of exactness equal to n. In (9.26) and (9.27), $\gamma_n = (n+2)$ if the formula is open while $\gamma_n = n$ if the formula is closed.

Proof. We only consider the case where n is even. Using (9.19), and noticing that M_n does not depend on the integration interval, we get

$$E_{n,m}(f) = \sum_{j=0}^{m-1}\left[I(f)|_{T_j} - I_n(f)|_{T_j}\right] = \frac{M_n}{(n+2)!}\sum_{j=0}^{m-1} h_j^{n+3} f^{(n+2)}(\xi_j),$$

where, for $j = 0,\ldots,(m-1)$, $h_j = |T_j|/(n+2) = (b-a)/(m(n+2))$; this time, ξ_j is a suitable point of T_j. Since $(b-a)/m = H$, we obtain

$$E_{n,m}(f) = \frac{M_n}{(n+2)!}\frac{b-a}{m(n+2)^{n+3}}H^{n+2}\sum_{j=0}^{m-1} f^{(n+2)}(\xi_j),$$

from which, applying Theorem 9.1 with $u(x) = f^{(n+2)}(x)$ and $\delta_j = 1$ for $j = 0,\ldots,m-1$, (9.26) immediately follows. A similar procedure can be followed to prove (9.27). ◇

We notice that, for n fixed, $E_{n,m}(f) \to 0$ as $m \to \infty$ (i.e., as $H \to 0$). This ensures the convergence of the numerical integral to the exact value $I(f)$. We notice also that the degree of exactness of composite formulae coincides with that of simple formulae, whereas its order of infinitesimal (with respect to H) is reduced by 1 with respect to the order of infinitesimal (in h) of simple formulae.

In practical computations, it is convenient to resort to a local interpolation of low degree (typically $n \leq 2$, as done in Section 9.2), this leads to composite quadrature rules with positive weights, with a minimization of the rounding errors.

Example 9.4 For the same integral (9.24) considered in Example 9.3, we show in Table 9.7 the behavior of the absolute error as a function of the number of subintervals m, in the case of the composite midpoint, trapezoidal and Cavalieri-Simpson formulae. Convergence of $I_{n,m}(f)$ to $I(f)$ as m increases can be clearly observed. Moreover, we notice that $E_{0,m}(f) \simeq E_{1,m}(f)/2$ for $m \geq 32$ (see Exercise 1). •

Convergence of $I_{n,m}(f)$ to $I(f)$ can be established under less stringent regularity assumptions on f than those required by Theorem 9.3. In this regard, the following result holds (see for the proof [IK66], pp. 341-343).

Table 9.7. Absolute error for composite quadratures in the computation of (9.24)

m	$\lvert E_{0,m}\rvert$	$\lvert E_{1,m}\rvert$	$\lvert E_{2,m}\rvert$
1	7.253	2.362	4.04
2	1.367	2.445	$9.65 \cdot 10^{-2}$
8	$3.90 \cdot 10^{-2}$	$3.77 \cdot 10^{-2}$	$1.35 \cdot 10^{-2}$
32	$1.20 \cdot 10^{-4}$	$2.40 \cdot 10^{-4}$	$4.55 \cdot 10^{-8}$
128	$7.52 \cdot 10^{-6}$	$1.50 \cdot 10^{-5}$	$1.63 \cdot 10^{-10}$
512	$4.70 \cdot 10^{-7}$	$9.40 \cdot 10^{-7}$	$6.36 \cdot 10^{-13}$

Property 9.1 *Let $f \in C^0([a,b])$ and assume that the weights $\alpha_k^{(j)}$ in (9.25) are nonnegative. Then*

$$\lim_{m\to\infty} I_{n,m}(f) = \int_a^b f(x)dx, \qquad \forall n \geq 0.$$

Moreover

$$\left| \int_a^b f(x)dx - I_{n,m}(f) \right| \leq 2(b-a)\Omega(f;H),$$

where

$$\Omega(f;H) = \sup\{|f(x) - f(y)|,\, x, y \in [a,b],\, x \neq y,\, |x-y| \leq H\}$$

is the module of continuity of function f.

9.5 Hermite Quadrature Formulae

Thus far we have considered quadrature formulae based on Lagrange interpolation (simple or composite). More accurate formulae can be devised by resorting to Hermite interpolation (see Section 8.5).

Suppose that $2(n+1)$ values $f(x_k)$, $f'(x_k)$ are available at $n+1$ distinct points $x_0, \ldots, x_n$, then the Hermite interpolating polynomial of f is given by

$$H_{2n+1}f(x) = \sum_{i=0}^{n} f(x_i)\mathcal{L}_i(x) + \sum_{i=0}^{n} f'(x_i)\mathcal{M}_i(x), \tag{9.28}$$

where the polynomials $\mathcal{L}_k, \mathcal{M}_k \in \mathbb{P}_{2n+1}$ are defined, for $k = 0, \ldots, n$, as

$$\mathcal{L}_k(x) = \left[1 - \frac{\omega''_{n+1}(x_k)}{\omega'_{n+1}(x_k)}(x - x_k)\right] l_k^2(x), \qquad \mathcal{M}_k(x) = (x - x_k) l_k^2(x).$$

Integrating (9.28) over $[a,b]$, we get the quadrature formula of type (9.4)

$$I_n(f) = \sum_{k=0}^{n} \alpha_k f(x_k) + \sum_{k=0}^{n} \beta_k f'(x_k), \tag{9.29}$$

where

$$\alpha_k = I(\mathcal{L}_k),\ \beta_k = I(\mathcal{M}_k),\ k = 0, \ldots, n.$$

Formula (9.29) has degree of exactness equal to $2n+1$. Taking $n = 1$, the so-called *corrected trapezoidal formula* is obtained

$$I_1^{corr}(f) = \frac{b-a}{2}\left[f(a) + f(b)\right] + \frac{(b-a)^2}{12}\left[f'(a) - f'(b)\right], \tag{9.30}$$

with weights $\alpha_0 = \alpha_1 = (b-a)/2$, $\beta_0 = (b-a)^2/12$ and $\beta_1 = -\beta_0$. Assuming $f \in C^4([a,b])$, the quadrature error associated with (9.30) is

$$E_1^{corr}(f) = \frac{h^5}{720} f^{(4)}(\xi), \qquad h = b - a, \tag{9.31}$$

with $\xi \in (a,b)$. Notice the increase of accuracy from $\mathcal{O}(h^3)$ to $\mathcal{O}(h^5)$ with respect to the corresponding expression (9.12) (of the same order as the Cavalieri-Simpson formula (9.15)). The composite formula can be generated in a similar manner

$$\begin{aligned} I_{1,m}^{corr}(f) = \frac{b-a}{m}\Big\{\frac{1}{2}\left[f(x_0) + f(x_m)\right] \\ + f(x_1) + \ldots + f(x_{m-1})\Big\} + \frac{(b-a)^2}{12m^2}\left[f'(a) - f'(b)\right], \end{aligned} \tag{9.32}$$

where the assumption that $f \in C^1([a,b])$ gives rise to the cancellation of the first derivatives at the nodes x_k, with $k = 1, \ldots, m-1$.

Example 9.5 Let us check experimentally the error estimate (9.31) in the simple ($m = 1$) and composite ($m > 1$) cases, running Program 75 for the approximate computation of integral (9.18). Table 9.8 reports the behavior of the module of the absolute error as H is halved (that is, m is doubled) and the ratio $\mathcal{R}_m$ between two consecutive errors. This ratio, as happens in the case of Cavalieri-Simpson formula, tends to 16, demonstrating that formula (9.32) has order of infinitesimal equal to 4. Comparing Table 9.8 with the corresponding Table 9.1, we can also notice that $|E_{1,m}^{corr}(f)| \simeq 4|E_{2,m}(f)|$ (see Exercise 9). •

The corrected composite trapezoidal quadrature is implemented in Program 75, where **dfun** contains the expression of the derivative of f.

Table 9.8. Absolute error for the corrected trapezoidal formula in the computation of $I(f) = \int_0^{2\pi} xe^{-x}\cos(2x)dx$

m	$E_{1,m}^{corr}(f)$	$\mathcal{R}_m$	m	$E_{1,m}^{corr}(f)$	$\mathcal{R}_m$	m	$E_{1,m}^{corr}(f)$	$\mathcal{R}_m$
1	3.4813		8	$4.4 \cdot 10^{-3}$	6.1	64	$1.1 \cdot 10^{-6}$	15.957
2	1.398	2.4	16	$2.9 \cdot 10^{-4}$	14.9	128	$7.3 \cdot 10^{-8}$	15.990
4	$2.72 \cdot 10^{-2}$	51.4	32	$1.8 \cdot 10^{-5}$	15.8	256	$4.5 \cdot 10^{-9}$	15.997

Program 75 - trapmodc : Composite corrected trapezoidal formula

```
function int = trapmodc(a,b,m,fun,dfun)
%TRAPMODC Composite corrected trapezoidal formula
% INT=TRAPMODC(A,B,M,FUN,DFUN) computes an approximation of the inte-
gral of the
% function FUN over (A,B) via the corrected trapezoidal method (with M equispaced
% intervals). FUN and DFUN accept a real vector input x and returns a real vector value.
h=(b-a)/m;
x=[a:h:b];
y=eval(fun);
x=a; f1a=eval(dfun);
x=b; f1b=eval(dfun);
int=h*(0.5*y(1)+sum(y(2:m))+0.5*y(m+1))+(h^2/12)*(f1a-f1b);
return
```

9.6 Richardson Extrapolation

The *Richardson extrapolation method* is a procedure which combines several approximations of a certain quantity α_0 in a smart way to yield a more accurate approximation of α_0. More precisely, assume that a method is available to approximate α_0 by a quantity $\mathcal{A}(h)$ that is computable for any value of the parameter $h \neq 0$. Moreover, assume that, for a suitable $k \geq 0$, $\mathcal{A}(h)$ can be expanded as follows

$$\mathcal{A}(h) = \alpha_0 + \alpha_1 h + \ldots + \alpha_k h^k + \mathcal{R}_{k+1}(h), \tag{9.33}$$

where $|\mathcal{R}_{k+1}(h)| \leq C_{k+1} h^{k+1}$. The constants C_{k+1} and the coefficients α_i, for $i = 0, \ldots, k$, are independent of h. Henceforth, $\alpha_0 = \lim_{h \to 0} \mathcal{A}(h)$.

Writing (9.33) with δh instead of h, for $0 < \delta < 1$ (typically, $\delta = 1/2$), we get

$$\mathcal{A}(\delta h) = \alpha_0 + \alpha_1(\delta h) + \ldots + \alpha_k(\delta h)^k + \mathcal{R}_{k+1}(\delta h).$$

Subtracting (9.33) multiplied by δ from this expression then yields

$$\mathcal{B}(h) = \frac{\mathcal{A}(\delta h) - \delta\mathcal{A}(h)}{1-\delta} = \alpha_0 + \widetilde{\alpha}_2 h^2 + \ldots + \widetilde{\alpha}_k h^k + \widetilde{\mathcal{R}}_{k+1}(h),$$

having defined, for $k \geq 2$, $\widetilde{\alpha}_i = \alpha_i(\delta^i - \delta)/(1-\delta)$, for $i = 2, \ldots, k$ and $\widetilde{\mathcal{R}}_{k+1}(h) = [\mathcal{R}_{k+1}(\delta h) - \delta\mathcal{R}_{k+1}(h)]/(1-\delta)$.

Notice that $\widetilde{\alpha}_i \neq 0$ iff $\alpha_i \neq 0$. In particular, if $\alpha_1 \neq 0$, then $\mathcal{A}(h)$ is a first-order approximation of α_0, while $\mathcal{B}(h)$ is at least second-order accurate. More generally, if $\mathcal{A}(h)$ is an approximation of α_0 of order p, then the quantity $\mathcal{B}(h) = [\mathcal{A}(\delta h) - \delta^p\mathcal{A}(h)]/(1-\delta^p)$ approximates α_0 up to order $p+1$ (at least).

Proceeding by induction, the following Richardson extrapolation algorithm is generated: setting $n \geq 0$, $h > 0$ and $\delta \in (0,1)$, we construct the sequences

$$\begin{array}{ll} \mathcal{A}_{m,0} = \mathcal{A}(\delta^m h), & m = 0, \ldots, n, \\ \mathcal{A}_{m,q+1} = \dfrac{\mathcal{A}_{m,q} - \delta^{q+1}\mathcal{A}_{m-1,q}}{1-\delta^{q+1}}, & q = 0, \ldots, n-1, \\ & m = q+1, \ldots, n, \end{array} \tag{9.34}$$

which can be represented by the diagram below

$$\begin{array}{ccccccccc} \mathcal{A}_{0,0} & & & & & & & & \\ & \searrow & & & & & & & \\ \mathcal{A}_{1,0} & \rightarrow & \mathcal{A}_{1,1} & & & & & & \\ & \searrow & & \searrow & & & & & \\ \mathcal{A}_{2,0} & \rightarrow & \mathcal{A}_{2,1} & \rightarrow & \mathcal{A}_{2,2} & & & & \\ & \searrow & & \searrow & & \searrow & & & \\ \mathcal{A}_{3,0} & \rightarrow & \mathcal{A}_{3,1} & \rightarrow & \mathcal{A}_{3,2} & \rightarrow & \mathcal{A}_{3,3} & & \\ & \searrow & & \searrow & & \searrow & & \searrow & \\ \vdots & \ddots & & \ddots & & \ddots & & \ddots & \\ & \searrow & & \searrow & & \searrow & & \searrow & \\ \mathcal{A}_{n,0} & \rightarrow & \mathcal{A}_{n,1} & \rightarrow & \mathcal{A}_{n,2} & \rightarrow & \mathcal{A}_{n,3} & \ldots \rightarrow & \mathcal{A}_{n,n} \end{array}$$

where the arrows indicate the way the terms which have been already computed contribute to the construction of the "new" ones.

The following result can be proved (see [Com95], Proposition 4.1).

Property 9.2 *For $n \geq 0$ and $\delta \in (0,1)$*

$$\mathcal{A}_{m,n} = \alpha_0 + \mathcal{O}((\delta^m h)^{n+1}), \qquad m = 0, \ldots, n. \tag{9.35}$$

In particular, for the terms in the first column ($n = 0$) the convergence rate to α_0 is $\mathcal{O}((\delta^m h))$, while for those of the last one it is $\mathcal{O}((\delta^m h)^{n+1})$, i.e., n times higher.

Example 9.6 Richardson extrapolation has been employed to approximate at $\overline{x} = 0$ the derivative of the function $f(x) = xe^{-x}\cos(2x)$, introduced in Example 9.1. For this purpose, algorithm (9.34) has been executed with $\mathcal{A}(h) = [f(\overline{x}+h) - f(\overline{x})]/h$, $\delta = 0.5$, $n = 5$ and $h = 0.1$. Table 9.9 reports the sequence of absolute errors $E_{m,k} = |\alpha_0 - \mathcal{A}_{m,k}|$. The results demonstrate that the error decays as predicted by (9.35). •

9.6.1 Romberg Integration

The *Romberg integration method* is an application of Richardson extrapolation to the composite trapezoidal rule. The following result, known as the Euler-MacLaurin formula, will be useful (for its proof see, e.g., [Ral65], pp. 131-133, and [DR75], pp. 106-111).

Table 9.9. Errors in the Richardson extrapolation for the approximate evaluation of $f'(0)$ where $f(x) = xe^{-x}\cos(2x)$

$E_{m,0}$	$E_{m,1}$	$E_{m,2}$	$E_{m,3}$	$E_{m,4}$	$E_{m,5}$
0.113	–	–	–	–	–
$5.3 \cdot 10^{-2}$	$6.1 \cdot 10^{-3}$	–	–	–	–
$2.6 \cdot 10^{-2}$	$1.7 \cdot 10^{-3}$	$2.2 \cdot 10^{-4}$	–	–	–
$1.3 \cdot 10^{-2}$	$4.5 \cdot 10^{-4}$	$2.8 \cdot 10^{-5}$	$5.5 \cdot 10^{-7}$	–	–
$6.3 \cdot 10^{-3}$	$1.1 \cdot 10^{-4}$	$3.5 \cdot 10^{-6}$	$3.1 \cdot 10^{-8}$	$3.0 \cdot 10^{-9}$	–
$3.1 \cdot 10^{-3}$	$2.9 \cdot 10^{-5}$	$4.5 \cdot 10^{-7}$	$1.9 \cdot 10^{-9}$	$9.9 \cdot 10^{-11}$	$4.9 \cdot 10^{-12}$

Property 9.3 *Let $f \in C^{2k+2}([a,b])$, for $k \geq 0$, and let us approximate $\alpha_0 = \int_a^b f(x)dx$ by the composite trapezoidal rule (9.14). Letting $h_m = (b-a)/m$ for $m \geq 1$,*

$$
\begin{aligned}
I_{1,m}(f) = \alpha_0 + \sum_{i=1}^{k} \frac{B_{2i}}{(2i)!} h_m^{2i} \left(f^{(2i-1)}(b) - f^{(2i-1)}(a) \right) \\
+ \frac{B_{2k+2}}{(2k+2)!} h_m^{2k+2} (b-a) f^{(2k+2)}(\eta),
\end{aligned}
\tag{9.36}
$$

where $\eta \in (a,b)$ and $B_{2j} = (-1)^{j-1}\left[\sum_{n=1}^{+\infty} 2/(2n\pi)^{2j}\right](2j)!$, for $j \geq 1$, are the Bernoulli numbers.

Equation (9.36) is a special case of (9.33) where $h = h_m^2$ and $\mathcal{A}(h) = I_{1,m}(f)$; notice that *only even powers* of the parameter h appear in the expansion.

The Richardson extrapolation algorithm (9.34) applied to (9.36) gives

$$
\begin{aligned}
\mathcal{A}_{m,0} &= \mathcal{A}(\delta^m h), & m &= 0, \ldots, n, \\
\mathcal{A}_{m,q+1} &= \frac{\mathcal{A}_{m,q} - \delta^{2(q+1)} \mathcal{A}_{m-1,q}}{1 - \delta^{2(q+1)}}, & q &= 0, \ldots, n-1, \\
& & m &= q+1, \ldots, n.
\end{aligned}
\tag{9.37}
$$

Setting $h = b - a$ and $\delta = 1/2$ into (9.37) and denoting by $T(h_s) = I_{1,s}(f)$ the composite trapezoidal formula (9.14) over $s = 2^m$ subintervals of width $h_s = (b-a)/2^m$, for $m \geq 0$, the algorithm (9.37) becomes

$$
\begin{aligned}
\mathcal{A}_{m,0} &= T((b-a)/2^m), & m &= 0, \ldots, n, \\
\mathcal{A}_{m,q+1} &= \frac{4^{q+1}\mathcal{A}_{m,q} - \mathcal{A}_{m-1,q}}{4^{q+1} - 1}, & q &= 0, \ldots, n-1, \\
& & m &= q+1, \ldots, n.
\end{aligned}
$$

This is the Romberg numerical integration algorithm. Recalling (9.35), the following convergence result holds for Romberg integration

$$\mathcal{A}_{m,n} = \int_a^b f(x)dx + \mathcal{O}(h_s^{2(n+1)}), \quad n \geq 0.$$

Example 9.7 Table 9.10 shows the results obtained by running Program 76 to compute the quantity α_0 in the two cases $\alpha_0^{(1)} = \int_0^\pi e^x \cos(x)dx = -(e^\pi + 1)/2$ and $\alpha_0^{(2)} = \int_0^1 \sqrt{x}dx = 2/3$.

The maximum size n has been set equal to 9. In the second and third columns we show the modules of the absolute errors $E_k^{(r)} = |\alpha_0^{(r)} - \mathcal{A}_{k+1,k+1}^{(r)}|$, for $r = 1, 2$ and $k = 0, \ldots, 6$.

The convergence to zero is much faster for $E_k^{(1)}$ than for $E_k^{(2)}$. Indeed, the first integrand function is infinitely differentiable whereas the second is only continuous. •

The Romberg algorithm is implemented in Program 76.

Program 76 - romberg : Romberg integration

```
function int = romberg(a,b,n,fun)
%ROMBERG Romberg integration
% INT=ROMBERG(A,B,N,FUN) computes an approximation of the integral of the
% function FUN over (A,B) via the Romberg method. FUN accepts a real vector
% input x and returns a real vector value.
for i=1:n+1
  A(i,1)=trapezc(a,b,2^(i-1),fun);
end
for j=2:n+1
  for i=j:n+1
    A(i,j)=(4^(j-1)*A(i,j-1)-A(i-1,j-1))/(4^(j-1)-1);
  end
end
int=A(n+1,n+1);
return
```

Table 9.10. Romberg integration for the approximate evaluation of $\int_0^\pi e^x \cos(x)dx$ (error $E_k^{(1)}$) and $\int_0^1 \sqrt{x}dx$ (error $E_k^{(2)}$)

k	$E_k^{(1)}$	$E_k^{(2)}$	k	$E_k^{(1)}$	$E_k^{(2)}$
0	22.71	0.1670	4	$8.923 \cdot 10^{-7}$	$1.074 \cdot 10^{-3}$
1	0.4775	$2.860 \cdot 10^{-2}$	5	$6.850 \cdot 10^{-11}$	$3.790 \cdot 10^{-4}$
2	$5.926 \cdot 10^{-2}$	$8.910 \cdot 10^{-3}$	6	$5.330 \cdot 10^{-14}$	$1.340 \cdot 10^{-4}$
3	$7.410 \cdot 10^{-5}$	$3.060 \cdot 10^{-3}$	7	0	$4.734 \cdot 10^{-5}$

9.7 Automatic Integration

An *automatic numerical integration* program, or *automatic integrator*, is a set of algorithms which yield an approximation of the integral $I(f) = \int_a^b f(x)dx$, within a given tolerance, ε_a, or relative tolerance, ε_r, prescribed by the user.

With this aim, the program generates a sequence $\{\mathcal{I}_k, \mathcal{E}_k\}$, for $k = 1, \ldots, N$, where $\mathcal{I}_k$ is the approximation of $I(f)$ at the k-th step of the computational process, $\mathcal{E}_k$ is an estimate of the error $I(f) - \mathcal{I}_k$, and is N a suitable fixed integer.

The sequence terminates at the s-th level, with $s \leq N$, such that the automatic integrator fulfills the following requirement on the accuracy

$$\max\left\{\varepsilon_a, \varepsilon_r|\widetilde{I}(f)|\right\} \geq |\mathcal{E}_s|(\simeq |I(f) - \mathcal{I}_s|), \tag{9.38}$$

where $\widetilde{I}(f)$ is a reasonable guess of the integral $I(f)$ provided as an input datum by the user. Otherwise, the integrator returns the last computed approximation $\mathcal{I}_N$, together with a suitable error message that warns the user of the algorithm's failure to converge.
Ideally, an automatic integrator should:

(a) provide a reliable criterion for determining $|\mathcal{E}_s|$ that allows for monitoring the convergence check (9.38);
(b) ensure an *efficient implementation*, which minimizes the number of functional evaluations for yielding the desired approximation $\mathcal{I}_s$.

In computational practice, for each $k \geq 1$, moving from level k to level $k+1$ of the automatic integration process can be done according to two different strategies, which we define as *nonadaptive* or *adaptive*.

In the nonadaptive case, the law of distribution of the quadrature nodes is fixed a priori and the quality of the estimate $\mathcal{I}_k$ is refined by increasing the number of nodes corresponding to each level of the computational process. An example of an automatic integrator that is based on such a procedure is provided by the composite Newton-Cotes formulae on m and $2m$ subintervals, respectively, at levels k and $k+1$, as described in Section 9.7.1.

In the adaptive case, the positions of the nodes is not set a priori, but at each level k of the process they depend on the information that has been stored during the previous $k-1$ levels. An adaptive automatic integration algorithm is performed by partitioning the interval $[a, b]$ into successive subdivisions which are characterized by a nonuniform density of the nodes, this density being typically higher in a neighborhood of strong gradients or singularities of f. An example of an adaptive integrator based on the Cavalieri-Simpson formula is described in Section 9.7.2.

9.7.1 Nonadaptive Integration Algorithms

In this section, we employ the composite Newton-Cotes formulae. Our aim is to devise a criterion for estimating the absolute error $|I(f) - \mathcal{I}_k|$ by using

Richardson extrapolation. From (9.26) and (9.27) it turns out that, for $m \geq 1$ and $n \geq 0$, $I_{n,m}(f)$ has order of infinitesimal equal to H^{n+p}, with $p = 2$ for n even and $p = 1$ for n odd, where m, n and $H = (b-a)/m$ are the number of partitions of $[a, b]$, the number of quadrature nodes over each subinterval and the constant length of each subinterval, respectively. By doubling the value of m (i.e., halving the stepsize H) and proceeding by extrapolation, we get

$$I(f) - I_{n,2m}(f) \simeq \frac{1}{2^{n+p}}\left[I(f) - I_{n,m}(f)\right]. \tag{9.39}$$

The use of the symbol $\simeq$ instead of $=$ is due to the fact that the point ξ or η, where the derivative in (9.26) and (9.27) must be evaluated, changes when passing from m to $2m$ subintervals. Solving (9.39) with respect to $I(f)$ yields the following *absolute error estimate* for $I_{n,2m}(f)$

$$I(f) - I_{n,2m}(f) \simeq \frac{I_{n,2m}(f) - I_{n,m}(f)}{2^{n+p} - 1}. \tag{9.40}$$

If the composite Simpson rule is considered (i.e., $n = 2$), (9.40) predicts a reduction of the absolute error by a factor of 15 when passing from m to $2m$ subintervals. Notice also that only 2^{m-1} extra functional evaluations are needed to compute the new approximation $I_{2,2m}(f)$ starting from $I_{2,m}(f)$. Relation (9.40) is an instance of an *a posteriori error estimate* (see Chapter 2, Section 2.3). It is based on the combined use of an *a priori estimate* (in this case, (9.26) or (9.27)) and of two evaluations of the quantity to be approximated (the integral $I(f)$) for two different values of the discretization parameter (that is, $H = (b-a)/m$).

Example 9.8 Let us employ the a posteriori estimate (9.40) in the case of the composite Simpson formula ($n = p = 2$), for the approximation of the integral

$$\int_0^{\pi} (e^{x/2} + \cos 4x)dx = 2(e^{\pi} - 1) \simeq 7.621,$$

where we require the absolute error to be less than 10^{-4}. For $k = 0, 1, \ldots$, set $h_k = (b-a)/2^k$ and denote by $I_{2,m(k)}(f)$ the integral of f which is computed using the composite Simpson formula on a grid of size h_k with $m(k) = 2^k$ intervals. We can thus assume as a conservative estimate of the quadrature error the following quantity

$$|E_k| = |I(f) - I_{2,m(k)}(f)| \simeq \frac{1}{10}|I_{2,2m(k)}(f) - I_{2,m(k)}(f)| = |\mathcal{E}_k|, \qquad k \geq 1. \tag{9.41}$$

Table 9.11 shows the sequence of the estimated errors $|\mathcal{E}_k|$ and of the corresponding absolute errors $|E_k|$ that have been *actually* made by the numerical integration process. Notice that, when convergence has been achieved, the error estimated by (9.41) is definitely higher than the actual error, due to the conservative choice above. •

Table 9.11. Nonadaptive automatic Simpson rule for the approximation of $\int_0^\pi (e^{x/2} + \cos 4x)dx$

k	$\lvert\mathcal{E}_k\rvert$	$\lvert E_k\rvert$	k	$\lvert\mathcal{E}_k\rvert$	$\lvert E_k\rvert$
0		3.156	2	0.10	$4.52 \cdot 10^{-5}$
1	0.42	1.047	3	$5.8 \cdot 10^{-6}$	$2 \cdot 10^{-9}$

An alternative approach for fulfilling the constraints (a) and (b) consists of employing a *nested sequence* of special Gaussian quadratures $I_k(f)$ (see Chapter 10), having increasing degree of exactness for $k = 1, \ldots, N$. These formulae are constructed in such a way that, denoting by $\mathcal{S}_{n_k} = \{x_1, \ldots, x_{n_k}\}$ the set of quadrature nodes relative to quadrature $I_k(f)$, $\mathcal{S}_{n_k} \subset \mathcal{S}_{n_{k+1}}$ for any $k = 1, \ldots, N-1$. As a result, for $k \geq 1$, the formula at the $k+1$-th level employs *all* the nodes of the formula at level k and this makes nested formulae quite effective for computer implementation.

As an example, we recall the Gauss-Kronrod formulae with 10, 21, 43 and 87 points, that are available in [PdKÜK83] (in this case, $N = 4$). The Gauss-Kronrod formulae have degree of exactness r_{n_k} (optimal) equal to $2n_k - 1$, where n_k is the number of nodes for each formula, with $n_1 = 10$ and $n_{k+1} = 2n_k + 1$ for $k = 1, 2, 3$. The criterion for devising an error estimate is based on comparing the results given by two successive formulae $I_{n_k}(f)$ and $I_{n_{k+1}}(f)$ with $k = 1, 2, 3$, and then terminating the computational process at the level k such that (see also [DR75], p. 321)

$$|\mathcal{I}_{k+1} - \mathcal{I}_k| \leq \max\{\varepsilon_a, \varepsilon_r |\mathcal{I}_{k+1}|\}.$$

9.7.2 Adaptive Integration Algorithms

The goal of an adaptive integrator is to yield an approximation of $I(f)$ within a fixed tolerance ε by a *nonuniform* distribution of the integration stepsize along the interval $[a, b]$. An optimal algorithm is able to adapt automatically the choice of the steplength according to the behavior of the integrand function, by increasing the density of the quadrature nodes where the function exhibits stronger variations.

In view of describing the method, it is convenient to restrict our attention to a generic subinterval $[\alpha, \beta] \subseteq [a, b]$. Recalling the error estimates for the Newton-Cotes formulae, it turns out that the evaluation of the derivatives of f, up to a certain order, is needed to set a stepsize h such that a fixed accuracy is ensured, say $\varepsilon(\beta - \alpha)/(b - a)$. This procedure, which is unfeasible in practical computations, is carried out by an automatic integrator as follows. We consider throughout this section the Cavalieri-Simpson formula (9.15), although the method can be extended to other quadrature rules.

Set $I_f(\alpha, \beta) = \int_\alpha^\beta f(x)dx$, $h = h_0 = (\beta - \alpha)/2$ and

$$S_f(\alpha, \beta) = (h_0/3)\left[f(\alpha) + 4f(\alpha + h_0) + f(\beta)\right].$$

From (9.16) we get

$$I_f(\alpha,\beta) - S_f(\alpha,\beta) = -\frac{h_0^5}{90} f^{(4)}(\xi), \tag{9.42}$$

where ξ is a point in (α,β). To estimate the error $I_f(\alpha,\beta) - S_f(\alpha,\beta)$ *without* using explicitly the function $f^{(4)}$ we employ again the Cavalieri-Simpson formula over the union of the two subintervals $[\alpha,(\alpha+\beta)/2]$ and $[(\alpha+\beta)/2,\beta]$, obtaining, for $h = h_0/2 = (\beta-\alpha)/4$

$$I_f(\alpha,\beta) - S_{f,2}(\alpha,\beta) = -\frac{(h_0/2)^5}{90}\left(f^{(4)}(\xi) + f^{(4)}(\eta)\right),$$

where $\xi \in (\alpha,(\alpha+\beta)/2)$, $\eta \in ((\alpha+\beta)/2,\beta)$ and $S_{f,2}(\alpha,\beta) = S_f(\alpha,(\alpha+\beta)/2) + S_f((\alpha+\beta)/2,\beta)$.

Let us now make the assumption that $f^{(4)}(\xi) \simeq f^{(4)}(\eta)$ (which is true, in general, only if the function $f^{(4)}$ does not vary "too much" on $[\alpha,\beta]$). Then,

$$I_f(\alpha,\beta) - S_{f,2}(\alpha,\beta) \simeq -\frac{1}{16}\frac{h_0^5}{90} f^{(4)}(\xi), \tag{9.43}$$

with a reduction of the error by a factor 16 with respect to (9.42), corresponding to the choice of a steplength of doubled size. Comparing (9.42) and (9.43), we get the estimate

$$\frac{h_0^5}{90} f^{(4)}(\xi) \simeq \frac{16}{15}\mathcal{E}_f(\alpha,\beta),$$

where $\mathcal{E}_f(\alpha,\beta) = S_f(\alpha,\beta) - S_{f,2}(\alpha,\beta)$. Then, from (9.43), we have

$$|I_f(\alpha,\beta) - S_{f,2}(\alpha,\beta)| \simeq \frac{|\mathcal{E}_f(\alpha,\beta)|}{15}. \tag{9.44}$$

We have thus obtained a formula that allows for easily *computing* the error made by using composite Cavalieri-Simpson numerical integration on the generic interval $[\alpha,\beta]$. Relation (9.44), as well as (9.40), is another instance of an *a posteriori error estimate*. It combines the use of an *a priori estimate* (in this case, (9.16)) and of two evaluations of the quantity to be approximated (the integral $I(f)$) for two different values of the discretization parameter h.

In the practice, it might be convenient to assume a more conservative error estimate, precisely

$$|I_f(\alpha,\beta) - S_{f,2}(\alpha,\beta)| \simeq |\mathcal{E}_f(\alpha,\beta)|/10.$$

Moreover, to ensure a global accuracy on $[a,b]$ equal to the fixed tolerance ε, it will suffice to enforce that the error $\mathcal{E}_f(\alpha,\beta)$ satisfies on each single subinterval $[\alpha,\beta] \subseteq [a,b]$ the following constraint

$$\frac{|\mathcal{E}_f(\alpha,\beta)|}{10} \le \varepsilon\frac{\beta-\alpha}{b-a}. \tag{9.45}$$

The adaptive automatic integration algorithm can be described as follows. Denote by:

1. A: the *active* integration interval, i.e., the interval where the integral is being computed;
2. S: the integration interval already examined, for which the error test (9.45) has been successfully passed;
3. N: the integration interval yet to be examined.

At the beginning of the integration process we have $N = [a, b]$, $A = N$ and $S = \emptyset$, while the situation at the generic step of the algorithm is depicted in Figure 9.3. Set $J_S(f) \simeq \int_a^\alpha f(x)dx$, with $J_S(f) = 0$ at the beginning of the process; if the algorithm successfully terminates, $J_S(f)$ yields the desired approximation of $I(f)$. We also denote by $J_{(\alpha,\beta)}(f)$ the approximate integral of f over the "active" interval $[\alpha, \beta]$. This interval is drawn in bold in Figure 9.3. At each step of the adaptive integration method the following decisions are taken:

1. if the local error test (9.45) is passed, then:
   (i) $J_S(f)$ is increased by $J_{(\alpha,\beta)}(f)$, that is, $J_S(f) \leftarrow J_S(f) + J_{(\alpha,\beta)}(f)$;
   (ii) we let $S \leftarrow S \cup A$, $A = N$ (corresponding to the path (I) in Figure 9.3), $\beta = b$;
2. if the local error test (9.45) fails, then:
   (j) A is halved, and the new active interval is set to $A = [\alpha, \alpha']$ with $\alpha' = (\alpha + \beta)/2$ (corresponding to the path (II) in Figure 9.3);
   (jj) we let $N \leftarrow N \cup [\alpha', \beta]$, $\beta \leftarrow \alpha'$;
   (jjj) a new error estimate is provided.

In order to prevent the algorithm from generating too small stepsizes, it is convenient to monitor the width of A and warn the user, in case of an excessive reduction of the steplength, about the presence of a possible singularity in the integrand function (see Section 9.8).

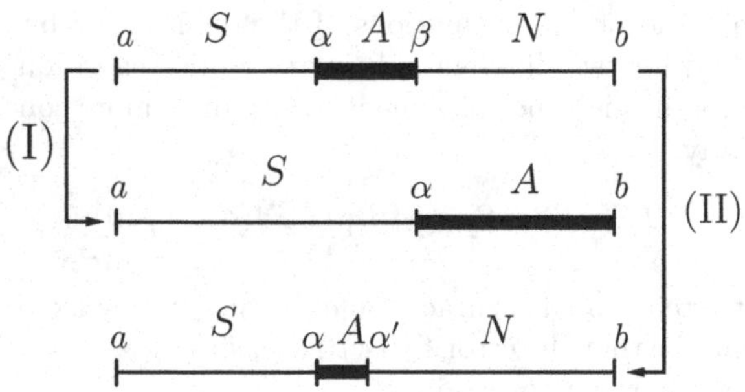

Fig. 9.3. Distribution of the integration intervals at the generic step of the adaptive algorithm and updating of the integration grid

Example 9.9 Let us employ Cavalieri-Simpson adaptive integration for computing the integral

$$I(f) = \int_{-3}^{4} \tan^{-1}(10x)dx$$
$$= 4\tan^{-1}(40) + 3\tan^{-1}(-30) - (1/20)\log(16/9) \simeq 1.54201193.$$

Running Program 77 with `tol` $= 10^{-4}$ and `hmin` $= 10^{-3}$ yields an approximation of the integral with an absolute error of $2.104 \cdot 10^{-5}$. The algorithm performs 77 functional evaluations, corresponding to partitioning the interval $[a, b]$ into 38 nonuniform subintervals. We notice that the corresponding composite formula with uniform stepsize would have required 128 subintervals with an absolute error of $2.413 \cdot 10^{-5}$.

In Figure 9.4 (*left*) we show, together with the plot of the integrand function, the distribution of the quadrature nodes as a function of x, while on the right the integration step density (piecewise constant) $\Delta_h(x)$ is shown, defined as the inverse of the step size h over each active interval A. Notice the high value attained by Δ_h at $x = 0$, where the derivative of the integrand function is maximum. •

The adaptive algorithm described above is implemented in Program 77. Among the input parameters, `hmin` is the minimum admissible value of the integration steplength. In output the program returns the approximate value of the integral `JSF` and the set of integration points `nodes`.

Program 77 - simpadpt : Adaptive Cavalieri-Simpson formula

```
function [JSf,nodes]=simpadpt(f,a,b,tol,hmin,varargin)
%SIMPADPT Adaptive Simpson quadrature.
% [JSF,NODES] = SIMPADPT(FUN,A,B,TOL,HMIN) tries to approximate the
% integral of function FUN over (A,B) to within an error of TOL
% using recursive adaptive Simpson quadrature. The inline function Y = FUN(V)
```

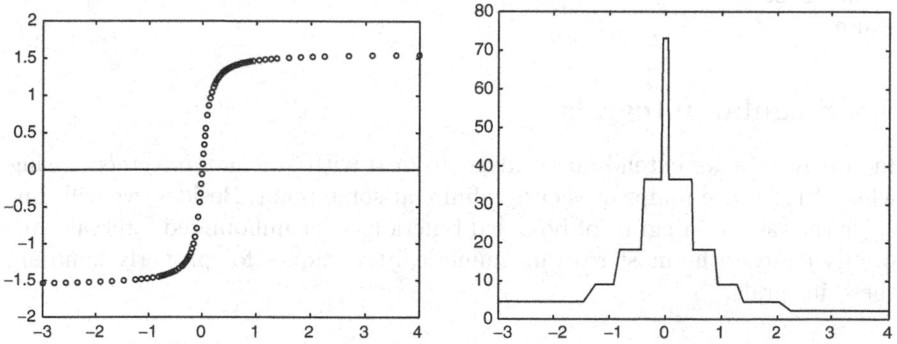

Fig. 9.4. Distribution of quadrature nodes (*left*); density of the integration stepsize in the approximation of the integral of Example 9.9 (*right*)

```
% should accept a vector argument V and return a vector result
% Y, the integrand evaluated at each element of X.
% JSF = SIMPADPT(FUN,A,B,TOL,HMIN,P1,P2,...) calls the function FUN passing
% the optional parameters P1,P2,... as FUN(X,P1,P2,...).
A=[a,b]; N=[]; S=[]; JSf = 0; ba = b - a; nodes=[];
while ~isempty(A),
  [deltaI,ISc]=caldeltai(A,f,varargin{:});
  if abs(deltaI) <= 15*tol*(A(2)-A(1))/ba;
     JSf = JSf + ISc;    S = union(S,A);
     nodes = [nodes, A(1) (A(1)+A(2))*0.5 A(2)];
     S = [S(1), S(end)]; A = N; N = [];
  elseif A(2)-A(1) < hmin
     JSf=JSf+ISc;         S = union(S,A);
     S = [S(1), S(end)]; A=N; N=[];
     warning('Too small integration-step');
  else
     Am = (A(1)+A(2))*0.5;
     A = [A(1) Am];
     N = [Am, b];
  end
end
nodes=unique(nodes);
return

function [deltaI,ISc]=caldeltai(A,f,varargin)
L=A(2)-A(1);
t=[0; 0.25; 0.5; 0.5; 0.75; 1];
x=L*t+A(1);
L=L/6;
w=[1; 4; 1];
fx=feval(f,x,varargin{:}).*ones(6,1);
IS=L*sum(fx([1 3 6]).*w);
ISc=0.5*L*sum(fx.*[w;w]);
deltaI=IS-ISc;
return
```

9.8 Singular Integrals

In this section we extend our analysis to deal with *singular integrals*, arising when f has finite jumps or is even infinite at some point. Besides, we will consider the case of integrals of bounded functions over unbounded intervals. We briefly address the most relevant numerical techniques for properly handling these integrals.

9.8.1 Integrals of Functions with Finite Jump Discontinuities

Let c be a *known* point within $[a, b]$ and assume that f is a continuous and bounded function in $[a, c)$ and $(c, b]$, with finite jump $f(c^+) - f(c^-)$. Since

$$I(f) = \int_a^b f(x)dx = \int_a^c f(x)dx + \int_c^b f(x)dx, \tag{9.46}$$

any integration formula of the previous sections can be used on $[a, c^-]$ and $[c^+, b]$ to furnish an approximation of $I(f)$. We proceed similarly if f admits a *finite* number of jump discontinuities within $[a, b]$.

When the position of the discontinuity points of f is *not* known a priori, a preliminary analysis of the graph of the function should be carried out. Alternatively, one can resort to an adaptive integrator that is able to detect the presence of discontinuities when the integration steplength falls below a given tolerance (see Section 9.7.2).

9.8.2 Integrals of Infinite Functions

Let us deal with the case in which $\lim_{x\to a^+} f(x) = \infty$; similar considerations hold when f is infinite as $x \to b^-$, while the case of a point of singularity c internal to the interval $[a, b]$ can be recast to one of the previous two cases owing to (9.46). Assume that the integrand function is of the form

$$f(x) = \frac{\phi(x)}{(x-a)^\mu}, \qquad 0 \le \mu < 1,$$

where ϕ is a function whose absolute value is bounded by M. Then

$$|I(f)| \le M \lim_{t\to a^+} \int_t^b \frac{1}{(x-a)^\mu} dx = M \frac{(b-a)^{1-\mu}}{1-\mu}.$$

Suppose we wish to approximate $I(f)$ up to a fixed tolerance δ. For this, let us describe the following two methods (for further details, see also [IK66], Section 7.6, and [DR75], Section 2.12 and Appendix 1).

Method 1. For any ε such that $0 < \varepsilon < (b-a)$, we write the singular integral as $I(f) = I_1 + I_2$, where

$$I_1 = \int_a^{a+\varepsilon} \frac{\phi(x)}{(x-a)^\mu} dx, \qquad I_2 = \int_{a+\varepsilon}^b \frac{\phi(x)}{(x-a)^\mu} dx.$$

The computation of I_2 is not troublesome. After replacing ϕ in I_1 by its p-th order Taylor's expansion around $x = a$, we obtain

$$\phi(x) = \Phi_p(x) + \frac{(x-a)^{p+1}}{(p+1)!} \phi^{(p+1)}(\xi(x)), \qquad p \ge 0, \tag{9.47}$$

where $\Phi_p(x) = \sum_{k=0}^{p} \phi^{(k)}(a)(x-a)^k/k!$. Then

$$I_1 = \varepsilon^{1-\mu}\sum_{k=0}^{p} \frac{\varepsilon^k \phi^{(k)}(a)}{k!(k+1-\mu)} + \frac{1}{(p+1)!}\int_a^{a+\varepsilon} (x-a)^{p+1-\mu}\phi^{(p+1)}(\xi(x))dx.$$

Replacing I_1 by the finite sum, the corresponding error E_1 can be bounded as

$$|E_1| \leq \frac{\varepsilon^{p+2-\mu}}{(p+1)!(p+2-\mu)} \max_{a\leq x\leq a+\varepsilon} |\phi^{(p+1)}(x)|, \qquad p \geq 0. \tag{9.48}$$

For fixed p, the right side of (9.48) is an increasing function of ε. On the other hand, taking $\varepsilon < 1$ and assuming that the successive derivatives of ϕ do not grow too much as p increases, the same function is decreasing as p grows.

Let us next approximate I_2 using a composite Newton-Cotes formula with m subintervals and $n+1$ quadrature nodes for each subinterval, n being an even integer. Recalling (9.26) and aiming at equidistributing the error δ between I_1 and I_2, it turns out that

$$|E_2| \leq \mathcal{M}^{(n+2)}(\varepsilon)\frac{b-a-\varepsilon}{(n+2)!}\frac{|M_n|}{n^{n+3}}\left(\frac{b-a-\varepsilon}{m}\right)^{n+2} = \delta/2, \tag{9.49}$$

where

$$\mathcal{M}^{(n+2)}(\varepsilon) = \max_{a+\varepsilon\leq x\leq b}\left|\frac{d^{n+2}}{dx^{n+2}}\left(\frac{\phi(x)}{(x-a)^\mu}\right)\right|.$$

The value of the constant $\mathcal{M}^{(n+2)}(\varepsilon)$ grows rapidly as ε tends to zero; as a consequence, (9.49) might require such a large number of subintervals $m_\varepsilon = m(\varepsilon)$ to make the method at hand of little practical use.

Example 9.10 Consider the singular integral (known as the *Fresnel integral*)

$$I(f) = \int_0^{\pi/2} \frac{\cos(x)}{\sqrt{x}} dx. \tag{9.50}$$

Expanding the integrand function in a Taylor's series around the origin and applying the theorem of integration by series, we get

$$I(f) = \sum_{k=0}^{\infty} \frac{(-1)^k}{(2k)!}\frac{1}{(2k+1/2)}(\pi/2)^{2k+1/2}.$$

Truncating the series at the first 10 terms, we obtain an approximate value of the integral equal to 1.9549.

Using the composite Cavalieri-Simpson formula, the a priori estimate (9.49) yields, as ε tends to zero and letting $n=2$, $|M_2| = 4/15$,

$$m_\varepsilon \simeq \left[\frac{0.018}{\delta}\left(\frac{\pi}{2}-\varepsilon\right)^5 \varepsilon^{-9/2}\right]^{1/4}.$$

For $\delta = 10^{-4}$, taking $\varepsilon = 10^{-2}$, it turns out that 1140 (uniform) subintervals are needed, while for $\varepsilon = 10^{-4}$ and $\varepsilon = 10^{-6}$ the number of subintervals is $2 \cdot 10^5$ and $3.6 \cdot 10^7$, respectively.

As a comparison, running Program 77 (adaptive integration with Cavalieri-Simpson formula) with `a` $= \varepsilon = 10^{-10}$, `hmin` $= 10^{-12}$ and `tol` $= 10^{-4}$, we get the approximate value 1.955 for the integral at the price of 1057 functional evaluations, which correspond to 528 nonuniform subdivisions of the interval $[0, \pi/2]$. •

Method 2. Using the Taylor expansion (9.47) we obtain

$$I(f) = \int_a^b \frac{\phi(x) - \Phi_p(x)}{(x-a)^\mu}dx + \int_a^b \frac{\Phi_p(x)}{(x-a)^\mu}dx = I_1 + I_2.$$

Exact computation of I_2 yields

$$I_2 = (b-a)^{1-\mu}\sum_{k=0}^{p}\frac{(b-a)^k\phi^{(k)}(a)}{k!(k+1-\mu)}. \tag{9.51}$$

For $p \geq 0$, the integral I_1 is equal to

$$I_1 = \int_a^b (x-a)^{p+1-\mu}\frac{\phi^{(p+1)}(\xi(x))}{(p+1)!}dx = \int_a^b g(x)dx. \tag{9.52}$$

Unlike the case of method 1, the integrand function g *does not* blow up at $x = a$, since its first p derivatives are finite at $x = a$. As a consequence, assuming we approximate I_1 using a composite Newton-Cotes formula, it is possible to give an estimate of the quadrature error, provided that $p \geq n+2$, for $n \geq 0$ even, or $p \geq n+1$, for n odd.

Example 9.11 Consider again the singular Fresnel integral (9.50), and assume we use the composite Cavalieri-Simpson formula for approximating I_1. We will take $p = 4$ in (9.51) and (9.52). Computing I_2 yields the value $(\pi/2)^{1/2}(2-(1/5)(\pi/2)^2+(1/108)(\pi/2)^4) \simeq 1.9588$. Applying the error estimate (9.26) with $n = 2$ shows that only 2 subdivisions of $[0, \pi/2]$ suffice for approximating I_1 up to an error $\delta = 10^{-4}$, obtaining the value $I_1 \simeq -0.0173$. As a whole, method 2 returns for (9.50) the approximate value 1.9415. •

9.8.3 Integrals over Unbounded Intervals

Let $f \in C^0([a, +\infty))$; should it exist and be finite, the following limit

$$\lim_{t\to+\infty}\int_a^t f(x)dx$$

is taken as being the value of the singular integral

$$I(f) = \int_a^{\infty} f(x)dx = \lim_{t \to +\infty} \int_a^t f(x)dx. \qquad (9.53)$$

An analogous definition holds if f is continuous over $(-\infty, b]$, while for a function $f : \mathbb{R} \to \mathbb{R}$, integrable over any bounded interval, we let

$$\int_{-\infty}^{\infty} f(x)dx = \int_{-\infty}^{c} f(x)dx + \int_{c}^{+\infty} f(x)dx \qquad (9.54)$$

if c is any real number and the two singular integrals on the right hand side of (9.54) are convergent. This definition is correct since the value of $I(f)$ *does not* depend on the choice of c.

A sufficient condition for f to be integrable over $[a, +\infty)$ is that

$$\exists \rho > 0, \text{ such that } \lim_{x \to +\infty} x^{1+\rho} f(x) = 0,$$

that is, we require f to be infinitesimal of order > 1 with respect to $1/x$ as $x \to \infty$. For the numerical approximation of (9.53) up to a tolerance δ, we consider the following methods, referring for further details to [DR75], Chapter 3.

Method 1. To compute (9.53), we can split $I(f)$ as $I(f) = I_1 + I_2$, where $I_1 = \int_a^c f(x)dx$ and $I_2 = \int_c^{\infty} f(x)dx$.

The end-point c, which can be taken arbitrarily, is chosen in such a way that the contribution of I_2 is negligible. Precisely, taking advantage of the asymptotic behavior of f, c is selected to guarantee that I_2 equals a fraction of the fixed tolerance, say, $I_2 = \delta/2$.

Then, I_1 will be computed up to an absolute error equal to $\delta/2$. This ensures that the global error in the computation of $I_1 + I_2$ is below the tolerance δ.

Example 9.12 Compute up to an error $\delta = 10^{-3}$ the integral

$$I(f) = \int_0^{\infty} \cos^2(x) e^{-x} dx = 3/5.$$

For any given $c > 0$, we have $I_2 = \int_c^{\infty} \cos^2(x) e^{-x} dx \leq \int_c^{\infty} e^{-x} dx = e^{-c}$; requiring that $e^{-c} = \delta/2$, one gets $c \simeq 7.6$. Then, assuming we use the composite trapezoidal formula for approximating I_1, thanks to (9.27) with $n = 1$ and $M = \max_{0 \leq x \leq c} |f''(x)| \simeq 1.04$, we obtain $m \geq \left(Mc^3/(6\delta)\right)^{1/2} = 277$.

Program 72 returns the value $\mathcal{I}_1 \simeq 0.599905$, instead of the exact value $I_1 = 3/5 - e^{-c}(\cos^2(c) - (\sin(2c) + 2\cos(2c))/5) \simeq 0.599842$, with an absolute error of about $6.27 \cdot 10^{-5}$. The global numerical outcome is thus $\mathcal{I}_1 + I_2 \simeq 0.600405$, with an absolute error with respect to $I(f)$ equal to $4.05 \cdot 10^{-4}$. •

Method 2. For any real number c, we let $I(f) = I_1 + I_2$, as for method 1, then we introduce the change of variable $x = 1/t$ in order to transform I_2 into an integral over the *bounded* interval $[0, 1/c]$

$$I_2 = \int_0^{1/c} f(t)t^{-2}dt = \int_0^{1/c} g(t)dt. \tag{9.55}$$

If $g(t)$ is not singular at $t = 0$, (9.55) can be treated by any quadrature formula introduced in this chapter. Otherwise, one can resort to the integration methods considered in Section 9.8.2.

Method 3. Gaussian interpolatory formulae are used, where the integration nodes are the zeros of Laguerre and Hermite orthogonal polynomials (see Section 10.5).

9.9 Multidimensional Numerical Integration

Let Ω be a bounded domain in $\mathbb{R}^2$ with a sufficiently smooth boundary. We consider the problem of approximating the integral $I(f) = \int_\Omega f(x,y)dxdy$, where f is a continuous function in $\overline{\Omega}$. For this purpose, in Sections 9.9.1 and 9.9.2 we address two methods.

The first method applies when Ω is a *normal* domain with respect to a coordinate axis. It is based on the reduction formula for double integrals and consists of using one-dimensional quadratures along both coordinate direction. The second method, which applies when Ω is a polygon, consists of employing composite quadratures of low degree on a triangular decomposition of the domain Ω. Section 9.9.3 briefly addresses the Monte Carlo method, which is particularly well-suited to integration in several dimensions.

9.9.1 The Method of Reduction Formula

Let Ω be a normal domain with respect to the x axis, as drawn in Figure 9.5, and assume for the sake of simplicity that $\phi_2(x) > \phi_1(x)$, $\forall x \in [a, b]$.
The reduction formula for double integrals gives (with obvious choice of notation)

$$I(f) = \int_a^b \int_{\phi_1(x)}^{\phi_2(x)} f(x,y)dydx = \int_a^b F_f(x)dx. \tag{9.56}$$

The integral $I(F_f) = \int_a^b F_f(x)dx$ can be approximated by a composite quadrature rule using M_x subintervals $\{J_k,\ k = 1, \dots, M_x\}$, of width $H = (b-a)/M_x$, and in each subinterval $n_x^{(k)} + 1$ nodes $\{x_i^k,\ i = 0, \dots, n_x^{(k)}\}$. Thus, in the x direction we can write

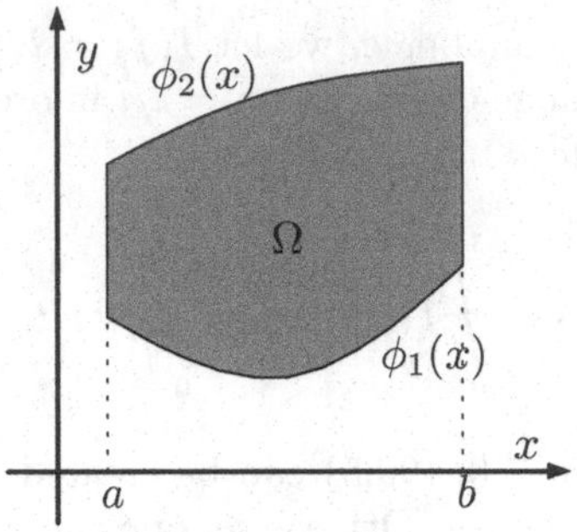

Fig. 9.5. Normal domain with respect to x axis

$$I^c_{n_x}(f) = \sum_{k=1}^{M_x}\sum_{i=0}^{n_x^{(k)}} \alpha_i^k F_f(x_i^k),$$

where the coefficients α_i^k are the quadrature weights on each subinterval J_k. For each node x_i^k, the approximate evaluation of the integral $F_f(x_i^k)$ is then carried out by a composite quadrature using M_y subintervals $\{J_m, m = 1, \ldots, M_y\}$, of width $h_i^k = (\phi_2(x_i^k) - \phi_1(x_i^k))/M_y$ and in each subinterval $n_y^{(m)} + 1$ nodes $\{y_{j,m}^{i,k}, j = 0, \ldots, n_y^{(m)}\}$.
In the particular case $M_x = M_y = M$, $n_x^{(k)} = n_y^{(m)} = 0$, for $k, m = 1, \ldots, M$, the resulting quadrature formula is the *midpoint reduction formula*

$$I^c_{0,0}(f) = H \sum_{k=1}^{M} h_0^k \sum_{m=1}^{M} f(x_0^k, y_{0,m}^{0,k}),$$

where $H = (b-a)/M$, $x_0^k = a + (k - 1/2)H$ for $k = 1, \ldots, M$ and $y_{0,m}^{0,k} = \phi_1(x_0^k) + (m - 1/2)h_0^k$ for $m = 1, \ldots, M$. With a similar procedure the *trapezoidal reduction formula* can be constructed along the coordinate directions (in that case, $n_x^{(k)} = n_y^{(m)} = 1$, for $k, m = 1, \ldots, M$).
The efficiency of the approach can obviously be increased by employing the adaptive method described in Section 9.7.2 to suitably allocate the quadrature nodes x_i^k and $y_{j,m}^{i,k}$ according to the variations of f over the domain Ω. The use of the reduction formulae above becomes less and less convenient as the dimension d of the domain $\Omega \subset \mathbb{R}^d$ gets larger, due to the large increase in the computational effort. Indeed, if any simple integral requires N functional evaluations, the overall cost would be equal to N^d.

The midpoint and trapezoidal reduction formulae for approximating the integral (9.56) are implemented in Programs 78 and 79. For the sake of simplicity, we set $M_x = M_y = M$. The variables `phi1` and `phi2` contain the expressions of the functions ϕ_1 and ϕ_2 which delimitate the integration domain.

Program 78 - redmidpt : Midpoint reduction formula

```
function int=redmidpt(a,b,phi1,phi2,m,fun)
%REDMIDPT Midpoint reduction quadrature formula
% INT=REDMIDPT(A,B,PHI1,PHI2,M,FUN) computes the integral of the function
% FUN over the 2D domain with X in (A,B) and Y delimited by the functions PHI1 and
% PHI2. FUN is a function of y.
H=(b-a)/m;
xx=[a+H/2:H:b];
dim=length(xx);
for i=1:dim
   x=xx(i); d=eval(phi2); c=eval(phi1); h=(d-c)/m;
   y=[c+h/2:h:d]; w=eval(fun); psi(i)=h*sum(w(1:m));
end
int=H*sum(psi(1:m));
return
```

Program 79 - redtrap : Trapezoidal reduction formula

```
function int=redtrap(a,b,phi1,phi2,m,fun)
%REDTRAP Trapezoidal reduction quadrature formula
% INT=REDTRAP(A,B,PHI1,PHI2,M,FUN) computes the integral of the function
% FUN over the 2D domain with X in (A,B) and Y delimited by the functions PHI1 and
% PHI2. FUN is a function of y.
H=(b-a)/m;
xx=[a:H:b];
dim=length(xx);
for i=1:dim
   x=xx(i); d=eval(phi2); c=eval(phi1); h=(d-c)/m;
   y=[c:h:d]; w=eval(fun); psi(i)=h*(0.5*w(1)+sum(w(2:m))+0.5*w(m+1));
end
int=H*(0.5*psi(1)+sum(psi(2:m))+0.5*psi(m+1));
return
```

9.9.2 Two-Dimensional Composite Quadratures

In this section we extend to the two-dimensional case the composite interpolatory quadratures that have been considered in Section 9.4. We assume that Ω is a convex polygon on which we introduce a *triangulation* $\mathcal{T}_h$ of N_T triangles or *elements*, such that $\overline{\Omega} = \bigcup_{T \in \mathcal{T}_h} T$, where the parameter $h > 0$ is the maximum edge-length in $\mathcal{T}_h$ (see Section 8.6.2).

Exactly as happens in the one-dimensional case, interpolatory composite quadrature rules on triangles can be devised by replacing $\int_\Omega f(x,y)dxdy$ with $\int_\Omega \Pi_h^k f(x,y)dxdy$, where, for $k \geq 0$, $\Pi_h^k f$ is the composite interpolating polynomial of f on the triangulation $\mathcal{T}_h$ introduced in Section 8.6.2.

For an efficient evaluation of this last integral, we employ the property of additivity which, combined with (8.42), leads to the following interpolatory composite rule

$$\begin{aligned} I_k^c(f) &= \int_\Omega \Pi_h^k f(x,y)dxdy = \sum_{T\in\mathcal{T}_h} \int_T \Pi_T^k f(x,y)dxdy = \sum_{T\in\mathcal{T}_h} I_k^T(f) \\ &= \sum_{T\in\mathcal{T}_h} \sum_{j=0}^{d_k-1} f(\tilde{\mathbf{z}}_j^T) \int_T l_j^T(x,y)dxdy = \sum_{T\in\mathcal{T}_h} \sum_{j=0}^{d_k-1} \alpha_j^T f(\tilde{\mathbf{z}}_j^T). \end{aligned} \tag{9.57}$$

The coefficients α_j^T and the points $\tilde{\mathbf{z}}_j^T$ are called the *local* weights and nodes of the quadrature formula (9.57), respectively.

The weights α_j^T can be computed on the reference triangle $\hat{T}$ of vertices $(0,0)$, $(1,0)$ and $(0,1)$, as follows

$$\alpha_j^T = \int_T l_j^T(x,y)dxdy = 2|T| \int_{\hat{T}} \hat{l}_j(\hat{x},\hat{y})d\hat{x}d\hat{y}, \qquad j = 0,\ldots,d_k-1,$$

where $|T|$ is the area of T. If $k=0$, we get $\alpha_0^T = |T|$, while if $k=1$ we have $\alpha_j^T = |T|/3$, for $j=0,1,2$.

Denoting respectively by $\mathbf{a}_j^T$ and $\mathbf{a}^T = \sum_{j=1}^3 \mathbf{a}_T^{(j)}/3$, for $j=1,2,3$, the vertices and the center of gravity of the triangle $T \in \mathcal{T}_h$, the following formulae are obtained.

Composite midpoint formula

$$I_0^c(f) = \sum_{T\in\mathcal{T}_h} |T| f(\mathbf{a}^T). \tag{9.58}$$

Composite trapezoidal formula

$$I_1^c(f) = \frac{1}{3} \sum_{T\in\mathcal{T}_h} |T| \sum_{j=1}^{3} f(\mathbf{a}_j^T). \tag{9.59}$$

In view of the analysis of the quadrature error $E_k^c(f) = I(f) - I_k^c(f)$, we introduce the following definition.

Definition 9.1 The quadrature formula (9.57) has *degree of exactness equal to* n, with $n \geq 0$, if $I_k^{\widehat{T}}(p) = \int_{\widehat{T}} p dxdy$ for any $p \in \mathbb{P}_n(\widehat{T})$, where $\mathbb{P}_n(\widehat{T})$ is defined in (8.39). ■

The following result can be proved (see [IK66], pp. 361–362).

Property 9.4 *Assume that the quadrature rule* (9.57) *has degree of exactness on* Ω *equal to* n*, with* $n \geq 0$*, and that its weights are all nonnegative. Then, there exists a positive constant* K_n*, independent of* h*, such that*

$$|E_k^c(f)| \leq K_n h^{n+1} |\Omega| M_{n+1},$$

for any function $f \in C^{n+1}(\Omega)$*, where* M_{n+1} *is the maximum value of the modules of the derivatives of order* $n+1$ *of* f *and* $|\Omega|$ *is the area of* Ω.

The composite formulae (9.58) and (9.59) both have degrees of exactness equal to 1; then, due to Property 9.4, their order of infinitesimal with respect to h is equal to 2.
An alternative family of quadrature rules on triangles is provided by the so-called *symmetric formulae.* These are Gaussian formulae with n nodes and high degree of exactness, and exhibit the feature that the quadrature nodes occupy symmetric positions with respect to all corners of the reference triangle $\widehat{T}$ or, as happens for Gauss-Radau formulae, with respect to the straight line $\widehat{y} = \widehat{x}$.
Considering the generic triangle $T \in \mathcal{T}_h$ and denoting by $\mathbf{a}^T_{(j)}$, $j = 1,2,3$, the midpoints of the edges of T, two examples of symmetric formulae, having degree of exactness equal to 2 and 3, respectively, are the following

$$I_3(f) = \frac{|T|}{3}\sum_{j=1}^{3} f(\mathbf{a}^T_{(j)}), \qquad n = 3,$$

$$I_7(f) = \frac{|T|}{60}\left(3\sum_{j=1}^{3} f(\mathbf{a}^T_j) + 8\sum_{j=1}^{3} f(\mathbf{a}^T_{(j)}) + 27 f(\mathbf{a}^T)\right), \qquad n = 7.$$

For a description and analysis of symmetric formulae for triangles, see [Dun85], while we refer to [Kea86] and [Dun86] for their extension to tetrahedra and cubes.

The composite quadrature rules (9.58) and (9.59) are implemented in Programs 80 and 81 for the approximate evaluation of the integral of $f(x, y)$ over a single triangle $T \in \mathcal{T}_h$. To compute the integral over Ω it suffices to sum the result provided by the program over each triangle of $\mathcal{T}_h$. The coordinates of the vertices of the triangle T are stored in the arrays xv and yv.

Program 80 - midptr2d : Midpoint rule on a triangle

```
function int=midptr2d(xv,yv,fun)
%MIDPTR2D Midpoint formula on a triangle.
% INT=MIDPTR2D(XV,YV,FUN) computes the integral of FUN on the triangle with
% vertices XV(K),YV(K), K=1,2,3. FUN is a function of x and y.
y12=yv(1)-yv(2);
y23=yv(2)-yv(3);
y31=yv(3)-yv(1);
areat=0.5*abs(xv(1)*y23+xv(2)*y31+xv(3)*y12);
x=sum(xv)/3; y=sum(yv)/3;
int=areat*eval(fun);
return
```

Program 81 - traptr2d : Trapezoidal rule on a triangle

```
function int=traptr2d(xv,yv,fun)
%TRAPTR2D Trapezoidal formula on a triangle.
```

```
% INT=TRAPTR2D(XV,YV,FUN) computes the integral of FUN on the
% triangle with vertices XV(K),YV(K), K=1,2,3. FUN is a function of x and y.
y12=yv(1)-yv(2);
y23=yv(2)-yv(3);
y31=yv(3)-yv(1);
areat=0.5*abs(xv(1)*y23+xv(2)*y31+xv(3)*y12);
int=0;
for i=1:3
  x=xv(i); y=yv(i); int=int+eval(fun);
end
int=int*areat/3;
return
```

9.9.3 Monte Carlo Methods for Numerical Integration

Numerical integration methods based on Monte Carlo techniques are a valid tool for approximating multidimensional integrals when the space dimension of $\mathbb{R}^n$ gets large. These methods differ from the approaches considered thus far, since the choice of quadrature nodes is done *statistically* according to the values attained by random variables having a known probability distribution.

The basic idea of the method is to interpret the integral as a *statistic mean value*

$$\int_{\Omega} f(\mathbf{x})d\mathbf{x} = |\Omega| \int_{\mathbb{R}^n} |\Omega|^{-1}\chi_{\Omega}(\mathbf{x}) f(\mathbf{x})d\mathbf{x} = |\Omega|\mu(f),$$

where $\mathbf{x} = (x_1, x_2, \ldots, x_n)^T$ and $|\Omega|$ denotes the n-dimensional volume of Ω, $\chi_{\Omega}(\mathbf{x})$ is the characteristic function of the set Ω, equal to 1 for $\mathbf{x} \in \Omega$ and to 0 elsewhere, while $\mu(f)$ is the mean value of the function $f(X)$, where X is a random variable with uniform probability density $|\Omega|^{-1}\chi_{\Omega}$ over $\mathbb{R}^n$.

We recall that the *random variable* $X \in \mathbb{R}^n$ (or, more properly, *random vector*) is an n-tuple of real numbers $X_1(\zeta), \ldots, X_n(\zeta)$ assigned to every outcome ζ of a random experiment (see [Pap87], Chapter 4).

Having fixed a vector $\mathbf{x} \in \mathbb{R}^n$, the probability $\mathcal{P}\{X \le \mathbf{x}\}$ of the random event $\{X_1 \le x_1, \ldots, X_n \le x_n\}$ is given by

$$\mathcal{P}\{X \le \mathbf{x}\} = \int_{-\infty}^{x_1} \cdots \int_{-\infty}^{x_n} f(X_1, \ldots, X_n)dX_1 \ldots dX_n,$$

where $f(X) = f(X_1, \ldots, X_n)$ is the *probability density* of the random variable $X \in \mathbb{R}^n$, such that

$$f(X_1, \ldots, X_n) \ge 0, \qquad \int_{\mathbb{R}^n} f(X_1, \ldots, X_n)dX = 1.$$

The numerical computation of the mean value $\mu(f)$ is carried out by taking N independent samples $\mathbf{x}_1, \ldots, \mathbf{x}_N \in \mathbb{R}^n$ with probability density $|\Omega|^{-1}\chi_{\Omega}$ and evaluating their *average*

$$\overline{f}_N = \frac{1}{N}\sum_{i=1}^{N} f(\mathbf{x}_i) = I_N(f). \tag{9.60}$$

From a statistical standpoint, the samples $\mathbf{x}_1, \ldots, \mathbf{x}_N$ can be regarded as the realizations of a sequence of N random variables $\{X_1, \ldots, X_N\}$, mutually independent and each with probability density $|\Omega|^{-1}\chi_\Omega$.

For such a sequence the strong law of large numbers ensures with probability 1 the convergence of the average $I_N(f) = \left(\sum_{i=1}^N f(X_i)\right)/N$ to the mean value $\mu(f)$ as $N \to \infty$. In computational practice the sequence of samples $\mathbf{x}_1, \ldots, \mathbf{x}_N$ is deterministically produced by a random-number generator, giving rise to the so-called *pseudo-random integration formulae*.
The quadrature error $E_N(f) = \mu(f) - I_N(f)$ as function of N can be characterized through the *variance*

$$\sigma(I_N(f)) = \sqrt{\mu\left(I_N(f) - \mu(f)\right)^2}.$$

Interpreting again f as a function of the random variable X, distributed with uniform probability density $|\Omega|^{-1}$ in $\Omega \subseteq \mathbb{R}^n$ and variance $\sigma(f)$, we have

$$\sigma(I_N(f)) = \frac{\sigma(f)}{\sqrt{N}}, \tag{9.61}$$

from which, as $N \to \infty$, a convergence rate of $\mathcal{O}(N^{-1/2})$ follows for the statistical estimate of the error $\sigma(I_N(f))$. Such convergence rate *does not* depend on the dimension n of the integration domain, and this is a most relevant feature of the Monte Carlo method. However, it is worth noting that the convergence rate is independent of the *regularity* of f; thus, unlike interpolatory quadratures, Monte Carlo methods *do not* yield more accurate results when dealing with smooth integrands.

The estimate (9.61) is extremely weak and in practice one does often obtain poorly accurate results. A more efficient implementation of Monte Carlo methods is based on composite approach or semi-analytical methods; an example of these techniques is provided in [NAG95], where a composite Monte Carlo method is employed for the computation of integrals over hypercubes in $\mathbb{R}^n$.

9.10 Applications

We consider in the next sections the computation of two integrals suggested by applications in geometry and the mechanics of rigid bodies.

9.10.1 Computation of an Ellipsoid Surface

Let E be the ellipsoid obtained by rotating the ellipse in Figure 9.6 around the x axis, where the radius ρ is described as a function of the axial coordinate by the equation

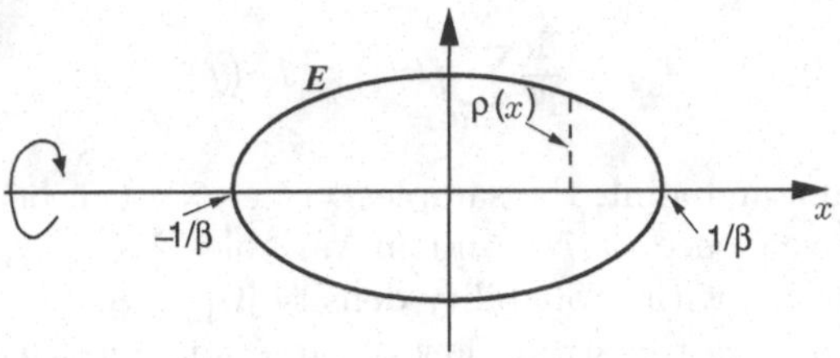

Fig. 9.6. Section of the ellipsoid

$$\rho^2(x) = \alpha^2(1 - \beta^2 x^2), \qquad -\frac{1}{\beta} \le x \le \frac{1}{\beta},$$

α and β being given constants such that $\alpha^2\beta^2 < 1$.
We set the following values for the parameters: $\alpha^2 = (3 - 2\sqrt{2})/100$ and $\beta^2 = 100$. Letting $K^2 = \beta^2\sqrt{1 - \alpha^2\beta^2}$, $f(x) = \sqrt{1 - K^2x^2}$ and $\theta = \cos^{-1}(K/\beta)$, the computation of the surface of E requires evaluating the integral

$$I(f) = 4\pi\alpha \int_0^{1/\beta} f(x)dx = \frac{2\pi\alpha}{K}\left[(\pi/2 - \theta) + \sin(2\theta)/2\right]. \qquad (9.62)$$

Notice that $f'(1/\beta) = -100$; this prompts us to use a numerical adaptive formula able to provide a nonuniform distribution of quadrature nodes, with a possible refinement around $x = 1/\beta$. With this aim, we have run Program 77 taking `hmin`$=10^{-5}$ and `tol`$=10^{-8}$.

In Figure 9.7 (*left*), we show, together with the graph of f, the nonuniform distribution of the quadrature nodes on the x axis, while in Figure 9.7 (*right*) we plot the logarithmic graph of the integration step density (piecewise constant) $\Delta_h(x)$, defined as the inverse of the value of the stepsize h on each active interval A (see Section 9.7.2).

Notice the high value of Δ_h at $x = 1/\beta$, where the derivative of the integrand function is maximum.

9.10.2 Computation of the Wind Action on a Sailboat Mast

Let us consider the sailboat schematically drawn in Figure 9.8 (*left*) and subject to the action of the wind force. The mast, of length L, is denoted by the straight line AB, while one of the two shrouds (strings for the side stiffening of the mast) is represented by the straight line BO. Any infinitesimal element of the sail transmits to the corresponding element of length dx of the mast a force of magnitude equal to $f(x)dx$. The change of f along with the height x, measured from the point A (basis of the mast), is expressed by the following law

$$f(x) = \frac{\alpha x}{x + \beta} e^{-\gamma x},$$

where α, β and γ are given constants.

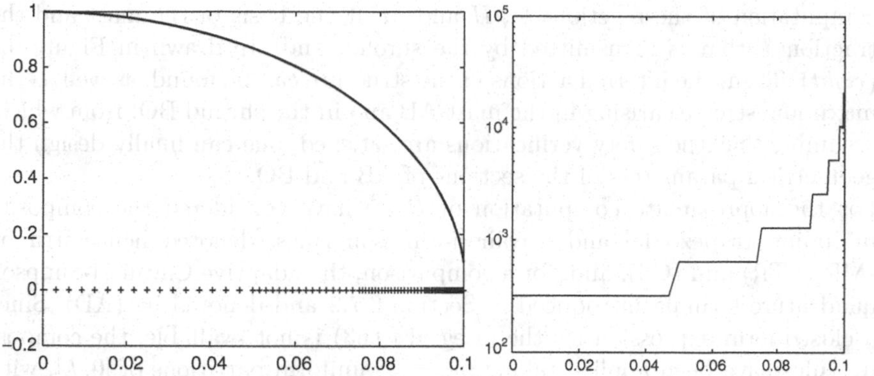

Fig. 9.7. Distribution of quadrature nodes (*left*); integration stepsize density in the approximation of integral (9.62) (*right*)

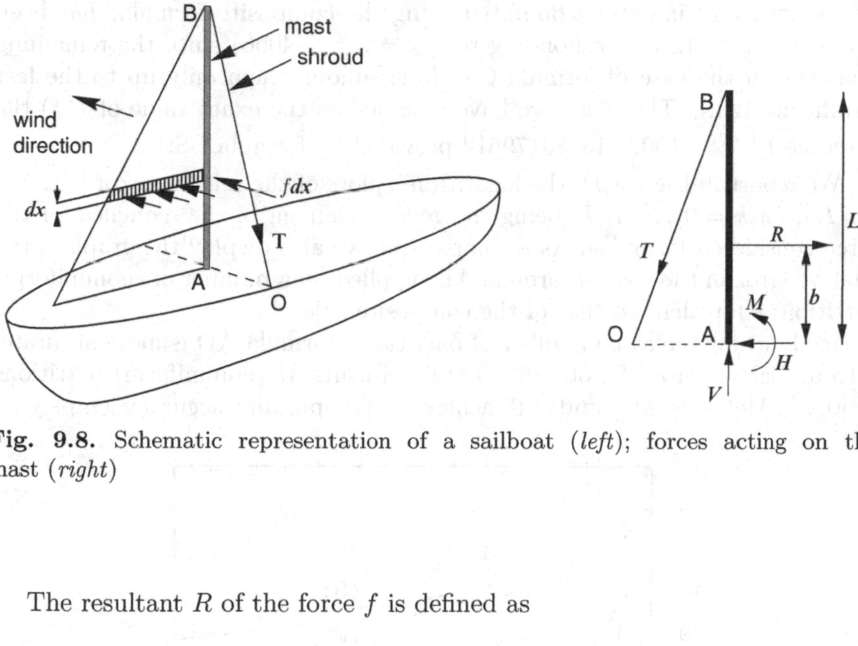

Fig. 9.8. Schematic representation of a sailboat (*left*); forces acting on the mast (*right*)

The resultant R of the force f is defined as

$$R = \int_0^L f(x)dx \equiv I(f), \tag{9.63}$$

and is applied at a point at distance equal to b (to be determined) from the basis of the mast.

Computing R and the distance b, given by $b = I(xf)/I(f)$, is crucial for the structural design of the mast and shroud sections. Indeed, once the values of R and b are known, it is possible to analyze the hyperstatic structure mast-shroud (using for instance the method of forces), thus allowing for the

computation of the reactions V, H and M at the basis of the mast and the traction T that is transmitted by the shroud, and are drawn in Figure 9.8 (*right*). Then, the internal actions in the structure can be found, as well as the maximum stresses arising in the mast AB and in the shroud BO, from which, assuming that the safety verifications are satisfied, one can finally design the geometrical parameters of the sections of AB and BO.
For the approximate computation of R we have considered the composite midpoint, trapezoidal and Cavalieri-Simpson rules, denoted henceforth by (MP), (TR) and (CS), and, for a comparison, the adaptive Cavalieri-Simpson quadrature formula introduced in Section 9.7.2 and denoted by (AD). Since a closed-form expression for the integral (9.63) is not available, the composite rules have been applied taking $m_k = 2^k$ uniform partitions of $[0, L]$, with $k = 0, \ldots, 15$.

We have assumed in the numerical experiments $\alpha = 50$, $\beta = 5/3$ and $\gamma = 1/4$ and we have run Program 77 taking $\texttt{tol}=10^{-4}$ and $\texttt{hmin}=10^{-3}$. The sequence of integrals computed using the composite formulae has been stopped at $k = 12$ (corresponding to $m_k = 2^{12} = 4096$) since the remaining elements, in the case of formula CS, differ among them only up to the last significant figure. Therefore, we have assumed as the exact value of $I(f)$ the outcome $I_{12}^{(CS)} = 100.0613683179612$ provided by formula CS.

We report in Figure 9.9 the logarithmic plots of the relative error $|I_{12}^{(CS)} - I_k|/I_{12}$, for $k = 0, \ldots, 7$, I_k being the generic element of the sequence for the three considered formulae. As a comparison, we also display the graph of the relative error in the case of formula AD, applied on a number of (nonuniform) partitions equivalent to that of the composite rules.
Notice how, for the same number of partitions, formula AD is more accurate, with a relative error of $2.06 \cdot 10^{-7}$ obtained using 37 (nonuniform) partitions of $[0, L]$. Methods MP and TR achieve a comparable accuracy employing

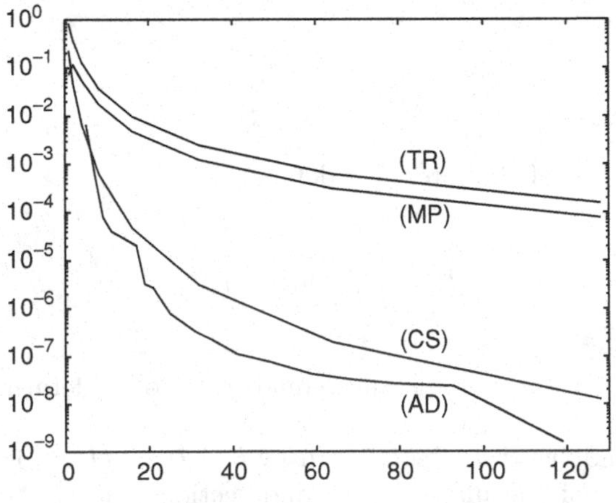

Fig. 9.9. Relative errors in the approximate computation of the integral $\int_0^L (\alpha x e^{-\gamma x})/(x + \beta) dx$

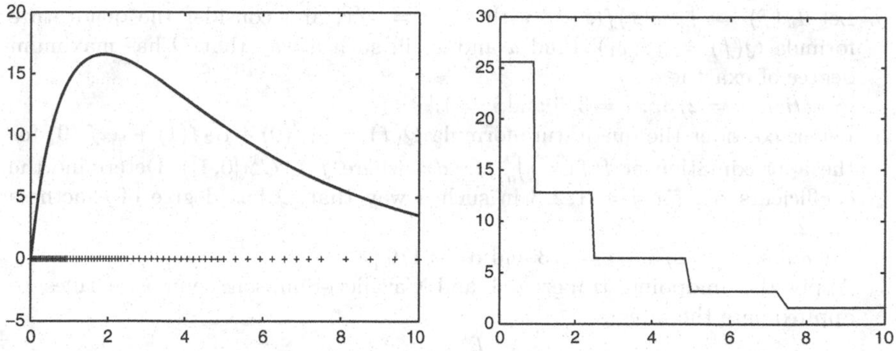

Fig. 9.10. Distribution of quadrature nodes (*left*); integration step density in the approximation of the integral $\int_0^L (\alpha x e^{-\gamma x})/(x+\beta)dx$ (*right*)

2048 and 4096 uniform subintervals, respectively, while formula CS requires about 64 partitions. The effectiveness of the adaptivity procedure is demonstrated by the plots in Figure 9.10, which show, together with the graph of f, the distribution of the quadrature nodes (*left*) and the function $\Delta_h(x)$ (*right*) that expresses the (piecewise constant) density of the integration stepsize h, defined as the inverse of the value of h over each active interval A (see Section 9.7.2).

Notice also the high value of Δ_h at $x = 0$, where the derivatives of f are maximum.

9.11 Exercises

1. Let $E_0(f)$ and $E_1(f)$ be the quadrature errors in (9.6) and (9.12). Prove that $|E_1(f)| \simeq 2|E_0(f)|$.
2. Check that the error estimates for the midpoint, trapezoidal and Cavalieri-Simpson formulae, given respectively by (9.6), (9.12) and (9.16), are special instances of (9.19) or (9.20). In particular, show that $M_0 = 2/3$, $K_1 = -1/6$ and $M_2 = -4/15$ and determine, using the definition, the degree of exactness r of each formula.
[*Hint*: find r such that $I_n(x^k) = \int_a^b x^k dx$, for $k = 0, \ldots, r$, and $I_n(x^j) \neq \int_a^b x^j dx$, for $j > r$.]
3. Let $I_n(f) = \sum_{k=0}^n \alpha_k f(x_k)$ be a Lagrange quadrature formula on $n+1$ nodes. Compute the degree of exactness r of the formulae:
(a) $I_2(f) = (2/3)[2f(-1/2) - f(0) + 2f(1/2)]$,
(b) $I_4(f) = (1/4)[f(-1) + 3f(-1/3) + 3f(1/3) + f(1)]$.
Which is the order of infinitesimal p for (a) and (b)?
[*Solution*: $r = 3$ and $p = 5$ for both $I_2(f)$ and $I_4(f)$.]
4. Compute $\mathrm{d}f[x_0, \ldots, x_n, x]/\mathrm{d}x$ by checking (9.22).
[*Hint*: proceed by computing directly the derivative at x as an incremental ratio, in the case where only one node x_0 exists, then upgrade progressively the order of the divided difference.]

5. Let $I_w(f) = \int_0^1 w(x)f(x)dx$ with $w(x) = \sqrt{x}$, and consider the quadrature formula $Q(f) = af(x_1)$. Find a and x_1 in such a way that Q has maximum degree of exactness r.
[*Solution*: $a = 2/3$, $x_1 = 3/5$ and $r = 1$.]
6. Let us consider the quadrature formula $Q(f) = \alpha_1 f(0) + \alpha_2 f(1) + \alpha_3 f'(0)$ for the approximation of $I(f) = \int_0^1 f(x)dx$, where $f \in C^1([0,1])$. Determine the coefficients α_j, for $j = 1,2,3$ in such a way that Q has degree of exactness $r = 2$.
[*Solution*: $\alpha_1 = 2/3$, $\alpha_2 = 1/3$ and $\alpha_3 = 1/6$.]
7. Apply the midpoint, trapezoidal and Cavalieri-Simpson composite rules to approximate the integral
$$\int_{-1}^{1} |x|e^x\,dx,$$
and discuss their convergence as a function of the size H of the subintervals.
8. Consider the integral $I(f) = \int_0^1 e^x\,dx$ and estimate the minimum number m of subintervals that is needed for computing $I(f)$ up to an absolute error $\leq 5 \cdot 10^{-4}$ using the composite trapezoidal (TR) and Cavalieri-Simpson (CS) rules. Evaluate in both cases the absolute error Err that is actually made.
[*Solution*: for TR, we have $m = 17$ and $Err = 4.95 \cdot 10^{-4}$, while for CS, $m = 2$ and $Err = 3.70 \cdot 10^{-5}$.]
9. Consider the corrected trapezoidal formula (9.30) and check that $|E_1^{corr}(f)| \simeq 4|E_2(f)|$, where $E_1^{corr}(f)$ and $E_2(f)$ are defined in (9.31) and (9.16), respectively.

10. Compute, with an error less than 10^{-4}, the following integrals:
$$(a) \int_0^\infty \sin(x)/(1+x^4)dx; \; (b) \int_0^\infty e^{-x}(1+x)^{-5}dx; \; (c) \int_{-\infty}^\infty \cos(x)e^{-x^2}dx.$$

11. Use the reduction midpoint and trapezoidal formulae for computing the double integral $I(f) = \int_\Omega \dfrac{y}{(1+xy)}dxdy$ over the domain $\Omega = (0,1)^2$. Run Programs 78 and 79 with $M = 2^i$, for $i = 0, \ldots, 10$ and plot in log-scale the absolute error in the two cases as a function of M. Which method is the most accurate? How many functional evaluations are needed to get an (absolute) accuracy of the order of 10^{-6}?
[*Solution*: the exact integral is $I(f) = \log(4) - 1$, and almost $200^2 = 40000$ functional evaluations are needed.]

Part IV

Transforms, Differentiation and Problem Discretization

10

Orthogonal Polynomials in Approximation Theory

Trigonometric polynomials, as well as other orthogonal polynomials like Legendre's and Chebyshev's, are widely employed in approximation theory.

This chapter addresses the most relevant properties of orthogonal polynomials, and introduces the transforms associated with them, in particular the discrete Fourier transform and the FFT, but also the Zeta and Wavelet transforms.

Application to interpolation, least-squares approximation, numerical differentiation and Gaussian integration are addressed.

10.1 Approximation of Functions by Generalized Fourier Series

Let $w = w(x)$ be a weight function on the interval $(-1, 1)$, i.e., a nonnegative integrable function in $(-1, 1)$. Let us denote by $\{p_k,\ k = 0, 1, \ldots\}$ a system of algebraic polynomials, with p_k of degree equal to k for each k, mutually orthogonal on the interval $(-1, 1)$ with respect to w. This means that

$$\int_{-1}^{1} p_k(x)p_m(x)w(x)dx = 0 \qquad \text{if } k \neq m.$$

Set $(f, g)_w = \int_{-1}^{1} f(x)g(x)w(x)dx$ and $\|f\|_w = (f, f)_w^{1/2}$; $(\cdot, \cdot)_w$ and $\|\cdot\|_w$ are respectively the scalar product and the norm for the function space

$$\mathrm{L}_w^2 = \mathrm{L}_w^2(-1, 1) = \left\{ f : (-1, 1) \to \mathbb{R},\ \int_{-1}^{1} f^2(x)w(x)dx < \infty \right\}. \qquad (10.1)$$

For any function $f \in \mathrm{L}_w^2$ the series

$$Sf = \sum_{k=0}^{+\infty} \widehat{f}_k p_k, \qquad \text{with } \widehat{f}_k = \frac{(f, p_k)_w}{\|p_k\|_w^2},$$

is called the *generalized Fourier series of* f, and $\widehat{f}_k$ is the k*-th Fourier coefficient.* As is well-known, Sf converges *in average* (or *in the sense of* L^2_w) to f. This means that, letting for any integer n

$$f_n(x) = \sum_{k=0}^{n} \widehat{f}_k p_k(x) \tag{10.2}$$

($f_n \in \mathbb{P}_n$ is the truncation of order n of the generalized Fourier series of f), the following convergence result holds

$$\lim_{n\to+\infty} \|f - f_n\|_w = 0.$$

Moreover, the following Parseval's equality holds

$$\|f\|_w^2 = \sum_{k=0}^{+\infty} \widehat{f}_k^2 \|p_k\|_w^2$$

and, for any n, $\|f - f_n\|_w^2 = \sum_{k=n+1}^{+\infty} \widehat{f}_k^2 \|p_k\|_w^2$ is the square of the remainder of the generalized Fourier series.

The polynomial $f_n \in \mathbb{P}_n$ satisfies the following minimization property

$$\|f - f_n\|_w = \min_{q\in\mathbb{P}_n} \|f - q\|_w. \tag{10.3}$$

Indeed, since $f - f_n = \sum_{k=n+1}^{+\infty} \widehat{f}_k p_k$, the property of orthogonality of polynomials $\{p_k\}$ implies $(f - f_n, q)_w = 0 \ \forall q \in \mathbb{P}_n$. Then, the Cauchy-Schwarz inequality (8.33) yields

$$\begin{aligned}\|f - f_n\|_w^2 &= (f - f_n, f - f_n)_w = (f - f_n, f - q)_w + (f - f_n, q - f_n)_w \\ &= (f - f_n, f - q)_w \le \|f - f_n\|_w \|f - q\|_w, \quad \forall q \in \mathbb{P}_n,\end{aligned}$$

and (10.3) follows since q is arbitrary in $\mathbb{P}_n$. In such a case, we say that f_n is the orthogonal projection of f over $\mathbb{P}_n$ in the sense of L^2_w. It is therefore interesting to compute the coefficients $\widehat{f}_k$ of f_n. As will be seen in later sections, this is done by approximating the integrals that appear in the definition of $\widehat{f}_k$ using Gaussian quadratures. By doing so, one gets the so-called *discrete coefficients* $\tilde{f}_k$ of f, and, as a consequence, the new polynomial

$$f_n^*(x) = \sum_{k=0}^{n} \tilde{f}_k p_k(x), \tag{10.4}$$

which is called the *discrete truncation of order* n of the Fourier series of f. Typically,

$$\tilde{f}_k = \frac{(f, p_k)_n}{\|p_k\|_n^2}, \tag{10.5}$$

where, for any pair of continuous functions f and g, $(f,g)_n$ is the approximation of the scalar product $(f,g)_w$ and $\|g\|_n = \sqrt{(g,g)_n}$ is the seminorm associated with $(\cdot,\cdot)_n$. In a manner analogous to what was done for f_n, it can be checked that

$$\|f - f_n^*\|_n = \min_{q\in\mathbb{P}_n} \|f - q\|_n \tag{10.6}$$

and we say that f_n^* is the approximation to f in $\mathbb{P}_n$ *in the least-squares sense* (the reason for using this name will be made clear later on).

We conclude this section by recalling that, for any family of monic orthogonal polynomials $\{p_k\}$, the following recursive three-term formula holds (for the proof, see for instance [Gau96])

$$\begin{cases} p_{k+1}(x) = (x - \alpha_k)p_k(x) - \beta_k p_{k-1}(x), & k \geq 0, \\ p_{-1}(x) = 0, \quad p_0(x) = 1, \end{cases} \tag{10.7}$$

where

$$\alpha_k = \frac{(xp_k, p_k)_w}{(p_k, p_k)_w}, \qquad \beta_{k+1} = \frac{(p_{k+1}, p_{k+1})_w}{(p_k, p_k)_w}, \qquad k \geq 0. \tag{10.8}$$

Since $p_{-1} = 0$, the coefficient β_0 is arbitrary and is chosen according to the particular family of orthogonal polynomials at hand. The recursive three-term relation is generally quite stable and can thus be conveniently employed in the numerical computation of orthogonal polynomials, as will be seen in Section 10.6.
In the forthcoming sections we introduce two relevant families of orthogonal polynomials.

10.1.1 The Chebyshev Polynomials

Consider the Chebyshev weight function $w(x) = (1-x^2)^{-1/2}$ on the interval $(-1,1)$, and, according to (10.1), introduce the space of square-integrable functions with respect to the weight w

$$\mathrm{L}^2_w(-1,1) = \left\{ f : (-1,1) \to \mathbb{R} : \int_{-1}^{1} f^2(x)(1-x^2)^{-1/2}dx < \infty \right\}.$$

A scalar product and a norm for this space are defined as

$$\begin{aligned} (f,g)_w &= \int_{-1}^{1} f(x)g(x)(1-x^2)^{-1/2}dx, \\ \|f\|_w &= \left\{ \int_{-1}^{1} f^2(x)(1-x^2)^{-1/2}dx \right\}^{1/2}. \end{aligned} \tag{10.9}$$

The Chebyshev polynomials are defined as follows

$$T_k(x) = \cos k\theta, \ \theta = \arccos x, \ k = 0, 1, 2, \ldots \tag{10.10}$$

They can be recursively generated by the following formula (a consequence of (10.7), see [DR75], pp. 25-26)

$$\begin{cases} T_{k+1}(x) = 2xT_k(x) - T_{k-1}(x) & k = 1, 2, \ldots, \\ T_0(x) = 1, \qquad T_1(x) = x. \end{cases} \tag{10.11}$$

In particular, for any $k \geq 0$, we notice that $T_k \in \mathbb{P}_k$, i.e., $T_k(x)$ is an algebraic polynomial of degree k with respect to x. Using well-known trigonometric relations, we have

$$(T_k, T_n)_w = 0 \text{ if } k \neq n, \ (T_n, T_n)_w = \begin{cases} c_0 = \pi & \text{if } n = 0, \\ c_n = \pi/2 & \text{if } n \neq 0, \end{cases}$$

which expresses the orthogonality of the Chebyshev polynomials with respect to the scalar product $(\cdot, \cdot)_w$. Therefore, the Chebyshev series of a function $f \in \mathrm{L}^2_w$ takes the form

$$Cf = \sum_{k=0}^{\infty} \widehat{f}_k T_k, \text{ with } \widehat{f}_k = \frac{1}{c_k} \int_{-1}^{1} f(x) T_k(x) (1 - x^2)^{-1/2} dx.$$

Notice that $\|T_n\|_\infty = 1$ for every n and the following *minimax* property holds

$$\|2^{1-n} T_n\|_\infty \leq \min_{p \in \mathbb{P}^1_n} \|p\|_\infty, \text{ if } n \geq 1,$$

where $\mathbb{P}^1_n = \{p(x) = \sum_{k=0}^n a_k x^k, \ a_n = 1\}$ denotes the subset of polynomials of degree n with leading coefficient equal to 1.

10.1.2 The Legendre Polynomials

The Legendre polynomials are orthogonal polynomials over the interval $(-1, 1)$ with respect to the weight function $w(x) = 1$. For these polynomials, L^2_w is the usual $\mathrm{L}^2(-1, 1)$ space introduced in (8.29), while $(\cdot, \cdot)_w$ and $\| \cdot \|_w$ coincide with the scalar product and norm in $\mathrm{L}^2(-1, 1)$, respectively given by

$$(f, g) = \int_{-1}^{1} f(x) g(x) \, dx, \ \|f\|_{\mathrm{L}^2(-1,1)} = \left(\int_{-1}^{1} f^2(x) \, dx \right)^{\frac{1}{2}}.$$

The Legendre polynomials are defined as

$$L_k(x) = \frac{1}{2^k}\sum_{l=0}^{[k/2]}(-1)^l \binom{k}{l}\binom{2k-2l}{k} x^{k-2l}, \qquad k = 0, 1, \ldots, \tag{10.12}$$

where $[k/2]$ is the integer part of $k/2$, or, recursively, through the three-term relation

$$\begin{cases} L_{k+1}(x) = \dfrac{2k+1}{k+1} x L_k(x) - \dfrac{k}{k+1} L_{k-1}(x), & k = 1, 2 \ldots, \\ \\ L_0(x) = 1, \qquad L_1(x) = x. \end{cases}$$

For every $k = 0, 1 \ldots$, $L_k \in \mathbb{P}_k$ and $(L_k, L_m) = \delta_{km}(k + 1/2)^{-1}$ for $k, m = 0, 1, 2, \ldots$. For any function $f \in \mathrm{L}^2(-1,1)$, its Legendre series takes the following form

$$Lf = \sum_{k=0}^{\infty} \widehat{f}_k L_k, \text{ with } \widehat{f}_k = \left(k + \frac{1}{2}\right) \int_{1}^{1} f(x) L_k(x) dx.$$

Remark 10.1 (The Jacobi polynomials) The polynomials previously introduced belong to the wider family of Jacobi polynomials $\{J_k^{\alpha\beta}, k = 0, \ldots, n\}$, that are orthogonal with respect to the weight $w(x) = (1-x)^{\alpha}(1+x)^{\beta}$, for $\alpha, \beta > -1$. Indeed, setting $\alpha = \beta = 0$ we recover the Legendre polynomials, while choosing $\alpha = \beta = -1/2$ gives the Chebyshev polynomials. ■

10.2 Gaussian Integration and Interpolation

Orthogonal polynomials play a crucial role in devising quadrature formulae with maximal degrees of exactness. Let $x_0, \ldots, x_n$ be $n+1$ given distinct points in the interval $[-1, 1]$. For the approximation of the weighted integral $I_w(f) = \int_{-1}^{1} f(x)w(x)dx$, being $f \in C^0([-1,1])$, we consider quadrature rules of the type

$$I_{n,w}(f) = \sum_{i=0}^{n} \alpha_i f(x_i), \tag{10.13}$$

where α_i are coefficients to be suitably determined. Obviously, both nodes and weights depend on n, however this dependence will be understood. Denoting by

$$E_{n,w}(f) = I_w(f) - I_{n,w}(f)$$

the error between the exact integral and its approximation (10.13), if $E_{n,w}(p) = 0$ for any $p \in \mathbb{P}_r$ (for a suitable $r \geq 0$) we shall say that formula (10.13) has degree of exactness r with respect to the weight w. This definition generalizes the one given for ordinary integration with weight $w = 1$.

Clearly, we can get a degree of exactness equal to (at least) n taking

$$I_{n,w}(f) = \int_{-1}^{1} \Pi_n f(x) w(x) dx,$$

where $\Pi_n f \in \mathbb{P}_n$ is the Lagrange interpolating polynomial of the function f at the nodes $\{x_i, i = 0, \ldots, n\}$, given by (8.4). Therefore, (10.13) has degree of exactness at least equal to n taking

$$\alpha_i = \int_{-1}^{1} l_i(x) w(x) dx, \qquad i = 0, \ldots, n, \tag{10.14}$$

where $l_i \in \mathbb{P}_n$ is the i-th characteristic Lagrange polynomial such that $l_i(x_j) = \delta_{ij}$, for $i, j = 0, \ldots, n$.

The question that arises is whether suitable choices of the nodes exist such that the degree of exactness is greater than n, say, equal to $r = n+m$ for some $m > 0$. The answer to this question is furnished by the following theorem, due to Jacobi [Jac26].

Theorem 10.1 *For a given $m > 0$, the quadrature formula* (10.13) *has degree of exactness $n+m$ iff it is of interpolatory type and the nodal polynomial ω_{n+1}* (8.6) *associated with the nodes $\{x_i\}$ is such that*

$$\int_{-1}^{1} \omega_{n+1}(x) p(x) w(x) dx = 0, \qquad \forall p \in \mathbb{P}_{m-1}. \tag{10.15}$$

Proof. Let us prove that these conditions are sufficient. If $f \in \mathbb{P}_{n+m}$ then there exist a quotient $\pi_{m-1} \in \mathbb{P}_{m-1}$ and a remainder $q_n \in \mathbb{P}_n$, such that $f = \omega_{n+1}\pi_{m-1} + q_n$. Since the degree of exactness of an interpolatory formula with $n+1$ nodes is equal to n (at least), we get

$$\sum_{i=0}^{n} \alpha_i q_n(x_i) = \int_{-1}^{1} q_n(x) w(x) dx = \int_{-1}^{1} f(x) w(x) dx - \int_{-1}^{1} \omega_{n+1}(x) \pi_{m-1}(x) w(x) dx.$$

As a consequence of (10.15), the last integral is null, thus

$$\int_{-1}^{1} f(x) w(x) dx = \sum_{i=0}^{n} \alpha_i q_n(x_i) = \sum_{i=0}^{n} \alpha_i f(x_i).$$

Since f is arbitrary, we conclude that $E_{n,w}(f) = 0$ for any $f \in \mathbb{P}_{n+m}$. Proving that the conditions are also necessary is an exercise left to the reader. ◇

Corollary 10.2 *The maximum degree of exactness of the quadrature formula* (10.13) *is $2n + 1$.*

Proof. If this would not be true, one could take $m \geq n+2$ in the previous theorem. This, in turn, would allow us to choose $p = \omega_{n+1}$ in (10.15) and come to the conclusion that ω_{n+1} is identically zero, which is absurd. ◇

Setting $m = n+1$ (the maximum admissible value), from (10.15) we get that the nodal polynomial ω_{n+1} satisfies the relation

$$\int_{-1}^{1} \omega_{n+1}(x)p(x)w(x)dx = 0, \qquad \forall p \in \mathbb{P}_n.$$

Since ω_{n+1} is a polynomial of degree $n+1$ orthogonal to all the polynomials of lower degree, we conclude that ω_{n+1} is the only monic polynomial multiple of p_{n+1} (recall that $\{p_k\}$ is the system of orthogonal polynomials introduced in Section 10.1). In particular, its roots $\{x_j\}$ coincide with those of p_{n+1}, that is

$$p_{n+1}(x_j) = 0, \qquad j = 0, \ldots, n. \tag{10.16}$$

The abscissae $\{x_j\}$ are the *Gauss nodes* associated with the weight function $w(x)$. We can thus conclude that the quadrature formula (10.13) with coefficients and nodes given by (10.14) and (10.16), respectively, has degree of exactness $2n+1$, the maximum value that can be achieved using interpolatory quadrature formulae with $n+1$ nodes, and is called the *Gauss quadrature formula.*

Its weights are all positive and the nodes are *internal* to the interval $(-1, 1)$ (see, for instance, [CHQZ06], p. 70). However, it is often useful to also include the end points of the interval among the quadrature nodes. By doing so, the Gauss formula with the highest degree of exactness is the one that employs as nodes the $n+1$ roots of the polynomial

$$\overline{\omega}_{n+1}(x) = p_{n+1}(x) + ap_n(x) + bp_{n-1}(x), \tag{10.17}$$

where the constants a and b are selected in such a way that $\overline{\omega}_{n+1}(-1) = \overline{\omega}_{n+1}(1) = 0$.

Denoting these roots by $\overline{x}_0 = -1, \overline{x}_1, \ldots, \overline{x}_n = 1$, the coefficients $\{\overline{\alpha}_i, i = 0, \ldots, n\}$ can then be obtained from the usual formulae (10.14), that is

$$\overline{\alpha}_i = \int_{-1}^{1} \overline{l}_i(x)w(x)dx, \qquad i = 0, \ldots, n,$$

where $\overline{l}_i \in \mathbb{P}_n$ is the i-th characteristic Lagrange polynomial such that $\overline{l}_i(\overline{x}_j) = \delta_{ij}$, for $i, j = 0, \ldots, n$. The quadrature formula

$$I_{n,w}^{GL}(f) = \sum_{i=0}^{n} \overline{\alpha}_i f(\overline{x}_i) \tag{10.18}$$

is called the *Gauss-Lobatto formula* with $n+1$ nodes, and has degree of exactness $2n-1$. Indeed, for any $f \in \mathbb{P}_{2n-1}$, there exist a polynomial $\pi_{n-2} \in \mathbb{P}_{n-2}$ and a remainder $q_n \in \mathbb{P}_n$ such that $f = \overline{\omega}_{n+1}\pi_{n-2} + q_n$.

The quadrature formula (10.18) has degree of exactness at least equal to n (being interpolatory with $n+1$ distinct nodes), thus we get

$$\sum_{j=0}^{n} \overline{\alpha}_j q_n(\overline{x}_j) = \int_{-1}^{1} q_n(x)w(x)dx = \int_{-1}^{1} f(x)w(x)dx - \int_{-1}^{1} \overline{\omega}_{n+1}(x)\pi_{n-2}(x)w(x)dx.$$

From (10.17) we conclude that $\bar{\omega}_{n+1}$ is orthogonal to all the polynomials of degree $\leq n-2$, so that the last integral is null. Moreover, since $f(\overline{x}_j) = q_n(\overline{x}_j)$ for $j = 0, \ldots, n$, we conclude that

$$\int_{-1}^{1} f(x)w(x)dx = \sum_{i=0}^{n} \overline{\alpha}_i f(\overline{x}_i), \qquad \forall f \in \mathbb{P}_{2n-1}.$$

Denoting by $\Pi_{n,w}^{GL} f$ the polynomial of degree n that interpolates f at the nodes $\{\overline{x}_j, j = 0, \ldots, n\}$, we get

$$\Pi_{n,w}^{GL} f(x) = \sum_{i=0}^{n} f(\overline{x}_i)\overline{l}_i(x) \tag{10.19}$$

and thus $I_{n,w}^{GL}(f) = \int_{-1}^{1} \Pi_{n,w}^{GL} f(x)w(x)dx$.

Remark 10.2 In the special case where the Gauss-Lobatto quadrature is considered with respect to the Jacobi weight $w(x) = (1-x)^{\alpha}(1+x)^{\beta}$, with $\alpha, \beta > -1$, the internal nodes $\overline{x}_1, \ldots, \overline{x}_{n-1}$ can be identified as the roots of the polynomial $(J_n^{(\alpha,\beta)})'$, that is, the extremants of the n-th Jacobi polynomial $J_n^{(\alpha,\beta)}$ (see [CHQZ06], pp. 71-72). ■

The following convergence result holds for Gaussian integration (see [Atk89], Chapter 5)

$$\lim_{n\to+\infty} \left| \int_{-1}^{1} f(x)w(x)dx - \sum_{j=0}^{n} \alpha_j f(x_j) \right| = 0, \qquad \forall f \in C^0([-1,1]).$$

A similar result also holds for Gauss-Lobatto integration. If the integrand function is not only continuous, but also differentiable up to the order $p \geq 1$, we shall see that Gaussian integration converges with an order of infinitesimal with respect to $1/n$ that is larger when p is greater. In the forthcoming sections, the previous results will be specified in the cases of the Chebyshev and Legendre polynomials.

Remark 10.3 (Integration over an arbitrary interval) A quadrature formula with nodes ξ_j and coefficients β_j, $j = 0, \ldots, n$ over the interval $[-1,1]$ can be mapped on any interval $[a,b]$. Indeed, let $\varphi : [-1,1] \to [a,b]$ be the affine map $x = \varphi(\xi) = \frac{b-a}{2}\xi + \frac{a+b}{2}$. Then

$$\int_a^b f(x)dx = \frac{b-a}{2}\int_{-1}^{1}(f \circ \varphi)(\xi)d\xi.$$

Therefore, we can employ on the interval $[a,b]$ the quadrature formula with nodes $x_j = \varphi(\xi_j)$ and weights $\alpha_j = \frac{b-a}{2}\beta_j$. Notice that this formula maintains on the interval $[a,b]$ the same degree of exactness of the generating formula over $[-1,1]$. Indeed, assuming that

$$\int_{-1}^{1} p(\xi)d\xi = \sum_{j=0}^{n} p(\xi_j)\beta_j$$

for any polynomial p of degree r over $[-1,1]$ (for a suitable integer r), for any polynomial q of the same degree on $[a,b]$ we get

$$\sum_{j=0}^{n} q(x_j)\alpha_j = \frac{b-a}{2}\sum_{j=0}^{n}(q \circ \varphi)(\xi_j)\beta_j = \frac{b-a}{2}\int_{-1}^{1}(q \circ \varphi)(\xi)d\xi = \int_a^b q(x)dx,$$

having recalled that $(q \circ \varphi)(\xi)$ is a polynomial of degree r on $[-1,1]$. ■

10.3 Chebyshev Integration and Interpolation

If Gaussian quadratures are considered with respect to the Chebyshev weight $w(x) = (1-x^2)^{-1/2}$, Gauss nodes and coefficients are given by

$$x_j = -\cos\frac{(2j+1)\pi}{2(n+1)}, \; \alpha_j = \frac{\pi}{n+1}, \; 0 \le j \le n, \qquad (10.20)$$

while Gauss-Lobatto nodes and weights are

$$\overline{x}_j = -\cos\frac{\pi j}{n}, \; \overline{\alpha}_j = \frac{\pi}{d_j n}, \; 0 \le j \le n, \, n \ge 1, \qquad (10.21)$$

where $d_0 = d_n = 2$ and $d_j = 1$ for $j = 1, \ldots, n-1$. Notice that the Gauss nodes (10.20) are, for a fixed $n \ge 0$, the zeros of the Chebyshev polynomial $T_{n+1} \in \mathbb{P}_{n+1}$, while, for $n \ge 1$, the internal nodes $\{\bar{x}_j, \; j = 1, \ldots, n-1\}$ are the zeros of T'_n, as anticipated in Remark 10.2.

Denoting by $\Pi^{GL}_{n,w} f$ the polynomial of degree n that interpolates f at the nodes (10.21), it can be shown that the interpolation error can be bounded as

$$\|f - \Pi_{n,w}^{GL} f\|_w \leq C n^{-s} \|f\|_{s,w}, \qquad \text{for } s \geq 1, \tag{10.22}$$

where $\|\cdot\|_w$ is the norm in L^2_w defined in (10.9), provided that for some $s \geq 1$ the function f has derivatives $f^{(k)}$ of order $k = 0, \ldots, s$ in L^2_w. In such a case

$$\|f\|_{s,w} = \left(\sum_{k=0}^{s} \|f^{(k)}\|_w^2 \right)^{\frac{1}{2}}. \tag{10.23}$$

Here and in the following, C is a constant independent of n that can assume different values at different places. In particular, for any continuous function f the following pointwise error estimate can be derived (see Exercise 3)

$$\|f - \Pi_{n,w}^{GL} f\|_\infty \leq C n^{1/2-s} \|f\|_{s,w}. \tag{10.24}$$

Thus, $\Pi_{n,w}^{GL} f$ converges pointwise to f as $n \to \infty$, for any $f \in C^1([-1,1])$. The same kind of results (10.22) and (10.24) hold if $\Pi_{n,w}^{GL} f$ is replaced with the polynomial $\Pi_n^G f$ of degree n that interpolates f at the $n+1$ Gauss nodes x_j in (10.20). (For the proof of these results see, for instance, [CHQZ06], p. 296, or [QV94], p. 112). We have also the following result (see [Riv74], p.13)

$$\|f - \Pi_n^G f\|_\infty \leq (1 + \Lambda_n) E_n^*(f), \quad \text{with } \Lambda_n \leq \frac{2}{\pi} \log(n+1) + 1, \tag{10.25}$$

where $\forall n$, $E_n^*(f) = \inf_{p \in \mathbb{P}_n} \|f - p\|_\infty$ is the best approximation error for f in $\mathbb{P}_n$ and Λ_n is the Lebesgue constant associated with the Chebyshev nodes (10.20).

Remark 10.4 (Barycentric weigths) When Gauss nodes (10.20) or Gauss-Lobatto nodes (10.21) are used, it is possible to compute in an explicit form the weigths of the barycentric formula (8.24). For Gauss nodes we have

$$w_j = (-1)^j \sin \frac{(2j+1)\pi}{2n+2},$$

while for Gauss-Lobatto nodes become

$$w_j = (-1)^j \gamma_j, \qquad \gamma_j = \begin{cases} \frac{1}{2}, \text{ if } j = 0 \text{ or } j = n, \\ 1, \text{ otherwise.} \end{cases}$$

We can done in this case a numerical example in order to appreciate the efficiency and the robustness of the barycentric formula for the Lagrange interpolant. We interpolate in 1001 Gauss-Lobatto nodes the function $f(x) = |x| + x/2 - x^2$ and we evaluate the Lagrange polynomial in 5000 uniform nodes in $[-1,1]$. On a PC this computation requires 0.16 s if one use the barycentric formula and 23.17 s if the Newton formula is used. Moreover, in this last case the evaluated polynomial is complete wrong since it is affected by *overflow* errors, while the barycentric formula produces an accurate evaluation of Π_n. ■

As far as the numerical integration error is concerned, let us consider, for instance, the Gauss-Lobatto quadrature rule (10.18) with nodes and weights given in (10.21). First of all, notice that

$$\int_{-1}^{1} f(x)(1-x^2)^{-1/2}dx = \lim_{n\to\infty} I_{n,w}^{GL}(f)$$

for any function f whose left integral is finite (see [Sze67], p. 342). If, moreover, $\|f\|_{s,w}$ is finite for some $s \geq 1$, we have

$$\left| \int_{-1}^{1} f(x)(1-x^2)^{-1/2}dx - I_{n,w}^{GL}(f) \right| \leq Cn^{-s}\|f\|_{s,w}. \tag{10.26}$$

This result follows from the more general one

$$|(f,v_n)_w - (f,v_n)_n| \leq Cn^{-s}\|f\|_{s,w}\|v_n\|_w, \qquad \forall v_n \in \mathbb{P}_n, \tag{10.27}$$

where the so-called *discrete scalar product* has been introduced

$$(f,g)_n = \sum_{j=0}^{n} \overline{\alpha}_j f(\overline{x}_j) g(\overline{x}_j) = I_{n,w}^{GL}(fg). \tag{10.28}$$

Actually, (10.26) follows from (10.27) setting $v_n \equiv 1$ and noticing that $\|v_n\|_w = \left(\int_{-1}^{1}(1-x^2)^{-1/2}dx\right)^{1/2} = \sqrt{\pi}$. Thanks to (10.26) we can thus conclude that the (Chebyshev) Gauss-Lobatto formula has degree of exactness $2n-1$ and order of accuracy (with respect to n^{-1}) equal to s, provided that $\|f\|_{s,w} < \infty$. Therefore, the order of accuracy is only limited by the regularity threshold s of the integrand function. Completely similar considerations can be drawn for (Chebyshev) Gauss formulae with $n+1$ nodes.
Let us finally determine the coefficients $\tilde{f}_k$, $k = 0, \ldots, n$, of the interpolating polynomial $\Pi_{n,w}^{GL} f$ at the $n+1$ Gauss-Lobatto nodes in the expansion with respect to the Chebyshev polynomials (10.10)

$$\Pi_{n,w}^{GL} f(x) = \sum_{k=0}^{n} \tilde{f}_k T_k(x). \tag{10.29}$$

Notice that $\Pi_{n,w}^{GL} f$ coincides with the discrete truncation of the Chebyshev series f_n^* defined in (10.4). Enforcing the equality $\Pi_{n,w}^{GL} f(\overline{x}_j) = f(\overline{x}_j)$, $j = 0, \ldots, n$, we find

$$f(\overline{x}_j) = \sum_{k=0}^{n} \cos\left(\frac{kj\pi}{n}\right) \tilde{f}_k, \qquad j = 0, \ldots, n. \tag{10.30}$$

Recalling the exactness of the Gauss-Lobatto quadrature, it can be checked that (see Exercise 2)

$$\tilde{f}_k = \frac{2}{nd_k} \sum_{j=0}^{n} \frac{1}{d_j} \cos\left(\frac{kj\pi}{n}\right) f(\overline{x}_j), \qquad k = 0, \ldots, n, \tag{10.31}$$

where $d_j = 2$ if $j = 0, n$, and $d_j = 1$ if $j = 1, \ldots, n-1$. Relation (10.31) yields the discrete coefficients $\{\tilde{f}_k, k = 0, \ldots, n\}$ in terms of the nodal values $\{f(\overline{x}_j), j = 0, \ldots, n\}$. For this reason it is called the *Chebyshev discrete transform* (CDT) and, thanks to its trigonometric structure, it can be efficiently computed using the FFT algorithm (Fast Fourier transform) with a number of floating-point operations of the order of $n \log_2 n$ (see Section 10.9.2). Of course, (10.30) is the *inverse* of the CDT, and can be computed using the FFT.

10.4 Legendre Integration and Interpolation

As previously noticed, the Legendre weight is $w(x) \equiv 1$. For $n \geq 0$, the Gauss nodes and the related coefficients are given by

$$x_j \text{ zeros of } L_{n+1}(x), \; \alpha_j = \frac{2}{(1 - x_j^2)[L'_{n+1}(x_j)]^2}, \; j = 0, \ldots, n, \tag{10.32}$$

while the Gauss-Lobatto ones are, for $n \geq 1$

$$\overline{x}_0 = -1, \; \overline{x}_n = 1, \; \overline{x}_j \text{ zeros of } L'_n(x), \; j = 1, \ldots, n-1, \tag{10.33}$$

$$\overline{\alpha}_j = \frac{2}{n(n+1)} \frac{1}{[L_n(x_j)]^2}, \qquad j = 0, \ldots, n, \tag{10.34}$$

where L_n is the n-th Legendre polynomial defined in (10.12). It can be checked that, for a suitable constant C independent of n,

$$\frac{2}{n(n+1)} \leq \overline{\alpha}_j \leq \frac{C}{n}, \qquad \forall j = 0, \ldots, n$$

(see [BM92], p. 76). Then, letting $\Pi_n^{GL} f$ be the polynomial of degree n that interpolates f at the $n+1$ nodes $\overline{x}_j$ given by (10.33), it can be proved that it fulfills the same error estimates as those reported in (10.22) and (10.24) in the case of the corresponding Chebyshev polynomial.

Of course, the norm $\|\cdot\|_w$ must here be replaced by the norm $\|\cdot\|_{L^2(-1,1)}$, while $\|f\|_{s,w}$ becomes

$$\|f\|_s = \left(\sum_{k=0}^{s} \|f^{(k)}\|^2_{L^2(-1,1)} \right)^{\frac{1}{2}}. \tag{10.35}$$

The same kinds of results are ensured if $\Pi_n^{GL} f$ is replaced by the polynomial of degree n that interpolates f at the $n+1$ nodes x_j given by (10.32).

Referring to the discrete scalar product defined in (10.28), but taking now the nodes and coefficients given by (10.33) and (10.34), we see that $(\cdot,\cdot)_n$ is an approximation of the usual scalar product $(\cdot,\cdot)$ of $L^2(-1,1)$. Actually, the equivalent relation to (10.27) now reads

$$|(f,v_n)-(f,v_n)_n| \leq Cn^{-s}\|f\|_s\|v_n\|_{L^2(-1,1)}, \qquad \forall v_n \in \mathbb{P}_n \tag{10.36}$$

and holds for any $s \geq 1$ such that $\|f\|_s < \infty$. In particular, setting $v_n \equiv 1$, we get $\|v_n\| = \sqrt{2}$, and from (10.36) it follows that

$$\left|\int_{-1}^{1} f(x)dx - I_n^{GL}(f)\right| \leq Cn^{-s}\|f\|_s, \tag{10.37}$$

which demonstrates a convergence of the Gauss-Legendre-Lobatto quadrature formula to the exact integral of f with order of accuracy s with respect to n^{-1} provided that $\|f\|_s < \infty$. A similar result holds for the Gauss-Legendre quadrature formulae.

Example 10.1 Consider the approximate evaluation of the integral of $f(x) = |x|^{\alpha+\frac{3}{5}}$ over $[-1,1]$ for $\alpha = 0,1,2$. Notice that f has "piecewise" derivatives up to order $s = s(\alpha) = \alpha+1$ in $L^2(-1,1)$. Figure 10.1 shows the behavior of the error as a function of n for the Gauss-Legendre quadrature formula. According to (10.37), the convergence rate of the formula increases by one when α increases by one. •

The interpolating polynomial at the nodes (10.33) is given by

$$\Pi_n^{GL} f(x) = \sum_{k=0}^{n} \tilde{f}_k L_k(x). \tag{10.38}$$

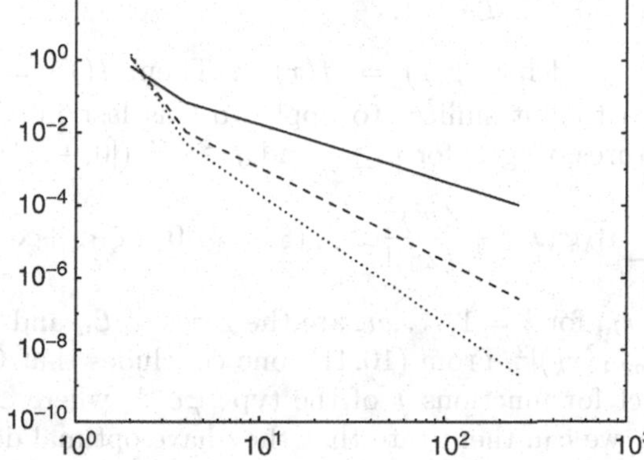

Fig. 10.1. The quadrature error in logarithmic scale as a function of n in the case of a function with the first s derivatives in $L^2(-1,1)$ for $s=1$ (*solid line*), $s=2$ (*dashed line*), $s=3$ (*dotted line*)

Notice that also in this case $\Pi_n^{GL} f$ coincides with the discrete truncation of the Legendre series f_n^* defined in (10.4). Proceeding as in the previous section, we get

$$f(\overline{x}_j) = \sum_{k=0}^{n} \tilde{f}_k L_k(\overline{x}_j), \qquad j = 0, \ldots, n, \tag{10.39}$$

and also

$$\tilde{f}_k = \begin{cases} \dfrac{2k+1}{n(n+1)} \displaystyle\sum_{j=0}^{n} L_k(\overline{x}_j) \frac{1}{L_n^2(\overline{x}_j)} f(\overline{x}_j), & k = 0, \ldots, n-1, \\ \dfrac{1}{n+1} \displaystyle\sum_{j=0}^{n} \frac{1}{L_n(\overline{x}_j)} f(\overline{x}_j), & k = n \end{cases} \tag{10.40}$$

(see Exercise 6). Formulae (10.40) and (10.39) provide, respectively, the *discrete Legendre transform* (DLT) and its inverse.

10.5 Gaussian Integration over Unbounded Intervals

We consider integration on both half and on the whole of real axis. In both cases we use interpolatory Gaussian formulae whose nodes are the zeros of Laguerre and Hermite orthogonal polynomials, respectively.

The Laguerre polynomials. These are algebraic polynomials, orthogonal on the interval $[0, +\infty)$ with respect to the weight function $w(x) = e^{-x}$. They are defined by

$$\mathcal{L}_n(x) = e^x \frac{d^n}{dx^n}(e^{-x} x^n), \qquad n \geq 0,$$

and satisfy the following three-term recursive relation

$$\begin{cases} \mathcal{L}_{n+1}(x) = (2n+1-x)\mathcal{L}_n(x) - n^2 \mathcal{L}_{n-1}(x), \; n \geq 0, \\ \mathcal{L}_{-1} = 0, \qquad \mathcal{L}_0 = 1. \end{cases}$$

For any function f, define $\varphi(x) = f(x)e^x$. Then, $I(f) = \int_0^\infty f(x)dx = \int_0^\infty e^{-x}\varphi(x)dx$, so that it suffices to apply to this last integral the Gauss-Laguerre quadratures, to get, for $n \geq 1$ and $f \in C^{2n}([0, +\infty))$

$$I(f) = \sum_{k=1}^{n} \alpha_k \varphi(x_k) + \frac{(n!)^2}{(2n)!} \varphi^{(2n)}(\xi), \qquad 0 < \xi < +\infty, \tag{10.41}$$

where the nodes x_k, for $k = 1, \ldots, n$, are the zeros of $\mathcal{L}_n$ and the weights are $\alpha_k = (n!)^2 x_k / [\mathcal{L}_{n+1}(x_k)]^2$. From (10.41), one concludes that Gauss-Laguerre formulae are exact for functions f of the type φe^{-x}, where $\varphi \in \mathbb{P}_{2n-1}$. In a generalized sense, we can then state that they have optimal degrees of exactness equal to $2n - 1$.

Example 10.2 Using a Gauss-Laguerre quadrature formula with $n = 12$ to compute the integral in Example 9.12 we obtain the value 0.5997 with an absolute error with respect to exact integration equal to $2.96 \cdot 10^{-4}$. For the sake of comparison, the composite trapezoidal formula would require 277 nodes to obtain the same accuracy. •

The Hermite polynomials. These are orthogonal polynomials on the real line with respect to the weight function $w(x) = e^{-x^2}$. They are defined by

$$\mathcal{H}_n(x) = (-1)^n e^{x^2} \frac{d^n}{dx^n}(e^{-x^2}), \qquad n \geq 0.$$

Hermite polynomials can be recursively generated as

$$\begin{cases} \mathcal{H}_{n+1}(x) = 2x\mathcal{H}_n(x) - 2n\mathcal{H}_{n-1}(x), \, n \geq 0, \\ \mathcal{H}_{-1} = 0, \qquad \mathcal{H}_0 = 1. \end{cases}$$

As in the previous case, letting $\varphi(x) = f(x)e^{x^2}$, we have $I(f) = \int_{-\infty}^{\infty} f(x)dx = \int_{-\infty}^{\infty} e^{-x^2}\varphi(x)dx$. Applying to this last integral the Gauss-Hermite quadratures we obtain, for $n \geq 1$ and $f \in C^{2n}(\mathbb{R})$

$$I(f) = \int_{-\infty}^{\infty} e^{-x^2}\varphi(x)dx = \sum_{k=1}^{n} \alpha_k \varphi(x_k) + \frac{(n!)\sqrt{\pi}}{2^n(2n)!}\varphi^{(2n)}(\xi), \qquad \xi \in \mathbb{R}, \quad (10.42)$$

where the nodes x_k, for $k = 1, \ldots, n$, are the zeros of $\mathcal{H}_n$ and the weights are $\alpha_k = 2^{n+1}n!\sqrt{\pi}/[\mathcal{H}_{n+1}(x_k)]^2$. As for Gauss-Laguerre quadratures, the Gauss-Hermite rules also are exact for functions f of the form φe^{-x^2}, where $\varphi \in \mathbb{P}_{2n-1}$; therefore, they have optimal degrees of exactness equal to $2n - 1$.

More details on the subject can be found in [DR75], pp. 173-174.

10.6 Programs for the Implementation of Gaussian Quadratures

Programs 82, 83 and 84 compute the coefficients $\{\alpha_k\}$ and $\{\beta_k\}$, introduced in (10.8), in the cases of the Legendre, Laguerre and Hermite polynomials. These programs are then called by Program 85 for the computation of nodes and weights (10.32), in the case of the Gauss-Legendre formulae, and by Programs 86, 87 for computing nodes and weights in the Gauss-Laguerre and Gauss-Hermite quadrature rules (10.41) and (10.42). All the codings reported in this section are excerpts from the library `ORTHPOL` [Gau94].

Program 82 - coeflege : Coefficients of Legendre polynomials

```
function [a,b]=coeflege(n)
%COEFLEGE Coefficients of Legendre polynomials.
%  [A,B]=COEFLEGE(N): A and B are the alpha(k) and beta(k) coefficients
```

```
%   for the Legendre polynomial of degree N.
if n<=1, error('n must be >1');end
a = zeros(n,1); b=a; b(1)=2;
k=[2:n]; b(k)=1./(4-1./(k-1).^2);
return
```

Program 83 - coeflagu : Coefficients of Laguerre polynomials

```
function [a,b]=coeflagu(n)
%COEFLAGU Coefficients of Laguerre polynomials.
%   [A,B]=COEFLAGU(N): A and B are the alpha(k) and beta(k) coefficients
%   for the Laguerre polynomial of degree N.
if n<=1, error('n must be >1 '); end
a=zeros(n,1); b=zeros(n,1); a(1)=1; b(1)=1;
k=[2:n]; a(k)=2*(k-1)+1; b(k)=(k-1).^2;
return
```

Program 84 - coefherm : Coefficients of Hermite polynomials

```
function [a,b]=coefherm(n)
%COEFHERM Coefficients of Hermite polynomials.
%   [A,B]=COEFHERM(N): A and B are the alpha(k) and beta(k) coefficients
%   for the Hermite polynomial of degree N.
if n<=1, error('n must be >1 '); end
a=zeros(n,1); b=zeros(n,1); b(1)=sqrt(4.*atan(1.));
k=[2:n]; b(k)=0.5*(k-1);
return
```

Program 85 - zplege : Coefficients of Gauss-Legendre formulae

```
function [x,w]=zplege(n)
%ZPLEGE Gauss-Legendre formula.
%   [X,W]=ZPLEGE(N) computes the nodes and the weights of the Gauss-Legendre
%   formula with N nodes.
if n<=1, error('n must be >1'); end
[a,b]=coeflege(n);
JacM=diag(a)+diag(sqrt(b(2:n)),1)+diag(sqrt(b(2:n)),-1);
[w,x]=eig(JacM); x=diag(x); scal=2; w=w(1,:)'.^2*scal;
[x,ind]=sort(x); w=w(ind);
return
```

Program 86 - zplagu : Coefficients of Gauss-Laguerre formulae

```
function [x,w]=zplagu(n)
%ZPLAGU Gauss-Laguerre formula.
%   [X,W]=ZPLAGU(N) computes the nodes and the weights of the Gauss-Laguerre
%   formula with N nodes.
if n<=1, error('n must be >1 '); end
```

```
[a,b]=coeflagu(n);
JacM=diag(a)+diag(sqrt(b(2:n)),1)+diag(sqrt(b(2:n)),-1);
[w,x]=eig(JacM); x=diag(x); w=w(1,:)'.^2;
return
```

Program 87 - zpherm : Coefficients of Gauss-Hermite formulae

```
function [x,w]=zpherm(n)
%ZPHERM Gauss-Hermite formula.
%  [X,W]=ZPHERM(N) computes the nodes and the weights of the Gauss-Hermite
%  formula with N nodes.
if n<=1, error('n must be >1 '); end
[a,b]=coefherm(n);
JacM=diag(a)+diag(sqrt(b(2:n)),1)+diag(sqrt(b(2:n)),-1);
[w,x]=eig(JacM); x=diag(x); scal=sqrt(pi); w=w(1,:)'.^2*scal;
[x,ind]=sort(x); w=w(ind);
return
```

10.7 Approximation of a Function in the Least-Squares Sense

Given a function $f \in \mathrm{L}^2_w(a,b)$, we look for a polynomial r_n of degree $\leq n$ that satisfies

$$\|f - r_n\|_w = \min_{p_n \in \mathbb{P}_n} \|f - p_n\|_w,$$

where w is a fixed weight function in (a,b). Should it exist, r_n is called a *least-squares polynomial*. The name derives from the fact that, if $w \equiv 1$, r_n is the polynomial that minimizes the mean-square error $E = \|f - r_n\|_{\mathrm{L}^2(a,b)}$ (see Exercise 8).

As seen in Section 10.1, r_n coincides with the truncation f_n of order n of the Fourier series (see (10.2) and (10.3)). Depending on the choice of the weight $w(x)$, different least-squares polynomials arise with different convergence properties.

Analogous to Section 10.1, we can introduce the discrete truncation f_n^* (10.4) of the Chebyshev series (setting $p_k = T_k$) or the Legendre series (setting $p_k = L_k$). If the discrete scalar product induced by the Gauss-Lobatto quadrature rule (10.28) is used in (10.5) then the $\tilde{f}_k$'s coincide with the coefficients of the expansion of the interpolating polynomial $\Pi^{GL}_{n,w} f$ (see (10.29) in the Chebyshev case, or (10.38) in the Legendre case).

Consequently, $f_n^* = \Pi^{GL}_{n,w} f$, i.e., the discrete truncation of the (Chebyshev or Legendre) series of f turns out to coincide with the interpolating polynomial at the $n+1$ Gauss-Lobatto nodes. In particular, in such a case (10.6) is trivially satisfied, since $\|f - f_n^*\|_n = 0$.

10.7.1 Discrete Least-Squares Approximation

Several applications require representing in a synthetic way, using elementary functions, a large set of data that are available at a discrete level, for instance, the results of experimental measurements. This approximation process, often referred to as *data fitting*, can be satisfactorily solved using the discrete least-squares technique that can be formulated as follows.

Assume we are given $m+1$ pairs of data

$$\{(x_i, y_i),\, i = 0, \ldots, m\}, \tag{10.43}$$

where y_i may represent, for instance, the value of a physical quantity measured at the position x_i. We assume that all the abscissae are distinct.

We look for a polynomial $p_n(x) = \sum_{i=0}^{n} a_i \varphi_i(x)$ such that

$$\sum_{j=0}^{m} w_j |p_n(x_j) - y_j|^2 \leq \sum_{j=0}^{m} w_j |q_n(x_j) - y_j|^2 \;\forall q_n \in \mathbb{P}_n, \tag{10.44}$$

for suitable coefficients $w_j > 0$. If $n = m$ the polynomial p_n clearly coincides with the interpolating polynomial of degree n at the nodes $\{x_i\}$. Problem (10.44) is called a *discrete least-squares problem* since a discrete scalar product is involved, and is the discrete counterpart of the continuous least-squares problem. The solution p_n is therefore referred to as a *least-squares polynomial.* Notice that

$$|||q||| = \left\{ \sum_{j=0}^{m} w_j [q(x_j)]^2 \right\}^{1/2} \tag{10.45}$$

is an *essentially strict* seminorm on $\mathbb{P}_n$ (see, Exercise 7). By definition a discrete norm (or seminorm) $\|\cdot\|_*$ is essentially strict if $\|f+g\|_* = \|f\|_* + \|g\|_*$ implies there exist nonnull α, β such that $\alpha f(x_i) + \beta g(x_i) = 0$ for $i = 0, \ldots, m$. Since $|||\cdot|||$ is an essentially strict seminorm, problem (10.44) admits a unique solution (see, [IK66], Section 3.5). Proceeding as in Section 3.13, we find

$$\sum_{k=0}^{n} a_k \sum_{j=0}^{m} w_j \varphi_k(x_j) \varphi_i(x_j) = \sum_{j=0}^{m} w_j y_j \varphi_i(x_j), \qquad \forall i = 0, \ldots, n,$$

which is called a *system of normal equations*, and can be conveniently written in the form

$$\mathrm{B}^T \mathrm{B}\mathbf{a} = \mathrm{B}^T \mathbf{y}, \tag{10.46}$$

where B is the rectangular matrix $(m+1) \times (n+1)$ of entries $b_{ij} = \varphi_j(x_i)$, $i = 0, \ldots, m$, $j = 0, \ldots, n$, $\mathbf{a} \in \mathbb{R}^{n+1}$ is the vector of the unknown coefficients and $\mathbf{y} \in \mathbb{R}^{m+1}$ is the vector of data.

Notice that the system of normal equations obtained in (10.46) is of the same nature as that introduced in Section 3.13 in the case of over-determined systems. Actually, if $w_j = 1$ for $j = 0, \ldots, m$, the above system can be regarded as the solution in the least-squares sense of the system

$$\sum_{k=0}^{n} a_k \varphi_k(x_i) = y_i, \qquad i = 0, 1, \ldots, m,$$

which would not admit a solution in the classical sense, since the number of rows is greater than the number of columns. In the case $n = 1$, the solution to (10.44) is a linear function, called *linear regression* for the data fitting of (10.43). The associated system of normal equations is

$$\sum_{k=0}^{1} \sum_{j=0}^{m} w_j \varphi_i(x_j) \varphi_k(x_j) a_k = \sum_{j=0}^{m} w_j \varphi_i(x_j) y_j, \qquad i = 0, 1.$$

Setting $(f, g)_m = \sum_{j=0}^{m} w_j f(x_j) g(x_j)$ the previous system becomes

$$\begin{cases} (\varphi_0, \varphi_0)_m a_0 + (\varphi_1, \varphi_0)_m a_1 = (y, \varphi_0)_m, \\ (\varphi_0, \varphi_1)_m a_0 + (\varphi_1, \varphi_1)_m a_1 = (y, \varphi_1)_m, \end{cases}$$

where $y(x)$ is a function that takes the value y_i at the nodes x_i, $i = 0, \ldots, m$. After some algebra, we get this explicit form for the coefficients

$$a_0 = \frac{(y, \varphi_0)_m (\varphi_1, \varphi_1)_m - (y, \varphi_1)_m (\varphi_1, \varphi_0)_m}{(\varphi_1, \varphi_1)_m (\varphi_0, \varphi_0)_m - (\varphi_0, \varphi_1)_m^2},$$

$$a_1 = \frac{(y, \varphi_1)_m (\varphi_0, \varphi_0)_m - (y, \varphi_0)_m (\varphi_1, \varphi_0)_m}{(\varphi_1, \varphi_1)_m (\varphi_0, \varphi_0)_m - (\varphi_0, \varphi_1)_m^2}.$$

Example 10.3 As already seen in Example 8.2, small changes in the data can give rise to large variations on the interpolating polynomial of a given function f. This doesn't happen for the least-squares polynomial where m is much larger than n. As an example, consider the function $f(x) = \sin(2\pi x)$ in $[-1, 1]$ and evaluate it at the 22 equally spaced nodes $x_i = -1 + 2i/21$, $i = 0, \ldots, 21$, setting $f_i = f(x_i)$. Then, suppose to add to the data f_i a random perturbation of the order of 10^{-3} and denote by p_5 and $\tilde{p}_5$ the least-squares polynomials of degree 5 approximating the data f_i and $\tilde{f}_i$, respectively. The maximum norm of the difference $p_5 - \tilde{p}_5$ over $[-1, 1]$ is of the order of 10^{-3}, i.e., it is of the same order as the perturbation on the data. For comparison, the same difference in the case of Lagrange interpolation is about equal to 2 as can be seen in Figure 10.2. •

10.8 The Polynomial of Best Approximation

Consider a function $f \in C^0([a, b])$. A polynomial $p_n^* \in \mathbb{P}_n$ is said to be the *polynomial of best approximation of* f if it satisfies

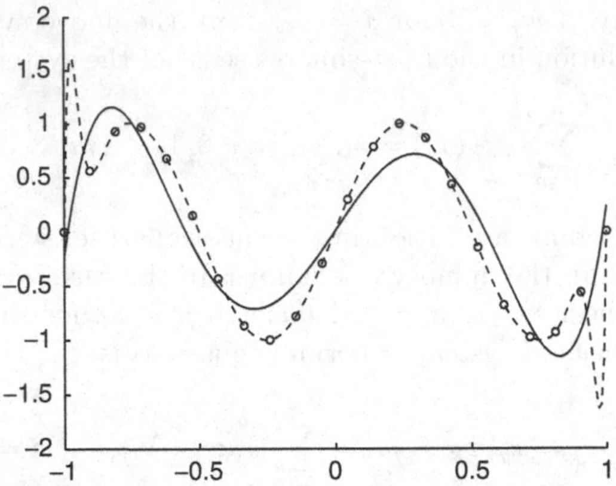

Fig. 10.2. The perturbed data (*circles*), the associated least-squares polynomial of degree 5 (*solid line*) and the Lagrange interpolating polynomial (*dashed line*)

$$\|f - p_n^*\|_\infty = \min_{p_n \in \mathbb{P}_n} \|f - p_n\|_\infty, \quad \forall p_n \in \mathbb{P}_n, \tag{10.47}$$

where $\|g\|_\infty = \max_{a \le x \le b} |g(x)|$. This problem is referred to as a *minimax approximation*, as we are looking for the minimum error measured in the maximum norm.

Property 10.1 (Chebyshev equioscillation theorem) *For any $n \geq 0$, the polynomial of best approximation p_n^* of f exists and is unique. Moreover, in $[a, b]$ there exist $n+2$ points $x_0 < x_1 < \ldots < x_{n+1}$ such that*

$$f(x_j) - p_n^*(x_j) = \sigma(-1)^j E_n^*(f), \qquad j = 0, \ldots, n+1,$$

with $\sigma = 1$ or $\sigma = -1$ depending on f and n, and $E_n^*(f) = \|f - p_n^*\|_\infty$.

(For the proof, see [Dav63], Chapter 7). As a consequence, there exist $n+1$ points $\tilde{x}_0 < \tilde{x}_1 < \ldots < \tilde{x}_n$, with $x_k < \tilde{x}_k < x_{k+1}$ for $k = 0, \ldots, n$, to be determined in $[a, b]$ such that

$$p_n^*(\tilde{x}_j) = f(\tilde{x}_j), \; j = 0, 1, \ldots, n,$$

so that the best approximation polynomial is a polynomial of degree n that interpolates f at $n+1$ unknown nodes.

The following result yields an estimate of $E_n^*(f)$ without explicitly computing p_n^* (we refer for the proof to [Atk89], Chapter 4).

Property 10.2 (de la Vallée-Poussin theorem) *Let $f \in C^0([a, b])$ and $n \geq 0$, and let $x_0 < x_1 < \ldots < x_{n+1}$ be $n+2$ points in $[a, b]$. If there exists a polynomial q_n of degree $\leq n$ such that*

$$f(x_j) - q_n(x_j) = (-1)^j e_j \; j = 0, 1, \ldots, n+1,$$

where all e_j have the same sign and are nonnull, then

$$\min_{0\le j\le n+1} |e_j| \le E_n^*(f).$$

We can now relate $E_n^*(f)$ with the interpolation error. Indeed,

$$\|f-\Pi_n f\|_\infty \le \|f-p_n^*\|_\infty + \|p_n^* - \Pi_n f\|_\infty.$$

On the other hand, using the Lagrange representation of p_n^* we get

$$\|p_n^* - \Pi_n f\|_\infty = \left\| \sum_{i=0}^{n} (p_n^*(x_i) - f(x_i)) l_i \right\|_\infty \le \|p_n^* - f\|_\infty \left\| \sum_{i=0}^{n} |l_i| \right\|_\infty ,$$

from which it follows

$$\|f-\Pi_n f\|_\infty \le (1+\Lambda_n) E_n^*(f),$$

where Λ_n is the Lebesgue constant (8.11) associated with the nodes $\{x_i\}$. Thanks to (10.25) we can conclude that the Lagrange interpolating polynomial on the Chebyshev nodes is a good approximation of p_n^*. The above results yield a characterization of the best approximation polynomial, but do not provide a constructive way for generating it. However, starting from the Chebyshev equioscillation theorem, it is possible to devise an algorithm, called the Remes algorithm, that is able to construct an arbitrarily good approximation of the polynomial p_n^* (see [Atk89], Section 4.7).

10.9 Fourier Trigonometric Polynomials

Let us apply the theory developed in the previous sections to a particular family of orthogonal polynomials which are no longer algebraic polynomials but rather trigonometric. The *Fourier polynomials* on $(0, 2\pi)$ are defined as

$$\varphi_k(x) = e^{ikx}, \qquad k = 0, \pm 1, \pm 2, \ldots,$$

where i is the imaginary unit. These are complex-valued periodic functions with period equal to 2π. We shall use the notation $\mathrm{L}^2(0, 2\pi)$ to denote the complex-valued functions that are square integrable over $(0, 2\pi)$. Therefore

$$\mathrm{L}^2(0,2\pi) = \left\{ f : (0,2\pi) \to \mathbb{C} \text{ such that } \int_0^{2\pi} |f(x)|^2 dx < \infty \right\}$$

with scalar product and norm defined respectively by

$$(f,g) = \textstyle\int_0^{2\pi} f(x)\overline{g(x)}dx, \; \|f\|_{\mathrm{L}^2(0,2\pi)} = \sqrt{(f,f)}.$$

If $f \in \mathrm{L}^2(0, 2\pi)$, its Fourier series is

$$Ff = \sum_{k=-\infty}^{\infty} \widehat{f}_k \varphi_k, \text{ with } \widehat{f}_k = \frac{1}{2\pi} \int_0^{2\pi} f(x) e^{-ikx} dx = \frac{1}{2\pi}(f, \varphi_k). \quad (10.48)$$

If f is complex-valued we set $f(x) = \alpha(x) + i\beta(x)$ for $x \in [0, 2\pi]$, where $\alpha(x)$ is the real part of f and $\beta(x)$ is the imaginary one. Recalling that $e^{-ikx} = \cos(kx) - i\sin(kx)$ and letting

$$a_k = \frac{1}{2\pi}\int_0^{2\pi} [\alpha(x)\cos(kx) + \beta(x)\sin(kx)]\,dx,$$

$$b_k = \frac{1}{2\pi}\int_0^{2\pi} [-\alpha(x)\sin(kx) + \beta(x)\cos(kx)]\,dx,$$

the *Fourier coefficients* of the function f can be written as

$$\widehat{f}_k = a_k + ib_k \qquad \forall k = 0, \pm 1, \pm 2, \ldots. \tag{10.49}$$

We shall assume henceforth that f is a real-valued function; in such a case $\widehat{f}_{-k} = \overline{\widehat{f}_k}$ for any k.

Let N be an even positive integer. Analogously to what was done in Section 10.1, we call the *truncation of order* N of the Fourier series the function

$$f_N^*(x) = \sum_{k=-\frac{N}{2}}^{\frac{N}{2}-1} \widehat{f}_k e^{ikx}.$$

The use of capital N instead of small n is to conform with the notation usually adopted in the analysis of discrete Fourier series (see [Bri74], [Wal91]). To simplify the notations we also introduce an index shift so that

$$f_N^*(x) = \sum_{k=0}^{N-1} \widehat{f}_k e^{i(k-\frac{N}{2})x},$$

where now

$$\widehat{f}_k = \frac{1}{2\pi}\int_0^{2\pi} f(x)e^{-i(k-N/2)x}dx = \frac{1}{2\pi}(f, \widetilde{\varphi}_k),\ k = 0, \ldots, N-1 \tag{10.50}$$

and $\widetilde{\varphi}_k = e^{i(k-N/2)x}$. Denoting by

$$S_N = \text{span}\{\widetilde{\varphi}_k,\ 0 \le k \le N-1\},$$

if $f \in \mathrm{L}^2(0, 2\pi)$ its truncation of order N satisfies the following optimal approximation property in the least-squares sense

$$\|f - f_N^*\|_{\mathrm{L}^2(0,2\pi)} = \min_{g \in S_N} \|f - g\|_{\mathrm{L}^2(0,2\pi)}.$$

Set $h = 2\pi/N$ and $x_j = jh$, for $j = 0, \ldots, N-1$, and introduce the following *discrete scalar product*

$$(f,g)_N = h\sum_{j=0}^{N-1} f(x_j)\overline{g(x_j)}. \tag{10.51}$$

Replacing $(f, \widetilde{\varphi}_k)$ in (10.50) with $(f, \widetilde{\varphi}_k)_N$, we get the *discrete Fourier coefficients* of the function f

$$\widetilde{f}_k = \frac{1}{N}\sum_{j=0}^{N-1} f(x_j)e^{-ikjh}e^{ij\pi} = \frac{1}{N}\sum_{j=0}^{N-1} f(x_j)W_N^{(k-\frac{N}{2})j} \tag{10.52}$$

for $k = 0, \ldots, N-1$, where

$$W_N = \exp\left(-i\frac{2\pi}{N}\right)$$

is the *principal root of order N* of unity. According to (10.4), the trigonometric polynomial

$$\Pi_N^F f(x) = \sum_{k=0}^{N-1} \widetilde{f}_k e^{i(k-\frac{N}{2})x} \tag{10.53}$$

is called the *discrete Fourier series of order N* of f.

Lemma 10.2 *The following property holds*

$$(\varphi_l, \varphi_j)_N = h\sum_{k=0}^{N-1} e^{-ik(l-j)h} = 2\pi\delta_{jl}, \qquad 0 \le l, j \le N-1, \tag{10.54}$$

where δ_{jl} is the Kronecker symbol.

Proof. For $l = j$ the result is immediate. Thus, assume $l \neq j$; we have that

$$\sum_{k=0}^{N-1} e^{-ik(l-j)h} = \frac{1-\left(e^{-i(l-j)h}\right)^N}{1-e^{-i(l-j)h}} = 0.$$

Indeed, the numerator is $1 - (\cos(2\pi(l-j)) - i\sin(2\pi(l-j))) = 1 - 1 = 0$, while the denominator cannot vanish. Actually, it vanishes iff $(j-l)h = 2\pi$, i.e., $j - l = N$, which is impossible. ◇

Thanks to Lemma 10.2, the trigonometric polynomial $\Pi_N^F f$ is the Fourier *interpolate* of f at the nodes x_j, that is

$$\Pi_N^F f(x_j) = f(x_j), \qquad j = 0, 1, \ldots, N-1.$$

Indeed, using (10.52) and (10.54) in (10.53) it follows that

$$\Pi_N^F f(x_j) = \sum_{k=0}^{N-1} \widetilde{f}_k e^{ikjh} e^{-ijh\frac{N}{2}} = \sum_{l=0}^{N-1} f(x_l)\left[\frac{1}{N}\sum_{k=0}^{N-1} e^{-ik(l-j)h}\right] = f(x_j).$$

Therefore, looking at the first and last equality, we get

$$f(x_j) = \sum_{k=0}^{N-1} \widetilde{f}_k e^{ik(j-\frac{N}{2})h} = \sum_{k=0}^{N-1} \widetilde{f}_k W_N^{-(j-\frac{N}{2})k}, \; j = 0, \ldots, N-1. \quad (10.55)$$

The mapping $\{f(x_j)\} \to \{\widetilde{f}_k\}$ described by (10.52) is called the *Discrete Fourier Transform* (DFT), while the mapping (10.55) from $\{\widetilde{f}_k\}$ to $\{f(x_j)\}$ is called the *inverse transform* (IDFT). Both DFT and IDFT can be written in matrix form as $\{\widetilde{f}_k\} = \mathrm{T}\{f(x_j)\}$ and $\{f(x_j)\} = \mathrm{C}\{\widetilde{f}_k\}$, where $\mathrm{T} \in \mathbb{C}^{N\times N}$, C denotes the inverse of T and

$$T_{kj} = \frac{1}{N} W_N^{(k-\frac{N}{2})j}, \; k, j = 0, \ldots, N-1,$$

$$C_{jk} = W_N^{-(j-\frac{N}{2})k}, \quad j, k = 0, \ldots, N-1.$$

A naive implementation of the matrix-vector computation in the DFT and IDFT would require N^2 operations. Using the FFT (*Fast Fourier Transform*) algorithm only $\mathcal{O}(N \log_2 N)$ flops are needed, provided that N is a power of 2, as will be explained in Section 10.9.2.

The function $\Pi_N^F f \in S_N$ introduced in (10.53) is the solution of the minimization problem $\|f - \Pi_N^F f\|_N \leq \|f - g\|_N$ for any $g \in S_N$, where $\|\cdot\|_N = (\cdot,\cdot)_N^{1/2}$ is a discrete norm for S_N. In the case where f is periodic with all its derivatives up to order s ($s \geq 1$), an error estimate analogous to that for Chebyshev and Legendre interpolation holds

$$\|f - \Pi_N^F f\|_{\mathrm{L}^2(0,2\pi)} \leq C N^{-s} \|f\|_s$$

and also

$$\max_{0 \leq x \leq 2\pi} |f(x) - \Pi_N^F f(x)| \leq C N^{1/2-s} \|f\|_s.$$

In a similar manner, we also have

$$|(f, v_N) - (f, v_N)_N| \leq C N^{-s} \|f\|_s \|v_N\|$$

for any $v_N \in S_N$, and in particular, setting $v_N = 1$ we have the following error for the quadrature formula (10.51)

$$\left| \int_0^{2\pi} f(x)dx - h \sum_{j=0}^{N-1} f(x_j) \right| \leq C N^{-s} \|f\|_s$$

(see for the proof [CHQZ06], Chapter 2).

Notice that $h \sum_{j=0}^{N-1} f(x_j)$ is nothing else than the composite trapezoidal rule for approximating the integral $\int_0^{2\pi} f(x)dx$. Therefore, such a formula turns out to be extremely accurate when dealing with periodic and smooth integrands.

Programs 88 and 89 provide an implementation of the DFT and IDFT. The input parameter `f` is a string containing the function f to be transformed while `fc` is a vector of size N containing the values $\widetilde{f}_k$.

Program 88 - dft : Discrete Fourier transform

```
function fc=dft(N,f)
%DFT Discrete Fourier transform.
%  FC=DFT(N,F) computes the coefficients of the discrete Fourier
%  transform of a function F.
h = 2*pi/N;
x=[0:h:2*pi*(1-1/N)]; fx = eval(f);
wn = exp(-i*h);
for k=0:N-1,
  s = 0;
  for j=0:N-1
    s = s + fx(j+1)*wn^((k-N/2)*j);
  end
  fc (k+1) = s/N;
end
return
```

Program 89 - idft : Inverse discrete Fourier transform

```
function fv = idft(N,fc)
%IDFT Inverse discrete Fourier transform.
%  FV=IDFT(N,F) computes the coefficients of the inverse discrete Fourier
%  transform of a function F.
h  = 2*pi/N; wn = exp(-i*h);
for k=0:N-1
  s = 0;
  for j=0:N-1
    s = s + fc(j+1)*wn^(-k*(j-N/2));
  end
  fv (k+1) = s;
end
return
```

10.9.1 The Gibbs Phenomenon

Consider the discontinuous function $f(x) = x/\pi$ for $x \in [0, \pi]$ and equal to $x/\pi - 2$ for $x \in (\pi, 2\pi]$, and compute its DFT using Program 88. The interpolate $\Pi_N^F f$ is shown in Figure 10.3 (above) for $N = 8, 16, 32$. Notice the spurious oscillations around the point of discontinuity of f whose maximum amplitude, however, tends to a finite limit. The arising of these oscillations is known as *Gibbs phenomenon* and is typical of functions with isolated jump discontinuities; it affects the behavior of the truncated Fourier series not only in the neighborhood of the discontinuity but also over the entire interval, as can be clearly seen in the figure. The convergence rate of the truncated series for functions with jump discontinuities is linear in N^{-1} at every given non-singular point of the interval of definition of the function (see [CHQZ06], Section 2.1.4).

Since the Gibbs phenomenon is related to the slow decay of the Fourier coefficients of a discontinuous function, smoothing procedures can be profitably employed to attenuate the higher-order Fourier coefficients. This can be done by multiplying each coefficient $\widetilde{f}_k$ by a factor σ_k such that σ_k is a decreasing function of k. An example is provided by the *Lanczos smoothing*

$$\sigma_k = \frac{\sin(2(k - N/2)(\pi/N))}{2(k - N/2)(\pi/N)}, \qquad k = 0, \dots, N-1. \tag{10.56}$$

The effect of applying the Lanczos smoothing to the computation of the DFT of the above function f is represented in Figure 10.3 (below), which shows that the oscillations have almost completely disappeared.

For a deeper analysis of this subject we refer to [CHQZ06], Chapter 2.

10.9.2 The Fast Fourier Transform

As pointed out in the previous section, computing the discrete Fourier transform (DFT) or its inverse (IDFT) as a matrix-vector product, would require N^2 operations. In this section we illustrate the basic steps of the Cooley-Tukey algorithm [CT65], commonly known as Fast Fourier Transform (FFT). The computation of a DFT of order N is split into DFTs of order $p_0, \dots, p_m$, where $\{p_i\}$ are the prime factors of N. If N is a power of 2, the computational cost has the order of $N \log_2 N$ flops.

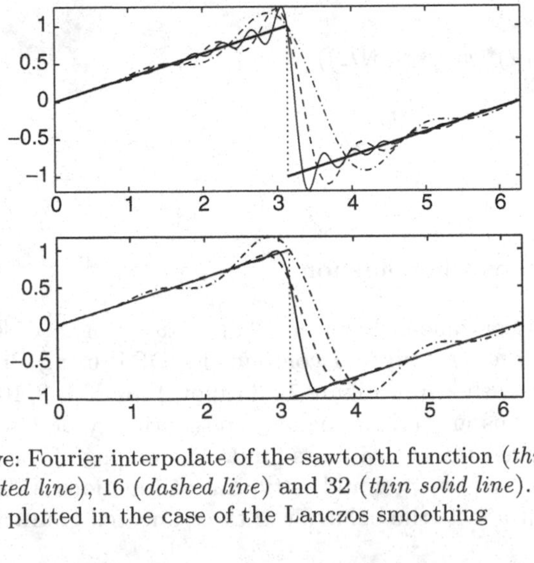

Fig. 10.3. Above: Fourier interpolate of the sawtooth function (*thick solid line*) for $N = 8$ (*dash-dotted line*), 16 (*dashed line*) and 32 (*thin solid line*). Below: the same informations are plotted in the case of the Lanczos smoothing

A recursive algorithm to compute the DFT when N is a power of 2 is described in the following. Let $\mathbf{f} = (f_i)^T$, $i = 0, \ldots, N-1$ and set $p(x) = \frac{1}{N}\sum_{j=0}^{N-1} f_j x^j$. Then, computing the DFT of the vector $\mathbf{f}$ amounts to evaluating $p(W_N^{k-\frac{N}{2}})$ for $k = 0, \ldots, N-1$. Let us introduce the polynomials

$$p_e(x) = \frac{1}{N}\left[f_0 + f_2 x + \ldots + f_{N-2} x^{\frac{N}{2}-1}\right],$$
$$p_o(x) = \frac{1}{N}\left[f_1 + f_3 x + \ldots + f_{N-1} x^{\frac{N}{2}-1}\right].$$

Notice that

$$p(x) = p_e(x^2) + x p_o(x^2)$$

from which it follows that the computation of the DFT of $\mathbf{f}$ can be carried out by evaluating the polynomials p_e and p_o at the points $W_N^{2(k-\frac{N}{2})}$, $k = 0, \ldots, N-1$. Since

$$W_N^{2(k-\frac{N}{2})} = W_N^{2k-N} = \exp\left(-i\frac{2\pi k}{N/2}\right)\exp(i2\pi) = W_{N/2}^k,$$

it turns out that we must evaluate p_e and p_o at the principal roots of unity of order $N/2$. In this manner the DFT of order N is rewritten in terms of two DFTs of order $N/2$; of course, we can recursively apply again this procedure to p_o and p_e. The process is terminated when the degree of the last generated polynomials is equal to one.
In Program 90 we propose a simple implementation of the FFT recursive algorithm. The input parameters are the vector `f` containing the `NN` values f_k, where `NN` is a power of 2.

Program 90 - fftrec : FFT algorithm in the recursive version

```
function [fftv]=fftrec(f,NN)
%FFTREC FFT algorithm in recursive form.
N = length(f);   w = exp(-2*pi*sqrt(-1)/N);
if N == 2
  fftv = f(1)+w.^[-NN/2:NN-1-NN/2]*f(2);
else
  a1 = f(1:2:N);   b1 = f(2:2:N);
  a2 = fftrec(a1,NN); b2 = fftrec(b1,NN);
  for k=-NN/2:NN-1-NN/2
    fftv(k+1+NN/2) = a2(k+1+NN/2) + b2(k+1+NN/2)*w^k;
  end
end
return
```

Remark 10.5 A FFT procedure can also be set up when N is not a power of 2. The simplest approach consists of adding some zero samples to the original sequence $\{f_i\}$ in such a way to obtain a total number of $\tilde{N} = 2^p$ values. This technique, however, does not necessarily yield the correct result. Therefore, an effective alternative is based on partitioning the Fourier matrix C into subblocks of smaller size. Practical FFT implementations can handle both strategies (see, for instance, the **fft** package available in MATLAB). ■

10.10 Approximation of Function Derivatives

A problem which is often encountered in numerical analysis is the approximation of the derivative of a function $f(x)$ on a given interval $[a, b]$. A natural approach to it consists of introducing in $[a, b]$ $n+1$ nodes $\{x_k,\ k = 0, \ldots, n\}$, with $x_0 = a$, $x_n = b$ and $x_{k+1} = x_k + h$, $k = 0, \ldots, n-1$ where $h = (b-a)/n$. Then, we approximate $f'(x_i)$ using the nodal values $f(x_k)$ as

$$h \sum_{k=-m}^{m} \alpha_k u_{i-k} = \sum_{k=-m'}^{m'} \beta_k f(x_{i-k}), \tag{10.57}$$

where $\{\alpha_k\}$, $\{\beta_k\} \in \mathbb{R}$ are $2(m + m' + 1)$ coefficients to be determined and u_k is the desired approximation to $f'(x_k)$.

A nonnegligible issue in the choice of scheme (10.57) is the computational efficiency. Regarding this concern, it is worth noting that, if $m \neq 0$, determining the values $\{u_i\}$ requires the solution of a linear system.

The set of nodes which are involved in constructing the derivative of f at a certain node, is called a *stencil*. The band of the matrix associated with system (10.57) increases as the stencil gets larger.

10.10.1 Classical Finite Difference Methods

The simplest way to generate a formula like (10.57) consists of resorting to the definition of the derivative. If $f'(x_i)$ exists, then

$$f'(x_i) = \lim_{h \to 0^+} \frac{f(x_i + h) - f(x_i)}{h}. \tag{10.58}$$

Replacing the limit with the incremental ratio, with h finite, yields the approximation

$$u_i^{FD} = \frac{f(x_{i+1}) - f(x_i)}{h}, \qquad 0 \le i \le n-1. \tag{10.59}$$

Relation (10.59) is a special instance of (10.57) setting $m = 0$, $\alpha_0 = 1$, $m' = 1$, $\beta_{-1} = 1$, $\beta_0 = -1$, $\beta_1 = 0$.

The right side of (10.59) is called the *forward finite difference* and the approximation that is being used corresponds to replacing $f'(x_i)$ with the slope of the straight line passing through the points $(x_i, f(x_i))$ and $(x_{i+1}, f(x_{i+1}))$, as shown in Figure 10.4.

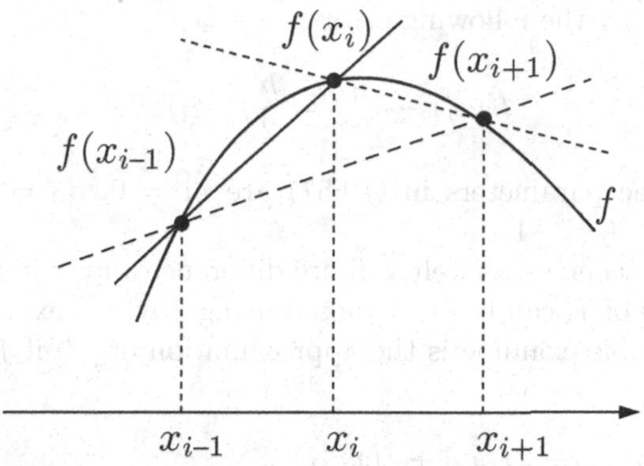

Fig. 10.4. Finite difference approximation of $f'(x_i)$: backward (*solid line*), forward (*dotted line*) and centered (*dashed line*)

To estimate the error that is made, it suffices to expand f in Taylor's series, obtaining

$$f(x_{i+1}) = f(x_i) + hf'(x_i) + \frac{h^2}{2} f''(\xi_i) \qquad \text{with } \xi_i \in (x_i, x_{i+1}).$$

We assume henceforth that f has the required regularity, so that

$$f'(x_i) - u_i^{FD} = -\frac{h}{2} f''(\xi_i). \tag{10.60}$$

Obviously, instead of (10.58) we could employ a centered incremental ratio, obtaining the following approximation

$$u_i^{CD} = \frac{f(x_{i+1}) - f(x_{i-1})}{2h}, \qquad 1 \le i \le n-1. \tag{10.61}$$

Scheme (10.61) is a special instance of (10.57) setting $m = 0$, $\alpha_0 = 1$, $m' = 1$, $\beta_{-1} = 1/2$, $\beta_0 = 0$, $\beta_1 = -1/2$.
The right side of (10.61) is called the *centered finite difference* and geometrically amounts to replacing $f'(x_i)$ with the slope of the straight line passing through the points $(x_{i-1}, f(x_{i-1}))$ and $(x_{i+1}, f(x_{i+1}))$ (see Figure 10.4). Resorting again to Taylor's series, we get

$$f'(x_i) - u_i^{CD} = -\frac{h^2}{6} f'''(\xi_i). \tag{10.62}$$

Formula (10.61) thus provides a second-order approximation to $f'(x_i)$ with respect to h.

Finally, with a similar procedure, we can derive a *backward finite difference* scheme, where

$$u_i^{BD} = \frac{f(x_i) - f(x_{i-1})}{h}, \qquad 1 \le i \le n, \tag{10.63}$$

which is affected by the following error

$$f'(x_i) - u_i^{BD} = \frac{h}{2} f''(\xi_i). \tag{10.64}$$

The values of the parameters in (10.57) are $m = 0$, $\alpha_0 = 1$, $m' = 1$ and $\beta_{-1} = 0$, $\beta_0 = 1$, $\beta_1 = -1$.

Higher-order schemes, as well as finite difference approximations of higher-order derivatives of f, can be constructed using Taylor's expansions of higher order. A remarkable example is the approximation of f''; if $f \in C^4([a,b])$ we easily get

$$\begin{aligned} f''(x_i) = & \frac{f(x_{i+1}) - 2f(x_i) + f(x_{i-1})}{h^2} \\ & - \frac{h^2}{24} \left(f^{(4)}(x_i + \theta_i h) + f^{(4)}(x_i - \omega_i h) \right), \qquad 0 < \theta_i, \omega_i < 1. \end{aligned}$$

The following *centered finite difference* scheme can thus be derived

$$u_i'' = \frac{f(x_{i+1}) - 2f(x_i) + f(x_{i-1})}{h^2}, \qquad 1 \le i \le n-1 \tag{10.65}$$

which is affected by the error

$$f''(x_i) - u_i'' = -\frac{h^2}{24} \left(f^{(4)}(x_i + \theta_i h) + f^{(4)}(x_i - \omega_i h) \right). \tag{10.66}$$

Formula (10.65) provides a second-order approximation to $f''(x_i)$ with respect to h.

10.10.2 Compact Finite Differences

More accurate approximations are provided by using the following formula (which we call *compact differences*)

$$\alpha u_{i-1} + u_i + \alpha u_{i+1} = \frac{\beta}{2h}(f_{i+1} - f_{i-1}) + \frac{\gamma}{4h}(f_{i+2} - f_{i-2}) \tag{10.67}$$

for $i = 2, \ldots, n-2$. We have set, for brevity, $f_i = f(x_i)$.

The coefficients α, β and γ are to be determined in such a way that the relations (10.67) yield values u_i that approximate $f'(x_i)$ up to the highest order with respect to h. For this purpose, the coefficients are selected in such a way as to minimize the *consistency error* (see Section 2.2)

$$\begin{aligned} \sigma_i(h) = & \alpha f_{i-1}^{(1)} + f_i^{(1)} + \alpha f_{i+1}^{(1)} \\ & - \left(\frac{\beta}{2h}(f_{i+1} - f_{i-1}) + \frac{\gamma}{4h}(f_{i+2} - f_{i-2}) \right), \end{aligned} \tag{10.68}$$

which comes from "forcing" f to satisfy the numerical scheme (10.67). For brevity, we set $f_i^{(k)} = f^{(k)}(x_i)$, $k = 1, 2, \ldots$.

Precisely, assuming that $f \in C^5([a,b])$ and expanding it in a Taylor's series around x_i, we find

$$f_{i\pm 1} = f_i \pm h f_i^{(1)} + \tfrac{h^2}{2} f_i^{(2)} \pm \tfrac{h^3}{6} f_i^{(3)} + \tfrac{h^4}{24} f_i^{(4)} \pm \tfrac{h^5}{120} f_i^{(5)} + \mathcal{O}(h^6),$$

$$f_{i\pm 1}^{(1)} = f_i^{(1)} \pm h f_i^{(2)} + \tfrac{h^2}{2} f_i^{(3)} \pm \tfrac{h^3}{6} f_i^{(4)} + \tfrac{h^4}{24} f_i^{(5)} + \mathcal{O}(h^5).$$

Substituting into (10.68) we get

$$\begin{aligned}\sigma_i(h) = (2\alpha+1) f_i^{(1)} + \alpha \frac{h^2}{2} f_i^{(3)} + \alpha \frac{h^4}{12} f_i^{(5)} - (\beta+\gamma) f_i^{(1)} \\ - \frac{h^2}{2}\left(\frac{\beta}{6} + \frac{2\gamma}{3}\right) f_i^{(3)} - \frac{h^4}{60}\left(\frac{\beta}{2} + 8\gamma\right) f_i^{(5)} + \mathcal{O}(h^6).\end{aligned}$$

Second-order methods are obtained by equating to zero the coefficient of $f_i^{(1)}$, i.e., if $2\alpha + 1 = \beta + \gamma$, while we obtain schemes of order 4 by equating to zero also the coefficient of $f_i^{(3)}$, yielding $6\alpha = \beta + 4\gamma$ and finally, methods of order 6 are obtained by setting to zero also the coefficient of $f_i^{(5)}$, i.e., $10\alpha = \beta + 16\gamma$. The linear system formed by these last three equations has a nonsingular matrix. Thus, there exists a unique scheme of order 6 that corresponds to the following choice of the parameters

$$\alpha = 1/3, \quad \beta = 14/9, \quad \gamma = 1/9, \tag{10.69}$$

while there exist infinitely many methods of second and fourth order. Among these infinite methods, a popular scheme has coefficients $\alpha = 1/4$, $\beta = 3/2$ and $\gamma = 0$. Schemes of higher order can be generated at the expense of furtherly expanding the computational stencil.

Traditional finite difference schemes correspond to setting $\alpha = 0$ and allow for computing explicitly the approximant of the first derivative of f at a node, in contrast with compact schemes which are required in any case to solve a linear system of the form $\mathbf{Au} = \mathbf{Bf}$ (where the notation has the obvious meaning). To make the system solvable, it is necessary to provide values to the variables u_i with $i < 0$ and $i > n$. A particularly favorable instance is that where f is a periodic function of period $b - a$, in which case $u_{i+n} = u_i$ for any $i \in \mathbb{Z}$. In the nonperiodic case, system (10.67) must be supplied by suitable relations at the nodes near the boundary of the approximation interval. For example, the first derivative at x_0 can be computed using the relation

$$u_0 + \alpha u_1 = \frac{1}{h}(\mathcal{A} f_1 + \mathcal{B} f_2 + \mathcal{C} f_3 + \mathcal{D} f_4),$$

and requiring that

$$\mathcal{A} = -\frac{3 + \alpha + 2\mathcal{D}}{2}, \; \mathcal{B} = 2 + 3\mathcal{D}, \; \mathcal{C} = -\frac{1 - \alpha + 6\mathcal{D}}{2},$$

in order for the scheme to be at least second-order accurate (see [Lel92] for the relations to enforce in the case of higher-order methods). Finally, we notice that, for any given order of accuracy, compact schemes have a stencil smaller than the one of standard finite differences.
Program 91 provides an implementation of the compact finite difference schemes (10.67) for the approximation of the derivative of a given function f which is assumed to be periodic on the interval $[a, b)$. The input parameters `alpha`, `beta` and `gamma` contain the coefficients of the scheme, `a` and `b` are the endpoints of the interval, `f` is a string containing the expression of f and `n` denotes the number of subintervals in which $[a, b]$ is partitioned. The output vectors `u` and `x` contain the computed approximate values u_i and the node coordinates. Notice that setting `alpha`=`gamma`=0 and `beta`=1 we recover the centered finite difference approximation (10.61).

Program 91 - compdiff : Compact difference schemes

```
function [u,x] = compdiff(alpha,beta,gamma,a,b,n,f)
%COMPDIFF Compact difference scheme.
%   [U,X]=COMPDIFF(ALPHA,BETA,GAMMA,A,B,N,F) computes the first
%   derivative of a function F over the interval (A,B) using a compact finite
%   difference scheme with coefficients ALPHA, BETA and GAMMA.
h=(b-a)/(n+1); x=[a:h:b]; fx = eval(f);
A=eye(n+2)+alpha*diag(ones(n+1,1),1)+alpha*diag(ones(n+1,1),-1);
rhs=0.5*beta/h*(fx(4:n+1)-fx(2:n-1))+0.25*gamma/h*(fx(5:n+2)-fx(1:n-2));
if gamma == 0
   rhs=[0.5*beta/h*(fx(3)-fx(1)), rhs, 0.5*beta/h*(fx(n+2)-fx(n))];
   A(1,1:n+2)=zeros(1,n+2);
   A(1,1)= 1; A(1,2)=alpha; A(1,n+1)=alpha;
   rhs=[0.5*beta/h*(fx(2)-fx(n+1)), rhs];
   A(n+2,1:n+2)=zeros(1,n+2);
   A(n+2,n+2)=1; A(n+2,n+1)=alpha; A(n+2,2)=alpha;
   rhs=[rhs, 0.5*beta/h*(fx(2)-fx(n+1))];
else
   rhs=[0.5*beta/h*(fx(3)-fx(1))+0.25*gamma/h*(fx(4)-fx(n+1)), rhs];
   A(1,1:n+2)=zeros(1,n+2);
   A(1,1)=1; A(1,2)=alpha; A(1,n+1)=alpha;
   rhs=[0.5*beta/h*(fx(2)-fx(n+1))+0.25*gamma/h*(fx(3)-fx(n)), rhs];
   rhs=[rhs,0.5*beta/h*(fx(n+2)-fx(n))+0.25*gamma/h*(fx(2)-fx(n-1))];
   A(n+2,1:n+2)=zeros(1,n+2);
   A(n+2,n+2)=1; A(n+2,n+1)=alpha; A(n+2,2)=alpha;
   rhs=[rhs,0.5*beta/h*(fx(2)-fx(n+1))+0.25*gamma/h*(fx(3)-fx(n))];
end
u = A\rhs';
return
```

Example 10.4 Let us consider the approximate evaluation of the derivative of the function $f(x) = \sin(x)$ on the interval $[0, 2\pi]$. Figure 10.5 shows the logarithm of the

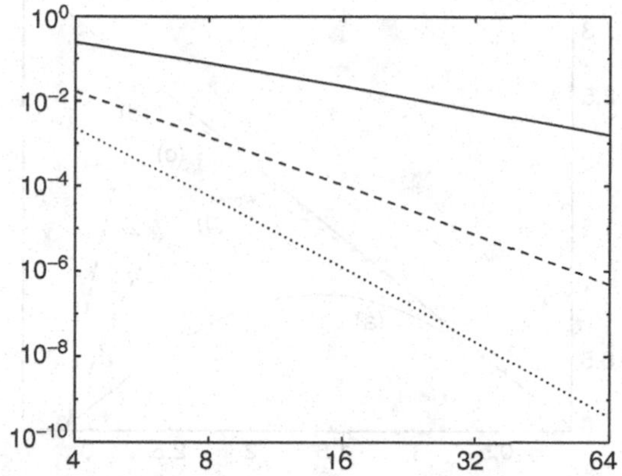

Fig. 10.5. Maximum nodal errors for the second-order centered finite difference scheme (*solid line*) and for the fourth (*dashed line*) and sixth-order (*dotted line*) compact difference schemes as functions of $p = \log(n)$

maximum nodal errors for the second-order centered finite difference scheme (10.61) and of the fourth and sixth-order compact difference schemes introduced above, as a function of $p = \log(n)$. •

Another nice feature of compact schemes is that they maximize the range of *well-resolved waves* as we are going to explain. Assume that f is a real and periodic function on $[0, 2\pi]$, that is, $f(0) = f(2\pi)$. Using the same notation as in Section 10.9, we let N be an even positive integer and set $h = 2\pi/N$. Then replace f by its truncated Fourier series

$$f_N^*(x) = \sum_{k=-N/2}^{N/2-1} \widehat{f}_k e^{ikx}.$$

Since the function f is real-valued, $\widehat{f}_k = \overline{\widehat{f}}_{-k}$ for $k = 1, \ldots, N/2$ and $\widehat{f}_0 = \overline{\widehat{f}}_0$. For sake of convenience, introduce the *normalized wave number* $w_k = kh = 2\pi k/N$ and perform a scaling of the coordinates setting $s = x/h$. As a consequence, we get

$$f_N^*(x(s)) = \sum_{k=-N/2}^{N/2-1} \widehat{f}_k e^{iksh} = \sum_{k=-N/2}^{N/2-1} \widehat{f}_k e^{iw_k s}. \tag{10.70}$$

Taking the first derivative of (10.70) with respect to s yields a function whose Fourier coefficients are $\widehat{f}'_k = i w_k \widehat{f}_k$. We can thus estimate the approximation error on $(f_N^*)'$ by comparing the exact coefficients $\widehat{f}'_k$ with the corresponding ones obtained by an approximate derivative, in particular, by comparing the exact wave number w_k with the approximate one, say $w_{k,app}$.

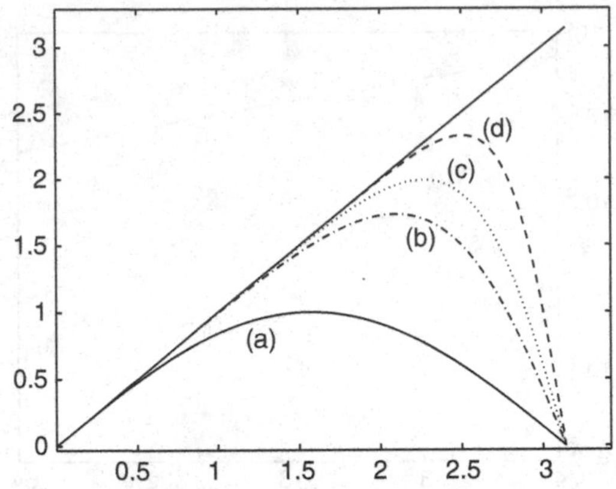

Fig. 10.6. Computed wave numbers for centered finite differences (10.61) (a) and for compact schemes of fourth (b), sixth (c) and tenth (d) order, compared with the exact wave number (*the straight line*). On the x axis the normalized coordinate s is represented

Let us neglect the subscript k and perform the comparison over the whole interval $[0, \pi)$ where w_k is varying. It is clear that methods based on the Fourier expansion have $w_{app} = w$ if $w \neq \pi$ ($w_{app} = 0$ if $w = \pi$). The family of schemes (10.67) is instead characterized by the wave number

$$w_{app}(z) = \frac{a\sin(z) + (b/2)\sin(2z) + (c/3)\sin(3z)}{1 + 2\alpha\cos(z) + 2\beta\cos(2z)}, \qquad z \in [0, \pi)$$

(see [Lel92]). Figure 10.6 displays a comparison among wave numbers of several schemes, of compact and noncompact type.

The range of values for which the wave number computed by the numerical scheme adequately approximates the exact wave number, is the set of *well-resolved* waves. As a consequence, if w_{min} is the smallest well-resolved wave, the difference $1 - w_{min}/\pi$ represents the fraction of waves that are unresolved by the numerical scheme. As can be seen in Figure 10.6, the standard finite difference schemes approximate correctly the exact wave number only for small wave numbers.

10.10.3 Pseudo-Spectral Derivative

An alternative way for numerical differentiation consists of approximating the first derivative of a function f with the exact first derivative of the polynomial $\Pi_n f$ interpolating f at the nodes $\{x_0, \ldots, x_n\}$.

Exactly as happens for Lagrange interpolation, using equally spaced nodes does not yield stable approximations to the first derivative of f for n large. For this reason, we limit ourselves to considering the case where the nodes are nonuniformly distributed according to the Gauss-Lobatto-Chebyshev formula.

For simplicity, assume that $I = [a, b] = [-1, 1]$ and for $n \geq 1$, take in I the Gauss-Lobatto-Chebyshev nodes as in (10.21). Then, consider the Lagrange interpolating polynomial $\Pi_{n,w}^{GL} f$, introduced in Section 10.3. We define the *pseudo-spectral derivative* of $f \in C^0(I)$ to be the derivative of the polynomial $\Pi_{n,w}^{GL} f$

$$\mathcal{D}_n f = (\Pi_{n,w}^{GL} f)' \in \mathbb{P}_{n-1}(I).$$

The error made in replacing f' with $\mathcal{D}_n f$ is of *exponential* type, that is, it only depends on the smoothness of the function f. More precisely, there exists a constant $C > 0$ independent of n such that

$$\|f' - \mathcal{D}_n f\|_w \leq C n^{1-m} \|f\|_{m,w}, \tag{10.71}$$

for any $m \geq 2$ such that the norm $\|f\|_{m,w}$, introduced in (10.23), is finite. Recalling (10.19)

$$(\mathcal{D}_n f)(\bar{x}_i) = \sum_{j=0}^{n} f(\bar{x}_j) \bar{l}'_j(\bar{x}_i), \qquad i = 0, \ldots, n, \tag{10.72}$$

so that the pseudo-spectral derivative at the interpolation nodes can be computed knowing only the nodal values of f and of $\bar{l}'_j$. These values can be computed once for all and stored in a matrix $D \in \mathbb{R}^{(n+1)\times(n+1)}$: $D_{ij} = \bar{l}'_j(\bar{x}_i)$ for $i, j = 0, ..., n$, called a *pseudo-spectral differentiation matrix.*

Relation (10.72) can thus be cast in matrix form as $\mathbf{f}' = D\mathbf{f}$, letting $\mathbf{f} = (f(\bar{x}_i))$ and $\mathbf{f}' = ((\mathcal{D}_n f)(\bar{x}_i))$ for $i = 0, ..., n$.
The entries of D have the following explicit form (see [CHQZ06], p. 89)

$$D_{lj} = \begin{cases} \dfrac{d_l}{d_j} \dfrac{(-1)^{l+j}}{\bar{x}_l - \bar{x}_j}, \, l \neq j, \\ \dfrac{-\bar{x}_j}{2(1 - \bar{x}_j^2)}, \quad 1 \leq l = j \leq n-1, \\ -\dfrac{2n^2+1}{6}, \quad l = j = 0, \\ \dfrac{2n^2+1}{6}, \quad l = j = n, \end{cases} \tag{10.73}$$

where the coefficients d_l have been defined in Section 10.3 (see also Example 5.13 concerning the approximation of the multiple eigenvalue $\lambda = 0$ of D). To compute the pseudo-spectral derivative of a function f over the generic interval $[a, b]$, we only have to resort to the change of variables considered in Remark 10.3.

The second-order pseudo-spectral derivative can be computed as the product of the matrix D and the vector $\mathbf{f}'$, that is, $\mathbf{f}'' = D\mathbf{f}'$, or by directly applying matrix D^2 to the vector $\mathbf{f}$.

10.11 Transforms and Their Applications

In this section we provide a short introduction to the most relevant integral transforms and discuss their basic analytical and numerical properties.

10.11.1 The Fourier Transform

Definition 10.1 Let $\mathrm{L}^1(\mathbb{R})$ denote the space of real or complex functions defined on the real line such that

$$\int_{-\infty}^{\infty} |f(t)|\, dt < +\infty.$$

For any $f \in \mathrm{L}^1(\mathbb{R})$ its Fourier transform is a complex-valued function $F = \mathcal{F}[f]$ defined as

$$F(\nu) = \int_{-\infty}^{\infty} f(t) e^{-i2\pi\nu t}\, dt.$$

■

Should the independent variable t denote time, then ν would have the meaning of frequency. Thus, the Fourier transform is a mapping that to a function of time (typically, real-valued) associates a complex-valued function of frequency. The following result provides the conditions under which an inversion formula exists that allows us to recover the function f from its Fourier transform F (for the proof see [Rud83], p. 199).

Property 10.3 (Inversion theorem) *Let f be a given function in $\mathrm{L}^1(\mathbb{R})$, $F \in \mathrm{L}^1(\mathbb{R})$ be its Fourier transform and g be the function defined by*

$$g(t) = \int_{-\infty}^{\infty} F(\nu) e^{i2\pi\nu t}\, d\nu, \qquad t \in \mathbb{R}. \tag{10.74}$$

Then $g \in C^0(\mathbb{R})$, with $\lim_{|x|\to\infty} g(x) = 0$, and $f(t) = g(t)$ almost everywhere in $\mathbb{R}$ (i.e., for any t unless possibly a set of zero measure).

The integral at right-hand side of (10.74) is to be meant in the *Cauchy principal value sense*, i.e., we let

$$\int_{-\infty}^{\infty} F(\nu) e^{i2\pi\nu t}\, d\nu = \lim_{a\to\infty} \int_{-a}^{a} F(\nu) e^{i2\pi\nu t}\, d\nu$$

and we call it the *inverse Fourier transform* or *inversion formula of the Fourier transform.* This mapping that associates to the complex function F the generating function f will be denoted by $\mathcal{F}^{-1}[F]$, i.e., $F = \mathcal{F}[f]$ iff $f = \mathcal{F}^{-1}[F]$. Let us briefly summarize the main properties of the Fourier transform and its inverse:

1. $\mathcal{F}$ and $\mathcal{F}^{-1}$ are *linear* operators, i.e.
$$\begin{aligned} &\mathcal{F}[\alpha f + \beta g] = \alpha\mathcal{F}[f] + \beta\mathcal{F}[g], \qquad \forall\alpha,\beta \in \mathbb{C}, \\ &\mathcal{F}^{-1}[\alpha F + \beta G] = \alpha\mathcal{F}^{-1}[F] + \beta\mathcal{F}^{-1}[G], \forall\alpha,\beta \in \mathbb{C}; \end{aligned} \tag{10.75}$$
2. *scaling*: if α is any nonzero real number and f_α is the function $f_\alpha(t) = f(\alpha t)$, then
$$\mathcal{F}[f_\alpha] = \frac{1}{|\alpha|}F_{\frac{1}{\alpha}},$$
where $F_{\frac{1}{\alpha}}(\nu) = F(\nu/\alpha)$;
3. *duality*: let $f(t)$ be a given function and $F(\nu)$ be its Fourier transform. Then the function $g(t) = F(-t)$ has a Fourier transform given by $f(\nu)$. Thus, once an associated function-transform pair is found, another dual pair is automatically generated. An application of this property is provided by the pair $r(t)$-$\mathcal{F}[r]$ in Example 10.5;
4. *parity*: if $f(t)$ is a real even function then $F(\nu)$ is real and even, while if $f(t)$ is a real and odd function then $F(\nu)$ is imaginary and odd. This property allows one to work only with nonnegative frequencies;
5. *convolution and product*: for any given functions $f, g \in \mathrm{L}^1(\mathbb{R})$, we have
$$\mathcal{F}[f * g] = \mathcal{F}[f]\mathcal{F}[g], \ \mathcal{F}[fg] = F * G, \tag{10.76}$$
where the *convolution integral* of two functions ϕ and ψ is given by
$$(\phi * \psi)(t) = \int_{-\infty}^{\infty} \phi(\tau)\psi(t-\tau)\,d\tau. \tag{10.77}$$

Example 10.5 We provide two examples of the computation of the Fourier transforms of functions that are typically encountered in signal processing.
Let us first consider the *square wave (or rectangular)* function $r(t)$ defined as
$$r(t) = \begin{cases} A \text{ if } -\frac{T}{2} \le t \le \frac{T}{2}, \\ 0 \text{ otherwise}, \end{cases}$$
where T and A are two given positive numbers. Its Fourier transform $\mathcal{F}[r]$ is the function
$$F(\nu) = \int_{-T/2}^{T/2} Ae^{-i2\pi\nu t}\,dt = AT\frac{\sin(\pi\nu T)}{\pi\nu T}, \qquad \nu \in \mathbb{R},$$
where AT is the area of the rectangular function.

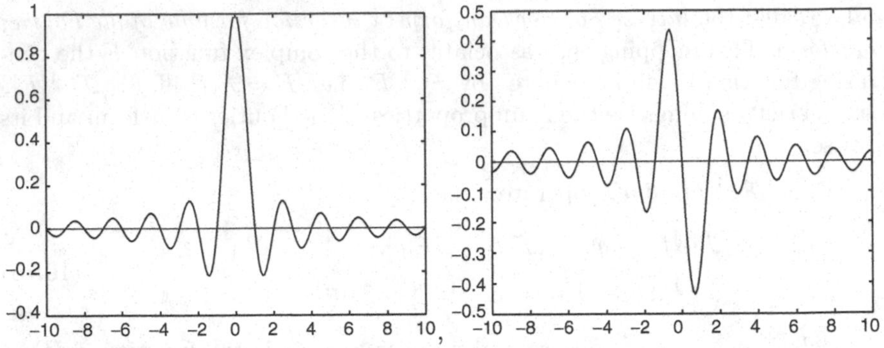

Fig. 10.7. Fourier transforms of the rectangular (*left*) and the sawtooth (*right*) functions

Let us consider the *sawtooth* function

$$s(t) = \begin{cases} \dfrac{2At}{T} & \text{if } -\frac{T}{2} \le t \le \frac{T}{2}, \\ 0 & \text{otherwise,} \end{cases}$$

whose DFT is shown in Figure 10.3 and whose Fourier transform $\mathcal{F}[s]$ is the function

$$F(\nu) = i\,\frac{AT}{\pi\nu T}\left[\cos(\pi\nu T) - \frac{\sin(\pi\nu T)}{\pi\nu T}\right], \qquad \nu \in \mathbb{R}$$

and is purely imaginary since s is an odd real function. Notice also that the functions r and s have a finite support whereas their transforms have an infinite support (see Figure 10.7). In signal theory this corresponds to saying that the transform has an infinite *bandwidth*. •

Example 10.6 The Fourier transform of a sinusoidal function is of paramount interest in signal and communication systems. To start with, consider the constant function $f(t) = A$ for a given $A \in \mathbb{R}$. Since it has an *infinite* time duration its Fourier transform $\mathcal{F}[A]$ is the function

$$F(\nu) = \lim_{a\to\infty} \int_{-a}^{a} Ae^{-i2\pi\nu t}\, dt = A \lim_{a\to\infty} \frac{\sin(2\pi\nu a)}{\pi\nu},$$

where the integral above is again the Cauchy principal value of the corresponding integral over $(-\infty, \infty)$. It can be proved that the limit exists and is unique *in the sense of distributions* (see Section 12.4) yielding

$$F(\nu) = A\delta(\nu), \tag{10.78}$$

where δ is the so-called Dirac mass, i.e., a distribution that satisfies

$$\int_{-\infty}^{\infty} \delta(\xi)\phi(\xi)\, d\xi = \phi(0) \tag{10.79}$$

for any function ϕ continuous at the origin. From (10.78) we see that the transform of a function with infinite time duration has a finite bandwidth.

Let us now consider the computation of the Fourier transform of the function $f(t) = A\cos(2\pi\nu_0 t)$, where ν_0 is a fixed frequency. Recalling Euler's formula

$$\cos(\theta) = \frac{e^{i\theta} + e^{-i\theta}}{2}, \qquad \theta \in \mathbb{R},$$

and applying (10.78) twice we get

$$\mathcal{F}[A\cos(2\pi\nu_0 t)] = \frac{A}{2}\delta(\nu - \nu_0) + \frac{A}{2}\delta(\nu + \nu_0),$$

which shows that the spectrum of a sinusoidal function with frequency ν_0 is centered around $\pm\nu_0$ (notice that the transform is even and real since the same holds for the function $f(t)$). •

It is worth noting that in real-life there do not exist functions (i.e. signals) with infinite duration or bandwidth. Actually, if $f(t)$ is a function whose value may be considered as "negligible" outside of some interval (t_a, t_b), then we can assume that the *effective duration* of f is the length $\Delta t = t_b - t_a$. In a similar manner, if $F(\nu)$ is the Fourier transform of f and it happens that $F(\nu)$ may be considered as "negligible" outside of some interval (ν_a, ν_b), then the *effective bandwidth* of f is $\Delta\nu = \nu_b - \nu_a$. Referring to Figure 10.7, we clearly see that the effective bandwidth of the rectangular function can be taken as $(-10, 10)$.

10.11.2 (Physical) Linear Systems and Fourier Transform

Mathematically speaking, a physical linear system (LS) can be regarded as a linear operator S that enjoys the linearity property (10.75). Denoting by $i(t)$ and $u(t)$ an admissible input function for S and its corresponding output function respectively, the LS can be represented as $u(t) = S(i(t))$ or $S : i \to u$. A special category of LS are the so-called *shift invariant* (or time-invariant) linear systems (ILS) which satisfy the property

$$S(i(t - t_0)) = u(t - t_0), \qquad \forall t_0 \in \mathbb{R}$$

and for any admissible input function i.

Let S be an ILS system and let f and g be two admissible input functions for S with $w = S(g)$. An immediate consequence of the linearity and shift-invariance is that

$$S((f * g)(t)) = (f * S(g))(t) = (f * w)(t), \tag{10.80}$$

where $*$ is the convolution operator defined in (10.77).

Assume we take as input function the *impulse function* $\delta(t)$ introduced in the previous section and denote by $h(t) = S(\delta(t))$ the corresponding output through S (usually referred to as the system *impulse response function*).

Property (10.79) implies that for any function ϕ, $(\phi * \delta)(t) = \phi(t)$, so that, recalling (10.80) and taking $\phi(t) = i(t)$ we have

$$u(t) = S(i(t)) = S(i * \delta)(t) = (i * S(\delta))(t) = (i * h)(t).$$

Thus, S can be completely described through its impulse response function. Equivalently, we can pass to the frequency domain by means of the first relation in (10.76) obtaining

$$U(\nu) = I(\nu)H(\nu), \tag{10.81}$$

where I, U and H are the Fourier transforms of $i(t)$, $u(t)$ and $h(t)$, respectively; H is the so-called system *transfer function.*

Relation (10.81) plays a central role in the analysis of linear time-invariant systems as it is simpler to deal with the system transfer function than with the corresponding impulse response function, as demonstrated in the following example.

Example 10.7 (Ideal low-pass filter) An ideal low-pass filter is an ILS characterized by the transfer function

$$H(\nu) = \begin{cases} 1, \text{ if } |\nu| \leq \nu_0/2, \\ 0, \text{ otherwise.} \end{cases}$$

Using the duality property, the impulse response function $\mathcal{F}^{-1}[H]$ is

$$h(t) = \nu_0 \frac{\sin(\pi \nu_0 t)}{\pi \nu_0 t}.$$

Given an input signal $i(t)$ with Fourier transform $I(\nu)$, the corresponding output $u(t)$ has a spectrum given by (10.81)

$$I(\nu)H(\nu) = \begin{cases} I(\nu), \text{ if } |\nu| \leq \nu_0/2, \\ 0, \quad \text{otherwise.} \end{cases}$$

The effect of the filter is to cut off the input frequencies that lie outside the *window* $|\nu| \leq \nu_0/2$. •

The input/output functions $i(t)$ and $u(t)$ usually denote *signals* and the linear system described by $H(\nu)$ is typically a communication system. Therefore, as pointed out at the end of Section 10.11.1, we are legitimated in assuming that both $i(t)$ and $u(t)$ have a finite effective duration. In particular, referring to $i(t)$ we suppose $i(t) = 0$ if $t \notin [0, T_0)$. Then, the computation of the Fourier transform of $i(t)$ yields

$$I(\nu) = \int_0^{T_0} i(t) e^{-i2\pi\nu t}\, dt.$$

Letting $\Delta t = T_0/n$ for $n \geq 1$ and approximating the integral above by the composite trapezoidal formula (9.14), we get

$$\tilde{I}(\nu) = \Delta t \sum_{k=0}^{n-1} i(k\Delta t) e^{-i2\pi\nu k\Delta t}.$$

It can be proved (see, e.g., [Pap62]) that $\tilde{I}(\nu)/\Delta t$ is the Fourier transform of the so-called *sampled signal*

$$i_s(t) = \sum_{k=-\infty}^{\infty} i(k\Delta t)\delta(t - k\Delta t),$$

where $\delta(t - k\Delta t)$ is the Dirac mass at $k\Delta t$. Then, using the convolution and the duality properties of the Fourier transform, we get

$$\tilde{I}(\nu) = \sum_{j=-\infty}^{\infty} I\left(\nu - \frac{j}{\Delta t}\right), \tag{10.82}$$

which amounts to replacing $I(\nu)$ by its periodic repetition with period $1/\Delta t$. Let $\mathcal{J}_{\Delta t} = [-\frac{1}{2\Delta t}, \frac{1}{2\Delta t}]$; then, it suffices to compute (10.82) for $\nu \in \mathcal{J}_{\Delta t}$. This can be done numerically by introducing a uniform discretization of $\mathcal{J}_{\Delta t}$ with frequency step $\nu_0 = 1/(m\Delta t)$ for $m \geq 1$. By doing so, the computation of $\tilde{I}(\nu)$ requires evaluating the following $m + 1$ discrete Fourier transforms (DFT)

$$\tilde{I}(j\nu_0) = \Delta t \sum_{k=0}^{n-1} i(k\Delta t) e^{-i2\pi j\nu_0 k\Delta t}, \; j = -\tfrac{m}{2}, \ldots, \tfrac{m}{2}.$$

For an efficient computation of each DFT in the formula above it is crucial to use the FFT algorithm described in Section 10.9.2.

10.11.3 The Laplace Transform

The Laplace transform can be employed to solve ordinary differential equations with constant coefficients as well as partial differential equations.

Definition 10.2 Let $f \in \mathrm{L}^1_{loc}([0, \infty))$ i.e., $f \in \mathrm{L}^1([0, T])$ for any $T > 0$. Let $s = \sigma + i\omega$ be a complex variable. The Laplace integral of f is defined as

$$\int_0^\infty f(t)e^{-st}\, dt = \lim_{T\to\infty} \int_0^T f(t)e^{-st}\, dt.$$

If this integral exists for some s, it turns out to be a function of s; then, the *Laplace transform* $\mathcal{L}[f]$ of f is the function

$$L(s) = \int_0^\infty f(t)e^{-st}\, dt.$$

■

The following relation between Laplace and Fourier transforms holds

$$L(s) = F(e^{-\sigma t}\tilde{f}(t)),$$

where $\tilde{f}(t) = f(t)$ if $t \geq 0$ while $\tilde{f}(t) = 0$ if $t < 0$.

Example 10.8 The Laplace transform of the unit step function $f(t) = 1$ if $t > 0$, $f(t) = 0$ otherwise, is given by

$$L(s) = \int_0^\infty e^{-st}\, dt = \frac{1}{s}.$$

We notice that the Laplace integral exists if $\sigma > 0$. •

In Example 10.8 the convergence region of the Laplace integral is the half-plane $\{\mathrm{Re}(s) > 0\}$ of the complex field. This property is quite general, as stated by the following result.

Property 10.4 *If the Laplace transform exists for $s = \bar{s}$ then it exists for all s with $\mathrm{Re}(s) > \mathrm{Re}(\bar{s})$. Moreover, let E be the set of the real parts of s such that the Laplace integral exists and denote by λ the infimum of E. If λ happens to be finite, the Laplace integral exists in the half-plane $\mathrm{Re}(s) > \lambda$. If $\lambda = -\infty$ then it exists for all $s \in \mathbb{C}$; λ is called the abscissa of convergence.*

We recall that the Laplace transform enjoys properties completely analogous to those of the Fourier transform. The inverse Laplace transform is denoted formally as $\mathcal{L}^{-1}$ and is such that

$$f(t) = \mathcal{L}^{-1}[L(s)].$$

Example 10.9 Let us consider the ordinary differential equation $y'(t)+ay(t) = g(t)$ with $y(0) = y_0$. Multiplying by e^{st}, integrating between 0 and ∞ and passing to the Laplace transform, yields

$$sY(s) - y_0 + aY(s) = G(s). \tag{10.83}$$

Should $G(s)$ be easily computable, (10.83) would furnish $Y(s)$ and then, by applying the inverse Laplace transform, the generating function $y(t)$. For instance, if $g(t)$ is the unit step function, we obtain

$$y(t) = \mathcal{L}^{-1}\left\{\frac{1}{a}\left[\frac{1}{s} - \frac{1}{s+a}\right] + \frac{y_0}{s+a}\right\} = \frac{1}{a}(1 - e^{-at}) + y_0 e^{-at}.$$

•

For an extensive presentation and analysis of the Laplace transform see, e.g., [Tit37]. In the next section we describe a discrete version of the Laplace transform, known as the Z-transform.

10.11.4 The Z-Transform

Definition 10.3 Let f be a given function, defined for any $t \geq 0$, and $\Delta t > 0$ be a given time step. The function

$$Z(z) = \sum_{n=0}^{\infty} f(n\Delta t) z^{-n}, \qquad z \in \mathbb{C}, \tag{10.84}$$

is called the Z-transform of the sequence $\{f(n\Delta t)\}$ and is denoted by $\mathcal{Z}[f(n\Delta t)]$. ■

The parameter Δt is the *sampling time step* of the sequence of samples $f(n\Delta t)$. The infinite sum (10.84) converges if

$$|z| > R = \limsup_{n \to \infty} \sqrt[n]{|f(n\Delta t)|}.$$

It is possible to deduce the Z-transform from the Laplace transform as follows. Denoting by $f_0(t)$ the piecewise constant function such that $f_0(t) = f(n\Delta t)$ for $t \in (n\Delta t, (n+1)\Delta t)$, the Laplace transform $\mathcal{L}[f_0]$ of f_0 is the function

$$\begin{aligned} L(s) &= \int_0^{\infty} f_0(t) e^{-st}\, dt = \sum_{n=0}^{\infty} \int_{n\Delta t}^{(n+1)\Delta t} e^{-st} f(n\Delta t)\, dt \\ &= \sum_{n=0}^{\infty} f(n\Delta t) \frac{e^{-ns\Delta t} - e^{-(n+1)s\Delta t}}{s} = \left(\frac{1 - e^{-s\Delta t}}{s}\right) \sum_{n=0}^{\infty} f(n\Delta t) e^{-ns\Delta t}. \end{aligned}$$

The *discrete Laplace transform* $\mathcal{Z}^d[f_0]$ of f_0 is the function

$$Z^d(s) = \sum_{n=0}^{\infty} f(n\Delta t) e^{-ns\Delta t}.$$

Then, the Z-transform of the sequence $\{f(n\Delta t),\ n = 0, \ldots, \infty\}$ coincides with the discrete Laplace transform of f_0 up to the change of variable $z = e^{-s\Delta t}$. The Z-transform enjoys similar properties (linearity, scaling, convolution and product) to those already seen in the continuous case.
The inverse Z-transform is denoted by $\mathcal{Z}^{-1}$ and is defined as

$$f(n\Delta t) = \mathcal{Z}^{-1}[Z(z)].$$

The practical computation of $\mathcal{Z}^{-1}$ can be carried out by resorting to classical techniques of complex analysis (for example, using the Laurent formula or the Cauchy theorem for residual integral evaluation) coupled with an extensive use of tables (see, e.g., [Pou96]).

10.12 The Wavelet Transform

This technique, originally developed in the area of signal processing, has successively been extended to many different branches of approximation theory, including the solution of differential equations. It is based on the so-called wavelets, which are functions generated by an elementary wavelet through traslations and dilations. We shall limit ourselves to a brief introduction of univariate wavelets and their transform in both the continuous and discrete cases referring to [DL92], [Dau88] and to the references cited therein for a detailed presentation and analysis.

10.12.1 The Continuous Wavelet Transform

Any function

$$h_{s,\tau}(t) = \frac{1}{\sqrt{s}} h\left(\frac{t-\tau}{s}\right), \qquad t \in \mathbb{R}, \tag{10.85}$$

that is obtained from a reference function $h \in \mathrm{L}^2(\mathbb{R})$ by means of traslations by a *traslation factor* τ and dilations by a positive *scaling factor* s is called a *wavelet*. The function h is called an *elementary wavelet*.
Its Fourier transform, written in terms of $\omega = 2\pi\nu$, is

$$H_{s,\tau}(\omega) = \sqrt{s} H(s\omega) e^{-i\omega\tau}, \tag{10.86}$$

where i denotes the imaginary unit and $H(\omega)$ is the Fourier transform of the elementary wavelet. A dilation t/s ($s > 1$) in the real domain produces therefore a contraction $s\omega$ in the frequency domain. Therefore, the factor $1/s$ plays the role of the frequency ν in the Fourier transform (see Section 10.11.1). In wavelets theory s is usually referred to as the *scale*. Formula (10.86) is known as the filter of the wavelet transform.

Definition 10.4 Given a function $f \in \mathrm{L}^2(\mathbb{R})$, its continuous *wavelet transform* $W_f = \mathcal{W}[f]$ is a decomposition of $f(t)$ onto a wavelet basis $\{h_{s,\tau}(t)\}$, that is

$$W_f(s,\tau) = \int_{-\infty}^{\infty} f(t) \bar{h}_{s,\tau}(t)\, dt, \tag{10.87}$$

where the overline bar denotes complex conjugate. ■

When t denotes the time-variable, the wavelet transform of $f(t)$ is a function of the two variables s (scale) and τ (time shift); as such, it is a representation of f in the time-scale space and is usually referred to as *time-scale joint representation* of f. The time-scale representation is the analogue of the time-frequency representation introduced in the Fourier analysis. This latter representation has an intrinsic limitation: the product of the resolution in time Δt and the resolution in frequency $\Delta\omega$ must satisfy the following constraint (Heisenberg inequality)

$$\Delta t \Delta\omega \geq \frac{1}{2}, \tag{10.88}$$

which is the counterpart of the Heisenberg uncertainty principle in quantum mechanics. This inequality states that a signal cannot be represented as a point in the time-frequency space. We can only determine its position within a rectangle of area $\Delta t\Delta\omega$ in the time-frequency space.

The wavelet transform (10.87) can be rewritten in terms of the Fourier transform $F(\omega)$ of f as

$$W_f(s,\tau) = \frac{\sqrt{s}}{2\pi}\int_{-\infty}^{\infty} F(\omega)\bar{H}(s\omega)e^{i\omega\tau}\, d\omega, \quad \forall s \neq 0,\ \forall\tau,$$

which shows that the wavelets transform is a bank of wavelet filters characterized by different scales. More precisely, if the scale is small the wavelet is concentrated in time and the wavelet transform provides a detailed description of $f(t)$ (which is the signal). Conversely, if the scale is large, the wavelet transform is able to resolve only the large-scale details of f. Thus, the wavelet transform can be regarded as a bank of *multiresolution filters.*

The theoretical properties of this transform do not depend on the particular elementary wavelet that is considered. Hence, specific bases of wavelets can be derived for specific applications. Some examples of elementary wavelets are reported below.

Example 10.10 (Haar wavelets) These functions can be obtained by choosing as the elementary wavelet the Haar function defined as

$$h(x) = \begin{cases} 1, & \text{if } x \in (0, \frac{1}{2}), \\ -1, & \text{if } x \in (\frac{1}{2}, 1), \\ 0, & \text{otherwise.} \end{cases}$$

Its Fourier transform is the complex-valued function

$$H(\omega) = 4ie^{-i\omega/2}\left(1 - \cos(\frac{\omega}{2})\right)/\omega,$$

which has symmetric module with respect to the origin (see Figure 10.8). The bases that are obtained from this wavelet are not used in practice due to their ineffective localization properties in the frequency domain. •

Example 10.11 (Morlet wavelets) The *Morlet wavelet* is defined as follows (see [MMG87])

$$h(x) = e^{i\omega_0 x}e^{-x^2/2}.$$

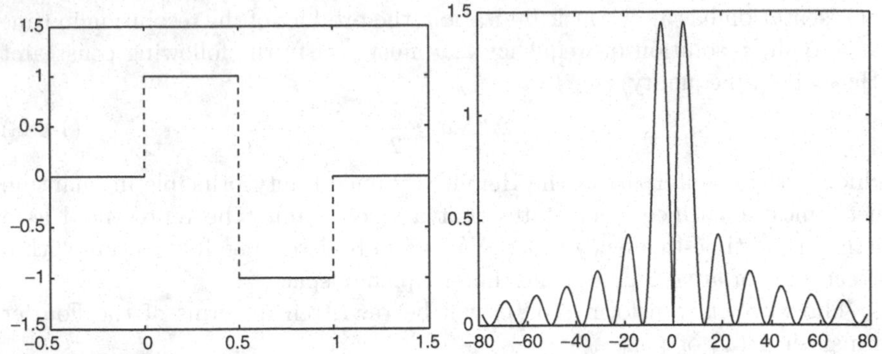

Fig. 10.8. The Haar wavelet (*left*) and the module of its Fourier transform (*right*)

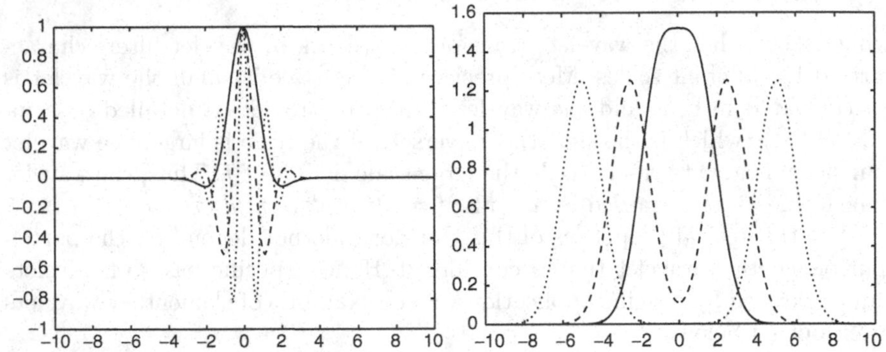

Fig. 10.9. The real part of the Morlet wavelet (*left*) and the real part of the corresponding Fourier transforms (*right*) for $\omega_0 = 1$ (*solid line*), $\omega_0 = 2.5$ (*dashed line*) and $\omega_0 = 5$ (*dotted line*)

Thus, it is a complex-valued function whose real part has a real positive Fourier transform, symmetric with respect to the origin, given by

$$H(\omega) = \sqrt{\pi}\left[e^{-(\omega-\omega_0)^2/2} + e^{-(\omega+\omega_0)^2/2}\right].$$

•

We point out that the presence of the dilation factor allows for the wavelets to easily handle possible discontinuities or singularities in f. Indeed, using the multi-resolution analysis, the signal, properly divided into frequency bandwidths, can be processed at each frequency by suitably tuning up the scale factor of the wavelets.
Recalling what was already pointed out in Section 10.11.1, the time localization of the wavelet gives rise to a filter with infinite bandwidth. In particular, defining the bandwidth $\Delta\omega$ of the wavelet filter as

$$\Delta\omega = \left(\int_{-\infty}^{\infty} \omega^2 |H(\omega)|^2 \, d\omega \Big/ \int_{-\infty}^{\infty} |H(\omega)|^2 \, d\omega \right)^2,$$

then the bandwidth of the wavelet filter with scale equal to s is

$$\Delta\omega_s = \left(\int_{-\infty}^{\infty} \omega^2 |H(s\omega)|^2 \, d\omega \Big/ \int_{-\infty}^{\infty} |H(s\omega)|^2 \, d\omega \right)^2 = \frac{1}{s}\Delta\omega.$$

Consequently, the *quality factor* Q of the wavelet filter, defined as the inverse of the bandwidth of the filter, is independent of s since

$$Q = \frac{1/s}{\Delta\omega_s} = \Delta\omega$$

provided that (10.88) holds. At low frequencies, corresponding to large values of s, the wavelet filter has a small bandwidth and a large temporal width (called *window*) with a low resolution. Conversely, at high frequencies the filter has a large bandwidth and a small temporal window with a high resolution. Thus, the resolution furnished by the wavelet analysis increases with the frequency of the signal. This property of *adaptivity* makes the wavelets a crucial tool in the analysis of unsteady signals or signals with fast transients for which the standard Fourier analysis turns out to be ineffective.

10.12.2 Discrete and Orthonormal Wavelets

The continuous wavelet transform maps a function of one variable into a bidimensional representation in the time-scale domain. In many applications this description is excessively rich. Resorting to the discrete wavelets is an attempt to represent a function using a finite (and small) number of parameters.

A *discrete wavelet* is a continuous wavelet that is generated by using discrete scale and translation factors. For $s_0 > 1$, denote by $s = s_0^j$ the scale factors; the dilation factors usually depend on the scale factors by setting $\tau = k\tau_0 s_0^j$, $\tau_0 \in \mathbb{R}$. The corresponding discrete wavelet is

$$h_{j,k}(t) = s_0^{-j/2} h(s_0^{-j}(t - k\tau_0 s_0^j)) = s_0^{-j/2} h(s_0^{-j} t - k\tau_0).$$

The scale factor s_0^j corresponds to the magnification or the resolution of the observation, while the translation factor τ_0 is the location where the observations are made. If one looks at very small details, the magnification must be large, which corresponds to large negative index j. In this case the step of translation is small and the wavelet is very concentrated around the observation point. For large and positive j, the wavelet is spread out and large translation steps are used.

The behavior of the discrete wavelets depends on the steps s_0 and τ_0. When s_0 is close to 1 and τ_0 is small, the discrete wavelets are close to the continuous ones. For a fixed scale s_0 the localization points of the discrete wavelets along the scale axis are logarithmic as $\log s = j \log s_0$. The choice $s_0 = 2$ corresponds to the dyadic sampling in frequency. The discrete time-step is $\tau_0 s_0^j$ and, typically, $\tau_0 = 1$. Hence, the time-sampling step is a function of the scale and along the time axis the localization points of the wavelet depend on the scale.

For a given function $f \in \mathrm{L}^1(\mathbb{R})$, the corresponding discrete wavelet transform is

$$W_f(j,k) = \int_{-\infty}^{\infty} f(t)\bar{h}_{j,k}(t)\, dt.$$

It is possible to introduce an orthonormal wavelet basis using discrete dilation and traslation factors, i.e.

$$\int_{-\infty}^{\infty} h_{i,j}\bar{h}_{k,l}(t)\, dt = \delta_{ik}\delta_{jl}, \qquad \forall i,j,k,l \in \mathbb{Z}.$$

With an orthogonal wavelet basis, an arbitrary function f can be reconstructed by the expansion

$$f(t) = A \sum_{j,k\in\mathbb{Z}} W_f(j,k) h_{j,k}(t),$$

where A is a constant that does not depend on f.

As of the computational standpoint, the wavelet discrete transform can be implemented at even a cheaper cost than the FFT algorithm for computing the Fourier transform.

10.13 Applications

In this section we apply the theory of orthogonal polynomials to solve two problems arising in quantum physics. In the first example we deal with Gauss-Laguerre quadratures, while in the second case the Fourier analysis and the FFT are considered.

10.13.1 Numerical Computation of Blackbody Radiation

The monochromatic energy density $\mathcal{E}(\nu)$ of blackbody radiation as a function of frequency ν is expressed by the following law

$$\mathcal{E}(\nu) = \frac{8\pi h}{c^3} \frac{\nu^3}{e^{h\nu/K_B T} - 1},$$

where h is the Planck constant, c is the speed of light, K_B is the Boltzmann constant and T is the absolute temperature of the blackbody (see, for instance, [AF83]).

To compute the total density of monochromatic energy that is emitted by the blackbody (that is, the emitted energy per unit volume) we must evaluate the integral

$$E = \int_0^\infty \mathcal{E}(\nu) d\nu = \alpha T^4 \int_0^\infty \frac{x^3}{e^x - 1} dx,$$

where $x = h\nu/K_B T$ and $\alpha = (8\pi K_B^4)/(ch)^3 \simeq 1.16 \cdot 10^{-16}$ $[J][K^{-4}][m^{-3}]$. We also let $f(x) = x^3/(e^x - 1)$ and $I(f) = \int_0^\infty f(x)dx$.
To approximate $I(f)$ up to a previously fixed absolute error $\leq \delta$, we compare method 1. introduced in Section 9.8.3 with Gauss-Laguerre quadratures.
In the case of method 1. we proceed as follows. For any $a > 0$ we let $I(f) = \int_0^a f(x)dx + \int_a^\infty f(x)dx$ and try to find a function ϕ such that

$$\int_a^\infty f(x)dx \leq \int_a^\infty \phi(x)dx \leq \frac{\delta}{2}, \qquad (10.89)$$

the integral $\int_a^\infty \phi(x)dx$ being "easy" to compute. Once the value of a has been found such that (10.89) is fulfilled, we compute the integral $I_1(f) = \int_0^a f(x)dx$ using for instance the adaptive Cavalieri-Simpson formula introduced in Section 9.7.2 and denoted in the following by AD.
A natural choice of a bounding function for f is $\phi(x) = Kx^3e^{-x}$, for a suitable constant $K > 1$. Thus, we have $K \geq e^x/(e^x - 1)$, for any $x > 0$, that is, letting $x = a$, $K = e^a/(e^a - 1)$. Substituting back into (10.89) yields

$$\int_a^\infty f(x)dx \leq \frac{a^3 + 3a^2 + 6a + 6}{e^a - 1} = \eta(a) \leq \frac{\delta}{2}.$$

Letting $\delta = 10^{-3}$, we see that (10.89) is satisfied by taking $a \simeq 16$. Program 77 for computing $I_1(f)$ with the AD method, setting `hmin`$=10^{-3}$ and `tol`$=5 \cdot 10^{-4}$, yields the approximate value $I_1 \simeq 6.4934$ with a number of (nonuniform) partitions equal to 25.

The distribution of the quadrature nodes produced by the adaptive algorithm is plotted in Figure 10.10. Globally, using method 1. yields an approximation of $I(f)$ equal to 6.4984. Table 10.1 shows, for sake of comparison, some approximate values of $I(f)$ obtained using the Gauss-Laguerre formulae with the number of nodes varying between 2 to 20. Notice that, taking $n = 4$ nodes, the accuracy of the two computational procedures is roughly equivalent.

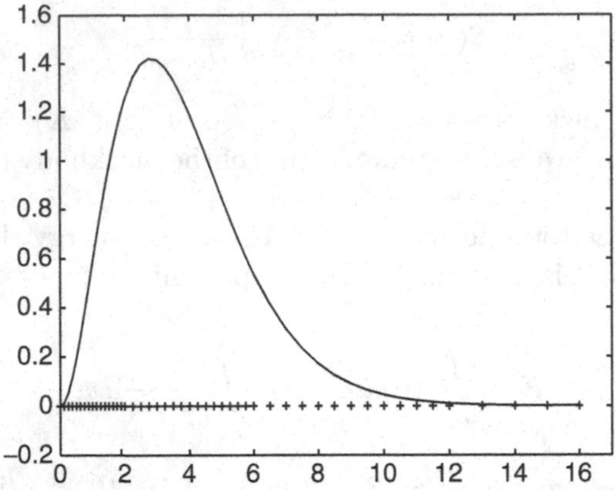

Fig. 10.10. Distribution of quadrature nodes and graph of the integrand function

Table 10.1. Approximate evaluation of $I(f) = \int_0^\infty x^3/(e^x - 1)dx$ with Gauss-Laguerre quadratures

n	$I_n(f)$
2	6.413727469517582
3	6.481130171540022
4	6.494535639802632
5	6.494313365790864
10	6.493939967652101
15	6.493939402671590
20	6.493939402219742

10.13.2 Numerical Solution of Schrödinger Equation

Let us consider the following differential equation arising in quantum mechanics known as the *Schrödinger equation*

$$i\frac{\partial \psi}{\partial t} = -\frac{\hbar}{2m}\frac{\partial^2 \psi}{\partial x^2}, \qquad x \in \mathbb{R} \qquad t > 0. \tag{10.90}$$

The symbols i and $\hbar$ denote the imaginary unit and the reduced Planck constant, respectively. The complex-valued function $\psi = \psi(x,t)$, the solution of (10.90), is called a *wave function* and the quantity $|\psi(x,t)|^2$ defines the probability density in the space x of a free electron of mass m at time t (see [FRL55]).

The corresponding Cauchy problem may represent a physical model for describing the motion of an electron in a cell of an infinite lattice (for more details see, e.g., [AF83]).

Consider the initial condition $\psi(x,0) = w(x)$, where w is the step function that takes the value $1/\sqrt{2b}$ for $|x| \leq b$ and is zero for $|x| > b$, with $b = a/5$, and where $2a$ represents the inter-ionic distance in the lattice. Therefore, we are searching for periodic solutions, with period equal to $2a$.

Solving problem (10.90) can be carried out using Fourier analysis as follows. We first write the Fourier series of w and ψ (for any $t > 0$)

$$\begin{aligned} w(x) &= \sum_{k=-N/2}^{N/2-1} \widehat{w}_k e^{i\pi kx/a}, \qquad \widehat{w}_k = \frac{1}{2a}\int_{-a}^{a} w(x)e^{-i\pi kx/a}dx, \\ \psi(x,t) &= \sum_{k=-N/2}^{N/2-1} \widehat{\psi}_k(t)e^{i\pi kx/a}, \ \widehat{\psi}_k(t) = \frac{1}{2a}\int_{-a}^{a} \psi(x,t)e^{-i\pi kx/a}dx. \end{aligned} \tag{10.91}$$

Then, we substitute back (10.91) into (10.90), obtaining the following Cauchy problem for the Fourier coefficients $\widehat{\psi}_k$, for $k = -N/2, \ldots, N/2-1$

$$\begin{cases} \widehat{\psi}_k'(t) = -i\dfrac{\hbar}{2m}\left(\dfrac{k\pi}{a}\right)^2 \widehat{\psi}_k(t), \\ \widehat{\psi}_k(0) = \widetilde{\widehat{w}}_k. \end{cases} \tag{10.92}$$

The coefficients $\{\widetilde{\widehat{w}}_k\}$ have been computed by regularizing the coefficients $\{\widehat{w}_k\}$ of the step function w using the Lanczos smoothing (10.56) in order to avoid the Gibbs phenomenon arising around the discontinuities of w (see Section 10.9.1).
After solving (10.92), we finally get, recalling (10.91), the following expression for the wave function

$$\psi_N(x,t) = \sum_{k=-N/2}^{N/2-1} \widetilde{\widehat{w}}_k e^{-iE_k t/\hbar} e^{i\pi kx/a}, \tag{10.93}$$

where the coefficients $E_k = (k^2\pi^2\hbar^2)/(2ma^2)$ represent, from the physical standpoint, the energy levels that the electron may assume in its motion within the potential well.
To compute the coefficients $\widehat{w}_k$ (and, as a consequence, $\widetilde{\widehat{w}}_k$), we have used the MATLAB intrinsic function `fft` (see Section 10.9.2), employing $N = 2^6 = 64$ points and letting $a = 10[\mathring{A}] = 10^{-9}[m]$. Time analysis has been carried out up to $T = 10\,[s]$, with time steps of $1\,[s]$; in all the reported graphs, the x-axis is measured in $[\mathring{A}]$, while the y-axes are respectively in units of $10^5\,[m^{-1/2}]$ and $10^{10}\,[m^{-1}]$.

In Figure 10.11 we draw the probability density $|\psi(x,t)|^2$ at $t = 0$, 2 and $5\,[s]$. The result obtained without the regularizing procedure above is shown on the left, while the same calculation with the "filtering" of the Fourier coefficients is reported on the right. The second plot demonstrates the smoothing effect on the solution by the regularization, at the price of a slight enlargement of the step-like initial probability distribution.

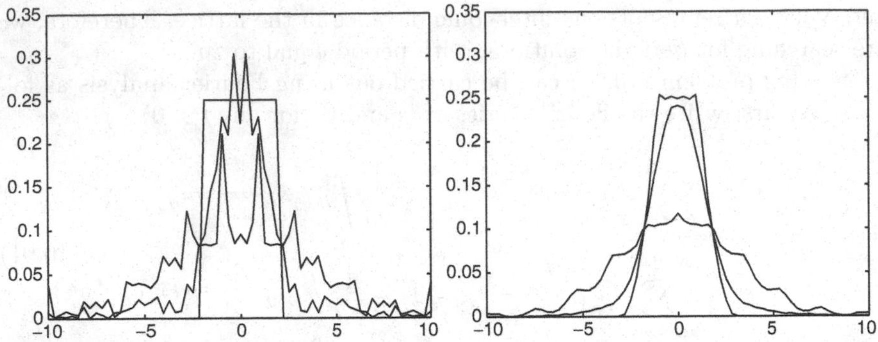

Fig. 10.11. Probability density $|\psi(x,t)|^2$ at $t = 0,\ 2,\ 5\,[s]$, corresponding to a step function as initial datum: solution without filtering (*left*), with Lanczos filtering (*right*)

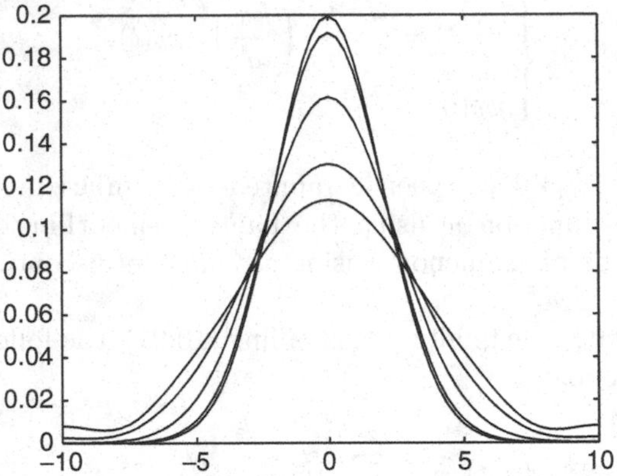

Fig. 10.12. Probability density $|\psi(x,t)|^2$ at $t = 0,\ 2,\ 5,\ 7,\ 9[s]$, corresponding to an initial datum with Gaussian form

Finally, it is interesting to apply Fourier analysis to solve problem (10.90) starting from a smooth initial datum. For this, we choose an initial probability density w of Gaussian form such that $\|w\|_2 = 1$. The solution $|\psi(x,t)|^2$, this time computed without regularization, is shown in Figure 10.12, at $t = 0,\ 2,\ 5,\ 7,\ 9[s]$. Notice the absence of spurious oscillations with respect to the previous case.

10.14 Exercises

1. Prove the three-term relation (10.11).
   [*Hint*: set $x = \cos(\theta)$, for $0 \le \theta \le \pi$.]
2. Prove (10.31).

[*Hint*: first prove that $\|v_n\|_n = (v_n, v_n)^{1/2}$, $\|T_k\|_n = \|T_k\|_w$ for $k < n$ and $\|T_n\|_n^2 = 2\|T_n\|_w^2$ (see [QV94], formula (4.3.16)). Then, the thesis follows from (10.29) multiplying by T_l ($l \neq k$) and taking $(\cdot,\cdot)_n$.]

3. Prove (10.24) after showing that $\|(f - \Pi_n^{GL} f)'\|_\omega \leq Cn^{1-s}\|f\|_{s,\omega}$.
[*Hint*: use the Gagliardo-Nirenberg inequality

$$\max_{-1\leq x\leq 1} |f(x)| \leq \|f\|^{1/2}\|f'\|^{1/2}$$

valid for any $f \in \mathrm{L}^2$ with $f' \in \mathrm{L}^2$. Next, use the relation that has been just shown to prove (10.24).]

4. Prove that the discrete seminorm $\|f\|_n = (f, f)_n^{1/2}$ is a norm for $\mathbb{P}_n$.
5. Compute weights and nodes of the following quadrature formulae

$$\int_a^b w(x) f(x) dx = \sum_{i=0}^{n} \omega_i f(x_i),$$

in such a way that the order is maximum, setting

$$\begin{array}{ll} \omega(x) = \sqrt{x}, & a = 0, \quad b = 1, n = 1; \\ \omega(x) = 2x^2 + 1, & a = -1, b = 1, n = 0; \\ \omega(x) = \begin{cases} 2 \text{ if } \ 0 < x \leq 1, \\ 1 \text{ if } \ -1 \leq x \leq 0 \end{cases} & a = -1, b = 1, n = 1. \end{array}$$

[*Solution*: for $\omega(x) = \sqrt{x}$, the nodes $x_1 = \frac{5}{9} + \frac{2}{9}\sqrt{10/7}$, $x_2 = \frac{5}{9} - \frac{2}{9}\sqrt{10/7}$ are obtained, from which the weights can be computed (order 3); for $\omega(x) = 2x^2 + 1$, we get $x_1 = 3/5$ and $\omega_1 = 5/3$ (order 1); for $\omega(x) = 2x^2 + 1$, we have $x_1 = \frac{1}{22} + \frac{1}{22}\sqrt{155}$, $x_2 = \frac{1}{22} - \frac{1}{22}\sqrt{155}$ (order 3).]

6. Prove (10.40).
[*Hint*: notice that $(\Pi_n^{GL} f, L_j)_n = \sum_k f_k^*(L_k, L_j)_n = \ldots$, distinguishing the case $j < n$ from the case $j = n$.]
7. Show that $|||\cdot|||$, defined in (10.45), is an essentially strict seminorm.
[*Solution* : use the Cauchy-Schwarz inequality (1.14) to check that the triangular inequality is satisfied. This proves that $|||\cdot|||$ is a seminorm. The second part of the exercise follows after a direct computation.]
8. Consider in an interval $[a, b]$ the nodes

$$x_j = a + \left(j - \frac{1}{2}\right)\left(\frac{b-a}{m}\right), \ j = 1, 2, \ldots, m$$

for $m \geq 1$. They are the midpoints of m equally spaced intervals in $[a, b]$. Let f be a given function; prove that the least-squares polynomial r_n with respect to the weight $w(x) = 1$ minimizes the error average, defined as

$$E = \lim_{m\to\infty} \left\{ \frac{1}{m} \sum_{j=1}^{m} [f(x_j) - r_n(x_j)]^2 \right\}^{1/2}.$$

9. Consider the function

$$F(a_0, a_1, \ldots, a_n) = \int_0^1 \left[f(x) - \sum_{j=0}^{n} a_j x^j \right]^2 dx$$

and determine the coefficients $a_0, a_1, \ldots, a_n$ in such a way that F is minimized. Which kind of linear system is obtained?
[*Hint*: enforce the conditions $\partial F / \partial a_i = 0$ with $i = 0, 1, \ldots, n$. The matrix of the final linear system is the Hilbert matrix (see Example 3.2, Chapter 3) which is strongly ill-conditioned.]

11

Numerical Solution of Ordinary Differential Equations

In this chapter we deal with the numerical solutions of the Cauchy problem for ordinary differential equations (henceforth abbreviated by ODEs). After a brief review of basic notions about ODEs, we introduce the most widely used techniques for the numerical approximation of scalar equations. The concepts of consistency, convergence, zero-stability and absolute stability will be addressed. Then, we extend our analysis to systems of ODEs, with emphasis on *stiff* problems.

11.1 The Cauchy Problem

The Cauchy problem (also known as the initial-value problem) consists of finding the solution of an ODE, in the scalar or vector case, given suitable initial conditions. In particular, in the scalar case, denoting by I an interval of $\mathbb{R}$ containing the point t_0, the Cauchy problem associated with a first order ODE reads:

find a real-valued function $y \in C^1(I)$, such that

$$\begin{cases} y'(t) = f(t, y(t)), & t \in I, \\ y(t_0) = y_0, \end{cases} \tag{11.1}$$

where $f(t, y)$ is a given real-valued function in the strip $S = I \times (-\infty, +\infty)$, which is continuous with respect to both variables. Should f depend on t only through y, the differential equation is called *autonomous*.

Most of our analysis will be concerned with one single differential equation (scalar case). The extension to the case of systems of first-order ODEs will be addressed in Section 11.9.

If f is continuous with respect to t, then the solution to (11.1) satisfies

$$y(t) - y_0 = \int_{t_0}^{t} f(\tau, y(\tau)) d\tau. \tag{11.2}$$

Conversely, if y is defined by (11.2), then it is continuous in I and $y(t_0) = y_0$. Moreover, since y is a primitive of the continuous function $f(\cdot, y(\cdot))$, $y \in C^1(I)$ and satisfies the differential equation $y'(t) = f(t, y(t))$.

Thus, if f is continuous the Cauchy problem (11.1) is equivalent to the integral equation (11.2). We shall see later on how to take advantage of this equivalence in the numerical methods.

Let us now recall two existence and uniqueness results for (11.1).

1. **Local existence and uniqueness.**
   Suppose that $f(t, y)$ is locally Lipschitz continuous at (t_0, y_0) with respect to y, that is, there exist two neighborhoods, $J \subseteq I$ of t_0 of width r_J, and Σ of y_0 of width r_Σ, and a constant $L > 0$, such that
   $$|f(t, y_1) - f(t, y_2)| \leq L|y_1 - y_2| \quad \forall t \in J,\ \forall y_1, y_2 \in \Sigma. \tag{11.3}$$
   Then, the Cauchy problem (11.1) admits a unique solution in a neighborhood of t_0 with radius r_0 with $0 < r_0 < \min(r_J, r_\Sigma/M, 1/L)$, where M is the maximum of $|f(t, y)|$ on $J \times \Sigma$. This solution is called the *local solution.*
   Notice that condition (11.3) is automatically satisfied if f has continuous derivative with respect to y: indeed, in such a case it suffices to choose L as the maximum of $|\partial f(t, y)/\partial y|$ in $\overline{J \times \Sigma}$.
2. **Global existence and uniqueness.** The problem admits a unique *global solution* if one can take $J = I$ and $\Sigma = \mathbb{R}$ in (11.3), that is, if f is *uniformly Lipschitz continuous* with respect to y.

In view of the stability analysis of the Cauchy problem, we consider the following problem

$$\begin{cases} z'(t) = f(t, z(t)) + \delta(t), & t \in I, \\ z(t_0) = y_0 + \delta_0, \end{cases} \tag{11.4}$$

where $\delta_0 \in \mathbb{R}$ and δ is a continuous function on I. Problem (11.4) is derived from (11.1) by perturbing both the initial datum y_0 and the function f. Let us now characterize the sensitivity of the solution z to those perturbations.

Definition 11.1 ([Hah67] or [Ste71]). Let I be a bounded set. The Cauchy problem (11.1) is *stable in the sense of Liapunov* (or *stable*) on I if, for any perturbation $(\delta_0, \delta(t))$ satisfying

$$|\delta_0| < \varepsilon, \qquad |\delta(t)| < \varepsilon \quad \forall t \in I,$$

with $\varepsilon > 0$ sufficiently small to guarantee that the solution to the perturbed problem (11.4) does exist, then

$$\exists C > 0 \text{ such that } |y(t) - z(t)| < C\varepsilon, \qquad \forall t \in I. \tag{11.5}$$

The constant C depends in general on problem data t_0, y_0 and f, but not on ε.

If I has no upper bound we say that (11.1) is *asymptotically stable* if, as well as being Liapunov stable in any bounded interval I, the following limit also holds

$$|y(t) - z(t)| \to 0, \qquad \text{for } t \to +\infty, \tag{11.6}$$

provided that $\lim_{t\to\infty} |\delta(t)| = 0$. ■

The requirement that the Cauchy problem is stable is equivalent to requiring that it is well-posed in the sense stated in Chapter 2.

The uniform Lipschitz-continuity of f with respect to y suffices to ensure the stability of the Cauchy problem. Indeed, letting $w(t) = z(t) - y(t)$, we have

$$w'(t) = f(t, z(t)) - f(t, y(t)) + \delta(t).$$

Therefore,

$$w(t) = \delta_0 + \int_{t_0}^{t} [f(s, z(s)) - f(s, y(s))]\, ds + \int_{t_0}^{t} \delta(s) ds, \qquad \forall t \in I.$$

Thanks to previous assumptions, it follows that

$$|w(t)| \le (1 + |t - t_0|)\, \varepsilon + L \int_{t_0}^{t} |w(s)| ds.$$

Applying the Gronwall lemma (which we include below for the reader's ease) yields

$$|w(t)| \le (1 + |t - t_0|)\, \varepsilon e^{L|t-t_0|}, \qquad \forall t \in I$$

and, thus, (11.5) with $C = (1 + K_I)e^{LK_I}$, where $K_I = \max_{t\in I} |t - t_0|$.

Lemma 11.1 (Gronwall) *Let p be an integrable function nonnegative on the interval (t_0, t_0+T), and let g and φ be two continuous functions on $[t_0, t_0+T]$, g being nondecreasing. If φ satisfies the inequality*

$$\varphi(t) \le g(t) + \int_{t_0}^{t} p(\tau)\varphi(\tau) d\tau, \qquad \forall t \in [t_0, t_0 + T],$$

then

$$\varphi(t) \le g(t) \exp\left(\int_{t_0}^{t} p(\tau) d\tau \right), \qquad \forall t \in [t_0, t_0 + T].$$

For the proof, see, for instance, [QV94], Lemma 1.4.1.

The constant C that appears in (11.5) could be very large and, in general, depends on the upper extreme of the interval I, as in the proof above. For that reason, the property of asymptotic stability is more suitable for describing the behavior of the *dynamical system* (11.1) as $t \to +\infty$ (see [Arn73]).

As is well-known, only a restricted number of nonlinear ODEs can be solved in closed form (see, for instance, [Arn73]). Moreover, even when this is possible, it is not always a straightforward task to find an explicit expression of the solution; for example, consider the (very simple) equation $y' = (y-t)/(y+t)$, whose solution is only implicitly defined by the relation $(1/2)\log(t^2+y^2)+\tan^{-1}(y/t) = C$, where C is a constant depending on the initial condition.

For this reason we are interested in numerical methods, since these can be applied to any ODE under the sole condition that it admits a unique solution.

11.2 One-Step Numerical Methods

Let us address the numerical approximation of the Cauchy problem (11.1). Fix $0 < T < +\infty$ and let $I = (t_0, t_0+T)$ be the integration interval and, correspondingly, for $h > 0$, let $t_n = t_0 + nh$, with $n = 0, 1, 2, \ldots, N_h$, be the sequence of discretization nodes of I into subintervals $I_n = [t_n, t_{n+1}]$. The width h of such subintervals is called the *discretization stepsize*. Notice that N_h is the maximum integer such that $t_{N_h} \leq t_0 + T$. Let u_j be the approximation at node t_j of the exact solution $y(t_j)$; this solution will be henceforth shortly denoted by y_j. Similarly, f_j denotes the value $f(t_j, u_j)$. We obviously set $u_0 = y_0$.

Definition 11.2 A numerical method for the approximation of problem (11.1) is called a *one-step* method if $\forall n \geq 0$, u_{n+1} depends only on u_n. Otherwise, the scheme is called a *multistep* method. ■

For now, we focus our attention on one-step methods. Here are some of them:

1. **forward Euler method**

$$u_{n+1} = u_n + hf_n; \tag{11.7}$$

2. **backward Euler method**

$$u_{n+1} = u_n + hf_{n+1}. \tag{11.8}$$

In both cases, y' is approximated through a finite difference: forward and backward differences are used in (11.7) and (11.8), respectively. Both finite differences are first-order approximations of the first derivative of y with respect to h (see Section 10.10.1).

3. **trapezoidal (or Crank-Nicolson) method**

$$u_{n+1} = u_n + \frac{h}{2}\left[f_n + f_{n+1}\right]. \tag{11.9}$$

This method stems from approximating the integral on the right side of (11.2) by the trapezoidal quadrature rule (9.11).

4. **Heun method**

$$u_{n+1} = u_n + \frac{h}{2}[f_n + f(t_{n+1}, u_n + hf_n)]. \tag{11.10}$$

This method can be derived from the trapezoidal method substituting $f(t_{n+1}, u_n + hf_n)$ for f_{n+1} in (11.9) (i.e., using the forward Euler method to compute u_{n+1}).

In this last case, we notice that the aim is to transform an *implicit* method into an *explicit* one. Addressing this concern, we recall the following.

Definition 11.3 (Explicit and implicit methods) A method is called *explicit* if u_{n+1} can be computed directly in terms of (some of) the previous values u_k, $k \leq n$, *implicit* if u_{n+1} depends implicitly on itself through f. ■

Methods (11.7) and (11.10) are explicit, while (11.8) and (11.9) are implicit. These latter require at each time step to solving a nonlinear problem if f depends nonlinearly on the second argument.

A remarkable example of one-step methods are the Runge-Kutta methods, which will be analyzed in Section 11.8.

11.3 Analysis of One-Step Methods

Any one-step explicit method for the approximation of (11.1) can be cast in the concise form

$$u_{n+1} = u_n + h\Phi(t_n, u_n, f_n; h), \quad 0 \leq n \leq N_h - 1, \quad u_0 = y_0, \tag{11.11}$$

where $\Phi(\cdot, \cdot, \cdot; \cdot)$ is called an *increment function.* Letting as usual $y_n = y(t_n)$, analogously to (11.11) we can write

$$y_{n+1} = y_n + h\Phi(t_n, y_n, f(t_n, y_n); h) + \varepsilon_{n+1}, \quad 0 \leq n \leq N_h - 1, \tag{11.12}$$

where ε_{n+1} is the residual arising at the point t_{n+1} when we pretend that the exact solution "satisfies" the numerical scheme. Let us write the residual as

$$\varepsilon_{n+1} = h\tau_{n+1}(h).$$

The quantity $\tau_{n+1}(h)$ is called the *local truncation error* (LTE) at the node t_{n+1}. We thus define the *global truncation error* to be the quantity

$$\tau(h) = \max_{0 \le n \le N_h - 1} |\tau_{n+1}(h)|.$$

Notice that $\tau(h)$ depends on the solution y of the Cauchy problem (11.1).

The forward Euler's method is a special instance of (11.11), where

$$\Phi(t_n, u_n, f_n; h) = f_n,$$

while to recover Heun's method we must set

$$\Phi(t_n, u_n, f_n; h) = \frac{1}{2}\left[f_n + f(t_n + h, u_n + hf_n)\right].$$

A one-step explicit scheme is fully characterized by its increment function Φ. This function, in all the cases considered thus far, is such that

$$\lim_{h \to 0} \Phi(t_n, y_n, f(t_n, y_n); h) = f(t_n, y_n), \quad \forall t_n \ge t_0. \tag{11.13}$$

Property (11.13), together with the obvious relation $y_{n+1} - y_n = hy'(t_n) + \mathcal{O}(h^2)$, $\forall n \ge 0$, allows one to obtain from (11.12) that $\lim_{h\to 0} \tau_{n+1}(h) = 0$, $0 \le n \le N_h - 1$. In turn, this condition ensures that

$$\lim_{h \to 0} \tau(h) = 0,$$

which expresses the *consistency* of the numerical method (11.11) with the Cauchy problem (11.1). In general, a method is said to be *consistent* if its LTE is infinitesimal with respect to h. Moreover, a scheme has *order* p if, $\forall t \in I$, the solution $y(t)$ of the Cauchy problem (11.1) fulfills the condition

$$\tau(h) = \mathcal{O}(h^p) \quad \text{for } h \to 0. \tag{11.14}$$

Using Taylor expansions, as was done in Section 11.2, it can be proved that the forward Euler method has order 1, while the Heun method has order 2 (see Exercises 1 and 2).

11.3.1 The Zero-Stability

Let us formulate a requirement analogous to the one for Liapunov stability (11.5), specifically for the numerical scheme. If (11.5) is satisfied with a constant C independent of h, we shall say that the numerical problem is *zero-stable.* Precisely:

Definition 11.4 (Zero-stability of one-step methods) The numerical method (11.11) for the approximation of problem (11.1) is *zero-stable* if $\exists h_0 > 0$, $\exists C > 0$ such that $\forall h \in (0, h_0]$, $\forall \varepsilon > 0$ sufficiently small, if $|\delta_n| \le \varepsilon$, $0 \le n \le N_h$, then

$$|z_n^{(h)} - u_n^{(h)}| \le C\varepsilon, \qquad 0 \le n \le N_h, \tag{11.15}$$

where $z_n^{(h)}$, $u_n^{(h)}$ are respectively the solutions of the problems

$$\begin{cases} z_{n+1}^{(h)} = z_n^{(h)} + h\left[\Phi(t_n, z_n^{(h)}, f(t_n, z_n^{(h)}); h) + \delta_{n+1}\right], \\ z_0^{(h)} \quad = y_0 + \delta_0, \end{cases} \tag{11.16}$$

$$\begin{cases} u_{n+1}^{(h)} = u_n^{(h)} + h\Phi(t_n, u_n^{(h)}, f(t_n, u_n^{(h)}); h), \\ u_0^{(h)} \quad = y_0, \end{cases} \tag{11.17}$$

for $0 \leq n \leq N_h - 1$. ■

Both constants C and h_0 may depend on problem's data t_0, T, y_0 and f. Zero-stability thus requires that, in a bounded interval, (11.15) holds for any value $h \leq h_0$. This property deals, in particular, with the behavior of the numerical method in the limit case $h \to 0$ and this justifies the name of *zero*-stability. This latter is therefore a distinguishing property of the numerical method itself, not of the Cauchy problem (which, indeed, is stable thanks to the uniform Lipschitz continuity of f). Property (11.15) ensures that the numerical method has a weak sensitivity with respect to small changes in the data and is thus stable in the sense of the general definition given in Chapter 2.

The request that a numerical method be stable arises, before anything else, from the need of keeping under control the (unavoidable) errors introduced by the finite arithmetic of the computer. Indeed, if the numerical method were not zero-stable, the rounding errors made on y_0 as well as in the process of computing $f(t_n, u_n)$ would make the computed solution useless.

Theorem 11.1 (Zero-stability) *Consider the explicit one-step method* (11.11) *for the numerical solution of the Cauchy problem* (11.1). *Assume that the increment function* Φ *is Lipschitz continuous with respect to the second argument, with constant* Λ *independent of* h *and of the nodes* $t_j \in [t_0, t_0 + T]$, *that is*

$$\begin{aligned} &\exists h_0 > 0, \exists \Lambda > 0 : \forall h \in (0, h_0] \\ &|\Phi(t_n, u_n^{(h)}, f(t_n, u_n^{(h)}); h) - \Phi(t_n, z_n^{(h)}, f(t_n, z_n^{(h)}); h)| \\ &\qquad\qquad \leq \Lambda |u_n^{(h)} - z_n^{(h)}|, \ 0 \leq n \leq N_h. \end{aligned} \tag{11.18}$$

Then, method (11.11) *is zero-stable.*

Proof. Setting $w_j^{(h)} = z_j^{(h)} - u_j^{(h)}$, by subtracting (11.17) from (11.16) we obtain, for $j = 0, \ldots, N_h - 1$,

$$w_{j+1}^{(h)} = w_j^{(h)} + h\left[\Phi(t_j, z_j^{(h)}, f(t_j, z_j^{(h)}); h) - \Phi(t_j, u_j^{(h)}, f(t_j, u_j^{(h)}); h)\right] + h\delta_{j+1}.$$

Summing over j gives, for $n = 1, \ldots, N_h$,

$$w_n^{(h)} = w_0^{(h)}$$
$$+h\sum_{j=0}^{n-1}\delta_{j+1} + h\sum_{j=0}^{n-1}\left(\Phi(t_j, z_j^{(h)}, f(t_j, z_j^{(h)}); h) - \Phi(t_j, u_j^{(h)}, f(t_j, u_j^{(h)}); h)\right),$$

so that, by (11.18)

$$|w_n^{(h)}| \leq |w_0| + h\sum_{j=0}^{n-1}|\delta_{j+1}| + h\Lambda\sum_{j=0}^{n-1}|w_j^{(h)}|, \qquad 1 \leq n \leq N_h. \tag{11.19}$$

Applying the discrete Gronwall lemma, given below, we obtain

$$|w_n^{(h)}| \leq (1 + hn)\,\varepsilon e^{nh\Lambda}, \qquad 1 \leq n \leq N_h.$$

Then (11.15) follows from noticing that $hn \leq T$ and setting $C = (1 + T)\,e^{\Lambda T}$. ◇

Notice that zero-stability implies the boundedness of the solution when f is linear with respect to the second argument.

Lemma 11.2 (Discrete Gronwall) *Let k_n be a nonnegative sequence and φ_n a sequence such that*

$$\begin{cases} \varphi_0 \leq g_0, \\ \varphi_n \leq g_0 + \displaystyle\sum_{s=0}^{n-1} p_s + \sum_{s=0}^{n-1} k_s\varphi_s, \qquad n \geq 1. \end{cases}$$

If $g_0 \geq 0$ and $p_n \geq 0$ for any $n \geq 0$, then

$$\varphi_n \leq \left(g_0 + \sum_{s=0}^{n-1} p_s\right)\exp\left(\sum_{s=0}^{n-1} k_s\right), \qquad n \geq 1.$$

For the proof, see, for instance, [QV94], Lemma 1.4.2. In the specific case of the Euler method, checking the property of zero-stability can be done directly using the Lipschitz continuity of f (we refer the reader to the end of Section 11.3.2). In the case of multistep methods, the analysis will lead to the verification of a purely algebraic property, the so-called *root condition* (see Section 11.6.3).

11.3.2 Convergence Analysis

Definition 11.5 A method is said to be *convergent* if

$$\forall n = 0, \ldots, N_h, \qquad |u_n - y_n| \leq C(h),$$

where $C(h)$ is an infinitesimal with respect to h. In that case, it is said to be *convergent with order* p if $\exists \mathcal{C} > 0$ such that $C(h) = \mathcal{C}h^p$. ■

We can prove the following theorem.

Theorem 11.2 (Convergence) *Under the same assumptions as in Theorem 11.1, we have*

$$|y_n - u_n| \leq (|y_0 - u_0| + nh\tau(h))\, e^{nh\Lambda}, \qquad 1 \leq n \leq N_h. \tag{11.20}$$

Therefore, if the consistency assumption (11.13) *holds and* $|y_0 - u_0| \to 0$ *as* $h \to 0$*, then the method is convergent. Moreover, if* $|y_0 - u_0| = \mathcal{O}(h^p)$ *and the method has order* p*, then it is also convergent with order* p*.*

Proof. Setting $w_j = y_j - u_j$, subtracting (11.11) from (11.12) and proceeding as in the proof of the previous theorem yields inequality (11.19), with the understanding that

$$w_0 = y_0 - u_0, \text{ and } \delta_{j+1} = \tau_{j+1}(h).$$

The estimate (11.20) is then obtained by applying again the discrete Gronwall lemma. From the fact that $nh \leq T$ and $\tau(h) = \mathcal{O}(h^p)$, we can conclude that $|y_n - u_n| \leq Ch^p$ with C depending on T and Λ but not on h. $\diamond$

A consistent and zero-stable method is thus convergent. This property is known as the *Lax-Richtmyer theorem* or *equivalence theorem* (the converse: "a convergent method is zero-stable" being obviously true). This theorem, which is proven in [IK66], was already advocated in Section 2.2.1 and is a central result in the analysis of numerical methods for ODEs (see [Dah56] or [Hen62] for linear multistep methods, [But66], [MNS74] for a wider classes of methods). It will be considered again in Section 11.5 for the analysis of multistep methods.

We carry out in detail the convergence analysis in the case of the forward Euler method, without resorting to the discrete Gronwall lemma. In the first part of the proof we assume that any operation is performed in exact arithmetic and that $u_0 = y_0$.

Denote by $e_{n+1} = y_{n+1} - u_{n+1}$ the error at node t_{n+1} with $n = 0, 1, \ldots$ and notice that

$$e_{n+1} = (y_{n+1} - u^*_{n+1}) + (u^*_{n+1} - u_{n+1}), \tag{11.21}$$

where $u^*_{n+1} = y_n + hf(t_n, y_n)$ is the solution obtained after one step of the forward Euler method starting from the initial datum y_n (see Figure 11.1). The first addendum in (11.21) accounts for the consistency error, the second one for the cumulation of these errors. Then

$$y_{n+1} - u^*_{n+1} = h\tau_{n+1}(h), \quad u^*_{n+1} - u_{n+1} = e_n + h\left[f(t_n, y_n) - f(t_n, u_n)\right].$$

As a consequence,

$$|e_{n+1}| \leq h|\tau_{n+1}(h)| + |e_n| + h|f(t_n, y_n) - f(t_n, u_n)| \leq h\tau(h) + (1 + hL)|e_n|,$$

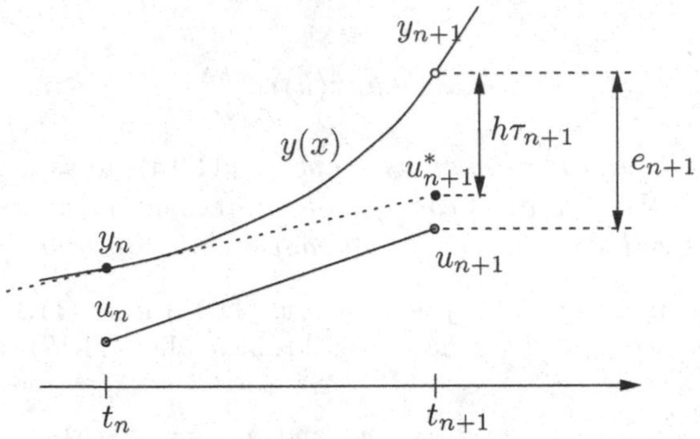

Fig. 11.1. Geometrical interpretation of the local truncation error τ_{n+1} and the true error e_{n+1} at node t_{n+1} for the forward Euler method

L being the Lipschitz constant of f. By recursion on n, we find

$$|e_{n+1}| \leq [1 + (1 + hL) + \ldots + (1 + hL)^n]\, h\tau(h)$$
$$= \frac{(1 + hL)^{n+1} - 1}{L}\tau(h) \leq \frac{e^{L(t_{n+1}-t_0)} - 1}{L}\tau(h).$$

The last inequality follows from noticing that $1 + hL \leq e^{hL}$ and $(n + 1)h = t_{n+1} - t_0$.

On the other hand, if $y \in C^2(I)$, the LTE for the forward Euler method is (see Section 10.10.1)

$$\tau_{n+1}(h) = \frac{h}{2}y''(\xi), \;\; \xi \in (t_n, t_{n+1}),$$

and thus, $\tau(h) \leq (M/2)h$, where $M = \max_{\xi \in I} |y''(\xi)|$. In conclusion,

$$|e_{n+1}| \leq \frac{e^{L(t_{n+1}-t_0)} - 1}{L}\frac{M}{2}h, \quad \forall n \geq 0, \tag{11.22}$$

from which it follows that the error tends to zero with the same order as the local truncation error.

If also the rounding errors are accounted for, we can assume that the solution $\bar{u}_{n+1}$, actually computed by the forward Euler method at time t_{n+1}, is such that

$$\bar{u}_0 = y_0 + \zeta_0, \;\; \bar{u}_{n+1} = \bar{u}_n + hf(t_n, \bar{u}_n) + \zeta_{n+1}, \tag{11.23}$$

having denoted the rounding error by ζ_j, for $j \geq 0$.
Problem (11.23) is an instance of (11.16), provided that we identify ζ_{n+1} and $\bar{u}_n$ with $h\delta_{n+1}$ and $z_n^{(h)}$ in (11.16), respectively. Combining Theorems 11.1 and 11.2 we get, instead of (11.22), the following error estimate

$$|y_{n+1} - \bar{u}_{n+1}| \le e^{L(t_{n+1}-t_0)} \left[|\zeta_0| + \frac{1}{L}\left(\frac{M}{2}h + \frac{\zeta}{h}\right)\right],$$

where $\zeta = \max_{1\le j\le n+1} |\zeta_j|$. The presence of rounding errors does not allow, therefore, to conclude that as $h \to 0$, the error goes to zero. Actually, there exists an optimal (nonnull) value of h, h_{opt}, for which the error is minimized. For $h < h_{opt}$, the rounding error dominates the truncation error and the error increases.

11.3.3 The Absolute Stability

The property of *absolute stability* is in some way specular to zero-stability, as far as the roles played by h and I are concerned. Heuristically, we say that a numerical method is absolutely stable if, *for h fixed*, u_n remains bounded as $t_n \to +\infty$. This property, thus, deals with the asymptotic behavior of u_n, as opposed to a zero-stable method for which, for a fixed integration interval, u_n remains bounded as $h \to 0$.
For a precise definition, consider the linear Cauchy problem (that from now on, we shall refer to as the *test problem*)

$$\begin{cases} y'(t) = \lambda y(t), & t > 0, \\ y(0) = 1, \end{cases} \tag{11.24}$$

with $\lambda \in \mathbb{C}$, whose solution is $y(t) = e^{\lambda t}$. Notice that $\lim_{t\to+\infty} |y(t)| = 0$ if $\mathrm{Re}(\lambda) < 0$.

Definition 11.6 A numerical method for approximating (11.24) is *absolutely stable* if

$$|u_n| \longrightarrow 0 \text{ as } t_n \longrightarrow +\infty. \tag{11.25}$$

Let h be the discretization stepsize. The numerical solution u_n of (11.24) obviously depends on h and λ. Therefore, a method will be absolutely stable for certain values of h and λ and not fot other values. More precisely, the *region of absolute stability* of the numerical method is defined as the subset of the complex plane

$$\mathcal{A} = \{z = h\lambda \in \mathbb{C} : \text{(11.25) is satisfied}\}. \tag{11.26}$$

Thus, $\mathcal{A}$ is the set of the values of the product $h\lambda$ for which the numerical method furnishes solutions that decay to zero as t_n tends to infinity. ■

Remark 11.1 Let us now consider the general case of the Cauchy problem (11.1) and assume that there exist two positive constant μ_{min} and μ_{max} such that

$$-\mu_{max} < \frac{\partial f}{\partial y}(t, y(t)) < -\mu_{min}, \forall t \in I.$$

Then, a suitable candidate to play the role of λ in the previous stability analysis is $-\mu_{max}$ (for more details, see [QS06]). ■

Let us check whether the one-step methods introduced previously are absolutely stable.

1. *Forward Euler method*: applying (11.7) to problem (11.24) yields $u_{n+1} = u_n + h\lambda u_n$ for $n \geq 0$, with $u_0 = 1$. Proceeding recursively on n we get

$$u_n = (1 + h\lambda)^n, \qquad n \geq 0.$$

Therefore, condition (11.25) is satisfied iff $|1 + h\lambda| < 1$, that is, if $h\lambda$ lies within the unit circle with center at $(-1, 0)$ (see Figure 11.3). This amounts to requiring that

$$h\lambda \in \mathbb{C}^- \text{ and } 0 < h < -\frac{2\text{Re}(\lambda)}{|\lambda|^2}, \tag{11.27}$$

where

$$\mathbb{C}^- = \{z \in \mathbb{C} : \ \text{Re}(z) < 0\}.$$

Example 11.1 For the Cauchy problem $y'(x) = -5y(x)$ for $x > 0$ and $y(0) = 1$, condition (11.27) implies $0 < h < 2/5$. Figure 11.2 on the left shows the behavior of the computed solution for two values of h which do not fulfill this condition, while on the right we show the solutions for two values of h that do. Notice that in this second case the oscillations, if present, damp out as t grows. •

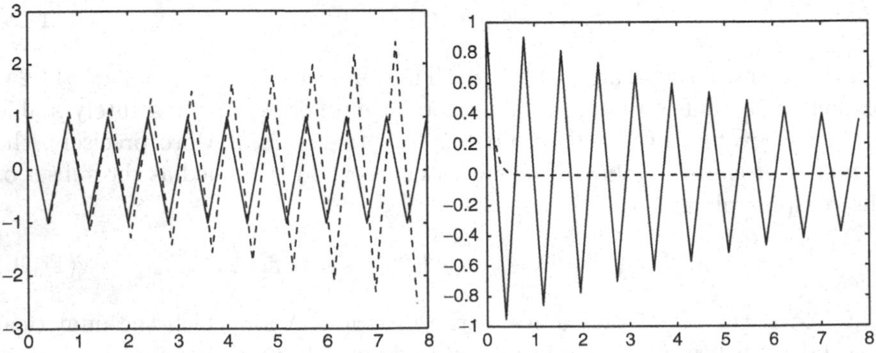

Fig. 11.2. Left: computed solutions for $h = 0.41 > 2/5$ (*dashed line*) and $h = 2/5$ (*solid line*). Notice how, in the limiting case $h = 2/5$, the oscillations remain unmodified as t grows. Right: two solutions are reported for $h = 0.39$ (*solid line*) and $h = 0.15$ (*dashed line*)

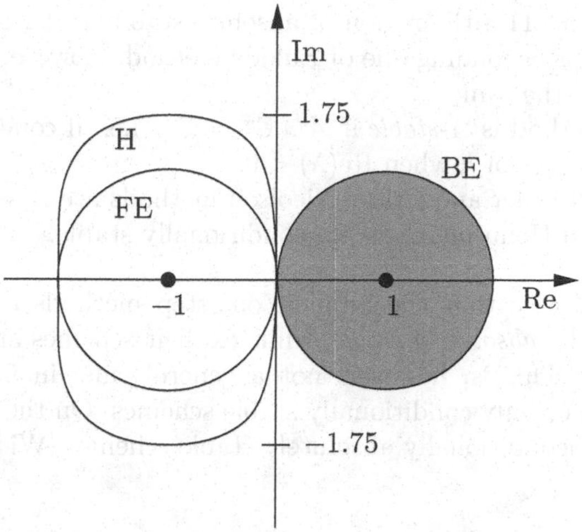

Fig. 11.3. Regions of absolute stability for the forward (FE) and backward Euler (BE) methods and for Heun's method (H). Notice that the region of absolute stability of the BE method lies outside the unit circle of center (1, 0) (shaded area)

2. *Backward Euler method*: proceeding as before, we get this time

$$u_n = \frac{1}{(1-h\lambda)^n}, \qquad n \geq 0.$$

The absolute stability property (11.25) is satisfied *for any value of* $h\lambda$ that does not belong to the unit circle of center (1, 0) (see Figure 11.3, on the right).

Example 11.2 The numerical solution given by the backward Euler method in the case of Example 11.1 does not exhibit any oscillation for any value of h. On the other hand, the same method, if applied to the problem $y'(t) = 5y(t)$ for $t > 0$ and with $y(0) = 1$, computes a solution that decays *anyway* to zero as $t \to \infty$ if $h > 2/5$, despite the fact that the exact solution of the Cauchy problem tends to infinity. •

3. *Trapezoidal (or Crank-Nicolson) method*: we get

$$u_n = \left[\left(1 + \frac{1}{2}\lambda h\right) / \left(1 - \frac{1}{2}\lambda h\right)\right]^n, \qquad n \geq 0,$$

hence (11.25) is fulfilled for any $h\lambda \in \mathbb{C}^-$.

4. *Heun's method*: applying (11.10) to problem (11.24) and proceeding by recursion on n, we obtain

$$u_n = \left[1 + h\lambda + \frac{(h\lambda)^2}{2}\right]^n, \qquad n \geq 0.$$

As shown in Figure 11.3 the region of absolute stability of Heun's method is larger than the corresponding one of Euler's method. However, its restriction to the real axis is the same.
We say that a method is *A-stable* if $\mathcal{A} \cap \mathbb{C}^- = \mathbb{C}^-$, i.e., if condition (11.25) is satisfied for all values of h when $\mathrm{Re}(\lambda) < 0$.

The backward Euler and Crank-Nicolson methods are A-stable, while the forward Euler and Heun methods are conditionally stable.

Remark 11.2 Notice that the implicit one-step methods examined so far are *unconditionally absolutely stable*, while explicit schemes are *conditionally absolutely stable*. This is, however, not a general rule: in fact, there exist implicit unstable or only conditionally stable schemes. On the contrary, there are no explicit unconditionally absolutely stable schemes [Wid67]. ■

11.4 Difference Equations

For any integer $k \geq 1$, an equation of the form

$$u_{n+k} + \alpha_{k-1} u_{n+k-1} + \ldots + \alpha_0 u_n = \varphi_{n+k}, \quad n = 0, 1, \ldots \tag{11.28}$$

is called a *linear difference equation* of order k. The coefficients $\alpha_0 \neq 0$, $\alpha_1, \ldots, \alpha_{k-1}$ may or may not depend on n. If, for any n, the right side φ_{n+k} is equal to zero, the equation is said *homogeneous*, while if the α_j's are independent of n it is called *linear difference equation with constant coefficients*. Difference equations arise for instance in the discretization of ordinary differential equations. Regarding this, we notice that all the numerical methods examined so far generate equations like (11.28). More generally, equations like (11.28) are encountered when quantities are defined through linear recursive relations. Another relevant application is concerned with the discretization of boundary value problems (see Chapter 12). For further details on the subject, we refer to Chapters 2 and 5 of [BO78] and to Chapter 6 of [Gau97].

Any sequence $\{u_n, n = 0, 1, \ldots\}$ of values that satisfy (11.28) is called a *solution* to the equation (11.28). Given k *initial values* $u_0, \ldots, u_{k-1}$, it is always possible to construct a solution of (11.28) by computing (sequentially)

$$u_{n+k} = \varphi_{n+k} - (\alpha_{k-1} u_{n+k-1} + \ldots + \alpha_0 u_n), \quad n = 0, 1, \ldots.$$

However, our interest is to find an expression of the solution u_{n+k} which depends only on the coefficients and on the initial values.

We start by considering the *homogeneous case with constant coefficients*,

$$u_{n+k} + \alpha_{k-1} u_{n+k-1} + \ldots + \alpha_0 u_n = 0, \quad n = 0, 1, \ldots \tag{11.29}$$

and associate with (11.29) the *characteristic polynomial* $\Pi \in \mathbb{P}_k$ defined as

$$\Pi(r) = r^k + \alpha_{k-1} r^{k-1} + \ldots + \alpha_1 r + \alpha_0. \tag{11.30}$$

Denoting its roots by r_j, $j = 0, \ldots, k-1$, any sequence of the form

$$\{r_j^n, n = 0, 1, \ldots\}, \qquad \text{for } j = 0, \ldots, k-1, \tag{11.31}$$

is a solution of (11.29), since

$$\begin{aligned} & r_j^{n+k} + \alpha_{k-1} r_j^{n+k-1} + \ldots + \alpha_0 r_j^n \\ & = r_j^n \left(r_j^k + \alpha_{k-1} r_j^{k-1} + \ldots + \alpha_0 \right) = r_j^n \Pi(r_j) = 0. \end{aligned}$$

We say that the k sequences defined in (11.31) are the *fundamental solutions* of the homogeneous equation (11.29). Any sequence of the form

$$u_n = \gamma_0 r_0^n + \gamma_1 r_1^n + \ldots + \gamma_{k-1} r_{k-1}^n, \qquad n = 0, 1, \ldots \tag{11.32}$$

is still a solution to (11.29), since it is a linear equation.
The coefficients $\gamma_0, \ldots, \gamma_{k-1}$ can be determined by imposing the k initial conditions $u_0, \ldots, u_{k-1}$. Moreover, it can be proved that if all the roots of Π are simple, then *all* the solutions of (11.29) can be cast in the form (11.32).

This last statement no longer holds if there are roots of Π with multiplicity greater than 1. If, for a certain j, the root r_j has multiplicity $m \geq 2$, in order to obtain a system of fundamental solutions that generate all the solutions of (11.29), it suffices to replace the corresponding fundamental solution $\{r_j^n, n = 0, 1, \ldots\}$ with the m sequences

$$\{r_j^n, n = 0, 1, \ldots\}, \ \{n r_j^n, n = 0, 1, \ldots\}, \ \ldots, \ \{n^{m-1} r_j^n, n = 0, 1, \ldots\}.$$

More generally, assuming that $r_0, \ldots, r_{k'}$ are distinct roots of Π, with multiplicities equal to $m_0, \ldots, m_{k'}$, respectively, we can write the solution of (11.29) as

$$u_n = \sum_{j=0}^{k'} \left(\sum_{s=0}^{m_j - 1} \gamma_{sj} n^s \right) r_j^n, \qquad n = 0, 1, \ldots. \tag{11.33}$$

Notice that even in presence of complex conjugate roots one can still obtain a real solution (see Exercise 3).

Example 11.3 For the difference equation $u_{n+2} - u_n = 0$, we have $\Pi(r) = r^2 - 1$, then $r_0 = -1$ and $r_1 = 1$, therefore the solution is given by $u_n = \gamma_{00}(-1)^n + \gamma_{01}$. In particular, enforcing the initial conditions u_0 and u_1 gives $\gamma_{00} = (u_0 - u_1)/2$, $\gamma_{01} = (u_0 + u_1)/2$. •

Example 11.4 For the difference equation $u_{n+3} - 2u_{n+2} - 7u_{n+1} - 4u_n = 0$, $\Pi(r) = r^3 - 2r^2 - 7r - 4$. Its roots are $r_0 = -1$ (with multiplicity 2), $r_1 = 4$ and the solution is $u_n = (\gamma_{00} + n\gamma_{10})(-1)^n + \gamma_{01} 4^n$. Enforcing the initial conditions we can compute the unknown coefficients as the solution of the following linear system

$$\begin{cases} \gamma_{00} + \gamma_{01} & = u_0, \\ -\gamma_{00} - \gamma_{10} + 4\gamma_{01} & = u_1, \\ \gamma_{00} + 2\gamma_{10} + 16\gamma_{01} & = u_2, \end{cases}$$

that yields $\gamma_{00} = (24u_0 - 2u_1 - u_2)/25$, $\gamma_{10} = (u_2 - 3u_1 - 4u_0)/5$ and $\gamma_{01} = (2u_1 + u_0 + u_2)/25$. •

The expression (11.33) is of little practical use since it does not outline the dependence of u_n on the k initial conditions. A more convenient representation is obtained by introducing a new set $\left\{\psi_j^{(n)},\ n = 0, 1, \ldots\right\}$ of fundamental solutions that satisfy

$$\psi_j^{(i)} = \delta_{ij}, \quad i, j = 0, 1, \ldots, k-1. \tag{11.34}$$

Then, the solution of (11.29) subject to the initial conditions $u_0, \ldots, u_{k-1}$ is given by

$$u_n = \sum_{j=0}^{k-1} u_j \psi_j^{(n)}, \qquad n = 0, 1, \ldots. \tag{11.35}$$

The new fundamental solutions $\left\{\psi_j^{(n)},\ n = 0, 1, \ldots\right\}$ can be represented in terms of those in (11.31) as follows

$$\psi_j^{(n)} = \sum_{m=0}^{k-1} \beta_{j,m} r_m^n \quad \text{for } j = 0, \ldots, k-1,\ n = 0, 1, \ldots. \tag{11.36}$$

By requiring (11.34), we obtain the k linear systems

$$\sum_{m=0}^{k-1} \beta_{j,m} r_m^i = \delta_{ij}, \qquad i, j = 0, \ldots, k-1,$$

whose matrix form is

$$\mathrm{R}\mathbf{b}_j = \mathbf{e}_j, \qquad j = 0, \ldots, k-1. \tag{11.37}$$

Here $\mathbf{e}_j$ denotes the unit vector of $\mathbb{R}^k$, $\mathrm{R} = (r_{im}) = (r_m^i)$ and $\mathbf{b}_j = (\beta_{j,0}, \ldots, \beta_{j,k-1})^T$. If all r_j's are simple roots of Π, the matrix R is nonsingular (see Exercise 5).

The general case where Π has $k'+1$ distinct roots $r_0, \ldots, r_{k'}$ with multiplicities $m_0, \ldots, m_{k'}$ respectively, can be dealt with by replacing in (11.36) $\{r_j^n,\ n = 0, 1, \ldots\}$ with $\{r_j^n n^s,\ n = 0, 1, \ldots\}$, where $j = 0, \ldots, k'$ and $s = 0, \ldots, m_j - 1$.

Example 11.5 We consider again the difference equation of Example 11.4. Here we have $\{r_0^n,\ n r_0^n,\ r_1^n,\ n = 0, 1, \ldots\}$ so that the matrix R becomes

$$\mathrm{R} = \begin{bmatrix} r_0^0 & 0 & r_1^0 \\ r_0^1 & r_0^1 & r_1^1 \\ r_0^2 & 2r_0^2 & r_1^2 \end{bmatrix} = \begin{bmatrix} 1 & 0 & 1 \\ -1 & -1 & 4 \\ 1 & 2 & 16 \end{bmatrix}.$$

Solving the three systems (11.37) yields

$$\begin{aligned} \psi_0^{(n)} &= \frac{24}{25}(-1)^n - \frac{4}{5}n(-1)^n + \frac{1}{25}4^n, \\ \psi_1^{(n)} &= -\frac{2}{25}(-1)^n - \frac{3}{5}n(-1)^n + \frac{2}{25}4^n, \\ \psi_2^{(n)} &= -\frac{1}{25}(-1)^n + \frac{1}{5}n(-1)^n + \frac{1}{25}4^n, \end{aligned}$$

from which it can be checked that the solution $u_n = \sum_{j=0}^{2} u_j \psi_j^{(n)}$ coincides with the one already found in Example 11.4. •

Now we return to the case of *nonconstant coefficients* and consider the following homogeneous equation

$$u_{n+k} + \sum_{j=1}^{k} \alpha_{k-j}(n) u_{n+k-j} = 0, \qquad n = 0, 1, \ldots. \tag{11.38}$$

The goal is to transform it into an ODE by means of a function F, called the *generating function* of the equation (11.38). F depends on the real variable t and is derived as follows. We require that the n-th coefficient of the Taylor series of F around $t = 0$ can be written as $\gamma_n u_n$, for some unknown constant γ_n, so that

$$F(t) = \sum_{n=0}^{\infty} \gamma_n u_n t^n. \tag{11.39}$$

The coefficients $\{\gamma_n\}$ are unknown and must be determined in such a way that

$$\sum_{j=0}^{k} c_j F^{(k-j)}(t) = \sum_{n=0}^{\infty} \left[u_{n+k} + \sum_{j=1}^{k} \alpha_{k-j}(n) u_{n+k-j} \right] t^n, \tag{11.40}$$

where c_j are suitable unknown constants not depending on n. Note that owing to (11.38) we obtain the ODE

$$\sum_{j=0}^{k} c_j F^{(k-j)}(t) = 0$$

to which we must add the initial conditions $F^{(j)}(0) = \gamma_j u_j$ for $j = 0, \ldots, k-1$. Once F is available, it is simple to recover u_n through the definition of F itself.

Example 11.6 Consider the difference equation

$$(n+2)(n+1)u_{n+2} - 2(n+1)u_{n+1} - 3u_n = 0, \quad n = 0, 1, \ldots \tag{11.41}$$

with the initial conditions $u_0 = u_1 = 2$. We look for a generating function of the form (11.39). By term-to-term derivation of the series, we get

$$F'(t) = \sum_{n=0}^{\infty} \gamma_n n u_n t^{n-1}, \quad F''(t) = \sum_{n=0}^{\infty} \gamma_n n(n-1) u_n t^{n-2},$$

and, after some algebra, we find

$$F'(t) = \sum_{n=0}^{\infty} \gamma_n n u_n t^{n-1} = \sum_{n=0}^{\infty} \gamma_{n+1}(n+1) u_{n+1} t^n,$$
$$F''(t) = \sum_{n=0}^{\infty} \gamma_n n(n-1) u_n t^{n-2} = \sum_{n=0}^{\infty} \gamma_{n+2}(n+2)(n+1) u_{n+2} t^n.$$

As a consequence, (11.40) becomes

$$\sum_{n=0}^{\infty}(n+1)(n+2)u_{n+2}t^n - 2\sum_{n=0}^{\infty}(n+1)u_{n+1}t^n - 3\sum_{n=0}^{\infty}u_n t^n$$
$$= c_0\sum_{n=0}^{\infty}\gamma_{n+2}(n+2)(n+1)u_{n+2}t^n + c_1\sum_{n=0}^{\infty}\gamma_{n+1}(n+1)u_{n+1}t^n + c_2\sum_{n=0}^{\infty}\gamma_n u_n t^n,$$

so that, equating both sides, we find

$$\gamma_n = 1 \ \forall n \geq 0, \ c_0 = 1, \ c_1 = -2, \ c_2 = -3.$$

We have thus associated with the difference equation the following ODE with constant coefficients

$$F''(t) - 2F'(t) - 3F(t) = 0,$$

with the initial condition $F(0) = F'(0) = 2$. The n-th coefficient of the solution $F(t) = e^{3t} + e^{-t}$ is

$$\frac{1}{n!}F^{(n)}(0) = \frac{1}{n!}\left[(-1)^n + 3^n\right],$$

so that $u_n = (1/n!)\left[(-1)^n + 3^n\right]$ is the solution of (11.41). •

The *nonhomogeneous case* (11.28) can be tackled by searching for solutions of the form

$$u_n = u_n^{(0)} + u_n^{(\varphi)},$$

where $u_n^{(0)}$ is the solution of the associated homogeneous equation and $u_n^{(\varphi)}$ is a particular solution of the nonhomogeneous equation. Once the solution of the homogeneous equation is available, a general technique to obtain the solution of the nonhomogeneous equation is based on the method of variation of parameters, combined with a reduction of the order of the difference equation (see [BO78]).

In the special case of difference equations with constant coefficients, with φ_{n+k} of the form $c^n Q(n)$, where c is a constant and Q is a polynomial of degree

p with respect to the variable n, a possible approach is that of *undetermined coefficients*, where one looks for a particular solution that depends on some undetermined constants and has a known form for some classes of right sides φ_{n+k}. It suffices to look for a particular solution of the form

$$u_n^{(\varphi)} = c^n(b_p n^p + b_{p-1}n^{p-1} + \ldots + b_0),$$

where $b_p, \ldots, b_0$ are constants to be determined in such a way that $u_n^{(\varphi)}$ is actually a solution of (11.28).

Example 11.7 Consider the difference equation $u_{n+3} - u_{n+2} + u_{n+1} - u_n = 2^n n^2$. The particular solution is of the form $u_n = 2^n(b_2 n^2 + b_1 n + b_0)$. Substituting this solution into the equation, we find $5b_2 n^2 + (36b_2 + 5b_1)n + (58b_2 + 18b_1 + 5b_0) = n^2$, from which, recalling the principle of identity for polynomials, one gets $b_2 = 1/5$, $b_1 = -36/25$ and $b_0 = 358/125$. •

Analogous to the homogeneous case, it is possible to express the solution of (11.28) as

$$u_n = \sum_{j=0}^{k-1} u_j \psi_j^{(n)} + \sum_{l=k}^{n} \varphi_l \psi_{k-1}^{(n-l+k-1)}, \qquad n = 0, 1, \ldots, \tag{11.42}$$

where we define $\psi_{k-1}^{(i)} \equiv 0$ for all $i < 0$ and $\varphi_j \equiv 0$ for all $j < k$.

11.5 Multistep Methods

Let us now introduce some examples of multistep methods (shortly, MS).

Definition 11.7 (q-steps methods) A q-step method ($q \geq 1$) is a method for which $\forall n \geq q-1$, u_{n+1} depends on u_{n+1-q}, but not on the values u_k with $k < n+1-q$. ■

A well-known *two-step* explicit method can be obtained by using the centered finite difference (10.61) to approximate the first order derivative in (11.1). This yields the *midpoint method*

$$u_{n+1} = u_{n-1} + 2hf_n, \qquad n \geq 1, \tag{11.43}$$

where $u_0 = y_0$, u_1 is to be determined and f_k denotes the value $f(t_k, u_k)$.

An example of an implicit two-step scheme is the *Simpson method*, obtained from (11.2) with $t_0 = t_{n-1}$ and $t = t_{n+1}$ and by using the Cavalieri-Simpson quadrature rule to approximate the integral of f

$$u_{n+1} = u_{n-1} + \frac{h}{3}[f_{n-1} + 4f_n + f_{n+1}], \qquad n \geq 1, \tag{11.44}$$

where $u_0 = y_0$, and u_1 is to be determined.

From these examples, it is clear that a multistep method requires q initial values $u_0, \ldots, u_{q-1}$ for "taking off". Since the Cauchy problem provides only one datum (u_0), one way to assign the remaining values consists of resorting to explicit one-step methods of high order. An example is given by Heun's method (11.10), other examples are provided by the Runge-Kutta methods, which will be introduced in Section 11.8.

In this section we deal with *linear multistep methods*

$$u_{n+1} = \sum_{j=0}^{p} a_j u_{n-j} + h\sum_{j=0}^{p} b_j f_{n-j} + hb_{-1}f_{n+1}, \quad n = p, p+1, \ldots \quad (11.45)$$

which are $p+1$-step methods, $p \geq 0$. For $p = 0$, we recover one-step methods.

The coefficients a_j, b_j are real and fully identify the scheme; they are such that $a_p \neq 0$ or $b_p \neq 0$. If $b_{-1} \neq 0$ the scheme is implicit, otherwise it is explicit. Also for MS methods we can characterize consistency in terms of the local truncation error, according to the following definition.

Definition 11.8 The *local truncation error* (LTE) $\tau_{n+1}(h)$ introduced by the multistep method (11.45) at t_{n+1} (for $n \geq p$) is defined through the following relation

$$h\tau_{n+1}(h) = y_{n+1} - \left[\sum_{j=0}^{p} a_j y_{n-j} + h \sum_{j=-1}^{p} b_j y'_{n-j}\right], \qquad n \geq p, \quad (11.46)$$

where $y_{n-j} = y(t_{n-j})$ and $y'_{n-j} = y'(t_{n-j})$ for $j = -1, \ldots, p$. ■

Analogous to one-step methods, the quantity $h\tau_{n+1}(h)$ is the residual generated at t_{n+1} if we pretend that the exact solution "satisfies" the numerical scheme. Letting $\tau(h) = \max_n |\tau_n(h)|$, we have the following definition.

Definition 11.9 (Consistency) The multistep method (11.45) is *consistent* if $\tau(h) \to 0$ as $h \to 0$. Moreover, if $\tau(h) = \mathcal{O}(h^q)$, for some $q \geq 1$, then the method is said to have *order* q. ■

A more precise characterization of the LTE can be given by introducing the following linear operator $\mathcal{L}$ associated with the linear MS method (11.45)

$$\mathcal{L}[w(t); h] = w(t+h) - \sum_{j=0}^{p} a_j w(t - jh) - h \sum_{j=-1}^{p} b_j w'(t - jh), \quad (11.47)$$

where $w \in C^1(I)$ is an arbitrary function. Notice that the LTE is exactly $\mathcal{L}[y(t_n); h]$. If we assume that w is sufficiently smooth and expand $w(t - jh)$ and $w'(t - jh)$ about $t - ph$, we obtain

$$\mathcal{L}[w(t);h] = C_0 w(t-ph) + C_1 h w^{(1)}(t-ph) + \ldots + C_k h^k w^{(k)}(t-ph) + \ldots$$

Consequently, if the MS method has order q and $y \in C^{q+1}(I)$, we obtain

$$\tau_{n+1}(h) = C_{q+1} h^{q+1} y^{(q+1)}(t_{n-p}) + \mathcal{O}(h^{q+2}).$$

The term $C_{q+1}h^{q+1}y^{(q+1)}(t_{n-p})$ is the so-called *principal local truncation error* (PLTE) while C_{q+1} is the *error constant.* The PLTE is widely employed in devising adaptive strategies for MS methods (see [Lam91], Chapter 3).

Program 92 provides an implementation of the multistep method in the form (11.45) for the solution of a Cauchy problem on the interval (t_0, T). The input parameters are: the column vector `a` containing the $p+1$ coefficients a_i; the column vector `b` containing the $p+2$ coefficients b_i; the discretization stepsize h; the vector of initial data `u0` at the corresponding time instants `t0`; the macros `fun` and `dfun` containing the functions f and $\partial f/\partial y$. If the MS method is implicit, a tolerance `tol` and a maximum number of admissible iterations `itmax` must be provided. These two parameters monitor the convergence of Newton's method that is employed to solve the nonlinear equation (11.45) associated with the MS method. In output the code returns the vectors `u` and `t` containing the computed solution at the time instants `t`.

Program 92 - multistep : Linear multistep methods

```
function [t,u]=multistep(a,b,tf,t0,u0,h,fun,dfun,tol,itmax)
%MULTISTEP Multistep method.
%  [T,U]=MULTISTEP(A,B,TF,T0,U0,H,FUN,DFUN,TOL,ITMAX) solves the
%  Cauchy problem Y'=FUN(T,Y) for T in (T0,TF) using a multistep method
%  with coefficients A and B. H specifies the time step. TOL specifies the
%  tolerance of the fixed-point iteration when the selected multistep method
%  is of implicit type.
y = u0;  t = t0;  f = eval (fun); p = length(a) - 1; u = u0;
nt = fix((tf - t0 (1))/h);
for k = 1:nt
   lu=length(u);
   G=a'*u(lu:-1:lu-p)+ h*b(2:p+2)'*f(lu:-1:lu-p);
   lt=length(t0);
   t0=[t0; t0(lt)+h];
   unew=u(lu);
   t=t0(lt+1); err=tol+1; it=0;
   while err>tol & it<=itmax
      y=unew;
      den=1-h*b(1)*eval(dfun);
      fnew=eval(fun);
      if den == 0
         it=itmax+1;
      else
         it=it+1;
         unew=unew-(unew-G-h*b(1)* fnew)/den;
```

```
      err=abs(unew-y);
    end
  end
  u=[u; unew]; f=[f; fnew];
end
t=t0;
return
```

In the forthcoming sections we examine some families of multistep methods.

11.5.1 Adams Methods

These methods are derived from the integral form (11.2) through an approximate evaluation of the integral of f between t_n and t_{n+1}. We suppose that the discretization nodes are equally spaced, i.e., $t_j = t_0 + jh$, with $h > 0$ and $j \geq 1$, and then we integrate, instead of f, its interpolating polynomial on $\tilde{p} + \theta$ distinct nodes, where $\theta = 1$ if the methods are explicit ($\tilde{p} \geq 0$ in this case) and $\theta = 2$ if the methods are implicit ($\tilde{p} \geq -1$). The resulting schemes are thus *consistent* by construction and have the following form

$$u_{n+1} = u_n + h \sum_{j=-1}^{\tilde{p}+\theta} b_j f_{n-j}. \tag{11.48}$$

The interpolation nodes can be either:

1. $t_n, t_{n-1}, \ldots, t_{n-\tilde{p}}$ (in this case $b_{-1} = 0$ and the resulting method is explicit);

   or
2. $t_{n+1}, t_n, \ldots, t_{n-\tilde{p}}$ (in this case $b_{-1} \neq 0$ and the scheme is implicit).

The *implicit* schemes are called *Adams-Moulton* methods, while the *explicit* ones are called *Adams-Bashforth* methods.

Adams-Bashforth methods (AB)

Taking $\tilde{p} = 0$ we recover the forward Euler method, since the interpolating polynomial of degree zero at node t_n is given by $\Pi_0 f = f_n$. For $\tilde{p} = 1$, the linear interpolating polynomial at the nodes t_{n-1} and t_n is

$$\Pi_1 f(t) = f_n + (t - t_n)\frac{f_{n-1} - f_n}{t_{n-1} - t_n}.$$

Since $\Pi_1 f(t_n) = f_n$ and $\Pi_1 f(t_{n+1}) = 2f_n - f_{n-1}$, we get

$$\int_{t_n}^{t_{n+1}} \Pi_1 f(t) = \frac{h}{2}\left[\Pi_1 f(t_n) + \Pi_1 f(t_{n+1})\right] = \frac{h}{2}\left[3f_n - f_{n-1}\right].$$

Therefore, the two-step AB method is

Table 11.1. Error constants for Adams-Bashforth methods (C^*_{q+1}) and Adams-Moulton methods (C_{q+1}) of order q

q	C^*_{q+1}	C_{q+1}	q	C^*_{q+1}	C_{q+1}
1	$\frac{1}{2}$	$-\frac{1}{2}$	3	$\frac{3}{8}$	$-\frac{1}{24}$
2	$\frac{5}{12}$	$-\frac{1}{12}$	4	$\frac{251}{720}$	$-\frac{19}{720}$

$$u_{n+1} = u_n + \frac{h}{2}\left[3f_n - f_{n-1}\right]. \tag{11.49}$$

With a similar procedure, if $\tilde{p} = 2$, we find the three-step AB method

$$u_{n+1} = u_n + \frac{h}{12}\left[23f_n - 16f_{n-1} + 5f_{n-2}\right],$$

while for $\tilde{p} = 3$ we get the four-step AB scheme

$$u_{n+1} = u_n + \frac{h}{24}\left(55f_n - 59f_{n-1} + 37f_{n-2} - 9f_{n-3}\right).$$

Note that the Adams-Bashforth methods use $\tilde{p}+1$ nodes and are $\tilde{p}+1$-step methods (with $\tilde{p} \geq 0$). In general, q-step Adams-Bashforth methods have order q. The error constants C^*_{q+1} of these methods are collected in Table 11.1.

Adams-Moulton methods (AM)

If $\tilde{p} = -1$, the Backward Euler scheme is recovered, while if $\tilde{p} = 0$, we construct the linear polynomial interpolating f at the nodes t_n and t_{n+1} to recover the Crank-Nicolson scheme (11.9). In the case of the two-step method ($\tilde{p} = 1$), the polynomial of degree 2 interpolating f at the nodes t_{n-1}, t_n, t_{n+1} is generated, yielding the following scheme

$$u_{n+1} = u_n + \frac{h}{12}\left[5f_{n+1} + 8f_n - f_{n-1}\right]. \tag{11.50}$$

The methods corresponding to $\tilde{p} = 2$ and $\tilde{p} = 3$ are respectively given by

$$u_{n+1} = u_n + \frac{h}{24}\left(9f_{n+1} + 19f_n - 5f_{n-1} + f_{n-2}\right),$$

$$u_{n+1} = u_n + \frac{h}{720}\left(251f_{n+1} + 646f_n - 264f_{n-1} + 106f_{n-2} - 19f_{n-3}\right).$$

The Adams-Moulton methods use $\tilde{p}+2$ nodes and are $\tilde{p}+1$-steps methods if $\tilde{p} \geq 0$, the only exception being the Backward Euler scheme (corresponding to $\tilde{p} = -1$) which uses one node and is a one-step method. In general, the q-steps Adams-Moulton methods have order $q+1$ (the only exception being again the Backward Euler scheme which is a one-step method of order one) and their error constants C_{q+1} are summarized in Table 11.1.

Table 11.2. Coefficients of zero-stable BDF methods for $p = 0, 1, \ldots, 5$

p	a_0	a_1	a_2	a_3	a_4	a_5	b_{-1}
0	1	0	0	0	0	0	1
1	$\frac{4}{3}$	$-\frac{1}{3}$	0	0	0	0	$\frac{2}{3}$
2	$\frac{18}{11}$	$-\frac{9}{11}$	$\frac{2}{11}$	0	0	0	$\frac{6}{11}$
3	$\frac{48}{25}$	$-\frac{36}{25}$	$\frac{16}{25}$	$-\frac{3}{25}$	0	0	$\frac{12}{25}$
4	$\frac{300}{137}$	$-\frac{300}{137}$	$\frac{200}{137}$	$-\frac{75}{137}$	$\frac{12}{137}$	0	$\frac{60}{137}$
5	$\frac{360}{147}$	$-\frac{450}{147}$	$\frac{400}{147}$	$-\frac{225}{147}$	$\frac{72}{147}$	$-\frac{10}{147}$	$\frac{60}{147}$

11.5.2 BDF Methods

The so-called *backward differentiation formulae* (henceforth denoted by BDF) are implicit MS methods derived from a complementary approach to the one followed for the Adams methods. In fact, for the Adams methods we have resorted to numerical integration for the source function f, whereas in BDF methods we directly approximate the value of the first derivative of y at node t_{n+1} through the first derivative of the polynomial of degree $p + 1$ interpolating y at the $p + 2$ nodes $t_{n+1}, t_n, \ldots, t_{n-p}$ with $p \geq 0$.

By doing so, we get schemes of the form

$$u_{n+1} = \sum_{j=0}^{p} a_j u_{n-j} + h b_{-1} f_{n+1},$$

with $b_{-1} \neq 0$. Method (11.8) represents the most elementary example, corresponding to the coefficients $a_0 = 1$ and $b_{-1} = 1$.

We summarize in Table 11.2 the coefficients of BDF methods that are zero-stable. In fact, we shall see in Section 11.6.3 that only for $p \leq 5$ are BDF methods zero-stable (see [Cry73]).

11.6 Analysis of Multistep Methods

Analogous to what has been done for one-step methods, in this section we provide algebraic conditions that ensure consistency and stability of multistep methods.

11.6.1 Consistency

The property of consistency of a multistep method introduced in Definition 11.9 can be verified by checking that the coefficients satisfy certain algebraic equations, as stated in the following theorem.

Theorem 11.3 *The multistep method* (11.45) *is consistent iff the following algebraic relations among the coefficients are satisfied*

$$\sum_{j=0}^{p} a_j = 1, \ -\sum_{j=0}^{p} j a_j + \sum_{j=-1}^{p} b_j = 1. \tag{11.51}$$

Moreover, if $y \in C^{q+1}(I)$ for some $q \geq 1$, where y is the solution of the Cauchy problem (11.1)*, then the method is of order q iff* (11.51) *holds and the following additional conditions are satisfied*

$$\sum_{j=0}^{p} (-j)^i a_j + i \sum_{j=-1}^{p} (-j)^{i-1} b_j = 1, \ i = 1, \ldots, q.$$

(Note that if $q = 1$ this reduces to the second condition of (11.51)*.)*

Proof. Expanding y and f in a Taylor series yields, for any $n \geq p$

$$y_{n-j} = y_n - jhy'_n + \mathcal{O}(h^2), \qquad f(t_{n-j}, y_{n-j}) = f(t_n, y_n) + \mathcal{O}(h). \tag{11.52}$$

Plugging these values back into the multistep scheme and neglecting the terms in h of order higher than 1 gives

$$\begin{aligned}
&y_{n+1} - \sum_{j=0}^{p} a_j y_{n-j} - h \sum_{j=-1}^{p} b_j f(t_{n-j}, y_{n-j}) \\
&= y_{n+1} - \sum_{j=0}^{p} a_j y_n + h \sum_{j=0}^{p} j a_j y'_n - h \sum_{j=-1}^{p} b_j f(t_n, y_n) - \mathcal{O}(h^2) \left(\sum_{j=0}^{p} a_j - \sum_{j=-1}^{p} b_j \right) \\
&= y_{n+1} - \sum_{j=0}^{p} a_j y_n - hy'_n \left(-\sum_{j=0}^{p} j a_j + \sum_{j=-1}^{p} b_j \right) - \mathcal{O}(h^2) \left(\sum_{j=0}^{p} a_j - \sum_{j=-1}^{p} b_j \right),
\end{aligned}$$

where we have replaced y'_n by $f(t_n, y_n)$. From the definition (11.46) we then obtain

$$h\tau_{n+1}(h) = y_{n+1} - \sum_{j=0}^{p} a_j y_n - hy'_n \left(-\sum_{j=0}^{p} j a_j + \sum_{j=-1}^{p} b_j \right) - \mathcal{O}(h^2) \left(\sum_{j=0}^{p} a_j - \sum_{j=-1}^{p} b_j \right),$$

from which the local truncation error is

$$\begin{aligned}
\tau_{n+1}(h) = {} & \frac{y_{n+1} - y_n}{h} + \frac{y_n}{h} \left(1 - \sum_{j=0}^{p} a_j \right) \\
& + y'_n \left(\sum_{j=0}^{p} j a_j - \sum_{j=-1}^{p} b_j \right) - \mathcal{O}(h) \left(\sum_{j=0}^{p} a_j - \sum_{j=-1}^{p} b_j \right).
\end{aligned}$$

Since, for any n, $(y_{n+1} - y_n)/h \to y'_n$, as $h \to 0$, it follows that $\tau_{n+1}(h)$ tends to 0 as h goes to 0 iff the algebraic conditions (11.51) are satisfied. The rest of the proof can be carried out in a similar manner, accounting for terms of progressively higher order in the expansions (11.52). ◇

11.6.2 The Root Conditions

Let us employ the multistep method (11.45) to solve the model problem (11.24). The numerical solution satisfies the linear difference equation

$$u_{n+1} = \sum_{j=0}^{p} a_j u_{n-j} + h\lambda \sum_{j=-1}^{p} b_j u_{n-j}, \tag{11.53}$$

which fits the form (11.29). We can therefore apply the theory developed in Section 11.4 and look for fundamental solutions of the form $u_k = [r_i(h\lambda)]^k$, $k = 0, 1, \ldots$, where $r_i(h\lambda)$, for $i = 0, \ldots, p$, are the roots of the polynomial $\Pi \in \mathbb{P}_{p+1}$

$$\Pi(r) = \rho(r) - h\lambda\sigma(r). \tag{11.54}$$

We have denoted by

$$\rho(r) = r^{p+1} - \sum_{j=0}^{p} a_j r^{p-j}, \quad \sigma(r) = b_{-1} r^{p+1} + \sum_{j=0}^{p} b_j r^{p-j}$$

the *first* and *second characteristic polynomials* of the multistep method (11.45), respectively. The polynomial $\Pi(r)$ is the *characteristic polynomial* associated with the difference equation (11.53), and $r_j(h\lambda)$ are its *characteristic roots.*

The roots of ρ are $r_i(0)$, $i = 0, \ldots, p$, and will be abbreviated henceforth by r_i. From the first condition in (11.51) it follows that if a multistep method is consistent then 1 is a root of ρ. We shall assume that such a root (the consistency root) is labelled as $r_0(0) = r_0$ and call the corresponding root $r_0(h\lambda)$ the *principal root.*

Definition 11.10 (Root condition) The multistep method (11.45) is said to satisfy the root condition if all roots r_i are contained within the unit circle centered at the origin of the complex plane, otherwise, if they fall on its boundary, they must be simple roots of ρ. Equivalently,

$$\begin{cases} |r_j| \leq 1, & j = 0, \ldots, p; \\ \text{furthermore, for those } j \text{ such that } |r_j| = 1, \text{ then } \rho'(r_j) \neq 0. \end{cases} \tag{11.55}$$

■

Definition 11.11 (Strong root condition) The multistep method (11.45) is said to satisfy the strong root condition if it satisfies the root condition and $r_0 = 1$ is the only root lying on the boundary of the unit circle. Equivalently,

$$|r_j| < 1, \qquad j = 1, \ldots, p. \tag{11.56}$$

■

Definition 11.12 (Absolute root condition) The multistep method (11.45) satisfies the absolute root condition if there exists $h_0 > 0$ such that

$$|r_j(h\lambda)| < 1, \qquad j = 0, \ldots, p, \quad \forall h \leq h_0.$$

■

11.6.3 Stability and Convergence Analysis for Multistep Methods

Let us now examine the relation between root conditions and the stability of multistep methods. Generalizing the Definition 11.4, we can get the following definition of zero-stability.

Definition 11.13 (Zero-stability of multistep methods) The multistep method (11.45) is zero-stable if $\exists h_0 > 0$, $\exists C > 0$ such that $\forall h \in (0, h_0]$, $\forall \varepsilon > 0$ sufficiently small, if $|\delta_k| \leq \varepsilon$, $0 \leq k \leq N_h$, then

$$|z_n^{(h)} - u_n^{(h)}| \leq C\varepsilon, \qquad 0 \leq n \leq N_h, \tag{11.57}$$

where $N_h = \max\{n : t_n \leq t_0 + T\}$ and $z_n^{(h)}$ and $u_n^{(h)}$ are, respectively, the solutions of problems

$$\begin{cases} z_{n+1}^{(h)} = \displaystyle\sum_{j=0}^{p} a_j z_{n-j}^{(h)} + h \sum_{j=-1}^{p} b_j f(t_{n-j}, z_{n-j}^{(h)}) + h\delta_{n+1}, \\ z_k^{(h)} = w_k^{(h)} + \delta_k, \qquad k = 0, \ldots, p \end{cases} \tag{11.58}$$

$$\begin{cases} u_{n+1}^{(h)} = \displaystyle\sum_{j=0}^{p} a_j u_{n-j}^{(h)} + h \sum_{j=-1}^{p} b_j f(t_{n-j}, u_{n-j}^{(h)}), \\ u_k^{(h)} = w_k^{(h)}, \qquad k = 0, \ldots, p \end{cases} \tag{11.59}$$

for $p \leq n \leq N_h - 1$, where $w_0^{(h)} = y_0$ and $w_k^{(h)}$, $k = 1, \ldots, p$, are p initial values generated by using another numerical scheme. ■

Theorem 11.4 (Equivalence of zero-stability and root condition) *For a consistent multistep method, the root condition is equivalent to zero-stability.*

Proof. Let us begin by proving that the root condition is *necessary* for the zero-stability to hold. We proceed by contradiction and assume that the method is zero-stable and there exists a root r_i which violates the root condition.

Since the method is zero-stable, condition (11.57) must be satisfied for *any* Cauchy problem, in particular for the problem $y'(t) = 0$ with $y(0) = 0$, whose solution is, clearly, the null function. Similarly, the solution $u_n^{(h)}$ of (11.59) with $f = 0$ and $w_k^{(h)} = 0$ for $k = 0, \ldots, p$ is identically zero.

Consider first the case $|r_i| > 1$. Then, define

$$\delta_n = \begin{cases} \varepsilon r_i^n & \text{if } r_i \in \mathbb{R}, \\ \varepsilon (r_i + \bar{r}_i)^n & \text{if } r_i \in \mathbb{C}, \end{cases}$$

for $\varepsilon > 0$. It is simple to check that the sequence $z_n^{(h)} = \delta_n$ for $n = 0, 1, \ldots$ is a solution of (11.58) with initial conditions $z_k^{(h)} = \delta_k$ and that $|\delta_k| \leq \varepsilon$ for $k = 0, 1, \ldots, p$. Let us now choose $\bar{t}$ in $(t_0, t_0 + T)$ and let x_n be the nearest grid node to $\bar{t}$. Clearly, n is the integral part of $\bar{t}/h$ and $\lim_{h\to 0} |z_n^{(h)}| = \lim_{h \to 0} |u_n^{(h)} - z_n^{(h)}| \to \infty$ as $h \to 0$. This proves that $|u_n^{(h)} - z_n^{(h)}|$ cannot be uniformly bounded with respect to h as $h \to 0$, which contradicts the assumption that the method is zero-stable.

A similar proof can be carried out if $|r_i| = 1$ but has multiplicity greater than 1, provided that one takes into account the form of the solution (11.33).

Let us now prove that the root condition is sufficient for method (11.45) to be zero-stable. With this aim, it is convenient to reformulate (11.45) as follows

$$\sum_{s=0}^{p+1} \alpha_s u_{n+s} = h \sum_{s=0}^{p+1} \beta_s f(t_{n+s}, u_{n+s}), \; n = 0, 1, \ldots, N_h - (p+1) \tag{11.60}$$

having set $\alpha_{p+1} = 1$, $\alpha_s = -a_{p-s}$ for $s = 0, \ldots, p$ and $\beta_s = b_{p-s}$ for $s = 0, \ldots, p+1$. Relation (11.60) is a special instance of the linear difference equation (11.28), where we set $k = p + 1$ and $\varphi_{n+j} = h\beta_j f(t_{n+j}, u_{n+j})$, for $j = 0, \ldots, p+1$. Using (11.60) and denoting by $z_{n+j}^{(h)}$ and $u_{n+j}^{(h)}$ the solutions to (11.58) and (11.59), respectively, for $j \geq 1$, it turns out that the function $w_{n+j}^{(h)} = z_{n+j}^{(h)} - u_{n+j}^{(h)}$ satisfies the following difference equation

$$\sum_{j=0}^{p+1} \alpha_j w_{n+j}^{(h)} = \varphi_{n+p+1}, \; n = 0, \ldots, N_h - (p+1), \tag{11.61}$$

having set

$$\varphi_{n+p+1} = h \sum_{j=0}^{p+1} \beta_j \left[f(t_{n+j}, z_{n+j}^{(h)}) - f(t_{n+j}, u_{n+j}^{(h)}) \right] + h\delta_{n+p+1}. \tag{11.62}$$

Denote by $\left\{ \psi_j^{(n)} \right\}$ a sequence of fundamental solutions to the homogeneous equation associated with (11.61). Recalling (11.42), the general solution of (11.61) is given by

$$w_n^{(h)} = \sum_{j=0}^{p} w_j^{(h)} \psi_j^{(n)} + \sum_{l=p+1}^{n} \psi_p^{(n-l+p)} \varphi_l, \; n = p+1, \ldots$$

The following result expresses the connection between the root condition and the boundedness of the solution of a difference equation (for the proof, see [Gau97], Theorem 6.3.2).

Lemma 11.3 *For any solution* $\{u_n\}$ *of the difference equation* (11.28) *there exists a constant* $M > 0$ *such that*

$$|u_n| \leq M \left\{ \max_{j=0,\ldots,k-1} |u_j| + \sum_{l=k}^{n} |\varphi_l| \right\}, \; n = 0, 1, \ldots \tag{11.63}$$

iff the root condition is satisfied for the polynomial (11.30), *i.e.*, (11.55) *holds for the zeros of the polynomial* (11.30).

Let us now recall that, for any j, $\{\psi_j^{(n)}\}$ is solution of a homogeneous difference equation whose initial data are $\psi_j^{(i)} = \delta_{ij}$, $i, j = 0, \ldots, p$. On the other hand, for any l, $\psi_p^{(n-l+p)}$ is solution of a difference equation with zero initial conditions and right-hand sides equal to zero except for the one corresponding to $n = l$ which is $\psi_p^{(p)} = 1$.

Therefore, Lemma 11.3 can be applied in both cases so we can conclude that there exists a constant $M > 0$ such that $|\psi_j^{(n)}| \leq M$ and $|\psi_p^{(n-l+p)}| \leq M$, uniformly with respect to n and l. The following estimate thus holds

$$|w_n^{(h)}| \leq M \left\{ (p+1) \max_{j=0,\ldots,p} |w_j^{(h)}| + \sum_{l=p+1}^{n} |\varphi_l| \right\}, \quad n = 0, 1, \ldots, N_h. \tag{11.64}$$

If L denotes the Lipschitz constant of f, from (11.62) we get

$$|\varphi_{n+p+1}| \leq h \max_{j=0,\ldots,p+1} |\beta_j| L \sum_{j=0}^{p+1} |w_{n+j}^{(h)}| + h|\delta_{n+p+1}|.$$

Let $\beta = \max_{j=0,\ldots,p+1} |\beta_j|$ and $\Delta_{[q,r]} = \max_{j=q,\ldots,r} |\delta_{j+q}|$, q and r being some integers with $q \leq r$. From (11.64), the following estimate is therefore obtained

$$\begin{aligned} |w_n^{(h)}| &\leq M \left\{ (p+1)\Delta_{[0,p]} + h\beta L \sum_{l=p+1}^{n} \sum_{j=0}^{p+1} |w_{l-p-1+j}^{(h)}| + N_h h \Delta_{[p+1,n]} \right\} \\ &\leq M \left\{ (p+1)\Delta_{[0,p]} + h\beta L(p+2) \sum_{m=0}^{n} |w_m^{(h)}| + T\Delta_{[p+1,n]} \right\}. \end{aligned}$$

Let $Q = 2(p+2)\beta LM$ and $h_0 = 1/Q$, so that $1 - h\frac{Q}{2} \geq \frac{1}{2}$ if $h \leq h_0$. Then

$$\begin{aligned} \frac{1}{2}|w_n^{(h)}| &\leq |w_n^{(h)}|(1 - h\tfrac{Q}{2}) \\ &\leq M \left\{ (p+1)\Delta_{[0,p]} + h\beta L(p+2) \sum_{m=0}^{n-1} |w_m^{(h)}| + T\Delta_{[p+1,n]} \right\}. \end{aligned}$$

Letting $R = 2M \left\{ (p+1)\Delta_{[0,p]} + T\Delta_{[p+1,n]} \right\}$, we finally obtain

$$|w_n^{(h)}| \leq hQ \sum_{m=0}^{n-1} |w_m^{(h)}| + R.$$

Applying Lemma 11.2 with the following identifications: $\varphi_n = |w_n^{(h)}|$, $g_0 = R$, $p_s = 0$ and $k_s = hQ$ for any $s = 0, \ldots, n-1$, yields

$$|w_n^{(h)}| \leq 2Me^{TQ} \left\{ (p+1)\Delta_{[0,p]} + T\Delta_{[p+1,n]} \right\}. \tag{11.65}$$

Method (11.45) is thus zero-stable for any $h \leq h_0$. ◇

Theorem 11.4 allows for characterizing the stability behavior of several families of discretization methods.

In the special case of consistent one-step methods, the polynomial ρ admits only the root $r_0 = 1$. They thus *automatically satisfy the root condition* and are zero-stable.

For the Adams methods (11.48), the polynomial ρ is always of the form $\rho(r) = r^{p+1} - r^p$. Thus, its roots are $r_0 = 1$ and $r_1 = 0$ (with multiplicity p) so that all Adams methods are zero-stable.

Also the midpoint method (11.43) and Simpson method (11.44) are zero-stable: for both of them, the first characteristic polynomial is $\rho(r) = r^2 - 1$, so that $r_0 = 1$ and $r_1 = -1$.

Finally, the BDF methods of Section 11.5.2 are zero-stable provided that $p \leq 5$, since in such a case the root condition is satisfied (see [Cry73]).
We are in position to give the following convergence result.

Theorem 11.5 (Convergence) *A consistent multistep method is convergent iff it satisfies the root condition and the error on the initial data tends to zero as $h \to 0$. Moreover, the method converges with order q if it has order q and the error on the initial data tends to zero as $\mathcal{O}(h^q)$.*

Proof. Suppose that the MS method is consistent and convergent. To prove that the root condition is satisfied, we refer to the problem $y'(t) = 0$ with $y(0) = 0$ and on the interval $I = (0, T)$. Convergence means that the numerical solution $\{u_n\}$ must tend to the exact solution $y(t) = 0$ for any converging set of initial data u_k, $k = 0, \ldots, p$, i.e. $\max_{k=0,\ldots,p} |u_k| \to 0$ as $h \to 0$. From this observation, the proof follows by contradiction along the same lines as the proof of Theorem 11.4, where the parameter ε is now replaced by h.

Let us now prove that consistency, together with the root condition, implies convergence under the assumption that the error on the initial data tends to zero as $h \to 0$. We can apply Theorem 11.4, setting $u_n^{(h)} = u_n$ (approximate solution of the Cauchy problem) and $z_n^{(h)} = y_n$ (exact solution), and from (11.46) it turns out that $\delta_m = \tau_m(h)$. Then, due to (11.65), for any $n \geq p+1$ we obtain

$$|u_n - y_n| \leq 2Me^{TQ} \left\{ (p+1) \max_{j=0,\ldots,p} |u_j - y_j| + T \max_{j=p+1,\ldots,n} |\tau_j(h)| \right\}.$$

Convergence holds by noticing that the right-hand side of this inequality tends to zero as $h \to 0$. ◇

A remarkable consequence of the above theorem is the following equivalence Lax-Richtmyer theorem.

Corollary 11.1 (Equivalence theorem) *A consistent multistep method is convergent iff it is zero-stable and if the error on the initial data tends to zero as h tends to zero.*

We conclude this section with the following result, which establishes an upper limit for the order of multistep methods (see [Dah63]).

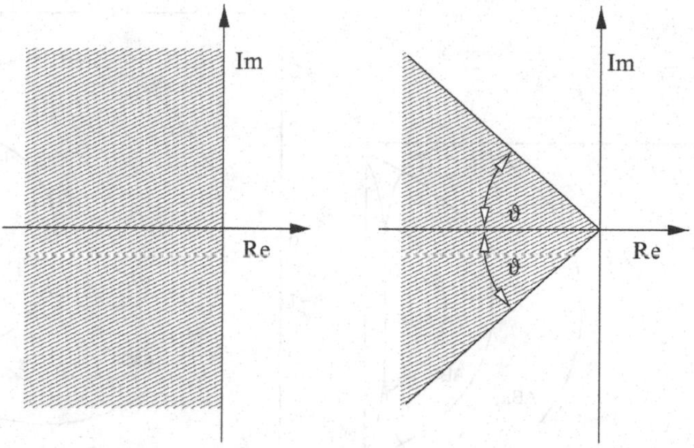

Fig. 11.4. Regions of absolute stability for A-stable (*left*) and ϑ-stable methods (*right*)

Property 11.1 (First Dahlquist barrier) *There isn't any zero-stable, q-step linear multistep method with order greater than $q+1$ if q is odd, $q+2$ if q is even.*

11.6.4 Absolute Stability of Multistep Methods

Consider again the difference equation (11.53), which was obtained by applying the MS method (11.45) to the model problem (11.24). According to (11.33), its solution takes the form

$$u_n = \sum_{j=1}^{k'} \left(\sum_{s=0}^{m_j-1} \gamma_{sj} n^s \right) [r_j(h\lambda)]^n, \qquad n = 0, 1, \ldots,$$

where $r_j(h\lambda)$, $j = 1, \ldots, k'$, are the distinct roots of the characteristic polynomial (11.54), and having denoted by m_j the multiplicity of $r_j(h\lambda)$. In view of (11.25), it is clear that the *absolute root condition* introduced by Definition 11.12 is necessary and sufficient to ensure that the multistep method (11.45) is absolutely stable as $h \le h_0$.

Among the methods enjoying the absolute stability property, the preference should go to those for which the region of absolute stability $\mathcal{A}$, introduced in (11.26), is as wide as possible or even unbounded. Among these are the *A-stable* methods introduced at the end of Section 11.3.3 and the *ϑ-stable* methods; the latter are those for which $\mathcal{A}$ contains the angular region defined by $z \in \mathbb{C}$ such that $-\vartheta < \pi - \arg(z) < \vartheta$, with $\vartheta \in (0, \pi/2)$. A-stable methods are of remarkable importance when solving *stiff* problems (see Section 11.10). The following result, whose proof is given in [Wid67], establishes a relation between the order of a multistep method, the number of its steps and its stability properties.

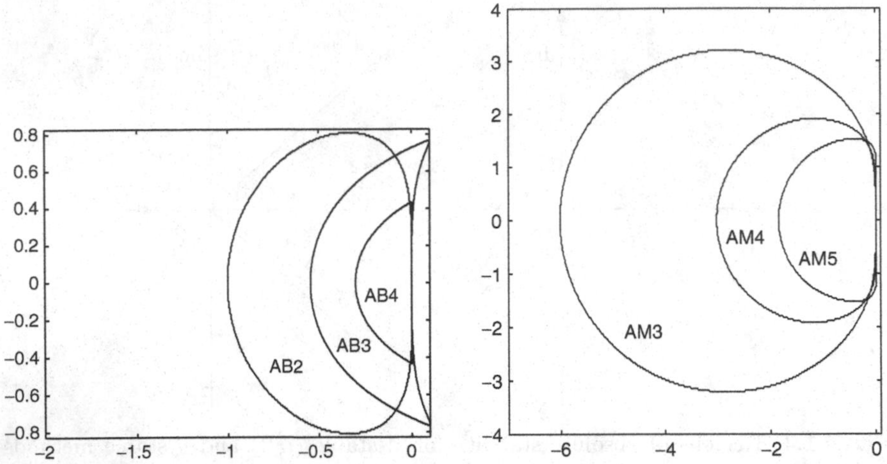

Fig. 11.5. Outer contours of the regions of absolute stability for Adams-Bashforth methods (*left*) ranging from second to fourth-order (AB2, AB3 and AB4) and for Adams-Moulton methods (*right*), from third to sixth-order (AM3, AM4 and AM5). Notice that the region of the AB3 method extends into the half-plane with positive real part. The region for the explicit Euler (AB1) method was drawn in Figure 11.3

Property 11.2 (Second Dahlquist barrier) *A linear explicit multistep method can be neither A-stable, nor ϑ-stable. Moreover, there is no A-stable linear multistep method with order greater than 2. Finally, for any $\vartheta \in (0, \pi/2)$, there only exist ϑ-stable q-step linear multistep methods of order q for $q = 3$ and $q = 4$.*

Let us now examine the region of absolute stability of several MS methods. The regions of absolute stability of both explicit and implicit Adams schemes reduce progressively as the order of the method increases. In Figure 11.5 (*left*) we show the regions of absolute stability for the AB methods examined in Section 11.5.1, with exception of the Forward Euler method whose region is shown in Figure 11.3.

The regions of absolute stability of the Adams-Moulton schemes, except for the Crank-Nicolson method which is A-stable, are represented in Figure 11.5 (*right*).

In Figure 11.6 the regions of absolute stability of some of the BDF methods introduced in Section 11.5.2 are drawn. They are unbounded and always contain the negative real numbers. These stability features make BDF methods quite attractive for solving *stiff* problems (see Section 11.10).

Remark 11.3 Some authors (see, e.g., [BD74]) adopt an alternative definition of absolute stability by replacing (11.25) with the milder property

$$\exists C > 0 \,:\, |u_n| \leq C, \text{ as } t_n \to +\infty.$$

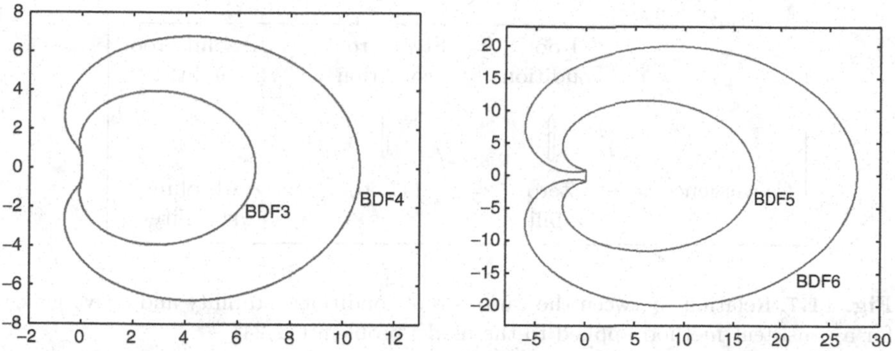

Fig. 11.6. Inner contours of regions of absolute stability for three- and four-step BDF methods (BDF3 and BDF4, *left*), and five- and six-step BDF methods (BDF5 and BDF6, *right*). Unlike Adams methods, these regions are unbounded and extend outside the limited portion that is shown in the figure

According to this new definition, the absolute stability of a numerical method should be regarded as the counterpart of the *asymptotic* stability (11.6) of the Cauchy problem. The new region of absolute stability $\mathcal{A}^*$ would be

$$\mathcal{A}^* = \{z \in \mathbb{C} : \exists C > 0, \ |u_n| \le C, \ \forall n \ge 0\}$$

and it would not necessarily coincide with $\mathcal{A}$. For example, in the case of the midpoint method $\mathcal{A}$ is empty (thus, it is unconditionally absolutely *unstable*), while $\mathcal{A}^* = \{z = \alpha i, \ \alpha \in [-1, 1]\}$.

In general, if $\mathcal{A}$ is nonempty, then $\mathcal{A}^*$ is its closure. We notice that *zero-stable methods are those for which the region $\mathcal{A}^*$ contains the origin $z = 0$ of the complex plane.* ■

To conclude, let us notice that the strong root condition (11.56) implies, for a linear problem, that

$$\forall h \le h_0, \ \exists C > 0 : |u_n| \le C(|u_0| + \ldots + |u_p|), \quad \forall n \ge p + 1. \quad (11.66)$$

We say that a method is *relatively stable* if it satisfies (11.66). Clearly, (11.66) implies zero-stability, but the converse does not hold.

Figure 11.7 summarizes the main conclusions drawn in this section about stability, convergence and root-conditions, in the particular case of a consistent method applied to the model problem (11.24).

11.7 Predictor-Corrector Methods

When solving a nonlinear Cauchy problem of the form (11.1), at each time step implicit schemes require dealing with a nonlinear equation. For instance,

$$
\begin{array}{ccccc}
\text{Root} & \Longleftarrow & \text{Strong root} & \Longleftarrow & \text{Absolute root} \\
\text{condition} & & \text{condition} & & \text{condition} \\
\Updownarrow & & \Downarrow & & \Updownarrow \\
\text{Zero} & \Longleftarrow & (11.66) & \Longleftarrow & \text{Absolute} \\
\text{stability} & & & & \text{stability}
\end{array}
$$

Convergence $\Longleftrightarrow$

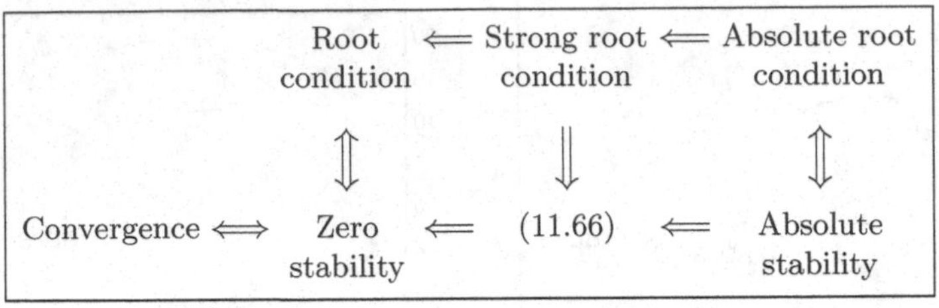

Fig. 11.7. Relations between the various root conditions, stability and convergence for a consistent method applied to the model problem (11.24)

if the Crank-Nicolson method is used, we get the nonlinear equation

$$
u_{n+1} = u_n + \frac{h}{2} [f_n + f_{n+1}] = \Psi(u_{n+1}),
$$

that can be cast in the form $\Phi(u_{n+1}) = 0$, where $\Phi(u_{n+1}) = u_{n+1} - \Psi(u_{n+1})$. To solve this equation the Newton method would give

$$
u_{n+1}^{(k+1)} = u_{n+1}^{(k)} - \Phi(u_{n+1}^{(k)})/\Phi'(u_{n+1}^{(k)}),
$$

for $k = 0, 1, \ldots$, until convergence and require an initial datum $u_{n+1}^{(0)}$ sufficiently close to u_{n+1}. Alternatively, one can resort to fixed-point iterations

$$
u_{n+1}^{(k+1)} = \Psi(u_{n+1}^{(k)}) \tag{11.67}
$$

for $k = 0, 1, \ldots$, until convergence. In such a case, the global convergence condition for the fixed-point method (see Theorem 6.1) sets a constraint on the discretization stepsize of the form

$$
h < \frac{1}{|b_{-1}|L}, \tag{11.68}
$$

where L is the Lipschitz constant of f with respect to y. In practice, except for the case of stiff problems (see Section 11.10), this restriction on h is not significant since considerations of accuracy put a much more restrictive constraint on h. However, each iteration of (11.67) requires one evaluation of the function f and the computational cost can be reduced by providing a good initial guess $u_{n+1}^{(0)}$. This can be done by taking one step of an explicit MS method and then iterating on (11.67) for a fixed number m of iterations. By doing so, the implicit MS method that is employed in the fixed-point scheme "corrects" the value of u_{n+1} "predicted" by the explicit MS method. A procedure of this type is called a *predictor-corrector method*, or PC method. There are many ways in which a predictor-corrector method can be implemented.

In its basic version, the value $u_{n+1}^{(0)}$ is computed by an explicit $\tilde{p}+1$-step method, called the *predictor* (here identified by the coefficients $\{\tilde{a}_j, \tilde{b}_j\}$)

$$[P]\ u_{n+1}^{(0)} = \sum_{j=0}^{\tilde{p}} \tilde{a}_j u_{n-j}^{(1)} + h\sum_{j=0}^{\tilde{p}} \tilde{b}_j f_{n-j}^{(0)},$$

where $f_k^{(0)} = f(t_k, u_k^{(0)})$ and $u_k^{(1)}$ are the solutions computed by the PC method at the previous steps or are the initial conditions. Then, we evaluate the function f at the new point $(t_{n+1}, u_{n+1}^{(0)})$ (*evaluation step*)

$$[E]\ f_{n+1}^{(0)} = f(t_{n+1}, u_{n+1}^{(0)}),$$

and finally, one single fixed-point iteration is carried out using an implicit MS scheme of the form (11.45)

$$[C]\ u_{n+1}^{(1)} = \sum_{j=0}^{p} a_j u_{n-j}^{(1)} + h\sum_{j=0}^{p} b_j f_{n-j}^{(0)} + hb_{-1} f_{n+1}^{(0)}.$$

This second step of the procedure, which is actually explicit, is called the *corrector*. The overall procedure is shortly denoted by PEC or $P(EC)^1$ method, in which P and C denote one application at time t_{n+1} of the predictor and the corrector methods, respectively, while E indicates one evaluation of the function f.

This strategy above can be generalized supposing to perform $m > 1$ iterations at each step t_{n+1}. The corresponding methods are called *predictor-multicorrector* schemes and compute $u_{n+1}^{(0)}$ at time step t_{n+1} using the predictor in the following form

$$[P]\ u_{n+1}^{(0)} = \sum_{j=0}^{\tilde{p}} \tilde{a}_j u_{n-j}^{(m)} + h\sum_{j=0}^{\tilde{p}} \tilde{b}_j f_{n-j}^{(m-1)}. \tag{11.69}$$

Here $m \geq 1$ denotes the (fixed) number of corrector iterations that are carried out in the following steps $[E]$, $[C]$: for $k = 0, 1, \ldots, m-1$

$$\begin{aligned}
[E] \quad f_{n+1}^{(k)} &= f(t_{n+1}, u_{n+1}^{(k)}),\\
[C] \quad u_{n+1}^{(k+1)} &= \sum_{j=0}^{p} a_j u_{n-j}^{(m)} + hb_{-1} f_{n+1}^{(k)} + h\sum_{j=0}^{p} b_j f_{n-j}^{(m-1)}.
\end{aligned}$$

These implementations of the predictor-corrector technique are referred to as $P(EC)^m$. Another implementation, denoted by $P(EC)^m E$, consists of updating at the end of the process also the function f and is given by

$$[P] \quad u_{n+1}^{(0)} = \sum_{j=0}^{\tilde{p}} \tilde{a}_j u_{n-j}^{(m)} + h\sum_{j=0}^{\tilde{p}} \tilde{b}_j f_{n-j}^{(m)},$$

and for $k = 0, 1, \ldots, m-1$,

$$[E] \quad f_{n+1}^{(k)} = f(t_{n+1}, u_{n+1}^{(k)}),$$

$$[C] \quad u_{n+1}^{(k+1)} = \sum_{j=0}^{p} a_j u_{n-j}^{(m)} + hb_{-1} f_{n+1}^{(k)} + h \sum_{j=0}^{p} b_j f_{n-j}^{(m)},$$

followed by

$$[E] \quad f_{n+1}^{(m)} = f(t_{n+1}, u_{n+1}^{(m)}).$$

Example 11.8 Heun's method (11.10) can be regarded as a predictor-corrector method whose predictor is the forward Euler method, while the corrector is the Crank-Nicolson method.

Another example is provided by the Adams-Bashforth method of order 2 (11.49) and the Adams-Moulton method of order 3 (11.50). Its corresponding *PEC* implementation is: given $u_0^{(0)} = u_0^{(1)} = u_0$, $u_1^{(0)} = u_1^{(1)} = u_1$ and $f_0^{(0)} = f(t_0, u_0^{(0)})$, $f_1^{(0)} = f(t_1, u_1^{(0)})$, compute for $n = 1, 2, \ldots$,

$$[P] \; u_{n+1}^{(0)} = u_n^{(1)} + \frac{h}{2}\left[3f_n^{(0)} - f_{n-1}^{(0)}\right],$$

$$[E] \; f_{n+1}^{(0)} = f(t_{n+1}, u_{n+1}^{(0)}),$$

$$[C] \; u_{n+1}^{(1)} = u_n^{(1)} + \frac{h}{12}\left[5f_{n+1}^{(0)} + 8f_n^{(0)} - f_{n-1}^{(0)}\right],$$

while the *PECE* implementation is: given $u_0^{(0)} = u_0^{(1)} = u_0$, $u_1^{(0)} = u_1^{(1)} = u_1$ and $f_0^{(1)} = f(t_0, u_0^{(1)})$, $f_1^{(1)} = f(t_1, u_1^{(1)})$, compute for $n = 1, 2, \ldots$,

$$[P] \; u_{n+1}^{(0)} = u_n^{(1)} + \frac{h}{2}\left[3f_n^{(1)} - f_{n-1}^{(1)}\right],$$

$$[E] \; f_{n+1}^{(0)} = f(t_{n+1}, u_{n+1}^{(0)}),$$

$$[C] \; u_{n+1}^{(1)} = u_n^{(1)} + \frac{h}{12}\left[5f_{n+1}^{(0)} + 8f_n^{(1)} - f_{n-1}^{(1)}\right],$$

$$[E] \; f_{n+1}^{(1)} = f(t_{n+1}, u_{n+1}^{(1)}).$$

•

Before studying the convergence of predictor-corrector methods, we introduce a simplification in the notation. Usually the number of steps of the predictor is greater than those of the corrector, so that we define the number of steps of the predictor-corrector pair as being equal to the number of steps of the predictor. This number will be denoted henceforth by p. Owing to this definition we no longer demand that the coefficients of the corrector satisfy $|a_p| + |b_p| \neq 0$. Consider for example the predictor-corrector pair

$$[P]\ u_{n+1}^{(0)} = u_n^{(1)} + hf(t_{n-1}, u_{n-1}^{(0)}),$$
$$[C]\ u_{n+1}^{(1)} = u_n^{(1)} + \tfrac{h}{2}\left[f(t_n, u_n^{(0)}) + f(t_{n+1}, u_{n+1}^{(0)})\right],$$

for which $p = 2$ (even though the corrector is a one-step method). Consequently, the first and the second characteristic polynomials of the corrector method will be $\rho(r) = r^2 - r$ and $\sigma(r) = (r^2 + r)/2$ instead of $\rho(r) = r - 1$ and $\sigma(r) = (r + 1)/2$.

In any predictor-corrector method, the truncation error of the *predictor* combines with the one of the *corrector*, generating a new truncation error which we are going to examine. Let $\tilde{q}$ and q be, respectively, the orders of the predictor and the corrector and assume that $y \in C^{\widehat{q}+1}$, where $\widehat{q} = \max(\tilde{q}, q)$. Then

$$y(t_{n+1}) - \sum_{j=0}^{p} \tilde{a}_j y(t_{n-j}) - h\sum_{j=0}^{p} \tilde{b}_j f(t_{n-j}, y_{n-j})$$
$$= \tilde{C}_{\tilde{q}+1} h^{\tilde{q}+1} y^{(\tilde{q}+1)}(t_{n-p}) + \mathcal{O}(h^{\tilde{q}+2}),$$

$$y(t_{n+1}) - \sum_{j=0}^{p} a_j y(t_{n-j}) - h\sum_{j=-1}^{p} b_j f(t_{n-j}, y_{n-j})$$
$$= C_{q+1} h^{q+1} y^{(q+1)}(t_{n-p}) + \mathcal{O}(h^{q+2}),$$

where $\tilde{C}_{\tilde{q}+1}, C_{q+1}$ are the error constants of the predictor and the corrector method respectively. The following result holds.

Property 11.3 *Let the predictor method have order $\tilde{q}$ and the corrector method have order q. Then:*

- *if $\tilde{q} \geq q$ (or $\tilde{q} < q$ with $m > q - \tilde{q}$), the predictor-corrector method has the same order and the same PLTE as the corrector;*
- *if $\tilde{q} < q$ and $m = q - \tilde{q}$, the predictor-corrector method has the same order as the corrector, but different PLTE;*
- *if $\tilde{q} < q$ and $m \leq q - \tilde{q} - 1$, the predictor-corrector method has order equal to $\tilde{q} + m$ (thus less than q).*

In particular, notice that if the predictor has order $q-1$ and the corrector has order q, the PEC suffices to get a method of order q. Moreover, the $P(EC)^m E$ and $P(EC)^m$ schemes have always the same order and the same PLTE.

Combining the Adams-Bashforth method of order q with the corresponding Adams-Moulton method of the same order we obtain the so-called ABM method of order q. It is possible to estimate its PLTE as

$$\frac{C_{q+1}}{C^*_{q+1} - C_{q+1}}\left(u_{n+1}^{(m)} - u_{n+1}^{(0)}\right),$$

where C_{q+1} and C^*_{q+1} are the error constants given in Table 11.1. Accordingly, the steplength h can be decreased if the estimate of the PLTE exceeds a given tolerance and increased otherwise (for the adaptivity of the step length in a predictor-corrector method, see [Lam91], pp.128–147).

Program 93 provides an implementation of the $P(EC)^mE$ methods. The input parameters `at`, `bt`, `a`, `b` contain the coefficients $\tilde{a}_j, \tilde{b}_j$ $(j = 0, \ldots, \tilde{p})$ of the predictor and the coefficients a_j $(j = 0, \ldots, p)$, b_j $(j = -1, \ldots, p)$ of the corrector. Moreover, `f` is a string containing the expression of $f(t, y)$, `h` is the stepsize, `t0` and `tf` are the end points of the time integration interval, `u0` is the vector of the initial data, `m` is the number of the corrector inner iterations. The input variable `pece` must be set equal to `'y'` if the $P(EC)^mE$ is selected, conversely the $P(EC)^m$ scheme is chosen.

Program 93 - predcor : Predictor-corrector scheme

```
function [t,u]=predcor(a,b,at,bt,h,f,t0,u0,tf,pece,m)
%PREDCOR Predictor-corrector method.
%   [T,U]=PREDCOR(A,B,AT,BT,TF,T0,U0,H,FUN,PECE,M) solves the Cauchy
%   problem Y'=FUN(T,Y) for T in (T0,TF) using a predictor method with
%   coefficients AT and BT for the predictor, A and B for the corrector. H specifies
%    the time step. If PECE=1, then the P(EC)^mE method is selected, other-
wise the P(EC)^m
%   is considered.
p  = max(length(a),length(b)-1);
pt = max(length(at),length(bt));
q  = max(p,pt); if length(u0)<q, break, end;
t  = [t0:h:t0+(q-1)*h]; u = u0; y = u0; fe = eval(f);
k  = q;
for t = t0+q*h:h:tf
    ut=sum(at.*u(k:-1:k-pt+1))+h*sum(bt.*fe(k:-1:k-pt+1));
    y=ut; foy=eval(f);
    uv=sum(a.*u(k:-1:k-p+1))+h*sum(b(2:p+1).*fe(k:-1:k-p+1));
    k = k+1;
    for j=1:m
       fy=foy; up=uv+h*b(1)*fy; y=up; foy=eval(f);
    end
    if pece=='y'|pece=='Y'
       fe=[fe,foy];
    else
      fe=[fe,fy];
    end
    u=[u,up];
end
t=[t0:h:tf];
return
```

Example 11.9 Let us check the performance of the $P(EC)^mE$ method on the Cauchy problem $y'(t) = e^{-y(t)}$ for $t \in [0, 1]$ with $y(0) = 0$. The exact solution is

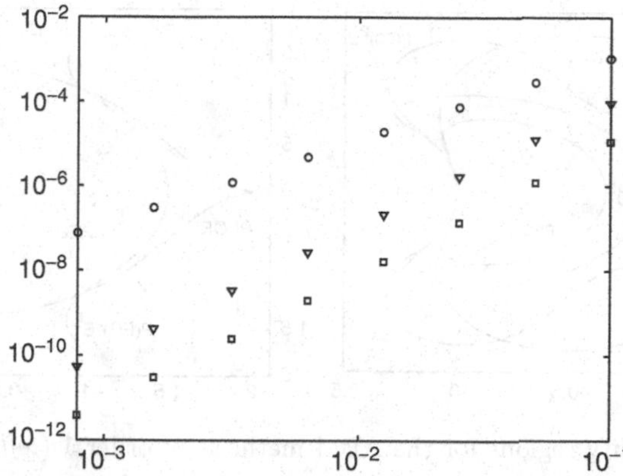

Fig. 11.8. Convergence rate for $P(EC)^m E$ methods as a function of $\log(h)$. The symbol ∇ refers to the AB2-AM3 method ($m = 1$), $\circ$ to AB1-AM3 ($m = 1$) and $\square$ to AB1-AM3 with $m = 2$

$y(t) = \log(1 + t)$. In all the numerical experiments, the corrector method is the Adams-Moulton third-order scheme (AM3), while the explicit Euler (AB1) and the Adams-Bashforth second-order (AB2) methods are used as predictors. Figure 11.8 shows that the pair AB2-AM3 ($m = 1$) yields third-order convergence rate, while AB1-AM3 ($m = 1$) has a second-order accuracy. Taking $m = 2$ allows to recover the third-order convergence rate of the corrector for the AB1-AM3 pair. •

As for the absolute stability, the characteristic polynomial of $P(EC)^m$ methods reads

$$\Pi_{P(EC)^m}(r) = b_{-1}r^p\left(\widehat{\rho}(r) - h\lambda\widehat{\sigma}(r)\right) + \frac{H^m(1-H)}{1-H^m}\left(\tilde{\rho}(r)\widehat{\sigma}(r) - \widehat{\rho}(r)\tilde{\sigma}(r)\right),$$

while for $P(EC)^m E$ we have

$$\Pi_{P(EC)^m E}(r) = \widehat{\rho}(r) - h\lambda\widehat{\sigma}(r) + \frac{H^m(1-H)}{1-H^m}\left(\tilde{\rho}(r) - h\lambda\tilde{\sigma}(r)\right).$$

We have set $H = h\lambda b_{-1}$ and denoted by $\tilde{\rho}$ and $\tilde{\sigma}$ the first and second characteristic polynomial of the *predictor* method, respectively. The polynomials $\widehat{\rho}$ and $\widehat{\sigma}$ are related to the first and second characteristic polynomials of the corrector, as previously explained after Example 11.8. Notice that in both cases the characteristic polynomial tends to the corresponding polynomial of the *corrector* method, since the function $H^m(1-H)/(1-H^m)$ tends to zero as m tends to infinity.

Example 11.10 If we consider the ABM methods with a number of steps p, the characteristic polynomials are $\widehat{\rho}(r) = \tilde{\rho}(r) = r(r^{p-1} - r^{p-2})$, while $\widehat{\sigma}(r) = r\sigma(r)$, where $\sigma(r)$ is the second characteristic polynomial of the corrector. In Figure 11.9 (*right*) the stability regions for the ABM methods of order 2 are plotted. In the case

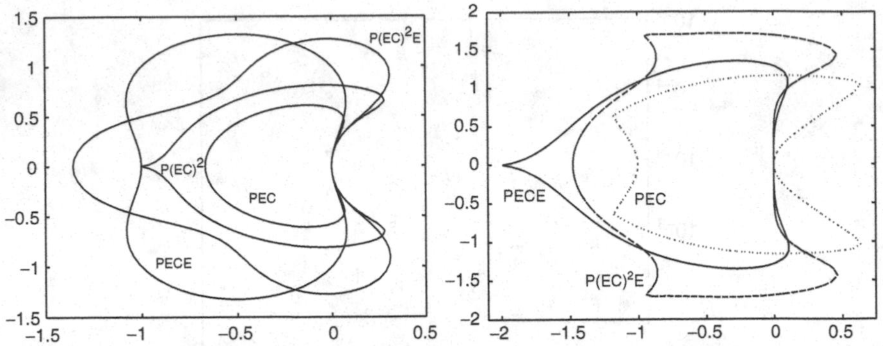

Fig. 11.9. Stability regions for the ABM methods of order 1 (*left*) and 2 (*right*)

of the ABM methods of order 2, 3 and 4, the corresponding stability regions can be ordered by size, namely, from the largest to the smallest one the regions of $PECE$, $P(EC)^2E$, the predictor and PEC methods are plotted in Figure 11.9, left. The one-step ABM method is an exception to the rule and the largest region is the one corresponding to the predictor method (see Figure 11.9, *left*). •

11.8 Runge-Kutta (RK) Methods

When moving from the forward Euler method (11.7) toward higher-order methods, linear multistep methods (MS) and Runge-Kutta methods (RK) adopt two opposite strategies.

Like the Euler method, MS schemes are linear with respect to both u_n and $f_n = f(t_n, u_n)$, require only one functional evaluation at each time step and their accuracy can be increased at the expense of increasing the number of steps. On the other hand, RK methods maintain the structure of one-step methods, and increase their accuracy at the price of an increase of functional evaluations at each time level, thus sacrifying linearity.

A consequence is that RK methods are more suitable than MS methods at adapting the stepsize, whereas estimating the local error for RK methods is more difficult than it is in the case of MS methods.
In its most general form, an RK method can be written as

$$u_{n+1} = u_n + hF(t_n, u_n, h; f), \qquad n \geq 0, \tag{11.70}$$

where F is the increment function defined as follows

$$\begin{aligned} &F(t_n, u_n, h; f) = \sum_{i=1}^{s} b_i K_i, \\ &K_i = f(t_n + c_i h, u_n + h\sum_{j=1}^{s} a_{ij} K_j),\ i = 1, 2, \ldots, s \end{aligned} \tag{11.71}$$

and s denotes the number of *stages* of the method. The coefficients $\{a_{ij}\}$, $\{c_i\}$ and $\{b_i\}$ fully characterize an RK method and are usually collected in the so-called *Butcher array*

$$\begin{array}{c|cccc} c_1 & a_{11} & a_{12} & \dots & a_{1s} \\ c_2 & a_{21} & a_{22} & & a_{2s} \\ \vdots & \vdots & & \ddots & \vdots \\ c_s & a_{s1} & a_{s2} & \dots & a_{ss} \\ \hline & b_1 & b_2 & \dots & b_s \end{array} \qquad \text{or} \qquad \begin{array}{c|c} \mathbf{c} & \mathrm{A} \\ \hline & \mathbf{b}^T \end{array},$$

where $\mathrm{A} = (a_{ij}) \in \mathbb{R}^{s\times s}$, $\mathbf{b} = (b_1, \dots, b_s)^T \in \mathbb{R}^s$ and $\mathbf{c} = (c_1, \dots, c_s)^T \in \mathbb{R}^s$. We shall henceforth assume that the following condition holds

$$c_i = \sum_{j=1}^{s} a_{ij} \; i = 1, \dots, s. \tag{11.72}$$

If the coefficients a_{ij} in A are equal to zero for $j \geq i$, with $i = 1, 2, \dots, s$, then each K_i can be explicitly computed in terms of the $i-1$ coefficients $K_1, \dots, K_{i-1}$ that have already been determined. In such a case the RK method is *explicit*. Otherwise, it is *implicit* and solving a nonlinear system of size s is necessary for computing the coefficients K_i.

The increase in the computational effort for implicit schemes makes their use quite expensive; an acceptable compromise is provided by RK *semi-implicit* methods, in which case $a_{ij} = 0$ for $j > i$ so that each K_i is the solution of the nonlinear equation

$$K_i = f\left(t_n + c_i h, u_n + h a_{ii} \boxed{K_i} + h \sum_{j=1}^{i-1} a_{ij} K_j\right).$$

A semi-implicit scheme thus requires s nonlinear independent equations to be solved.

The local truncation error $\tau_{n+1}(h)$ at node t_{n+1} of the RK method (11.70) is defined through the residual equation

$$h\tau_{n+1}(h) = y_{n+1} - y_n - hF(t_n, y_n, h; f),$$

where $y(t)$ is the exact solution to the Cauchy problem (11.1). Method (11.70) is *consistent* if $\tau(h) = \max_n |\tau_n(h)| \to 0$ as $h \to 0$. It can be shown (see [Lam91]) that this happens iff

$$\sum_{i=1}^{s} b_i = 1.$$

As usual, we say that (11.70) is a method of order p (≥ 1) with respect to h if $\tau(h) = \mathcal{O}(h^p)$ as $h \to 0$.

As for *convergence*, since RK methods are one-step methods, consistency implies stability and, in turn, convergence. As happens for MS methods, estimates of $\tau(h)$ can be derived; however, these estimates are often too involved to be profitably used. We only mention that, as for MS methods, if a RK scheme has a local truncation error $\tau_n(h) = \mathcal{O}(h^p)$, for any n, then also the convergence order will be equal to p.

The following result establishes a relation between order and number of stages of explicit RK methods.

Property 11.4 *The order of an s-stage explicit RK method cannot be greater than s. Also, there do not exist s-stage explicit RK methods with order s if $s \geq 5$.*

We refer the reader to [But87] for the proofs of this result and the results we give below. In particular, for orders ranging between 1 and 8, the minimum number of stages s_{min} required to get a method of corresponding order is shown below

order	1	2	3	4	5	6	7	8
s_{min}	1	2	3	4	6	7	9	11

Notice that 4 is the maximum number of stages for which the order of the method is not less than the number of stages itself. An example of a fourth-order RK method is provided by the following explicit 4-stage method

$$
\begin{aligned}
u_{n+1} &= u_n + \frac{h}{6}(K_1 + 2K_2 + 2K_3 + K_4), \\
K_1 &= f_n, \\
K_2 &= f(t_n + \tfrac{h}{2}, u_n + \tfrac{h}{2}K_1), \\
K_3 &= f(t_n + \tfrac{h}{2}, u_n + \tfrac{h}{2}K_2), \\
K_4 &= f(t_{n+1}, u_n + hK_3).
\end{aligned}
\qquad (11.73)
$$

As far as implicit schemes are concerned, the maximum achievable order using s stages is equal to $2s$.

Remark 11.4 (The case of systems of ODEs) A RK method can be readily extended to systems of ODEs. However, the order of a RK method in the scalar case does not necessarily coincide with that in the vector case. In particular, for $p \geq 4$, a method having order p in the case of the autonomous system $\mathbf{y}' = \mathbf{f}(\mathbf{y})$, with $\mathbf{f} : \mathbb{R}^m \to \mathbb{R}^n$, maintains order p even when applied to an autonomous scalar equation $y' = f(y)$, but the converse is not true, see [Lam91], Section 5.8. ■

11.8.1 Derivation of an Explicit RK Method

The standard technique for deriving an explicit RK method consists of enforcing that the highest number of terms in Taylor's expansion of the exact solution y_{n+1} about t_n coincide with those of the approximate solution u_{n+1}, assuming that we take one step of the RK method starting from the exact solution y_n. We provide an example of this technique in the case of an explicit 2-stage RK method.

Let us consider a 2-stage explicit RK method and assume to dispose at the n-th step of the exact solution y_n. Then

$$u_{n+1} = y_n + hF(t_n, y_n, h; f) = y_n + h(b_1K_1 + b_2K_2),$$

$$K_1 = f(t_n, y_n), \qquad K_2 = f(t_n + hc_2, y_n + hc_2K_1),$$

having assumed that (11.72) is satisfied. Expanding K_2 in a Taylor series in a neighborhood of t_n and truncating the expansion at the second order, we get

$$K_2 = f_n + hc_2(f_{n,t} + K_1 f_{n,y}) + \mathcal{O}(h^2).$$

We have denoted by $f_{n,z}$ (for $z = t$ or $z = y$) the partial derivative of f with respect to z evaluated at (t_n, y_n). Then

$$u_{n+1} = y_n + hf_n(b_1 + b_2) + h^2c_2b_2(f_{n,t} + f_nf_{n,y}) + \mathcal{O}(h^3).$$

If we perform the same expansion on the exact solution, we find

$$y_{n+1} = y_n + hy_n' + \frac{h^2}{2}y_n'' + \mathcal{O}(h^3) = y_n + hf_n + \frac{h^2}{2}(f_{n,t} + f_nf_{n,y}) + \mathcal{O}(h^3).$$

Forcing the coefficients in the two expansions above to agree, up to higher-order terms, we obtain that the coefficients of the RK method must satisfy $b_1 + b_2 = 1$, $c_2b_2 = \frac{1}{2}$.

Thus, there are infinitely many 2-stage explicit RK methods with second-order accuracy. Two examples are the Heun method (11.10) and the modified Euler method (11.91). Of course, with similar (and cumbersome) computations in the case of higher-stage methods, and accounting for a higher number of terms in Taylor's expansion, one can generate higher-order RK methods. For instance, retaining all the terms up to the fifth one, we get scheme (11.73).

11.8.2 Stepsize Adaptivity for RK Methods

Since RK schemes are one-step methods, they are well-suited to adapting the stepsize h, provided that an efficient estimator of the local error is available. Usually, a tool of this kind is an a posteriori error estimator, since the a priori local error estimates are too complicated to be used in practice. The error estimator can be constructed in two ways:

- using the same RK method, but with two different stepsizes (typically $2h$ and h);
- using two RK methods of different order, but with the same number s of stages.

In the first case, if a RK method of order p is being used, one pretends that, starting from an exact datum $u_n = y_n$ (which would not be available if $n \geq 1$), the local error at t_{n+1} is less than a fixed tolerance. The following relation holds

$$y_{n+1} - u_{n+1} = \Phi(y_n)h^{p+1} + \mathcal{O}(h^{p+2}), \tag{11.74}$$

where Φ is an unknown function evaluated at y_n. (Notice that, in this special case, $y_{n+1} - u_{n+1} = h\tau_{n+1}(h)$.)

Carrying out the same computation with a stepsize of $2h$, starting from t_{n-1}, and denoting by $\widehat{u}_{n+1}$ the computed solution, yields

$$\begin{aligned} y_{n+1} - \widehat{u}_{n+1} &= \Phi(y_{n-1})(2h)^{p+1} + \mathcal{O}(h^{p+2}) \\ &= \Phi(y_n)(2h)^{p+1} + \mathcal{O}(h^{p+2}), \end{aligned} \tag{11.75}$$

having expanded also y_{n-1} with respect to t_n. Subtracting (11.74) from (11.75), we get

$$(2^{p+1} - 1)h^{p+1}\Phi(y_n) = u_{n+1} - \widehat{u}_{n+1} + \mathcal{O}(h^{p+2}),$$

from which

$$y_{n+1} - u_{n+1} \simeq \frac{u_{n+1} - \widehat{u}_{n+1}}{(2^{p+1} - 1)} = \mathcal{E}.$$

If $|\mathcal{E}|$ is less than the fixed tolerance ε, the scheme moves to the next time step, otherwise the estimate is repeated with a halved stepsize. In general, the stepsize is doubled whenever $|\mathcal{E}|$ is less than $\varepsilon/2^{p+1}$.

This approach yields a considerable increase in the computational effort, due to the $s-1$ extra functional evaluations needed to generate the value $\widehat{u}_{n+1}$. Moreover, if one needs to half the stepsize, the value u_n must also be computed again.

An alternative that does not require extra functional evaluations consists of using simultaneously two different RK methods with s stages, of order p and $p+1$, respectively, which share the same set of values K_i. These methods are synthetically represented by the modified Butcher array

$$\begin{array}{c|c} \mathbf{c} & \mathrm{A} \\ \hline & \mathbf{b}^T \\ & \widehat{\mathbf{b}}^T \\ \hline & \mathbf{E}^T \end{array}, \tag{11.76}$$

where the method of order p is identified by the coefficients $\mathbf{c}$, A and $\mathbf{b}$, while that of order $p+1$ is identified by $\mathbf{c}$, A and $\widehat{\mathbf{b}}$, and where $\mathbf{E} = \mathbf{b} - \widehat{\mathbf{b}}$.

Taking the difference between the approximate solutions at t_{n+1} produced by the two methods provides an estimate of the local truncation error for the scheme of lower order. On the other hand, since the coefficients K_i coincide, this difference is given by $h\sum_{i=1}^{s} E_i K_i$ and thus it does not require extra functional evaluations.

Notice that, if the solution u_{n+1} computed by the scheme of order p is used to initialize the scheme at time step $n+2$, the method will have order p, as a whole. If, conversely, the solution computed by the scheme of order $p+1$ is employed, the resulting scheme would still have order $p+1$ (exactly as happens with predictor-corrector methods).

The Runge-Kutta Fehlberg method of fourth-order is one of the most popular schemes of the form (11.76) and consists of a fourth-order RK scheme coupled with a fifth-order RK method (for this reason, it is known as the RK45 method). The modified Butcher array for this method is shown below

$$
\begin{array}{c|cccccc}
0 & 0 & 0 & 0 & 0 & 0 & 0 \\
\frac{1}{4} & \frac{1}{4} & 0 & 0 & 0 & 0 & 0 \\
\frac{3}{8} & \frac{3}{32} & \frac{9}{32} & 0 & 0 & 0 & 0 \\
\frac{12}{13} & \frac{1932}{2197} & -\frac{7200}{2197} & \frac{7296}{2197} & 0 & 0 & 0 \\
1 & \frac{439}{216} & -8 & \frac{3680}{513} & -\frac{845}{4104} & 0 & 0 \\
\frac{1}{2} & -\frac{8}{27} & 2 & -\frac{3544}{2565} & \frac{1859}{4104} & -\frac{11}{40} & 0 \\
\hline
 & \frac{25}{216} & 0 & \frac{1408}{2565} & \frac{2197}{4104} & -\frac{1}{5} & 0 \\
 & \frac{16}{135} & 0 & \frac{6656}{12825} & \frac{28561}{56430} & -\frac{9}{50} & \frac{2}{55} \\
\hline
 & \frac{1}{360} & 0 & -\frac{128}{4275} & -\frac{2197}{75240} & \frac{1}{50} & \frac{2}{55}
\end{array}
$$

This method tends to underestimate the error. As such, its use is not completely reliable when the stepsize h is large.

Remark 11.5 MATLAB provides a package tool `funfun`, which, besides the two classical Runge-Kutta Fehlberg methods, RK23 (second-order and third-order pair) and RK45 (fourth-order and fifth-order pair), also implements other methods suitable for solving *stiff* problems. These methods are derived from BDF methods (see [SR97]) and are included in the MATLAB program `ode15s`. ■

11.8.3 Implicit RK Methods

Implicit RK methods can be derived from the integral formulation (11.2) of the Cauchy problem. In fact, if a quadrature formula with s nodes in (t_n, t_{n+1})

is employed to approximate the integral of f (which we assume, for simplicity, to depend only on t), we get

$$\int_{t_n}^{t_{n+1}} f(\tau)\, d\tau \simeq h \sum_{j=1}^{s} b_j f(t_n + c_j h),$$

having denoted by b_j the weights and by $t_n + c_j h$ the quadrature nodes. It can be proved (see [But64]) that for any RK formula (11.70)-(11.71), there exists a correspondence between the coefficients b_j, c_j of the formula and the weights and nodes of a Gauss quadrature rule (see, [Lam91], Section 5.11).

Once the s coefficients c_j have been found, we can construct RK methods of order $2s$, by determining the coefficients a_{ij} and b_j as being the solutions of the linear systems

$$\sum_{j=1}^{s} c_j^{k-1} a_{ij} = (1/k) c_i^k, \quad k = 1, 2, \ldots, s, \quad i = 1, \ldots, s,$$

$$\sum_{j=1}^{s} c_j^{k-1} b_j = 1/k, \qquad k = 1, 2, \ldots, s.$$

The following families can be derived:

1. *Gauss-Legendre RK methods*, if Gauss-Legendre quadrature nodes are used. These methods, for a fixed number of stages s, attain the maximum possible order $2s$. Remarkable examples are the one-stage method (*implicit midpoint method*) of order 2

$$u_{n+1} = u_n + hf\left(t_n + \tfrac{1}{2}h, \tfrac{1}{2}(u_n + u_{n+1})\right), \quad \begin{array}{c|c} \frac{1}{2} & \frac{1}{2} \\ \hline & 1 \end{array}$$

and the 2-stage method of order 4, described by the following Butcher array

$$\begin{array}{c|cc} \frac{3-\sqrt{3}}{6} & \frac{1}{4} & \frac{3-2\sqrt{3}}{12} \\ \frac{3+\sqrt{3}}{6} & \frac{3+2\sqrt{3}}{12} & \frac{1}{4} \\ \hline & \frac{1}{2} & \frac{1}{2} \end{array};$$

2. *Gauss-Radau methods*, which are characterized by the fact that the quadrature nodes include one of the two endpoints of the interval (t_n, t_{n+1}). The maximum order that can be achieved by these methods is $2s - 1$, when s stages are used. Elementary examples correspond to the following Butcher arrays

$$\begin{array}{c|c} 0 & 1 \\ \hline & 1 \end{array}, \quad \begin{array}{c|c} 1 & 1 \\ \hline & 1 \end{array}, \quad \begin{array}{c|cc} \frac{1}{3} & \frac{5}{12} & -\frac{1}{12} \\ 1 & \frac{3}{4} & \frac{1}{4} \\ \hline & \frac{3}{4} & \frac{1}{4} \end{array}$$

and have order 1, 1 and 3, respectively. The Butcher array in the middle represents the backward Euler method.

3. *Gauss-Lobatto methods*, where both the endpoints t_n and t_{n+1} are quadrature nodes. The maximum order that can be achieved using s stages is $2s-2$. We recall the methods of the family corresponding to the following Butcher arrays

$$\begin{array}{c|cc} 0 & 0 & 0 \\ 1 & \frac{1}{2} & \frac{1}{2} \\ \hline & \frac{1}{2} & \frac{1}{2} \end{array}, \quad \begin{array}{c|cc} 0 & \frac{1}{2} & 0 \\ 1 & \frac{1}{2} & 0 \\ \hline & \frac{1}{2} & \frac{1}{2} \end{array}, \quad \begin{array}{c|ccc} 0 & \frac{1}{6} & -\frac{1}{3} & \frac{1}{6} \\ \frac{1}{2} & \frac{1}{6} & \frac{5}{12} & -\frac{1}{12} \\ 1 & \frac{1}{6} & \frac{2}{3} & \frac{1}{6} \\ \hline & \frac{1}{6} & \frac{2}{3} & \frac{1}{6} \end{array},$$

which have order 2, 2 and 3, respectively. The first array represents the Crank-Nicolson method.

As for semi-implicit RK methods, we limit ourselves to mentioning the case of DIRK methods (*diagonally implicit RK*), which, for $s = 3$, are represented by the following Butcher array

$$\begin{array}{c|ccc} \frac{1+\mu}{2} & \frac{1+\mu}{2} & 0 & 0 \\ \frac{1}{2} & -\frac{\mu}{2} & \frac{1+\mu}{2} & 0 \\ \frac{1-\mu}{2} & 1+\mu & -1-2\mu & \frac{1+\mu}{2} \\ \hline & \frac{1}{6\mu^2} & 1-\frac{1}{3\mu^2} & \frac{1}{6\mu^2} \end{array}.$$

The parameter μ represents one of the three roots of $3\mu^3 - 3\mu - 1 = 0$ (i.e., $(2/\sqrt{3})\cos(10°)$, $-(2/\sqrt{3})\cos(50°)$, $-(2/\sqrt{3})\cos(70°)$). The maximum order that has been determined in the literature for these methods is 4.

11.8.4 Regions of Absolute Stability for RK Methods

Applying an s-stage RK method to the model problem (11.24) yields

$$K_i = \lambda\left(u_n + h\sum_{j=1}^{s} a_{ij}K_j\right), \quad u_{n+1} = u_n + h\sum_{i=1}^{s} b_iK_i, \qquad (11.77)$$

that is, a first-order difference equation. If $\mathbf{K}$ and $\mathbf{1}$ are the vectors of components $(K_1, \ldots, K_s)^T$ and $(1, \ldots, 1)^T$, respectively, then (11.77) becomes

$$\mathbf{K} = \lambda(u_n \mathbf{1} + h\mathrm{A}\mathbf{K}),\; u_{n+1} = u_n + h\mathbf{b}^T\mathbf{K},$$

from which, $\mathbf{K} = (\mathrm{I} - h\lambda\mathrm{A})^{-1}\mathbf{1}\lambda u_n$ and thus

$$u_{n+1} = \left[1 + h\lambda\mathbf{b}^T(\mathrm{I} - h\lambda\mathrm{A})^{-1}\mathbf{1}\right] u_n = R(h\lambda)u_n,$$

where $R(h\lambda)$ is the so-called *stability function.*

The RK method is absolutely stable, i.e., the sequence $\{u_n\}$ satisfies (11.25), iff $|R(h\lambda)| < 1$. Its region of absolute stability is given by

$$\mathcal{A} = \{z = h\lambda \in \mathbb{C} \text{ such that } |R(h\lambda)| < 1\}.$$

If the method is explicit, A is strictly lower triangular and the function R can be written in the following form (see [DV84])

$$R(h\lambda) = \frac{\det(\mathrm{I} - h\lambda\mathrm{A} + h\lambda\mathbf{1}\mathbf{b}^T)}{\det(\mathrm{I} - h\lambda\mathrm{A})}.$$

Thus since $\det(\mathrm{I} - h\lambda\mathrm{A}) = 1$, $R(h\lambda)$ is a polynomial function in the variable $h\lambda$, $|R(h\lambda)|$ can never be less than 1 for all values of $h\lambda$. Consequently, $\mathcal{A}$ can never be unbounded for an explicit RK method.
In the special case of an explicit RK of order $s = 1, \ldots, 4$, one gets (see [Lam91])

$$R(h\lambda) = \sum_{k=0}^{s} \frac{1}{k!}(h\lambda)^k.$$

The corresponding regions of absolute stability are drawn in Figure 11.10. Notice that, unlike MS methods, the regions of absolute stability of RK methods increase in size as the order grows.
We finally notice that the regions of absolute stability for explicit RK methods can fail to be connected; an example is given in Exercise 14.

11.9 Systems of ODEs

Let us consider the system of first-order ODEs

$$\mathbf{y}' = \mathbf{F}(t, \mathbf{y}), \tag{11.78}$$

where $\mathbf{F} : \mathbb{R} \times \mathbb{R}^n \to \mathbb{R}^n$ is a given vector function and $\mathbf{y} \in \mathbb{R}^n$ is the solution vector which depends on n arbitrary constants set by the n initial conditions

$$\mathbf{y}(t_0) = \mathbf{y}_0. \tag{11.79}$$

Let us recall the following property (see [Lam91], Theorem 1.1).

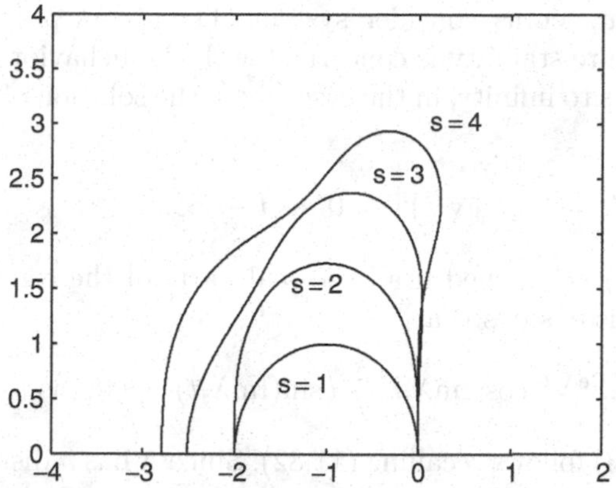

Fig. 11.10. Regions of absolute stability for s-stage explicit RK methods, with $s = 1,\dots,4$. The plot only shows the portion $\mathrm{Im}(h\lambda) \geq 0$ since the regions are symmetric about the real axis

Property 11.5 *Let* $\mathbf{F} : \mathbb{R} \times \mathbb{R}^n \to \mathbb{R}^n$ *be a continuous function on* $D = [t_0, T] \times \mathbb{R}^n$, *with* t_0 *and* T *finite. Then, if there exists a positive constant* L *such that*

$$\|\mathbf{F}(t, \mathbf{y}) - \mathbf{F}(t, \bar{\mathbf{y}})\| \leq L\|\mathbf{y} - \bar{\mathbf{y}}\| \qquad (11.80)$$

holds for any $(t, \mathbf{y})$ *and* $(t, \bar{\mathbf{y}}) \in D$, *then, for any* $\mathbf{y}_0 \in \mathbb{R}^n$ *there exists a unique* $\mathbf{y}$, *continuous and differentiable with respect to* t *for any* $(t, \mathbf{y}) \in D$, *which is a solution of the Cauchy problem* (11.78)-(11.79).

Condition (11.80) expresses the fact that $\mathbf{F}$ is *Lipschitz continuous* with respect to the second argument.

It is seldom possible to write the solution to system (11.78) in closed form. A special case is where the system is linear and autonomous, that is

$$\mathbf{y}'(t) = \mathrm{A}\mathbf{y}(t), \qquad (11.81)$$

with $\mathrm{A} \in \mathbb{R}^{n\times n}$. Assume that A has n distinct eigenvalues λ_j, $j = 1,\dots,n$; therefore, the solution $\mathbf{y}$ can be written as

$$\mathbf{y}(t) = \sum_{j=1}^{n} C_j e^{\lambda_j t} \mathbf{v}_j, \qquad (11.82)$$

where $C_1,\dots,C_n$ are some constants and $\{\mathbf{v}_j\}$ is a basis formed by the eigenvectors of A, associated with the eigenvalues λ_j for $j = 1,\dots,n$. The solution is determined by setting n initial conditions.

From the numerical standpoint, the methods introduced in the scalar case can be extended to systems. A delicate matter is how to generalize the theory developed about absolute stability.

With this aim, let us consider system (11.81). As previously seen, the property of absolute stability is concerned with the behavior of the numerical solution as t grows to infinity, in the case where the solution of problem (11.78) satisfies

$$\|\mathbf{y}(t)\| \to 0 \quad \text{as } t \to \infty. \tag{11.83}$$

Condition (11.83) is satisfied if all the real parts of the eigenvalues of A are negative since this ensures that

$$e^{\lambda_j t} = e^{\mathrm{Re}\lambda_j t}(\cos(\mathrm{Im}\lambda_j t) + i \sin(\mathrm{Im}\lambda_j t)) \to 0, \quad \text{as } t \to \infty, \tag{11.84}$$

from which (11.83) follows recalling (11.82). Since A has n distinct eigenvalues, there exists a nonsingular matrix Q such that $\Lambda = \mathrm{Q}^{-1}\mathrm{AQ}$, Λ being the diagonal matrix whose entries are the eigenvalues of A (see Section 1.8).

Introducing the auxiliary variable $\mathbf{z} = \mathrm{Q}^{-1}\mathbf{y}$, the initial system can therefore be transformed into

$$\mathbf{z}' = \Lambda \mathbf{z}. \tag{11.85}$$

Since Λ is a diagonal matrix, the results holding in the scalar case immediately apply to the vector case as well, provided that the analysis is repeated on all the (scalar) equations of system (11.85).

11.10 Stiff Problems

Consider a nonhomogeneous linear system of ODEs with constant coefficients

$$\mathbf{y}'(t) = \mathrm{A}\mathbf{y}(t) + \boldsymbol{\varphi}(t), \qquad \text{with } \mathrm{A} \in \mathbb{R}^{n\times n}, \quad \boldsymbol{\varphi}(t) \in \mathbb{R}^n,$$

and assume that A has n distinct eigenvalues λ_j, $j = 1, \ldots, n$. Then

$$\mathbf{y}(t) = \sum_{j=1}^{n} C_j e^{\lambda_j t}\mathbf{v}_j + \boldsymbol{\psi}(t) = \mathbf{y}_{hom}(t) + \boldsymbol{\psi}(t),$$

where $C_1, \ldots, C_n$, are n constants, $\{\mathbf{v}_j\}$ is a basis formed by the eigenvectors of A and $\boldsymbol{\psi}(t)$ is a particular solution of the ODE at hand. Throughout the section, we assume that $\mathrm{Re}\lambda_j < 0$ for all j.

As $t \to \infty$, the solution $\mathbf{y}$ tends to the particular solution $\boldsymbol{\psi}$. We can therefore interpret $\boldsymbol{\psi}$ as the *steady-state solution* (that is, after an infinite time) and $\mathbf{y}_{hom}$ as being the *transient solution* (that is, for t finite). Assume that we are interested only in the steady-state. If we employ a numerical scheme with a bounded region of absolute stability, the stepsize h is subject to a constraint that depends on the maximum module eigenvalue of A. On the other hand, the greater this module, the shorter the time interval where the corresponding component in the solution is meaningful. We are thus faced with a sort of paradox: the scheme is forced to employ a small integration stepsize to track a component of the solution that is virtually flat for large values of t.

Precisely, if we assume that

$$\sigma \leq \text{Re}\lambda_j \leq \tau < 0, \qquad \forall j = 1, \ldots, n \tag{11.86}$$

and introduce the *stiffness quotient* $r_s = \sigma/\tau$, we say that a linear system of ODEs with constant coefficients is *stiff* if the eigenvalues of matrix A all have negative real parts and $r_s \gg 1$.
However, referring only to the spectrum of A to characterize the *stiffness* of a problem might have some drawbacks. For instance, when $\tau \simeq 0$, the stiffness quotient can be very large while the problem appears to be "genuinely" stiff only if $|\sigma|$ is very large. Moreover, enforcing suitable initial conditions can affect the stiffness of the problem (for example, selecting the data in such a way that the constants multiplying the "stiff" components of the solution vanish).

For this reason, several authors find the previous definition of a stiff problem unsatisfactory, and, on the other hand, they agree on the fact that it is not possible to exactly state what it is meant by a stiff problem. We limit ourselves to quoting only one alternative definition, which is of some interest since it focuses on what is observed in practice to be a stiff problem.

Definition 11.14 (from [Lam91], p. 220) A system of ODEs is stiff if, when approximated by a numerical scheme characterized by a region of absolute stability with finite size, it forces the method, for any initial condition for which the problem admits a solution, to employ a discretization stepsize excessively small with respect to the smoothness of the exact solution. ■

From this definition, it is clear that no conditionally absolute stable method is suitable for approximating a stiff problem. This prompts resorting to implicit methods, such as MS or RK, which are more expensive than explicit schemes, but have regions of absolute stability of infinite size. However, it is worth recalling that, for nonlinear problems, implicit methods lead to nonlinear equations, for which it is thus crucial to select iterative numerical methods free of limitations on h for convergence.

For instance, in the case of MS methods, we have seen that using fixed-point iterations sets the constraint (11.68) on h in terms of the Lipschitz constant L of f. In the case of a linear system this constraint is

$$L \geq \max_{i=1,\ldots,n} |\lambda_i|,$$

so that (11.68) would imply a strong limitation on h (which could even be more stringent than those required for an explicit scheme to be stable). One way of circumventing this drawback consists of resorting to Newton's method or some variants. The presence of Dahlquist barriers imposes a strong limitation on the use of MS methods, the only exception being BDF methods, which, as already seen, are θ-stable for $p \leq 5$ (for a larger number of steps they are even not zero-stable). The situation becomes definitely more favorable if implicit RK methods are considered, as observed at the end of Section 11.8.4.

The theory developed so far holds rigorously if the system is linear. In the nonlinear case, let us consider the Cauchy problem (11.78), where the function $\mathbf{F} : \mathbb{R} \times \mathbb{R}^n \to \mathbb{R}^n$ is assumed to be differentiable. To study its stability a possible strategy consists of linearizing the system as

$$\mathbf{y}'(t) = \mathbf{F}(\tau, \mathbf{y}(\tau)) + \mathrm{J}_{\mathbf{F}}(\tau, \mathbf{y}(\tau)) \left[\mathbf{y}(t) - \mathbf{y}(\tau)\right],$$

in a neighborhood $(\tau, \mathbf{y}(\tau))$, where τ is an arbitrarily chosen value of t within the time integration interval.

The above technique might be dangerous since the eigenvalues of $\mathrm{J}_{\mathbf{F}}$ do not suffice in general to describe the behavior of the exact solution of the original problem. Actually, some counterexamples can be found where:

1. $\mathrm{J}_{\mathbf{F}}$ has complex conjugate eigenvalues, while the solution of (11.78) does not exhibit oscillatory behavior;
2. $\mathrm{J}_{\mathbf{F}}$ has real nonnegative eigenvalues, while the solution of (11.78) does not grow monotonically with t;
3. $\mathrm{J}_{\mathbf{F}}$ has eigenvalues with negative real parts, but the solution of (11.78) does not decay monotonically with t.

As an example of the case at item 3. let us consider the system of ODEs

$$\mathbf{y}' = \begin{bmatrix} -\dfrac{1}{2t} & \dfrac{2}{t^3} \\ -\dfrac{t}{2} & -\dfrac{1}{2t} \end{bmatrix} \mathbf{y} = \mathrm{A}(t)\mathbf{y}.$$

For $t \geq 1$ its solution is

$$\mathbf{y}(t) = C_1 \begin{bmatrix} t^{-3/2} \\ -\frac{1}{2}t^{1/2} \end{bmatrix} + C_2 \begin{bmatrix} 2t^{-3/2}\log t \\ t^{1/2}(1 - \log t) \end{bmatrix},$$

whose Euclidean norm diverges monotonically for $t > (12)^{1/4} \simeq 1.86$ when $C_1 = 1$, $C_2 = 0$, whilst the eigenvalues of $\mathrm{A}(t)$, equal to $(-1 \pm 2i)/(2t)$, have negative real parts.

Therefore, the nonlinear case must be tackled using *ad hoc* techniques, by suitably reformulating the concept of stability itself (see [Lam91], Chapter 7).

11.11 Applications

We consider two examples of dynamical systems that are well-suited to checking the performances of several numerical methods introduced in the previous sections.

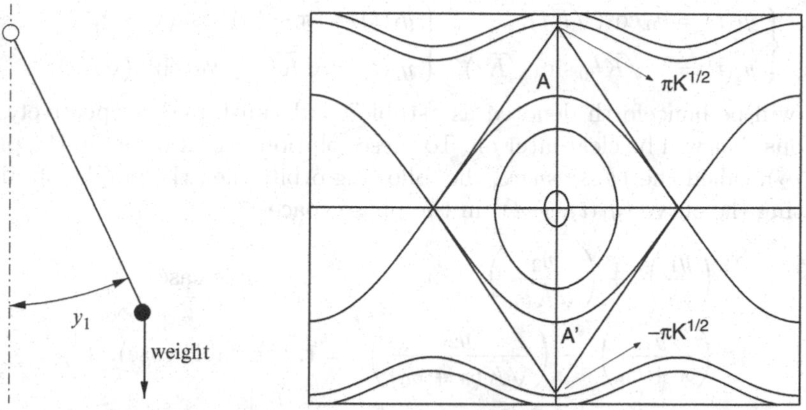

Fig. 11.11. Left: frictionless pendulum; right: orbits of system (11.87) in the phase space

11.11.1 Analysis of the Motion of a Frictionless Pendulum

Let us consider the frictionless pendulum in Figure 11.11 (*left*), whose motion is governed by the following system of ODEs

$$\begin{cases} y_1' = y_2, \\ y_2' = -K\sin(y_1), \end{cases} \tag{11.87}$$

for $t > 0$, where $y_1(t)$ and $y_2(t)$ represent the position and angular velocity of the pendulum at time t, respectively, while K is a positive constant depending on the geometrical-mechanical parameters of the pendulum. We consider the initial conditions: $y_1(0) = \theta_0$, $y_2(0) = 0$.

Denoting by $\mathbf{y} = (y_1, y_2)^T$ the solution to system (11.87), this admits infinitely many equilibrium conditions of the form $\mathbf{y} = (n\pi, 0)^T$ for $n \in \mathbb{Z}$, corresponding to the situations where the pendulum is vertical with zero velocity. For n even, the equilibrium is stable, while for n odd it is unstable. These conclusions can be drawn by analyzing the linearized system

$$\mathbf{y}' = \mathrm{A}_e\mathbf{y} = \begin{bmatrix} 0 & 1 \\ -K & 0 \end{bmatrix}\mathbf{y}, \qquad \mathbf{y}' = \mathrm{A}_o\mathbf{y} = \begin{bmatrix} 0 & 1 \\ K & 0 \end{bmatrix}\mathbf{y}.$$

If n is even, matrix A_e has complex conjugate eigenvalues $\lambda_{1,2} = \pm i\sqrt{K}$ and associated eigenvectors $\mathbf{y}_{1,2} = (\mp i/\sqrt{K}, 1)^T$, while for n odd, A_o has real and opposite eigenvalues $\lambda_{1,2} = \pm\sqrt{K}$ and eigenvectors $\mathbf{y}_{1,2} = (1/\sqrt{K}, \mp 1)^T$.

Let us consider two different sets of initial data: $\mathbf{y}^{(0)} = (\theta_0, 0)^T$ and $\mathbf{y}^{(0)} = (\pi + \theta_0, 0)^T$, where $|\theta_0| \ll 1$. The solutions of the corresponding linearized system are, respectively,

$$\begin{cases} y_1(t) = \theta_0 \cos(\sqrt{K}t), \\ y_2(t) = -\sqrt{K}\theta_0 \sin(\sqrt{K}t), \end{cases} \qquad \begin{cases} y_1(t) = (\pi + \theta_0)\cosh(\sqrt{K}t), \\ y_2(t) = \sqrt{K}(\pi + \theta_0)\sinh(\sqrt{K}t), \end{cases}$$

and will be henceforth denoted as "stable" and "unstable", respectively, for reasons that will be clear later on. To these solutions we associate in the plane (y_1, y_2), called the *phase space*, the following orbits (i.e., the graphs obtained plotting the curve $(y_1(t), y_2(t))$ in the phase space).

$$\left(\frac{y_1}{\theta_0}\right)^2 + \left(\frac{y_2}{\sqrt{K}\theta_0}\right)^2 = 1, \qquad \text{(stable case)},$$

$$\left(\frac{y_1}{\pi + \theta_0}\right)^2 - \left(\frac{y_2}{\sqrt{K}(\pi + \theta_0)}\right)^2 = 1, \text{ (unstable case)}.$$

In the stable case, the orbits are ellipses with period $2\pi/\sqrt{K}$ and are centered at $(0,0)^T$, while in the unstable case they are hyperbolae centered at $(0,0)^T$ and asymptotic to the straight lines $y_2 = \pm\sqrt{K}y_1$.

The complete picture of the motion of the pendulum in the phase space is shown in Figure 11.11 (*right*). Notice that, letting $v = |y_2|$ and fixing the initial position $y_1(0) = 0$, there exists a limit value $v_L = 2\sqrt{K}$ which corresponds in the figure to the points A and A'. For $v(0) < v_L$, the orbits are closed, while for $v(0) > v_L$ they are open, corresponding to a continuous revolution of the pendulum, with infinite passages (with periodic and non null velocity) through the two equilibrium positions $y_1 = 0$ and $y_1 = \pi$. The limit case $v(0) = v_L$ yields a solution such that, thanks to the total energy conservation principle, $y_2 = 0$ when $y_1 = \pi$. Actually, these two values are attained asymptotically only as $t \to \infty$.

The first-order nonlinear differential system (11.87) has been numerically solved using the forward Euler method (FE), the midpoint method (MP) and the Adams-Bashforth second-order scheme (AB). In Figure 11.12 we show the orbits in the phase space that have been computed by the two methods on the time interval $(0, 30)$ and taking $K = 1$ and $h = 0.1$. The crosses denote initial conditions.

As can be noticed, the orbits generated by FE do not close. This kind of instability is due to the fact that the region of absolute stability of the FE method completely excludes the imaginary axis. On the contrary, the MP method describes accurately the closed system orbits due to the fact that its region of asymptotic stability (see Section 11.6.4) includes pure imaginary eigenvalues in the neighborhood of the origin of the complex plane. It must also be noticed that the MP scheme gives rise to oscillating solutions as v_0 gets larger. The second-order AB method, instead, describes correctly all kinds of orbits.

11.11.2 Compliance of Arterial Walls

An arterial wall subject to blood flow can be modelled by a compliant circular cylinder of length L and radius R_0 with walls made by an incompressible,

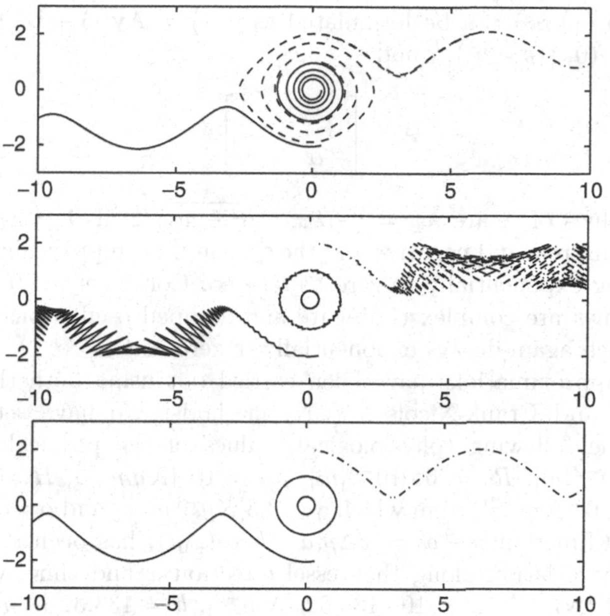

Fig. 11.12. Orbits for system (11.87) in the case $K = 1$ and $h = 0.1$, computed using the FE method (*upper plot*), the MP method (*central plot*) and the AB method (*lower plot*), respectively. The initial conditions are $\theta_0 = \pi/10$ and $v_0 = 0$ (*thin solid line*), $v_0 = 1$ (*dashed line*), $v_0 = 2$ (*dash-dotted line*) and $v_0 = -2$ (*thick solid line*)

homogeneous, isotropic, elastic tissue of thickness H. A simple model describing the mechanical behavior of the walls interacting with the blood flow is the so called "independent-rings" model according to which the vessel wall is regarded as an assembly of rings which are not influenced one by the others.

This amounts to neglecting the longitudinal (or axial) inner actions along the vessel, and to assuming that the walls can deform only in the radial direction. Thus, the vessel radius R is given by $R(t) = R_0 + y(t)$, where y is the radial deformation of the ring with respect to a reference radius R_0 and t is the time variable. The application of Newton's law to the independent-ring system yields the following equation modeling the time mechanical behavior of the wall

$$y''(t) + \beta y'(t) + \alpha y(t) = \gamma(p(t) - p_0), \tag{11.88}$$

where $\alpha = E/(\rho_w R_0^2)$, $\gamma = 1/(\rho_w H)$ and β is a positive constant. The physical parameters ρ_w and E denote the vascular wall density and the Young modulus of the vascular tissue, respectively. The function $p - p_0$ is the forcing term acting on the wall due to the pressure drop between the inner part of the vessel (where the blood flows) and its outer part (surrounding organs). At rest, if $p = p_0$, the vessel configuration coincides with the undeformed circular cylinder having radius equal exactly to R_0 ($y = 0$).

Equation (11.88) can be formulated as $\mathbf{y}'(t) = \mathrm{A}\mathbf{y}(t) + \mathbf{b}(t)$ where $\mathbf{y} = (y, y')^T$, $\mathbf{b} = (0, \gamma(p - p_0))^T$ and

$$\mathrm{A} = \begin{bmatrix} 0 & 1 \\ -\alpha & -\beta \end{bmatrix}. \tag{11.89}$$

The eigenvalues of A are $\lambda_\pm = (-\beta \pm \sqrt{\beta^2 - 4\alpha})/2$; if $\beta \geq 2\sqrt{\alpha}$ both the eigenvalues are real and negative and the system is asymptotically stable with $\mathbf{y}(t)$ decaying exponentially to zero as $t \to \infty$. Conversely, if $0 < \beta < 2\sqrt{\alpha}$ the eigenvalues are complex conjugate and damped oscillations arise in the solution which again decays exponentially to zero as $t \to \infty$.
Numerical approximations have been carried out using both the backward Euler (BE) and Crank-Nicolson (CN) methods. We have set $\mathbf{y}(t) = \mathbf{0}$ and used the following (physiological) values of the physical parameters: $L = 5 \cdot 10^{-2}[m]$, $R_0 = 5 \cdot 10^{-3}[m]$, $\rho_w = 10^3[Kgm^{-3}]$, $H = 3 \cdot 10^{-4}[m]$ and $E = 9 \cdot 10^5[Nm^{-2}]$, from which $\gamma \simeq 3.3[Kg^{-1}m^{-2}]$ and $\alpha = 36 \cdot 10^6[s^{-2}]$. A sinusoidal function $p - p_0 = x\Delta p(a + b\cos(\omega_0 t))$ has been used to model the pressure variation along the vessel direction x and time, where $\Delta p = 0.25 \cdot 133.32\ [Nm^{-2}]$, $a = 10 \cdot 133.32\ [Nm^{-2}]$, $b = 133.32\ [Nm^{-2}]$ and the pulsation $\omega_0 = 2\pi/0.8\ [rad\, s^{-1}]$ corresponds to a heart beat.

The results reported below refer to the ring coordinate $x = L/2$. The two (very different) cases (1) $\beta = \sqrt{\alpha}\,[s^{-1}]$ and (2) $\beta = \alpha\,[s^{-1}]$ have been analyzed; it is easily seen that in case (2) the stiffness quotient (see Section 11.10) is almost equal to α, thus the problem is highly stiff. We notice also that in both cases the real parts of the eigenvalues of A are very large, so that an appropriately small time step should be taken to accurately describe the fast transient of the problem.
In case (1) the differential system has been studied on the time interval $[0, 2.5 \cdot 10^{-3}]$ with a time step $h = 10^{-4}$. We notice that the two eigenvalues of A have modules equal to 6000, thus our choice of h is compatible with the use of an explicit method as well.

Figure 11.13 (*left*) shows the numerical solutions as functions of time. The solid (thin) line is the exact solution while the thick dashed and solid lines are the solutions given by the CN and BE methods, respectively. A far better accuracy of the CN method over the BE is clearly demonstrated; this is confirmed by the plot in Figure 11.13 (*right*) which shows the trajectories of the computed solutions in the phase space. In this case the differential system has been integrated on the time interval $[0, 0.25]$ with a time step $h = 2.5 \cdot 10^{-4}$. The dashed line is the trajectory of the CN method while the solid line is the corresponding one obtained using the BE scheme. A strong dissipation is clearly introduced by the BE method with respect to the CN scheme; the plot also shows that both methods converge to a limit cycle which corresponds to the cosine component of the forcing term.
In the second case (2) the differential system has been integrated on the time interval $[0, 10]$ with a time step $h = 0.1$. The stiffness of the problem

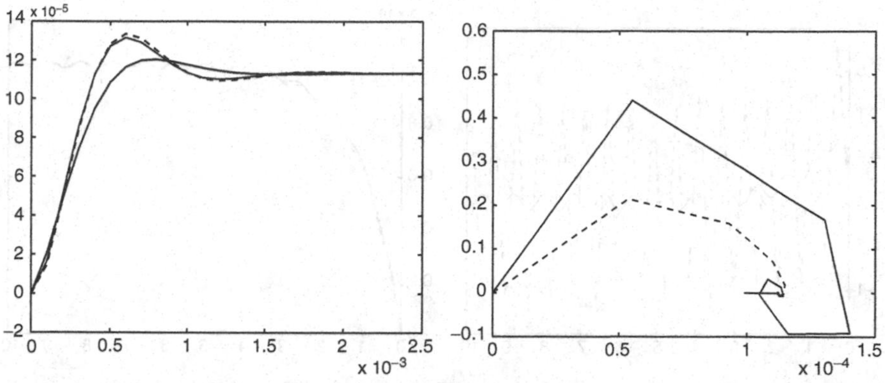

Fig. 11.13. Transient simulation (*left*) and phase space trajectories (*right*)

is demonstrated by the plot of the deformation velocities z shown in Figure 11.14 (left). The solid line is the solution computed by the BE method while the dashed line is the corresponding one given by the CN scheme; for the sake of graphical clarity, only one third of the nodal values have been plotted for the CN method. Strong oscillations arise since the eigenvalues of matrix A are $\lambda_1 = -1$, $\lambda_2 = -36 \cdot 10^6$ so that the CN method approximates the first component y of the solution $\mathbf{y}$ as

$$y_k^{CN} = \left(\frac{1 + (h\lambda_1)/2}{1 - (h\lambda_1)/2}\right)^k \simeq (0.9048)^k, \qquad k \geq 0,$$

which is clearly stable, while the approximate second component $z(= y')$ is

$$z_k^{CN} = \left(\frac{1 + (h\lambda_2)/2}{1 - (h\lambda_2)/2}\right)^k \simeq (-0.9999)^k, \qquad k \geq 0,$$

which is obviously oscillating. On the contrary, the BE method yields

$$y_k^{BE} = \left(\frac{1}{1 - h\lambda_1}\right)^k \simeq (0.9090)^k, \qquad k \geq 0,$$

and

$$z_k^{BE} = \left(\frac{1}{1 - h\lambda_2}\right)^k \simeq (0.2777)^k, \qquad k \geq 0,$$

which are both stable for every $h > 0$. According to these conclusions the first component y of the vector solution $\mathbf{y}$ is correctly approximated by both the methods as can be seen in Figure 11.14 (*right*) where the solid line refers to the BE scheme while the dashed line is the solution computed by the CN method.

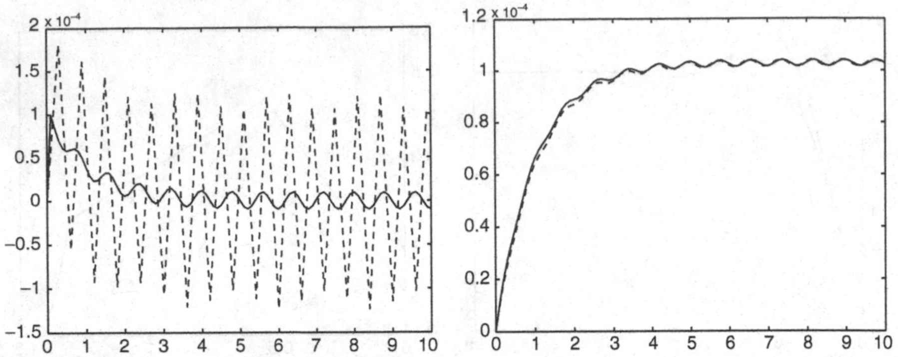

Fig. 11.14. Long-time behavior of the solution: velocities (*left*) and displacements (*right*)

11.12 Exercises

1. Prove that Heun's method has order 2 with respect to h.
[*Hint*: notice that $h\tau_{n+1} = y_{n+1} - y_n - h\Phi(t_n, y_n; h) = E_1 + E_2$, where

$$E_1 = \int_{t_n}^{t_{n+1}} f(s, y(s))ds - \frac{h}{2}[f(t_n, y_n) + f(t_{n+1}, y_{n+1})]$$

and

$$E_2 = \frac{h}{2}\{[f(t_{n+1}, y_{n+1}) - f(t_{n+1}, y_n + hf(t_n, y_n))]\},$$

where E_1 is the error due to numerical integration with the trapezoidal method and E_2 can be bounded by the error due to using the forward Euler method.]
2. Prove that the Crank-Nicoloson method has order 2 with respect to h.
[*Solution*: using (9.12) we get, for a suitable ξ_n in (t_n, t_{n+1})

$$y_{n+1} = y_n + \frac{h}{2}[f(t_n, y_n) + f(t_{n+1}, y_{n+1})] - \frac{h^3}{12} f''(\xi_n, y(\xi_n))$$

or, equivalently,

$$\frac{y_{n+1} - y_n}{h} = \frac{1}{2}[f(t_n, y_n) + f(t_{n+1}, y_{n+1})] - \frac{h^2}{12} f''(\xi_n, y(\xi_n)). \qquad (11.90)$$

Therefore, relation (11.9) coincides with (11.90) up to an infinitesimal of order 2 with respect to h, provided that $f \in C^2(I)$.]
3. Solve the difference equation $u_{n+4} - 6u_{n+3} + 14u_{n+2} - 16u_{n+1} + 8u_n = n$ subject to the initial conditions $u_0 = 1$, $u_1 = 2$, $u_2 = 3$ and $u_3 = 4$.
[*Solution*: $u_n = 2^n(n/4 - 1) + 2^{(n-2)/2}\sin(\pi/4) + n + 2$.]
4. Prove that if the characteristic polynomial Π defined in (11.30) has simple roots, then any solution of the associated difference equation can be written in the form (11.32).
[*Hint*: notice that a generic solution u_{n+k} is completely determined by the initial values $u_0, \ldots, u_{k-1}$. Moreover, if the roots r_i of Π are distinct, there exist unique k coefficients α_i such that $\alpha_1 r_1^j + \ldots + \alpha_k r_k^j = u_j$ with $j = 0, \ldots, k-1$.]

5. Prove that if the characteristic polynomial Π has simple roots, the matrix R defined in (11.37) is not singular.
[*Hint*: it coincides with the transpose of the Vandermonde matrix where x_i^j is replaced by r_j^i (see Exercise 2, Chapter 8).]
6. The Legendre polynomials L_i satisfy the difference equation

$$(n+1)L_{n+1}(x) - (2n+1)xL_n(x) + nL_{n-1}(x) = 0,$$

with $L_0(x) = 1$ and $L_1(x) = x$ (see Section 10.1.2). Defining the generating function $F(z,x) = \sum_{n=0}^{\infty} P_n(x)z^n$, prove that $F(z,x) = (1-2zx+z^2)^{-1/2}$.
7. Prove that the *gamma function*

$$\Gamma(z) = \int_0^{\infty} e^{-t}t^{z-1}dt, \qquad z \in \mathbb{C}, \quad \mathrm{Re} z > 0,$$

is the solution of the difference equation $\Gamma(z+1) = z\Gamma(z)$
[*Hint*: integrate by parts.]
8. Study, as functions of $\alpha \in \mathbb{R}$, stability and order of the family of linear multistep methods

$$u_{n+1} = \alpha u_n + (1-\alpha)u_{n-1} + 2hf_n + \frac{h\alpha}{2}\left[f_{n-1} - 3f_n\right].$$

9. Consider the following family of one-step methods depending on the real parameter α

$$u_{n+1} = u_n + h[(1-\frac{\alpha}{2})f(x_n, u_n) + \frac{\alpha}{2}f(x_{n+1}, u_{n+1})].$$

Study their consistency as a function of α; then, take $\alpha = 1$ and use the corresponding method to solve the Cauchy problem

$$\begin{cases} y'(x) = -10y(x), & x > 0, \\ y(0) = 1. \end{cases}$$

Determine the values of h in correspondance of which the method is absolutely stable.
[*Solution*: the family of methods is consistent for any value of α. The method of highest order (equal to two) is obtained for $\alpha = 1$ and coincides with the Crank-Nicolson method.]
10. Consider the family of linear multistep methods

$$u_{n+1} = \alpha u_n + \frac{h}{2}\left(2(1-\alpha)f_{n+1} + 3\alpha f_n - \alpha f_{n-1}\right),$$

where α is a real parameter.
   a) Analyze consistency and order of the methods as functions of α, determining the value α^* for which the resulting method has maximal order.
   b) Study the zero-stability of the method with $\alpha = \alpha^*$, write its characteristic polynomial $\Pi(r; h\lambda)$ and, using MATLAB, draw its region of absolute stability in the complex plane.

11. Adams methods can be easily generalized, integrating between t_{n-r} and t_{n+1} with $r \geq 1$. Show that, by doing so, we get methods of the form

$$u_{n+1} = u_{n-r} + h \sum_{j=-1}^{p} b_j f_{n-j}$$

and prove that for $r = 1$ the midpoint method introduced in (11.43) is recovered (the methods of this family are called *Nystron methods.*)

12. Check that Heun's method (11.10) is an explicit two-stage RK method and write the Butcher arrays of the method. Then, do the same for the *modified Euler method*, given by

$$u_{n+1} = u_n + hf(t_n + \frac{h}{2}, \quad u_n + \frac{h}{2} f_n), \; n \geq 0. \tag{11.91}$$

[*Solution*: the methods have the following Butcher arrays

$$\begin{array}{c|cc} 0 & 0 & 0 \\ 1 & 1 & 0 \\ \hline & \frac{1}{2} & \frac{1}{2} \end{array} \qquad \begin{array}{c|cc} 0 & 0 & 0 \\ \frac{1}{2} & \frac{1}{2} & 0 \\ \hline & 0 & 1 \end{array}$$
.]

13. Check that the Butcher array for method (11.73) is given by

$$\begin{array}{c|cccc} 0 & 0 & 0 & 0 & 0 \\ \frac{1}{2} & \frac{1}{2} & 0 & 0 & 0 \\ \frac{1}{2} & 0 & \frac{1}{2} & 0 & 0 \\ 1 & 0 & 0 & 1 & 0 \\ \hline & \frac{1}{6} & \frac{1}{3} & \frac{1}{3} & \frac{1}{6} \end{array} .$$

14. Write a MATLAB program to draw the regions of absolute stability for a RK method, for which the function $R(h\lambda)$ is available. Check the code in the special case of

$$R(h\lambda) = 1 + h\lambda + (h\lambda)^2/2 + (h\lambda)^3/6 + (h\lambda)^4/24 + (h\lambda)^5/120 + (h\lambda)^6/600$$

and verify that such a region is not connected.

15. Determine the function $R(h\lambda)$ associated with the *Merson method*, whose Butcher array is

$$\begin{array}{c|ccccc} 0 & 0 & 0 & 0 & 0 & 0 \\ \frac{1}{3} & \frac{1}{3} & 0 & 0 & 0 & 0 \\ \frac{1}{3} & \frac{1}{6} & \frac{1}{6} & 0 & 0 & 0 \\ \frac{1}{2} & \frac{1}{8} & 0 & \frac{3}{8} & 0 & 0 \\ 1 & \frac{1}{2} & 0 & -\frac{3}{2} & 2 & 0 \\ \hline & \frac{1}{6} & 0 & 0 & \frac{2}{3} & \frac{1}{6} \end{array} .$$

[*Solution*: one gets $R(h\lambda) = 1 + \sum_{i=1}^{4} (h\lambda)^i / i! + (h\lambda)^5/144$.]

12

Two-Point Boundary Value Problems

This chapter is devoted to the analysis of approximation methods for two-point boundary value problems for differential equations of elliptic type. Finite differences, finite elements and spectral methods will be considered. A short account is also given on the extension to elliptic boundary value problems in two-dimensional regions.

12.1 A Model Problem

To start with, let us consider the two-point boundary value problem

$$-u''(x) = f(x),\ 0 < x < 1, \tag{12.1}$$

$$u(0) = u(1) = 0. \tag{12.2}$$

From the fundamental theorem of calculus, if $u \in C^2([0,1])$ and satisfies the differential equation (12.1) then

$$u(x) = c_1 + c_2 x - \int_0^x F(s)\, ds,$$

where c_1 and c_2 are arbitrary constants and $F(s) = \int_0^s f(t)\, dt$. Using integration by parts one has

$$\int_0^x F(s)\, ds = [sF(s)]_0^x - \int_0^x sF'(s)\, ds = \int_0^x (x-s)f(s)\, ds.$$

The constants c_1 and c_2 can be determined by enforcing the boundary conditions. The condition $u(0) = 0$ implies that $c_1 = 0$, and then $u(1) = 0$ yields

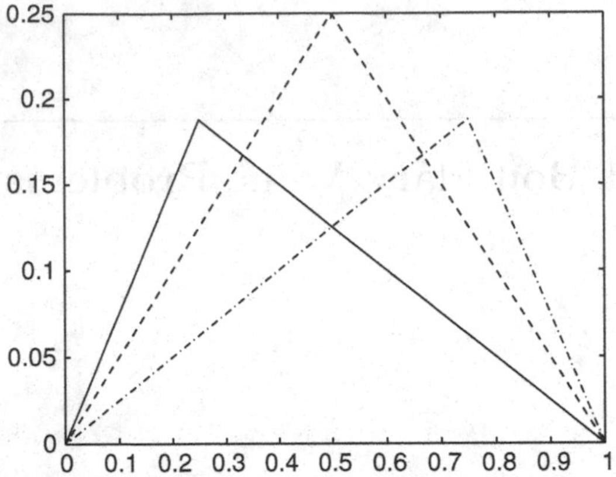

Fig. 12.1. Green's function for three different values of x: $x = 1/4$ (*solid line*), $x = 1/2$ (*dashed line*), $x = 3/4$ (*dash-dotted line*)

$c_2 = \int_0^1 (1-s)f(s)\,ds$. Consequently, the solution of (12.1)-(12.2) can be written in the following form

$$u(x) = x\int_0^1 (1-s)f(s)\,ds - \int_0^x (x-s)f(s)\,ds$$

or, more compactly,

$$u(x) = \int_0^1 G(x,s)f(s)\,ds, \tag{12.3}$$

where, for any fixed x, we have defined

$$G(x,s) = \begin{cases} s(1-x) \text{ if } 0 \le s \le x, \\ x(1-s) \text{ if } x \le s \le 1. \end{cases} \tag{12.4}$$

The function G is called *Green's function* for the boundary value problem (12.1)-(12.2). It is a piecewise linear function of x for fixed s, and vice versa. It is continuous, symmetric (i.e., $G(x,s) = G(s,x)$ for all $x, s \in [0,1]$), nonnegative, null if x or s are equal to 0 or 1, and $\int_0^1 G(x,s)\,ds = \frac{1}{2}x(1-x)$. The function is plotted in Figure 12.1.

We can therefore conclude that for every $f \in C^0([0,1])$ there is a unique solution $u \in C^2([0,1])$ of the boundary value problem (12.1)-(12.2) which admits the representation (12.3). Further smoothness of u can be derived by (12.1); indeed, if $f \in C^m([0,1])$ for some $m \ge 0$ then $u \in C^{m+2}([0,1])$.
An interesting property of the solution u is that if $f \in C^0([0,1])$ is a nonnegative function, then u is also nonnegative. This is referred to as the

monotonicity property, and follows directly from (12.3), since $G(x,s) \geq 0$ for all $x, s \in [0,1]$. The next property is called the *maximum principle* and states that if $f \in C^0([0,1])$,

$$\|u\|_\infty \leq \frac{1}{8}\|f\|_\infty, \tag{12.5}$$

where $\|u\|_\infty = \max_{0\leq x\leq 1} |u(x)|$ is the maximum norm. Indeed, since G is nonnegative,

$$|u(x)| \leq \int_0^1 G(x,s)|f(s)|\, ds \leq \|f\|_\infty \int_0^1 G(x,s)\, ds = \frac{1}{2}x(1-x)\|f\|_\infty$$

from which the inequality (12.5) follows.

12.2 Finite Difference Approximation

We introduce on $[0,1]$ the grid points $\{x_j\}_{j=0}^n$ given by $x_j = jh$, where $n \geq 2$ is an integer and $h = 1/n$ is the grid spacing. The approximation to the solution u is a finite sequence $\{u_j\}_{j=0}^n$ defined only at the grid points (with the understanding that u_j approximates $u(x_j)$) by requiring that

$$-\frac{u_{j+1} - 2u_j + u_{j-1}}{h^2} = f(x_j), \qquad \text{for } j = 1, \ldots, n-1 \tag{12.6}$$

and $u_0 = u_n = 0$. This corresponds to having replaced $u''(x_j)$ by its second order centered finite difference (10.65) (see Section 10.10.1).

If we set $\mathbf{u} = (u_1, \ldots, u_{n-1})^T$ and $\mathbf{f} = (f_1, \ldots, f_{n-1})^T$, with $f_i = f(x_i)$, it is a simple matter to see that (12.6) can be written in the more compact form

$$\mathrm{A}_{\mathrm{fd}}\mathbf{u} = \mathbf{f}, \tag{12.7}$$

where A_{fd} is the symmetric $(n-1)\times(n-1)$ finite difference matrix defined as

$$\mathrm{A}_{\mathrm{fd}} = h^{-2}\mathrm{tridiag}_{n-1}(-1, 2, -1). \tag{12.8}$$

This matrix is diagonally dominant by rows; moreover, it is positive definite since for any vector $\mathbf{x} \in \mathbb{R}^{n-1}$

$$\mathbf{x}^T\mathrm{A}_{\mathrm{fd}}\mathbf{x} = h^{-2}\left[x_1^2 + x_{n-1}^2 + \sum_{i=2}^{n-1}(x_i - x_{i-1})^2\right].$$

This implies that (12.7) admits a unique solution. Another interesting property is that A_{fd} is an M-matrix (see Definition 1.25 and Exercise 2), which guarantees that the finite difference solution enjoys the same monotonicity property as the exact solution $u(x)$, namely $\mathbf{u}$ is nonnegative if $\mathbf{f}$ is nonnegative. This property is called *discrete maximum principle.*

In order to rewrite (12.6) in operator form, let V_h be a collection of discrete functions defined at the grid points x_j for $j = 0, \ldots, n$. If $v_h \in V_h$, then $v_h(x_j)$ is defined for all j and we sometimes use the shorthand notation v_j instead of $v_h(x_j)$. Next, we let V_h^0 be the subset of V_h containing discrete functions that are zero at the endpoints x_0 and x_n. For a function w_h we define the operator L_h by

$$(L_h w_h)(x_j) = -\frac{w_{j+1} - 2w_j + w_{j-1}}{h^2}, \qquad j = 1, \ldots, n-1, \tag{12.9}$$

and reformulate the finite difference problem (12.6) equivalently as: find $u_h \in V_h^0$ such that

$$(L_h u_h)(x_j) = f(x_j) \qquad \text{for } j = 1, \ldots, n-1. \tag{12.10}$$

Notice that, in this formulation, the boundary conditions are taken care of by the requirement that $u_h \in V_h^0$.

Finite differences can be used to provide approximations of higher-order differential operators than the one considered in this section. An example is given in Section 4.7.2 where the finite difference centered discretization of the fourth-order derivative $-u^{(iv)}(x)$ is carried out by applying twice the discrete operator L_h (see also Exercise 11). Again, extra care is needed to properly handle the boundary conditions.

12.2.1 Stability Analysis by the Energy Method

For two discrete functions $w_h, v_h \in V_h$ we define the *discrete inner product*

$$(w_h, v_h)_h = h \sum_{k=0}^{n} c_k w_k v_k,$$

with $c_0 = c_n = 1/2$ and $c_k = 1$ for $k = 1, \ldots, n-1$. This is nothing but the composite trapezoidal rule (9.13) which is here used to evaluate the inner product $(w, v) = \int_0^1 w(x)v(x)dx$. Clearly,

$$\|v_h\|_h = (v_h, v_h)_h^{1/2}$$

is a norm on V_h.

Lemma 12.1 *The operator L_h is symmetric, i.e.*

$$(L_h w_h, v_h)_h = (w_h, L_h v_h)_h \qquad \forall\, w_h, v_h \in V_h^0,$$

and is positive definite, i.e.

$$(L_h v_h, v_h)_h \geq 0 \qquad \forall v_h \in V_h^0,$$

with equality only if $v_h \equiv 0$.

Proof. From the identity

$$w_{j+1}v_{j+1} - w_j v_j = (w_{j+1} - w_j)v_j + (v_{j+1} - v_j)w_{j+1},$$

upon summation over j from 0 to $n-1$ we obtain the following relation for all $w_h, v_h \in V_h$

$$\sum_{j=0}^{n-1}(w_{j+1} - w_j)v_j = w_n v_n - w_0 v_0 - \sum_{j=0}^{n-1}(v_{j+1} - v_j)w_{j+1},$$

which is referred to as *summation by parts.* Using summation by parts twice, and setting for ease of notation $w_{-1} = v_{-1} = 0$, for all $w_h, v_h \in V_h^0$ we obtain

$$\begin{aligned}(L_h w_h, v_h)_h &= -h^{-1}\sum_{j=0}^{n-1}\left[(w_{j+1} - w_j) - (w_j - w_{j-1})\right] v_j \\ &= h^{-1}\sum_{j=0}^{n-1}(w_{j+1} - w_j)(v_{j+1} - v_j).\end{aligned}$$

From this relation we deduce that $(L_h w_h, v_h)_h = (w_h, L_h v_h)_h$; moreover, taking $w_h = v_h$ we obtain

$$(L_h v_h, v_h)_h = h^{-1}\sum_{j=0}^{n-1}(v_{j+1} - v_j)^2. \tag{12.11}$$

This quantity is always positive, unless $v_{j+1} = v_j$ for $j = 0, \ldots, n-1$, in which case $v_j = 0$ for $j = 0, \ldots, n$ since $v_0 = 0$. ◇

For any grid function $v_h \in V_h^0$ we define the following norm

$$|||v_h|||_h = \left\{ h\sum_{j=0}^{n-1}\left(\frac{v_{j+1} - v_j}{h}\right)^2 \right\}^{1/2}. \tag{12.12}$$

Thus, (12.11) is equivalent to

$$(L_h v_h, v_h)_h = |||v_h|||_h^2 \qquad \text{for all } v_h \in V_h^0. \tag{12.13}$$

Lemma 12.2 *The following inequality holds for any function $v_h \in V_h^0$*

$$\|v_h\|_h \leq \frac{1}{\sqrt{2}}|||v_h|||_h. \tag{12.14}$$

Proof. Since $v_0 = 0$, we have

$$v_j = h\sum_{k=0}^{j-1}\frac{v_{k+1} - v_k}{h} \qquad \text{for all } j = 1, \ldots, n-1.$$

Then,

$$v_j^2 = h^2 \left[\sum_{k=0}^{j-1} \left(\frac{v_{k+1} - v_k}{h} \right) \right]^2.$$

Using the Minkowski inequality

$$\left(\sum_{k=1}^{m} p_k \right)^2 \leq m \left(\sum_{k=1}^{m} p_k^2 \right), \tag{12.15}$$

which holds for every integer $m \geq 1$ and every sequence $\{p_1, \ldots, p_m\}$ of real numbers (see Exercise 4), we obtain

$$\sum_{j=1}^{n-1} v_j^2 \leq h^2 \sum_{j=1}^{n-1} j \sum_{k=0}^{j-1} \left(\frac{v_{k+1} - v_k}{h} \right)^2.$$

Then for every $v_h \in V_h^0$ we get

$$\|v_h\|_h^2 = h \sum_{j=1}^{n-1} v_j^2 \leq h^2 \sum_{j=1}^{n-1} jh \sum_{k=0}^{n-1} \left(\frac{v_{k+1} - v_k}{h} \right)^2 = h^2 \frac{(n-1)n}{2} |||v_h|||_h^2.$$

Inequality (12.14) follows since $h = 1/n$. ◇

Remark 12.1 For every $v_h \in V_h^0$, the grid function $v_h^{(1)}$ whose grid values are $(v_{j+1} - v_j)/h$, $j = 0, \ldots, n-1$, can be regarded as a discrete derivative of v_h (see Section 10.10.1). Inequality (12.14) can thus be rewritten as

$$\|v_h\|_h \leq \frac{1}{\sqrt{2}} \|v_h^{(1)}\|_h \qquad \forall v_h \in V_h^0.$$

It can be regarded as the discrete counterpart in $[0, 1]$ of the following *Poincaré inequality*: for every interval $[a, b]$ there exists a constant $C_P > 0$ such that

$$\|v\|_{L^2(a,b)} \leq C_P \|v^{(1)}\|_{L^2(a,b)} \tag{12.16}$$

for all $v \in C^1([a, b])$ such that $v(a) = v(b) = 0$ and where $\| \cdot \|_{L^2(a,b)}$ is the norm in $L^2(a, b)$ (see (8.29)). ■

Inequality (12.14) has an interesting consequence. If we multiply every equation of (12.10) by u_j and then sum for j from 1 on $n-1$, we obtain

$$(L_h u_h, u_h)_h = (f, u_h)_h.$$

Applying to (12.13) the Cauchy-Schwarz inequality (1.14) (valid in the finite dimensional case), we obtain

$$|||u_h|||_h^2 \leq \|f_h\|_h \|u_h\|_h,$$

where $f_h \in V_h$ is the grid function such that $f_h(x_j) = f(x_j)$ for all $j = 0, \ldots, n$. Owing to (12.14) we conclude that

$$\|u_h\|_h \leq \frac{1}{2}\|f_h\|_h, \tag{12.17}$$

from which we deduce that the finite difference problem (12.6) has a unique solution (equivalently, the only solution corresponding to $f_h = 0$ is $u_h = 0$). Moreover, (12.17) is a *stability* result, as it states that the finite difference solution is bounded by the given datum f_h.

To prove convergence, we first introduce the notion of consistency. According to our general definition (2.13), if $f \in C^0([0,1])$ and $u \in C^2([0,1])$ is the corresponding solution of (12.1)-(12.2), the local truncation error is the grid function τ_h defined by

$$\tau_h(x_j) = (L_h u)(x_j) - f(x_j), \qquad j = 1, \ldots, n-1. \tag{12.18}$$

By Taylor series expansion and recalling (10.66), one obtains

$$\begin{aligned}
\tau_h(x_j) &= -h^{-2}\left[u(x_{j-1}) - 2u(x_j) + u(x_{j+1})\right] - f(x_j) \\
&= -u''(x_j) - f(x_j) + \frac{h^2}{24}(u^{(iv)}(\xi_j) + u^{(iv)}(\eta_j)) \\
&= \frac{h^2}{24}(u^{(iv)}(\xi_j) + u^{(iv)}(\eta_j))
\end{aligned} \tag{12.19}$$

for suitable $\xi_j \in (x_{j-1}, x_j)$ and $\eta_j \in (x_j, x_{j+1})$. Upon defining the *discrete maximum norm* as

$$\|v_h\|_{h,\infty} = \max_{0 \leq j \leq n} |v_h(x_j)|,$$

we obtain from (12.19)

$$\|\tau_h\|_{h,\infty} \leq \frac{\|f''\|_\infty}{12} h^2 \tag{12.20}$$

provided that $f \in C^2([0,1])$. In particular, $\lim_{h \to 0} \|\tau_h\|_{h,\infty} = 0$ and therefore the finite difference scheme is consistent with the differential problem (12.1)-(12.2).

Remark 12.2 Taylor's expansion of u around x_j can also be written as

$$u(x_j \pm h) = u(x_j) \pm h u'(x_j) + \frac{h^2}{2} u''(x_j) \pm \frac{h^3}{6} u'''(x_j) + \mathcal{R}_4(x_j \pm h)$$

with the following integral form of the remainder

$$\mathcal{R}_4(x_j + h) = \int_{x_j}^{x_j + h} (u'''(t) - u'''(x_j)) \frac{(x_j + h - t)^2}{2} dt,$$

$$\mathcal{R}_4(x_j - h) = -\int_{x_j - h}^{x_j} (u'''(t) - u'''(x_j)) \frac{(x_j - h - t)^2}{2} dt.$$

Using the two formulae above, by inspection on (12.18) it is easy to see that

$$\tau_h(x_j) = \frac{1}{h^2}\left(\mathcal{R}_4(x_j + h) + \mathcal{R}_4(x_j - h)\right). \tag{12.21}$$

For any integer $m \geq 0$, we denote by $C^{m,1}(0,1)$ the space of all functions in $C^m(0,1)$ whose m-th derivative is Lipschitz continuous, i.e.

$$\max_{x,y\in(0,1),x\neq y} \frac{|v^{(m)}(x) - v^{(m)}(y)|}{|x-y|} \leq M < \infty.$$

Looking at (12.21) we see that it suffices to assuming that $u \in C^{3,1}(0,1)$ to conclude that

$$\|\tau_h\|_{h,\infty} \leq Mh^2,$$

which shows that the finite difference scheme is consistent with the differential problem (12.1)-(12.2) even under a slightly weaker regularity of the exact solution u. ■

Remark 12.3 Let $e = u - u_h$ be the *discretization error* grid function. Then,

$$L_h e = L_h u - L_h u_h = L_h u - f_h = \tau_h. \tag{12.22}$$

It can be shown (see Exercise 5) that

$$\|\tau_h\|_h^2 \leq 3\left(\|f\|_h^2 + \|f\|_{L^2(0,1)}^2\right) \tag{12.23}$$

from which it follows that the norm of the discrete second-order derivative of the discretization error is bounded, provided that the norms of f at the right-hand side of (12.23) are also bounded. ■

12.2.2 Convergence Analysis

The finite difference solution u_h can be characterized by a discrete Green's function as follows. For a given grid point x_k define a grid function $G^k \in V_h^0$ as the solution to the following problem

$$L_h G^k = e^k, \tag{12.24}$$

where $e^k \in V_h^0$ satisfies $e^k(x_j) = \delta_{kj}$, $1 \leq j \leq n-1$. It is easy to see that $G^k(x_j) = hG(x_j, x_k)$, where G is the Green's function introduced in (12.4) (see Exercise 6). For any grid function $g \in V_h^0$ we can define the grid function

$$w_h = T_h g, \qquad w_h = \sum_{k=1}^{n-1} g(x_k) G^k. \tag{12.25}$$

Then

$$L_h w_h = \sum_{k=1}^{n-1} g(x_k) L_h G^k = \sum_{k=1}^{n-1} g(x_k) e^k = g.$$

In particular, the solution u_h of (12.10) satisfies $u_h = T_h f$, therefore

$$u_h = \sum_{k=1}^{n-1} f(x_k) G^k, \text{ and } u_h(x_j) = h \sum_{k=1}^{n-1} G(x_j, x_k) f(x_k). \tag{12.26}$$

Theorem 12.1 *Assume that $f \in C^2([0,1])$. Then, the nodal error $e(x_j) = u(x_j) - u_h(x_j)$ satisfies*

$$\|u - u_h\|_{h,\infty} \le \frac{h^2}{96} \|f''\|_\infty, \tag{12.27}$$

i.e. u_h converges to u (in the discrete maximum norm) with second order with respect to h.

Proof. We start by noticing that, thanks to the representation (12.26), the following discrete counterpart of (12.5) holds

$$\|u_h\|_{h,\infty} \le \frac{1}{8} \|f\|_{h,\infty}. \tag{12.28}$$

Indeed, we have

$$\begin{aligned} |u_h(x_j)| &\le h \sum_{k=1}^{n-1} G(x_j, x_k) |f(x_k)| \le \|f\|_{h,\infty} \left(h \sum_{k=1}^{n-1} G(x_j, x_k) \right) \\ &= \|f\|_{h,\infty} \frac{1}{2} x_j (1 - x_j) \le \frac{1}{8} \|f\|_{h,\infty} \end{aligned}$$

since, if $g = 1$, then $T_h g$ is such that $T_h g(x_j) = \frac{1}{2} x_j (1 - x_j)$ (see Exercise 7).

Inequality (12.28) provides a result of stability in the discrete maximum norm for the finite difference solution u_h. Using (12.22), by the same argument used to prove (12.28) we obtain

$$\|e\|_{h,\infty} \le \frac{1}{8} \|\tau_h\|_{h,\infty}.$$

Finally, the thesis (12.27) follows owing to (12.20). ◇

Observe that for the derivation of the convergence result (12.27) we have used both stability and consistency. In particular, the discretization error is of the same order (with respect to h) as the consistency error τ_h.

12.2.3 Finite Differences for Two-Point Boundary Value Problems with Variable Coefficients

A two-point boundary value problem more general than (12.1)-(12.2) is the following one

$$\begin{aligned} &Lu(x) = -(J(u)(x))' + \gamma(x)u(x) = f(x), 0 < x < 1, \\ &u(0) = d_0, \quad u(1) = d_1, \end{aligned} \tag{12.29}$$

where

$$J(u)(x) = \alpha(x)u'(x), \tag{12.30}$$

d_0 and d_1 are assigned constants and α, γ and f are given functions that are continuous in $[0, 1]$. Finally, $\gamma(x) \geq 0$ in $[0, 1]$ and $\alpha(x) \geq \alpha_0 > 0$ for a suitable α_0. The auxiliary variable $J(u)$ is the *flux* associated with u and very often has a precise physical meaning.

For the approximation, it is convenient to introduce on $[0, 1]$ a new grid made by the midpoints $x_{j+1/2} = (x_j + x_{j+1})/2$ of the intervals $[x_j, x_{j+1}]$ for $j = 0, \ldots, n-1$. Then, a finite difference approximation of (12.29) is given by: find $u_h \in V_h$ such that

$$\begin{aligned} &L_h u_h(x_j) = f(x_j) \text{ for all } j = 1, \ldots, n-1, \\ &u_h(x_0) = d_0, \qquad u_h(x_n) = d_1, \end{aligned} \tag{12.31}$$

where L_h is defined for $j = 1, \ldots, n-1$ as

$$L_h w_h(x_j) = -\frac{J_{j+1/2}(w_h) - J_{j-1/2}(w_h)}{h} + \gamma_j w_j. \tag{12.32}$$

We have defined $\gamma_j = \gamma(x_j)$ and, for $j = 0, \ldots, n-1$, the *approximate fluxes* are given by

$$J_{j+1/2}(w_h) = \alpha_{j+1/2}\frac{w_{j+1} - w_j}{h}, \tag{12.33}$$

with $\alpha_{j+1/2} = \alpha(x_{j+1/2})$.
The finite difference scheme (12.31)-(12.32) with the approximate fluxes (12.33) can still be cast in the form (12.7) by setting

$$\mathrm{A}_{\mathrm{fd}} = h^{-2}\mathrm{tridiag}_{n-1}(\mathbf{a}, \mathbf{d}, \mathbf{a}) + \mathrm{diag}_{n-1}(\mathbf{c}), \tag{12.34}$$

where

$$\begin{aligned} &\mathbf{a} = -\left(\alpha_{3/2}, \alpha_{5/2}, \ldots, \alpha_{n-3/2}\right)^T \in \mathbb{R}^{n-2}, \\ &\mathbf{d} = \left(\alpha_{1/2} + \alpha_{3/2}, \ldots, \alpha_{n-3/2} + \alpha_{n-1/2}\right)^T \in \mathbb{R}^{n-1}, \\ &\mathbf{c} = (\gamma_1, \ldots, \gamma_{n-1})^T \in \mathbb{R}^{n-1}. \end{aligned}$$

The matrix (12.34) is symmetric positive definite and is also strictly diagonally dominant if $\gamma > 0$.

The convergence analysis of the scheme (12.31)-(12.32) can be carried out by extending straightforwardly the techniques developed in Sections 12.2.1 and 12.2.2.

We conclude this section by addressing boundary conditions that are more general than those considered in (12.29). For this purpose we assume that

$$u(0) = d_0,\ J(u)(1) = g_1,$$

where d_0 and g_1 are two given data. The boundary condition at $x = 1$ is called a *Neumann* condition while the condition at $x = 0$ (where the value of u is assigned) is a *Dirichlet* boundary condition. The finite difference discretization of the Neumann boundary condition can be performed by using the *mirror imaging* technique. For any sufficiently smooth function ψ we write its truncated Taylor's expansion at x_n as

$$\psi_n = \frac{\psi_{n-1/2} + \psi_{n+1/2}}{2} - \frac{h^2}{16}(\psi''(\eta_n) + \psi''(\xi_n))$$

for suitable $\eta_n \in (x_{n-1/2}, x_n)$, $\xi_n \in (x_n, x_{n+1/2})$. Taking $\psi = J(u)$ and neglecting the term containing h^2 yields

$$J_{n+1/2}(u_h) = 2g_1 - J_{n-1/2}(u_h). \tag{12.35}$$

Notice that the point $x_{n+1/2} = x_n + h/2$ and the corresponding flux $J_{n+1/2}$ do not really exist (indeed, $x_{n+1/2}$ is called a "ghost" point), but it is generated by linear extrapolation of the flux at the nearby nodes $x_{n-1/2}$ and x_n. The finite difference equation (12.32) at the node x_n reads

$$\frac{J_{n-1/2}(u_h) - J_{n+1/2}(u_h)}{h} + \gamma_n u_n = f_n.$$

Using (12.35) to obtain $J_{n+1/2}(u_h)$ we finally get the second-order accurate approximation

$$-\alpha_{n-1/2}\frac{u_{n-1}}{h^2} + \left(\frac{\alpha_{n-1/2}}{h^2} + \frac{\gamma_n}{2}\right) u_n = \frac{g_1}{h} + \frac{f_n}{2}.$$

This formula suggests easy modification of the matrix and right-hand side entries in the finite difference system (12.7).

For a further generalization of the boundary conditions in (12.29) and their discretization using finite differences we refer to Exercise 10 where boundary conditions of the form $\lambda u + \mu u' = g$ at both the endpoints of $(0, 1)$ are considered for u (*Robin* boundary conditions).

For a thorough presentation and analysis of finite difference approximations of two-point boundary value problems, see, e.g., [Str89] and [HGR96].

12.3 The Spectral Collocation Method

Other discretization schemes can be derived which exhibit the same structure as the finite difference problem (12.10), with a discrete operator L_h being defined in a different manner, though.

Actually, numerical approximations of the second derivative other than the centered finite difference one can be set up, as described in Section 10.10.3. A noticeable instance is provided by the spectral collocation method. In that case we assume the differential equation (12.1) to be set on the interval $(-1,1)$ and choose the nodes $\{x_0,\dots,x_n\}$ to coincide with the $n+1$ Legendre-Gauss-Lobatto nodes introduced in Section 10.4. Besides, u_h is a polynomial of degree n. For coherence, we will use throughout the section the index n instead of h.

The spectral collocation problem reads

$$\text{find } u_n \in \mathbb{P}_n^0: \ L_n u_n(x_j) = f(x_j), \quad j=1,\dots,n-1, \tag{12.36}$$

where $\mathbb{P}_n^0$ is the set of polynomials $p \in \mathbb{P}_n([-1,1])$ such that $p(-1)=p(1)=0$. Besides, $L_n v = L I_n v$ for any continuous function v where $I_n v \in \mathbb{P}_n$ is the interpolant of v at the nodes $\{x_0,\dots,x_n\}$ and L denotes the differential operator at hand, which, in the case of equation (12.1), coincides with $-d^2/dx^2$. Clearly, if $v \in \mathbb{P}_n$ then $L_n v = Lv$.

The algebraic form of (12.36) becomes

$$\mathrm{A}_{\mathrm{sp}}\mathbf{u} = \mathbf{f},$$

where $u_j = u_n(x_j)$, $f_j = f(x_j)$ $j=1,\dots,n-1$ and the spectral collocation matrix $\mathrm{A}_{\mathrm{sp}} \in \mathbb{R}^{(n-1)\times(n-1)}$ is equal to $\tilde{D}^2$, where $\tilde{D}$ is the matrix obtained from the pseudo-spectral differentiation matrix (10.73) by eliminating the first and the $n+1$-th rows and columns.

For the analysis of (12.36) we can introduce the following discrete scalar product

$$(u,v)_n = \sum_{j=0}^{n} u(x_j)v(x_j)w_j, \tag{12.37}$$

where w_j are the weights of the Legendre-Gauss-Lobatto quadrature formula (see Section 10.4). Then (12.36) is equivalent to

$$(L_n u_n, v_n)_n = (f, v_n)_n \ \forall v_n \in \mathbb{P}_n^0. \tag{12.38}$$

Since (12.37) is exact for u, v such that $uv \in \mathbb{P}_{2n-1}$ (see Section 10.2) then

$$(L_n v_n, v_n)_n = (L_n v_n, v_n) = \|v_n'\|^2_{\mathrm{L}^2(-1,1)}, \qquad \forall v_n \in \mathbb{P}_n^0.$$

Besides,

$$(f, v_n)_n \le \|f\|_n \|v_n\|_n \le \sqrt{6}\|f\|_\infty \ \|v_n\|_{\mathrm{L}^2(-1,1)},$$

where $\|f\|_\infty$ denotes the maximum of f in $[-1,1]$ and we have used the fact that $\|f\|_n \leq \sqrt{2}\|f\|_\infty$ and the result of equivalence

$$\|v_n\|_{\mathrm{L}^2(-1,1)} \leq \|v_n\|_n \leq \sqrt{3}\|v_n\|_{\mathrm{L}^2(-1,1)}, \qquad \forall v_n \in \mathbb{P}_n$$

(see [CHQZ06], p. 280).

Taking $v_n = u_n$ in (12.38) and using the Poincaré inequality (12.16) we finally obtain

$$\|u_n'\|_{\mathrm{L}^2(-1,1)} \leq \sqrt{6}C_P\|f\|_\infty,$$

which ensures that problem (12.36) has a unique solution which is stable. As for consistency, we can notice that

$$\tau_n(x_j) = (L_n u - f)(x_j) = (-(I_n u)'' - f)(x_j) = (u - I_n u)''(x_j)$$

and this right-hand side tends to zero as $n \to \infty$ provided that $u \in C^2([-1,1])$.

Let us now establish a convergence result for the spectral collocation approximation of (12.1). In the following, C is a constant independent of n that can assume different values at different places.

Moreover, we denote by $\mathrm{H}^s(a,b)$, for $s \geq 1$, the space of the functions $f \in C^{s-1}(a,b)$ such that $f^{(s-1)}$ is continuous and piecewise differentiable, so that $f^{(s)}$ exists unless for a finite number of points and belongs to $\mathrm{L}^2(a,b)$. The space $\mathrm{H}^s(a,b)$ is known as the *Sobolev* function space of order s and is endowed with the norm $\|\cdot\|_{\mathrm{H}^s(a,b)}$ defined in (10.35).

Theorem 12.2 *Let $f \in \mathrm{H}^s(-1,1)$ for some $s \geq 1$. Then*

$$\|u' - u_n'\|_{\mathrm{L}^2(-1,1)} \leq Cn^{-s}\left(\|f\|_{\mathrm{H}^s(-1,1)} + \|u\|_{\mathrm{H}^{s+1}(-1,1)}\right). \tag{12.39}$$

Proof. Note that u_n satisfies

$$(u_n', v_n') = (f, v_n)_n,$$

where $(u,v) = \int_{-1}^1 uv\,dx$ is the scalar product of $\mathrm{L}^2(-1,1)$. Similarly, u satisfies

$$(u', v') = (f, v) \qquad \forall v \in C^1([0,1]) \text{ such that } v(0) = v(1) = 0$$

(see (12.43) of Section 12.4). Then

$$((u-u_n)', v_n') = (f, v_n) - (f, v_n)_n =: E(f, v_n), \qquad \forall v_n \in \mathbb{P}_n^0.$$

It follows that

$$\begin{aligned}((u-u_n)', (u-u_n)') &= ((u-u_n)', (u - I_n u)') + ((u-u_n)', (I_n u - u_n)') \\ &= ((u-u_n)', (u - I_n u)') + E(f, I_n u - u_n).\end{aligned}$$

We recall the following result (see (10.36))

$$|E(f, v_n)| \leq Cn^{-s}\|f\|_{\mathrm{H}^s(-1,1)}\|v_n\|_{\mathrm{L}^2(-1,1)}.$$

Then

$$|E(f, I_n u - u_n)| \leq C n^{-s} \|f\|_{H^s(-1,1)} \left(\|I_n u - u\|_{L^2(-1,1)} + \|u - u_n\|_{L^2(-1,1)}\right).$$

We recall now the following Young's inequality (see Exercise 8)

$$ab \leq \varepsilon a^2 + \frac{1}{4\varepsilon} b^2, \qquad \forall a, b \in \mathbb{R}, \quad \forall \varepsilon > 0. \tag{12.40}$$

Using this inequality we obtain

$$\left((u - u_n)', (u - I_n u)'\right) \leq \frac{1}{4} \|(u - u_n)'\|^2_{L^2(-1,1)} + \|(u - I_n u)'\|^2_{L^2(-1,1)},$$

and also (using the Poincaré inequality (12.16))

$$\begin{aligned} C n^{-s} \|f\|_{H^s(-1,1)} \|u - u_n\|_{L^2(-1,1)} &\leq C C_P n^{-s} \|f\|_{H^s(-1,1)} \|(u - u_n)'\|_{L^2(-1,1)} \\ &\leq (C C_P)^2 n^{-2s} \|f\|^2_{H^s(-1,1)} + \frac{1}{4} \|(u - u_n)'\|^2_{L^2(-1,1)}. \end{aligned}$$

Finally,

$$C n^{-s} \|f\|_{H^s(-1,1)} \|I_n u - u\|_{L^2(-1,1)} \leq \frac{1}{2} C^2 n^{-2s} \|f\|^2_{H^s(-1,1)} + \frac{1}{2} \|I_n u - u\|^2_{L^2(-1,1)}.$$

Using the interpolation error estimate (10.22) for $u - I_n u$ we finally obtain the desired error estimate (12.39). ◇

12.4 The Galerkin Method

We now derive the Galerkin approximation of problem (12.1)-(12.2), which is the basic ingredient of the finite element method and the spectral method, widely employed in the numerical approximation of boundary value problems.

12.4.1 Integral Formulation of Boundary Value Problems

We consider a problem which is slightly more general than (12.1), namely

$$-(\alpha u')'(x) + (\beta u')(x) + (\gamma u)(x) = f(x) \; 0 < x < 1, \tag{12.41}$$

with $u(0) = u(1) = 0$, where α, β and γ are continuous functions on $[0, 1]$ with $\alpha(x) \geq \alpha_0 > 0$ for any $x \in [0, 1]$. Let us now multiply (12.41) by a function $v \in C^1([0, 1])$, hereafter called a "test function", and integrate over the interval $[0, 1]$

$$\int_0^1 \alpha u' v' \, dx + \int_0^1 \beta u' v \, dx + \int_0^1 \gamma u v \, dx = \int_0^1 f v \, dx + [\alpha u' v]_0^1,$$

where we have used integration by parts on the first integral. If the function v is required to vanish at $x = 0$ and $x = 1$ we obtain

$$\int_0^1 \alpha u'v'\,dx + \int_0^1 \beta u'v\,dx + \int_0^1 \gamma uv\,dx = \int_0^1 fv\,dx.$$

We will denote by V the test function space. This consists of all functions v that are continuous, vanish at $x = 0$ and $x = 1$ and whose first derivative is *piecewise continuous*, i.e., continuous everywhere except at a finite number of points in $[0,1]$ where the left and right limits v'_- and v'_+ exist but do not necessarily coincide.

V is actually a vector space which is denoted by $\mathrm{H}_0^1(0,1)$. Precisely,

$$\mathrm{H}_0^1(0,1) = \left\{v \in \mathrm{L}^2(0,1) :\ v' \in \mathrm{L}^2(0,1),\ v(0) = v(1) = 0\right\}, \qquad (12.42)$$

where v' is the *distributional derivative* of v whose definition is given in Section 12.4.2.

We have therefore shown that if a function $u \in C^2([0,1])$ satisfies (12.41), then u is also a solution of the following problem

$$\text{find } u \in V :\ a(u,v) = (f,v) \ \text{ for all } v \in V, \qquad (12.43)$$

where now $(f,v) = \int_0^1 fv\,dx$ denotes the scalar product of $\mathrm{L}^2(0,1)$ and

$$a(u,v) = \int_0^1 \alpha u'v'\,dx + \int_0^1 \beta u'v\,dx + \int_0^1 \gamma uv\,dx \qquad (12.44)$$

is a bilinear form, i.e. it is linear with respect to both arguments u and v. Problem (12.43) is called the *weak formulation* of problem (12.1). Since (12.43) contains only the first derivative of u it might cover cases in which a classical solution $u \in C^2([0,1])$ of (12.41) does not exist although the physical problem is well defined.

If for instance, $\alpha = 1$, $\beta = \gamma = 0$, the solution $u(x)$ denotes of the displacement at point x of an elastic cord having linear density equal to f, whose position at rest is $u(x) = 0$ for all $x \in [0,1]$ and which remains fixed at the endpoints $x = 0$ and $x = 1$. Figure 12.2 (*right*) shows the solution $u(x)$ corresponding to a function f which is discontinuous (see Figure 12.2, *left*). Clearly, u'' does not exist at the points $x = 0.4$ and $x = 0.6$ where f is discontinuous.

If (12.41) is supplied with nonhomogeneous boundary conditions, say $u(0) = u_0$, $u(1) = u_1$, we can still obtain a formulation like (12.43) by proceeding as follows. Let $\bar{u}(x) = xu_1 + (1-x)u_0$ be the straight line that interpolates the data at the endpoints, and set $\overset{0}{u} = u(x) - \bar{u}(x)$. Then $\overset{0}{u} \in V$ satisfies the following problem

$$\text{find } \overset{0}{u} \in V :\ a(\overset{0}{u}, v) = (f,v) - a(\bar{u}, v) \ \text{ for all } v \in V.$$

A similar problem is obtained in the case of Neumann boundary conditions, say $u'(0) = u'(1) = 0$. Proceeding as we did to obtain (12.43), we see that the

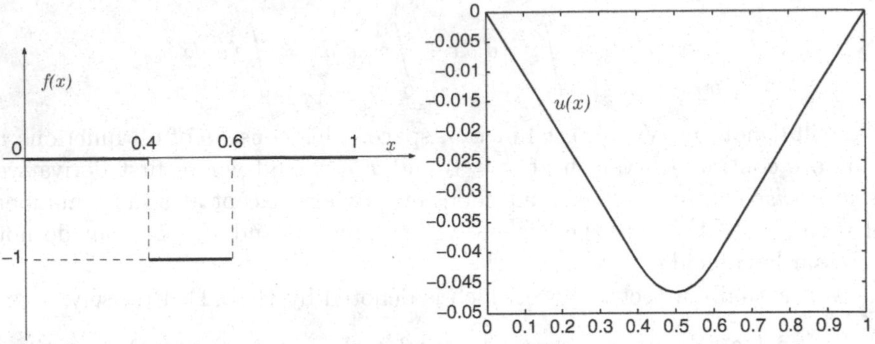

Fig. 12.2. Elastic cord fixed at the endpoints and subject to a discontinuous load f (*left*). The vertical displacement u is shown on the right

solution u of this homogeneous Neumann problem satisfies the same problem (12.43) provided the space V is now $\mathrm{H}^1(0,1)$. More general boundary conditions of mixed type can be considered as well (see Exercise 12).

12.4.2 A Quick Introduction to Distributions

Let X be a Banach space, i.e., a normed and complete vector space. We say that a functional $T : X \to \mathbb{R}$ is *continuous* if $\lim_{x\to x_0} T(x) = T(x_0)$ for all $x_0 \in X$ and *linear* if $T(x+y) = T(x) + T(y)$ for any $x, y \in X$ and $T(\lambda x) = \lambda T(x)$ for any $x \in X$ and $\lambda \in \mathbb{R}$.

Usually, a linear continuous functional is denoted by $\langle T, x\rangle$ and the symbol $\langle \cdot, \cdot \rangle$ is called *duality*. As an example, let $X = C^0([0,1])$ be endowed with the maximum norm $\|\cdot\|_\infty$ and consider on X the two functionals defined as

$$\langle T, x\rangle = x(0), \qquad \langle S, x\rangle = \int_0^1 x(t)\sin(t)dt.$$

It is easy to check that both T and S are linear and continuous functionals on X. The set of all linear continuous functionals on X identifies an abstract space which is called the *dual space* of X and is denoted by X'.

We then introduce the space $C_0^\infty(0,1)$ (or $\mathcal{D}(0,1)$) of infinitely differentiable functions having compact support in $[0,1]$, i.e., vanishing outside a bounded open set $(a,b) \subset (0,1)$ with $0 < a < b < 1$. We say that $v_n \in \mathcal{D}(0,1)$ converges to $v \in \mathcal{D}(0,1)$ if there exists a closed bounded set $K \subset (0,1)$ such that v_n vanishes outside K for each n and for any $k \geq 0$ the derivative $v_n^{(k)}$ converges to $v^{(k)}$ uniformly in $(0,1)$.

The space of linear functionals on $\mathcal{D}(0,1)$ which are continuous with respect to the convergence introduced above is denoted by $\mathcal{D}'(0,1)$ (the *dual space* of $\mathcal{D}(0,1)$) and its elements are called *distributions*.

We are now in position to introduce the *derivative of a distribution.* Let T be a distribution, i.e. an element of $\mathcal{D}'(0,1)$. Then, for any $k \geq 0$, $T^{(k)}$ is also a distribution, defined as

$$\langle T^{(k)}, \varphi\rangle = (-1)^k \langle T, \varphi^{(k)}\rangle, \qquad \forall \varphi \in \mathcal{D}(0,1). \tag{12.45}$$

As an example, consider the Heaviside function

$$H(x) = \begin{cases} 1 \; x \geq 0, \\ 0 \; x < 0. \end{cases}$$

The distributional derivative of H is the *Dirac mass* δ at the origin, defined as

$$v \rightarrow \delta(v) = v(0), \qquad v \in \mathcal{D}(\mathbb{R}).$$

From the definition (12.45), it turns out that any distribution is infinitely differentiable; moreover, if T is a differentiable function its distributional derivative coincides with the usual one.

12.4.3 Formulation and Properties of the Galerkin Method

Unlike the finite difference method which stems directly from the differential (or *strong*) form (12.41), the Galerkin method is based on the weak formulation (12.43). If V_h is a finite dimensional vector subspace of V, the Galerkin method consists of approximating (12.43) by the problem

$$\text{find } u_h \in V_h: \; a(u_h, v_h) = (f, v_h) \quad \forall v_h \in V_h. \tag{12.46}$$

This is a finite dimensional problem. Actually, let $\{\varphi_1, \ldots, \varphi_N\}$ denote a basis of V_h, i.e. a set of N linearly independent functions of V_h. Then we can write

$$u_h(x) = \sum_{j=1}^{N} u_j \varphi_j(x).$$

The integer N denotes the dimension of the vector space V_h. Taking $v_h = \varphi_i$ in (12.46), it turns out that the Galerkin problem (12.46) is equivalent to seeking N unknown coefficients $\{u_1, \ldots, u_N\}$ such that

$$\sum_{j=1}^{N} u_j a(\varphi_j, \varphi_i) = (f, \varphi_i) \qquad \forall i = 1, \ldots, N. \tag{12.47}$$

We have used the linearity of $a(\cdot,\cdot)$ with respect to its first argument, i.e.

$$a(\sum_{j=1}^{N} u_j \varphi_j, \varphi_i) = \sum_{j=1}^{N} u_j a(\varphi_j, \varphi_i).$$

If we introduce the matrix $\mathrm{A_G} = (a_{ij})$, $a_{ij} = a(\varphi_j, \varphi_i)$ (called the *stiffness matrix*), the unknown vector $\mathbf{u} = [u_1, \ldots, u_N]^T$ and the right-hand side vector $\mathbf{f}_\mathrm{G} = [f_1, \ldots, f_N]^T$, with $f_i = (f, \varphi_i)$, we see that (12.47) is equivalent to the linear system

$$\mathrm{A_G}\mathbf{u} = \mathbf{f}_\mathrm{G}. \tag{12.48}$$

The structure of $\mathrm{A_G}$, as well as the degree of accuracy of u_h, depends on the form of the basis functions $\{\varphi_i\}$, and therefore on the choice of V_h.

We will see two remarkable instances, the *finite element method*, where V_h is a space of piecewise polynomials over subintervals of $[0, 1]$ of length not greater than h which are continuous and vanish at the endpoints $x = 0$ and 1, and the *spectral method* in which V_h is a space of algebraic polynomials still vanishing at the endpoints $x = 0, 1$.

However, before specifically addressing those cases, we state a couple of general results that hold for any Galerkin problem (12.46).

12.4.4 Analysis of the Galerkin Method

We endow the space $\mathrm{H}_0^1(0, 1)$ with the following norm

$$|v|_{\mathrm{H}^1(0,1)} = \left\{ \int_0^1 |v'(x)|^2 \, dx \right\}^{1/2}. \tag{12.49}$$

We will address the special case where $\beta = 0$ and $\gamma(x) \geq 0$. In the most general case given by the differential problem (12.41) we shall assume that the coefficients satisfy

$$-\frac{1}{2}\beta' + \gamma \geq 0, \qquad \forall x \in [0, 1]. \tag{12.50}$$

This ensures that the Galerkin problem (12.46) admits a unique solution depending continuously on the data. Taking $v_h = u_h$ in (12.46) we obtain

$$\alpha_0 |u_h|^2_{\mathrm{H}^1(0,1)} \leq \int_0^1 \alpha u_h' u_h' \, dx + \int_0^1 \gamma u_h u_h \, dx = (f, u_h) \leq \|f\|_{\mathrm{L}^2(0,1)} \|u_h\|_{\mathrm{L}^2(0,1)},$$

where we have used the Cauchy-Schwarz inequality (8.33) to set the right-hand side inequality. Owing to the Poincaré inequality (12.16) we conclude that

$$|u_h|_{\mathrm{H}^1(0,1)} \leq \frac{C_P}{\alpha_0} \|f\|_{\mathrm{L}^2(0,1)}. \tag{12.51}$$

Thus, the norm of the Galerkin solution remains bounded (uniformly with respect to the dimension of the subspace V_h) provided that $f \in \mathrm{L}^2(0, 1)$.

Inequality (12.51) therefore represents a stability result for the solution of the Galerkin problem.

As for convergence, we can prove the following result.

Theorem 12.3 *Let $C = \alpha_0^{-1}(\|\alpha\|_\infty + C_P^2\|\gamma\|_\infty)$; then, we have*

$$|u - u_h|_{\mathrm{H}^1(0,1)} \leq C \min_{w_h \in V_h} |u - w_h|_{\mathrm{H}^1(0,1)}. \tag{12.52}$$

Proof. Subtracting (12.46) from (12.43) (where we use $v_h \in V_h \subset V$), owing to the bilinearity of the form $a(\cdot, \cdot)$ we obtain

$$a(u - u_h, v_h) = 0 \qquad \forall v_h \in V_h. \tag{12.53}$$

Then, setting $e(x) = u(x) - u_h(x)$, we deduce

$$\alpha_0 |e|^2_{\mathrm{H}^1(0,1)} \leq a(e, e) = a(e, u - w_h) + a(e, w_h - u_h) \qquad \forall w_h \in V_h.$$

The last term is null due to (12.53). On the other hand, still by the Cauchy-Schwarz inequality we obtain

$$\begin{aligned} a(e, u - w_h) &= \int_0^1 \alpha e'(u - w_h)' \, dx + \int_0^1 \gamma e(u - w_h) \, dx \\ &\leq \|\alpha\|_\infty |e|_{\mathrm{H}^1(0,1)} |u - w_h|_{\mathrm{H}^1(0,1)} + \|\gamma\|_\infty \|e\|_{\mathrm{L}^2(0,1)} \|u - w_h\|_{\mathrm{L}^2(0,1)}. \end{aligned}$$

The desired result (12.52) now follows by using again the Poincaré inequality for both $\|e\|_{\mathrm{L}^2(0,1)}$ and $\|u - w_h\|_{\mathrm{L}^2(0,1)}$. ◇

The previous results can be obtained under more general hypotheses on problems (12.43) and (12.46). Precisely, we can assume that V is a Hilbert space, endowed with norm $\|\cdot\|_V$, and that the bilinear form $a : V \times V \to \mathbb{R}$ satisfies the following properties:

$$\exists \alpha_0 > 0 : \ a(v, v) \geq \alpha_0 \|v\|_V^2 \quad \forall v \in V \ (\textit{coercivity}), \tag{12.54}$$

$$\exists M > 0 : \ |a(u, v)| \leq M \|u\|_V \|v\|_V \quad \forall u, v \in V \ (\textit{continuity}). \tag{12.55}$$

Moreover, the right hand side (f, v) satisfies the following inequality

$$|(f, v)| \leq K \|v\|_V \qquad \forall v \in V.$$

Then both problems (12.43) and (12.46) admit unique solutions that satisfy

$$\|u\|_V \leq \frac{K}{\alpha_0}, \ \|u_h\|_V \leq \frac{K}{\alpha_0}.$$

This is a celebrated result which is known as the Lax-Milgram Lemma (for its proof see, e.g., [QV94]). Besides, the following error inequality holds

$$\|u - u_h\|_V \leq \frac{M}{\alpha_0} \min_{w_h \in V_h} \|u - w_h\|_V. \tag{12.56}$$

The proof of this last result, which is known as Céa's Lemma, is very similar to that of (12.52) and is left to the reader.
We now wish to notice that, under the assumption (12.54), the matrix introduced in (12.48) is positive definite. To show this, we must check that $\mathbf{v}^T \mathrm{A_G} \mathbf{v} \geq 0 \; \forall \mathbf{v} \in \mathbb{R}^N$ and that $\mathbf{v}^T \mathrm{A_G} \mathbf{v} = 0 \Leftrightarrow \mathbf{v} = \mathbf{0}$ (see Section 1.12).

Let us associate with a generic vector $\mathbf{v} = (v_i)$ of $\mathbb{R}^N$ the function $v_h = \sum_{j=1}^N v_j \varphi_j \in V_h$. Since the form $a(\cdot,\cdot)$ is bilinear and coercive we get

$$\begin{aligned}
\mathbf{v}^T \mathrm{A_G} \mathbf{v} &= \sum_{j=1}^N \sum_{i=1}^N v_i a_{ij} v_j = \sum_{j=1}^N \sum_{i=1}^N v_i a(\varphi_j, \varphi_i) v_j \\
&= \sum_{j=1}^N \sum_{i=1}^N a(v_j \varphi_j, v_i \varphi_i) = a\left(\sum_{j=1}^N v_j \varphi_j, \sum_{i=1}^N v_i \varphi_i \right) \\
&= a(v_h, v_h) \geq \alpha_0 \|v_h\|_V^2 \geq 0.
\end{aligned}$$

Moreover, if $\mathbf{v}^T \mathrm{A_G} \mathbf{v} = 0$ then also $\|v_h\|_V^2 = 0$ which implies $v_h = 0$ and thus $\mathbf{v} = \mathbf{0}$.
It is also easy to check that the matrix $\mathrm{A_G}$ is symmetric iff the bilinear form $a(\cdot,\cdot)$ is symmetric.

For example, in the case of problem (12.41) with $\beta = \gamma = 0$ the matrix $\mathrm{A_G}$ is symmetric and positive definite (s.p.d.) while if β and γ are nonvanishing, $\mathrm{A_G}$ is positive definite only under the assumption (12.50). If $\mathrm{A_G}$ is s.p.d. the numerical solution of the linear system (12.48) can be efficiently carried out using direct methods like the Cholesky factorization (see Section 3.4.2) as well as iterative methods like the conjugate gradient method (see Section 4.3.4). This is of particular interest in the solution of boundary value problems in more than one space dimension (see Section 12.6).

12.4.5 The Finite Element Method

The finite element method (FEM) is a special technique for constructing a subspace V_h in (12.46) based on the piecewise polynomial interpolation considered in Section 8.4. With this aim, we introduce a partition $\mathcal{T}_h$ of [0,1] into n subintervals $I_j = [x_j, x_{j+1}]$, $n \geq 2$, of width $h_j = x_{j+1} - x_j$, $j = 0, \ldots, n-1$, with

$$0 = x_0 < x_1 < \ldots < x_{n-1} < x_n = 1$$

and let $h = \max_{\mathcal{T}_h}(h_j)$. Since functions in $\mathrm{H}_0^1(0,1)$ are continuous it makes sense to consider for $k \geq 1$ the family of piecewise polynomials X_h^k introduced in (8.26) (where now $[a,b]$ must be replaced by $[0,1]$). Any function $v_h \in X_h^k$ is a continuous piecewise polynomial over $[0,1]$ and its restriction over each interval $I_j \in \mathcal{T}_h$ is a polynomial of degree $\leq k$. In the following we shall mainly deal with the cases $k = 1$ and $k = 2$.

Then, we set

$$V_h = X_h^{k,0} = \left\{ v_h \in X_h^k : v_h(0) = v_h(1) = 0 \right\}. \tag{12.57}$$

The dimension N of the finite element space V_h is equal to $nk - 1$.

To assess the accuracy of the Galerkin FEM we first notice that, thanks to Céa's lemma (12.56), we have

$$\min_{w_h \in V_h} \|u - w_h\|_{\mathrm{H}_0^1(0,1)} \leq \|u - \Pi_h^k u\|_{\mathrm{H}_0^1(0,1)}, \tag{12.58}$$

where $\Pi_h^k u$ is the interpolant of the exact solution $u \in V$ of (12.43) (see Section 8.4). From inequality (12.58) we conclude that the matter of estimating the Galerkin *approximation error* $\|u - u_h\|_{\mathrm{H}_0^1(0,1)}$ is turned into the estimate of the *interpolation error* $\|u - \Pi_h^k u\|_{\mathrm{H}_0^1(0,1)}$. When $k = 1$, using (12.56) and (8.31) we obtain

$$\|u - u_h\|_{\mathrm{H}_0^1(0,1)} \leq \frac{M}{\alpha_0} Ch \|u\|_{\mathrm{H}^2(0,1)},$$

provided that $u \in \mathrm{H}^2(0,1)$. This estimate can be extended to the case $k > 1$ as stated in the following convergence result (for its proof we refer, e.g., to [QV94], Theorem 6.2.1).

Property 12.1 *Let* $u \in \mathrm{H}_0^1(0,1)$ *be the exact solution of* (12.43) *and* $u_h \in V_h$ *its finite element approximation using continuous piecewise polynomials of degree* $k \geq 1$. *Assume also that* $u \in \mathrm{H}^s(0,1)$ *for some* $s \geq 2$. *Then the following error estimate holds*

$$\|u - u_h\|_{\mathrm{H}_0^1(0,1)} \leq \frac{M}{\alpha_0} Ch^l \|u\|_{\mathrm{H}^{l+1}(0,1)}, \tag{12.59}$$

where $l = \min(k, s-1)$. *Under the same assumptions, one can also prove that*

$$\|u - u_h\|_{\mathrm{L}^2(0,1)} \leq Ch^{l+1} \|u\|_{\mathrm{H}^{l+1}(0,1)}. \tag{12.60}$$

The estimate (12.59) shows that the Galerkin method is *convergent*, i.e. the approximation error tends to zero as $h \to 0$ and the order of convergence is l. We also see that there is no convenience in increasing the degree k of the finite element approximation if the solution u is not sufficiently smooth. In this respect l is called a *regularity threshold.* The obvious alternative to gain accuracy in such a case is to reduce the stepzise h. Spectral methods, which will be considered in Section 12.4.7, instead pursue the opposite strategy (i.e. increasing the degree k) and are thus ideally suited to approximating problems with highly smooth solutions.

An interesting situation is that where the exact solution u has the *minimum* regularity ($s = 1$). In such a case, Céa's lemma ensures that the Galerkin FEM is still convergent since as $h \to 0$ the subspace V_h becomes dense into

Table 12.1. Order of convergence of the FEM as a function of k (the degree of interpolation) and s (the Sobolev regularity of the solution u)

k	$s=1$	$s=2$	$s=3$	$s=4$	$s=5$
1	only convergence	$\boxed{h^1}$	h^1	h^1	h^1
2	only convergence	h^1	$\boxed{h^2}$	h^2	h^2
3	only convergence	h^1	h^2	$\boxed{h^3}$	h^3
4	only convergence	h^1	h^2	h^3	$\boxed{h^4}$

V. However, the estimate (12.59) is no longer valid so that it is not possible to establish the order of convergence of the numerical method. Table 12.1 summarizes the orders of convergence of the FEM for $k = 1, \ldots, 4$ and $s = 1, \ldots, 5$.

Let us now focus on how to generate a suitable basis $\{\varphi_j\}$ for the finite element space X_h^k in the special cases $k = 1$ and $k = 2$. The basic point is to choose appropriately a set of *degrees of freedom* for each element I_j of the partition $\mathcal{T}_h$ (i.e., the parameters which permit uniquely identifying a function in X_h^k). The generic function v_h in X_h^k can therefore be written as

$$v_h(x) = \sum_{i=0}^{nk} v_i \varphi_i(x),$$

where $\{v_i\}$ denote the set of the degrees of freedom of v_h and the basis functions φ_i (which are also called *shape functions*) are assumed to satisfy the Lagrange interpolation property $\varphi_i(x_j) = \delta_{ij}$, $i, j = 0, \ldots, n$, where δ_{ij} is the Kronecker symbol.

The space X_h^1

This space consists of all continuous and piecewise linear functions over the partition $\mathcal{T}_h$. Since a unique straight line passes through two distinct nodes the number of degrees of freedom for v_h is equal to the number $n+1$ of nodes in the partition. As a consequence, $n+1$ shape functions φ_i, $i = 0, \ldots, n$, are needed to completely span the space X_h^1. The most natural choice for φ_i, $i = 1, \ldots, n-1$, is

$$\varphi_i(x) = \begin{cases} \dfrac{x - x_{i-1}}{x_i - x_{i-1}} & \text{for } x_{i-1} \leq x \leq x_i, \\[2ex] \dfrac{x_{i+1} - x}{x_{i+1} - x_i} & \text{for } x_i \leq x \leq x_{i+1}, \\[2ex] 0 & \text{elsewhere.} \end{cases} \tag{12.61}$$

The shape function φ_i is thus piecewise linear over $\mathcal{T}_h$, its value is 1 at the node x_i and 0 at all the other nodes of the partition. Its support (i.e., the subset of

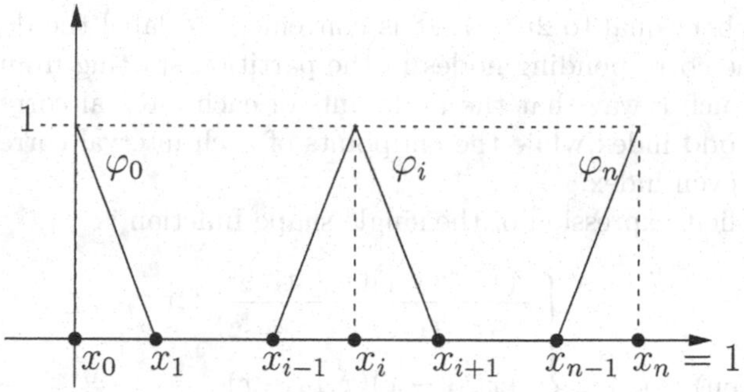

Fig. 12.3. Shape functions of X_h^1 associated with internal and boundary nodes

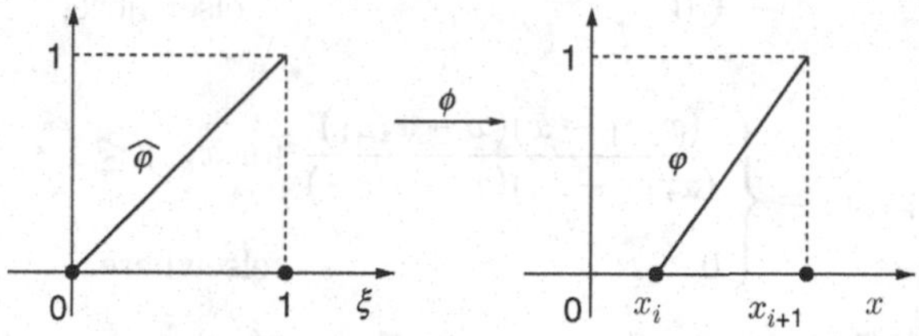

Fig. 12.4. Linear affine mapping ϕ from the reference interval to the generic interval of the partition

$[0, 1]$ where φ_i is nonvanishing) consists of the union of the intervals I_{i-1} and I_i if $1 \leq i \leq n-1$ while it coincides with the interval I_0 (respectively I_{n-1}) if $i = 0$ (resp., $i = n$). The plots of φ_i, φ_0 and φ_n are shown in Figure 12.3.
For any interval $I_i = [x_i, x_{i+1}]$, $i = 0, \ldots, n-1$, the two basis functions φ_i and φ_{i+1} can be regarded as the images of two "reference" shape functions $\widehat{\varphi}_0$ and $\widehat{\varphi}_1$ (defined over the *reference interval* $[0, 1]$) through the linear affine mapping $\phi : [0, 1] \to I_i$

$$x = \phi(\xi) = x_i + \xi(x_{i+1} - x_i), \qquad i = 0, \ldots, n-1. \tag{12.62}$$

Defining $\widehat{\varphi}_0(\xi) = 1 - \xi$, $\widehat{\varphi}_1(\xi) = \xi$, the two shape functions φ_i and φ_{i+1} can be constructed over the interval I_i as

$$\varphi_i(x) = \widehat{\varphi}_0(\xi(x)), \qquad \varphi_{i+1}(x) = \widehat{\varphi}_1(\xi(x)),$$

where $\xi(x) = (x - x_i)/(x_{i+1} - x_i)$ (see Figure 12.4).

The space X_h^2

The generic function $v_h \in X_h^2$ is a piecewise polynomial of degree 2 over each interval I_i. As such, it can be uniquely determined once three values of it at three distinct points of I_i are assigned. To ensure continuity of v_h over $[0, 1]$ the degrees of freedom are chosen as the function values at the nodes x_i of $\mathcal{T}_h$, $i = 0, \ldots, n$, and at the midpoints of each interval I_i, $i = 0, \ldots, n-1$, for a total number equal to $2n+1$. It is convenient to label the degrees of freedom and the corresponding nodes in the partition starting from $x_0 = 0$ until $x_{2n} = 1$ in such a way that the midpoints of each interval correspond to the nodes with odd index while the endpoints of each interval correspond to the nodes with even index.

The explicit expression of the single shape function is

$$(i\,\text{even})\ \varphi_i(x) = \begin{cases} \dfrac{(x-x_{i-1})(x-x_{i-2})}{(x_i-x_{i-1})(x_i-x_{i-2})} & \text{for } x_{i-2} \le x \le x_i, \\ \dfrac{(x_{i+1}-x)(x_{i+2}-x)}{(x_{i+1}-x_i)(x_{i+2}-x_i)} & \text{for } x_i \le x \le x_{i+2}, \\ 0 & \text{elsewhere,} \end{cases} \tag{12.63}$$

$$(i\,\text{odd})\ \varphi_i(x) = \begin{cases} \dfrac{(x_{i+1}-x)(x-x_{i-1})}{(x_{i+1}-x_i)(x_i-x_{i-1})} & \text{for } x_{i-1} \le x \le x_{i+1}, \\ 0 & \text{elsewhere.} \end{cases} \tag{12.64}$$

Each basis function enjoys the property that $\varphi_i(x_j) = \delta_{ij}$, $i, j = 0, \ldots, 2n$. The shape functions for X_h^2 on the reference interval $[0,1]$ are

$$\widehat{\varphi}_0(\xi) = (1-\xi)(1-2\xi),\ \widehat{\varphi}_1(\xi) = 4(1-\xi)\xi,\ \widehat{\varphi}_2(\xi) = \xi(2\xi-1) \tag{12.65}$$

and are shown in Figure 12.5. As in the case of piecewise linear finite elements of X_h^1 the shape functions (12.63) and (12.64) are the images of (12.65) through the affine mapping (12.62). Notice that the support of the basis function φ_{2i+1} associated with the midpoint x_{2i+1} coincides with the interval to which the midpoint belongs. Due to its shape φ_{2i+1} is usually referred to as *bubble function.*

So far, we have considered only lagrangian-type shape functions. If this constraint is removed other kind of bases can be derived. A notable example (on the reference interval) is given by

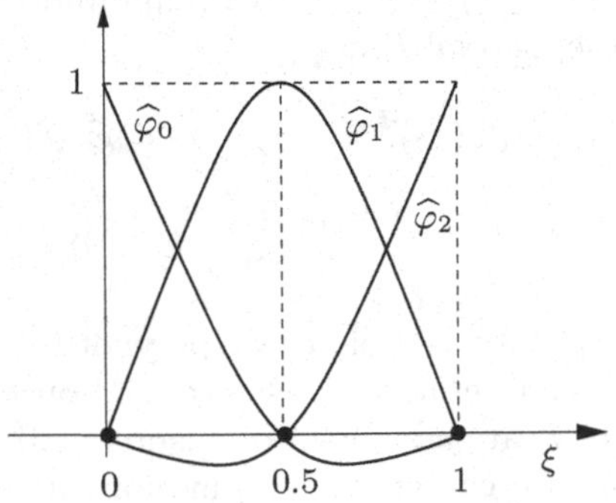

Fig. 12.5. Basis functions of X_h^2 on the reference interval

$$\widehat{\psi}_0(\xi) = 1 - \xi,\ \widehat{\psi}_1(\xi) = (1-\xi)\xi,\ \widehat{\psi}_2(\xi) = \xi. \tag{12.66}$$

This basis is called *hierarchical* since it is generated using the shape functions of the subspace having an immediately lower dimension than X_h^2 (i.e. X_h^1). Precisely, the bubble function $\widehat{\psi}_1 \in X_h^2$ is added to the shape functions $\widehat{\psi}_0$ and $\widehat{\psi}_2$ which belong to X_h^1. Hierarchical bases can be of some interest in numerical computations if the local degree of the interpolation is adaptively increased (*p-type adaptivity*).

To check that (12.66) forms a basis for X_h^2 we must verify that its functions are linearly independent, i.e.

$$\alpha_0\widehat{\psi}_0(\xi) + \alpha_1\widehat{\psi}_1(\xi) + \alpha_2\widehat{\psi}_2(\xi) = 0, \quad \forall \xi \in [0,1] \Leftrightarrow \alpha_0 = \alpha_1 = \alpha_2 = 0.$$

In our case this holds true since if

$$\sum_{i=0}^{2} \alpha_i \widehat{\psi}_i(\xi) = \alpha_0 + \xi(\alpha_1 - \alpha_0 + \alpha_2) - \alpha_1\xi^2 = 0, \qquad \forall \xi \in [0,1],$$

then necessarily $\alpha_0 = 0$, $\alpha_1 = 0$ and thus $\alpha_2 = 0$.
A procedure analogous to that examined in the sections above can be used in principle to construct a basis for every subspace X_h^k with k being arbitrary. However, it is important to remember that an increase in the degree k of the polynomial approximation gives rise to an increase of the number of degrees of freedom of the FEM and, as a consequence, of the computational cost required for the solution of the linear system (12.48).
Let us now examine the structure and the basic properties of the stiffness matrix associated with system (12.48) in the case of the finite element method ($\mathrm{A_G} = \mathrm{A_{fe}}$).

Since the finite element basis functions for X_h^k have a local support, $\mathrm{A_{fe}}$ is *sparse*. In the particular case $k = 1$, the support of the shape function φ_i is the union of the intervals I_{i-1} and I_i if $1 \leq i \leq n-1$, and it coincides with the interval I_0 (respectively I_{n-1}) if $i = 0$ (resp., $i = n$). As a consequence, for a fixed $i = 1, \ldots, n-1$, only the shape functions φ_{i-1} and φ_{i+1} have a nonvanishing support intersection with that of φ_i, which implies that $\mathrm{A_{fe}}$ is *tridiagonal* since $a_{ij} = 0$ if $j \notin \{i-1, i, i+1\}$. In the case $k = 2$ one concludes with an analogous argument that $\mathrm{A_{fe}}$ is a *pentadiagonal* matrix.

The *condition number* of $\mathrm{A_{fe}}$ is a function of the grid size h; indeed,

$$K_2(\mathrm{A_{fe}}) = \|\mathrm{A_{fe}}\|_2 \|\mathrm{A_{fe}^{-1}}\|_2 = \mathcal{O}(h^{-2})$$

(for the proof, see [QV94], Section 6.3.2), which demonstrates that the conditioning of the finite element system (12.48) grows rapidly as $h \to 0$. This is clearly conflicting with the need of increasing the accuracy of the approximation and, in multidimensional problems, demands suitable preconditioning techniques if iterative solvers are used (see Section 4.3.2).

Remark 12.4 (Elliptic problems of higher order) The Galerkin method in general, and the finite element method in particular, can also be applied to other type of elliptic equations, for instance to those of fourth order. In that case, the numerical solution (as well as the test functions) should be continuous together with their first derivative. An example has been illustrated in Section 8.9.1. ■

12.4.6 Implementation Issues

In this section we implement the finite element (FE) approximation with piecewise linear elements ($k = 1$) of the boundary value problem (12.41) (shortly, BVP) with non homogeneous Dirichlet boundary conditions.
Here is the list of the input parameters of Program 94: `Nx` is the number of grid subintervals; `I` is the interval $[a, b]$, `alpha`, `beta`, `gamma` and `f` are the macros corresponding to the coefficients in the equation, `bc=[ua,ub]` is a vector containing the Dirichlet boundary conditions for u at $x = a$ and $x = b$ and `stabfun` is an optional string variable. It can assume different values, allowing the user to select the desired type of artificial viscosity that may be needed for dealing with the problems addressed in Section 12.5.

Program 94 - ellfem : Linear FE for two-point BVPs

```
function [uh,x] = ellfem(Nx,I,alpha,beta,gamma,f,bc,stabfun)
%ELLFEM Finite element solver.
%  [UH,X]=ELLFEM(NX,I,ALPHA,BETA,GAMMA,F,BC,STABFUN) solves the
%  boundary-value problem:
%  -ALPHA*U''+BETA*U'+GAMMA=F in (I(1),I(2))
%  U(I(1))=BC(1),       U(I(2))=BC(2)
%  with linear finite elements. If STABFUN=1, then a stabilized finite element
%  method is considered.
a=I(1); b=I(2); h=(b-a)/Nx; x=[a+h/2:h:b-h/2];
alpha=eval(alpha); beta=eval(beta); gamma=eval(gamma);
f=eval(f);
rhs=0.5*h*(f(1:Nx-1)+f(2:Nx));
if nargin == 8
   [Afe,rhsbc]=femmatr(Nx,h,alpha,beta,gamma,stabfun);
else
   [Afe,rhsbc]=femmatr(Nx,h,alpha,beta,gamma);
end
[L,U,P]=lu(Afe);
rhs(1)=rhs(1)-bc(1)*(-alpha(1)/h-beta(1)/2+h*gamma(1)/3+rhsbc(1));
rhs(Nx-1)=rhs(Nx-1)-bc(2)*(-alpha(Nx)/h+beta(Nx)/2+h*gamma(Nx)/3+rhsbc(2));
rhs=P*rhs';
z=L\rhs;
w=U\z;
uh=[bc(1), w', bc(2)]; x=[a:h:b];
return
```

Program 95 computes the stiffness matrix A_{fe}; with this aim, the coefficients α, β and γ and the forcing term f are replaced by piecewise constant functions on each mesh subinterval and the remaining integrals in (12.41), involving the basis functions and their derivatives, are evaluated exactly.

Program 95 - femmatr : Construction of the stiffness matrix

```
function [Afe,rhsbc] = femmatr(Nx,h,alpha,beta,gamma,stabfun)
%FEMMATR Stiffness matrix and right-hand side.
for i=2:Nx
   dd(i-1)=(alpha(i-1)+alpha(i))/h; dc(i-1)=-(beta(i)-beta(i-1))/2;
   dr(i-1)=h*(gamma(i-1)+gamma(i))/3;
   if i>2
      ld(i-2)=-alpha(i-1)/h; lc(i-2)=-beta(i-1)/2;
      lr(i-2)=h*gamma(i-1)/6;
   end
   if i<Nx
      ud(i-1)=-alpha(i)/h;
      uc(i-1)=beta(i)/2;
      ur(i-1)=h*gamma(i)/6;
   end
end
Kd=spdiags([[ld 0]',dd',[0 ud]'],-1:1,Nx-1,Nx-1);
Kc=spdiags([[lc 0]',dc',[0 uc]'],-1:1,Nx-1,Nx-1);
Kr=spdiags([[lr 0]',dr',[0 ur]'],-1:1,Nx-1,Nx-1);
Afe=Kd+Kc+Kr;
if nargin == 6
   s=['[Ks,rhsbc]=',stabfun,'(Nx,h,alpha,beta);']; eval(s)
   Afe = Afe + Ks;
else
   rhsbc = [0, 0];
end
return
```

The H^1-norm of the error can be computed by calling Program 96, which must be supplied by the macros `u` and `ux` containing the expression of the exact solution u and of u'. The computed numerical solution is stored in the output vector `uh`, while the vector `coord` contains the grid coordinates and `h` is the mesh size. The integrals involved in the computation of the H^1-norm of the error are evaluated using the composite Simpson formula (9.17).

Program 96 - H1error : Computation of the H^1-norm of the error

```
function [L2err,H1err]=H1error(coord,h,uh,u,udx)
%H1ERROR Computes the error in the H1-norm.
nvert=max(size(coord)); x=[]; k=0;
for i = 1:nvert-1
   xm=(coord(i+1)+coord(i))*0.5;
   x=[x, coord(i),xm];
```

```
  k=k+2;
end
ndof=k+1; x(ndof)=coord(nvert);
uq=eval(u); uxq=eval(udx);
L2err=0; H1err=0;
for i=1:nvert-1
   L2err = L2err + (h/6)*((uh(i)-uq(2*i-1))^2+...
      4*(0.5*uh(i)+0.5*uh(i+1)-uq(2*i))^2+(uh(i+1)-uq(2*i+1))^2);
   H1err = H1err + (1/(6*h))*((uh(i+1)-uh(i)-h*uxq(2*i-1))^2+...
      4*(uh(i+1)-uh(i)-h*uxq(2*i))^2+(uh(i+1)-uh(i)-h*uxq(2*i+1))^2);
end
H1err = sqrt(H1err + L2err); L2err = sqrt(L2err);
return
```

Example 12.1 We assess the accuracy of the finite element solution of the following problem. Consider a thin rod of length L whose temperature at $x = 0$ is fixed to t_0 while the other endpoint $x = L$ is thermally isolated. Assume that the rod has a cross-section with constant area equal to A and that the perimeter of A is p.

The temperature u of the rod at a generic point $x \in (0, L)$ is governed by the following boundary value problem with mixed Dirichlet-Neumann conditions

$$\begin{cases} -\mu A u'' + \sigma p u = 0 \; x \in (0, L), \\ u(0) = u_0, \qquad\qquad u'(L) = 0, \end{cases} \tag{12.67}$$

where μ denotes the thermal conductivity and σ is the convective transfer coefficient. The exact solution of the problem is the (smooth) function

$$u(x) = u_0 \frac{\cosh[m(L-x)]}{\cosh(mL)},$$

where $m = \sqrt{\sigma p / \mu A}$. We solve the problem by using linear and quadratic finite elements ($k = 1$ and $k = 2$) on a grid with uniform size. In the numerical computations we assume that the length of the rod is $L = 100[cm]$ and that the rod has a circular cross-section of radius $2[cm]$ (and thus, $A = 4\pi[cm^2]$, $p = 4\pi[cm]$). We also set $u_0 = 10[°C]$, $\sigma = 2$ and $\mu = 200$.

Figure 12.6 (*left*) shows the behavior of the error in the L^2 and H^1 norms for the linear and quadratic elements, respectively. Notice the excellent agreement between the numerical results and the expected theoretical estimates (12.59) and (12.60), i.e., the orders of convergence in the L^2 norm and the H^1 norm tend respectively to $k + 1$ and k if finite elements of degree k are employed, since the exact solution is smooth. •

12.4.7 Spectral Methods

It turns out that the spectral collocation method of Section 12.3 can be regarded as a Galerkin method where the subspace is $\mathbb{P}_n^0$ and the integrals

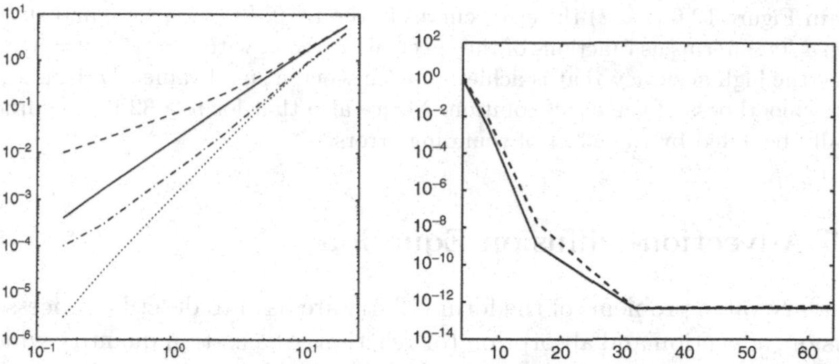

Fig. 12.6. Left: error curves for linear and quadratic elements. The dashed and solid lines denote the $\mathrm{H}^1(0, L)$ and $\mathrm{L}^2(0, L)$ norms of the error in the case $k = 1$, while the dot-line and dotted line denote the corresponding norms in the case $k = 2$. Right: error curves for the spectral collocation method. The dashed and solid lines denote the $\mathrm{H}^1(0, L)$ and $\mathrm{L}^2(0, L)$ norms of the error, respectively

are approximated by the Gauss-Lobatto quadrature formula. As a matter of fact, the approximation of problem (12.38) is

$$\text{find } u_n \in \mathbb{P}_n^0 : \ a_n(u_n, v_n) = (f, v_n)_n \ \forall v_n \in \mathbb{P}_n^0, \tag{12.68}$$

where a_n is the bilinear form that is obtained from the bilinear form a by replacing exact integrals by the Gauss-Lobatto formula (12.37). For problem (12.41) the associated bilinear form a was introduced in (12.44). We would therefore obtain

$$a_n(u_n, v_n) = (\alpha u_n', v_n')_n + (\beta u_n', v_n)_n + (\gamma u_n, v_n)_n. \tag{12.69}$$

This is no longer a Galerkin method, but is called a *generalized Galerkin approximation*. Its analysis requires more care than for Galerkin methods, as already seen in Section 12.3 and depends on the Strang lemma (see [QV94]). However, the same kind of error estimate (12.39) can be proven in this case as well.

A further generalization combining the finite element approach with piecewise polynomials of high degree and Gauss-Lobatto integration on each element yields the so-called *spectral element method* and the $h - p$ version of the finite element method (here p stands for the polynomial degree that we have denoted with n). In these cases convergence is achieved letting simultaneously (or independently) h go to zero and p go to infinity. (See, e.g., [BM92], [Sch98], [CHQZ06]).

Example 12.2 We consider again the two-point boundary value problem (12.67) and employ the spectral collocation method for its numerical approximation. We show in Figure 12.6 (*right*) the error curves in the $L^2(0, L)$ (*solid line*) and $H^1(0, L)$ (*dashed line*) norms as functions of the spectral degree n, with $n = 4^{-k}$, $k = 1, \ldots, 5$. Notice the high accuracy that is achieved, even when a small value of n is used, due to the smoothness of the exact solution. Notice also that for $n \geq 32$ the accuracy is actually bounded by the effect of rounding errors. •

12.5 Advection-Diffusion Equations

Boundary value problems of the form (12.41) are used to describe processes of diffusion, advection and absorption (or reaction) of a certain quantity which is identified with $u(x)$. The term $-(\alpha u')'$ is responsible for the diffusion, $\beta u'$ for the advection (or transport), γu for the absorption (if $\gamma > 0$). In this section we focus on the case where α is small compared with β (or γ). In these cases, the Galerkin method that we introduced earlier might be unsuitable for providing accurate numerical results. A heuristic explanation can be drawn from the inequality (12.56), noticing that in this case the constant M/α_0 can be very large, hence the error estimate can be meaningless unless h is much smaller than $(M/\alpha_0)^{-1}$. For instance, if $\alpha = \varepsilon$, $\gamma = 0$ and $\beta = const \gg 1$, then $\alpha_0 = \varepsilon$ and $M = \varepsilon + C_P\beta$. Similarly, if $\alpha = \varepsilon$, $\beta = 0$ and $\gamma = const \gg 1$ then $\alpha_0 = \varepsilon$ and $M = \varepsilon + C_P^2\gamma$.

To keep our analysis at the simplest possible level, we will consider the following elementary two-point boundary value problem

$$\begin{cases} -\varepsilon u'' + \beta u' = 0, \qquad 0 < x < 1, \\ u(0) = 0, \quad u(1) = 1, \end{cases} \tag{12.70}$$

where ε and β are two positive constants such that $\varepsilon/\beta \ll 1$. Despite its simplicity, (12.70) provides an interesting paradigm of an advection-diffusion problem in which advection dominates diffusion.

We define the *global Péclet number* as

$$\mathbb{P}\mathrm{e}_{gl} = \frac{|\beta| L}{2\varepsilon}, \tag{12.71}$$

where L is the size of the domain (equal to 1 in our case). The global Péclet number measures the dominance of the advective term over the diffusive one.

Let us first compute the exact solution of problem (12.70). tHE characteristic equation associated to the differential equation is $-\varepsilon\lambda^2 + \beta\lambda = 0$ and admits the roots $\lambda_1 = 0$ and $\lambda_2 = \beta/\varepsilon$. Then

$$u(x) = C_1 e^{\lambda_1 x} + C_2 e^{\lambda_2 x} = C_1 + C_2 e^{\frac{\beta}{\varepsilon} x},$$

where C_1 and C_2 are arbitrary constants. Imposing the boundary conditions yields $C_1 = -1/(e^{\beta/\varepsilon} - 1) = -C_2$, therefore

$$u(x) = (\exp(\beta x/\varepsilon) - 1) \, / \, (\exp(\beta/\varepsilon) - 1) \, .$$

If $\beta/\varepsilon \ll 1$ we can expand the exponentials up to first order obtaining

$$u(x) = (1 + \frac{\beta}{\varepsilon}x + \ldots - 1)/(1 + \frac{\beta}{\varepsilon} + \ldots - 1) \simeq (\beta\, x/\varepsilon)/(\beta/\varepsilon) = x,$$

thus the solution is close to the solution of the limit problem $-\varepsilon u'' = 0$, which is a straight line interpolating the boundary data.
However, if $\beta/\varepsilon \gg 1$ the exponentials attain big values so that

$$u(x) \simeq \frac{\exp(\beta/\varepsilon x)}{\exp(\beta/\varepsilon)} = \exp\left[-\frac{\beta}{\varepsilon}(1-x)\right].$$

Since the exponent is big and negative the solution is almost equal to zero everywhere unless a small neighborhood of the point $x = 1$ where the term $1-x$ becomes very small and the solution joins the value 1 with an exponential behaviour. The width of the neighbourhood is of the order of ε/β and thus it is quite small: in such an event, we say that the solution exhibits a *boundary layer* of width $\mathcal{O}(\varepsilon/\beta)$ at $x = 1$.

12.5.1 Galerkin Finite Element Approximation

Let us discretize problem (12.70) using the Galerkin finite element method introduced in Section 12.4.5 with $k = 1$ (piecewise linear finite elements). The approximation to the problem is: find $u_h \in X_h^1$ such that

$$\begin{cases} a(u_h, v_h) = 0 & \forall v_h \in X_h^{1,0}, \\ u_h(0) = 0,\ u_h(1) = 1, \end{cases} \tag{12.72}$$

where the finite element spaces X_h^1 and $X_h^{1,0}$ have been introduced in (8.26) and (12.57) and the bilinear form $a(\cdot,\cdot)$ is

$$a(u_h, v_h) = \int_0^1 (\varepsilon u_h' v_h' + \beta u_h' v_h)\, dx. \tag{12.73}$$

Remark 12.5 (Advection-diffusion problems in conservation form) Sometimes, the advection-diffusion problem (12.70) is written in the following *conservation form*

$$\begin{cases} -(J(u))' = 0, & 0 < x < 1, \\ u(0) = 0, \quad u(1) = 1, \end{cases} \tag{12.74}$$

where $J(u) = \varepsilon u' - \beta u$ is the *flux* (already introduced in the finite difference context in Section 12.2.3), ε and β are given functions with $\varepsilon(x) \geq \varepsilon_0 > 0$ for all $x \in [0,1]$. The Galerkin approximation of (12.74) using piecewise linear finite elements reads: find $u_h \in X_h^1$ such that

$$b(u_h, v_h) = 0, \qquad \forall v_h \in X_h^{1,0},$$

where $b(u_h, v_h) = \int_0^1 (\varepsilon u_h' - \beta u_h) v_h' \, dx$. The bilinear form $b(\cdot,\cdot)$ coincides with the corresponding one in (12.73) when ε and β are constant. ■

Taking v_h as a test function the generic basis function φ_i, (12.72) yields

$$\int_0^1 \varepsilon u_h' \varphi_i' \, dx + \int_0^1 \beta u_h' \varphi_i \, dx = 0, \qquad i = 1, \ldots, n-1.$$

Setting $u_h(x) = \sum_{j=0}^n u_j \varphi_j(x)$, and noting that $supp(\varphi_i) = [x_{i-1}, x_{i+1}]$ the above integral, for $i = 1, \ldots, n-1$, reduces to

$$\varepsilon \left[u_{i-1} \int_{x_{i-1}}^{x_i} \varphi_{i-1}' \varphi_i' \, dx + u_i \int_{x_{i-1}}^{x_{i+1}} (\varphi_i')^2 \, dx + u_{i+1} \int_{x_i}^{x_{i+1}} \varphi_i' \varphi_{i+1}' \, dx \right]$$
$$+\beta \left[u_{i-1} \int_{x_{i-1}}^{x_i} \varphi_{i-1}' \varphi_i \, dx + u_i \int_{x_{i-1}}^{x_{i+1}} \varphi_i' \varphi_i \, dx + u_{i+1} \int_{x_i}^{x_{i+1}} \varphi_{i+1}' \varphi_i \, dx \right] = 0.$$

Assuming a uniform partition of $[0,1]$ with $x_i = x_{i-1} + h$ for $i = 1, \ldots, n$, $h = 1/n$, and noting that $\varphi_j'(x) = \frac{1}{h}$ if $x_{j-1} \leq x \leq x_j$, $\varphi_j'(x) = -\frac{1}{h}$ if $x_j \leq x \leq x_{j+1}$, we deduce that

$$\frac{\varepsilon}{h}(-u_{i-1} + 2u_i - u_{i+1}) + \frac{1}{2}\beta(u_{i+1} - u_{i-1}) = 0, \qquad i = 1, \ldots, n-1. \quad (12.75)$$

Multiplying by h/ε and defining the *local Péclet number* to be

$$\mathbb{P}\mathsf{e} = \frac{|\beta| h}{2\varepsilon},$$

we finally obtain

$$(\mathbb{P}\mathsf{e} - 1) u_{i+1} + 2u_i - (\mathbb{P}\mathsf{e} + 1) u_{i-1} = 0, \qquad i = 1, \ldots, n-1. \quad (12.76)$$

This is a linear difference equation which admits a solution of the form $u_i = A_1 \rho_1^i + A_2 \rho_2^i$ for suitable constants A_1, A_2 (see Section 11.4), where ρ_1 and ρ_2 are the two roots of the following characteristic equation

$$(\mathbb{P}\mathsf{e} - 1)\rho^2 + 2\rho - (\mathbb{P}\mathsf{e} + 1) = 0.$$

Thus

$$\rho_{1,2} = \frac{-1 \pm \sqrt{1+\mathbb{P}\mathrm{e}^2 - 1}}{\mathbb{P}\mathrm{e} - 1} = \begin{cases} \dfrac{1+\mathbb{P}\mathrm{e}}{1-\mathbb{P}\mathrm{e}}, \\ 1. \end{cases}$$

Imposing the boundary conditions at $x = 0$ and $x = 1$ gives

$$A_1 = 1/(1 - \left(\tfrac{1+\mathbb{P}\mathrm{e}}{1-\mathbb{P}\mathrm{e}}\right)^n), \; A_2 = -A_1,$$

so that the solution of (12.76) is

$$u_i = \left(1 - \left(\frac{1+\mathbb{P}\mathrm{e}}{1-\mathbb{P}\mathrm{e}}\right)^i\right) / \left(1 - \left(\frac{1+\mathbb{P}\mathrm{e}}{1-\mathbb{P}\mathrm{e}}\right)^n\right), \qquad i = 0, \ldots, n.$$

We notice that if $\mathbb{P}\mathrm{e} > 1$, a power with a negative base appears at the numerator which gives rise to an oscillating solution. This is clearly visible in Figure 12.7 where for several values of the local Péclet number, the solution of (12.76) is compared with the exact solution (sampled at the mesh nodes) corresponding to a value of the global Péclet equal to 50.

The simplest remedy for preventing the oscillations consists of choosing a sufficiently small mesh stepsize h in such a way that $\mathbb{P}\mathrm{e} < 1$. However this approach is often impractical: for example, if $\beta = 1$ and $\varepsilon = 5 \cdot 10^{-5}$ one should take $h < 10^{-4}$ which amounts to dividing $[0, 1]$ into 10000 subintervals, a strategy that becomes unfeasible when dealing with multidimensional problems. Other strategies can be pursued, as will be addressed in the next sections.

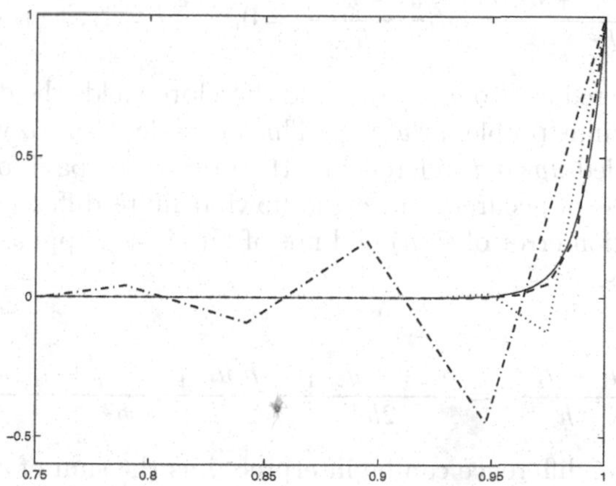

Fig. 12.7. Finite difference solution of the advection-diffusion problem (12.70) (with $\mathbb{P}\mathrm{e}_{gl} = 50$) for several values of the local Péclet number. Solid line: exact solution, dot-dashed line: $\mathbb{P}\mathrm{e} = 2.63$, dotted line: $\mathbb{P}\mathrm{e} = 1.28$, dashed line: $\mathbb{P}\mathrm{e} = 0.63$

12.5.2 The Relationship between Finite Elements and Finite Differences; the Numerical Viscosity

To examine the behaviour of the finite difference method (FD) when applied to the solution of advection-diffusion problems and its relationship with the finite element method (FE), we again consider the one-dimensional problem (12.70) with a uniform meshsize h.

To ensure that the local discretization error is of second order we approximate $u'(x_i)$ and $u''(x_i)$, $i = 1, \ldots, n-1$, by the centered finite differences (10.61) and (10.65) respectively (see Section 10.10.1). We obtain the following FD problem

$$\begin{cases} -\varepsilon \dfrac{u_{i+1} - 2u_i + u_{i-1}}{h^2} + \beta \dfrac{u_{i+1} - u_{i-1}}{2h} = 0, \qquad i = 1, \ldots, n-1, \\ u_0 = 0, \qquad u_n = 1. \end{cases} \tag{12.77}$$

If we multiply by h for every $i = 1, \ldots, n-1$, we obtain exactly the same equation (12.75) that was obtained using piecewise linear finite elements.

The equivalence between FD and FE can be profitably employed to devise a cure for the oscillations arising in the approximate solution of (12.75) when the local Péclet number is larger than 1. The important observation here is that the instability in the FD solution is due to the fact that the discretization scheme is a *centered* one. A possible remedy consists of approximating the first derivative by a one-sided finite difference according to the direction of the transport field. Precisely, we use the backward difference if the convective coefficient β is positive and the forward difference otherwise. The resulting scheme when $\beta > 0$ is

$$-\varepsilon \frac{u_{i+1} - 2u_i + u_{i-1}}{h^2} + \beta \frac{u_i - u_{i-1}}{h} = 0, \qquad i = 1, \ldots, n-1, \tag{12.78}$$

which, for $\varepsilon = 0$ reduces to $u_i = u_{i-1}$ and therefore yields the desired constant solution of the limit problem $\beta u' = 0$. This one-side discretization of the first derivative is called *upwind* differencing: the price to be paid for the enhanced stability is a loss of accuracy since the upwind finite difference introduces a local discretization error of $\mathcal{O}(h)$ and not of $\mathcal{O}(h^2)$ as happens using centered finite differences.

Noting that

$$\frac{u_i - u_{i-1}}{h} = \frac{u_{i+1} - u_{i-1}}{2h} - \frac{h}{2} \frac{u_{i+1} - 2u_i + u_{i-1}}{h^2},$$

the upwind finite difference can be interpreted as the sum of a centered finite difference approximating the first derivative and of a term proportional to the discretization of the second-order derivative. Consequently, (12.78) is equivalent to

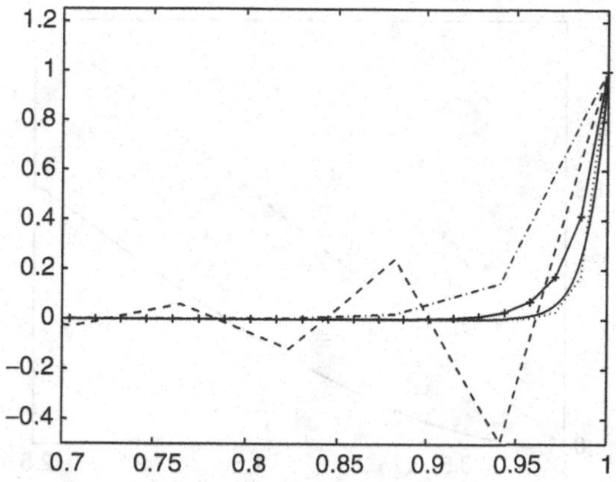

Fig. 12.8. Finite difference solution of (12.70) (with $\varepsilon = 1/100$) using a centered discretization (*dashed and dotted lines*) and the artificial viscosity (12.81) (*dot-dashed and starred lines*). The solid line denotes the exact solution. Notice the effect of eliminating the oscillations when the local Péclet number is large; conversely, notice also the corresponding loss of accuracy for low values of the local Péclet number

$$-\varepsilon_h \frac{u_{i+1} - 2u_i + u_{i-1}}{h^2} + \beta \frac{u_{i+1} - u_{i-1}}{2h} = 0 \qquad i = 1, \ldots, n-1, \qquad (12.79)$$

where $\varepsilon_h = \varepsilon(1 + \mathbb{P}\mathrm{e})$. This amounts to having replaced the differential equation (12.70) with the perturbed one

$$-\varepsilon_h u'' + \beta u' = 0, \qquad (12.80)$$

then using centered finite differences to approximate both u' and u''. The perturbation

$$-\varepsilon\, \mathbb{P}\mathrm{e}\, u'' = -\frac{\beta h}{2} u'' \qquad (12.81)$$

is called the *numerical viscosity* (or *artificial diffusion*). In Figure 12.8 a comparison between centered and upwinded discretizations of problem (12.72) is shown.

More generally, we can resort to a centered scheme of the form (12.78) with the following viscosity

$$\varepsilon_h = \varepsilon(1 + \phi(\mathbb{P}\mathrm{e})), \qquad (12.82)$$

where ϕ is a suitable function of the local Péclet number satisfying

$$\lim_{t \to 0+} \phi(t) = 0.$$

Notice that when $\phi(t) = 0$ for all t, one recovers the centered finite difference method (12.77), while if $\phi(t) = t$ the upwind finite difference scheme (12.78) (or, equivalently, (12.79)) is obtained. Other choices are admissible as well. For instance, taking

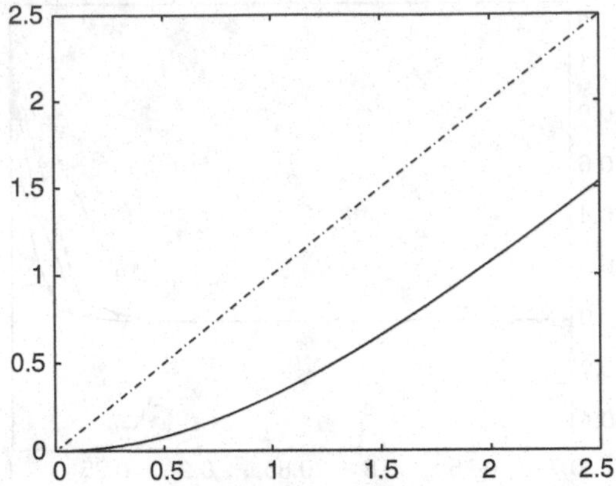

Fig. 12.9. The functions ϕ^{UP} (*dash-dotted line*) and ϕ^{SG} (*solid line*)

$$\phi(t) = t - 1 + B(2t), \tag{12.83}$$

where $B(t)$ is the inverse of the Bernoulli function defined as $B(t) = t/(e^t - 1)$ for $t \neq 0$ and $B(0) = 1$, yields the so called *exponential fitting* finite difference scheme which is also well known as the Scharfetter-Gummel (SG) method [SG69].

Remark 12.6 Denoting respectively by ϕ^C, ϕ^{UP} and ϕ^{SG} the previous three functions, i.e. $\phi^C = 0$, $\phi^{UP}(t) = t$ and $\phi^{SG}(t) = t - 1 + B(2t)$ (see Figure 12.9), we notice that $\phi^{SG} \simeq \phi^{UP}$ as $\mathbb{P}\mathsf{e} \to +\infty$ while $\phi^{SG} = \mathcal{O}(h^2)$ and $\phi^{UP} = \mathcal{O}(h)$ if $\mathbb{P}\mathsf{e} \to 0^+$. Therefore, the SG method is second-order accurate with respect to h and for this reason it is an *optimal viscosity* upwind method. Actually, one can show (see [HGR96], pp. 44-45) that if f is piecewise constant over the grid partition the SG scheme yields a numerical solution u_h^{SG} which is *nodally exact*, i.e., $u_h^{SG}(x_i) = u(x_i)$ for each node x_i, irrespectively of h (and, thus, of the size of the local Péclet number). This is demonstrated in Figure 12.10. ■

The new local Péclet number associated with the scheme (12.79)-(12.82) is defined as

$$\mathbb{P}\mathsf{e}^* = \frac{|\beta| h}{2\varepsilon_h} = \frac{\mathbb{P}\mathsf{e}}{(1 + \phi(\mathbb{P}\mathsf{e}))}.$$

For both the upwind and the SG schemes we have $\mathbb{P}\mathsf{e}^* < 1$ for any value of h. This implies that the matrix associated with these methods is a M-matrix for any h (see Definition 1.25 and Exercise 13), and, in turn, that the numerical solution u_h satisfies a discrete maximum principle (see Section 12.2.2).

12.5.3 Stabilized Finite Element Methods

In this section we extend the use of numerical viscosity introduced in the previous section for finite differences to the Galerkin method using finite elements

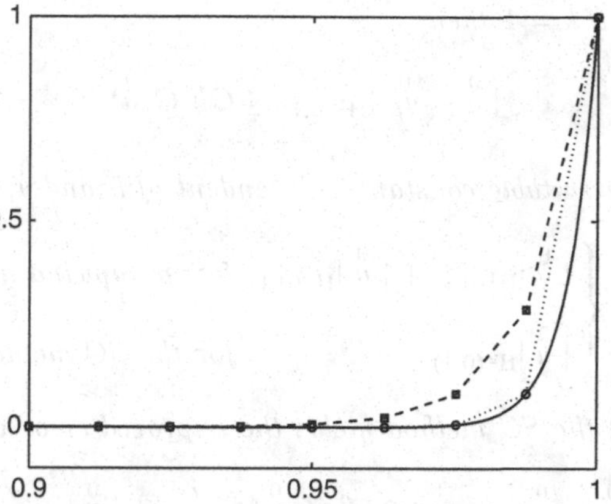

Fig. 12.10. Comparison between the numerical solutions of problem (12.70) (with $\varepsilon = 1/200$) obtained using the artificial viscosity (12.81) (dashed line where the symbol ■ denotes the nodal values) and with the optimal viscosity (12.83) (dotted line where the symbol • denotes the nodal values) in the case where $\mathbb{P}\mathrm{e} = 1.25$. The solid line denotes the exact solution

of arbitrary degree $k \geq 1$. For this purpose we consider the advection-diffusion problem (12.70) where the viscosity coefficient ε is replaced by (12.82). This yields the following modification of the original Galerkin problem (12.72):

find $\overset{0}{u}_h \in X_h^{k,0} = \left\{ v_h \in X_h^k : \ v_h(0) = v_h(1) = 0 \right\}$ such that

$$a_h(\overset{0}{u}_h, v_h) = -\int_0^1 \beta v_h \, dx \qquad \forall v_h \in X_h^{k,0}, \tag{12.84}$$

where

$$a_h(u, v) = a(u, v) + b(u, v),$$

and

$$b(u, v) = \varepsilon \, \phi(\mathbb{P}\mathrm{e}) \int_0^1 u' v' \, dx$$

is called the *stabilization* term. Since $a_h(v, v) = \varepsilon_h |v|_1^2$ for all $v \in \mathrm{H}_0^1(0,1)$ and $\varepsilon_h/\varepsilon = (1 + \phi(\mathbb{P}\mathrm{e})) \geq 1$, the modified problem (12.84) enjoys more favorable monotonicity properties than the corresponding nonstabilized Galerkin formulation (12.75).

To prove convergence, it is sufficient to show that $\overset{0}{u}_h$ tends to $\overset{0}{u}$ as $h \to 0$, where $\overset{0}{u}(x) = u(x) - x$. This is done in the following theorem, where we assume that $\overset{0}{u}$ (and henceforth u) has the required regularity.

Theorem 12.4 *If $k = 1$ then*

$$| \overset{0}{u} - \overset{0}{u}_h |_{H^1(0,1)} \leq Ch\, G(\overset{0}{u}), \tag{12.85}$$

where $C > 0$ is a suitable constant independent of h and $\overset{0}{u}$, and

$$G(\overset{0}{u}) = \begin{cases} | \overset{0}{u} |_{H^1(0,1)} + | \overset{0}{u} |_{H^2(0,1)} & \textit{for the upwind method,} \\ | \overset{0}{u} |_{H^2(0,1)} & \textit{for the SG method.} \end{cases}$$

Further, if $k = 2$ the SG method yields the improved error estimate

$$| \overset{0}{u} - \overset{0}{u}_h |_{H^1(0,1)} \leq Ch^2(| \overset{0}{u} |_{H^1(0,1)} + | \overset{0}{u} |_{H^3(0,1)}). \tag{12.86}$$

Proof. From (12.70) we obtain

$$a(\overset{0}{u}, v_h) = -\int_0^1 \beta v_h dx, \qquad \forall v_h \in X_h^{k,0}.$$

By comparison with (12.84) we get

$$a_h(\overset{0}{u} - \overset{0}{u}_h, v_h) = b(\overset{0}{u}, v_h), \qquad \forall v_h \in X_h^{k,0}. \tag{12.87}$$

Denote by $E_h = \overset{0}{u} - \overset{0}{u}_h$ the discretization error and recall that the space $H_0^1(0,1)$ is endowed with the norm (12.49). Then,

$$\begin{aligned} \varepsilon_h |E_h|^2_{H^1(0,1)} &= a_h(E_h, E_h) = a_h(E_h, \overset{0}{u} - \Pi_h^k \overset{0}{u}) + a_h(E_h, \Pi_h^k \overset{0}{u} - \overset{0}{u}_h) \\ &= a_h(E_h, \overset{0}{u} - \Pi_h^k \overset{0}{u}) + b(\overset{0}{u}, \Pi_h^k \overset{0}{u} - \overset{0}{u}_h), \end{aligned}$$

where we have applied (12.87) with $v_h = \Pi_h^k \overset{0}{u} - \overset{0}{u}_h$. Using the Cauchy-Schwarz inequality we get

$$\begin{aligned} \varepsilon_h |E_h|^2_{H^1(0,1)} &\leq M_h |E_h|_{H^1(0,1)} | \overset{0}{u} - \Pi_h^k \overset{0}{u} |_{H^1(0,1)} \\ &\quad + \varepsilon \phi(\mathbb{P}\mathrm{e}) \int_0^1 (\overset{0}{u})'(\Pi_h^k \overset{0}{u} - \overset{0}{u}_h)' dx, \end{aligned} \tag{12.88}$$

where $M_h = \varepsilon_h + |\beta| C_P$ is the continuity constant of the bilinear form $a_h(\cdot, \cdot)$ and C_P is the Poincaré constant introduced in (12.16).

Notice that if $k = 1$ (corresponding to piecewise linear finite elements) and $\phi = \phi^{SG}$ (SG optimal viscosity) the quantity in the second integral is identically zero since $\overset{0}{u}_h = \Pi_h^1 \overset{0}{u}$, as pointed out in Remark 12.6. Then, from (12.88) we get

$$|E_h|_{H^1(0,1)} \leq \left(1 + \frac{|\beta| C_P}{\varepsilon_h}\right) | \overset{0}{u} - \Pi_h^1 \overset{0}{u} |_{H^1(0,1)}.$$

Noting that $\varepsilon_h > \varepsilon$, using (12.71) and the interpolation estimate (8.31), we finally obtain the error bound

$$|E_h|_{\mathrm{H}^1(0,1)} \leq C(1 + 2\mathbb{P}\mathrm{e}_{gl} C_P)h| \overset{0}{u} |_{\mathrm{H}^2(0,1)}.$$

In the general case the error inequality (12.88) can be further manipulated. Using the Cauchy-Schwarz and triangular inequalities we obtain

$$\int_0^1 (\overset{0}{u})'(\Pi_h^k \overset{0}{u} - \overset{0}{u}_h)'dx \leq | \overset{0}{u} |_{\mathrm{H}^1(0,1)}(|\Pi_h^k \overset{0}{u} - \overset{0}{u} |_{\mathrm{H}^1(0,1)} + |E_h|_{\mathrm{H}^1(0,1)})$$

from which

$$\begin{aligned}\varepsilon_h|E_h|^2_{\mathrm{H}^1(0,1)} \leq |E_h|_{\mathrm{H}^1(0,1)} &\Big(M_h| \overset{0}{u} - \Pi_h^k \overset{0}{u} |_{\mathrm{H}^1(0,1)} \\ &+\varepsilon\phi(\mathbb{P}\mathrm{e})| \overset{0}{u} |_{\mathrm{H}^1(0,1)}\Big) + \varepsilon\phi(\mathbb{P}\mathrm{e})| \overset{0}{u} |_{\mathrm{H}^1(0,1)}| \overset{0}{u} - \Pi_h^k \overset{0}{u} |_{\mathrm{H}^1(0,1)}.\end{aligned}$$

Using again the interpolation estimate (8.31) yields

$$\begin{aligned}\varepsilon_h|E_h|^2_{\mathrm{H}^1(0,1)} \leq |E_h|_{\mathrm{H}^1(0,1)} &\Big(M_h C h^k| \overset{0}{u} |_{\mathrm{H}^{k+1}(0,1)} + \varepsilon\phi(\mathbb{P}\mathrm{e})| \overset{0}{u} |_{\mathrm{H}^1(0,1)}\Big) \\ &+C\varepsilon\phi(\mathbb{P}\mathrm{e})| \overset{0}{u} |_{\mathrm{H}^1(0,1)} h^k| \overset{0}{u} |_{\mathrm{H}^{k+1}(0,1)}.\end{aligned}$$

Using Young's inequality (12.40) gives

$$\begin{aligned}\varepsilon_h|E_h|^2_{\mathrm{H}^1(0,1)} \leq &\frac{\varepsilon_h|E_h|^2_{\mathrm{H}^1(0,1)}}{2} \\ &+\frac{3}{4\varepsilon_h}\left[(M_h C h^k| \overset{0}{u} |_{\mathrm{H}^{k+1}(0,1)})^2 + (\varepsilon\phi(\mathbb{P}\mathrm{e})| \overset{0}{u} |_{\mathrm{H}^1(0,1)})^2\right],\end{aligned}$$

from which it follows that

$$\begin{aligned}|E_h|^2_{\mathrm{H}^1(0,1)} \leq &\frac{3}{2}\left(\frac{M_h}{\varepsilon_h}\right)^2 C^2 h^{2k}| \overset{0}{u} |^2_{\mathrm{H}^{k+1}(0,1)} + \frac{3}{2}\left(\frac{\varepsilon}{\varepsilon_h}\right)^2 \phi(\mathbb{P}\mathrm{e})^2| \overset{0}{u} |^2_{\mathrm{H}^1(0,1)} \\ &+\frac{2\varepsilon}{\varepsilon_h}\phi(\mathbb{P}\mathrm{e})| \overset{0}{u} |_{\mathrm{H}^1(0,1)} C h^k| \overset{0}{u} |_{\mathrm{H}^{k+1}(0,1)}.\end{aligned}$$

Again using the fact that $\varepsilon_h > \varepsilon$ and the definition (12.71) we get $(M_h/\varepsilon_h) \leq (1 + 2C_P\mathbb{P}\mathrm{e}_{gl})$ and then

$$\begin{aligned}|E_h|^2_{\mathrm{H}^1(0,1)} \leq &\frac{3}{2}C^2(1 + 2C_P\mathbb{P}\mathrm{e}_{gl})^2 h^{2k}| \overset{0}{u} |^2_{\mathrm{H}^{k+1}(0,1)} \\ &+2\phi(\mathbb{P}\mathrm{e}) C h^k| \overset{0}{u} |_{\mathrm{H}^1(0,1)}| \overset{0}{u} |_{\mathrm{H}^{k+1}(0,1)} + \frac{3}{2}\phi(\mathbb{P}\mathrm{e})^2| \overset{0}{u} |^2_{\mathrm{H}^1(0,1)},\end{aligned}$$

which can be bounded further as

$$\begin{aligned}|E_h|^2_{\mathrm{H}^1(0,1)} \leq \mathcal{M}\Big[&h^{2k}| \overset{0}{u} |^2_{\mathrm{H}^{k+1}(0,1)} \\ &+ \phi(\mathbb{P}\mathrm{e}) h^k| \overset{0}{u} |_{\mathrm{H}^1(0,1)}| \overset{0}{u} |_{\mathrm{H}^{k+1}(0,1)} + \phi(\mathbb{P}\mathrm{e})^2| \overset{0}{u} |^2_{\mathrm{H}^1(0,1)}\Big]\end{aligned} \tag{12.89}$$

for a suitable positive constant $\mathcal{M}$.

If $\phi^{UP} = \mathcal{C}_\varepsilon h$, where $\mathcal{C}_\varepsilon = \beta/\varepsilon$, we obtain

$$|E_h|^2_{\mathrm{H}^1(0,1)} \le Ch^2 \Big[h^{2k-2} | \overset{0}{u} |^2_{\mathrm{H}^{k+1}(0,1)} + h^{k-1} | \overset{0}{u} |_{\mathrm{H}^1(0,1)} | \overset{0}{u} |_{\mathrm{H}^{k+1}(0,1)} + | \overset{0}{u} |^2_{\mathrm{H}^1(0,1)} \Big],$$

which shows that using piecewise linear finite elements (i.e., $k = 1$) plus the upwind artificial viscosity gives the linear convergence estimate (12.85).
In the case $\phi = \phi^{SG}$, assuming that for h sufficiently small $\phi^{SG} \le Kh^2$, for a suitable positive constant K, we get

$$|E_h|^2_{\mathrm{H}^1(0,1)} \le Ch^4 \Big[h^{2(k-2)} | \overset{0}{u} |^2_{\mathrm{H}^{k+1}(0,1)} + h^{k-2} | \overset{0}{u} |_{\mathrm{H}^1(0,1)} | \overset{0}{u} |_{\mathrm{H}^{k+1}(0,1)} + | \overset{0}{u} |^2_{\mathrm{H}^1(0,1)} \Big],$$

which shows that using quadratic finite elements (i.e., $k = 2$) plus the optimal artificial viscosity gives the second-order convergence estimate (12.86). ◇

Programs 97 and 98 implement the computation of the artificial and optimal artificial viscosities (12.81) and (12.83), respectively. These viscosities can be selected by the user setting the input parameter `stabfun` in Program 94 equal to `artvisc` or `sgvisc`. The function `sgvisc` employs the function `bern` to evaluate the inverse of the Bernoulli function in (12.83).

Program 97 - artvisc : Artificial viscosity

```
function [Kupw,rhsbc] = artvisc(Nx,h,nu,beta)
%ARTVISC Artificial viscosity: stiffness matrix and right-hand side.
Peclet=0.5*h*abs(beta);
for i=2:Nx
   dd(i-1)=(Peclet(i-1)+Peclet(i))/h;
   if i>2
      ld(i-2)=-Peclet(i-1)/h;
   end
   if i<Nx
      ud(i-1)=-Peclet(i)/h;
   end
end
Kupw=spdiags([[ld 0]',dd',[0 ud]'],-1:1,Nx-1,Nx-1);
rhsbc = - [Peclet(1)/h, Peclet(Nx)/h];
return
```

Program 98 - sgvisc : Optimal artificial viscosity

```
function [Ksg,rhsbc] = sgvisc(Nx, h, nu, beta)
%SGVISC Optimal artificial viscosity: stiffness matrix and right-hand side.
Peclet=0.5*h*abs(beta)./nu;
[bp, bn]=bern(2*Peclet);
```

```
Peclet=Peclet-1+bp;
for i=2:Nx
   dd(i-1)=(nu(i-1)*Peclet(i-1)+nu(i)*Peclet(i))/h;
   if i>2
      ld(i-2)=-nu(i-1)*Peclet(i-1)/h;
   end
   if i<Nx
      ud(i-1)=-nu(i)*Peclet(i)/h;
   end
end
Ksg=spdiags([[ld 0]',dd',[0 ud]'],-1:1,Nx-1,Nx-1);
rhsbc = - [nu(1)*Peclet(1)/h, nu(Nx)*Peclet(Nx)/h];
return
```

Program 99 - bern : Evaluation of the Bernoulli function

```
function [bp,bn]=bern(x)
%BERN Evaluation of the Bernoulli function
xlim=1e-2; ax=abs(x);
if ax==0,
   bp=1; bn=1; return
end
if ax>80
   if x>0
      bp=0.; bn=x;
      return
   else
      bp=-x; bn=0;
      return
   end
end
if ax>xlim
   bp=x/(exp(x)-1); bn=x+bp; return
else
   ii=1; fp=1.;fn=1.; df=1.; s=1.;
   while abs(df)>eps
      ii=ii+1;  s=-s; df=df*x/ii;
      fp=fp+df; fn=fn+s*df;
      bp=1./fp; bn=1./fn;
   end
   return
end
return
```

Example 12.3 We use Program 94 supplied with Programs 97 and 98 for the numerical approximation of problem (12.70) in the case $\varepsilon = 10^{-2}$. Figure 12.11 shows the convergence behavior as a function of $\log(h)$ of the Galerkin method without (G) and with numerical viscosity (upwind (UP) and SG methods are employed).

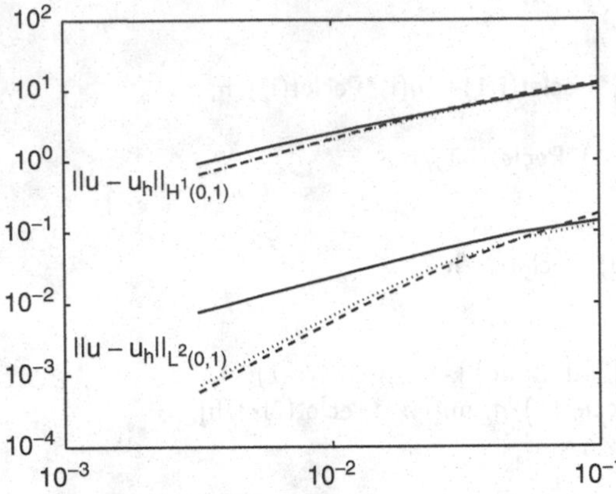

Fig. 12.11. Convergence analysis for an advection-diffusion problem

The figure shows the logarithmic plots of the $L^2(0,1)$ and $H^1(0,1)$-norms, where the solid line denotes the UP method and the dashed and dotted lines denote the G and SG methods, respectively. It is interesting to notice that the UP and SG schemes exhibit the same (linear) convergence rate as the pure Galerkin method in the H^1-norm, while the accuracy of the UP scheme in the L^2-norm deteriorates dramatically because of the effect of the artificial viscosity which is $\mathcal{O}(h)$. Conversely, the SG converges quadratically since the introduced numerical viscosity is in this case $\mathcal{O}(h^2)$ as h tends to zero. •

12.6 A Quick Glance at the Two-Dimensional Case

The game that we want to play is to extend (in a few pages) the basic ideas illustrated so far to the two-dimensional case. The obvious generalization of problem (12.1)-(12.2) is the celebrated *Poisson problem* with homogeneous Dirichlet boundary condition

$$\begin{cases} -\Delta u = f \text{ in } \Omega, \\ u = 0 \qquad \text{on } \partial\Omega, \end{cases} \tag{12.90}$$

where $\Delta u = \partial^2 u/\partial x^2 + \partial^2 u/\partial y^2$ is the Laplace operator and Ω is a two-dimensional bounded domain whose boundary is $\partial\Omega$. If we allow Ω to be the unit square $\Omega = (0,1)^2$, the finite difference approximation of (12.90) that mimics (12.10) is

$$\begin{cases} L_h u_h(x_{i,j}) = f(x_{i,j}) \text{ for } i,j = 1,\ldots,N-1, \\ u_h(x_{i,j}) = 0 \qquad \text{if } i = 0 \text{ or } N, \qquad j = 0 \text{ or } N, \end{cases} \tag{12.91}$$

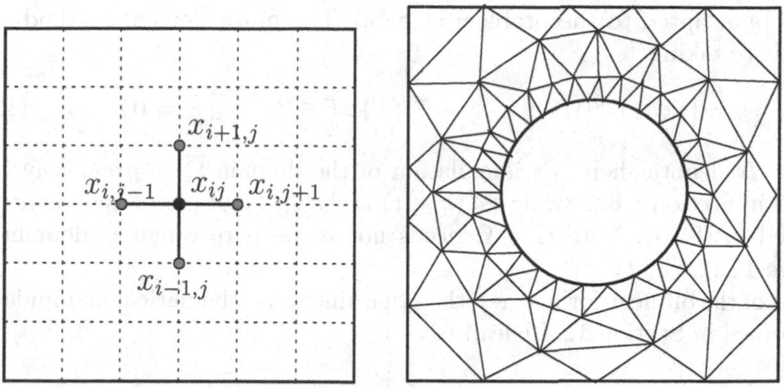

Fig. 12.12. Left: finite difference grid and stencil for a squared domain. Right: finite element triangulation of a region around a hole

where $x_{i,j} = (ih, jh)$ $(h = 1/N > 0)$ are the grid points and u_h is a grid function. Finally, L_h denotes any consistent approximation of the operator $L = -\triangle$. The classical choice is

$$L_h u_h(x_{i,j}) = \frac{1}{h^2}\left(4u_{i,j} - u_{i+1,j} - u_{i-1,j} - u_{i,j+1} - u_{i,j-1}\right), \quad (12.92)$$

where $u_{i,j} = u_h(x_{i,j})$, which amounts to adopting the second-order centered discretization of the second derivative (10.65) in both x and y directions (see Figure 12.12, *left*).

The resulting scheme is known as the *five-point discretization* of the Laplacian. It is an easy (and useful) exercise for the reader to check that the associated matrix A_{fd} has $(N-1)^2$ rows and is pentadiagonal, with the i-th row given by

$$(a_{\mathrm{fd}})_{ij} = \frac{1}{h^2}\begin{cases} 4 & \text{if } j = i, \\ -1 & \text{if } j = i-N-1,\ i-1,\ i+1,\ i+N+1, \\ 0 & \text{otherwise.} \end{cases} \quad (12.93)$$

Moreover, A_{fd} is symmetric positive definite and it is also an M-matrix (see Exercise 14). As expected, the consistency error associated with (12.92) is second-order with respect to h, and the same holds for the discretization error $\|u - u_h\|_{h,\infty}$ of the method. More general boundary conditions than the one considered in (12.90) can be dealt with by properly extending to the two-dimensional case the mirror imaging technique described in Section 12.2.3 and in Exercise 10 (for a thorough discussion of this subject, see, e.g., [Smi85]).

The extension of the Galerkin method is (formally speaking) even more straightforward and actually is still readable as (12.46) with, however, the implicit understanding that both the function space V_h and the bilinear form

$a(\cdot,\cdot)$ be adapted to the problem at hand. The finite element method corresponds to taking

$$V_h = \left\{ v_h \in C^0(\overline{\Omega}) \,:\, v_{h|T} \in \mathbb{P}_k(T)\, \forall T \in \mathcal{T}_h,\ v_{h|\partial\Omega} = 0 \right\}, \tag{12.94}$$

where $\mathcal{T}_h$ denotes here a triangulation of the domain $\overline{\Omega}$ as previously introduced in Section 8.6.2, while $\mathbb{P}_k$ $(k \geq 1)$ is the space of piecewise polynomials defined in (8.39). Note that Ω needs not to be a rectangular domain (see Figure 12.12, right).

As of the bilinear form $a(\cdot,\cdot)$, the same kind of mathematical manipulations performed in Section 12.4.1 lead to

$$a(u_h, v_h) = \int_\Omega \nabla u_h \cdot \nabla v_h \, dxdy,$$

where we have used the following Green's formula that generalizes the formula of integration by parts

$$\int_\Omega -\Delta u \, v \, dxdy = \int_\Omega \nabla u \cdot \nabla v \, dxdy - \int_{\partial\Omega} \nabla u \cdot \mathbf{n}\, v \, d\gamma, \tag{12.95}$$

for any u, v smooth enough and where $\mathbf{n}$ is the outward normal unit vector on $\partial\Omega$ (see Exercise 15).

The error analysis for the two-dimensional finite element approximation of (12.90) can still be performed through the combined use of Ceà's lemma and interpolation error estimates as in Section 12.4.5 and is summarized in the following result, which is the two-dimensional counterpart of Property 12.1 (for its proof we refer, e.g., to [QV94], Theorem 6.2.1).

Property 12.2 *Let $u \in \mathrm{H}^1_0(\Omega)$ be the exact weak solution of* (12.90) *and $u_h \in V_h$ be its finite element approximation using continuous piecewise polynomials of degree $k \geq 1$. Assume also that $u \in \mathrm{H}^s(\Omega)$ for some $s \geq 2$. Then the following error estimate holds*

$$\|u - u_h\|_{\mathrm{H}^1_0(\Omega)} \leq \frac{M}{\alpha_0} C h^l \|u\|_{\mathrm{H}^{l+1}(\Omega)}, \tag{12.96}$$

where $l = \min(k, s-1)$. Under the same assumptions, one can also prove that

$$\|u - u_h\|_{\mathrm{L}^2(\Omega)} \leq C h^{l+1} \|u\|_{\mathrm{H}^{l+1}(\Omega)}. \tag{12.97}$$

We notice that, for any integer $s \geq 0$, the Sobolev space $\mathrm{H}^s(\Omega)$ introduced above is defined as the space of functions with the first s partial derivatives (in the distributional sense) belonging to $\mathrm{L}^2(\Omega)$. Moreover, $\mathrm{H}^1_0(\Omega)$ is the space of functions of $\mathrm{H}^1(\Omega)$ such that $u = 0$ on $\partial\Omega$. The precise mathematical meaning of this latter statement has to be carefully addressed since, for instance, a function belonging to $H_0^1(\Omega)$ does not necessarily mean to be continuous everywhere. For a comprehensive presentation and analysis of Sobolev spaces we refer to [Ada75] and [LM68].

Following the same procedure as in Section 12.4.3, we can write the finite element solution u_h as

$$u_h(x,y) = \sum_{j=1}^{N} u_j \varphi_j(x,y),$$

where $\{\varphi_j\}_{j=1}^N$ is a basis for V_h. An example of such a basis in the case $k=1$ is provided by the so-called *hat functions* introduced in Section 8.6.2 (see Figure 8.7, right). The Galerkin finite element method leads to the solution of the linear system $A_{fe}\mathbf{u} = \mathbf{f}$, where $(a_{fe})_{ij} = a(\varphi_j, \varphi_i)$.

Exactly as happens in the one-dimensional case, the matrix A_{fe} is symmetric positive definite and, in general, sparse, the sparsity pattern being strongly dependent on the topology of $\mathcal{T}_h$ and the numbering of its nodes. Moreover, the spectral condition number of A_{fe} is still $\mathcal{O}(h^{-2})$, which implies that solving iteratively the linear system demands for the use of a suitable preconditioner (see Section 4.3.2). If, instead, a direct solver is used, one should resort to a suitable renumbering procedure, as explained in Section 3.9.

12.7 Applications

In this section we employ the finite element method for the numerical approximation of two problems arising in fluid mechanics and in the simulation of biological systems.

12.7.1 Lubrication of a Slider

Let us consider a rigid slider moving in the direction x along a physical support from which it is separated by a thin layer of a viscous fluid (which is the lubricant). Suppose that the slider, of length L, moves with a velocity U with respect to the plane support which is supposed to have an infinite length. The surface of the slider that is faced towards the support is described by the function s (see Figure 12.13, left).

Denoting by μ the viscosity of the lubricant, the pressure p acting on the slider can be modeled by the following Dirichlet problem

$$\begin{cases} -\left(\dfrac{s^3}{6\mu}p'\right)' = -(Us)' & x \in (0,L), \\ p(0) = 0, \qquad p(L) = 0. \end{cases}$$

Assume in the numerical computations that we are working with a convergent-divergent slider of unit length, whose surface is $s(x) = 1 - 3/2x + 9/8x^2$ with $\mu = 1$.

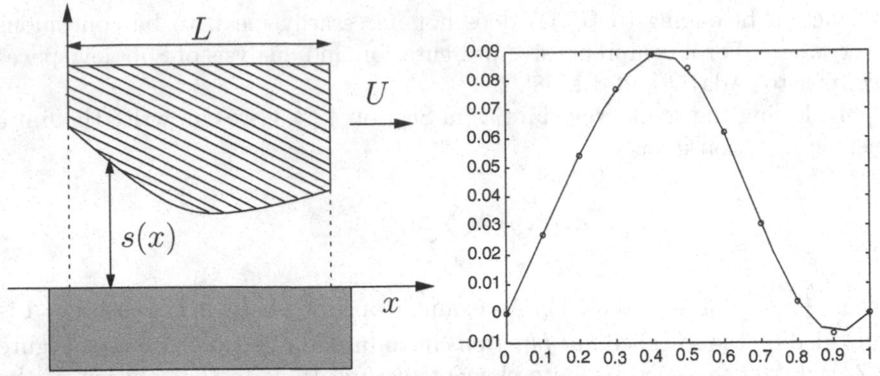

Fig. 12.13. Left: geometrical parameters of the slider considered in Section 12.7.1; right: pressure on a converging-diverging slider. The solid line denotes the solution obtained using quadratic finite elements, while the symbols ∘ denotes the nodal values of the solution obtained using linear finite elements

Table 12.2. Condition number of the stiffness matrix for linear and quadratic finite elements

h	$K_2(\mathrm{A}_1)$	p_1	$K_2(\mathrm{A}_2)$	p_2
0.10000	63.951	–	455.24	
0.05000	348.21	2.444	2225.7	2.28
0.02500	1703.7	2.290	10173.3	2.19
0.01250	7744.6	2.184	44329.6	2.12
0.00625	33579	2.116	187195.2	2.07

Figure 12.13 (right) shows the solution obtained using linear and quadratic finite elements with an uniform grid size $h = 0.2$. The linear system has been solved by the nonpreconditioned conjugate gradient method. To reduce the Euclidean norm of the residual below 10^{-10}, 4 iterations are needed in the case of linear finite elements while 9 are required for quadratic finite elements.

Table 12.2 reports a numerical study of the condition number $K_2(\mathrm{A}_{fe})$ as a function of h. In the case of linear finite elements we have denoted the matrix by A_1, while A_2 is the corresponding matrix for quadratic elements. Here we assume that the condition number approaches h^{-p} as h tends to zero; the numbers p_i are the estimated values of p. As can be seen, in both cases the condition number grows like h^{-2}, however, for every fixed h, $K_2(\mathrm{A}_2)$ is much bigger than $K_2(\mathrm{A}_1)$.

12.7.2 Vertical Distribution of Spore Concentration over Wide Regions

In this section we are concerned with diffusion and transport processes of spores in the air, such as the endspores of bacteria and the pollen of flowering plants. In particular, we study the vertical concentration distribution of spores and pollen grains over a wide area. These spores, in addition to settling under the influence of gravity, diffuse passively in the atmosphere.

The basic model assumes the *diffusivity* ν and the *settling velocity* β to be given constants, the averaging procedure incorporating various physical processes such as small-scale convection and horizontal advection-diffusion which can be neglected over a wide horizontal area. Denoting by $x \geq 0$ the vertical direction, the steady-state distribution $u(x)$ of the spore concentration is the solution of

$$\begin{cases} -\nu u'' + \beta u' = 0 & 0 < x < H, \\ u(0) = u_0, & -\nu u'(H) + \beta u(H) = 0, \end{cases} \tag{12.98}$$

where H is a fixed height at which we assume a vanishing Neumann condition for the *advective-diffusive flux* $-\nu u' + \beta u$ (see Section 12.4.1). Realistic values of the coefficients are $\nu = 10\ m^2 s^{-1}$ and $\beta = -0.03\ ms^{-1}$; as for u_0, a reference concentration of 1 pollen grain per m^3 has been used in the numerical experiments, while the height H has been set equal to 10 km. The global Péclet number is therefore $\mathbb{P}\mathrm{e}_{gl} = 15$.

The Galerkin finite element method with piecewise linear finite elements has been used for the approximation of (12.98). Figure 12.14 (*left*) shows the solution computed by running Program 94 on a uniform grid with stepsize $h = H/10$. The solution obtained using the (non stabilized) Galerkin formulation (G) is denoted by the solid line, while the dash-dotted and dashed lines refer to the Scharfetter-Gummel (SG) and upwind (UP) stabilized methods. Spurious oscillations can be noticed in the G solution while the one obtained using UP is clearly overdiffused with respect to the SG solution that is nodally exact. The local Péclet number is equal to 1.5 in this case. Taking $h = H/100$

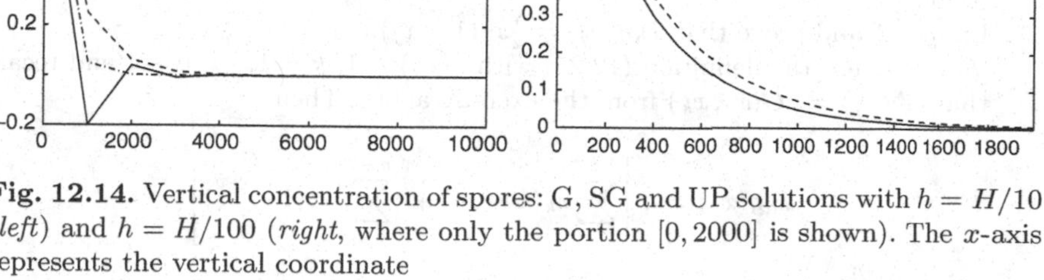

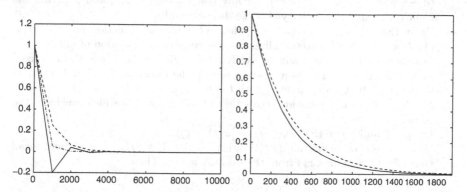

Fig. 12.14. Vertical concentration of spores: G, SG and UP solutions with $h = H/10$ (*left*) and $h = H/100$ (*right*, where only the portion $[0, 2000]$ is shown). The x-axis represents the vertical coordinate

yields for the pure Galerkin scheme a stable result as shown in Figure 12.14 (*right*) where the solutions furnished by G (*solid line*) and UP (*dashed line*) are compared.

12.8 Exercises

1. Consider the boundary value problem (12.1)-(12.2) with $f(x) = 1/x$. Using (12.3) prove that $u(x) = -x\log(x)$. This shows that $u \in C^2(0,1)$ but $u(0)$ is not defined and u', u'' do not exist at $x = 0$ ($\Rightarrow$: if $f \in C^0(0,1)$, but not $f \in C^0([0,1])$, then u does not belong to $C^0([0,1])$).
2. Prove that the matrix $\mathrm{A_{fd}}$ introduced in (12.8) is an M-matrix.
[*Hint*: check that $\mathrm{A_{fd}}\mathbf{x} \geq 0 \Rightarrow \mathbf{x} \geq 0$. To do this, for any $\alpha > 0$ set $\mathrm{A_{fd,\alpha}} = \mathrm{A_{fd}} + \alpha \mathrm{I}_{n-1}$. Then, compute $\mathbf{w} = \mathrm{A_{fd,\alpha}}\mathbf{x}$ and prove that $\min_{1\leq i\leq (n-1)} w_i \geq 0$. Finally, since the matrix $\mathrm{A_{fd,\alpha}}$ is invertible, being symmetric and positive definite, and since the entries of $\mathrm{A_{fd,\alpha}^{-1}}$ are continuous functions of $\alpha \geq 0$, one concludes that $\mathrm{A_{fd,\alpha}^{-1}}$ is a nonnegative matrix as $\alpha \to 0$.]
3. Prove that (12.13) defines a norm for V_h^0.
4. Prove (12.15) by induction on m.
5. Prove the estimate (12.23).
[*Hint*: for each internal node x_j, $j = 1, \ldots, n-1$, integrate by parts (12.21) to get

$$\tau_h(x_j) = -u''(x_j) - \frac{1}{h^2}\left[\int_{x_j-h}^{x_j} u''(t)(x_j - h - t)^2\, dt - \int_{x_j}^{x_j+h} u''(t)(x_j + h - t)^2\, dt\right].$$

Then, pass to the squares and sum $\tau_h(x_j)^2$ for $j = 1, \ldots, n-1$. On noting that $(a+b+c)^2 \leq 3(a^2+b^2+c^2)$, for any real numbers a, b, c, and applying the Cauchy-Schwarz inequality yields the desired result.]
6. Prove that $G^k(x_j) = hG(x_j, x_k)$, where G is Green's function introduced in (12.4) and G^k is its corresponding discrete counterpart solution of (12.4).
[*Solution*: we prove the result by verifying that $L_h G = he^k$. Indeed, for a fixed x_k the function $G(x_k, s)$ is a straight line on the intervals $[0, x_k]$ and $[x_k, 1]$ so that $L_h G = 0$ at every node x_l with $l = 0, \ldots, k-1$ and $l = k+1, \ldots, n+1$. Finally, a direct computation shows that $(L_h G)(x_k) = 1/h$ which concludes the proof.]
7. Let $g = 1$ and prove that $T_h g(x_j) = \frac{1}{2}x_j(1 - x_j)$.
[*Solution*: use the definition (12.25) with $g(x_k) = 1$, $k = 1, \ldots, n-1$ and recall that $G^k(x_j) = hG(x_j, x_k)$ from the exercise above. Then

$$T_h g(x_j) = h\left[\sum_{k=1}^{j} x_k(1 - x_j) + \sum_{k=j+1}^{n-1} x_j(1 - x_k)\right]$$

from which, after straightforward computations, one gets the desired result.]

8. Prove Young's inequality (12.40).
9. Show that $\|v_h\|_h \leq \|v_h\|_{h,\infty}\ \forall v_h \in V_h$.
10. Consider the two-point boundary value problem (12.29) with the following boundary conditions

$$\lambda_0 u(0) + \mu_0 u(0) = g_0, \qquad \lambda_1 u(1) + \mu_1 u(1) = g_1,$$

where λ_j, μ_j and g_j are given data ($j = 0, 1$). Using the mirror imaging technique described in Section 12.2.3 write the finite difference discretization of the equations corresponding to the nodes x_0 and x_n.
[*Solution*:

$$\text{node } x_0: \left(\frac{\alpha_{1/2}}{h^2} + \frac{\gamma_0}{2} + \frac{\alpha_0\lambda_0}{\mu_0 h}\right) u_0 - \alpha_{1/2}\frac{u_1}{h^2} = \frac{\alpha_0 g_0}{\mu_0 h} + \frac{f_0}{2},$$

$$\text{node } x_n: \left(\frac{\alpha_{n-1/2}}{h^2} + \frac{\gamma_n}{2} + \frac{\alpha_n\lambda_1}{\mu_1 h}\right) u_n - \alpha_{n-1/2}\frac{u_{n-1}}{h^2} = \frac{\alpha_n g_1}{\mu_1 h} + \frac{f_n}{2}.]$$

11. Discretize the fourth-order differential operator $Lu(x) = -u^{(iv)}(x)$ using centered finite differences.
[*Solution*: apply twice the second order centered finite difference operator L_h defined in (12.9).]
12. Consider problem (12.41) with nonhomogeneous Neumann boundary conditions $\alpha u'(0) = w_0$, $\alpha u'(1) = w_1$. Show that the solution satisfies problem (12.43) where $V = \mathrm{H}^1(0,1)$ and the right-hand side is replaced by $(f, v) + w_1 v(1) - w_0 v(0)$. Derive the formulation in the case of mixed boundary conditions $\alpha u'(0) = w_0$, $u(1) = u_1$.
13. Using Property 1.19 prove that the matrices corresponding to the stabilized finite element method (12.79) using the upwind and SG artificial viscosities ϕ^{UP} and ϕ^{SG} (see Section 12.5.2) are M-matrices irrespective of h.
[*Hint*: let us denote respectively by A^{UP} and A^{SG} the two stiffness matrices corrsponding to ϕ^{UP} and ϕ^{SG}. Take $v(x) = 1 + x$ and set $v_i = 1 + x_i$, $i = 0, \ldots, n$, being $x_i = ih$, $h = 1/n$. Then, by a direct computation check that $(\mathrm{A}^{UP}\mathbf{v})_i \geq \beta > 0$. As for the matrix A^{SG} the same result can be proved by noting that $B(-t) = t + B(t)$ for any $t \in \mathbb{R}$.]
14. Prove that the matrix A_{fd} with entries given by (12.93) is symmetric positive definite and it is also an M-matrix.
[*Solution*: to show that A_{fd} is positive definite, proceed as in the corresponding proof in Section 12.2, then proceed as in Exercise 2.]
15. Prove the Green's formula (12.95).
[*Solution*: first, notice that for any u, v sufficiently smooth, $\mathrm{div}(v\nabla u) = v\Delta u + \nabla u \cdot \nabla v$. Then, integrate this relation over Ω and use the divergence theorem

$$\int_{\Omega} \mathrm{div}(v\nabla u)\, dx dy = \int_{\partial\Omega} \frac{\partial u}{\partial n} v\, d\gamma.]$$

13

Parabolic and Hyperbolic Initial Boundary Value Problems

The final chapter of this book is devoted to the approximation of time-dependent partial differential equations. Parabolic and hyperbolic initial-boundary value problems will be addressed and either finite differences and finite elements will be considered for their discretization.

13.1 The Heat Equation

The problem we are considering is how to find a function $u = u(x,t)$ for $x \in [0,1]$ and $t > 0$ that satisfies the partial differential equation

$$\frac{\partial u}{\partial t} + Lu = f, \quad 0 < x < 1,\ t > 0, \tag{13.1}$$

subject to the boundary conditions

$$u(0,t) = u(1,t) = 0, \qquad t > 0, \tag{13.2}$$

and the initial condition

$$u(x,0) = u_0(x) \qquad 0 \leq x \leq 1. \tag{13.3}$$

The differential operator L is defined as

$$Lu = -\nu \frac{\partial^2 u}{\partial x^2}. \tag{13.4}$$

Equation (13.1) is called the heat equation. In fact, $u(x,t)$ describes the temperature at the point x and time t of a metallic bar of unit length that occupies the interval $[0,1]$, under the following conditions. Its thermal conductivity is constant and equal to $\nu > 0$, its extrema are kept at a constant temperature of zero degrees, at time $t = 0$ its temperature at point x is described by $u_0(x)$, and $f(x,t)$ represents the heat production per unit length supplied at point x

at time t. Here we are supposing that the volumetric density ρ and the specific heat per unit mass c_p are both constant and unitary. Otherwise, the temporal derivative $\partial u/\partial t$ should be multiplied by the product ρc_p in (13.1).

A solution of problem (13.1)-(13.3) is provided by a *Fourier series*. For instance, if $\nu = 1$ and $f \equiv 0$, it is given by

$$u(x,t) = \sum_{n=1}^{\infty} c_n e^{-(n\pi)^2 t} \sin(n\pi x), \tag{13.5}$$

where the coefficients c_n are the Fourier sine coefficients of the initial datum $u_0(x)$, i.e.

$$c_n = 2 \int_0^1 u_0(x) \sin(n\pi x)\, dx, \qquad n = 1, 2 \ldots .$$

If instead of (13.2) we consider the *Neumann conditions*

$$u_x(0,t) = u_x(1,t) = 0, \qquad t > 0, \tag{13.6}$$

the corresponding solution (still in the case where $\nu = 1$ and $f = 0$) would be

$$u(x,t) = \frac{d_0}{2} + \sum_{n=1}^{\infty} d_n e^{-(n\pi)^2 t} \cos(n\pi x),$$

where the coefficients d_n are the Fourier cosine coefficients of $u_0(x)$, i.e.

$$d_n = 2 \int_0^1 u_0(x) \cos(n\pi x)\, dx, \qquad n = 1, 2 \ldots .$$

These expressions show that the solution decays exponentially fast in time. A more general result can be stated concerning the behavior in time of the *energy*

$$E(t) = \int_0^1 u^2(x,t)\, dx.$$

Indeed, if we multiply (13.1) by u and integrate with respect to x over the interval $[0,1]$, we obtain

$$\begin{aligned}
&\int_0^1 \frac{\partial u}{\partial t}(x,t) u(x,t)\, dx - \nu \int_0^1 \frac{\partial^2 u}{\partial x^2}(x,t) u(x,t)\, dx \\
&= \frac{1}{2} \int_0^1 \frac{\partial u^2}{\partial t}(x,t)\, dx + \nu \int_0^1 \left(\frac{\partial u}{\partial x}(x,t)\right)^2 dx - \nu \left[\frac{\partial u}{\partial x}(x,t) u(x,t)\right]_{x=0}^{x=1} \\
&= \frac{1}{2} E'(t) + \nu \int_0^1 \left(\frac{\partial u}{\partial x}(x,t)\right)^2 dx,
\end{aligned}$$

having used integration by parts, the boundary conditions (13.2) or (13.6), and interchanged differentiation and integration.
Using the Cauchy-Schwarz inequality (8.33) yields

$$\int_0^1 f(x,t)u(x,t)\,dx \le \sqrt{F(t)}\sqrt{E(t)},$$

where $F(t) = \int_0^1 f^2(x,t)\,dx$. Then

$$E'(t) + 2\nu \int_0^1 \left(\frac{\partial u}{\partial x}(x,t)\right)^2 dx \le 2\sqrt{F(t)}\sqrt{E(t)}.$$

Owing to the Poincaré inequality (12.16) with $(a,b) = (0,1)$ we obtain

$$E'(t) + 2\frac{\nu}{(C_P)^2}E(t) \le 2\sqrt{F(t)}\sqrt{E(t)}.$$

By Young's inequality (12.40) we have

$$2\sqrt{F(t)}\sqrt{E(t)} \le \gamma E(t) + \frac{1}{\gamma}F(t),$$

having set $\gamma = \nu/C_P^2$. Therefore, $E'(t) + \gamma E(t) \le \frac{1}{\gamma}F(t)$, or, equivalently, $(e^{\gamma t}E(t))' \le \frac{1}{\gamma}e^{\gamma t}F(t)$. Then, integrating from 0 to t we get

$$E(t) \le e^{-\gamma t}E(0) + \frac{1}{\gamma}\int_0^t e^{\gamma(s-t)}F(s)ds. \qquad (13.7)$$

In particular, when $f \equiv 0$, (13.7) shows that the energy $E(t)$ decays exponentially fast in time.

13.2 Finite Difference Approximation of the Heat Equation

To solve the heat equation numerically we have to discretize both the x and t variables. We can start by dealing with the x-variable, following the same approach as in Section 12.2. We denote by $u_i(t)$ an approximation of $u(x_i,t)$, $i = 0,\ldots,n$, and approximate the Dirichlet problem (13.1)-(13.3) by the scheme

$$\begin{aligned}
&\dot{u}_i(t) - \frac{\nu}{h^2}(u_{i-1}(t) - 2u_i(t) + u_{i+1}(t)) = f_i(t), && i = 1,\ldots,n-1, \forall t > 0,\\
&u_0(t) = u_n(t) = 0, && \forall t > 0,\\
&u_i(0) = u_0(x_i), && i = 0,\ldots,n,
\end{aligned}$$

where the upper dot indicates derivation with respect to time, and $f_i(t) = f(x_i, t)$. This is actually a *semi-discretization* of problem (13.1)-(13.3), and is a system of ordinary differential equations of the following form

$$\begin{cases} \dot{\mathbf{u}}(t) = -\nu \mathrm{A}_{\mathrm{fd}}\mathbf{u}(t) + \mathbf{f}(t), \quad \forall t > 0, \\ \mathbf{u}(0) = \mathbf{u}_0, \end{cases} \tag{13.8}$$

where $\mathbf{u}(t) = [u_1(t), \ldots, u_{n-1}(t)]^T$ is the vector of unknowns, $\mathbf{f}(t) = [f_1(t), \ldots, f_{n-1}(t)]^T$, $\mathbf{u}_0 = [u_0(x_1), \ldots, u_0(x_{n-1})]^T$ and A_{fd} is the tridiagonal matrix introduced in (12.8). Note that for the derivation of (13.8) we have assumed that $u_0(x_0) = u_0(x_n) = 0$, which is coherent with the boundary condition (13.2).

A popular scheme for the integration of (13.8) with respect to time is the so-called *θ−method.* To construct the scheme, we denote by v^k the value of the variable v at time $t^k = k\Delta t$, for $\Delta t > 0$; then, the θ-method for the time-integration of (13.8) is

$$\begin{cases} \dfrac{\mathbf{u}^{k+1} - \mathbf{u}^k}{\Delta t} = -\nu \mathrm{A}_{\mathrm{fd}}(\theta \mathbf{u}^{k+1} + (1-\theta)\mathbf{u}^k) + \theta \mathbf{f}^{k+1} + (1-\theta)\mathbf{f}^k, \\ \hfill k = 0, 1, \ldots \\ \mathbf{u}^0 = \mathbf{u}_0, \end{cases} \tag{13.9}$$

or, equivalently,

$$(\mathrm{I} + \nu\theta\Delta t \mathrm{A}_{\mathrm{fd}})\, \mathbf{u}^{k+1} = (\mathrm{I} - \nu(1-\theta)\Delta t \mathrm{A}_{\mathrm{fd}})\, \mathbf{u}^k + \mathbf{g}^{k+1}, \tag{13.10}$$

where $\mathbf{g}^{k+1} = \Delta t(\theta \mathbf{f}^{k+1} + (1-\theta)\mathbf{f}^k)$ and I is the identity matrix of order $n-1$.

For suitable values of the parameter θ, from (13.10) we can recover some familiar methods that have been introduced in Chapter 11. For example, if $\theta = 0$ the method (13.10) coincides with the forward Euler scheme and we can get $\mathbf{u}^{k+1}$ explicitly; otherwise, a linear system (with constant matrix $\mathrm{I} + \nu\theta\Delta t \mathrm{A}_{\mathrm{fd}}$) needs be solved at each time-step.

Regarding stability, assume that $f \equiv 0$ (henceforth $\mathbf{g}^k = \mathbf{0}\ \forall k > 0$), so that from (13.5) the exact solution $u(x,t)$ tends to zero for every x as $t \to \infty$. Then we would expect the discrete solution to have the same behaviour, in which case we would call our scheme (13.10) *asymptotically stable*, this being coherent with what we did in Chapter 11, Section 11.1 for ordinary differential equations.

If $\theta = 0$, from (13.10) it follows that

$$\mathbf{u}^k = (\mathrm{I} - \nu\Delta t \mathrm{A}_{\mathrm{fd}})^k \mathbf{u}^0, \quad k = 1, 2, \ldots.$$

From the analysis of convergent matrices (see Section 1.11.2) we deduce that $\mathbf{u}^k \to \mathbf{0}$ as $k \to \infty$ iff

$$\rho(\mathrm{I} - \nu\Delta t \mathrm{A}_{\mathrm{fd}}) < 1. \tag{13.11}$$

On the other hand, the eigenvalues of A_{fd} are given by (see Exercise 3)

$$\mu_i = \frac{4}{h^2} \sin^2(i\pi h/2), \qquad i = 1, \dots, n-1.$$

Then (13.11) is satisfied iff

$$\Delta t < \frac{1}{2\nu} h^2.$$

As expected, the forward Euler method is conditionally stable, and the time-step Δt should decay as the square of the grid spacing h.

In the case of the backward Euler method ($\theta = 1$), we would have from (13.10)

$$\mathbf{u}^k = \left[(\mathrm{I} + \nu \Delta t \mathrm{A}_{\mathrm{fd}})^{-1} \right]^k \mathbf{u}^0, \qquad k = 1, 2, \dots.$$

Since all the eigenvalues of the matrix $(\mathrm{I} + \nu\Delta t \mathrm{A}_{\mathrm{fd}})^{-1}$ are real, positive and strictly less than 1 for every value of Δt, this scheme is unconditionally stable. More generally, the θ-scheme is unconditionally stable for all the values $1/2 \leq \theta \leq 1$, and conditionally stable if $0 \leq \theta < 1/2$ (see Section 13.3.1).

As far as the accuracy of the θ-method is concerned, its local truncation error is of the order of $\Delta t + h^2$ if $\theta \neq \frac{1}{2}$, while it is of the order of $\Delta t^2 + h^2$ if $\theta = \frac{1}{2}$. The method corresponding to $\theta = 1/2$ is frequently called the *Crank-Nicolson scheme* and is therefore unconditionally stable and second-order accurate with respect to both Δt and h.

13.3 Finite Element Approximation of the Heat Equation

The space discretization of (13.1)-(13.3) can also be accomplished using the Galerkin finite element method by proceeding as in Chapter 12 in the elliptic case. First, for all $t > 0$ we multiply (13.1) by a test function $v = v(x)$ and integrate over $(0,1)$. Then, we let $V = \mathrm{H}_0^1(0,1)$ and $\forall t > 0$ we look for a function $t \to u(x,t) \in V$ (briefly, $u(t) \in V$) such that

$$\int_0^1 \frac{\partial u(t)}{\partial t} v\,dx + a(u(t), v) = F(v) \qquad \forall v \in V, \tag{13.12}$$

with $u(0) = u_0$. Here, $a(u(t), v) = \int_0^1 \nu (\partial u(t)/\partial x)\,(\partial v/\partial x)\,dx$ and $F(v) = \int_0^1 f(t) v\,dx$ are the bilinear form and the linear functional respectively, associated with the elliptic operator L and the right-hand side f. Notice that $a(\cdot,\cdot)$ is a special case of (12.44) and that the dependence of u and f on the space variable x will be understood henceforth.

Let V_h be a suitable finite dimensional subspace of V. We consider the following Galerkin formulation: $\forall t > 0$, find $u_h(t) \in V_h$ such that

$$\int_0^1 \frac{\partial u_h(t)}{\partial t} v_h dx + a(u_h(t), v_h) = F(v_h) \quad \forall v_h \in V_h, \tag{13.13}$$

where $u_h(0) = u_{0h}$ and $u_{0h} \in V_h$ is a convenient approximation of u_0. Problem (13.13) is referred to as a *semi-discretization* of (13.12) since it is only a space discretization of the heat equation.

Proceeding in a similar manner to that used to obtain the energy estimate (13.7), we get the following a priori estimate for the discrete solution $u_h(t)$ of (13.13)

$$E_h(t) \le e^{-\gamma t} E_h(0) + \frac{1}{\gamma} \int_0^t e^{\gamma(s-t)} F(s) ds,$$

where $E_h(t) = \int_0^1 u_h^2(x,t)\, dx$.

As for the finite element discretization of (13.13), we introduce the finite element space V_h defined in (12.57) and consequently a basis $\{\varphi_j\}$ for V_h as already done in Section 12.4.5. Then, the solution u_h of (13.13) can be sought under the form

$$u_h(t) = \sum_{j=1}^{N_h} u_j(t) \varphi_j,$$

where $\{u_j(t)\}$ are the unknown coefficients and N_h is the dimension of V_h. Then, from (13.13) we obtain

$$\int_0^1 \sum_{j=1}^{N_h} \dot{u}_j(t) \varphi_j \varphi_i dx + a\left(\sum_{j=1}^{N_h} u_j(t) \varphi_j, \varphi_i \right) = F(\varphi_i), \qquad i = 1, \ldots, N_h$$

that is,

$$\sum_{j=1}^{N_h} \dot{u}_j(t) \int_0^1 \varphi_j \varphi_i dx + \sum_{j=1}^{N_h} u_j(t) a(\varphi_j, \varphi_i) = F(\varphi_i), \qquad i = 1, \ldots, N_h.$$

Using the same notation as in (13.8) we obtain

$$\mathrm{M}\dot{\mathbf{u}}(t) + \mathrm{A}_{\mathrm{fe}} \mathbf{u}(t) = \mathbf{f}_{\mathrm{fe}}(t), \tag{13.14}$$

where $\mathrm{A}_{\mathrm{fe}} = (a(\varphi_j, \varphi_i))$, $\mathbf{f}_{\mathrm{fe}}(t) = (F(\varphi_i))$ and $\mathrm{M} = (m_{ij}) = (\int_0^1 \varphi_j \varphi_i dx)$ for $i, j = 1, \ldots, N_h$. M is called the *mass matrix*. Since it is nonsingular, the system of ODEs (13.14) can be written in normal form as

$$\dot{\mathbf{u}}(t) = -\mathrm{M}^{-1} \mathrm{A}_{\mathrm{fe}} \mathbf{u}(t) + \mathrm{M}^{-1} \mathbf{f}_{\mathrm{fe}}(t). \tag{13.15}$$

To solve (13.15) approximately we can still apply the θ-method and obtain

$$\mathrm{M}\frac{\mathbf{u}^{k+1}-\mathbf{u}^{k}}{\Delta t}+\mathrm{A}_{\mathrm{fe}}\left[\theta\mathbf{u}^{k+1}+(1-\theta)\mathbf{u}^{k}\right]=\theta\mathbf{f}_{\mathrm{fe}}^{k+1}+(1-\theta)\mathbf{f}_{\mathrm{fe}}^{k}. \qquad (13.16)$$

As usual, the upper index k means that the quantity at hand is computed at time t^k. As in the finite difference case, for $\theta = 0, 1$ and $1/2$ we respectively obtain the forward Euler, backward Euler and Crank-Nicolson methods, where the Crank-Nicolson method is the only one which is second-order accurate with respect to Δt.

For each k, (13.16) is a linear system whose matrix is

$$\mathrm{K}=\frac{1}{\Delta t}\mathrm{M}+\theta\mathrm{A}_{\mathrm{fe}}.$$

Since M and A_{fe} are symmetric and positive definite, the matrix K is also symmetric and positive definite. Thus, its Cholesky decomposition $\mathrm{K}=\mathrm{H}^T\mathrm{H}$ where H is upper triangular (see Section 3.4.2) can be carried out at $t = 0$. Consequently, at each time step the following two linear triangular systems, each of size equal to N_h, must be solved, with a computational cost of $N_h^2/2$ flops

$$\begin{cases}\mathrm{H}^T\mathbf{y}=\left[\dfrac{1}{\Delta t}\mathrm{M}-(1-\theta)\mathrm{A}_{\mathrm{fe}}\right]\mathbf{u}^k+\theta\mathbf{f}_{\mathrm{fe}}^{k+1}+(1-\theta)\mathbf{f}_{\mathrm{fe}}^{k},\\ \mathrm{H}\mathbf{u}^{k+1}=\mathbf{y}.\end{cases}$$

When $\theta = 0$, a suitable diagonalization of M would allow to decouple the system equations (13.16). The procedure is carried out by the so-called *mass-lumping* in which we approximate M by a nonsingular diagonal matrix $\widetilde{\mathrm{M}}$. In the case of piecewise linear finite elements $\widetilde{\mathrm{M}}$ can be obtained using the composite trapezoidal formula over the nodes $\{x_i\}$ to evaluate the integrals $\int_0^1 \varphi_j\varphi_i\,dx$, obtaining $\tilde{m}_{ij} = h\delta_{ij}$, $i,j = 1,\ldots,N_h$ (see Exercise 2).

13.3.1 Stability Analysis of the θ-Method

Applying the θ-method to the Galerkin problem (13.13) yields

$$\begin{aligned}&\left(\frac{u_h^{k+1}-u_h^k}{\Delta t},v_h\right)+a\left(\theta u_h^{k+1}+(1-\theta)u_h^k,v_h\right)\\&\qquad=\theta F^{k+1}(v_h)+(1-\theta)F^k(v_h)\qquad\forall v_h\in V_h\end{aligned}\qquad(13.17)$$

for $k \geq 0$ and with $u_h^0 = u_{0h}$, $F^k(v_h) = \int_0^1 f(t^k)v_h(x)dx$. Since we are interested in the stability analysis, we can consider the special case where $F = 0$; moreover, for the time being, we focus on the case $\theta = 1$ (implicit Euler scheme), i.e.

$$\left(\frac{u_h^{k+1} - u_h^k}{\Delta t}, v_h\right) + a\left(u_h^{k+1}, v_h\right) = 0 \qquad \forall v_h \in V_h.$$

Letting $v_h = u_h^{k+1}$, we get

$$\left(\frac{u_h^{k+1} - u_h^k}{\Delta t}, u_h^{k+1}\right) + a(u_h^{k+1}, u_h^{k+1}) = 0.$$

From the definition of $a(\cdot, \cdot)$, it follows that

$$a\left(u_h^{k+1}, u_h^{k+1}\right) = \nu \left\| \frac{\partial u_h^{k+1}}{\partial x} \right\|^2_{\mathrm{L}^2(0,1)}. \tag{13.18}$$

Moreover, we remark that (see Exercise 3 for the proof of this result)

$$\|u_h^{k+1}\|^2_{\mathrm{L}^2(0,1)} + 2\nu\Delta t \left\| \frac{\partial u_h^{k+1}}{\partial x} \right\|^2_{\mathrm{L}^2(0,1)} \leq \|u_h^k\|^2_{\mathrm{L}^2(0,1)}. \tag{13.19}$$

It follows that, $\forall n \geq 1$

$$\sum_{k=0}^{n-1} \|u_h^{k+1}\|^2_{\mathrm{L}^2(0,1)} + 2\nu\Delta t \sum_{k=0}^{n-1} \left\| \frac{\partial u_h^{k+1}}{\partial x} \right\|^2_{\mathrm{L}^2(0,1)} \leq \sum_{k=0}^{n-1} \|u_h^k\|^2_{\mathrm{L}^2(0,1)}.$$

Since these are telescopic sums, we get

$$\|u_h^n\|^2_{\mathrm{L}^2(0,1)} + 2\nu\Delta t \sum_{k=0}^{n-1} \left\| \frac{\partial u_h^{k+1}}{\partial x} \right\|^2_{\mathrm{L}^2(0,1)} \leq \|u_{0h}\|^2_{\mathrm{L}^2(0,1)}, \tag{13.20}$$

which shows that the scheme is unconditionally stable. Proceeding similarly if $f \neq 0$, it can be shown that

$$\begin{aligned} &\|u_h^n\|^2_{\mathrm{L}^2(0,1)} + 2\nu\Delta t \sum_{k=0}^{n-1} \left\| \frac{\partial u_h^{k+1}}{\partial x} \right\|^2_{\mathrm{L}^2(0,1)} \\ &\leq C(n) \left(\|u_{0h}\|^2_{\mathrm{L}^2(0,1)} + \sum_{k=1}^{n} \Delta t \|f^k\|^2_{\mathrm{L}^2(0,1)} \right), \end{aligned} \tag{13.21}$$

where $C(n)$ is a constant independent of both h and Δt.

Remark 13.1 The same kind of stability inequalities (13.20) and (13.21) can be obtained if $a(\cdot, \cdot)$ is a more general bilinear form provided that it is continuous and coercive (see Exercise 4). ■

To carry out the stability analysis of the θ-method for every $\theta \in [0,1]$ we need defining the *eigenvalues* and *eigenvectors* of a bilinear form.

Definition 13.1 We say that λ is an eigenvalue and $w \in V$ is the associated eigenvector for the bilinear form $a(\cdot,\cdot) : V \times V \mapsto \mathbb{R}$ if

$$a(w,v) = \lambda(w,v) \qquad \forall v \in V,$$

where $(\cdot,\cdot)$ denotes the usual scalar product in $\mathrm{L}^2(0,1)$. ■

If the bilinear form $a(\cdot,\cdot)$ is symmetric and coercive, it has infinitely many real positive eigenvalues that form an unbounded sequence; moreover, its eigenvectors (called also *eigenfunctions*) form a basis for the space V.

At a discrete level the corresponding pair $\lambda_h \in \mathbb{R}$, $w_h \in V_h$ satisfies

$$a(w_h, v_h) = \lambda_h(w_h, v_h) \quad \forall v_h \in V_h. \tag{13.22}$$

From the algebraic standpoint, problem (13.22) can be formulated as

$$\mathrm{A_{fe}}\mathbf{w} = \lambda_h \mathrm{M}\mathbf{w}$$

(where $\mathbf{w}$ is the vector of the gridvalues of w_h) and can be regarded as a *generalized eigenvalue problem* (see Section 5.9). All the eigenvalues $\lambda_h^1, \ldots, \lambda_h^{N_h}$ are positive. The corresponding eigenvectors $w_h^1, \ldots, w_h^{N_h}$ form a basis for the subspace V_h and can be chosen in such a way as to be *orthonormal*, i.e., such that $(w_h^i, w_h^j) = \delta_{ij}$, $\forall i, j = 1, \ldots, N_h$. In particular, any function $v_h \in V_h$ can be represented as

$$v_h(x) = \sum_{j=1}^{N_h} v_j w_h^j(x).$$

Let us now assume that $\theta \in [0,1]$ and focus on the case where the bilinear form $a(\cdot,\cdot)$ is symmetric. Although the final stability result still holds in the nonsymmetric case, the proof that follows cannot apply since in that case the eigenvectors would no longer form a basis for V_h. Let $\{w_h^i\}$ be the eigenvectors of $a(\cdot,\cdot)$ whose span forms an orthonormal basis for V_h. Since at each time step $u_h^k \in V_h$, we can express u_h^k as

$$u_h^k(x) = \sum_{j=1}^{N_h} u_j^k w_h^j(x).$$

Letting $F = 0$ in (13.17) and taking $v_h = w_h^i$, we find

$$\begin{aligned} &\frac{1}{\Delta t}\sum_{j=1}^{N_h} \left[u_j^{k+1} - u_j^k\right] \left(w_h^j, w_h^i\right) \\ &\quad + \sum_{j=1}^{N_h} \left[\theta u_j^{k+1} + (1-\theta) u_j^k\right] a(w_h^j, w_h^i) = 0, \qquad i = 1, \ldots, N_h. \end{aligned}$$

Since w_h^j are eigenfunctions of $a(\cdot,\cdot)$ we obtain

$$a(w_h^j, w_h^i) = \lambda_h^j (w_h^j, w_h^i) = \lambda_h^j \delta_{ij} = \lambda_h^i,$$

so that

$$\frac{u_i^{k+1} - u_i^k}{\Delta t} + \left[\theta u_i^{k+1} + (1-\theta) u_i^k\right] \lambda_h^i = 0.$$

Solving this equation with respect to u_i^{k+1} gives

$$u_i^{k+1} = u_i^k \frac{\left[1 - (1-\theta)\lambda_h^i \Delta t\right]}{\left[1 + \theta \lambda_h^i \Delta t\right]}.$$

In order for the method to be unconditionally stable we must have (see Chapter 11)

$$\left| \frac{1 - (1-\theta)\lambda_h^i \Delta t}{1 + \theta \lambda_h^i \Delta t} \right| < 1,$$

that is

$$2\theta - 1 > -\frac{2}{\lambda_h^i \Delta t}.$$

If $\theta \geq 1/2$, this inequality is satisfied for any value of Δt. Conversely, if $\theta < 1/2$ we must have

$$\Delta t < \frac{2}{(1-2\theta)\lambda_h^i}.$$

Since this relation must hold for all the eigenvalues λ_h^i of the bilinear form, it suffices requiring that it is satisfied for the largest of them, which we assume to be $\lambda_h^{N_h}$.

We therefore conclude that if $\theta \geq 1/2$ the θ-method is unconditionally stable (i.e., it is stable $\forall \Delta t$), whereas if $0 \leq \theta < 1/2$ the θ-method is stable only if

$$\Delta t \leq \frac{2}{(1-2\theta)\lambda_h^{N_h}}.$$

It can be shown that there exist two positive constants c_1 and c_2, independent of h, such that

$$c_1 h^{-2} \leq \lambda_h^{N_h} = c_2 h^{-2}$$

(see for the proof, [QV94], Section 6.3.2). Accounting for this, we obtain that if $0 \leq \theta < 1/2$ the method is stable only if

$$\Delta t \leq C_1(\theta) h^2, \qquad (13.23)$$

for a suitable constant $C_1(\theta)$ independent of both h and Δt.

With an analogous proof, it can be shown that if a pseudo-spectral Galerkin approximation is used for problem (13.12), the $\theta-$method is unconditionally stable if $\theta \geq \frac{1}{2}$, while for $0 \leq \theta < \frac{1}{2}$ stability holds only if

$$\Delta t \leq C_2(\theta) N^{-4}, \qquad (13.24)$$

for a suitable constant $C_2(\theta)$ independent of both N and Δt. The difference between (13.23) and (13.24) is due to the fact that the largest eigenvalue of the spectral stiffness matrix grows like $\mathcal{O}(N^4)$ with respect to the degree of the approximating polynomial.

Comparing the solution of the globally discretized problem (13.17) with that of the semi-discrete problem (13.13), by a suitable use of the stability result (13.21) and of the truncation time discretization error, the following *convergence result* can be proved

$$\|u(t^k) - u_h^k\|_{\mathrm{L}^2(0,1)} \leq C(u_0, f, u)(\Delta t^{p(\theta)} + h^{r+1}), \qquad \forall k \geq 1,$$

where r denotes the piecewise polynomial degree of the finite element space V_h, $p(\theta) = 1$ if $\theta \neq 1/2$ while $p(1/2) = 2$ and C is a constant that depends on its arguments (assuming that they are sufficiently smooth) but not on h and Δt. In particular, if $f \equiv 0$ on can obtain the following improved estimates

$$\|u(t^k) - u_h^k\|_{\mathrm{L}^2(0,1)} \leq C \left[\left(\frac{h}{\sqrt{t^k}} \right)^{r+1} + \left(\frac{\Delta t}{t^k} \right)^{p(\theta)} \right] \|u_0\|_{\mathrm{L}^2(0,1)},$$

for $k \geq 1$, $\theta = 1$ or $\theta = 1/2$. (For the proof of these results, see [QV94], pp. 394-395).

Program 100 provides an implementation of the θ-method for the solution of the heat equation on the space-time domain $(a, b) \times (t_0, T)$. The discretization in space is based on piecewise-linear finite elements. The input parameters are: the column vector `I` containing the endpoints of the space interval ($a =$ `I(1)`, $b =$ `I(2)`) and of the time interval ($t_0 =$ `I(3)`, $T =$ `I(4)`); the column vector `n` containing the number of steps in space and time; the macros `u0` and `f` containing the functions u_{0h} and f, the constant viscosity `nu`, the Dirichlet boundary conditions `bc(1)` and `bc(2)`, and the value of the parameter `theta`.

Program 100 - thetameth : θ-method for the heat equation

```
function [u,x] = thetameth(I,n,u0,f,bc,nu,theta)
%THETAMETH Theta-method.
% [U,X]=THETAMETH(I,N,U0,F,BC,NU,THETA) solves the heat equation
% with the THETA-method.
nx=n(1); h=(I(2)-I(1))/nx; x=[I(1):h:I(2)];
t=I(3);
uold=(eval(u0))';
nt=n(2); k=(I(4)-I(3))/nt;
e=ones(nx+1,1);
K=spdiags([(h/(6*k)-nu*theta/h)*e, (2*h/(3*k)+2*nu*theta/h)*e, ...
   (h/(6*k)-nu*theta/h)*e],-1:1,nx+1,nx+1);
B=spdiags([(h/(6*k)+nu*(1-theta)/h)*e, (2*h/(3*k)-nu*2*(1-theta)/h)*e, ...
   (h/(6*k)+nu*(1-theta)/h)*e],-1:1,nx+1,nx+1);
K(1,1)=1;     K(1,2)=0;    B(1,1)= 0;     B(1,2)=0;
K(nx+1,nx+1)=1; K(nx+1,nx)=0; B(nx+1,nx+1)=0; B(nx+1,nx)=0;
```

```
[L,U]=lu(K);
t=I(3);
x=[I(1)+h:h:I(2)-h];
fold=(eval(f))';
fold=h*fold;
fold=[bc(1); fold; bc(2)];
for time=I(3)+k:k:I(4)
   t=time;
   fnew=(eval(f))';
   fnew=h*fnew;
   fnew=[bc(1); fnew; bc(2)];
   b=theta*fnew+(1-theta)*fold+B*uold;
   y=L\b;
   u=U\y;
   uold=u;
end
x=[I(1):h:I(2)];
return
```

Example 13.1 Let us assess the time-accuracy of the θ-method in the solution of the heat equation (13.1) on the space-time domain $(0,1)\times(0,1)$, where f is chosen in such a way that the exact solution is $u = \sin(2\pi x)\cos(2\pi t)$. A fixed spatial grid size $h = 1/500$ has been used while the time step Δt is equal to $(10k)^{-1}$, $k = 1,\ldots,4$. Finally, piecewise finite elements are used for the space discretization. Figure 13.1 shows the convergence behavior in the $L^2(0,1)$ norm (evaluated at time $t = 1$), as Δt tends to zero, of the backward Euler method (BE) ($\theta = 1$, solid line) and of the Crank-Nicolson scheme (CN) ($\theta = 1/2$, dashed line). As expected, the CN method is far more accurate than the BE method. •

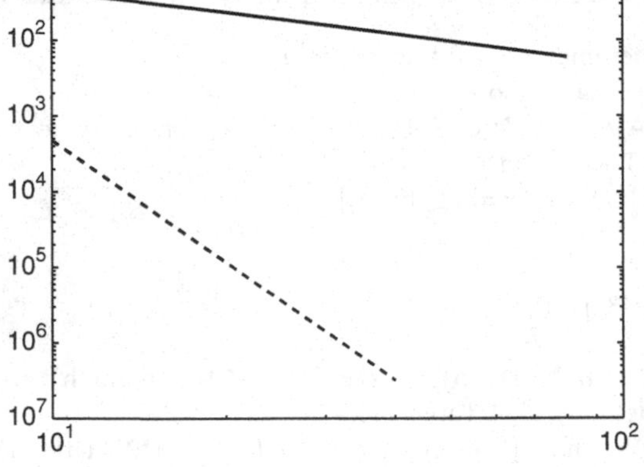

Fig. 13.1. Convergence analysis of the θ-method as a function of the number $1/\Delta t$ of time steps (represented on the x-axis): $\theta = 1$ (*solid line*) and $\theta = 0.5$ (*dashed line*)

13.4 Space-Time Finite Element Methods for the Heat Equation

An alternative approach for time discretization is based on the use of a Galerkin method to discretize both space and time variables.

Suppose to solve the heat equation for $x \in [0,1]$ and $t \in [0,T]$. Let us denote by $I_k = [t^{k-1}, t^k]$ the k-th time interval for $k = 1, \ldots, n$ with $\Delta t^k = t^k - t^{k-1}$; moreover, we let $\Delta t = \max_k \Delta t^k$; the rectangle $S_k = [0,1] \times I_k$ is the so called *space-time slab*. At each time level t^k, we consider a partition $\mathcal{T}_{h_k}$ of $(0,1)$ into m^k subintervals $K_j^k = [x_j^k, x_{j+1}^k]$, $j = 0, \ldots, m^k - 1$. We let $h_j^k = x_{j+1}^k - x_j^k$ and denote by $h_k = \max_j h_j^k$ and by $h = \max_k h_k$.

Let us now associate with S_k a space-time partition $\mathcal{S}_k = \cup_{j=1}^{m_k} R_j^k$ where $R_j^k = K_j^k \times I_k$ and $K_j^k \in \mathcal{T}_{h_k}$. The space-time slab S_k is thus decomposed into rectangles R_j^k (see Figure 13.2).

For each time slab S_k we introduce the space-time finite element space

$$\mathbb{Q}_q(S_k) = \left\{ v \in C^0(S_k),\ v_{|R_j^k} \in \mathbb{P}_1(K_j^k) \times \mathbb{P}_q(I_k),\ j = 0, \ldots, m^k - 1 \right\}$$

where, usually, $q = 0$ or $q = 1$. Then, the space-time finite element space over $[0,1] \times [0,T]$ is defined as follows

$$V_{h,\Delta t} = \left\{ v : [0,1] \times [0,T] \to \mathbb{R} :\ v_{|S_k} \in Y_{h,k},\ k = 1, \ldots, n \right\},$$

where

$$Y_{h,k} = \{ v \in \mathbb{Q}_q(S_k) :\ v(0,t) = v(1,t) = 0\ \forall t \in I_k \}.$$

The number of degrees of freedom of $V_{h,\Delta t}$ is equal to $(q+1)(m^k - 1)$. The functions in $V_{h,\Delta t}$ are linear and continuous in space while they are piecewise

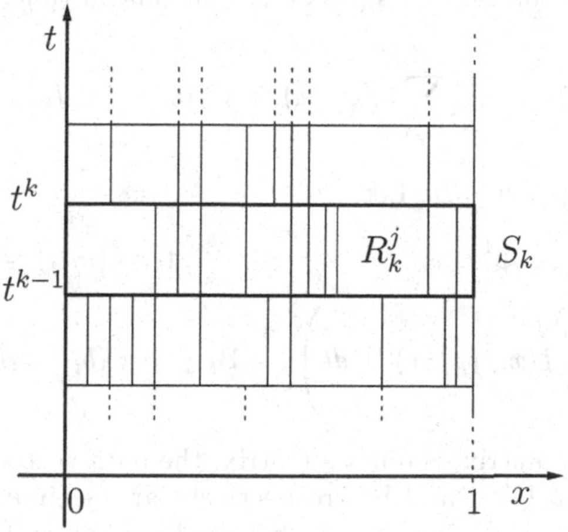

Fig. 13.2. Space-time finite element discretization

polynomials of degree q in time. These functions are in general discontinuous across the time levels t^k and the partitions $\mathcal{T}_h^k$ do not match at the interface between contiguous time levels (see Figure 13.2). For this reason, we adopt henceforth the following notation

$$v_\pm^k = \lim_{\tau \to 0} v(t^k \pm \tau),\ [v^k] = v_+^k - v_-^k.$$

The discretization of problem (13.12) using continuous finite elements in space of degree 1 and discontinuous finite elements of degree q in time (abbreviated by cG(1)dG(q) method) is: find $U \in V_{h,\Delta t}$ such that

$$\sum_{k=1}^{n} \int_{I_k} \left[\left(\frac{\partial U}{\partial t}, V \right) + a(U,V) \right] dt + \sum_{k=1}^{n-1} ([U^k], V_+^k)$$
$$+ (U_+^0, V_+^0) = \int_0^T (f,V)\, dt, \qquad \forall V \in \overset{0}{V}_{h,\Delta t},$$

where

$$\overset{0}{V}_{h,\Delta t} = \{ v \in V_{h,\Delta t} :\ v(0,t) = v(1,t) = 0\ \forall t \in [0,T] \},$$

$U_-^0 = u_{0h}$, $U^k = U(x,t^k)$ and $(u,v) = \int_0^1 uv\, dx$ denotes the scalar product of $\mathrm{L}^2(0,1)$. The continuity of U at each point t^k is therefore imposed only in a weak sense.

To construct the algebraic equations for the unknown U we need expanding it over a basis in time and in space. The single space-time basis function $\varphi_{jl}^k(x,t)$ can be written as $\varphi_{jl}^k(x,t) = \varphi_j^k(x)\psi_l(t)$, $j = 1, \ldots, m^k - 1$, $l = 0, \ldots, q$, where φ_j^k is the usual piecewise linear basis function and ψ_l is the l-th basis function of $\mathbb{P}_q(I_k)$.

When $q = 0$ the solution U is piecewise constant in time. In that case

$$U^k(x,t) = \sum_{j=1}^{N_h^k} U_j^k \varphi_j^k(x),\ x \in [0,1],\ t \in I_k,$$

where $U_j^k = U^k(x_j,t)\ \forall t \in I_k$. Let

$$\mathrm{A}_k = (a_{ij}) = (a(\varphi_j^k, \varphi_i^k)), \qquad \mathrm{M}_k = (m_{ij}) = ((\varphi_j^k, \varphi_i^k)),$$

$$\mathbf{f}_k = (f_i) = \left(\int_{S_k} f(x,t)\varphi_i^k(x) dx\, dt \right), \quad \mathrm{B}_{k,k-1} = (b_{ij}) = ((\varphi_j^k, \varphi_i^{k-1})),$$

denote the stiffness matrix, the mass matrix, the data vector and the projection matrix between V_h^{k-1} and V_h^k, respectively, at the time level t^k.
Then, letting $\mathbf{U}^k = (U_j^k)$, at each k-th time level the cG(1)dG(0) method requires solving the following linear system

$$\left(\mathrm{M}_k + \Delta t^k \mathrm{A}_k\right) \mathbf{U}^k = \mathrm{B}_{k,k-1}\mathbf{U}^{k-1} + \mathbf{f}_k,$$

which is nothing else than the Euler backward discretization scheme with a modified right-hand side.

When $q = 1$, the solution is piecewise linear in time. For ease of notation we let $U^k(x) = U_-(x, t^k)$ and $U^{k-1}(x) = U_+(x, t^{k-1})$. Moreover, we assume that the spatial partition $\mathcal{T}_{h_k}$ does not change with the time level and we let $m^k = m$ for every $k = 0, \ldots, n$. Then, we can write

$$U_{|S_k} = U^{k-1}(x)\frac{t^k - t}{\Delta t^k} + U^k(x)\frac{t - t^{k-1}}{\Delta t^k}.$$

Thus the cG(1)dG(1) method leads to the solution of the following 2×2 block-system in the unknowns $\mathbf{U}^k = (U_i^k)$ and $\mathbf{U}^{k-1} = (U_i^{k-1})$, $i = 1, \ldots, m-1$,

$$\begin{cases} \left(-\frac{1}{2}\mathrm{M}_k + \frac{\Delta t^k}{3}\mathrm{A}_k\right)\mathbf{U}^{k-1} + \left(\frac{1}{2}\mathrm{M}_k + \frac{\Delta t^k}{6}\mathrm{A}_k\right)\mathbf{U}^k = \mathbf{f}_{k-1} + \mathrm{B}_{k,k-1}\mathbf{U}_-^{k-1}, \\ \left(\frac{1}{2}\mathrm{M}_k + \frac{\Delta t^k}{6}\mathrm{A}_k\right)\mathbf{U}^{k-1} + \left(\frac{1}{2}\mathrm{M}_k + \frac{\Delta t^k}{3}\mathrm{A}_k\right)\mathbf{U}^k = \mathbf{f}_k, \end{cases}$$

where

$$\mathbf{f}_{k-1} = \int_{S_k} f(x,t)\varphi_i^k(x)\psi_1^k(t)dx\, dt, \quad \mathbf{f}_k = \int_{S_k} f(x,t)\varphi_i^k(x)\psi_2^k(t)dx\, dt$$

and $\psi_1^k(t) = (t^k - t)/\Delta t^k$, $\psi_2^k(t)(t - t^{k-1})/\Delta t^k$ are the two basis functions of $\mathbb{P}_1(I_k)$.

Assuming that $V_{h,k-1} \not\subset V_{h,k}$, it is possible to prove that (see for the proof [EEHJ96])

$$\|u(t^n) - U^n\|_{\mathrm{L}^2(0,1)} \leq C(u_{0h}, f, u, n)(\Delta t^2 + h^2), \tag{13.25}$$

where C is a constant that depends on its arguments (assuming that they are sufficiently smooth) but not on h and Δt.

An advantage in using space-time finite elements is the possibility to perform a space-time grid adaptivity on each time-slab based on a posteriori error estimates (the interested reader is referred to [EEHJ96] where the analysis of this method is carried out in detail).

Program 101 provides an implementation of the dG(1)cG(1) method for the solution of the heat equation on the space-time domain $(a, b) \times (t_0, T)$. The input parameters are the same as in Program 100.

Program 101 - pardg1cg1 : dG(1)cG(1) method for the heat equation

```
function [u,x]=pardg1cg1(I,n,u0,f,nu,bc)
%PARDG1CG1 dG(1)cG(1) scheme for the heat equation.
nx=n(1); h=(I(2)-I(1))/nx; x=[I(1):h:I(2)];
```

```
t=l(3); um=(eval(u0))';
nt=n(2); k=(l(4)-l(3))/nt;
e=ones(nx+1,1);
Add=spdiags([(h/12-k*nu/(3*h))*e, (h/3+2*k*nu/(3*h))*e, ...
    (h/12-k*nu/(3*h))*e],-1:1,nx+1,nx+1);
Aud=spdiags([(h/12-k*nu/(6*h))*e, (h/3+k*nu/(3*h))*e, ...
    (h/12-k*nu/(6*h))*e],-1:1,nx+1,nx+1);
Ald=spdiags([(-h/12-k*nu/(6*h))*e, (-h/3+k*nu/(3*h))*e, ...
    (-h/12-k*nu/(6*h))*e],-1:1,nx+1,nx+1);
B=spdiags([h*e/6, 2*h*e/3, h*e/6],-1:1,nx+1,nx+1);
Add(1,1)=1; Add(1,2)=0; B(1,1)=0;  B(1,2)=0;
Aud(1,1)=0; Aud(1,2)=0; Ald(1,1)=0; Ald(1,2)=0;
Add(nx+1,nx+1)=1; Add(nx+1,nx)=0;
B(nx+1,nx+1)=0;  B(nx+1,nx)=0;
Ald(nx+1,nx+1)=0; Ald(nx+1,nx)=0;
Aud(nx+1,nx+1)=0; Aud(nx+1,nx)=0;
[L,U]=lu([Add Aud; Ald Add]);
x=[l(1)+h:h:l(2)-h]; xx=[l(1),x,l(2)];
for time=l(3)+k:k:l(4)
  t=time;    fq1=0.5*k*h*eval(f);
  t=time-k;  fq0=0.5*k*h*eval(f);
  rhs0=[bc(1), fq0, bc(2)];
  rhs1=[bc(1), fq1, bc(2)];
  b=[rhs0'; rhs1'] + [B*um; zeros(nx+1,1)];
  y=L\b;
  u=U\y;
  um=u(nx+2:2*nx+2,1);
end
x=[l(1):h:l(2)]; u=um;
return
```

Example 13.2 We assess the accuracy of the dG(1)cG(1) method on the same test problem considered in Example 13.1. In order to neatly identify both spatial and temporal contributions in the error estimate (13.25) we have performed the numerical computations using Program 101 by varying either the time step or the space discretization step only, having chosen in each case the discretization step in the other variable sufficiently small that the corresponding error can be neglected. The convergence behavior in Figure 13.3 shows perfect agreement with the theoretical results (second-order accuracy in both space and time). •

13.5 Hyperbolic Equations: A Scalar Transport Problem

Let us consider the following scalar hyperbolic problem

$$\begin{cases} \dfrac{\partial u}{\partial t} + a\dfrac{\partial u}{\partial x} = 0, & x \in \mathbb{R},\ t > 0, \\ u(x,0) = u_0(x), & x \in \mathbb{R}, \end{cases} \tag{13.26}$$

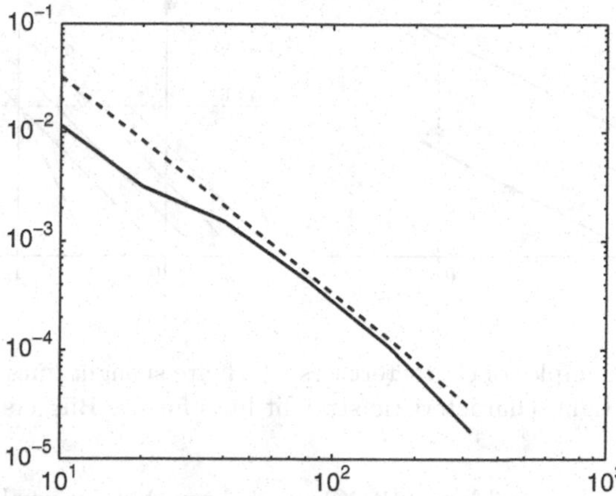

Fig. 13.3. Convergence analysis for the dG(1)cG(1) method. The solid line is the time discretization error while the dashed line is the space discretization error. In the first case the x-axis denotes the number of time steps while in second case it represents the number of space subintervals

where a is a positive real number. Its solution is given by

$$u(x,t) = u_0(x - at), \quad t \geq 0,$$

and represents a travelling wave with velocity a. The curves $(x(t), t)$ in the plane (x, t), that satisfy the following scalar ordinary differential equation

$$\begin{cases} \dfrac{dx(t)}{dt} = a, & t > 0, \\ x(0) = x_0, \end{cases} \tag{13.27}$$

are called *characteristic curves*. They are the straight lines $x(t) = x_0 + at$, $t > 0$. The solution of (13.26) remains constant along them since

$$\frac{du}{dt} = \frac{\partial u}{\partial t} + \frac{\partial u}{\partial x}\frac{dx}{dt} = 0 \qquad \text{on } (x(t), t).$$

For the more general problem

$$\begin{cases} \dfrac{\partial u}{\partial t} + a\dfrac{\partial u}{\partial x} + a_0 u = f, & x \in \mathbb{R}, t > 0, \\ u(x,0) = u_0(x), & x \in \mathbb{R}, \end{cases} \tag{13.28}$$

where a, a_0 and f are given functions of the variables (x, t), the characteristic curves are still defined as in (13.27). In this case, the solutions of (13.28) satisfy along the characteristics the following differential equation

$$\frac{du}{dt} = f - a_0 u \text{ on } (x(t), t).$$

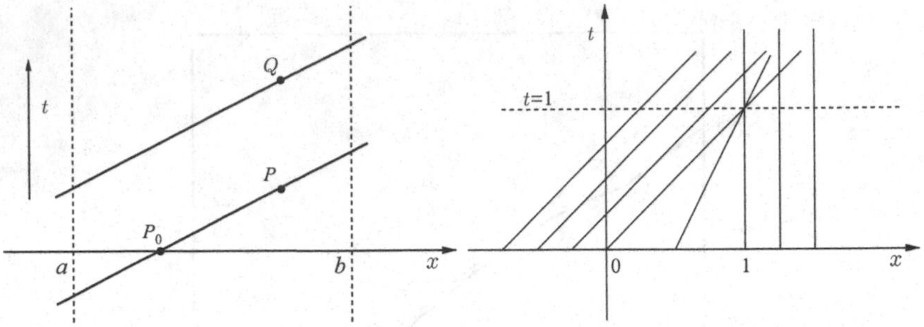

Fig. 13.4. Left: examples of characteristics which are straight lines issuing from the points P and Q. Right: characteristic straight lines for the Burgers equation

Let us now consider problem (13.26) on a bounded interval. For example, assume that $x \in [\alpha, \beta]$ and $a > 0$. Since u is constant along the characteristics, from Figure 13.4 (*left*) we deduce that the value of the solution at P attains the value of u_0 at P_0, the foot of the characteristic issuing from P. On the other hand, the characteristic issuing from Q intersects the straight line $x(t) = \alpha$ at a certain time $t = \bar{t} > 0$. Thus, the point $x = \alpha$ is an *inflow* point and it is necessary to assign a boundary value for u at $x = \alpha$ for every $t > 0$. Notice that if $a < 0$ then the inflow point is $x = \beta$.

Referring to problem (13.26) it is worth noting that if u_0 is discontinuous at a point x_0, then such a discontinuity propagates along the characteristics issuing from x_0. This process can be rigorously formalized by introducing the concept of *weak solutions* of hyperbolic problems, see e.g. [GR96]. Another reason for introducing weak solutions is that in the case of nonlinear hyperbolic problems the characteristic lines can intersect: in this case the solution cannot be continuous and no classical solution does exist.

Example 13.3 (Burgers equation) Let us consider the Burgers equation

$$\frac{\partial u}{\partial t} + u\frac{\partial u}{\partial x} = 0, \qquad x \in \mathbb{R}, \tag{13.29}$$

which is perhaps the simplest nontrivial example of a nonlinear hyperbolic equation. Taking as initial condition

$$u(x,0) = u_0(x) = \begin{cases} 1, & x \leq 0, \\ 1-x, & 0 \leq x \leq 1, \\ 0, & x \geq 1, \end{cases}$$

the characteristic line issuing from the point $(x_0, 0)$ is given by

$$x(t) = x_0 + tu_0(x_0) = \begin{cases} x_0 + t, & x_0 \leq 0, \\ x_0 + t(1-x_0), & 0 \leq x_0 \leq 1, \\ x_0, & x_0 \geq 1. \end{cases}$$

Notice that the characteristic lines do not intersect only if $t < 1$ (see Figure 13.4, right). •

13.6 Systems of Linear Hyperbolic Equations

Consider the linear hyperbolic systems of the form

$$\frac{\partial \mathbf{u}}{\partial t} + \mathrm{A}\frac{\partial \mathbf{u}}{\partial x} = \mathbf{0}, \quad x \in \mathbb{R},\, t > 0, \tag{13.30}$$

where $\mathbf{u} : \mathbb{R}\times[0,\infty) \to \mathbb{R}^p$ and $\mathrm{A} \in \mathbb{R}^{p\times p}$ is a matrix with constant coefficients.

This system is said *hyperbolic* if A is diagonalizable and has real eigenvalues, that is, if there exists a nonsingular matrix $\mathrm{T} \in \mathbb{R}^{p\times p}$ such that

$$\mathrm{A} = \mathrm{T}\Lambda\mathrm{T}^{-1},$$

where $\Lambda = \mathrm{diag}(\lambda_1, ..., \lambda_p)$ is the diagonal matrix of the real eigenvalues of A, while $\mathrm{T} = [\boldsymbol{\omega}^1, \boldsymbol{\omega}^2, \ldots, \boldsymbol{\omega}^p]$ is the matrix whose column vectors are the right eigenvectors of A (see Section 1.7). The system is said to be *strictly hyperbolic* if it is hyperbolic with distinct eigenvalues. Thus

$$\mathrm{A}\boldsymbol{\omega}^k = \lambda_k \boldsymbol{\omega}^k, \quad k = 1, \ldots, p.$$

Introducing the *characteristic variables* $\mathbf{w} = \mathrm{T}^{-1}\mathbf{u}$, system (13.30) becomes

$$\frac{\partial \mathbf{w}}{\partial t} + \Lambda\frac{\partial \mathbf{w}}{\partial x} = \mathbf{0}.$$

This is a system of p independent scalar equations of the form

$$\frac{\partial w_k}{\partial t} + \lambda_k \frac{\partial w_k}{\partial x} = 0, \quad k = 1, \ldots, p.$$

Proceeding as in Section 13.5 we obtain $w_k(x,t) = w_k(x - \lambda_k t, 0)$, and thus the solution $\mathbf{u} = \mathrm{T}\mathbf{w}$ of problem (13.30) can be written as

$$\mathbf{u}(x,t) = \sum_{k=1}^{p} w_k(x - \lambda_k t, 0)\boldsymbol{\omega}^k.$$

The curve $(x_k(t), t)$ in the plane (x,t) that satisfies $x_k'(t) = \lambda_k$ is the k-th characteristic curve and w_k is constant along it. A strictly hyperbolic system enjoys the property that p distinct characteristic curves pass through any point of the plane (x,t), for any fixed $\overline{x}$ and $\overline{t}$. Then $u(\overline{x}, \overline{t})$ depends only on the initial datum at the points $\overline{x} - \lambda_k \overline{t}$. For this reason, the set of p points that form the feet of the characteristics issuing from the point $(\overline{x}, \overline{t})$

$$D(\overline{t}, \overline{x}) = \left\{x \in \mathbb{R} \ : \ x = \overline{x} - \lambda_k \overline{t}\ ,\ k = 1, ..., p\right\}, \tag{13.31}$$

is called the *domain of dependence* of the solution $\mathbf{u}(\overline{x}, \overline{t})$.

If (13.30) is set on a bounded interval (α, β) instead of on the whole real line, the inflow point for each characteristic variable w_k is determined by the sign of λ_k. Correspondingly, the number of positive eigenvalues determines the number of boundary conditions that can be assigned at $x = \alpha$, whereas at $x = \beta$ it is admissible to assign a number of conditions which equals the number of negative eigenvalues. An example is discussed in Section 13.6.1.

Remark 13.2 (The nonlinear case) Let us consider the following nonlinear system of first-order equations

$$\frac{\partial \mathbf{u}}{\partial t} + \frac{\partial}{\partial x}\mathbf{g}(\mathbf{u}) = \mathbf{0}, \tag{13.32}$$

where $\mathbf{g} = [g_1, \ldots, g_p]^T$ is called the *flux function*. The system is hyperbolic if the jacobian matrix $\mathrm{A}(\mathbf{u})$ whose elements are $a_{ij} = \partial g_i(\mathbf{u})/\partial u_j$, $i, j = 1, \ldots, p$, is diagonalizable and has p real eigenvalues. ■

13.6.1 The Wave Equation

Consider the second-order hyperbolic equation

$$\frac{\partial^2 u}{\partial t^2} - \gamma^2 \frac{\partial^2 u}{\partial x^2} = f \quad x \in (\alpha, \beta), \quad t > 0, \tag{13.33}$$

with initial data

$$u(x, 0) = u_0(x) \quad \text{and} \quad \frac{\partial u}{\partial t}(x, 0) = v_0(x), \quad x \in (\alpha, \beta),$$

and boundary data

$$u(\alpha, t) = 0 \quad \text{and} \quad u(\beta, t) = 0, \quad t > 0. \tag{13.34}$$

In this case, u may represent the transverse displacement of an elastic vibrating string of length $\beta - \alpha$, fixed at the endpoints, and γ is a coefficient depending on the specific mass of the string and on its tension. The string is subject to a vertical force of density f.

The functions $u_0(x)$ and $v_0(x)$ denote respectively the initial displacement and the initial velocity of the string.
The change of variables

$$\omega_1 = \frac{\partial u}{\partial x}, \quad \omega_2 = \frac{\partial u}{\partial t},$$

transforms (13.33) into the following first-order system

$$\frac{\partial \boldsymbol{\omega}}{\partial t} + \mathrm{A}\frac{\partial \boldsymbol{\omega}}{\partial x} = \mathbf{f}, \quad x \in (\alpha, \beta), \quad t > 0, \tag{13.35}$$

where

$$\boldsymbol{\omega} = \begin{bmatrix} \omega_1 \\ \omega_2 \end{bmatrix}, \quad \mathrm{A} = \begin{bmatrix} 0 & -1 \\ -\gamma^2 & 0 \end{bmatrix}, \quad \mathbf{f} = \begin{bmatrix} 0 \\ f \end{bmatrix},$$

and the initial conditions are $\omega_1(x, 0) = u_0'(x)$ and $\omega_2(x, 0) = v_0(x)$.

Since the eigenvalues of A are the two distinct real numbers $\pm\gamma$ (representing the propagation velocities of the wave) we conclude that system (13.35) is hyperbolic. Moreover, one boundary condition needs to be prescribed at every end-point, as in (13.34). Notice that, also in this case, smooth solutions correspond to smooth initial data, while any discontinuity that is present in the initial data will propagate along the characteristics.

Remark 13.3 Notice that replacing $\frac{\partial^2 u}{\partial t^2}$ by t^2, $\frac{\partial^2 u}{\partial x^2}$ by x^2 and f by 1, the wave equation becomes

$$t^2 - \gamma^2 x^2 = 1,$$

which represents an hyperbola in the (x, t) plane. Proceeding analogously in the case of the heat equation (13.1), we end up with

$$t - \nu x^2 = 1,$$

which represents a parabola in the (x, t) plane. Finally, for the Poisson equation (12.90), replacing $\frac{\partial^2 u}{\partial x^2}$ by x^2, $\frac{\partial^2 u}{\partial y^2}$ by y^2 and f by 1, we get

$$x^2 + y^2 = 1,$$

which represents an ellipse in the (x, y) plane.

Due to the geometric interpretation above, the corresponding differential operators are classified as hyperbolic, parabolic and elliptic. ■

13.7 The Finite Difference Method for Hyperbolic Equations

Let us discretize the hyperbolic problem (13.26) by space-time finite differences. With this aim, the half-plane $\{(x, t) : -\infty < x < \infty,\ t > 0\}$ is discretized by choosing a spatial grid size Δx, a temporal step Δt and the grid points (x_j, t^n) as follows

$$x_j = j\Delta x, \quad j \in \mathbb{Z}, \quad t^n = n\Delta t, \quad n \in \mathbb{N}.$$

Let us set

$$\lambda = \Delta t/\Delta x,$$

and define $x_{j+1/2} = x_j + \Delta x/2$. We look for discrete solutions u_j^n which approximate the values $u(x_j, t^n)$ of the exact solution for any j, n.

Quite often, explicit methods are employed for advancing in time in hyperbolic initial-value problems, even though they require restrictions on the value of λ, unlike what typically happens with implicit methods.
Let us focus our attention on problem (13.26). Any explicit finite-difference method can be written in the form

$$u_j^{n+1} = u_j^n - \lambda(h_{j+1/2}^n - h_{j-1/2}^n), \tag{13.36}$$

where $h_{j+1/2}^n = h(u_j^n, u_{j+1}^n)$ for every j and $h(\cdot, \cdot)$ is a particular function that is called the *numerical flux*.

13.7.1 Discretization of the Scalar Equation

We illustrate several instances of explicit methods, and provide the corresponding numerical flux.

1. *Forward Euler/centered*

$$u_j^{n+1} = u_j^n - \frac{\lambda}{2} a (u_{j+1}^n - u_{j-1}^n) \tag{13.37}$$

which can be cast in the form (13.36) by setting

$$h_{j+1/2}^n = \frac{1}{2} a (u_{j+1}^n + u_j^n). \tag{13.38}$$

2. *Lax-Friedrichs*

$$u_j^{n+1} = \frac{1}{2}(u_{j+1}^n + u_{j-1}^n) - \frac{\lambda}{2} a (u_{j+1}^n - u_{j-1}^n) \tag{13.39}$$

which is of the form (13.36) with

$$h_{j+1/2}^n = \frac{1}{2}[a(u_{j+1}^n + u_j^n) - \lambda^{-1}(u_{j+1}^n - u_j^n)].$$

3. *Lax-Wendroff*

$$u_j^{n+1} = u_j^n - \frac{\lambda}{2} a (u_{j+1}^n - u_{j-1}^n) + \frac{\lambda^2}{2} a^2 (u_{j+1}^n - 2u_j^n + u_{j-1}^n) \tag{13.40}$$

which can be written in the form (13.36) provided that

$$h_{j+1/2}^n = \frac{1}{2}[a(u_{j+1}^n + u_j^n) - \lambda a^2 (u_{j+1}^n - u_j^n)].$$

4. *Upwind (or forward Euler/uncentered)*

$$u_j^{n+1} = u_j^n - \frac{\lambda}{2} a (u_{j+1}^n - u_{j-1}^n) + \frac{\lambda}{2} |a| (u_{j+1}^n - 2u_j^n + u_{j-1}^n) \tag{13.41}$$

which fits the form (13.36) when the numerical flux is defined to be

$$h_{j+1/2}^n = \frac{1}{2}[a(u_{j+1}^n + u_j^n) - |a|(u_{j+1}^n - u_j^n)].$$

The last three methods can be obtained from the forward Euler/centered method by adding a term proportional to a numerical diffusion, so that they can be written in the equivalent form

$$u_j^{n+1} = u_j^n - \frac{\lambda}{2} a (u_{j+1}^n - u_{j-1}^n) + \frac{1}{2} k \frac{u_{j+1}^n - 2u_j^n + u_{j-1}^n}{(\Delta x)^2}, \tag{13.42}$$

where the artificial viscosity k is given for the three cases in Table 13.1.

Table 13.1. Artificial viscosity, artificial flux and truncation error for Lax-Friedrichs, Lax-Wendroff and Upwind methods

methods	k	$h^{diff}_{j+1/2}$	$\tau(\Delta t, \Delta x)$
Lax-Friedrichs	Δx^2	$-\frac{1}{2\lambda}(u_{j+1} - u_j)$	$\mathcal{O}\left(\frac{\Delta x^2}{\Delta t} + \Delta t + \Delta x\right)$
Lax-Wendroff	$a^2\Delta t^2$	$-\frac{\lambda a^2}{2}(u_{j+1} - u_j)$	$\mathcal{O}\left(\Delta t^2 + \Delta x^2\right)$
Upwind	$\|a\|\Delta x\Delta t$	$-\frac{\|a\|}{2}(u_{j+1} - u_j)$	$\mathcal{O}(\Delta t + \Delta x)$

As a consequence, the numerical flux for each scheme can be written equivalently as

$$h_{j+1/2} = h^{FE}_{j+1/2} + h^{diff}_{j+1/2},$$

where $h^{FE}_{j+1/2}$ is the numerical flux of the forward Euler/centered scheme (which is given in (13.38)) and the *artificial diffusion flux* $h^{diff}_{j+1/2}$ is given for the three cases in Table 13.1.

An example of an implicit method is the *backward Euler/centered* scheme

$$u_j^{n+1} + \frac{\lambda}{2}a(u_{j+1}^{n+1} - u_{j-1}^{n+1}) = u_j^n. \tag{13.43}$$

It can still be written in the form (13.36) provided that h^n is replaced by h^{n+1}. In the example at hand, the numerical flux is the same as for the Forward Euler/centered method, and so is the artificial viscosity.

Finally, we report the following schemes for the approximation of the second-order wave equation (13.33):

1. *Leap-Frog*

$$u_j^{n+1} - 2u_j^n + u_j^{n-1} = (\gamma\lambda)^2(u_{j+1}^n - 2u_j^n + u_{j-1}^n); \tag{13.44}$$

2. *Newmark*

$$\begin{aligned} u_j^{n+1} - u_j^n &= \Delta t v_j^n + (\gamma\lambda)^2\left[\beta w_j^{n+1} + \left(\tfrac{1}{2} - \beta\right) w_j^n\right], \\ v_j^{n+1} - v_j^n &= \frac{(\gamma\lambda)^2}{\Delta t}\left[\theta w_j^{n+1} + (1-\theta) w_j^n\right], \end{aligned} \tag{13.45}$$

with $w_j = u_{j+1} - 2u_j + u_{j-1}$ and where the parameters β and θ satisfy $0 \leq \beta \leq \frac{1}{2}$, $0 \leq \theta \leq 1$.

13.8 Analysis of Finite Difference Methods

Let us analyze the properties of consistency, stability and convergence, as well as those of dissipation and dispersion, of the finite difference methods introduced above.

13.8.1 Consistency

As illustrated in Section 11.3, the local truncation error of a numerical scheme is the residual that is generated by pretending the exact solution to satisfy the numerical method itself.

Denoting by u the solution of the exact problem (13.26), in the case of method (13.37) the *local truncation error* at (x_j, t^n) is defined as follows

$$\tau_j^n = \frac{u(x_j, t^{n+1}) - u(x_j, t^n)}{\Delta t} - a\frac{u(x_{j+1}, t^n) - u(x_{j-1}, t^n)}{2\Delta x}.$$

The *truncation error* is

$$\tau(\Delta t, \Delta x) = \max_{j,n} |\tau_j^n|.$$

When $\tau(\Delta t, \Delta x)$ goes to zero as Δt and Δx tend to zero independently, the numerical scheme is said to be *consistent*.

Moreover, we say that it is of *order* p in time and of *order* q in space (for suitable integers p and q), if, for a sufficiently smooth solution of the exact problem, we have

$$\tau(\Delta t, \Delta x) = \mathcal{O}(\Delta t^p + \Delta x^q).$$

Using Taylor's expansion conveniently we can characterize the truncation error of the methods previously introduced as indicated in Table 13.1. The Leap-frog and Newmark methods are both second order accurate if $\Delta t = \Delta x$, while the forward (or backward) Euler centered method is $\mathcal{O}(\Delta t + \Delta x^2)$.

Finally, we say that a numerical scheme is *convergent* if

$$\lim_{\Delta t, \Delta x \to 0} \max_{j,n} |u(x_j, t^n) - u_j^n| = 0.$$

13.8.2 Stability

A numerical method for a hyperbolic problem (linear or nonlinear) is said to be *stable* if, for any time T, there exist two constants $C_T > 0$ (possibly depending on T) and $\delta_0 > 0$, such that

$$\|\mathbf{u}^n\|_\Delta \le C_T \|\mathbf{u}^0\|_\Delta, \tag{13.46}$$

for any n such that $n\Delta t \le T$ and for any Δt, Δx such that $0 < \Delta t \le \delta_0$, $0 < \Delta x \le \delta_0$. We have denoted by $\|\cdot\|_\Delta$ a suitable discrete norm, for instance one of those indicated below

$$\|\mathbf{v}\|_{\Delta,p} = \left(\Delta x \sum_{j=-\infty}^{\infty} |v_j|^p\right)^{\frac{1}{p}} \quad \text{for } p = 1, 2, \quad \|\mathbf{v}\|_{\Delta,\infty} = \sup_j |v_j|. \tag{13.47}$$

Note that $\|\cdot\|_{\Delta,p}$ is an approximation of the norm of $\mathrm{L}^p(\mathbb{R})$. For instance, the implicit Backward Euler/centered scheme (13.43) is unconditionally stable with respect to the norm $\|\cdot\|_{\Delta,2}$ (see Exercise 7).

13.8.3 The CFL Condition

Courant, Friedrichs and Lewy [CFL28] have shown that a necessary and sufficient condition for any explicit scheme of the form (13.36) to be stable is that the time and space discretization steps must obey the following condition

$$|a\lambda| = \left| a\frac{\Delta t}{\Delta x} \right| \leq 1, \tag{13.48}$$

which is known as the *CFL condition*. The number $a\lambda$, which is an adimensional number since a is a velocity, is commonly referred to as the *CFL number*. If a is not constant the CFL condition becomes

$$\Delta t \leq \frac{\Delta x}{\sup\limits_{x\in\mathbb{R},\ t>0} |a(x,t)|},$$

while, in the case of the hyperbolic system (13.30), the stability condition becomes

$$\left| \lambda_k \frac{\Delta t}{\Delta x} \right| \leq 1 \quad k = 1, \ldots, p,$$

where $\{\lambda_k,\ k = 1\ldots, p\}$ are the eigenvalues of A.

The CFL stability condition has the following geometric interpretation. In a finite difference scheme the value u_j^{n+1} depends, in general, on the values of u^n at the three points x_{j+i}, $i = -1, 0, 1$. Thus, at the time $t = 0$ the solution u_j^{n+1} will depend only on the initial data at the points x_{j+i}, for $i = -(n+1), \ldots, (n+1)$ (see Figure 13.5).

Let us define *numerical domain of dependence* $D_{\Delta t}(x_j, t^n)$ to be the set of values at time $t = 0$ the numerical solution u_j^n depends on, that is

$$D_{\Delta t}(x_j, t^n) \subset \left\{ x \in \mathbb{R} : \ |x - x_j| \leq n\Delta x = \frac{t^n}{\lambda} \right\}.$$

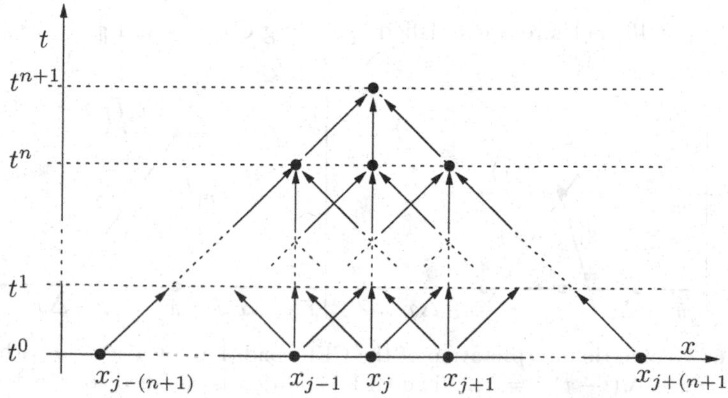

Fig. 13.5. The numerical domain of dependence $D_{\Delta t}(x_j, t^{n+1})$

Consequently, for any fixed point $(\overline{x},\overline{t})$ we have

$$D_{\Delta t}(\overline{x},\overline{t}) \subset \left\{ x \in \mathbb{R}: \; |x-\overline{x}| \leq \frac{\overline{t}}{\lambda} \right\}.$$

In particular, taking the limit as $\Delta t \to 0$ for a fixed λ, the numerical domain of dependence becomes

$$D_0(\overline{x},\overline{t}) = \left\{ x \in \mathbb{R}: \; |x-\overline{x}| \leq \frac{\overline{t}}{\lambda} \right\}.$$

The condition (13.48) is thus equivalent to the inclusion

$$D(\overline{x},\overline{t}) \subset D_0(\overline{x},\overline{t}), \tag{13.49}$$

where $D(\overline{x},\overline{t})$ is the domain of dependence defined in (13.31).

In the case of an hyperbolic system, thanks to (13.49), we can conclude that the CFL condition requires that any straight line $x = \overline{x} - \lambda_k(\overline{t}-t)$, $k = 1,\ldots,p$, must intersect the temporal straight line $t = \overline{t} - \Delta t$ at some point x belonging to the domain of dependence (see Figure 13.6).

Let us analyze the stability properties of some of the methods introduced in the previous section.

Assuming that $a > 0$, the upwind scheme (13.41) can be reformulated as

$$u_j^{n+1} = u_j^n - \lambda a(u_j^n - u_{j-1}^n). \tag{13.50}$$

Therefore

$$\|\mathbf{u}^{n+1}\|_{\Delta,1} \leq \Delta x \sum_j |(1-\lambda a)u_j^n| + \Delta x \sum_j |\lambda a u_{j-1}^n|.$$

Both λa and $1-\lambda a$ are nonnegative if (13.48) holds. Thus

$$\|\mathbf{u}^{n+1}\|_{\Delta,1} \leq \Delta x(1-\lambda a)\sum_j |u_j^n| + \Delta x \lambda a \sum_j |u_{j-1}^n| = \|\mathbf{u}^n\|_{\Delta,1}.$$

Inequality (13.46) is therefore satisfied by taking $C_T = 1$ and $\|\cdot\|_\Delta = \|\cdot\|_{\Delta,1}$.

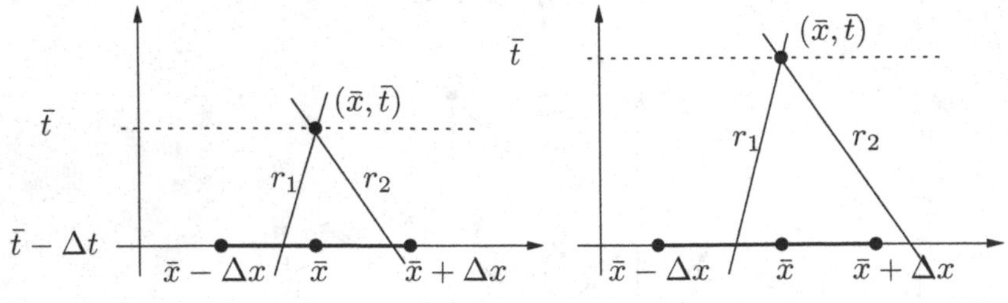

Fig. 13.6. Geometric interpretation of the CFL condition for a system with $p = 2$, where $r_i = \bar{x} - \lambda_i(t-\bar{t})$ $i = 1,2$. The CFL condition is satisfied for the left-hand case, while it is violated for the right-hand case

The Lax-Friedrichs scheme is also stable, upon assuming (13.48). Indeed, from (13.39) we get

$$u_j^{n+1} = \frac{1}{2}(1-\lambda a)u_{j+1}^n + \frac{1}{2}(1+\lambda a)u_{j-1}^n.$$

Therefore,

$$\begin{aligned}\|\mathbf{u}^{n+1}\|_{\Delta,1} &\leq \frac{1}{2}\Delta x \left[\sum_j |(1-\lambda a)u_{j+1}^n| + \sum_j |(1+\lambda a)u_{j-1}^n|\right] \\ &\leq \frac{1}{2}(1-\lambda a)\|\mathbf{u}^n\|_{\Delta,1} + \frac{1}{2}(1+\lambda a)\|\mathbf{u}^n\|_{\Delta,1} = \|\mathbf{u}^n\|_{\Delta,1}.\end{aligned}$$

Also the Lax-Wendroff scheme is stable under the usual assumption (13.48) on Δt (for the proof see, e.g., [QV94] Chapter 14).

13.8.4 Von Neumann Stability Analysis

Let us now show that the condition (13.48) is not sufficient to ensure that the forward Euler/centered scheme (13.37) is stable. For this purpose, we make the assumption that the function $u_0(x)$ is 2π-periodic so that it can be expanded in a Fourier series as

$$u_0(x) = \sum_{k=-\infty}^{\infty} \alpha_k e^{ikx}, \tag{13.51}$$

where

$$\alpha_k = \frac{1}{2\pi}\int_0^{2\pi} u_0(x)\, e^{-ikx}\, dx$$

is the k-th Fourier coefficient of u_0 (see Section 10.9). Therefore,

$$u_j^0 = u_0(x_j) = \sum_{k=-\infty}^{\infty} \alpha_k e^{ikjh} \quad j = 0, \pm 1, \pm 2, \dots,$$

where we have set $h = \Delta x$ for ease of notation. Applying (13.37) with $n = 0$ we get

$$\begin{aligned}u_j^1 &= \sum_{k=-\infty}^{\infty} \alpha_k e^{ikjh}\left(1 - \frac{a\Delta t}{2h}(e^{ikh} - e^{-ikh})\right) \\ &= \sum_{k=-\infty}^{\infty} \alpha_k e^{ikjh}\left(1 - \frac{a\Delta t}{h} i \sin(kh)\right).\end{aligned}$$

Setting

$$\gamma_k = 1 - \frac{a\Delta t}{h} i \sin(kh),$$

and proceeding recursively on n yields

$$u_j^n = \sum_{k=-\infty}^{\infty} \alpha_k e^{ikjh} \gamma_k^n \; j = 0, \pm 1, \pm 2, \ldots, \qquad n \geq 1. \tag{13.52}$$

The number $\gamma_k \in \mathbb{C}$ is said to be the *amplification coefficient* of the k-th frequency (or harmonic) at each time step. Since

$$|\gamma_k| = \left\{ 1 + \left(\frac{a\Delta t}{h} \sin(kh) \right)^2 \right\}^{\frac{1}{2}},$$

we deduce that

$$|\gamma_k| > 1 \text{ if } \; a \neq 0 \; \text{ and } \; k \neq \frac{m\pi}{h}, \qquad m = 0, \pm 1, \pm 2, \ldots .$$

Correspondingly, the nodal values $|u_j^n|$ continue to grow as $n \to \infty$ and the numerical solution "blows-up" whereas the exact solution satisfies

$$|u(x,t)| = |u_0(x - at)| \leq \max_{s \in \mathbb{R}} |u_0(s)| \quad \forall x \in \mathbb{R}, \;\; \forall t > 0.$$

The centered discretization scheme (13.37) is thus *unconditionally unstable*, i.e., it is unstable for any choice of the parameters Δt and Δx.

The previous analysis is based on the Fourier series expansion and is called *von Neumann analysis.* It can be applied to studying the stability of any numerical scheme with respect to the norm $\| \cdot \|_{\Delta,2}$ and for establishing the dissipation and dispersion of the method.

Any explicit finite difference numerical scheme for problem (13.26) satisfies a recursive relation analogous to (13.52), where γ_k depends a priori on Δt and h and is called the *k-th amplification coefficient* of the numerical scheme at hand.

Theorem 13.1 *Assume that for a suitable choice of Δt and h, $|\gamma_k| \leq 1 \; \forall k$; then, the numerical scheme is stable with respect to the $\| \cdot \|_{\Delta,2}$ norm.*

Proof. Take an initial datum with a finite Fourier expansion

$$u_0(x) = \sum_{k=-\frac{N}{2}}^{\frac{N}{2}-1} \alpha_k e^{ikx}$$

where N is a positive integer. Without loss of generality we can assume that problem (13.26) is well-posed on $[0, 2\pi]$ since u_0 is a 2π-periodic function. Take in this interval N equally spaced nodes

$$x_j = jh, \; j = 0, \ldots, N-1, \text{ with } h = \frac{2\pi}{N},$$

at which the numerical scheme (13.36) is applied. We get

$$u_j^0 = u_0(x_j) = \sum_{k=-\frac{N}{2}}^{\frac{N}{2}-1} \alpha_k e^{ikjh}, \quad u_j^n = \sum_{k=-\frac{N}{2}}^{\frac{N}{2}-1} \alpha_k \gamma_k^n e^{ikjh}.$$

Notice that

$$\|\mathbf{u}^n\|^2_{\Delta,2} = h\sum_{j=0}^{N-1}\sum_{k,m=-\frac{N}{2}}^{\frac{N}{2}-1} \alpha_k\overline{\alpha}_m(\gamma_k\overline{\gamma}_m)^n e^{i(k-m)jh}.$$

Recalling Lemma 10.2 we have

$$h\sum_{j=0}^{N-1} e^{i(k-m)jh} = 2\pi\delta_{km}, \quad -\frac{N}{2} \leq k, m \leq \frac{N}{2} - 1,$$

which yields

$$\|\mathbf{u}^n\|^2_{\Delta,2} = 2\pi \sum_{k=-\frac{N}{2}}^{\frac{N}{2}-1} |\alpha_k|^2|\gamma_k|^{2n}.$$

As a consequence, since $|\gamma_k| \leq 1 \; \forall k$, it turns out that

$$\|\mathbf{u}^n\|^2_{\Delta,2} \leq 2\pi \sum_{k=-\frac{N}{2}}^{\frac{N}{2}-1} |\alpha_k|^2 = \|\mathbf{u}^0\|^2_{\Delta,2}, \quad \forall n \geq 0,$$

which proves that the scheme is stable with respect to the $\|\cdot\|_{\Delta,2}$ norm. ◇

In the case of the upwind scheme (13.41), proceeding as was done for the centered scheme, we find the following amplification coefficients (see Exercise 6)

$$\gamma_k = \begin{cases} 1 - a\dfrac{\Delta t}{h}(1 - e^{-ikh}) & \text{if } a > 0, \\ 1 - a\dfrac{\Delta t}{h}(e^{-ikh} - 1) & \text{if } a < 0. \end{cases}$$

Therefore

$$\forall k, \quad |\gamma_k| \leq 1 \text{ if } \quad \Delta t \leq \frac{h}{|a|},$$

which is nothing but the CFL condition.

Thanks to Theorem 13.1, if the CFL condition is satisfied, the upwind scheme is stable with respect to the $\|\cdot\|_{\Delta,2}$ norm.

We conclude by noting that the upwind scheme (13.50) satisfies

$$u_j^{n+1} = (1 - \lambda a)u_j^n + \lambda a u_{j-1}^n.$$

Owing to (13.48), either λa or $1 - \lambda a$ are nonnegative, thus

$$\min(u_j^n, u_{j-1}^n) \leq u_j^{n+1} \leq \max(u_j^n, u_{j-1}^n).$$

It follows that

$$\inf_{l\in\mathbb{Z}} \{u_l^0\} \leq u_j^n \leq \sup_{l\in\mathbb{Z}} \{u_l^0\} \quad \forall j \in \mathbb{Z}, \; \forall n \geq 0,$$

that is,

$$\|\mathbf{u}^n\|_{\Delta,\infty} \leq \|\mathbf{u}^0\|_{\Delta,\infty} \quad \forall n \geq 0, \tag{13.53}$$

which proves that if (13.48) is satisfied, the upwind scheme is stable in the norm $\|\cdot\|_{\Delta,\infty}$. The relation (13.53) is called the *discrete maximum principle* (see also Section 12.2.2).

Remark 13.4 For the approximation of the wave equation (13.33) the Leap-Frog method (13.44) is stable under the CFL restriction $\Delta t \leq \Delta x/|\gamma|$, while the Newmark method (13.45) is unconditionally stable if $2\beta \geq \theta \geq \frac{1}{2}$ (see [Joh90]). ■

13.9 Dissipation and Dispersion

The von Neumann analysis on the amplification coefficients enlightens the study of the stability and *dissipation* of a numerical scheme.

Consider the exact solution to problem (13.26); the following relation holds

$$u(x, t^n) = u_0(x - an\Delta t), \quad \forall n \geq 0, \quad \forall x \in \mathbb{R}.$$

In particular, from applying (13.51) it follows that

$$u(x_j, t^n) = \sum_{k=-\infty}^{\infty} \alpha_k e^{ikjh} g_k^n, \quad \text{where} \quad g_k = e^{-iak\Delta t}. \tag{13.54}$$

Letting

$$\varphi_k = k\Delta x,$$

we have $k\Delta t = \lambda\varphi_k$ and thus

$$g_k = e^{-ia\lambda\varphi_k}. \tag{13.55}$$

The real number φ_k, here expressed in radians, is called the *phase angle* of the k-th harmonic. Comparing (13.54) with (13.52) we can see that γ_k is the counterpart of g_k which is generated by the specific numerical method at hand. Moreover, $|g_k| = 1$, whereas $|\gamma_k| \leq 1$, in order to ensure stability.

Thus, γ_k is a dissipation coefficient; the smaller $|\gamma_k|$, the higher the reduction of the amplitude α_k, and, as a consequence, the higher the numerical dissipation.

The ratio $\epsilon_a(k) = \frac{|\gamma_k|}{|g_k|}$ is called the *amplification error* of the k-th harmonic associated with the numerical scheme (in our case it coincides with the amplification coefficient). On the other hand, writing

$$\gamma_k = |\gamma_k| e^{-i\omega\Delta t} = |\gamma_k| e^{-i\frac{\omega}{k}\lambda\varphi_k},$$

and comparing this relation with (13.55), we can identify the *velocity of propagation* of the numerical solution, relative to its k-th harmonic, as being $\frac{\omega}{k}$. The ratio between this velocity and the velocity a of the exact solution is called the *dispersion error* ϵ_d relative to the k-th harmonic

$$\epsilon_d(k) = \frac{\omega}{ka} = \frac{\omega \Delta x}{\varphi_k a}.$$

The amplification and dispersion errors for the numerical schemes examined so far are functions of the phase angle φ_k and the CFL number $a\lambda$. This is shown in Figure 13.7 where we have only considered the interval $0 \leq \varphi_k \leq \pi$ and we have used degrees instead of radians to denote the values of φ_k.

In Figure 13.8 the numerical solutions of equation (13.26) with $a = 1$ and the initial datum u_0 given by a packet of two sinusoidal waves of equal wavelength l and centered at the origin $x = 0$ are shown. The three plots from the top of the figure refer to the case $l = 10\Delta x$ while from the bottom we have $l = 4\Delta x$. Since $k = (2\pi)/l$ we get $\varphi_k = ((2\pi)/l)\Delta x$, so that $\varphi_k = \pi/10$ in the left-side pictures and $\varphi_k = \pi/4$ in the right-side ones. All numerical solutions have been computed for a CFL number equal to 0.75, using the schemes introduced above. Notice that the dissipation effect is quite relevant at high frequencies ($\varphi_k = \pi/4$), especially for first-order methods (such as the upwind and the Lax-Friedrichs methods).

In order to highlight the effects of the dispersion, the same computations have been repeated for $\varphi_k = \pi/3$ and different values of the CFL number. The numerical solutions after 5 time steps are shown in Figure 13.9. The Lax-Wendroff method is the least dissipative for all the considered CFL numbers. Moreover, a comparison of the positions of the peaks of the numerical solutions with respect to the corresponding ones in the exact solution shows that the Lax-Friedrichs scheme is affected by a positive dispersion error, since the "numerical" wave advances faster than the exact one. Also, the upwind scheme exhibits a slight dispersion error for a CFL number of 0.75 which is absent for a CFL number of 0.5. The peaks are well aligned with those of the numerical solution, although they have been reduced in amplitude due to numerical dissipation. Finally, the Lax-Wendroff method exhibits a small negative dispersion error; the numerical solution is indeed slightly late with respect to the exact one.

13.9.1 Equivalent Equations

Using Taylor's expansion to the third order to represent the truncation error, it is possible to associate with any of the numerical schemes introduced so far an equivalent differential equation of the form

$$v_t + av_x = \mu v_{xx} + \nu v_{xxx}, \tag{13.56}$$

where the terms μv_{xx} and νv_{xxx} represent dissipation and dispersion, respectively. Table 13.2 shows the values of μ and ν for the various methods.

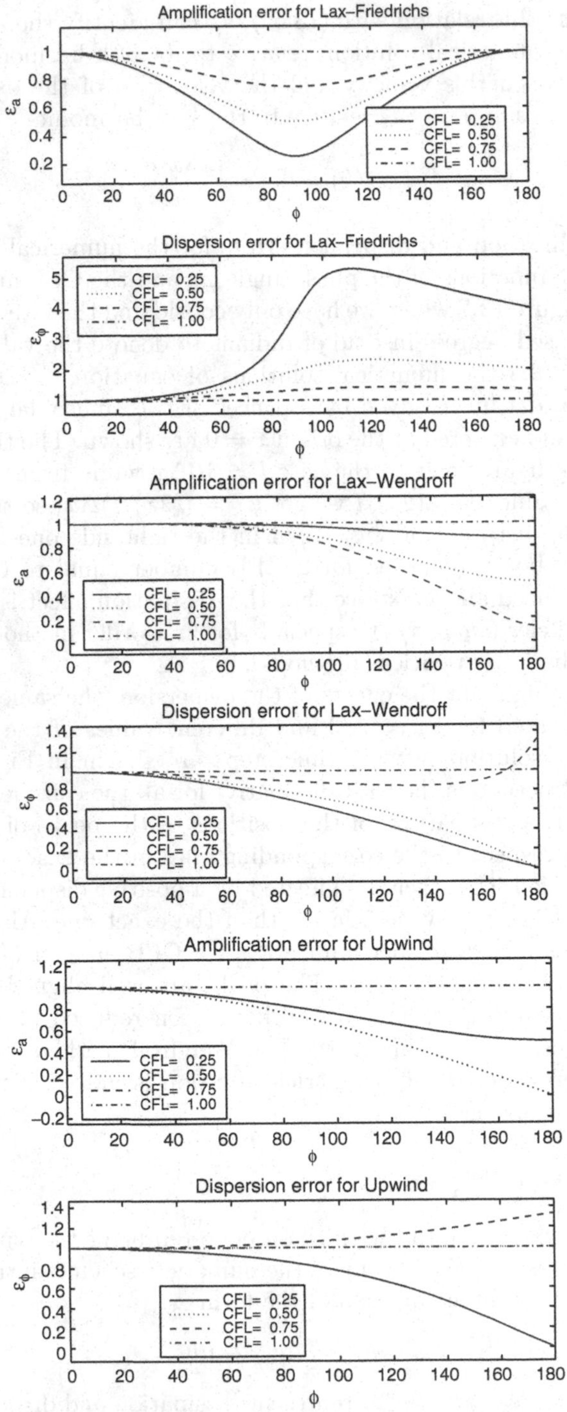

Fig. 13.7. Amplification and dispersion errors for several numerical schemes

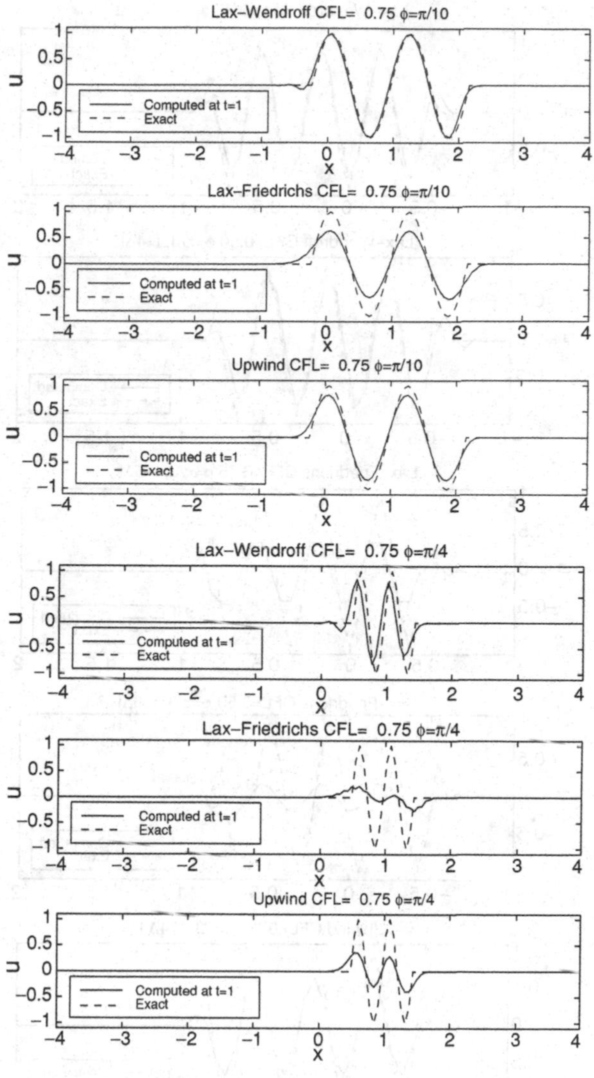

Fig. 13.8. Numerical solutions corresponding to the transport of a sinusoidal wave packet with different wavelengths

Let us give a proof of this procedure in the case of the upwind scheme. Let $v(x,t)$ be a smooth function which satisfies the difference equation (13.41); then, assuming that $a > 0$, we have

$$\frac{v(x,t+\Delta t)-v(x,t)}{\Delta t}+a\frac{v(x,t)-v(x-\Delta x,t)}{\Delta x}=0.$$

Truncating the Taylor expansions of v around (x,t) at the first and second order, respectively, we obtain

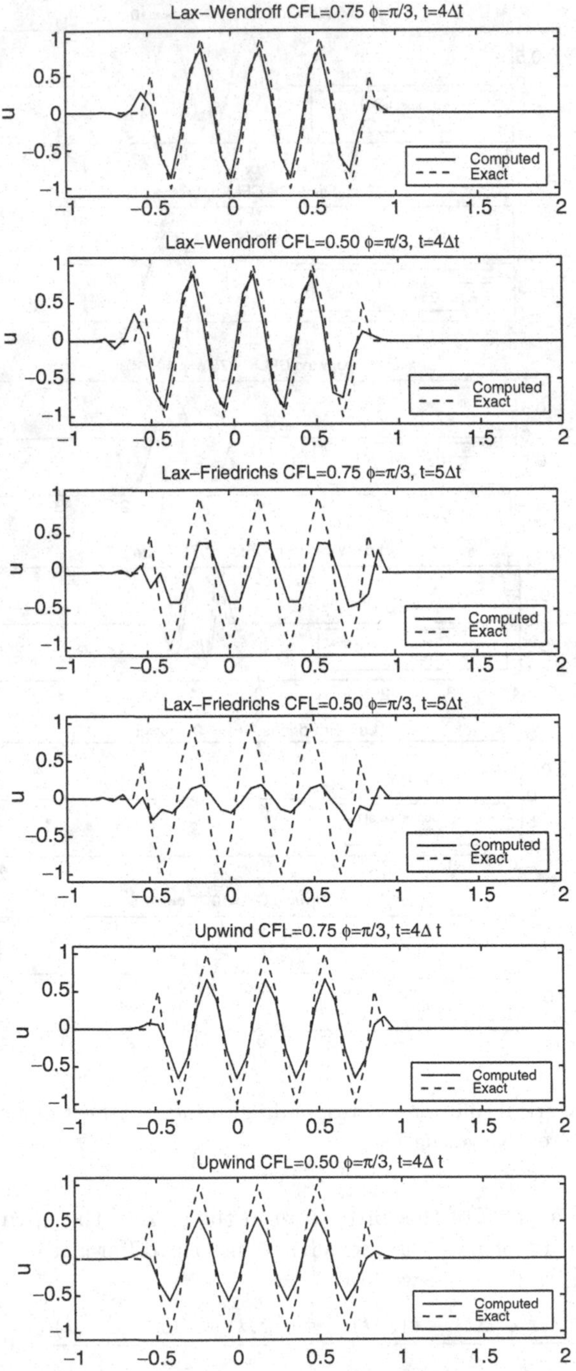

Fig. 13.9. Numerical solutions corresponding to the transport of a sinusoidal wave packet, for different CFL numbers

Table 13.2. Values of dissipation and dispersion coefficients for several numerical methods

Method	μ	ν
Upwind	$\frac{a\Delta x}{2} - \frac{a^2\Delta t}{2}$	$-\frac{a}{6}\left(\Delta x^2 - 3a\Delta x\Delta t + 2a^2\Delta t^2\right)$
Lax-Friedrichs	$\frac{\Delta x^2}{2\Delta t}\left(1-(a\lambda)^2\right)$	$\frac{a\Delta x^2}{3}\left(1-(a\lambda)^2\right)$
Lax-Wendroff	0	$\frac{a\Delta x^2}{6}\left((a\lambda)^2-1\right)$

$$v_t + O(\Delta t) + av_x + \mathcal{O}(\Delta x) = 0 \tag{13.57}$$

and

$$v_t + \frac{\Delta t}{2}v_{tt} + \mathcal{O}(\Delta t^2) + a\left(v_x + \frac{\Delta x}{2}v_{xx} + \mathcal{O}(\Delta x^2)\right) = 0, \tag{13.58}$$

where $v_t = \frac{\partial v}{\partial t}$ and $v_x = \frac{\partial v}{\partial x}$.

Differentiating (13.57) with respect to t and then with respect to x, we get

$$v_{tt} + av_{xt} = \mathcal{O}(\Delta x + \Delta t),$$

and

$$v_{tx} + av_{xx} = \mathcal{O}(\Delta x + \Delta t).$$

Thus, it follows that

$$v_{tt} = a^2 v_{xx} + \mathcal{O}(\Delta x + \Delta t),$$

which, after substituting into (13.58), yields the following equation

$$v_t + av_x = \mu v_{xx} \tag{13.59}$$

where

$$\mu = \frac{a\Delta x}{2} - \frac{a^2\Delta t}{2},$$

and having neglected the term $\mathcal{O}(\Delta x^2 + \Delta t^2)$. Relation (13.59) is the *equivalent differential equation* up to second order of the upwind scheme.

Following the same procedure and truncating the Taylor expansion at third order, yields

$$v_t + av_x = \mu v_{xx} + \nu v_{xxx} \tag{13.60}$$

where

$$\nu = -\frac{a}{6}\left(\Delta x^2 - 3a\Delta x\Delta t + 2a^2\Delta t^2\right).$$

We can give a heuristic explanation of the meaning of the dissipative and dispersive terms in the equivalent equation (13.56) by studying the following problem

$$\begin{cases} v_t + av_x = \mu v_{xx} + \nu v_{xxx} & x \in \mathbb{R},\ t > 0 \\ v(x,0) = e^{ikx}, & (k \in \mathbb{Z}) \end{cases} \tag{13.61}$$

Applying the Fourier transform yields, if $\mu = \nu = 0$,

$$v(x,t) = e^{ik(x-at)}, \tag{13.62}$$

while if μ and ν are arbitrary real numbers (with $\mu > 0$) we get

$$v(x,t) = e^{-\mu k^2 t} e^{ik[x-(a+\nu k^2)t]}. \tag{13.63}$$

Comparing (13.62) with (13.63) we can see that the module of the solution diminishes as μ grows and this becomes more relevant if the frequency k gets larger. Therefore, the term μv_{xx} in (13.61) has a dissipative effect on the solution. A further comparison between (13.62) and (13.63) shows that the presence of the term ν modifies the velocity of the propagation of the solution; the velocity is increased if $\nu > 0$ whereas it is diminuished if $\nu < 0$. Even in this case the effect is amplified at high frequencies. Therefore, the third-order differential term νv_{xxx} introduces a *dispersive* effect.

Generally speaking, even-order derivatives in the equivalent equation represent diffusive terms, while odd-order derivatives mean dispersive effects. In the case of first-order schemes (like the upwind method) the dispersive effect is often only slightly visible since it is hidden by the dissipative one. Actually, taking Δt and Δx of the same order, we have that $\nu \ll \mu$ as $\Delta x \to 0$, since $\nu = O(\Delta x^2)$ and $\mu = O(\Delta x)$. In particular, for a CFL number of $\frac{1}{2}$, the equivalent equation of the upwind method exhibits null dispersion, truncated at second order, according to the results of the previous section.

On the other hand, the dispersive effect is strikingly visible in the Lax-Friedrichs and in the Lax-Wendroff schemes; the latter, being second-order accurate, does not exhibit a dissipative term of the form μv_{xx}. However, it ought to be dissipative in order to be stable; actually, the equivalent equation (truncated at fourth order) for the Lax-Wendroff scheme reads

$$v_t + av_x = \frac{a\Delta x^2}{6}[(a\lambda)^2 - 1]v_{xxx} - \frac{a\Delta x^3}{6} a\lambda[1 - (a\lambda)^2]v_{xxxx},$$

where the last term is dissipative if $|a\lambda| < 1$. We thus recover the CFL condition for the Lax-Wendroff method.

13.10 Finite Element Approximation of Hyperbolic Equations

Let us consider the following first-order linear, scalar hyperbolic problem in the interval $(\alpha, \beta) \subset \mathbb{R}$

$$\begin{cases} \dfrac{\partial u}{\partial t} + a\dfrac{\partial u}{\partial x} + a_0 u = f & \text{in } Q_T = (\alpha, \beta) \times (0, T), \\ u(\alpha, t) = \varphi(t), & t \in (0, T), \\ u(x, 0) = u_0(x), & x \in \Omega, \end{cases} \tag{13.64}$$

where $a = a(x)$, $a_0 = a_0(x,t)$, $f = f(x,t)$, $\varphi = \varphi(t)$ and $u_0 = u_0(x)$ are given functions.

We assume that $a(x) > 0 \ \forall x \in [\alpha, \beta]$. In particular, this implies that the point $x = \alpha$ is the *inflow boundary*, and the boundary value has to be specified there.

13.10.1 Space Discretization with Continuous and Discontinuous Finite Elements

A semi-discrete approximation of problem (13.64) can be carried out by means of the Galerkin method (see Section 12.4). Define the spaces

$$V_h = X_h^r = \{v_h \in C^0([\alpha, \beta]) : \ v_{h|I_j} \in \mathbb{P}_r(I_j), \ \forall \, I_j \in \mathcal{T}_h\}$$

and

$$V_h^{in} = \{v_h \in V_h : \ v_h(\alpha) = 0\},$$

where $\mathcal{T}_h$ is a partition of Ω (see Section 12.4.5) into $n \geq 2$ subintervals $I_j = [x_j, x_{j+1}]$, for $j = 0, \ldots, n-1$.

Let $u_{0,h}$ be a suitable finite element approximation of u_0 and consider the problem: for any $t \in (0,T)$ find $u_h(t) \in V_h$ such that

$$\begin{cases} \displaystyle\int_\alpha^\beta \frac{\partial u_h(t)}{\partial t} v_h \, dx + \int_\alpha^\beta \left(a\frac{\partial u_h(t)}{\partial x} + a_0(t) u_h(t) \right) v_h \, dx \\ \displaystyle\qquad = \int_\alpha^\beta f(t) v_h \, dx \qquad \forall \, v_h \in V_h^{in}, \\ u_h(t) = \varphi_h(t) \quad \text{at } x = \alpha, \end{cases} \tag{13.65}$$

with $u_h(0) = u_{0,h} \in V_h$.

If φ is equal to zero, $u_h(t) \in V_h^{in}$, and we are allowed to taking $v_h = u_h(t)$ and get the following inequality

$$\|u_h(t)\|^2_{\mathrm{L}^2(\alpha,\beta)} + \int_0^t \mu_0 \|u_h(\tau)\|^2_{\mathrm{L}^2(\alpha,\beta)} \, d\tau + a(\beta) \int_0^t u_h^2(\tau, \beta) \, d\tau$$

$$\leq \|u_{0,h}\|^2_{\mathrm{L}^2(\alpha,\beta)} + \int_0^t \frac{1}{\mu_0} \|f(\tau)\|^2_{\mathrm{L}^2(\alpha,\beta)} d\tau \, ,$$

for any $t \in [0,T]$, where we have assumed that

$$0 < \mu_0 \leq a_0(x,t) - \frac{1}{2}a'(x). \tag{13.66}$$

Notice that in the special case in which both f and a_0 are identically zero, we obtain

$$\|u_h(t)\|_{\mathrm{L}^2(\alpha,\beta)} \leq \|u_{0,h}\|_{\mathrm{L}^2(\alpha,\beta)},$$

which expresses the conservation of the energy of the system. When (13.66) does not hold (for example, if a is a constant convective term and $a_0 = 0$), then an application of Gronwall's lemma 11.1 yields

$$\begin{aligned}
&\|u_h(t)\|^2_{\mathrm{L}^2(\alpha,\beta)} + a(\beta)\int_0^t u_h^2(\tau,\beta)\,d\tau \\
&\leq \left(\|u_{0,h}\|^2_{\mathrm{L}^2(\alpha,\beta)} + \int_0^t \|f(\tau)\|^2_{\mathrm{L}^2(\alpha,\beta)}\,d\tau\right)\exp\int_0^t [1+2\mu^*(\tau)]\,d\tau,
\end{aligned} \tag{13.67}$$

where $\mu^*(t) = \max_{x\in[\alpha,\beta]} |\mu(x,t)|$.

An alternative approach to the semi-discrete approximation of problem (13.64) is based on the use of *discontinuous* finite elements. This choice is motivated by the fact that, as previously pointed out, the solutions of hyperbolic problems (even in the linear case) may exhibit discontinuities.

The finite element space can be defined as follows

$$W_h = Y_h^r = \left\{v_h \in \mathrm{L}^2(\alpha,\beta) : v_{h|I_j} \in \mathbb{P}_r(I_j),\ \forall\, I_j \in \mathcal{T}_h\right\},$$

i.e., the space of piecewise polynomials of degree less than or equal to r, which are not necessarily continuous at the finite element nodes.

Then, the Galerkin discontinuous finite element space discretization reads: for any $t \in (0,T)$ find $u_h(t) \in W_h$ such that

$$\begin{cases}
\displaystyle\int_\alpha^\beta \frac{\partial u_h(t)}{\partial t} v_h\,dx \\
\displaystyle+\sum_{i=0}^{n-1}\left\{\int_{x_i}^{x_{i+1}}\left(a\frac{\partial u_h(t)}{\partial x} + a_0(x)u_h(t)\right)v_h\,dx + a(u_h^+ - U_h^-)(x_i,t)v_h^+(x_i)\right\} \\
\displaystyle= \int_\alpha^\beta f(t)v_h\,dx \quad \forall v_h \in W_h,
\end{cases} \tag{13.68}$$

where $\{x_i\}$ are the nodes of $\mathcal{T}_h$ with $x_0 = \alpha$ and $x_n = \beta$, and for each node x_i, $v_h^+(x_i)$ denotes the right-value of v_h at x_i while $v_h^-(x_i)$ is its left-value. Finally, $U_h^-(x_i,t) = u_h^-(x_i,t)$ if $i = 1,\ldots,n-1$, while $U_h^-(x_0,t) = \varphi(t)$ $\forall t > 0$.

If a is positive, x_j is the *inflow boundary* of I_j for every j and we set

$$[u]_j = u^+(x_j) - u^-(x_j),\ u^\pm(x_j) = \lim_{s\to 0^\pm} u(x_j + sa),\ j = 1,\dots,n-1.$$

Then, for any $t \in [0,T]$ the stability estimate for problem (13.68) reads

$$\begin{aligned}
&\|u_h(t)\|^2_{L^2(\alpha,\beta)} + \int_0^t \left(\|u_h(\tau)\|^2_{L^2(\alpha,\beta)} + \sum_{j=0}^{n-1} a(x_j)[u_h(\tau)]_j^2 \right) d\tau \\
&\quad \le C\left[\|u_{0,h}\|^2_{L^2(\alpha,\beta)} + \int_0^t \left(\|f(\tau)\|^2_{L^2(\alpha,\beta)} + a\varphi^2(\tau) \right) d\tau \right].
\end{aligned} \tag{13.69}$$

As for the convergence analysis, the following error estimate can be proved for continuous finite elements of degree r, $r \ge 1$ (see [QV94], Section 14.3.1)

$$\begin{aligned}
&\max_{t\in[0,T]} \|u(t) - u_h(t)\|_{L^2(\alpha,\beta)} + \left(\int_0^T a|u(\alpha,\tau) - u_h(\alpha,\tau)|^2\, d\tau \right)^{1/2} \\
&\quad = \mathcal{O}(\|u_0 - u_{0,h}\|_{L^2(\alpha,\beta)} + h^r),
\end{aligned}$$

If, instead, discontinuous finite elements of degree r are used, $r \ge 0$, the convergence estimate becomes (see [QV94], Section 14.3.3 and the references therein)

$$\begin{aligned}
&\max_{t\in[0,T]} \|u(t) - u_h(t)\|_{L^2(\alpha,\beta)} \\
&\quad + \left(\int_0^T \|u(t) - u_h(t)\|^2_{L^2(\alpha,\beta)}\, dt + \int_0^T \sum_{j=0}^{n-1} a(x_j)\,[u(t) - u_h(t)]_j^2\, dt \right)^{1/2} \\
&\quad = \mathcal{O}(\|u_0 - u_{0,h}\|_{L^2(\alpha,\beta)} + h^{r+1/2}).
\end{aligned}$$

13.10.2 Time Discretization

The time discretization of the finite element schemes introduced in the previous section can be carried out by resorting either to finite differences or finite elements. If an implicit finite difference scheme is adopted, both method (13.65) and (13.68) are unconditionally stable.

As an example, let us use the backward Euler method for the time discretization of problem (13.65). We obtain for each $n \ge 0$: find $u_h^{n+1} \in V_h$ such that

$$\begin{aligned}
&\frac{1}{\Delta t}\int_\alpha^\beta (u_h^{n+1} - u_h^n)v_h\, dx + \int_\alpha^\beta a\frac{\partial u_h^{n+1}}{\partial x} v_h\, dx \\
&\quad + \int_\alpha^\beta a_0^{n+1} u_h^{n+1} v_h\, dx = \int_\alpha^\beta f^{n+1} v_h\, dx \qquad \forall v_h \in V_h^{in},
\end{aligned} \tag{13.70}$$

with $u_h^{n+1}(\alpha) = \varphi^{n+1}$ and $u_h^0 = u_{0h}$. If $f \equiv 0$ and $\varphi \equiv 0$, taking $v_h = u_h^{n+1}$ in (13.70) we can obtain

$$\frac{1}{2\Delta t}\left(\|u_h^{n+1}\|^2_{\mathrm{L}^2(\alpha,\beta)} - \|u_h^n\|^2_{\mathrm{L}^2(\alpha,\beta)}\right) + a(\beta)(u_h^{n+1}(\beta))^2 + \mu_0\|u_h^{n+1}\|^2_{\mathrm{L}^2(\alpha,\beta)} \leq 0$$

$\forall n \geq 0$. Summing for n from 0 to $m-1$ yields, for $m \geq 1$,

$$\|u_h^m\|^2_{\mathrm{L}^2(\alpha,\beta)} + 2\Delta t\left(\sum_{j=1}^{m}\|u_h^j\|^2_{\mathrm{L}^2(\alpha,\beta)} + \sum_{j=1}^{m} a(\beta)(u_h^{j+1}(\beta))^2\right) \leq \|u_h^0\|^2_{\mathrm{L}^2(\alpha,\beta)}.$$

In particular, we conclude that

$$\|u_h^m\|_{\mathrm{L}^2(\alpha,\beta)} \leq \|u_h^0\|_{\mathrm{L}^2(\alpha,\beta)} \quad \forall m \geq 0.$$

On the other hand, explicit schemes for the hyperbolic equations are subject to a stability condition: for example, in the case of the forward Euler method the stability condition is $\Delta t = \mathcal{O}(\Delta x)$. In practice, this restriction is not as severe as happens in the case of parabolic equations and for this reason explicit schemes are often used in the approximation of hyperbolic equations.

Programs 102 and 103 provide an implementation of the discontinous Galerkin-finite element method of degree 0 (dG(0)) and 1 (dG(1)) in space coupled with the backward Euler method in time for the solution of (13.26) on the space-time domain $(\alpha, \beta) \times (t_0, T)$.

Program 102 - ipeidg0 : dG(0) implicit Euler

```
function [u,x]=ipeidg0(I,n,a,u0,bc)
%IPEIDG0 dG(0) implicit Euler for a scalar transport equation.
%  [U,X]=IPEIDG0(I,N,A,U0,BC) solves the equation
%    DU/DT+A+DU/DX=0   X in (I(1),I(2)), T in (I(3),I(4))
%  with a space-time finite element approximation.
nx=n(1); h=(I(2)-I(1))/nx; x=[I(1)+h/2:h:I(2)];
t=I(3);  u=(eval(u0))';
nt=n(2); k=(I(4)-I(3))/nt;
lambda=k/h;
e=ones(nx,1);
A=spdiags([-a*lambda*e, (1+a*lambda)*e],-1:0,nx,nx);
[L,U]=lu(A);
for t = I(3)+k:k:I(4)
   f = u;
   if a > 0
      f(1) = a*bc(1)+f(1);
   elseif a <= 0
      f(nx) = a*bc(2)+f(nx);
   end
```

```
    y = L \ f; u = U \ y;
end
return
```

Program 103 - ipeidg1 : dG(1) implicit Euler

```
function [u,x]=ipeidg1(I,n,a,u0,bc)
%IPEIDG0 dG(1) implicit Euler for a scalar transport equation.
%  [U,X]=IPEIDG1(I,N,A,U0,BC) solves the equation
%     DU/DT+A+DU/DX=0  X in (I(1),I(2)), T in (I(3),I(4))
%  with a space-time finite element approximation.
nx=n(1); h=(I(2)-I(1))/nx; x=[I(1):h:I(2)];
t=I(3);  um=(eval(u0))';
u=[]; xx=[];
for i=1:nx+1
   u=[u, um(i), um(i)];
   xx=[xx, x(i), x(i)];
end
u=u'; nt=n(2); k=(I(4)-I(3))/nt;
lambda=k/h;
e=ones(2*nx+2,1);
B=spdiags([1/6*e,1/3*e,1/6*e],-1:1,2*nx+2,2*nx+2);
dd=1/3+0.5*a*lambda;
du=1/6+0.5*a*lambda;
dl=1/6-0.5*a*lambda;
A=sparse([]);
A(1,1)=dd; A(1,2)=du; A(2,1)=dl; A(2,2)=dd;
for i=3:2:2*nx+2
   A(i,i-1)=-a*lambda;
   A(i,i)=dd;
   A(i,i+1)=du;
   A(i+1,i)= dl;
   A(i+1,i+1)=A(i,i);
end
[L,U]=lu(A);
for t = I(3)+k:k:I(4)
   f = B*u;
   if a>0
      f(1)=a*bc(1)+f(1);
   elseif a<=0
      f(nx)=a*bc(2)+f(nx);
   end
   y=L\f;
   u=U\y;
end
x=xx;
return
```

13.11 Applications

13.11.1 Heat Conduction in a Bar

Consider a homogeneous bar of unit length with thermal conductivity ν, which is connected at the endpoints to an external thermal source at a fixed temperature, say $u = 0$. Let $u_0(x)$ be the temperature distribution along the bar at time $t = 0$ and $f = f(x,t)$ be a given heat production term. Then, the initial-boundary value problem (13.1)-(13.4) provides a model of the time evolution of the temperature $u = u(x,t)$ throughout the bar.
In the following, we study the case where $f \equiv 0$ and the temperature of the bar is suddenly raised at the points around 1/2. A rough mathematical model for this situation is provided, for instance, by taking $u_0 = K$ in a certain subinterval $[a,b] \subseteq [0,1]$ and equal to 0 outside, where K is a given positive constant. The initial condition is therefore a discontinuous function.

We have used the θ-method with $\theta = 0.5$ (Crank-Nicolson method, CN) and $\theta = 1$ (Backward Euler method, BE). Program 100 has been run with $h = 1/20$, $\Delta t = 1/40$ and the obtained solutions at time $t = 2$ are shown in Figure 13.10. The results show that the CN method suffers a clear instability due to the low smoothness of the initial datum (about this point, see also [QV94], Chapter 11). On the contrary, the BE method provides a stable solution which decays correctly to zero as t grows since the source term f is null.

13.11.2 A Hyperbolic Model for Blood Flow Interaction with Arterial Walls

Let us consider again the problem of the fluid-structure interaction in a cylindrical artery considered in Section 11.11.2, where the simple independent rings model (11.88) was adopted.

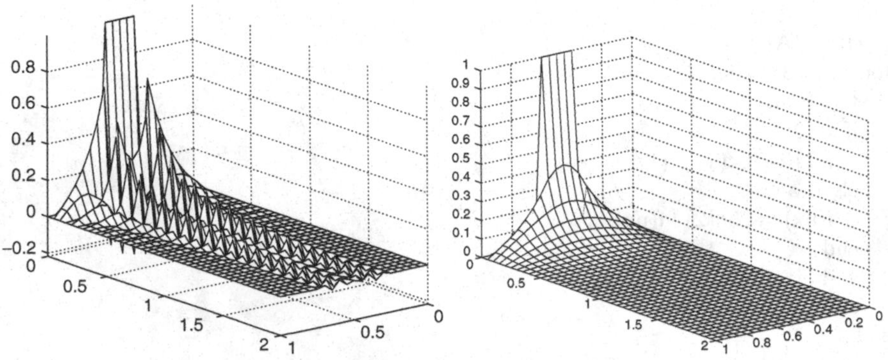

Fig. 13.10. Solutions for a parabolic problem with discontinuous initial datum: CN method (*left*) and BE method (*right*)

If the axial action due to the tension between the different rings is no longer neglected, denoting by z the longitudinal coordinate, equation (11.88) modifies into

$$\rho_w H \frac{\partial^2 \eta}{\partial t^2} - \sigma_z \frac{\partial^2 \eta}{\partial z^2} + \frac{HE}{R_0^2}\eta = P - P_0, \quad t > 0, \quad 0 < z < L, \qquad (13.71)$$

where σ_z is the radial component of the axial stress and L is the length of the cylindrical arterial district which is considered. In particular, neglecting the third term on the left-hand side and letting $\gamma^2 = \sigma_z/(\rho_w H)$, $f = (P - P_0)/(\rho_w H)$, we recover the wave equation (13.33).

We have performed two sets of numerical experiments using the Leap-Frog (LF) and Newmark (NW) methods. In the first example the space-time domain of integration is the cylinder $(0,1) \times (0,1)$ and the source term is $f = (1 + \pi^2\gamma^2)e^{-t}\sin(\pi x)$ in such a way that the exact solution is $u(x,t) = e^{-t}\sin(\pi x)$. Table 13.3 shows the estimated orders of convergence of the two methods, denoted by p_{LF} and p_{NW} respectively.

To compute these quantities we have first solved the wave equation on four grids with sizes $\Delta x = \Delta t = 1/(2^k \cdot 10)$, $k = 0, \ldots, 3$. Let $u_h^{(k)}$ denote the numerical solution corresponding to the space-time grid at the k-th refining level. Moreover, for $j = 1, \ldots, 10$, let $t_j^{(0)} = j/10$ be the time discretization nodes of the grid at the coarsest level $k = 0$. Then, for each level k, the maximum nodal errors e_j^k on the k-th spatial grid have been evaluated at each time $t_j^{(0)}$ in such a way that the convergence order $p_j^{(k)}$ can be estimated as

$$p_j^{(k)} = \frac{\log(e_j^0/e_j^k)}{\log(2^k)}, \qquad k = 1, 2, 3.$$

The results clearly show second-order convergence for both the methods, as theoretically expected.

Table 13.3. Estimated orders of convergence for the Leap-Frog (LF) and Newmark (NW) methods

$t_j^{(0)}$	$p_{LF}^{(1)}$	$p_{LF}^{(2)}$	$p_{LF}^{(3)}$	$t_j^{(0)}$	$p_{NW}^{(1)}$	$p_{NW}^{(2)}$	$p_{NW}^{(3)}$
0.1	2.0344	2.0215	2.0151	0.1	1.9549	1.9718	1.9803
0.2	2.0223	2.0139	2.0097	0.2	1.9701	1.9813	1.9869
0.3	2.0170	2.0106	2.0074	0.3	1.9754	1.9846	1.9892
0.4	2.0139	2.0087	2.0061	0.4	1.9791	1.9869	1.9909
0.5	2.0117	2.0073	2.0051	0.5	1.9827	1.9892	1.9924
0.6	2.0101	2.0063	2.0044	0.6	1.9865	1.9916	1.9941
0.7	2.0086	2.0054	2.0038	0.7	1.9910	1.9944	1.9961
0.8	2.0073	2.0046	2.0032	0.8	1.9965	1.9979	1.9985
0.9	2.0059	2.0037	2.0026	0.9	2.0034	2.0022	2.0015
1.0	2.0044	2.0028	2.0019	1.0	2.0125	2.0079	2.0055

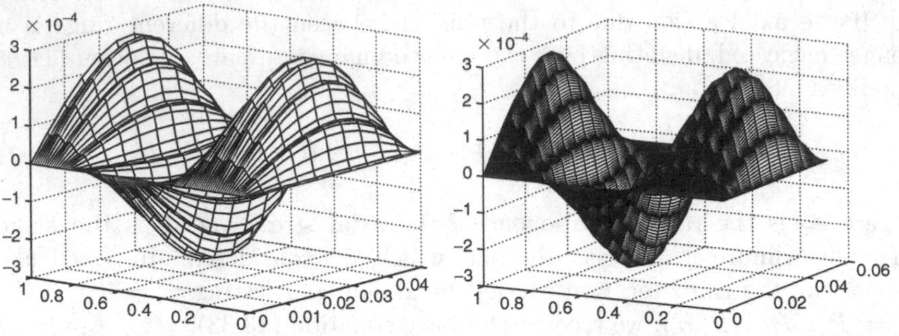

Fig. 13.11. Computed solutions using the NM method on a space-time grid with $\Delta t = T/100$ and $\Delta x = L/10$ (*left*) and the LF scheme on a space-time grid with the same value of Δx but with $\Delta t = T/400$ (*right*)

In the second example we have taken the following expressions for the coefficient and source term: $\gamma^2 = \sigma_z/(\rho_w H)$, with $\sigma_z = 1$ $[Kgs^{-2}]$, $f = (x\Delta p \cdot \sin(\omega_0 t))/(\rho_w H)$. The parameters ρ_w, H and the length L of the vessel are as in Section 11.11.2. The space-time computational domain is $(0, L) \times (0, T)$, with $T = 1$ $[s]$.

The Newmark method has been first used with $\Delta x = L/10$ and $\Delta t = T/100$; the corresponding value of $\gamma\lambda$ is 3.6515, where $\lambda = \Delta t/\Delta x$. Since the Newmark method is unconditionally stable, no spurious oscillations arise as is confirmed by Figure 13.11, left. Notice the correct periodical behaviour of the solution with a period corresponding to one heart beat; notice also that with the present values of Δt and Δx the Leap-Frog method cannot be employed since the CFL condition is not satisfied. To overcome this problem, we have therefore chosen a much smaller time-step $\Delta t = T/400$, in such a way that $\gamma\lambda \simeq 0.9129$ and the Leap-Frog scheme can be applied. The obtained result is shown in Figure 13.11, right; a similar solution has been computed by using the Newmark method with the same values of the discretization parameters.

13.12 Exercises

1. Apply the θ-method (13.9) to the approximate solution of the scalar Cauchy problem (11.1) and using the analysis of Section 11.3 prove that the local truncation error is of the order of $\Delta t + h^2$ if $\theta \neq \frac{1}{2}$ while it is of the order of $\Delta t^2 + h^2$ if $\theta = \frac{1}{2}$.
2. Prove that in the case of piecewise linear finite elements, the mass-lumping process described in Section 13.3 is equivalent to computing the integrals $m_{ij} = \int_0^1 \varphi_j \varphi_i \, dx$ by the trapezoidal quadrature formula (9.11). This, in particular, shows that the diagonal matrix $\widetilde{\mathrm{M}}$ is nonsingular.

[*Hint*: first, verify that exact integration yields

$$m_{ij} = \frac{h}{6}\begin{cases} \frac{1}{2} & i \neq j, \\ 1 & i = j. \end{cases}$$

Then, apply the trapezoidal rule to compute m_{ij} recalling that $\varphi_i(x_j) = \delta_{ij}$.]

3. Prove inequality (13.19).
   [*Hint*: using the Cauchy-Schwarz and Young inequalities, prove first that

$$\int_0^1 (u-v)u\,dx \geq \frac{1}{2}\left(\|u\|^2_{L^2(0,1)} - \|v\|^2_{L^2(0,1)}\right), \qquad \forall\, u, v \in L^2(0,1).$$

   Then, use (13.18).]
4. Assume that the bilinear form $a(\cdot,\cdot)$ in problem (13.12) is continuous and coercive over the function space V (see (12.54)-(12.55)) with continuity and coercivity constants M and α, respectively. Then, prove that the stability inequalities (13.20) and (13.21) still hold provided that ν is replaced by α.
5. Show that the methods (13.39), (13.40) and (13.41) can be written in the form (13.42). Then, show that the corresponding expressions of the artificial viscosity K and artificial diffusion flux $h^{diff}_{j+1/2}$ are as in Table (13.1).
6. Determine the CFL condition for the upwind scheme.
7. Show that for the scheme (13.43) one has $\|\mathbf{u}^{n+1}\|_{\Delta,2} \leq \|\mathbf{u}^n\|_{\Delta,2}$ for all $n \geq 0$.
   [*Hint*: multiply equation (13.43) by u_j^{n+1}, and notice that

$$(u_j^{n+1} - u_j^n)u_j^{n+1} \geq \frac{1}{2}\left(|u_j^{n+1}|^2 - |u_j^n|^2\right).$$

   Then, sum on j the resulting inequalities, and note that

$$\frac{\lambda a}{2}\sum_{j=-\infty}^{\infty}\left(u_{j+1}^{n+1} - u_{j-1}^{n+1}\right)u_j^{n+1} = 0$$

   since this sum is telescopic.]
8. Show how to find the values μ and ν in Table 13.2 for Lax-Friedrichs and Lax-Wendroff methods.
9. Prove (13.67).
10. Prove (13.69) when $f = 0$.
    [*Hint*: take $\forall t > 0$, $v_h = u_h(t)$ in (13.68).]

References

[Aas71] Aasen J. (1971) On the Reduction of a Symmetric Matrix to Tridiagonal Form. *BIT* 11: 233–242

[ABB+92] Anderson E., Bai Z., Bischof C., Demmel J., Dongarra J., Croz J. D., Greenbaum A., Hammarling S., McKenney A., Oustrouchov S., and Sorensen D. (1992) *LAPACK User's Guide, Release 1.0.* SIAM, Philadelphia

[Ada75] Adams D. (1975) *Sobolev Spaces.* Academic Press, New York

[ADR92] Arioli M., Duff I., and Ruiz D. (1992) Stopping Criteria for Iterative Solvers. *SIAM J. Matrix Anal. Appl.* 1(13)

[AF83] Alonso M. and Finn E. (1983) *Fundamental University Physics,* volume 3. Addison-Wesley, Reading, Massachusetts

[Arm66] Armijo L. (1966) Minimization of Functions Having Continuous Partial Derivatives. *Pacific Jour. Math.* 16: 1–3

[Arn73] Arnold V. I. (1973) *Ordinary Differential Equations.* The MIT Press, Cambridge, Massachusetts

[Atk89] Atkinson K. E. (1989) *An Introduction to Numerical Analysis.* John Wiley, New York

[Avr76] Avriel M. (1976) *Non Linear Programming: Analysis and Methods.* Prentice-Hall, Englewood Cliffs, New Jersey

[Axe94] Axelsson O. (1994) *Iterative Solution Methods.* Cambridge University Press, New York

[Bar89] Barnett S. (1989) Leverrier's Algorithm: A New Proof and Extensions. *Numer. Math.* 7: 338–352

[Bat90] Batterson S. (1990) Convergence of the Shifted QR Algorithm on 3 by 3 Normal Matrices. *Numer. Math.* 58: 341–352

[BBC+94] Barrett R., Berry M., Chan T., Demmel J., Donato J., Dongarra J., Eijkhout V., Pozo V., Romine C., and van der Vorst H. (1994) *Templates for the Solution of Linear Systems: Building Blocks for Iterative Methods.* SIAM, Philadelphia

[BD74] Björck A. and Dahlquist G. (1974) *Numerical Methods.* Prentice-Hall, Englewood Cliffs, N.J.

[BDMS79] Bunch J., Dongarra J., Moler C., and Stewart G. (1979) *LINPACK User's Guide.* SIAM, Philadelphia

[Ber82] Bertsekas D. P. (1982) *Constrained Optimization and Lagrange Multiplier Methods.* Academic Press. Inc., San Diego, California

[Bjö88] Björck A. (1988) *Least Squares Methods: Handbook of Numerical Analysis Vol. 1 Solution of Equations in* $\mathbb{R}^N$. Elsevier North Holland

[BM92] Bernardi C. and Maday Y. (1992) *Approximations Spectrales des Problémes aux Limites Elliptiques.* Springer-Verlag, Paris

[BM97] Berrut J. and Mittelmann H. (1997) Lebesgue constant minimizing linear rational interpolation of continuous functions over the interval. *Comput. Math.* 33: 77–86

[BMW67] Barth W., Martin R. S., and Wilkinson J. H. (1967) Calculation of the Eigenvalues of a Symmetric Tridiagonal Matrix by the Method of Bisection. *Numer. Math.* 9: 386–393

[BO78] Bender C. M. and Orszag S. A. (1978) *Advanced Mathematical Methods for Scientists and Engineers.* McGraw-Hill, New York

[Boe80] Boehm W. (1980) Inserting New Knots into B-spline Curves. *Computer Aided Design* 12: 199–201

[Bos93] Bossavit A. (1993) *Electromagnetisme, en vue de la modelisation.* Springer-Verlag, Paris

[BR81] Bank R. E. and Rose D. J. (1981) Global Approximate Newton Methods. *Numer. Math.* 37: 279–295

[Bra75] Bradley G. (1975) *A Primer of Linear Algebra.* Prentice-Hall, Englewood Cliffs, New York

[Bre73] Brent R. (1973) *Algorithms for Minimization Without Derivatives.* Prentice-Hall, Englewood Cliffs, New York

[Bri74] Brigham E. O. (1974) *The Fast Fourier Transform.* Prentice-Hall, Englewood Cliffs, New York

[BS90] Brown P. and Saad Y. (1990) Hybrid Krylov Methods for Nonlinear Systems of equations. *SIAM J. Sci. and Stat. Comput.* 11(3): 450–481

[BSG96] B. Smith P. B. and Gropp P. (1996) *Domain Decomposition, Parallel Multilevel Methods for Elliptic Partial Differential Equations.* Univ. Cambridge Press, Cambridge

[BT04] Berrut J. and Trefethen L. (2004) Barycentric Lagrange Interpolation. *SIAM Review* 46(3): 501–517

[But64] Butcher J. C. (1964) Implicit Runge-Kutta Processes. *Math. Comp.* 18: 233–244

[But66] Butcher J. C. (1966) On the Convergence of Numerical Solutions to Ordinary Differential Equations. *Math. Comp.* 20: 1–10

[But87] Butcher J. (1987) *The Numerical Analysis of Ordinary Differential Equations: Runge-Kutta and General Linear Methods.* Wiley, Chichester

[CCP70] Cannon M., Cullum C., and Polak E. (1970) *Theory and Optimal Control and Mathematical Programming.* McGraw-Hill, New York

[CFL28] Courant R., Friedrichs K., and Lewy H. (1928) Über die partiellen differenzengleichungen der mathematischen physik. *Math. Ann.* 100: 32–74

[CHQZ06] Canuto C., Hussaini M. Y., Quarteroni A., and Zang T. A. (2006) *Spectral Methods: Fundamentals in Single Domains.* Springer-Verlag, Berlin Heidelberg

[CI95] Chandrasekaren S. and Ipsen I. (1995) On the Sensitivity of Solution Components in Linear Systems of equations. *SIAM J. Matrix Anal. Appl.* 16: 93–112

[CL91] Ciarlet P. G. and Lions J. L. (1991) *Handbook of Numerical Analysis: Finite Element Methods (Part 1)*. North-Holland, Amsterdam

[CM94] Chan T. and Mathew T. (1994) Domain Decomposition Algorithms. *Acta Numerica* pages 61–143

[CMSW79] Cline A., Moler C., Stewart G., and Wilkinson J. (1979) An Estimate for the Condition Number of a Matrix. *SIAM J. Sci. and Stat. Comput.* 16: 368–375

[Col66] Collin R. E. (1966) *Foundations for Microwave Engineering*. McGraw-Hill Book Co., Singapore

[Com95] Comincioli V. (1995) *Analisi Numerica Metodi Modelli Applicazioni*. McGraw-Hill Libri Italia, Milano

[Cox72] Cox M. (1972) The Numerical Evaluation of B-splines. *Journal of the Inst. of Mathematics and its Applications* 10: 134–149

[Cry73] Cryer C. W. (1973) On the Instability of High Order Backward-Difference Multistep Methods. *BIT* 13: 153–159

[CT65] Cooley J. and Tukey J. (1965) An Algorithm for the Machine Calculation of Complex Fourier Series. *Math. Comp.* 19: 297–301

[Dah56] Dahlquist G. (1956) Convergence and Stability in the Numerical Integration of Ordinary Differential Equations. *Math. Scand.* 4: 33–53

[Dah63] Dahlquist G. (1963) A Special Stability Problem for Linear Multistep Methods. *BIT* 3: 27–43

[Dat95] Datta B. (1995) *Numerical Linear Algebra and Applications*. Brooks/Cole Publishing, Pacific Grove, CA

[Dau88] Daubechies I. (1988) Orthonormal bases of compactly supported wavelets. *Commun. on Pure and Appl. Math.* XLI

[Dav63] Davis P. (1963) *Interpolation and Approximation*. Blaisdell Pub., New York

[Day96] Day D. (1996) How the QR algorithm Fails to Converge and How to Fix It. Technical Report 96-0913J, Sandia National Laboratory, Albuquerque

[dB72] de Boor C. (1972) On Calculating with B-splines. *Journal of Approximation Theory* 6: 50–62

[dB83] de Boor C. (1983) A Practical Guide to Splines. In *Applied Mathematical Sciences*. (27), Springer-Verlag, New York

[dB90] de Boor C. (1990) *SPLINE TOOLBOX for use with MATLAB*. The Math Works, Inc., South Natick

[DD95] Davis T. and Duff I. (1995) A combined unifrontal/multifrontal method for unsymmetric sparse matrices. Technical Report TR-95-020, Computer and Information Sciences Department, University of Florida

[Dek69] Dekker T. (1969) Finding a Zero by means of Successive Linear Interpolation. In Dejon B. and Henrici P. (eds) *Constructive Aspects of the Fundamental Theorem of Algebra*, pages 37–51. Wiley, New York

[Dek71] Dekker T. (1971) A Floating-Point Technique for Extending the Available Precision. *Numer. Math.* 18: 224–242

[Dem97] Demmel J. (1997) *Applied Numerical Linear Algebra*. SIAM, Philadelphia

[Deu04] Deuflhard P. (2004) *Newton methods for nonlinear problems. Affine invariance and adaptive algorithms*, volume 35 of *Springer Series in Computational Mathematics*. Springer-Verlag, Berlin

[DGK84] Dongarra J., Gustavson F., and Karp A. (1984) Implementing Linear Algebra Algorithms for Dense Matrices on a Vector Pipeline Machine. *SIAM Review* 26(1): 91–112

[Die87a] Dierckx P. (1987) *FITPACK User Guide part 1: Curve Fitting Routines.* TW Report, Dept. of Computer Science, Katholieke Universiteit, Leuven, Belgium

[Die87b] Dierckx P. (1987) *FITPACK User Guide part 2: Surface Fitting Routines.* TW Report, Dept. of Computer Science, Katholieke Universiteit, Leuven, Belgium

[Die93] Dierckx P. (1993) *Curve and Surface Fitting with Splines.* Claredon Press, New York

[DL92] DeVore R. and Lucier J. (1992) Wavelets. *Acta Numerica* pages 1–56

[DR75] Davis P. and Rabinowitz P. (1975) *Methods of Numerical Integration.* Academic Press, New York

[DS83] Dennis J. and Schnabel R. (1983) *Numerical Methods for Unconstrained Optimization and Nonlinear Equations.* Prentice-Hall, Englewood Cliffs, New York

[Dun85] Dunavant D. (1985) High Degree Efficient Symmetrical Gaussian Quadrature Rules for the Triangle. *Internat. J. Numer. Meth. Engrg.* 21: 1129–1148

[Dun86] Dunavant D. (1986) Efficient Symmetrical Cubature Rules for Complete Polynomials of High Degree over the Unit Cube. *Internat. J. Numer. Meth. Engrg.* 23: 397–407

[DV84] Dekker K. and Verwer J. (1984) *Stability of Runge-Kutta Methods for Stiff Nonlinear Differential Equations.* North-Holland, Amsterdam

[dV89] der Vorst H. V. (1989) High Performance Preconditioning. *SIAM J. Sci. Stat. Comput.* 10: 1174–1185

[EEHJ96] Eriksson K., Estep D., Hansbo P., and Johnson C. (1996) *Computational Differential Equations.* Cambridge Univ. Press, Cambridge

[Elm86] Elman H. (1986) A Stability Analisys of Incomplete LU Factorization. *Math. Comp.* 47: 191–218

[Erd61] Erdös P. (1961) Problems and Results on the Theory of Interpolation. *Acta Math. Acad. Sci. Hungar.* 44: 235–244

[Erh97] Erhel J. (1997) About Newton-Krylov Methods. In Periaux J. and al. (eds) *Computational Science for* 21st *Century*, pages 53–61. Wiley, New York

[Fab14] Faber G. (1914) Über die interpolatorische Darstellung stetiger Funktionem. *Jber. Deutsch. Math. Verein.* 23: 192–210

[FF63] Faddeev D. K. and Faddeeva V. N. (1963) *Computational Methods of Linear Algebra.* Freeman, San Francisco and London

[Fle75] Fletcher R. (1975) Conjugate gradient methods for indefinite systems. In Springer-Verlag (ed) *Numerical Analysis*, pages 73–89. New York

[FM67] Forsythe G. E. and Moler C. B. (1967) *Computer Solution of Linear Algebraic Systems.* Prentice-Hall, Englewood Cliffs, New York

[Fra61] Francis J. G. F. (1961) The QR Transformation: A Unitary Analogue to the LR Transformation. Parts I and II. *Comp. J.* pages 265–272, 332–334

[FRL55] F. Richtmyer E. K. and Lauritsen T. (1955) *Introduction to Modern Physics.* McGraw-Hill, New York

[Gas83] Gastinel N. (1983) *Linear Numerical Analysis.* Kershaw Publishing, London

[Gau94] Gautschi W. (1994) Algorithm 726: ORTHPOL - A Package of Routines for Generating Orthogonal Polynomials and Gauss-type Quadrature Rules. *ACM Trans. Math. Software* 20: 21–62

[Gau96] Gautschi W. (1996) Orthogonal Polynomials: Applications and Computation. *Acta Numerica* pages 45–119

[Gau97] Gautschi W. (1997) *Numerical Analysis. An Introduction.* Birkhäuser, Berlin

[Geo73] George A. (1973) Nested Dissection of a Regular Finite Element Mesh. *SIAM J. Num. Anal.* 10: 345–363

[Giv54] Givens W. (1954) Numerical Computation of the Characteristic Values of a Real Symmetric Matrix. *Oak Ridge National Laboratory* ORNL-1574

[GL81] George A. and Liu J. (1981) *Computer Solution of Large Sparse Positive Definite Systems.* Prentice-Hall, Englewood Cliffs, New York

[GL89] Golub G. and Loan C. V. (1989) *Matrix Computations.* The John Hopkins Univ. Press, Baltimore and London

[GM83] Golub G. and Meurant G. (1983) *Resolution Numerique des Grands Systemes Lineaires.* Eyrolles, Paris

[GMW81] Gill P., Murray W., and Wright M. (1981) *Practical Optimization.* Academic Press, London

[God66] Godeman R. (1966) *Algebra.* Kershaw, London

[Gol91] Goldberg D. (1991) What Every Computer Scientist Should Know about Floating-point Arithmetic. *ACM Computing Surveys* 23(1): 5–48

[GP67] Goldstein A. A. and Price J. B. (1967) An Effective Algorithm for Minimization. *Numer. Math* 10: 184–189

[GR96] Godlewski E. and Raviart P. (1996) *Numerical Approximation of Hyperbolic System of Conservation Laws*, volume 118 of *Applied Mathematical Sciences.* Springer-Verlag, New York

[Hac94] Hackbush W. (1994) *Iterative Solution of Large Sparse Systems of Equations.* Springer-Verlag, New York

[Hah67] Hahn W. (1967) *Stability of Motion.* Springer-Verlag, Berlin

[Hal58] Halmos P. (1958) *Finite-Dimensional Vector Spaces.* Van Nostrand, Princeton, New York

[Hen62] Henrici P. (1962) *Discrete Variable Methods in Ordinary Differential Equations.* Wiley, New York

[Hen74] Henrici P. (1974) *Applied and Computational Complex Analysis*, volume 1. Wiley, New York

[Hen79] Henrici P. (1979) Barycentric formulas for interpolating trigonometric polynomials and their conjugates. *Numer. Math.* 33: 225–234

[HGR96] H-G. Roos M. Stynes L. T. (1996) *Numerical Methods for Singularly Perturbed Differential Equations.* Springer-Verlag, Berlin Heidelberg

[Hig88] Higham N. (1988) The Accuracy of Solutions to Triangular Systems. *University of Manchester, Dep. of Mathematics* 158: 91–112

[Hig89] Higham N. (1989) The Accuracy of Solutions to Triangular Systems. *SIAM J. Numer. Anal.* 26(5): 1252–1265

[Hig96] Higham N. (1996) *Accuracy and Stability of Numerical Algorithms.* SIAM Publications, Philadelphia, PA

[Hil87] Hildebrand F. (1987) *Introduction to Numerical Analysis.* McGraw-Hill, New York

[Hou75] Householder A. (1975) *The Theory of Matrices in Numerical Analysis.* Dover Publications, New York

[HP94] Hennessy J. and Patterson D. (1994) *Computer Organization and Design - The Hardware/Software Interface.* Morgan Kaufmann, San Mateo

[HW76] Hammarling S. and Wilkinson J. (1976) The Practical Behaviour of Linear Iterative Methods with Particular Reference to S.O.R. Technical Report Report NAC 69, National Physical Laboratory, Teddington, UK

[IK66] Isaacson E. and Keller H. (1966) *Analysis of Numerical Methods.* Wiley, New York

[Inm94] Inman D. (1994) *Engineering Vibration.* Prentice-Hall, Englewood Cliffs, NJ

[Iro70] Irons B. (1970) A Frontal Solution Program for Finite Element Analysis. *Int. J. for Numer. Meth. in Engng.* 2: 5–32

[Jac26] Jacobi C. (1826) Uber Gauβ neue Methode, die Werthe der Integrale näherungsweise zu finden. *J. Reine Angew. Math.* 30: 127–156

[Jer96] Jerome J. J. (1996) *Analysis of Charge Transport. A Mathematical Study of Semiconductor Devices.* Springer, Berlin Heidelberg

[Jia95] Jia Z. (1995) The Convergence of Generalized Lanczos Methods for Large Unsymmetric Eigenproblems. *SIAM J. Matrix Anal. Applic.* 16: 543–562

[JM92] Jennings A. and McKeown J. (1992) *Matrix Computation.* Wiley, Chichester

[Joh90] Johnson C. (1990) *Numerical Solution of Partial Differential Equations by the Finite Element Method.* Cambridge Univ. Press

[JW77] Jankowski M. and Wozniakowski M. (1977) Iterative Refinement Implies Numerical Stability. *BIT* 17: 303–311

[Kah66] Kahan W. (1966) Numerical Linear Algebra. *Canadian Math. Bull.* 9: 757–801

[Kan66] Kaniel S. (1966) Estimates for Some Computational Techniques in Linear Algebra. *Math. Comp.* 20: 369–378

[Kea86] Keast P. (1986) Moderate-Degree Tetrahedral Quadrature Formulas. *Comp. Meth. Appl. Mech. Engrg.* 55: 339–348

[Kel99] Kelley C. (1999) *Iterative Methods for Optimization*, volume 18 of *Frontiers in Applied Mathematics.* SIAM, Philadelphia

[KT51] Kuhn H. and Tucker A. (1951) Nonlinear Programming. In *Second Berkeley Symposium on Mathematical Statistics and Probability*, pages 481–492. Univ. of California Press, Berkeley and Los Angeles

[Lam91] Lambert J. (1991) *Numerical Methods for Ordinary Differential Systems.* John Wiley and Sons, Chichester

[Lan50] Lanczos C. (1950) An Iteration Method for the Solution of the Eigenvalue Problem of Linear Differential and Integral Operator. *J. Res. Nat. Bur. Stand.* 45: 255–282

[Lax65] Lax P. (1965) Numerical Solution of Partial Differential Equations. *Amer. Math. Monthly* 72(2): 74–84

[Lel92] Lele S. (1992) Compact Finite Difference Schemes with Spectral-like Resolution. *Journ. of Comp. Physics* 103(1): 16–42

[Lem89] Lemarechal C. (1989) Nondifferentiable Optimization. In Nemhauser G., Kan A. R., and Todd M. (eds) *Handbooks Oper. Res. Management Sci.*, volume 1. Optimization, pages 529–572. North-Holland, Amsterdam

[LH74] Lawson C. and Hanson R. (1974) *Solving Least Squares Problems*. Prentice-Hall, Englewood Cliffs, New York

[LM68] Lions J. L. and Magenes E. (1968) *Problemes aux limitès non-homogènes et applications*. Dunod, Paris

[LS96] Lehoucq R. and Sorensen D. (1996) Deflation Techniques for an Implicitly Restarted Iteration. *SIAM J. Matrix Anal. Applic.* 17(4): 789–821

[Lue73] Luenberger D. (1973) *Introduction to Linear and Non Linear Programming*. Addison-Wesley, Reading, Massachusetts

[Man69] Mangasarian O. (1969) *Non Linear Programming*. Prentice-Hall, Englewood Cliffs, New Jersey

[Man80] Manteuffel T. (1980) An Incomplete Factorization Technique for Positive Definite Linear Systems. *Math. Comp.* 150(34): 473–497

[Mar86] Markowich P. (1986) *The Stationary Semiconductor Device Equations*. Springer-Verlag, Wien and New York

[McK62] McKeeman W. (1962) Crout with Equilibration and Iteration. *Comm. ACM* 5: 553–555

[MdV77] Meijerink J. and der Vorst H. V. (1977) An Iterative Solution Method for Linear Systems of Which the Coefficient Matrix is a Symmetric M-matrix. *Math. Comp.* 137(31): 148–162

[MM71] Maxfield J. and Maxfield M. (1971) *Abstract Algebra and Solution by Radicals*. Saunders, Philadelphia

[MMG87] Martinet R., Morlet J., and Grossmann A. (1987) Analysis of sound patterns through wavelet transforms. *Int. J. of Pattern Recogn. and Artificial Intellig.* 1(2): 273–302

[MNS74] Mäkela M., Nevanlinna O., and Sipilä A. (1974) On the Concept of Convergence, Consistency and Stability in Connection with Some Numerical Methods. *Numer. Math.* 22: 261–274

[Mor84] Morozov V. (1984) *Methods for Solving Incorrectly Posed Problems*. Springer-Verlag, New York

[Mul56] Muller D. (1956) A Method for Solving Algebraic Equations using an Automatic Computer. *Math. Tables Aids Comput.* 10: 208–215

[NAG95] NAG (1995) *NAG Fortran Library Manual - Mark 17*. NAG Ltd., Oxford

[Nat65] Natanson I. (1965) *Constructive Function Theory*, volume III. Ungar, New York

[NM65] Nelder J. and Mead R. (1965) A simplex method for function minimization. *The Computer Journal* 7: 308–313

[Nob69] Noble B. (1969) *Applied Linear Algebra*. Prentice-Hall, Englewood Cliffs, New York

[OR70] Ortega J. and Rheinboldt W. (1970) *Iterative Solution of Nonlinear Equations in Several Variables*. Academic Press, New York and London

[Pap62] Papoulis A. (1962) *The Fourier Integral and its Application*. McGraw-Hill, New York

[Pap87] Papoulis A. (1987) *Probability, Random Variables, and Stochastic Processes*. McGraw-Hill, New York

[Par80] Parlett B. (1980) *The Symmetric Eigenvalue Problem.* Prentice-Hall, Englewood Cliffs, NJ

[PdKÜK83] Piessens R., deDoncker Kapenga E., Überhuber C. W., and Kahaner D. K. (1983) *QUADPACK: A Subroutine Package for Automatic Integration.* Springer-Verlag, Berlin and Heidelberg

[PJ55] Peaceman D. and Jr. H. R. (1955) The numerical solution of parabolic and elliptic differential equations. *J. Soc. Ind. Appl. Math.* 3: 28–41

[Pou96] Poularikas A. (1996) *The Transforms and Applications Handbook.* CRC Press, Inc., Boca Raton, Florida

[PR70] Parlett B. and Reid J. (1970) On the Solution of a System of Linear Equations Whose Matrix is Symmetric but not Definite. *BIT* 10: 386–397

[PW79] Peters G. and Wilkinson J. (1979) Inverse iteration, ill-conditioned equations, and newton's method. *SIAM Review* 21: 339–360

[QS06] Quarteroni A. and Saleri F. (2006) *Scientific Computing with Matlab and Octave.* Springer-Verlag, Berlin Heidelberg

[QV94] Quarteroni A. and Valli A. (1994) *Numerical Approximation of Partial Differential Equations.* Springer, Berlin and Heidelberg

[QV99] Quarteroni A. and Valli A. (1999) *Domain Decomposition Methods for Partial Differential Equations.* Oxford Science Publications, New York

[Ral65] Ralston A. (1965) *A First Course in Numerical Analysis.* McGraw-Hill, New York

[Red86] Reddy B. D. (1986) *Applied Functional Analysis and Variational Methods in Engineering.* McGraw-Hill, New York

[Ric81] Rice J. (1981) *Matrix Computations and Mathematical Software.* McGraw-Hill, New York

[Riv74] Rivlin T. (1974) *The Chebyshev Polynomials.* John Wiley and Sons, New York

[RM67] Richtmyer R. and Morton K. (1967) *Difference Methods for Initial Value Problems.* Wiley, New York

[RR78] Ralston A. and Rabinowitz P. (1978) *A First Course in Numerical Analysis.* McGraw-Hill, New York

[Rud83] Rudin W. (1983) *Real and Complex Analysis.* Tata McGraw-Hill, New Delhi

[Rut58] Rutishauser H. (1958) Solution of Eigenvalue Problems with the LR Transformation. *Nat. Bur. Stand. Appl. Math. Ser.* 49: 47–81

[Rut90] Rutishauser H. (1990) *Lectures on Numerical Mathematics.* Birkh auser, Boston

[Saa90] Saad Y. (1990) Sparskit: A basic tool kit for sparse matrix computations. Technical Report 90-20, Research Institute for Advanced Computer Science, NASA Ames Research Center, Moffet Field, CA

[Saa92] Saad Y. (1992) *Numerical Methods for Large Eigenvalue Problems.* Halstead Press, New York

[Saa96] Saad Y. (1996) *Iterative Methods for Sparse Linear Systems.* PWS Publishing Company, Boston

[Sch67] Schoenberg I. (1967) On Spline functions. In Shisha O. (ed) *Inequalities,* pages 255–291. Academic Press, New York

[Sch81] Schumaker L. (1981) *Splines Functions: Basic Theory.* Wiley, New York

[Sch98] Schwab C. (1998) *p- and hp-finite element methods. Theory and applications in solid and fluid mechanics.* Numerical Mathematics and Scientific Computation. The Clarendon Press, Oxford University Press

[Sel84] Selberherr S. (1984) *Analysis and Simulation of Semiconductor Devices.* Springer-Verlag, Wien and New York

[SG69] Scharfetter D. and Gummel H. (1969) Large-signal analysis of a silicon Read diode oscillator. *IEEE Trans. on Electr. Dev.* 16: 64–77

[Ske79] Skeel R. (1979) Scaling for Numerical Stability in Gaussian Elimination. *J. Assoc. Comput. Mach.* 26: 494–526

[Ske80] Skeel R. (1980) Iterative Refinement Implies Numerical Stability for Gaussian Elimination. *Math. Comp.* 35: 817–832

[SL89] Su B. and Liu D. (1989) *Computational Geometry: Curve and Surface Modeling.* Academic Press, New York

[Sla63] Slater J. (1963) *Introduction to Chemical Physics.* McGraw-Hill Book Co

[SM03] Suli E. and Mayers D. (2003) *An Introduction to Numerical Analysis.* Cambridge University Press, Cambridge

[Smi85] Smith G. (1985) *Numerical Solution of Partial Differential Equations: Finite Difference Methods.* Oxford University Press, Oxford

[Son89] Sonneveld P. (1989) Cgs, a fast lanczos-type solver for nonsymmetric linear systems. *SIAM Journal on Scientific and Statistical Computing* 10(1): 36–52

[SR97] Shampine L. F. and Reichelt M. W. (1997) The MATLAB ODE Suite. *SIAM J. Sci. Comput.* 18: 1–22

[SS90] Stewart G. and Sun J. (1990) *Matrix Perturbation Theory.* Academic Press, New York

[Ste71] Stetter H. (1971) Stability of discretization on infinite intervals. In Morris J. (ed) *Conf. on Applications of Numerical Analysis*, pages 207–222. Springer-Verlag, Berlin

[Ste73] Stewart G. (1973) *Introduction to Matrix Computations.* Academic Press, New York

[Str69] Strassen V. (1969) Gaussian Elimination is Not Optimal. *Numer. Math.* 13: 727–764

[Str80] Strang G. (1980) *Linear Algebra and Its Applications.* Academic Press, New York

[Str89] Strikwerda J. (1989) *Finite Difference Schemes and Partial Differential Equations.* Wadsworth and Brooks/Cole, Pacific Grove

[Sze67] Szegö G. (1967) *Orthogonal Polynomials.* AMS, Providence, R.I.

[Tit37] Titchmarsh E. (1937) *Introduction to the Theory of Fourier Integrals.* Oxford

[Var62] Varga R. (1962) *Matrix Iterative Analysis.* Prentice-Hall, Englewood Cliffs, New York

[vdV92] van der Vorst H. (1992) Bi-cgstab: a fast and smoothly converging variant of bi-cg for the solution of non-symmetric linear systems. *SIAM Jour. on Sci. and Stat. Comp.* 12: 631–644

[vdV03] van der Vorst H. (2003) *Iterative Krylov Methods for Large Linear systems.* Cambridge University Press, Cambridge

[Ver96] Verfürth R. (1996) *A Review of a Posteriori Error Estimation and Adaptive Mesh Refinement Techniques.* Wiley, Teubner, Germany

[Wac66] Wachspress E. (1966) *Iterative Solutions of Elliptic Systems.* Prentice-Hall, Englewood Cliffs, New York

[Wal75] Walsh G. (1975) *Methods of Optimization.* Wiley

[Wal91] Walker J. (1991) *Fast Fourier Transforms.* CRC Press, Boca Raton

[Wen66] Wendroff B. (1966) *Theoretical Numerical Analysis.* Academic Press, New York

[Wid67] Widlund O. (1967) A Note on Unconditionally Stable Linear Multistep Methods. *BIT* 7: 65–70

[Wil62] Wilkinson J. (1962) Note on the Quadratic Convergence of the Cyclic Jacobi Process. *Numer. Math.* 6: 296–300

[Wil63] Wilkinson J. (1963) *Rounding Errors in Algebraic Processes.* Prentice-Hall, Englewood Cliffs, New York

[Wil65] Wilkinson J. (1965) *The Algebraic Eigenvalue Problem.* Clarendon Press, Oxford

[Wil68] Wilkinson J. (1968) A priori Error Analysis of Algebraic Processes. In *Intern. Congress Math.*, volume 19, pages 629–639. Izdat. Mir, Moscow

[Wol69] Wolfe P. (1969) Convergence Conditions for Ascent Methods. *SIAM Review* 11: 226–235

[Wol71] Wolfe P. (1971) Convergence Conditions for Ascent Methods. II: Some Corrections. *SIAM Review* 13: 185–188

[Wol78] Wolfe M. (1978) *Numerical Methods for Unconstrained Optimization.* Van Nostrand Reinhold Company, New York

[You71] Young D. (1971) *Iterative Solution of Large Linear Systems.* Academic Press, New York

[Zie77] Zienkiewicz O. C. (1977) *The Finite Element Method (Third Edition).* McGraw Hill, London

Index of MATLAB Programs

Index

Texts in Applied Mathematics

(*continued from page ii*)

29. *Tveito/Winther:* Introduction to Partial Differential Equations: A Computational Approach
30. *Gasquet/Witomski:* Fourier Analysis and Applications: Filtering, Numerical Computation, Wavelets
31. *Brémaud:* Markov Chains: Gibbs Fields, Monte Carlo Simulation, and Queues
32. *Durran:* Numerical Methods for Fluid Dynamics with Applications in Geophysics
33. *Thomas:* Numerical Partial Differential Equations: Conservation Laws and Elliptic Equations
34. *Chicone:* Ordinary Differential Equations with Applications, 2nd ed.
35. *Kevorkian:* Partial Differential Equations: Analytical Solution Techniques, 2nd ed.
36. *Dullerud/Paganini:* A Course in Robust Control Theory: A Convex Approach
37. *Quarteroni/Sacco/Saleri:* Numerical Mathematics, 2nd ed.
38. *Gallier:* Geometric Methods and Applications: For Computer Science and Engineering
39. *Atkinson/Han:* Theoretical Numerical Analysis: A Functional Analysis Framework, 2nd ed.
40. *Brauer/Castillo-Chávez:* Mathematical Models in Population Biology and Epidemiology
41. *Davies:* Integral Transforms and Their Applications, 3rd ed.
42. *Deuflhard/Bornemann:* Scientific Computing with Ordinary Differential Equations
43. *Deuflhard/Hohmann:* Numerical Analysis in Modern Scientific Computing: An Introduction, 2nd ed.
44. *Knabner/Angerman:* Numerical Methods for Elliptic and Parabolic Partial Differential Equations
45. *Larsson/Thomée:* Partial Differential Equations with Numerical Methods
46. *Pedregal:* Introduction to Optimization
47. *Ockendon/Ockendon:* Waves and Compressible Flow
48. *Hinrichsen/Pritchard:* Mathematical Systems Theory I: Modelling, State Space Analysis, Stability and Robustness
49. *Bullo/Lewis:* Geometric Control of Mechanical Systems: Modeling, Analysis, and Design for Simple Mechanical Control Systems
50. *Verhulst:* Methods and Applications of Singular Perturbations: Boundary Layers and Multiple Timescale Dynamics
51. *Bondeson/Rylander/Ingelström:* Computational Electromagnetics
52. *Holmes:* Introduction to Numerical Methods in Differential Equations